T0190264

Lecture Notes in Computer Science 11165

Commenced Publication in 1973
Founding and Former Series Editors:
Gerhard Goos, Juris Hartmanis, and Jan van Leeuwen

Editorial Board

David Hutchison
Lancaster University, Lancaster, UK
Takeo Kanade
Carnegie Mellon University, Pittsburgh, PA, USA
Josef Kittler
University of Surrey, Guildford, UK
Jon M. Kleinberg
Cornell University, Ithaca, NY, USA
Friedemann Mattern
ETH Zurich, Zurich, Switzerland
John C. Mitchell
Stanford University, Stanford, CA, USA
Moni Naor
Weizmann Institute of Science, Rehovot, Israel
C. Pandu Rangan
Indian Institute of Technology Madras, Chennai, India
Bernhard Steffen
TU Dortmund University, Dortmund, Germany
Demetri Terzopoulos
University of California, Los Angeles, CA, USA
Doug Tygar
University of California, Berkeley, CA, USA
Gerhard Weikum
Max Planck Institute for Informatics, Saarbrücken, Germany

More information about this series at http://www.springer.com/series/7409

Richang Hong · Wen-Huang Cheng
Toshihiko Yamasaki · Meng Wang
Chong-Wah Ngo (Eds.)

Advances in Multimedia Information Processing – PCM 2018

19th Pacific-Rim Conference on Multimedia
Hefei, China, September 21–22, 2018
Proceedings, Part II

 Springer

Editors
Richang Hong
Hefei University of Technology
Hefei
China

Wen-Huang Cheng
National Chiao Tung University
Hsinchu
Taiwan

Meng Wang
Hefei University of Technology
Hefei
China

Chong-Wah Ngo
City University of Hong Kong
Hong Kong
Hong Kong, China

Toshihiko Yamasaki
University of Tokyo
Tokyo
Japan

ISSN 0302-9743   ISSN 1611-3349 (electronic)
Lecture Notes in Computer Science
ISBN 978-3-030-00766-9   ISBN 978-3-030-00767-6 (eBook)
https://doi.org/10.1007/978-3-030-00767-6

Library of Congress Control Number: 2018954671

LNCS Sublibrary: SL3 – Information Systems and Applications, incl. Internet/Web, and HCI

This Springer imprint is published by the registered company Springer Nature Switzerland AG
The registered company address is: Gewerbestrasse 11, 6330 Cham, Switzerland

Preface

The 19th Pacific-Rim Conference on Multimedia (PCM 2018) was held in Hefei, China, during September 21–22, 2018, and hosted by the Hefei University of Technology (HFUT). PCM is a major annual international conference for multimedia researchers and practitioners across academia and industry to demonstrate their scientific achievements and industrial innovations in the field of multimedia.

It is a great honor for HFUT to host PCM 2018, one of the most longstanding multimedia conferences, in Hefei, China. Hefei University of Technology, located in the capital of Anhui province, is one of the key universities administrated by the Ministry of Education, China. Recently its multimedia-related research has attracted more and more attentions from local and international multimedia community. Hefei is the capital city of Anhui Province, and is located in the center of Anhui between the Yangtze and Huaihe rivers. Well known both as a historic site famous from the Three Kingdoms Period and the hometown of Lord Bao, Hefei is a city with a history of more than 2000 years. In modern times, as an important base for science and education in China, Hefei is the first and sole Science and Technology Innovation Pilot City in China, and a member city of WTA (World Technopolis Association). We hope that PCM 2018 is a memorable experience for all participants.

PCM 2018 featured a comprehensive program. We received 422 submissions to the main conference by authors from more than ten countries. These submissions included a large number of high-quality papers in multimedia content analysis, multimedia signal processing and communications, and multimedia applications and services. We thank our Technical Program Committee with 178 members, who spent much time reviewing papers and providing valuable feedback to the authors. From the total of 422 submissions, the program chairs decided to accept 209 regular papers (49.5%) based on at least three reviews per submission. In total, 30 papers were received for four special sessions, while 20 of them were accepted. The volumes of the conference proceedings contain all the regular and special session papers.

We are also heavily indebted to many individuals for their significant contributions. We wish to acknowledge and express our deepest appreciation to general chairs, Meng Wang and Chong-Wah Ngo; program chairs, Richang Hong, Wen-Huang Cheng and Toshihiko Yamasaki; organizing chairs, Xueliang Liu, Yun Tie and Hanwang Zhang; publicity chairs, Jingdong Wang, Min Xu, Wei-Ta Chu and Yi Yu, special session chairs, Zhengjun Zha and Liqiang Nie. Without their efforts and enthusiasm, PCM 2018 would not have become a reality. Moreover, we want to thank our sponsors: Springer, Anhui Association for Artificial Intelligence, Shandong Artificial Intelligence Institute, Kuaishou Co. Ltd., and Zhongke Leinao Co. Ltd. Finally, we wish to thank all committee members, reviewers, session chairs, student volunteers, and supporters. Their contributions are much appreciated.

September 2018

Richang Hong
Wen-Huang Cheng
Toshihiko Yamasaki
Meng Wang
Chong-Wah Ngo

Organization

General Chairs

Meng Wang   Hefei University of Technology, China
Chong-Wah Ngo   City University of Hong Kong, Hong Kong, China

Technical Program Chairs

Richang Hong   Hefei University of Technology, China
Wen-Huang Cheng   National Chiao Tung University, Taiwan, China
Toshihiko Yamasaki   University of Tokyo, Japan

Organizing Chairs

Xueliang Liu   Hefei University of Technology, China
Yun Tie   Zhengzhou University, China
Hanwang Zhang   Nanyang Technological University, Singapore

Publicity Chairs

Jingdong Wang   Microsoft Research Asia, China
Min Xu   University of Technology Sydney, Australia
Wei-Ta Chu   National Chung Cheng University, Taiwan, China
Yi Yu   National Institute of Informatics, Japan

Special Session Chairs

Zhengjun Zha   University of Science and Technology of China, China
Liqiang Nie   Shandong University, China

Contents – Part II

Poster Papers

Poster Papers

Enhancing Low-Light Images with JPEG Artifact Based on Image Decomposition

Chenmin Xu, Shijie Hao$^{(\boxtimes)}$, Yanrong Guo, and Richang Hong

Hefei University of Technology, Hefei, China
hfut.hsj@gmail.com

Abstract. Images shared on the Internet are often compressed into a small size, and thus have the JPEG artifact. This issue becomes challenging for task of low-light image enhancement, as the artifacts hidden in dark image regions can be further boosted by traditional enhancement models. We use a divide-and-conquer strategy to tackle this problem. Specifically, we decompose an input image into an illumination layer and a reflectance layer, which decouple the issues of low lightness and JPEG artifacts. Therefore we can deal with them separately with the off-the-shelf enhancing and deblocking techniques. Qualitative and quantitative comparisons validate the effectiveness of our method.

Keywords: Image enhancement · Low light · JPEG artifact · Retinex model

1 Introduction

With the advance of photographing devices, people always prefer images with clear content details and few artifacts. Nevertheless, this is not always the case in various real-world situations. For example, people nowadays enjoy taking photographs and share them on the Internet. In this process, the quality of an image can be affected by many factors. In the stage of photograph shooting, poor lightness conditions (e.g. nighttime) and amateur shooting skills (e.g. back light) often bring in a dark visual appearance ((e.g. Fig. 1(a))). In the stage of photograph distribution, images are often unintentionally compressed or resized into a smaller size by social network software like WeChat or QQ. Many high-frequency image details can be filtered out in this process and JPEG artifacts (or called block artifacts) are therefore introduced. In this context, for these compressed low-light images, an image enhancement model is expected to have the abilities of lightness enhancement and block removal [1].

However, conventional image enhancement methods [2–5] are limited due to the following reasons. First, these methods are not equipped with the artifact-removing ability. When we directly apply the off-the-shelf low-light enhancement methods on them, the JPEG artifacts can be unnecessarily amplified (e.g. Fig. 1(b)). Furthermore, JPEG artifacts usually hide in the image regions of low contrast, which makes the enhancement task more challenging. For example, compared with the original clean image, the hidden JPEG artifacts are not visually obvious in the compressed images (e.g. Fig. 1(a)).

© Springer Nature Switzerland AG 2018
R. Hong et al. (Eds.): PCM 2018, LNCS 11165, pp. 3–12, 2018.
https://doi.org/10.1007/978-3-030-00767-6_1

(a) A visual comparison between low-light images without (first row) and with (second row) JPEG artifacts (Q=60). We use these six compressed low-light images as our experimental data.

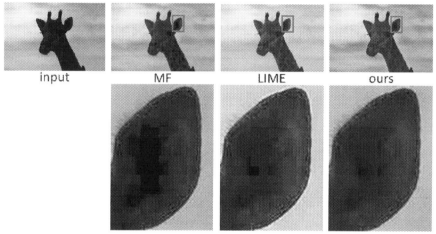

(b) Enhanced results directly based on state-of-the-art low-light enhancement methods (e.g. MF [12] and LIME [5]), and our method (better with a bright screen display).

(c) Enhanced results based on the technical roadmaps of deblock-enhance, enhance-deblock, and our method (better with a bright screen display).

Fig. 1. An illustration of our research background.

Intuitively, we can simply adopt the technical roadmap of deblock-enhance or alternatively enhance-deblock to tackle this problem. However, both of them are still limited. As shown in Fig. 1(c), the deblock-enhance roadmap tends to produce over-smoothed final results, while the enhance-deblock roadmap has difficulties in removing the unnecessarily boosted JPEG artifact. In this paper, we propose a novel framework that simultaneously enhances the lightness and removes the JPEG blocks well (Fig. 2). The key element is to decompose the input image into the illumination layer and the reflectance layer. In this way, the low lightness and JPEG artifact can be well separated, which avoids the risk of artifact-boosting. Our research is highlighted in the proposed framework that well enhances compressed low-light images. Experimental results demonstrate the superior performance over other related methods both qualitatively and quantitatively.

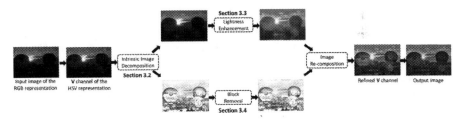

Fig. 2. The flowchart of our method.

2 Related Works

In recent years, many low-light image enhancement methods have been proposed. We can divide the enhancing methods into the single-source group and the multi-source group. The former refers to the methods with a single input image for enhancement, while the later uses multiple images as inputs.

As for single-source low-light image enhancement methods, we further divided into the histogram-based ones and the Retinex-based ones. As the histogram of a low-light image is heavily tailed at low intensities, the histogram-based methods are targeted on reshaping the histogram distribution [2, 6]. Since an image histogram is a global descriptor and drops all the spatial information, these methods are prone to produce over-enhanced [2] or under-enhanced [6] results for local regions. The Retinex-based models assume that an image is composed of an illumination layer representing the light intensity of an object, and a reflectance layer representing the physical characteristics of the object's surface. Based on this image representation, the low-light enhancement can be achieved by adjusting the illumination layer [3, 7, 8]. Differently, Guo et al. [5] propose a simplified Retinex model for the task of low-light image enhancement. Instead of the intrinsic decomposition, they directly estimate a piece-smooth map as the illumination layer. In general, the visual appearance of these single-source methods' results heavily depends on a properly chosen enhancing strength.

Low-light image enhancement methods based on multiple sources can relief the issue of choosing the enhancing parameter, as they adopt the technical roadmap of multi-source fusion. In principle, larger dynamic range can be captured for an imaging scene with multiple images of different exposures. Then these images are fused based on a multi-scale image pyramid [9] or a patch-based image composition [10]. In many cases, however, multiple sources are not available and only one low-quality input image is at hand. To address this challenge, a feasible way is to artificially produce multiple initial enhancements and then fuse them. Ying et al. [12] propose a method simulating the camera response model and generate multiple intermediate images of different exposing levels. Different from [11], multiple initial results with different enhancing models are generated and taken as the fusion sources in [12].

Although all the above methods well address the problem of dark image appearance, they are not equipped with the function of artifact removal. To the best of our knowledge, there are few works that concentrate on the enhancement of compressed low-light images. Li et al. [1] propose to decompose an image into the structure layer and a texture layer. The former layer is for the contrast enhancement; while the later one is for the JPEG block removal. Our method resembles the backbone of [1] but distinguishes itself in the technical details. First, we use a totally different image representation model (illumination-reflectance vs. structure-texture). Second, our enhancement model is also different from the one adopted in [1] (fusion-based vs. histogram-based).

3   Proposed Method

3.1   Overall Framework

Our technical roadmap is to separately solve the low lightness issue and the JPEG artifact issue. The proposed framework is shown in Fig. 2. We first convert the RGB input image into the HSV space. Then the V channel $\mathbf{V}$ of the input image $\mathbf{S}$ is firstly decomposed into two components (Sect. 3.2), i.e. the illumination layer $\mathbf{I}$ and the reflectance layer $\mathbf{R}$. Then we perform low-light enhancement on the illumination layer (Sect. 3.3), and perform the JPEG artifact removal on the reflectance layer (Sect. 3.4). Finally, the output image is obtained by re-combining the refined $\mathbf{I}'$ and $\mathbf{R}'$: $\mathbf{V}_{output} = \mathbf{I}' \cdot \mathbf{R}'$. By replacing the refined $\mathbf{V}_{output}$ with the original $\mathbf{V}$, the final output image $\mathbf{S}_{output}$ can be obtained by converting the HSV representation back into the RGB representation. In another word, we keep the color information of $\mathbf{S}$ unchanged during the whole process.

3.2   Image Decomposition

We decompose the input image $\mathbf{V}$ into the illumination layer $\mathbf{I}$ and the reflectance layer $\mathbf{R}$. Since we only have an observed image $\mathbf{S}$ at hand in real-world applications, the decomposition is an ill-posed task. Additionally, for our task, we further aim to decouple the low lightness and the JPEG artifact.

To meet the above demands, we use an image decomposition model by jointly considering shape, texture and illumination priors [8]. We minimize the following target function:

$$E(\mathbf{I}, \mathbf{R}) = \|\mathbf{V} - \mathbf{I} \cdot \mathbf{R}\|_2^2 + \eta_1 E_s(\mathbf{I}) + \eta_2 E_t(\mathbf{R}) + \eta_3 E_l(\mathbf{I}) \tag{1}$$

Here the first term represents the data fidelity, and the rest terms encode the three priors. The first prior is constructed as:

$$E_s(\mathbf{I}) = u_x \|\nabla_x \mathbf{I}\|_2^2 + u_y \|\nabla_y \mathbf{I}\|_2^2 \tag{2}$$

$$u_x = \left(\left|\frac{1}{\Omega} \sum_\Omega \nabla_x \mathbf{I}\right| |\nabla_x \mathbf{I}| + \varepsilon\right)^{-1}, \quad u_y = \left(\left|\frac{1}{\Omega} \sum_\Omega \nabla_y \mathbf{I}\right| |\nabla_y \mathbf{I}| + \varepsilon\right)^{-1} \tag{3}$$

The minimization of E_s leads to the extracted $\mathbf{I}$ that is consistent with the image structures of $\mathbf{V}$. The second prior is constructed as:

$$E_t(\mathbf{R}) = v_x \|\nabla_x \mathbf{R}\|_2^2 + v_y \|\nabla_y \mathbf{R}\|_2^2 \tag{4}$$

$$v_x = (|\nabla_x \mathbf{R}| + \varepsilon)^{-1}, \quad v_y = (|\nabla_y \mathbf{R}| + \varepsilon)^{-1} \tag{5}$$

The minimization of E_t preserves fine details of $\mathbf{V}$ for the extracted $\mathbf{R}$. The third prior is constructed as

$$E_l(\mathbf{I}) = \|\mathbf{I} - \mathbf{B}\|_2^2 \tag{6}$$

where $\mathbf{B}$ is the maxRGB matrix:

$$B(p) = \max_{c \in \{R,G,B\}} S_c(p) \tag{7}$$

The element in $\mathbf{B}$ represents the possible maximum brightness of a pixel. Through the optimization, the obtained $\mathbf{I}$ is forced to be consistent with the brightness distribution. An alternative optimization strategy is then applied to solve the optimal $\mathbf{I}$ and $\mathbf{R}$.

In summary, the first prior E_s and the third prior E_t enforce the illumination layer to be structure–aware and illumination-aware; the second prior E_t focuses on preserving as many texture details and block artifacts as possible for the reflectance layer. The examples in Fig. 3 validate the effectiveness of the chosen decomposition model. Specifically, we observe that the decomposed $\mathbf{R}$ contains almost all the block artifacts.

3.3   Low-Light Enhancement

We enhance the illumination layer $\mathbf{I}$ by using a fusion framework proposed in [8]. As we only have $\mathbf{I}$ at hand, we have to artificially produce several fusion sources. By noticing that the extracted $\mathbf{I}$ has already been piece-wisely smooth and detail-free,

Fig. 3. Examples of the image decomposition results.

we can simply apply the global contrast enhancement models. For producing the first fusion source $\mathbf{I}_1$, we first use a nonlinear intensity mapping:

$$\mathbf{I}_1(p) = \frac{2}{\pi}\arctan(\lambda\mathbf{I}(p)) \tag{8}$$

$$\lambda = \frac{1 - \mathrm{mean}(\mathbf{I})}{\mathrm{mean}(\mathbf{I})} + 10 \tag{9}$$

Then the enhancement result based on the well-known CLAHE method [3] is taken as the second fusion source $\mathbf{I}_2$. In case of over- or under-enhancement, we choose the original $\mathbf{I}$ as the third fusion source $\mathbf{I}_3$, which plays the role of regularization in the fusion process.

We construct pixel-level weight matrices. We first consider the brightness of $\mathbf{I}_k(k = 1, 2, 3)$:

$$\mathrm{W}_{\mathrm{B}}^k(p) = \exp{-\frac{(\mathbf{I}_k(p) - \beta)^2}{2\sigma^2}} \tag{10}$$

where β and σ represent the mean and standard deviation of the brightness in a natural image in a broadly statistical sense. They are empirically set as 0.5 and 0.25, respectively. When the pixel intensities are far from β, it means that they are possibly over- or under exposed, and the weights should be small. Second, we consider the chromatic contrast by incorporating the H and S channels of $\mathbf{S}$:

$$\mathrm{W}_{\mathrm{C}}^k(p) = \mathbf{I}_k(p) \cdot (1 + \cos(\alpha\mathbf{H}(p) + \varphi) \cdot \mathbf{S}(p)) \tag{11}$$

where α and φ are parameters to preserve the color consistency, and empirically set as 2 and 1.39 π. This weight emphasizes the image regions of high contrast and good colors.

By combining these two weights together and normalizing it, we can obtain the weights for each fusion source:

$$\overline{W}_k(p) = \frac{W_k(p)}{\sum_k W_k(p)} \tag{12}$$

$$W_k(p) = W_B^k(p) \cdot W_C^k(p) \tag{13}$$

To ensure a seamless fusion, we use multi-scale technique based on image pyramids. We first build Laplacian pyramids $L_l\{I_k\}$ for all the fusion sources. By building Gaussian pyramids $G_l\{\overline{W}_k\}$ for the weight matrices, we can fuse $\{I_k\}$ at various scales:

$$L_l' = \sum_k G_l\{\overline{W}_k\}L_l\{I_k\} \tag{14}$$

At last, the enhanced illumination layer I' can be obtained by collapsing the fused pyramid.

3.4 JPEG Artifact Removal

Since the JPEG artifact is divided from the illumination layer I, we can conduct the deblocking in the reflectance layer R. We adopt a simple but effective deblocking model in [1]:

$$\min_{R'(p)} \sum_p \left(R'(p) - R(p)\right)^2 + \mu \sum_{p' \in \mathcal{N}(p)} \left(\nabla R'(p')\right)^2 \tag{15}$$

In this model, the first fidelity term restricts the refined R' from going too far from the original R. The second term is specifically designed for eliminating block edges that are introduced by JPEG compression. By considering the specific pattern of JPEG blocks, we choose a specific neighboring system $\mathcal{N}(p)$ for each location p: $p' \in \mathcal{N}(p)$ refers to pixel positions along the border edges of each 8×8 image patch. By this setting, the optimization process of Eq. 15 concentrates on the JPEG blocks, and tries to preserve the original image structures. The parameter μ acts as a balancing weight between the two terms, and is empirically set as 0.5.

4   Experiments

In this section, we validate our method with qualitative and quantitative comparisons. The experimental images are shown in Fig. 1(a), of which the compression strength is set as $Q = 60$. The MATLAB codes were run on a PC with 8G RAM and 2.6G Hz CPU.

As our task is to enhance the low lightness and suppress the JPEG artifact, we first use two baseline models for comparison. **Baseline 1**: remove the artifact at first and then enhance the low lightness. **Baseline 2**: enhance the low lightness at first and then remove the artifact. For the fairness of the comparison, we use the methods mentioned in Sects. 3.3 and 3.4 to achieve the lightness enhancement and the artifact suppression, respectively. We also compare our method with the most related one proposed in [1] and term it as **ECCV14**. Its parameters are empirically set as the default values as in [1]. As for our methods, since the framework of our method is open for the choice of image enhancement models, we choose two kinds of them. The first is the one mentioned in Sect. 3.3, and we term it as **Ours-MF**. The second is the simple gamma correction process, and we term it as **Ours-GC**. Here we simply set the enhancement parameter γ as $1/2.2$ according to [8]. As for the image decomposition stage of **Ours-GC** and **Ours-MF**, we follow [8] to set $\eta_1 = 0.001$, $\eta_2 = 0.0001$, $\eta_3 = 0.25$, and $\varepsilon = 0.01$.

We first present the visual comparison in Fig. 4. From the two examples, we have the following observations. First, we observe the lost image details in the results of **Baseline 1** and **ECCV14**. The reason is that the image details hidden in the darkness are vulnerable to the JPEG removal process, and many of them are unnecessarily removed. In contrary, the artifact removal of our method is applied on the decomposed reflectance layer that extracts the image contents at various scales in advance. Second, **Baseline 2** well preserves the image details, but removes much fewer block artifacts than **Ours-MF** and **Ours-GC**. Since **Baseline 2** firstly enhances the input image, the originally weak block edges hidden in the darkness are unnecessarily boosted, which brings difficulty for the following suppression step. Our methods do not have this

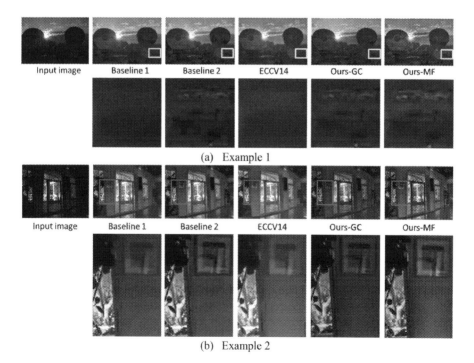

(a) Example 1

(b) Example 2

Fig. 4. Qualitative comparison between our methods and the counterparts.

problem due to the well-separated image layers. Third, by comparing the two versions based on our framework, we can see that **Ours-MF** achieves better results than **Ours-GC** in terms of the lightness condition.

We also quantitatively evaluate all the above methods with two metrics proposed in [13, 14]. The results are shown in Tables 1 and 2, in which the bold/underline numbers indicate the best/second best results among all the methods, respectively. From Table 1, we can see that **Ours-MF** and **Ours-GC** generally achieve better performance than other three methods in terms of removing JPEG artifacts. From Table 2, **Ours-MF** generally has the best performance in terms of image contrast. Differently, the performance of **Ours-GC** is less competitive than **ECCV14**. The reason is that the gamma correction model only imposes a global non-linear transform on the lightness layer. Based on the qualitative and quantitative comparison, **Ours-MF** achieves the best performance, and validates the effectiveness of our idea of decoupling the low lightness issue and the JPEG artifact issue at the beginning. In a word, all the above results validate the effectiveness of each part of our proposed framework.

Table 1. Quantitative comparison based on the metric measuring the block effect [13]

	Input	Baseline1	Baseline2	ECCV14	Ours-MF	Ours-GC
1	0.2109	0.0383	0.0296	0.0276	**0.0119**	0.0366
2	0.1975	0.0307	0.0216	0.0165	**0.0037**	0.0045
3	0.1900	0.0735	0.0472	0.0180	**0.0064**	0.0074
4	0.1941	0.0503	0.0407	0.0232	0.0150	**0.0133**
5	0.1894	0.0284	0.0171	0.0110	**0.0021**	0.0034
6	0.1268	0.0445	0.0195	0.0171	0.0043	**0.0038**
Average	0.1848	0.0443	0.0293	0.0189	**0.0072**	0.0115

The bold font means the best result

Table 2. Quantitative comparison based on the metric measuring the image contrast [14]

	Input	Baseline1	Baseline2	ECCV14	Ours-MF	Ours-GC
1	7.6885	7.1351	7.0683	6.0436	**5.8170**	7.6696
2	6.3567	4.7280	4.0690	4.6849	**3.3641**	4.3921
3	7.1700	5.3822	3.8854	3.5243	**3.1429**	3.7367
4	6.3027	5.0812	4.8454	4.3957	**4.2165**	4.9111
5	6.4311	2.9499	2.9322	**2.5457**	3.1349	3.1649
6	7.2752	3.8666	3.9664	**3.5832**	4.1252	5.3496
Average	6.8707	4.8571	4.4611	4.1296	**3.9668**	4.8707

The underline font means the second best result

5   Conclusions

In this paper, we propose an image enhancement method for compressed low-light images. Based on image decomposition, low lightness and JPEG artifact are separated into two decomposed layers, which can be well addressed by off-the-shelf enhancing techniques. Visual and quantitative comparisons demonstrate the effectiveness of our method. We plan to improve our method by introducing saliency detection [15, 16].

Acknowledgements. The research was supported by the National Nature Science Foundation of China (Grant Nos. 61632007, 61772171, and 61702156), Anhui Provincial Natural Science Foundation (Grant No. 1808085QF188)

References

1. Li, Y., Guo, F., Tan, R.T., Brown, M.S.: A contrast enhancement framework with JPEG artifacts suppression. In: Fleet, D., Pajdla, T., Schiele, B., Tuytelaars, T. (eds.) ECCV 2014. LNCS, vol. 8690, pp. 174–188. Springer, Cham (2014). https://doi.org/10.1007/978-3-319-10605-2_12
2. Reza, A.M.: Realization of the contrast limited adaptive histogram equalization (CLAHE) for real-time image enhancement. J. VLSI Signal Process. Syst. **38**(1), 35–44 (2004)
3. Yue, H., Yang, J., Sun, X., Wu, F., Hou, C.: Contrast enhancement based on intrinsic image decomposition. IEEE TIP **26**(8), 3981–3994 (2017)
4. Li, M., Liu, J., Yang, W., Sun, X., Guo, Z.: Structure-revealing low-light image enhancement via robust Retinex model. IEEE TIP **27**(6), 2828–2841 (2018)
5. Guo, X., Li, Y., Ling, H.: LIME: low-light image enhancement via illumination map estimation. IEEE TIP **26**(2), 982–993 (2017)
6. Lee, C., Lee, C., Kim, C.: Contrast enhancement based on layered difference representation of 2D histograms. IEEE TIP **22**(12), 5372–5384 (2013)
7. Fu, X., Zeng, D., Huang, Y., Zhang, X., Ding, X.: A weighted variational model for simultaneous reflectance and illumination estimation. In: Proceedings of CVPR (2016)
8. Cai, B., Xu, X., Guo, K., Jia, K., Hu, B., Tao, D.: A joint intrinsic-extrinsic prior model for retinex. In: Proceedings of ICCV (2017)
9. Kou, F., Wei, Z., Chen, W., Wu, X., Wen, C., Li, Z.: Intelligent detail enhancement for exposure fusion. IEEE TMM **20**(2), 484–495 (2018)
10. Ma, K., Li, H., Yong, H., Wang, Z., Meng, D., Zhang, L.: Robust multi-exposure image fusion: a structural patch decomposition approach. IEEE TIP **26**(5), 2519–2532 (2017)
11. Ying, Z., Li, G., Gao, W.: A bio-inspired multi-exposure fusion framework for low-light image enhancement. ArXiv, abs/1711.00591 (2017)
12. Fu, X., Zeng, D., Huang, Y., Liao, Y., Ding, X., Paisley, J.: A fusion-based enhancing method for weakly illuminated images. Sig. Process. **129**, 82–96 (2016)
13. Golestaneh, S.A., Chandler, D.M.: No-reference quality assessment of JPEG images via a quality relevance map. IEEE SPL **21**(2), 155–158 (2014)
14. Gu, K., Wang, S., Zhai, G.: Blind quality assessment of tone-mapped images via analysis of information, naturalness, and structure. IEEE TMM **18**(3), 432–443 (2016)
15. Jian, M., Qi, Q., Dong, J., Sun, X., Sun, Y., Lam, K.: Saliency detection using quaternionic distance based weber local descriptor & level priors. MTAP **77**(11), 14343–14360 (2018)
16. Jian, M., Lam, K., Dong, Y., Shen, L.: Visual-patch-attention-aware saliency detection. IEEE Trans. Cybernetics **45**(8), 1575–1586 (2015)

Depth Estimation from Monocular Images Using Dilated Convolution and Uncertainty Learning

Haojie Ma[1,2], Yinzhang Ding[1,2], Lianghao Wang[1,2,3(✉)], Ming Zhang[1,2], and Dongxiao Li[1,2]

[1] College of Information Science and Electronic Engineering, Zhejiang University, Hangzhou 310027, China
wanglianghao@zju.edu.cn
[2] Zhejiang Provincial Key Laboratory of Information Processing, Communication and Networking, Hangzhou 310027, China
[3] State Key Laboratory for Novel Software Technology, Nanjing University, Nanjing, People's Republic of China

Abstract. Depth cues are vital in many challenging computer vision tasks. In this paper, we address the problem of dense depth prediction from a single RGB image. Compared with stereo depth estimation, sensing the depth of a scene from monocular images is much more difficult and ambiguous because the epipolar geometry constraints cannot be exploited. In addition, the value of the scale is often unknown in monocular depth prediction. To facilitate an accurate single-view depth prediction, we introduce dilated convolution to capture multi-scale contextual information and then present a deep convolutional neural network. To improve the robustness of the system, we estimate the uncertainty of noisy data by modelling such uncertainty in a new loss function. The experiment results show that the proposed approach outperforms the previous state-of-the-art methods in depth estimation tasks.

Keywords: Depth estimation · Dilated convolution
Convolutional neural network · Uncertainty

1 Introduction

Depth estimation has been investigated for a long time because of its vital role in computer vision. Some studies have proven that accurate depth information is useful for various existing challenging tasks, such as image segmentation [1], 3D reconstruction [2], human pose estimation [3], and counter detection [5]. Humans can effectively predict monocular depth by using their past visual experiences to structurally understand their world and may even utilize such knowledge in unfamiliar environments. However, monocular depth prediction remains a difficult problem for computer vision systems due to the lack of reliable depth cues.

© Springer Nature Switzerland AG 2018
R. Hong et al. (Eds.): PCM 2018, LNCS 11165, pp. 13–23, 2018.
https://doi.org/10.1007/978-3-030-00767-6_2

Many studies have investigated the use of image correspondences that are included in stereo image pairs [6]. In the case of stereo images, depth estimation can be addressed when the correspondence between the points in the left and right parts of images is established. Many studies have also explored the method of motion [7], which initially estimates the camera pose based on the change in motion in video sequences and then recovers the depth via triangulation. Obtaining a sufficient number of point correspondences plays a key role in the aforementioned methods. These correspondences are often found by using the local feature selection and matching techniques. However, the feature-based method usually fails in the absence of texture and the occurrence of occlusion. Owing to the recent advancements in depth sensors, directly measuring depth has recently become affordable and achievable, but these sensors have their own limitations in practice. For instance, Microsoft Kinect is widely used indoors for acquiring RGB-D images but is limited by short measurement distance and large power consumption. When working outdoors, LiDAR and relatively cheaper millimeter wave radars are mainly used to capture depth data. However, these collected data are always sparse and noisy. Accordingly, there has always been a strong interest in accurate depth estimation from a single image. Recently, CNNs [8] with powerful feature representation capabilities have been widely used for single-view depth estimation by learning the implicit relationship between an RGB image and the depth map. Despite increasing the complexity of the task, the outputs of deep-learning-based approaches [13–17] showed significant improvements over those of traditional techniques [10–12] on public datasets.

In this work, we exploit CNNs to learn strong features for recovering the depth map from a single still image. Given that applying downsampling, upsampling, or deconvolution in a fully convolution network may result in the loss of many cues in the image boundary for pixel-level regression tasks, we introduce the dilated convolution [9] to learn multi-scale information from a single scale image input. Long skip connections are also applied to combine the abstract features with image features. To achieve more robust prediction, we further model the uncertainty in computer vision and proposed a novel loss function to measure such uncertainty during training without labels. The experimental results demonstrate that our proposed method outperform state-of-the-art approaches on standard benchmark datasets [2,21].

2   Related Work

Previous studies have often used probabilistic graphic model and have usually relied on hand-crafted features. Saxena et al. [10] proposed a superpixel-based method for inferring depth from a single still image and applied a multi-scale MRF to incorporate local and global features. Ladicky et al. [11] introduced semantic labels on the depth to learn a highly discriminative classifier. Karsch et al. [12] proposed a non-parametric approach for automatically generating depth. In this approach, the GIST features were initially extracted for the input image and other images in the database, then several candidate depths that correspond to the candidate images were selected before conducting warping and optimization procedures.

Recent studies have employed CNNs to solve the depth prediction problem. Eigen et al. [13] utilized two network stacks to regress depth. The first local network makes a coarse depth prediction for the global contents, while the other network refines the prediction locally. Liu et al. [15] combined CNN with continuous CRF in a unified framework. Wang et al. [16] jointly addressed depth prediction and semantic segmentation by using a common CNN. They proposed a two-layer hierarchical CRF model to refine the coarse network output. Laina et al. [17] proposed a fully convolutional network and introduced a robust loss function called berHu loss. Some unsupervised approaches have also been introduced recently to address the challenges in obtaining a large number of dense and reliable depth labels, especially in outdoor scenes. Garg et al. [18] treated depth estimation as an image reconstruction problem and proposed the use of photometric loss. Given that the loss is not completely differentiable, they performed a first-order Taylor expansion to linearize the results in warp images. Based on [18], Godard et al. [19] considered the left-right disparity consistency constraint and dealt with the warp image by bilinear interpolation.

In this paper, we construct a fully convolutional network for monocular depth prediction. To maintain additional feature information, we apply dilated convolution to enlarge the receptive field without reducing the resolution of the feature maps. Afterward, we implement low-level and high-level information fusion by using long skip connections. Unlike the previous CNNs that are unable to represent, or model uncertainty as probability distributions by using CRF, our network can accurately estimate uncertainty as the model attenuation.

3   Approach

3.1   Network Architecture

We adopt a deep fully convolutional network with an encoder-decoder architecture for single-view depth estimation (see Fig. 1). This network is constructed based on ResNet [4], which performs well in image classification. We remove the last average pooling layer and fully connected layer of the original version. In this way, we discard most of the network parameters and successfully train our model on the current hardware. As low-resolution feature maps contain less boundary information, we employ dilated convolution to expand the receptive field while maintaining the features within an appropriate size. The key components of the decoder part are the two up-sampling layers that are used to recover image resolution. To achieve a higher depth accuracy, we choose the up-projection module proposed in [17] as our up-sampling layer. This module comprises an unpooling layer (which increases the spatial resolution of the feature map) and two convolution layers with residual learning. We concatenate the corresponding feature maps from the encoder and decoder parts by skip connections. Eventually, a convolution is applied to generate the depth prediction.

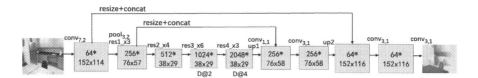

Fig. 1. Model architecture. We use conv$_{n,s}$ to denote a n × n convolution with stride s, and same notation is employed to max pooling pool$_{n,s}$. Let k* be feature maps and @r be the dilation rate. Residual blocks (res1, res2, res3, res4) consist of three convolutions with kernel size 1 × 1, 3 × 3, 1 × 1 and xm is the number of blocks. We replace the convolution by dilated convolution in res3_x6 and res4_x3.

Dilated Convolution. Dilated convolution has been recently proposed to overcome the reduced feature resolution problem caused by the successive pooling and down-sampling layers. Dilated convolution is a regular convolution with a kernel that is dilated by inserting zeros between non-zero values. Compared with standard convolution, dilated convolution can effectively increase the receptive field without increasing the number of parameters. Multi-scale contextual information is also extracted from the original resolution. A dilated convolution is defined as

$$(F *_n k)(\mathbf{p}) = \sum_{s+nt=p} F(\mathbf{s})k(\mathbf{t}), \tag{1}$$

where $F : \mathbb{Z}^2 \to \mathbb{R}$ is a discrete function, n is the dilation rate, $*_n$ is an n-dilated convolution, and $\Omega_r = [-r, r]^2 \cap \mathbb{Z}^2$. Let k : $\Omega_r \to \mathbb{R}$ be a discrete filter of size $(2r + 1)^2$.

3.2   Loss Function with Uncertainty Learning

Two major types of uncertainty can be modeled in deep learning. First, epistemic uncertainty, also known as systematic uncertainty, describes the uncertainty over the model parameters. Second, aleatoric uncertainty, also called statistical uncertainty, represents the inherent noise in the inputs and cannot be decreased no matter how much data are provided. We specifically focus on modelling aleatoric uncertainty, since epistemic uncertainty can be mostly eliminated by using large amounts of data.

To learn aleatoric uncertainty, we measure the variance of noise from the input RGB images. Compared with previous CNNs for depth prediction where the noise parameter σ is replaced by a fixed weight decay, our scheme assumes that the noise is variable for different inputs, since the depth for textureless regions is highly ambiguous. For a predicted depth map $\tilde{y}$ and the corresponding ground truth y, the variance is learned as loss attenuation and we define the new loss as

$$\mathcal{L} = \frac{1}{n} \sum_i^n \tilde{\sigma}_i^{-2} \|y_i - \tilde{y}_i\|_2^2 + \log \tilde{\sigma}_i^2, \tag{2}$$

where i indexes the n pixels over the image, and $\widetilde{\sigma}_i^2$ denotes the variance for pixel i. This loss consists of two components: a residual regression term and an uncertainty regularization term.

In practice, we predict $\widetilde{s}_i := \log \widetilde{\sigma}_i^2$ and

$$\mathcal{L} = \frac{1}{n}\sum_i^n \exp\left(-\widetilde{s}_i\right)\|y_i - \widetilde{y}_i\|_2^2 + \widetilde{s}_i. \tag{3}$$

The loss in Eq. 3 has a better numerical stability than that in Eq. 2 by avoiding division by zero.

The $\mathcal{L}_1 = \|y - \widetilde{y}\|_1$, $\mathcal{L}_2 = \|y - \widetilde{y}\|_2^2$, and berHu loss [17] were separately tested in the experiment, and the results revealed that $\mathcal{L}_1$ outperformed the others in estimating monocular depth. An explicit quantitative analysis is shown in Sect. 4. Therefore, we adopt the $\mathcal{L}_1$ norm instead of $\mathcal{L}_2$ norm as the residual term described in Eq. 3 during training.

4   Experiments

In this section, we perform with a quantitative analysis to test our proposed method. and then compare the performance of this method with other start-of-the-art models on two popular datasets, namely, NYU Depth v2 [21] and Make3D [2].

4.1   Experimental Setup

For the following analysis and evaluation, we implement our architecture by using Tensorflow, and train on a single NVIDIA GeForce GTX 1080Ti with 11GB memory. The weights of the network are initialed by the ResNet-50 model that is pre-trained on ImageNet data [23]. In all experiments, batch size and weight decay are set to 16 and 10^{-4}, respectively. We train the network for approximately 15 epochs on NYU Depth v2 and 20 epochs on Make3D. The starting learning rate is 0.001 and halved every 5 epochs. As for the initial values of the log variances, we set $s = 0.0$.

4.2   Datasets

The NYU Depth v2 dataset [21] contains 120 K unique RGB-D images taken from 464 different indoor scenes with a Kinect camera. We use 249 scenes for training and the other 215 scenes for testing according to the official train/test split. We sample equally spaced frames from each raw training sequence and obtain approximately 12 K RGB-D pairs. The missing depth values are filled in by using the toolbox provided by Silberman et al. [21]. To increase the size and variability of the training set, we perform a data augmentation similar to that in [13], and get roughly 96 K pairs. Following [17], the original frames are downsampled by half and then center-cropped to 304×228. For testing, we use

Table 1. Quantitative analysis of proposed architectures on the official test set of NYU Depth v2. For rel, rms, and $\log_{10}$, a lower is better, for δ_1, δ_2 and δ_3, a higher is better. Results in bold are best.

Architecture	Loss	rel	rms	$\log_{10}$	δ_1	δ_2	δ_3
Baseline	berHu	0.128	0.573	0.055	0.801	0.950	0.985
Ours (dilated convolution)	berHu	0.122	0.565	0.052	0.805	0.953	0.986
Ours (long skip connections)	berHu	0.118	0.560	0.050	0.814	0.955	0.988
Ours (full)	berHu	0.115	0.556	0.049	0.816	0.956	0.988
Ours (full)	$\mathcal{L}_2$	0.130	0.572	0.054	0.799	0.950	0.985
Ours (full)	$\mathcal{L}_1$	0.113	0.553	0.049	0.817	0.956	0.988
Ours (full)	$\mathcal{L}_1$+uncertainty	**0.110**	**0.552**	**0.048**	**0.820**	**0.958**	**0.989**

Table 2. Performance comparison with state-of-the-art methods on the NYU Depth v2 dataset. The values are originally reported by the authors in their respective papers

Method	rel	rms	$\log_{10}$	rms(log)	δ_1	δ_2	δ_3
Li et al. [26]	0.232	0.821	0.094	-	0.621	0.886	0.968
Liu et al. [15]	0.230	0.824	0.095	-	0.614	0.883	0.971
Wang et al. [16]	0.220	0.745	0.094	0.262	0.605	0.890	0.970
Eigen et al. [13]	0.215	0.907	-	0.285	0.611	0.887	0.971
Roy et al. [24]	0.187	0.744	0.078	-	-	-	-
Eigen and Fergus [14]	0.158	0.641	-	0.214	0.769	0.950	0.988
Laina et al. [17]	0.127	0.573	0.055	0.195	0.811	0.953	0.988
Xu et al. [25]	0.121	0.586	0.052	-	0.811	0.954	0.987
Ours	**0.110**	**0.550**	**0.048**	**0.173**	**0.820**	**0.958**	**0.989**

Table 3. Performance comparison with state-of-the-art methods on the Make3D dataset. The values are originally reported by the authors in their respective papers

Method	rel	rms	$\log_{10}$
Liu et al. [20]	0.335	9.49	0.137
Liu et al. [15]	0.314	8.60	0.119
Li et al. [26]	0.278	7.19	0.092
Laina et al. [17]	0.176	4.46	0.072
Xu et al. [25]	0.184	4.38	0.065
Ours	**0.165**	**4.35**	**0.063**

654 images from the labeled part of the dataset. The predictions are resized to 640×480 via bilinear interpolation to evaluate the performance of the model.

Make3D [2] is an outdoor scene dataset that consists of 534 RGB-D pairs, which are separated into 400 training images and 134 test images. Due to the limitations of the hardware, we resize the original images from 1704×2272 to 345×460. During training, RGB images are halved again. We also augment the training data with offline transformations and obtain about 15k samples. Given that the laser scanner has a maximum range of 81 m, we only compute the error for those pixels with a ground-truth depth less than 70 m.

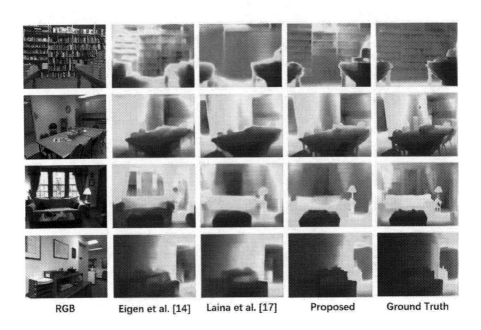

| RGB | Eigen et al. [14] | Laina et al. [17] | Proposed | Ground Truth |

Fig. 2. Qualitative results on the NYU Depth v2 dataset. For fair comparison, all depth predictions shown in color are scaled equally (blue is close and red is far). (Color figure online)

4.3 Evaluation Metrics

To objectively evaluate the performance of our depth estimation model, we use the following metrics:

- mean absolute relative error (rel): $\frac{1}{N}\sum_i \frac{|y_i - \widetilde{y}_i|}{y_i}$;
- root mean square error (rms): $\sqrt{\frac{1}{N}\sum_i (y_i - \widetilde{y}_i)^2}$;
- root mean square log error (rms(log)): $\sqrt{\frac{1}{N}\sum_i (\log y_i - \log \widetilde{y}_i)^2}$;

- mean $\log_{10}$ error ($\log_{10}$): $\frac{1}{N} \sum_i |\log y_i - \log \tilde{y}_i|$;
- δ_j: percentage of $\tilde{y}_i$ s.t. $\max(\frac{\tilde{y}_i}{y_i}, \frac{y_i}{\tilde{y}_i}) < 1.25^j$.

Where y_i and $\tilde{y}_i$ are the ground-truth depth and predicted depth at pixel indexed by i, and N is the number of pixels.

Fig. 3. Qualitative results on the Make3D dataset. We estimate depth for all pixels in the color maps.

4.4   Results

Architecture Evaluation. In this section, we analyze the effects of different architectures and loss functions on depth estimation performance. The quantitative results are shown in Table 1. For an ablation study, we train a baseline network composed of ResNet and up-sampling layers on the NYU Depth v2 dataset (row 1 in Table 1). To demonstrate the effectiveness of dilated convolution, we replace the last two down-sampling regular convolutions with 3×3 dilated convolutions and obtain better results (row 2 in Table 1). Table 1 also shows that long skip connections added to the baseline network can significantly improve the depth estimation performance (row 3, 4). Obviously, multi-scale contextual information fusion is beneficial to depth regression. Moreover, we compare the

$\mathcal{L}_1$, $\mathcal{L}_2$, and berHu loss functions with our proposed architecture. Rows 4 to 6 in Table 1 show that $\mathcal{L}_1$ is greater than both $\mathcal{L}_2$ and berHu. We further combine the $\mathcal{L}_1$ loss with uncertainty learning and achieve the best result as shown in row 7.

Comparison with the State-of-the-Art. Table 2 compares the performance of our method and that of several state-of-the-art approaches on the NYU Depth v2 dataset. Due to the use of dilated convolution, long skip connections, and heteroscedastic uncertainty, our method outperforms other approaches on all metrics. The quantitative results in Fig. 2 illustrate that the proposed model accurately estimates the depth in textureless regions (e.g., walls) and image edges. To demonstrate the generalization ability of the proposed model, we also compare its performance with that of previous related works on the Make3d dataset. Table 3 shows that our model outperforms the other state-of-the-art methods. Additional quantitative examples are provided in Fig. 3.

5   Conclusion

In this paper, we propose a novel approach for solving the monocular depth estimation problem. We introduce a deep residual network with dilated convolution and long skip connections that can aggregate multi-scale contextual information and generate a detailed depth map. By modelling the input-dependent aleatoric uncertainty as learned attenuation, we reduce the effect of noisy data and improve the accuracy of the depth estimation. The experimental results on two benchmark datasets demonstrate that our proposed method outperforms the other state-of-the-art approaches.

Depth information is beneficial for addressing various computer vision problems. In our future work, we plan to examine the application of our depth model to other tasks, such as object detection, semantic segmentation, and simultaneous localization and mapping.

Acknowledgments. This work was supported in part by Zhejiang Provincial Natural Science Foundation of China (Grant No. LY18F010004).

References

1. Long, J., Shelhamer, E., Darrell, T.: Fully convolutional networks for semantic segmentation. In: Proceedings of the IEEE Conference on Computer Vision and Pattern Recognition, pp. 3431–3440. IEEE, Boston (2015)
2. Saxena, A., Sun, M., Ng, A.Y.: Make3D: learning 3D scene structure from a single still image. IEEE Trans. Pattern Anal. Mach. Intell. **31**(5), 824–840 (2009)
3. Taylor, J., Shotton, J., Sharp, T., Fitzgibbon, A.: The Vitruvian manifold: inferring dense correspondences for one-shot human pose estimation. In: Proceedings of the IEEE Conference on Computer Vision and Pattern Recognition, pp. 103–110. IEEE, Providence (2012)

4. He, K., Zhang, X., Ren, S., Sun, J.: Deep residual learning for image recognition. In: Proceedings of the IEEE Conference on Computer Vision and Pattern Recognition, pp. 770–778. IEEE, Las Vegas (2016)

5. Bertasius, G., Shi, J., Torresani, L.: DeepEdge: a multi-scale bifurcated deep network for top-down contour detection. In: Proceedings of the IEEE Conference on Computer Vision and Pattern Recognition, pp. 4380–4389. IEEE, Boston (2015)

6. Scharstein, D., Szeliski, R.: A taxonomy and evaluation of dense two-frame stereo correspondence algorithms. Int. J. Comput. Vis. **47**(1–3), 7–42 (2002)

7. Szeliski, R.: Structure from motion. Computer Vision. Texts in Computer Science, pp. 303–334. Springer, London (2011). https://doi.org/10.1007/978-1-84882-935-7

8. LeCun, Y., et al.: Backpropagation applied to handwritten zip code recognition. Neural Comput. **1**(4), 541–551 (1989)

9. Yu, F., Koltun, V.: Multi-scale context aggregation by dilated convolutions. In: International Conference on Learning Representations, Caribe Hilton, Puerto Rico (2016)

10. Saxena, A., Chung, S.H., Ng, A.Y.: Learning depth from single monocular images. In: International Conference on Neural Information Processing Systems, pp. 1161–1168. MIT Press, Vancouver (2005)

11. Ladicky, L., Shi, J., Pollefeys, M.: Pulling things out of perspective. In: Proceedings of the IEEE Conference on Computer Vision and Pattern Recognition, pp. 89–96. IEEE, Columbus (2014)

12. Karsch, K., Liu, C., Kang, S.B.: Depth extraction from video using non-parametric sampling. In: Fitzgibbon, A., et al. (eds.) ECCV 2012. LNCS, vol. 7576, pp. 775–788. Springer, Heidelberg (2012). https://doi.org/10.1007/978-3-642-33715-4_56

13. Eigen, D., Puhrsch, C., Fergus, R.: Prediction from a single image using a multi-scale deep network. In: International Conference on Neural Information Processing Systems, pp. 2366–2374. MIT Press, Montreal (2014)

14. Eigen, D., Fergus, R.: Predicting depth, surface normals and semantic labels with a common multi-scale convolutional architecture. In: Proceedings of the IEEE International Conference on Computer Vision, pp. 2650–2658. IEEE, Santiago (2015)

15. Liu, F., Shen, C., Lin, G.: Deep convolutional neural fields for depth estimation from a single image. In: Proceedings of the IEEE Conference on Computer Vision and Pattern Recognition, pp. 5162–5170. IEEE, Boston (2015)

16. Wang, P., Shen, X., Lin, Z., Cohen, S., Price, B., Yuille, A.L.: Towards unified depth and semantic prediction from a single image. In: Proceedings of the IEEE Conference on Computer Vision and Pattern Recognition, pp. 2800–2809. IEEE, Boston (2015)

17. Laina, I., Rupprecht, C., Belagiannis, V., Tombari, F., Navab, N.: Deeper depth prediction with fully convolutional residual networks. In: 2016 Fourth International Conference on 3D Vision (3DV), pp. 239–248. IEEE, Stanford (2016)

18. Garg, R., Vijay Kumar, B.G., Carneiro, G., Reid, I.: Unsupervised CNN for single view depth estimation: geometry to the rescue. In: Leibe, B., Matas, J., Sebe, N., Welling, M. (eds.) ECCV 2016. LNCS, vol. 9912, pp. 740–756. Springer, Cham (2016). https://doi.org/10.1007/978-3-319-46484-8_45

19. Godard, C., Mac Aodha, O., Brostow, G.J.: Unsupervised monocular depth estimation with left-right consistency. In: Proceedings of the IEEE Conference on Computer Vision and Pattern Recognition, pp. 6602–6611. IEEE, Honolulu (2017)

20. Liu, M., Salzmann, M., He, X.: Discrete-continuous depth estimation from a single image. In: Proceedings of the IEEE Conference on Computer Vision and Pattern Recognition, pp. 716–723. IEEE, Columbus (2014)

21. Silberman, N., Hoiem, D., Kohli, P., Fergus, R.: Indoor segmentation and support inference from RGBD images. In: Fitzgibbon, A., Lazebnik, S., Perona, P., Sato, Y., Schmid, C. (eds.) ECCV 2012. LNCS, vol. 7576, pp. 746–760. Springer, Heidelberg (2012). https://doi.org/10.1007/978-3-642-33715-4_54

22. Paszke, A., Chaurasia, A., Kim, S., Culurciello, E.: ENet: a deep neural network architecture for real-time semantic segmentation. CoRR abs/1606.02147 (2016). http://arxiv.org/abs/1606.02147

23. Russakovsky, O., et al.: Imagenet large scale visual recognition challenge. Int. J. Comput. Vis. **115**(3), 211–252 (2015)

24. Roy, A., Todorovic, S.: Monocular depth estimation using neural regression forest. In: Proceedings of the IEEE Conference on Computer Vision and Pattern Recognition, pp. 5506–5514. IEEE, Las Vegas (2016)

25. Xu, D., Ricci, E., Ouyang, W., Wang, X., Sebe, N.: Multi-scale continuous crfs as sequential deep networks for monocular depth estimation. In: Proceedings of the IEEE Conference on Computer Vision and Pattern Recognition, pp. 161–169. IEEE, Honolulu (2017)

26. Li, B., Shen, C., Dai, Y., van den Hengel, A., He, M.: Depth and surface normal estimation from monocular images using regression on deep features and hierarchical CRFs. In: Proceedings of the IEEE Conference on Computer Vision and Pattern Recognition, pp. 1119–1127. IEEE, Boston (2015)

Enhancing Feature Correlation
for Bi-Modal Group Emotion Recognition

Ningjie Liu[1], Yuchun Fang[1(✉)], and Yike Guo[1,2]

[1] School of Computer Engineering and Science, Shanghai University, Shanghai, China
{liuningjie,ycfang}@shu.edu.cn
[2] Department of Computing, Imperial College London, London, UK
y.guo@imperial.ac.uk

Abstract. Group emotion recognition in the wild has received much attention in computer vision community. It is a very challenge issue, due to interactions taking place between various numbers of people, different occlusions. According to human cognitive and behavioral researches, background and facial expression play a dominating role in the perception of group's mood. Hence, in this paper, we propose a novel approach that combined these two features for image-based group emotion recognition with feature correlation enhancement. The feature enhancement is mainly reflected in two parts. For facial expression feature extraction, we plug non-local blocks into Xception network to enhance the feature correlation of different positions in low-level, which can avoid the fast loss of position information of the traditional CNNs and effectively enhance the network's feature representation capability. For global scene information, we build a bilinear convolutional neural network (B-CNN) consisting of VGG16 networks to model local pairwise feature interactions in a translationally invariant manner. The experimental results show that the fused feature could effectively improve the performance.

Keywords: Group emotion recognition · B-CNN · Non-local block

1 Introduction

Automatic analysis of group-level image in the wild has received much attention in computer vision in recent years. Several research fields, including emotion recognition, have started to shift their focus from individual to group-level [10, 11]. It has a variety of applications, for example, a car can monitor the emotion of all occupants and engage in additional safety measures. However, it is a very challenging task, due to interactions taking place between various numbers of people, different occlusions and variable illumination conditions in the image.

Researchers have built numerous robust approaches to solve this problem. For features, most researchers only utilized human face information in the image, especially expressions, to predict emotion of the entire group [21]. In [13], the algorithm includes detecting faces using classical Viola-Jones cascades and HOG

© Springer Nature Switzerland AG 2018
R. Hong et al. (Eds.): PCM 2018, LNCS 11165, pp. 24–34, 2018.
https://doi.org/10.1007/978-3-030-00767-6_3

features, detecting facial landmarks, extracting facial features using convolutional neural networks (CNNs) trained for face identification task. Recently, other researchers tried to extract multi-modal features from face, body and scene, such as in [7], and obtained promising performance. Based on human cognitive and behavioral researches [4,8], background/situation and facial expression play a dominating role in the perception of the mood of a group. It is also found that background may provide complementary information to group emotion recognition, especially when there is no face detected in the image. Therefore, in this paper, we mainly extract facial expression features and global scene features for group emotion recognition.

For feature extraction approaches, it can be seen that the approaches using CNNs, have been very successful at image-related tasks in recent years, for its ability to extract good representations from images. In [21], the researchers treated facial expressions as a regression problem in the Valence-Arousal space. And a two-stage fine-tuning was applied on deep CNN while doing transfer learning in [12]. For the EmotiW challenge, several winners utilized deep learning networks for emotion prediction. In [20], multiple deep network training layers are used for emotion prediction. However, convolutional operations build blocks that process one local neighborhood at a time, which may loss correlation information between different positions features and limit classification accuracy. Therefore, it is meaningful to enhance the feature correlation information representation ability for CNNs.

In this paper, we propose a novel bi-modal approach with feature correlation enhancement to address the group emotion recognition problem. For facial expression feature extraction, we plug non-local blocks [18] into Xception network [2] to enhance the feature correlation of different positions in low-level, which can avoid the fast loss of position information of the traditional CNNs and effectively enhance the network's feature representation capability. For global scene information, we extract the global features from the entire image through the novel bilinear embedding [9] VGG16 networks [15]. This architecture can model local pairwise feature interactions in a translationally invariant manner which is useful for global scene feature extraction. The bilinear form simplifies gradient computation and introduces no extra trained parameters, which is more suitable for large-size scene images than non-local blocks. Finally, we combine these features and put them into different classifiers.

2   Proposed Method

In this paper, we mainly extract facial expression features and global scene features for group emotion recognition. The overview of the proposed approach is shown in Fig. 1. We build a bilinear convolutional neural network (B-CNN) consisting of VGG16 networks to extract global scene features and plug non-local blocks into Xception network for facial expression feature extraction.

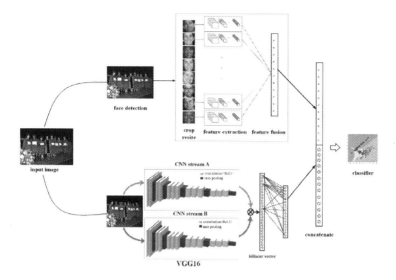

Fig. 1. The overview of the proposed approach. There are two branches: one for facial expression feature extraction and another for global scene feature extraction. All detected faces are cropped from the images and put into the novel Xception network with non-local blocks to extract facial expression features. At the same time, the whole images are put into the B-CNN model consisting of VGG16 networks to extract global scene features. After feature extraction, the features are fused and put into classifiers for 3-categorial classification.

2.1 The Novel Xeption Network with Non-local Blocks

The Xception network is used as our facial expression feature extractor, for its advantage of efficient use of model parameters and excellent performance in image-related tasks. The non-local blocks [18] is proposed as an efficient, simple, and generic component for capturing long-range dependencies with deep neural networks. A non-local operation computes the response at a position as a weighted sum of the features at all positions in the input feature maps, which are applicable for image, sequence, and video problems. The general definition of non-local operations is shown in Eq. (1), where i is the index of an output position whose response is to be computed and j is the index that enumerates all possible positions. x is the input image and y is the output signal of the same size as x. A pairwise function f computes a scalar between i and all j. The function g computes a representation of the input image at the position j. The response is normalized by a factor $C(x)$. And the non-local models are not sensitive to several presented versions of functions f and g in Ref. [18]. The non-local block is defined in Eq. (2), where y_i is given in Eq. (1) and $+x_i$ denotes a residual connection.

$$y_i = \frac{1}{C(x)} \sum_{\forall j} f(x_i, x_j) g(x_i) \tag{1}$$

$$z_i = W_z y_i + x_i \tag{2}$$

The non-local blocks compute the correlation of different positions, which is the important correlation information in expression. Enhancing the feature correlation may improve the model's robustness on the illumination and pose influence for facial expression feature extraction. Thus, we plug non-local blocks into the Xception network to enhance the CNN's ability of correlation information representation between low-level local facial features at different positions. The novel Xception network architecture relies on the small size kernels to capture the local information while using non-local blocks to capture the feature correlation information, complementing each other for better results. The second blocks in the novel Xception network is shown as an example in Fig. 2.

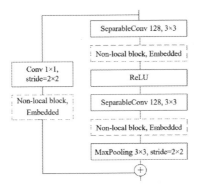

Fig. 2. The second block in the novel Xception network. *Embedded* represents the embedded Gaussian version non-local blocks. The layer marked with red-dotted lines represents adding non-local block after the separable convolutional layers in Xception blocks while the blue-dotted lines represents adding after convolutional layers in the residual connections.

2.2 The Bilinear Embedding VGG16 Networks

The bilinear model consists of two feature extractors whose outputs are multiplied using outer product at each location of the image and pooled to obtain an image descriptor. It is related to the two stream hypothesis of visual processing in the human brain [5]. This architecture can model local pairwise feature interactions in a translationally invariant manner which may provide complementary global information for group emotion recognition. The detail calculation process is as follows.

$$x = \sum_{l \in L} f_A(l, I)^T f_B(l, I) \tag{3}$$

$$y = sign(x)\sqrt{|x|} \tag{4}$$

$$z = \frac{y}{\|y\|_2} \tag{5}$$

Here f_A and f_B are mappings that takes an image I and a location L and outputs a feature of size. In Eq. (3), the feature outputs are combined at each location using the matrix outer product and then the pooling operation is to simply sum all the bilinear features. Finally, Eqs. (4) and (5) are used to normalize features and improve performance. The bilinear form simplifies gradient computation and introduces no extra trained parameters, which is more suitable for large-size scene images than non-local blocks.

3   Experiments

In previous section, we propose two important parts for group emotion recognition, including the novel Xception network to obtain facial expression features of multiple persons and our B-CNN model to obtain global scene information. In this section, we will thoroughly evaluate them on the GAF 2.0 dataset [4] and analyze the experimental results.

3.1   Dataset

The GAF 2.0 dataset [4] was provided as a part of 5th EmotiW challenge [3]. It was collected by searching images related to keywords that describe groups and events from Internet. The dataset was developed to understand attributes which affect the perception of affective state of a group. It consists of three emotion type of group-level images: Positive, Neutral and Negative. The distribution of images in each of train/validation sets are presented in Table 1.

Table 1. The GAF 2.0 dataset.

Emotion	Positive	Neutral	Negative	Total
Train	1272	1199	1159	3630
Validation	773	728	564	2065

3.2   Facial Expression Feature

In order to predict the emotion of a group-level image, we need to detect the emotion of each face in the image at first. The emotions of individual faces in the image would aid us in predicting the final emotion of the image. The face detection library from Yu [19] is used for this purpose. Then all detected faces cropped from 3630 images are resized to 72 × 72 pixels. To get better prediction for every face, we use FER2013 [1] and GENKI-4K dataset [17] to pre-train our novel Xception network firstly. After pre-training, our model can learn facial expression features better. Then all cropped and resized faces are fed into the pre-trained model to fine-tune parameters. And the emotion label of image is used for all the detected faces in this image for training.

After training, both faces from train set and validation set extract features using this model except the last fully-connected layer. And then the facial features for images are obtained by averaging the features of all detected faces in the same image, due to different number of faces in different images. If there is no face in the image, the facial features are all set to 0. Finally, we obtain the 128-dimensional facial expression features for the group-level images.

Without adding non-local block, the model achieves 52.78% for individual face emotion recognition on validation set but achieves 57.47% with 4 non-local blocks adding after the first 4 convolutional/separable convolutional layers of Xception blocks respectively. The accuracy improvement may prove that the facial correlation information of different positions plays a complementary role in emotion recognition. And after averaging all facial features in the same image, the model could achieve 63.73% accuracy. The average features get better results because of that persons have different emotions even in the same situation and humans mainly consider the emotion of the main characters in images to judge the group emotion. The confusion matrices for facial expression feature extraction model are shown in Fig. 3.

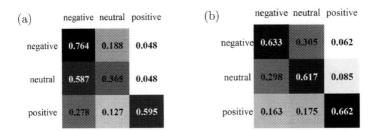

Fig. 3. Confusion matrices for (a) individual face (b) entire image.

Moreover, the adding position and the number of non-local blocks in the Xception network are discussed in our experiments. For positions, only the first 6 convolutional/separable convolutional layers in Xception blocks and the first 4 convolutional layers in residual connections are taking into consideration because too much position relevant information lost at high levels. (FC-n is used to present adding non-local blocks after the first n convolutional/separable convolutional layers in Xception blocks while FR-n presents adding non-local blocks after the first n convolutional layers in residual connections) The results on validation set are shown in Tables 2 and 3. In Table 2, it can be seen that the model achieves the best classification accuracy of FC-4, where the accuracies of all 3 classes are all close to 60% and balanced. In Table 3, the model achieves the best classification accuracy of FR-3. Moreover, in Table 2, the results of FC-5 and FC-6 are worse than FC-4, and the accuracy of FR-4 is worse than FR-3 in Table 3. The reasonable explanation is that gradients may disappear through adding too many non-local blocks, and noise may be introduced to affect the performance. Compared Table 2 with Table 3, the accuracy of applied the same number of non-local blocks to residual connections is almost better than to Xception blocks. It may because that the features before FR-1 are through 3 convolutional layers obtaining more useful information while only 1 convolutional layer before FC-1. Hence, it can be seen that adding non-local block after several convolutional layers at a time is more beneficial to performance than single convolutional layers. For better performance, we select the architecture of FC-4 for facial expression feature extraction.

Table 2. The results for plugged the non-local blocks into the first n convolutional/separable convolutional layers in Xception blocks.

Accuracy(%)	FC-1	FC-2	FC-3	FC-4	FC-5	FC-6
Negative	85.82	41.49	66.67	63.3	95.21	87.06
Neutral	20.52	68.73	42.15	61.71	0	3.31
Positive	60.41	66.75	71.28	66.24	67.27	61.06
Average	55.58	58.99	60.03	**63.75**	54.16	47.84

Table 3. The results for plugged the non-local blocks into the first n convolutional layers in residual connections.

Accuracy (%)	FR-1	FR-2	FR-3	FR-4
Negative	62.06	70.92	64.72	85.11
Neutral	53.58	50.41	54.41	0
Positive	68.69	61.97	70.76	72.96
Average	61.56	60.35	63.35	50.61

3.3   Global Scene Feature

In this section, we aim to see the benefit of scene for group emotion recognition. The CNNs, such as VGG16, VGG19 and ResNet50 [6], pre-trained by ImageNet dataset is used as two all-parameters-shared feature extractors for global scene feature extraction. Then, all 3630 images are fed into the model to fine-tune the parameters in fully-connected layers. The results are shown in Fig. 4.

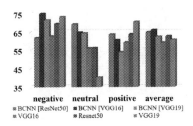

Fig. 4. The classification accuracy (%) for B-CNN consisting of different CNNs.

In comparison a fine-tuned bilinear model consisting of the two same VGG16 networks obtains 66.02% accuracy, outperforming the 59.09% accuracy of the existing VGG16 networks. VGG19 networks and ResNet50 networks are the same as VGG16. For all these three neural networks, the bilinear models perform at least 2.15% better than non-bilinear models. A reasonable explanation is that the local pairwise feature interactions strengthen the model's representation ability, especially for the global information representation. After training by train set data, both images from train set and validation set extract features using the model except the last classify layer. Finally, we obtain the 256-dimensional feature vectors as global scene features for group emotion recognition.

3.4   Feature Fusion

After facial and scene feature extraction, we combine the features and put them into several classifiers. The results for different classifiers are shown in Fig. 5. From Fig. 5, it is obvious that KNN obtains the best accuracy for positive classification and the best average accuracy using the combined features. Moreover, the average accuracies of other classifiers are all better than 58%. Then, we discuss the performance for different k value in KNN classifier. When $k = 119$, our model achieves the best accuracy 71.83% for group emotion recognition, which is 18.86% better than baseline on validation set. And the value of k influence little on classification performance, even the worst accuracy is 7.92% better than the baseline when $k = 3$.

In addition, our proposed method also compares with other methods on the same dataset. The comparison results are shown in Table 4. From Table 4, it can be seen that our average accuracy is 71.83% which is about 18.86% and 13.50% higher than in Ref. [4] with the accuracy (52.97%) and the accuracy

(58.33%) of Ref. [7]. For deep learning methods, the accuracy of our proposed method is about 4.08% and 1.19% higher than in Ref. [16] with the accuracy (67.75%) and the accuracy (70.64%) of Ref. [14]. Above all, it is obvious that our proposed approach evidently outperforms the compared deep learning methods on classification performance. In addition, the proposed method in this paper is simple in model structure and easy to rebuild.

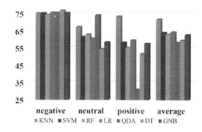

Fig. 5. The classification accuracy (%) using different classifiers. The classifiers include k-nearest neighbor (KNN), support vector machines (SVM), logical regression (LR), random forest (RF), Gaussian naive bayesian (GNB), quadratic discriminant analysis (QDA) and decision tree (DT).

Table 4. The comparison with other methods on GAF 2.0 dataset.

Study	Method	Average accuracy (%) on validation set
Baseline [4]	CENTRIST + SVM	52.97
Huang et al. [7]	INFA	58.33
Surace et al. [16]	CNNs + Bayesian	67.75
Shamsi et al. [14]	ResNet18 + Alexnet	70.64
Our proposed method	**Xception with non-local blocks + B-CNN + KNN**	**71.83**

4   Conclusions

We propose a novel bi-modal approach to address the group emotion recognition problem with feature correlation enhancement. For the facial expression feature extraction, we respectively plug non-local blocks into the Xception network to enhance the correlation information representation between facial features at different position, and obtain 4.69% better performance than without non-local blocks. For the scene information, we extract the global features through our B-CNN model, and obtain 6.93% better performance than without bilinear form, for it models local pairwise feature interactions in a translationally invariant manner. It indicates that the enhanced feature correlation improves the classification performance in group emotion recognition. Finally, our approach achieves 71.83% accuracy by fused these features. In addition, the adding number and position of non-local blocks on Xception networks for image processing are analyzed and there are 2 final conclusions: 1) adding non-local block after several low-level convolutional layers at a time is more beneficial to performance than after single low-level convolutional layers, and 2) adding more non-local blocks may limit the performance rather than fewer.

Acknowledgment. The work is funded by the National Natural Science Foundation of China (No. 61170155) and Shanghai Innovation Action Plan Project (No. 16511101200).

References

1. Arriaga, O., Valdenegro-Toro, M., Plöger, P.: Real-time convolutional neural networks for emotion and gender classification. arXiv preprint arXiv:1710.07557 (2017)
2. Chollet, F.: Xception: deep learning with depthwise separable convolutions. arXiv preprint arXiv:1610.02357 (2016)
3. Dhall, A., Goecke, R., Ghosh, S., Joshi, J., Hoey, J., Gedeon, T.: From individual to group-level emotion recognition: Emotiw 5.0. In: Proceedings of the 19th ACM International Conference on Multimodal Interaction (2017)
4. Dhall, A., Joshi, J., Sikka, K., Goecke, R., Sebe, N.: The more the merrier: analysing the affect of a group of people in images. In: 2015 11th IEEE International Conference and Workshops on Automatic Face and Gesture Recognition (FG), vol. 1, pp. 1–8. IEEE (2015)
5. Goodale, M.A., Milner, A.D.: Separate visual pathways for perception and action. Trends Neurosci. **15**(1), 20–25 (1992)
6. He, K., Zhang, X., Ren, S., Sun, J.: Deep residual learning for image recognition. In: Proceedings of the IEEE Conference on Computer Vision and Pattern Recognition, pp. 770–778 (2016)
7. Huang, X., et al.: Analyzing the affect of a group of people using multi-modal framework. arXiv preprint arXiv:1610.03640 (2016)
8. Kelly, J.R., Barsade, S.G.: Mood and emotions in small groups and work teams. Organ. Behav. Hum. Decis. Process. **86**(1), 99–130 (2001)
9. Lin, T.Y., RoyChowdhury, A., Maji, S.: Bilinear CNN models for fine-grained visual recognition. In: Proceedings of the IEEE International Conference on Computer Vision, pp. 1449–1457 (2015)
10. Mou, W., Celiktutan, O., Gunes, H.: Group-level arousal and valence recognition in static images: face, body and context. In: 2015 11th IEEE International Conference and Workshops on Automatic Face and Gesture Recognition (FG), vol. 5, pp. 1–6. IEEE (2015)
11. Mou, W., Gunes, H., Patras, I.: Alone versus in-a-group: a comparative analysis of facial affect recognition. In: Proceedings of the 2016 ACM on Multimedia Conference, pp. 521–525. ACM (2016)

12. Ng, H.W., Nguyen, V.D., Vonikakis, V., Winkler, S.: Deep learning for emotion recognition on small datasets using transfer learning. In: Proceedings of the 2015 ACM on International Conference on Multimodal Interaction, pp. 443–449. ACM (2015)

13. Rassadin, A.G., Gruzdev, A.S., Savchenko, A.V.: Group-level emotion recognition using transfer learning from face identification. arXiv preprint arXiv:1709.01688 (2017)

14. Shamsi, S., Rawat, B.P.S., Wadhwa, M.: Group affect prediction using emotion heatmaps and scene information. arXiv preprint arXiv:1710.01216 (2017)

15. Simonyan, K., Zisserman, A.: Very deep convolutional networks for large-scale image recognition. arXiv preprint arXiv:1409.1556 (2014)

16. Surace, L., Patacchiola, M., Sönmez, E.B., Spataro, W., Cangelosi, A.: Emotion recognition in the wild using deep neural networks and bayesian classifiers. arXiv preprint arXiv:1709.03820 (2017)

17. The MPLab GENKI Database, GENKI-4K Subset. http://mplab.ucsd.edu

18. Wang, X., Girshick, R., Gupta, A., He, K.: Non-local neural networks. arXiv preprint arXiv:1711.07971 (2017)

19. Yu, S.: Libfacedetection. https://github.com/ShiqiYu/libfacedetection

20. Yu, Z., Zhang, C.: Image based static facial expression recognition with multiple deep network learning. In: Proceedings of the 2015 ACM on International Conference on Multimodal Interaction, pp. 435–442. ACM (2015)

21. Zhang, L., Tjondronegoro, D., Chandran, V.: Representation of facial expression categories in continuous arousal-valence space: feature and correlation. Image Vis. Comput. **32**(12), 1067–1079 (2014)

Frame Rate Conversion Based High Efficient Compression Method for Video Satellite

Xu Wang[1,2], Ruimin Hu[1,2(✉)], and Jing Xiao[1,3]

[1] National Engineering Research Center for Multimedia Software,
School of Computer Science, Wuhan University, Wuhan, China
wangxu9191@gmail.com,hurm1964@gmail.com,jing@whu.edu.cn
[2] Hubei Key Laboratory of Multimedia and Network Communication Engineering,
Wuhan University, Wuhan, China
[3] Collaborative Innovation Center of Geospatial Technology, Wuhan, China

Abstract. Video transmission from satellites to ground devices usually requires a large amount of channel resource due to the huge size of satellite video. Subject to limited computation capability and transmission bandwidth in space environment, the video encoder for video satellite calls for higher coding efficiency. In this paper, we propose a high efficiency satellite video compression method based on frame rate conversion. We firstly down-sample frame rate of satellite video prior to encoding, and then adopt frame interpolation to recover its original frame rate after decoding in the ground. Furthermore, we raise a novel frame interpolation method via fusion of phase-based and region-based method to retain the naturalness of interpolated frames. Experiments show that our proposed coding scheme achieves higher efficiency than H.264 and HEVC in terms of rate-distortion performance. The proposed frame interpolation method is also verified to be more accurate than state-of-the-art methods.

Keywords: Video satellite · Video coding · Frame interpolation
Motion estimation

1 Introduction

Video satellite has been developed rapidly in recent years, whose staring video mode is widely used to provide continuous dynamic video. The fact that frame rate and image size of video captured by satellite have been becoming higher than ever leads to increasing amount of data. Due to the constraint of rare computation capability and bandwidth between satellite and ground, the video

The research was supported by the National Nature Science Foundation of China under Contracts 61671336, 61671332, U1736206, and the Open Research Fund of State Key Laboratory of Information Engineering in Surveying, Mapping and Remote Sensing, Wuhan University under Contract 17E03.

R. Hong et al. (Eds.): PCM 2018, LNCS 11165, pp. 35–44, 2018.
https://doi.org/10.1007/978-3-030-00767-6_4

compression is confronted with a huge challenge. At present, specific coding framework tuned for staring satellite video has not been exploited. Most of video satellites are using regular video encoders, like H.264 [11], HEVC [8], to compress data in the space.

However, the employment of common video encoder in satellite may cause some problems. On one hand, due to the complicated coding tools adopted by the common video encoders, the video encoding task is becoming more and more time-consuming. But computation capability in the satellite is constricted because of limited power, radiation, heat dissipation and so on. On the other hand, videos captured in the special scene will have its own characteristics of temporal and spatial redundancy, which cannot be fully exploited by common encoder designed for general purpose.

To promote the coding efficiency while maintain the computational complexity, we propose a video compression method for satellite based on temporal frame interpolation (TFI). The TFI algorithm performs the task of inserting frames to up convert frame rate. Currently, the state-of-the-art TFI methods can be roughly divided into space domain and frequency domain. In the space domain, the most advanced method is motion-compensated frame interpolation (MCFI), employing motion estimation (ME). A variety of ME methods arise from block-matching algorithm [5]. Jeon [3] proposed an unidirectional method to estimate motion vector. Kang [4] proposed a weighted index-based bidirectional motion compensated method to reduce artifacts. Tai [9] improved the accuracy of motion estimation by using multi-pass ME strategy along with variable block-sizes. However, because regular blocks inevitably contain pixels of different objects with different motion vectors, it will cause inaccuracy of estimated results. Lim [5] proposed a motion estimation method by segmenting frames into arbitrary-shaped regions. The frequency-domain methods are based on the assumption that subtle motions in the space correspond to the phase shift of phase spectrum. Following this idea, Wadhwa [10] proposed a phase-based method to magnify motions by using complex steerable pyramids. Meyer [6] proposed a frame interpolation method by interpolating phase of pixels.

In this work, we proposed a high efficient compression method for satellite video based on frame rate conversion. We first down-convert frame rate in the encoder to reduce the quantity of encoded frames. And then, we reconstruct lost frames using the proposed TFI method after videos are received and decompressed by the ground decoders. Therefore, the computation complexity will be shifted from satellite where the computation resource is expensive to ground devices. Meanwhile, the bit rate is saved greatly under the condition that the frame rate is still recovered.

Moreover, to get an accurate frame interpolation, we propose a novel TFI method via fusion of region-based and phase-based method. Considering the characteristic of staring satellite video that motionless background takes a large proportion of a satellite imagery, there is no need to segment background to get tinier regions. Therefore, we employ background model method to obtain more accurate motion foreground regions, and then carry out motion estimation and interpolation for extracted regions. Due to the fact that subtle motions of tiny regions like cars are hard to be detected by region-based method and large motions will cause serious blur in phase-based method, we further perform fusion on both results to obtain an increasing accuracy.

The rest of the paper is organized as follows. Section 2 presents the proposed satellite video compression method and frame interpolation method. Experimental results are shown in Sect. 3. Finally, we conclude in Sect. 4.

2   Proposed Method

2.1   Compression Framework

Figure 1 gives an outline of the proposed high efficient satellite video compression method. We first down sample original video into a low frame rate video. And the sampled frames in low frame rate video are extracted by equal intervals. Next the low frame rate video is compressed and transmit to the ground receiver. In the ground, the sampled frames are reconstructed by video decoder. Lastly, the frames lost in the down sample process are reconstructed by proposed TFI method to recover original frame rate. The proposed TFI method is described in Sect. 2.2.

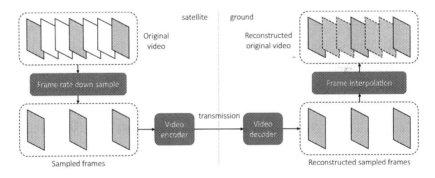

Fig. 1. Outline of the proposed compression framework for satellite video.

2.2   Frame Interpolation

The proposed TFI algorithm combine phase-based and region-based method to interpolate middle frames. Figure 2 shows the framework of proposed method. The input is a pair of adjacent frames in down sampled frame sequence. The left flow is phase-based method to handle small regions with subtle motions. And the right process is region-based method which is suitable for large motion regions. In the end, the interpolation results will be fused according to the motion scale of regions.

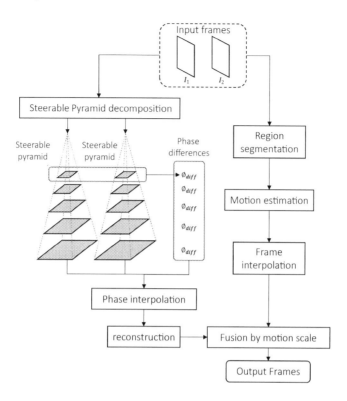

Fig. 2. Framework of the proposed TFI method based on fusion of phase-based and region-based method.

Phase-Based Method. Phase-based method is based on the insight that the motion of certain signals can be represented as phase-shift. Figure 2 left side describe the process of phase-based method.

Firstly, steerable pyramid decomposition is applied to obtain complex steerable pyramids [7] of input images. The complex-valued pyramid include local amplitude and phase which we exploit to process motion in frequency domain. The frequency domain transfer function is described as

$$
\begin{aligned}
\tilde{R}_{\omega,\theta}(x,y) &= (\tilde{I} * \Psi_{\omega,\theta})(x,y) \\
&= A_{\omega,\theta}(x,y)e^{i\Phi_{\omega,\theta}(x,y)} \\
&= C_{\omega,\theta}(x,y) + iS_{\omega,\theta}(x,y).
\end{aligned}
\tag{1}
$$

The $\Psi_{\omega,\theta}$ are scaled and rotated copies of a basic steerable filter. The $\tilde{I}$ are the image signals in different band level. And the $\tilde{R}_{\omega,\theta}$ is complex-valued response contains amplitude and phase.

Then the local phase is obtained by real and imaginary parts in Eq. 1. The calculation process of local phase is described as

$$
\phi_{\omega,\theta}(x,y) = \arctan(S_{\omega,\theta}(x,y)/C_{\omega,\theta}(x,y)),
\tag{2}
$$

where $C_{\omega,\theta}$ is a cosine function and $S_{\omega,\theta}$ is a sine function.

Next, phase difference between phase of two input images in each layer of pyramids is obtained by

$$\phi_{diff} = \text{atan2}(\sin(\phi_1 - \phi_2), \cos(\phi_1 - \phi_2)), \tag{3}$$

where atan2 is the four-quadrant inverse tangent function.

After phase difference is obtained, we can use a weight coefficient $\alpha \in (0,1)$ to interpolate phase between two input images. The interpolated phase is represented as

$$\phi_\alpha = \phi_1 + \alpha\phi_{diff}. \tag{4}$$

The α describes an intermediate position between two frames. In the video compression framework, α corresponding to the down sample times of frame rate.

Finally, the intermediate frame is reconstructed from a new steerable pyramid comprised by interpolated phase and amplitude. And the amplitude is obtained by linear interpolation.

Region-Based Method. This section introduce a region-based method to interpolate middle frames. Generally, a region-based interpolation method include three steps: region segmentation, motion estimation and frame interpolation. Then we describe the method by following these steps.

Region Segmentation. For staring satellite video, the moving regions are foreground in a short time, and the background are motionless. Therefore, it's no need to segment background to get small regions. Thus, we use GMM [12] background model method to segment background and moving foreground regions. Figure 3 shows the example.

Motion Estimation. In this step, the proposed method estimates motion vectors for foreground regions as

$$\mathbf{v}(\mathbf{r}) = \arg \min_{\mathbf{v}} \text{SAD}(\mathbf{r}, \mathbf{v}), \tag{5}$$

where $\mathbf{r}$ is a foreground region in current frame, and $\mathbf{v} = (v_x, v_y)$ means the motion vector. The $\text{SAD}(\mathbf{r}, \mathbf{v})$ is the sum of absolute difference which is defined as

$$\text{SAD}(\mathbf{r}, \mathbf{v}) = \sum_{(x,y)\in\mathbf{r}} |I_t(x, y) - I_{t-1}(x + v_x, y + v_y)|. \tag{6}$$

Frame Interpolation. This step is to interpolate middle frames between two frames. Firstly, background regions of frame I_{t-1} and I_t are fused to obtain a new background region for middle frame $I_{t-1+\alpha}$. And for the foreground regions the value of pixel is interpolated as

$$I_{t-1+\alpha}(x, y) = (1 - \alpha)I_{t-1}(x + \alpha v_x, y + \alpha v_y) \\ + \alpha I_t(x - (1 - \alpha)v_x, y - (1 - \alpha)v_y). \tag{7}$$

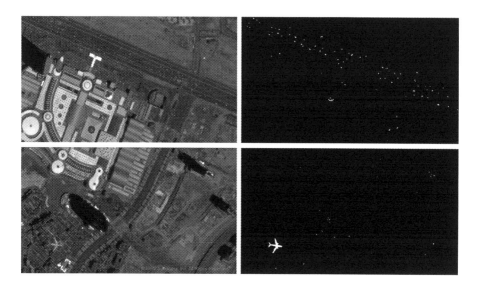

Fig. 3. Example of segmented foreground regions.

Figure 4 gives a visualized description of this process, where the red arrow represent a motion vector of region from I_t to I_{t-1}. And the interpolated frame $I_{t-1+\alpha}$ is between them donated by coefficient α. Correspondingly, the blue and yellow arrows represent motion vectors of interpolated region to regions in I_{t-1} and I_t. Therefore, the pixel value of foreground region in frame $I_{t-1+\alpha}$ is obtain by Eq. 7.

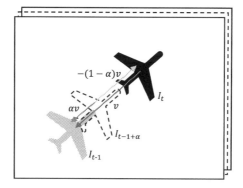

Fig. 4. Interpolation process of foreground regions.

Fusion. Lastly, the proposed method fuses phase-based and region-based method together to get a more accurate and stable result. The value of foreground regions are fused by

$$r_i(x,y) = e^{-\lambda|\mathbf{v}_i|}r_i^P(x,y) + (1 - e^{-\lambda|\mathbf{v}_r|})r_i^R(x,y), \tag{8}$$

where r_i^P and r_i^R are interpolation results of the i-th foreground region by phase-based and region-based method. $\mathbf{v}_i$ is the motion vector of the region, and λ is a constant coefficient in the experiment.

3 Experiments Evaluation

In the experiment, we employ four staring satellite video sequences to make comparison. The sequences include two captured by Skybox and two captured by China Jilin-1. Each sequence has 300 frames, and the sequences from each satellite include one 1080p and one 720p in resolution. Four sequences are named as S1, S2, J1, J2. Example frames are shown in Fig. 5.

3.1 Satellite Video Coding Comparisons

Two state-of-the-art coding scheme H.264 and HEVC are selected for comparison with proposed method. The compared codecs of H.264 and HEVC are x264 and x265, which are most advanced codecs. In the down sample step of the proposed

Fig. 5. Example frames of four staring satellite videos. The top two are captured by Skybox, the bottom two are captured by Jilin-1. And the left column are 1080p resolutions, right column are 720p resolutions.

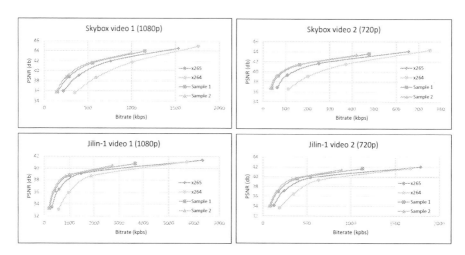

Fig. 6. Rate-distortion performance comparison for 4 test sequences. The sample 1 is the result by interval 1 in down sample, and the sample 2 is the result by interval 2 in down sample.

method, we use difference sample intervals, 1 and 2, to make compression for comparison. And the experiment of the proposed method is based on x265.

Figure 6 shows the rate distortion (R-D) of 4 test sequences. Compared with x264 and x265 codec, the proposed method gains a higher PSNR under the same bitrate, and the bitrate decrease obviously under the same PSNR. In the proposed method, sample 2 gains a better performance than proposed 1, Although the enhancement is not significant. Specifically, coding efficiency enhancement is evaluated by BD-Rate and BD-PSNR [1,2], as Table 1 shows.

Table 1. The BD-Rate (%) and BD-PSNR (dB) of four sequences compared with HEVC and H.264. And our method samples sequence in interval 1 and interval 2.

	S1		S2		J1		J2		Average	
Sample 1 vs.	dB	%	dB	%	dB	%	dB	%	dB	%
HEVC	1.01	−22.12	0.90	−29.94	0.66	−22.92	0.84	−24.89	0.85	−24.97
H.264	2.96	−48.77	2.46	−59.38	2.00	−55.20	2.15	−50.30	2.39	−53.41
Sample 2 vs.	dB	%	dB	%	dB	%	dB	%	dB	%
HEVC	1.77	−33.10	1.42	−40.08	1.06	−32.18	1.08	−29.22	1.33	−33.65
H.264	3.99	−54.74	3.16	−63.94	2.69	−60.21	2.62	−53.44	3.12	−58.08

3.2   Frame Interpolation Comparison

In this section, the proposed TFI method is compared with Meyer [6], Kang [4] and linear method by using objective and subjective quality evaluations. Table 2 shows the average PNSR of interpolated frame in four video sequences by different methods, and the proposed method gains a better performance that the others. Figure 7 shows the subjective comparison of methods. Through the detail views of interpolated frames, the proposed method introduce the least artifacts.

Table 2. PSNR comparison of methods (dB)

	Meyer [6]	Kang [4]	Linear	Proposed
Skybox 1	50.77	50.58	48.46	50.96
Skybox 2	51.65	51.14	50.82	51.74
Jilin-1 1	40.43	40.80	39.71	41.05
Jilin-1 2	42.38	42.78	41.62	43.12

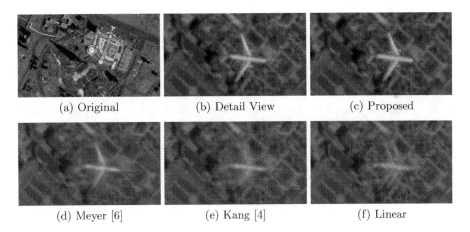

(a) Original   (b) Detail View   (c) Proposed

(d) Meyer [6]   (e) Kang [4]   (f) Linear

Fig. 7. Subjective comparison of interpolated frames by different methods. (a) is original frame. (b) is amplified view of red rectangle in (a). (c) is result of proposed method, and (d) (e) (f) are results of comparison methods.

4   Conclusion and Future Work

This paper has proposed a novel coding framework specific for satellite staring video by incorporating the frame subsampling and interpolation modules into video encoder. Particularly, we boost frame rate interpolation performance by fusing region-based and phase-based method. Experimentally, this method turns out to promote coding efficiency compared with standard HEVC, and meanwhile transfer computation complexity from satellite to ground successfully. In the future work, we will further explore unequal interval down-sampling method as motion redundancy varies among frames. Large interval and more accuracy interpolation method also need to be developed to improve coding efficiency.

References

1. Bjøntegaard, G.: Improvements of the BD-PSNR model (document ITU-T SC16/Q6, DocVCEG-AI11 2008)
2. Bjøntegaard, G.: Calculation of average PSNR differences between RD-curves (ITU-T SG16/Q6, Doc VCEG-M033, Austin, TX, April 2001)
3. Jeon, B.W., Lee, G.I., Lee, S.H., Park, R.H.: Coarse-to-fine frame interpolation for frame rate up-conversion using pyramid structure. IEEE Trans. Consum. Electron. **49**(3), 499–508 (2003)
4. Kang, S.J., Cho, K.R., Kim, Y.H.: Motion compensated frame rate up-conversion using extended bilateral motion estimation. IEEE Trans. Consum. Electron. **53**(4), 1759–1767 (2008)
5. Lim, H., Park, H.W.: A region-based motion-compensated frame interpolation method using a variance-distortion curve. IEEE Trans. Circuits Syst. Video Technol. **25**(3), 518–524 (2015)
6. Meyer, S., Wang, O., Zimmer, H., Grosse, M., Sorkinehornung, A.: Phase-based frame interpolation for video. In: IEEE Conference on Computer Vision and Pattern Recognition, pp. 1410–1418 (2015)
7. Simoncelli, E.P., Freeman, W.T.: The steerable pyramid: a flexible architecture for multi-scale derivative computation. In: International Conference on Image Processing, p. 3444 (1995)
8. Sullivan, G.J., Ohm, J.R., Han, W.J., Wiegand, T.: Overview of the high efficiency video coding (HEVC) standard. IEEE Trans. Circuits Syst. Video Technol. **22**(12), 1649–1668 (2013)
9. Tai, S.C., Chen, Y.R., Huang, Z.B., Wang, C.C.: A multi-pass true motion estimation scheme with motion vector propagation for frame rate up-conversion applications. J. Disp. Technol. **4**(2), 188–197 (2008)
10. Wadhwa, N., Rubinstein, M., Freeman, W.T.: Phase-based video motion processing. ACM Trans. Graph. **32**(4), 1–10 (2013)
11. Wiegand, T., Sullivan, G.J., Bjontegaard, G., Luthra, A.: Overview of the H.264/AVC video coding standard. IEEE Trans. Circuits Syst. Video Technol. **13**(7), 560–576 (2003)
12. Zivkovic, Z.: Improved adaptive Gaussian mixture model for background subtraction. In: International Conference on Pattern Recognition, pp. 28–31 (2004)

Spectral-Spatial Hyperspectral Image Classification via Adaptive Total Variation Filtering

Bing Tu[✉], Jinping Wang, Xiaofei Zhang, Siyuan Huang, and Guoyun Zhang

School of Information Science and Technology, Hunan Institute of Science and Technology, Yueyang, China
tubing@hnist.edu.cn

Abstract. It is unavoidable that existing noise interference in hyperspectral image (HSI). In order to reduce the noise in HSI and obtain a higher classification result, a spectral-spatial HSI classification via adaptive total variation filtering (ATVF) is proposed in this paper, which consists of the following steps: first, the principal component analysis (PCA) method is used for dimension reduction of HSI. Then, the adaptive total variation filtering is performed on the principal components so as to reduce the sensitiveness of noise and obtain a coarse contour feature. Next, the ensemble empirical mode decomposition is used to decompose each spectrum band into serial components, the characteristics of HSI can be further integrated in a transform domain. Finally, a pixel-level classifier (such as SVM) is used for classification of the processed image. The paper analyzes the effect of different parameters of ATVF method on the classification performance in detail, tests the proposed algorithm on the real hyperspectral data sets, and finally verifies the superiority of the proposed algorithm based on a contrastive analysis of different algorithms.

Keywords: Hyperspectral image · PCA · Adaptive total variation
Ensemble empirical mode decomposition · SVM

1 Introduction

Hyperspectral imaging greatly improve human's cognitive ability toward land cover objects with the high spectral resolution and wide spectral coverage of hyperspectral images. Hyperspectral images contain hundreds of spectral bands, which help in finding detailed spectrum characteristics hiding within a narrow

This work was supported the National Natural Science Foundation of China under Grant 51704115, by the Key Laboratory Open Fund Project of Hunan Province University under Grant 17K040 and 15K051, and by the Science and Technology Program of Hunan Province under Grant 2016TP1021.

R. Hong et al. (Eds.): PCM 2018, LNCS 11165, pp. 45–56, 2018.
https://doi.org/10.1007/978-3-030-00767-6_5

spectrum. Due to this advantage, Hyperspectral remote sensing plays an important role in various applications such as ocean exploration, forest monitoring, geological prospecting, modern military and modern agriculture.

Although HSI can provide abundant spectral information, the "Hughes" phenomenon is introduced unavoidably because of its' high spectral dimension. To solve this problem, many different types of advanced machine learning methods such as SVM [1] method have been proposed. However, the computing cost is dramatically increased due to the high spectral dimension. In recent years, Spectral-spatial feature extraction have been well researched. Benedkitsson et al. proposed a hyperspectral feature extraction method based on morphological filtering [2]. Mura et al. proposed the extended morphological attribute profiles and extended profiles with morphological attribute filters for the analysis of hyperspectral [3]. Prasad et al. [4] applied redundant discrete wavelet transform (RDWT) to extract image features. Compared with wavelet transform, a nonlinear signal processing method named empirical mode decomposition (EMD) [5] become more effective. However, if the physical criteria for the differences between two signals are not met, the sifting process derives an IMF with single tone modulated in amplitude instead of a superposition of two unimodular tones. Thus, the modulated signal would no longer encompass the characteristics of the original signals. To overcome the problem of mode mixing, Wu and Huang proposed ensemble empirical mode decomposition (EEMD) [6], which can also obtain a interesting performance as well.

Recently, many Total Variation (TV) methods [7] have been using in digital image processing which showed great effect to image denoising. Meanwhile, the solution of TV model always attracted the attention by researchers [8]. On this basis, Liu et al. [9] proposed adaptive total variation (ATV) model for image denoising with fast solving algorithm, which used the edge detection filters to choose the parameters adaptively. And then, this model combined shock filter with anisotropic diffusion to preprocess the noisy images. Due to the reason that the ATV model could keep the balance between noises smoothing and edges preserving adaptively, it has been widely used in low-dosed X-ray cone beam computed tomography (CBCT), compression perception (CP) and sparse reconstruction.

Considering that ATV has a fast solving convergence rate, which can effectively reduce the sensitivity about noise to denoise smoothly. And thus, the ATV is chosen to obtain a coarse contour feature. However, in the aspect of hyperspectral image feature selection, ATV image just selects the high frequency part which reflects the details of the image intuitively, and the selected image features do not contain low frequency grayscale change information. To solve this problem, a spectral-spatial hyperspectral classification method named adaptive total variation filtering (ATVF) is proposed in this paper. First, the PCA is used to reduce the spectral dimension of a HSI. Then, adaptive total variation is performed on the principal components obtained above. Next, the EEMD is used to decompose the spectrum into serial components and employ these components to improve the performance of spectral discrimination. Finally, a widely used pixel-level classifier, i.e., the SVM classifier, is used to classify the data. Experimental results are performed on several real hyperspectral data sets, in which we compare the proposed method with several widely used hyperspectral classification methods. The proposed method shows outstanding classification performances in terms of several quality indexes, such as overall accuracy and Kappa coefficient.

2   Related Work

2.1   Ensemble Empirical Mode Decomposition

Empirical mode decomposition (EMD) is an new adaptive signal processing method, which makes significant contributions in many applications. After that, a more sufficient method named the ensemble empirical mode decomposition (EEMD) is proposed, which adds a uniform distribution of white noise to signals before decomposition to reduce the effect of the mode mixing in the EMD process. For EEMD, the ratio p of the added white noise and the number of signals in the ensemble must be predetermined. The EEMD is composed of different intrinsic mode function (IMF) and all the IMFs must satisfy the following conditions:

- The local extreme points and the number of zero points of the function must be equal in the whole time range, or the maximum difference is one.
- The average of the upper envelope determined by the local maximum and the average of the lower envelope determined by the local minimum must be zero at any time.

Suppose $X_{l,m}^n(i, j)$ represents value about the nth iteration of the lth band of the mth IMF. To achieve the IMFs, the iterative process is begin from the first dimension data, and the process of screening is summarized as follows:

(1) Search all the local maxima and minimum points of each input band.
(2) Obtain the upper envelope $E_{max}(i, j)$ and the lower envelope $E_{min}(i, j)$ by the interpolating the maxima and minimum respectively.
(3) Calculate the mean of the upper and lower envelopes $Z_M^{(N)}(i, j)$ as follows:

$$Z_M^{(N)}(i, j) = \frac{(E_{max}(i, j) + E_{min}(i, j))}{2} \tag{1}$$

(4) Subtract the envelope mean from the input signal $D_M^{(N)}(i, j) = X_{l,m}^{(n)}(i, j) - Z_m^{(n)}(i, j)$
(5) Repeat (1)–(4) until the envelope signal fulfills the current IMF, i.e., $SD < \tau$ and IMF converges.

$$SD = \frac{max(abs(Z_m^{(n)}(i, j)))}{max(abs(R_m(i, j)))} < \tau \tag{2}$$

(6) Repeat (1)–(5) to generate a residual $R_m(i, j) = R_m(i, j) - IMF_{l,m}(i, j)$, this EMD program will terminate if the residual amount does not contain an extreme value.

To sum up, the spectral original band image $X_l(i,j)$ is obtained. The reconstruction of the IMF function and the residual quantity is represented as follows:

$$X_l(i,j) = \sum_{m=1}^{M} IMF_{l,m}(i,j) + R_m(i,j) \tag{3}$$

2.2 Fast ATV Algorithm

Using the proposed rapid solution algorithm to solve the Fmodel. Set $u^1 = f$, $b_x^0 = b_y^0 = 0$ ($k = 1,2,...$), a fast iterative algorithm (FIA) is shown as follows:

$$cut(c, \frac{1}{k}) := c - \frac{c}{|c|} \bullet max(|c| - \frac{1}{k}, 0) \tag{4}$$

$$b_x^k := cut(C\nabla_x u^k + b_x^{k-1}, \frac{1}{\lambda}) \tag{5}$$

$$b_y^k := cut(C\nabla_y u^k + b_y^{k-1}, \frac{1}{\lambda}) \tag{6}$$

$$u^{k+1} := f - \frac{\lambda}{u}(\nabla_x^T b_x^k + \nabla_y^T b_y^k) \tag{7}$$

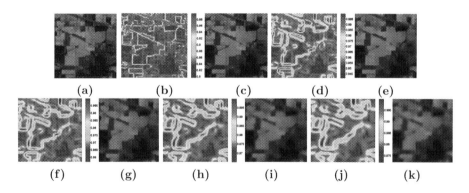

Fig. 1. ATV filtering results in different iterations number in $\lambda = 0.004$ and $\mu = 0.032$: (a) Input image; (b)–(k) the calculated parameter χ_f and filtered result in different iterations; (b, c) $In = 20$; (d, e) $In = 30$; (f, g) $In = 40$; (h, i) $In = 50$; (j, k) $In = 60$;

For $k = 0,1,...$, let b_x^k, b_y^k, u^{k+1} be given by the iteration scheme (5)–(7), if $0 < \lambda/\widehat{\mu} < 8$, then $\lim_{k \to \infty} u^k = u^*$. Because of $\widehat{\mu} = \mu/C$, $C \in (0,1]$, we can obtain $0 < C \bullet \lambda/\mu < 1/8$ from $0 < \lambda/\widehat{\mu} < 8$. To make any $C \in (0,1]$ meet $0 < C \bullet \lambda/\mu < 1/8$, we only need to guarantee $0 < \lambda/\mu < 1/8$ so that the FIA can convergence to the optimal solution of ATV. Figure 2 shows different number of iterations (In) with same λ and μ.

3   Spectral-Spatial Classification Framework

Figure 2 shows the proposed classification framework of HSI based on ATVF classification method. Unlike the single-band gray or three-band color image, the HSI usually has hundreds of spectral bands. Because PCA can reduce the spectral dimension of the hyperspectral data while well preserve the information in the mean square sense. This paper firstly use the PCA method to perform dimensionality reduction and obtain the first principal components for the following processing:

$$\mathbf{O}^K = PCA(\mathbf{I}) \tag{8}$$

where $\mathbf{I}$ presents as the original image, after dimensionality reduction into K wave bands, as shown in (8), and $\mathbf{O}^K$ is the principal components image.

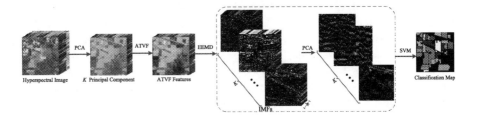

Fig. 2. Schematic of the proposed ATVF classification method.

Next, by using the ATV method, feature extraction is performed on the principal components. The resulting features used for classification are obtained as follows:

$$\mathbf{N}^K = ATV_{\mu,\lambda,In}(\mathbf{O}^K) \tag{9}$$

in which, μ, λ, In represent the filtering parameters respectively, K represents the data of various wave bands following PCA feature extraction, $\mathbf{N} = (\mathbf{N}^1, \mathbf{N}^2, ..., \mathbf{N}^K)$ is the characteristic data extracted from the HSI.

In the aspect of hyperspectral image feature selection, the features image just selects the high frequency part which reflects the details of the image intuitively, so the selected image features do not contain low frequency grayscale change information. Therefore, in order to further improve the characteristics of the hyperspectral image, the EEMD method is used to extract the intrinsic mode component of a distinct spatial structure from image $\mathbf{N}$. For each spectrum band of hyperspectral image, it can be decomposed of a series of internal solid mode function (IMFs) through EEMD. After that, the first principle component of IMFs can be obtained by PCA method to further integrate the spatial-spectral features into one component. Next, according to the relevant information between the IMFs and the residuals R, the hyperspectral image after EEMD can be reconstructed as follows:

$$\mathbf{X}^K = \sum_{m=1}^{h} IMF_{K,m} + R_m \tag{10}$$

where $\mathbf{X}^K \subset \mathbf{X}$ represents the contour feature of the Kth band by EEMD method. h is the total number of the IMFs in each band.

Finally, a widely used pixel-level classifier, i.e., the SVM classifier, is used to classify all pixel points for the feature data in hyperspectral image. SVM is one of the most commonly-used pixel-by-pixel spectral classifier. In addition, one of the major advantages of SVM is its robust performance for processing high dimensional data.

4   Experimental

4.1   Indian Pines Data Set

The Indian Pines HSI shows the Indian Pines Test Field in the northwest of Indiana, which is captured by Airborne Visible/Infrared Imaging Spectrometer (AVIRS) remote sensing device. The image is of a size 145×145 pixels with a spatial resolution of 20 m per pixel and 220 wave bands. With 20 water absorption wave bands removed, the 200 bands are used in the experiment. As the scene is captured in June, some crops such as corn and soybean etc, are still in an early stage of growth. In the reference classification map obtained from site exploration, the scene is divided into 16 different objectives.

4.2   Salinas Data Set

The Salinas image is acquired by the AVIRIS sensor over Salinas Valley, California, which contains 224 bands of size 512×217 pixel points with a spatial resolution of 3.7 m, and the no. 108–112, no. 154–167 and no. 224 water absorption bands are discarded before classification.

4.3   Parameter Settings

This paper first analyzes the influence of parameters to the performance of the proposed method. There are three common standards to evaluate classification accuracy of the HSI: The overall classification accuracy (OA), the average classification accuracy (AA), and Kappa coefficient. ATV filtering extraction steps includes two parameters μ and λ mentioned before. As shown in the Fig. 3, it is obvious that the accuracies of the OA, AA, and Kappa coefficients of the ATVF and SVM algorithms are much lower in a low feature dimension K. With the increasing of feature dimension K, the three common standards are also increase. Especially, when reaching to 21, the classification accuracies of surface features tend to be stable, which means that low feature dimension may result in missing abundant of meaningful information. Therefore, to obtain perfect classification performance, we choose the feature dimension K as 21.

Then, an analysis is conducted for the effect of different parameters λ, μ, p, K on the classification accuracy. In Figs. 4 and 5, 10% and 2% of ground reference data (Indian Pines and Salinas) are randomly selected as training samples. According to Control Variable Method, we only vary one parameter and the remaining

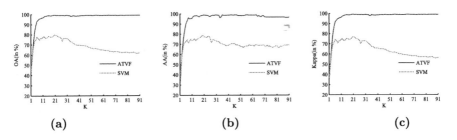

Fig. 3. Analysis of the influence of the parameter K to ATVF and SVM method: (a) $\mu = 0.032, \lambda = 0.004, p = 0.31$; (b) $\mu = 0.032, \lambda = 0.004, p = 0.31$; (c) $\mu = 0.032, \lambda = 0.004, p = 0.31$

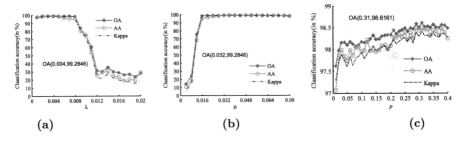

Fig. 4. Indian Pines image: analysis of the influence of the parameters μ, λ, h, p: (a) $\mu = 0.032, p = 0.31, K = 21$; (b) $\lambda = 0.004, p = 0.31, K = 21$; (c) $\lambda = 0.001, \mu = 0.032, K = 21$.

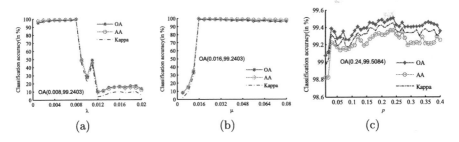

Fig. 5. Salinas image: analysis of the influence of the parameters μ, λ, h, p: (a) $\mu = 0.032, p = 0.24, K = 21$; (b) $\lambda = 0.004, p = 0.24, K = 21$; (c) $\lambda = 0.004, \mu = 0.032, K = 21$.

parameters are constant to illustrate the tendency of OA. Figure 4(a) analyzes the functional relationships between λ and the classification accuracy OA when $\mu = 0.032, p = 0.31, K = 21$. As can be seen, the classification accuracy OA could reach to 98.6161% when $\lambda=0.004$. Figure 4(b) shows the effect of μ on the classification accuracy when $\lambda = 0.004, p = 0.31, K = 21$. Figure 4(c) indicates the influence of p on the classification accuracy when $\lambda = 0.001, \mu = 0.032, K = 21$. In the same way, the Salinas data set is shown in Fig. 5. Therefore, $\mu = 0.032, \lambda = 0.004, p = 0.31, K = 21, In = 33$ are set as the default parameters.

4.4   Experiment with Indian Pines Data Set

This paper compares the ATVF method with several typical spectral and spectral-spatial classification algorithms, including the support vector machine (SVM) [1], joint sparse representation classification (JSRC) [10], extended morphological profiles (EMP) classification [2], edge preserving filtering (EPF) method [11], logistic regression and multi-level logistic (LMLL) method [12].

Table 1. Classification accuracies of the SVM, JSRC, EMP, EPF, LMLL, ATVF methods for the Indian Pines data set with 10% training samples.

Name	Train	Test	SVM	JSRC	EMP	EPF	LMLL	ATVF
Alfalfa	8	38	76.94	90.88	95.53	73.08	94.26	90.53
Corn_N	137	1291	79.49	81.33	87.38	93.34	97.16	97.59
Corn_M	83	747	80.55	61.04	92.68	96.24	90.41	98.11
Corn	24	213	67.20	68.54	83.99	88.29	98.42	95.96
Grass_M	48	435	89.27	92.87	92.74	99.01	98.39	99.43
Grass_T	73	657	89.86	59.51	98.31	92.12	99.22	99.09
Grass_p	8	20	88.62	89.15	91.00	95.00	93.18	94.00
Hay_W	48	430	97.24	83.26	99.86	97.00	98.87	98.55
Oats	8	12	48.59	41.67	97.50	96.40	97.51	81.45
Soybean_N	97	875	77.37	67.54	86.51	98.98	84.55	93.83
Soybean_M	239	2216	81.12	83.26	96.25	94.13	97.86	98.53
Soybean_C	59	534	77.99	55.81	87.40	96.73	98.55	97.88
Wheat	21	184	92.68	91.85	98.21	100.0	99.85	100.0
Woods	122	1143	92.56	95.98	99.50	99.29	98.37	99.96
Buildings	39	347	72.96	78.39	96.08	92.40	88.66	99.57
Stone	47	46	98.59	90.36	93.01	92.50	98.25	94.94
OA			83.28	82.65	93.54	95.52	95.97	98.53
AA			81.94	71.96	93.50	94.76	96.12	97.31
Kappa			80.86	80.22	92.63	94.88	95.38	98.32

The Indian Pines Data Set of experiments uses the data set from the Indian Pines HSI. Table 1 shows the number of the training samples and testing samples in the experiment (the training samples account for 10% of the reference data; selected randomly). The classification map and classification accuracy from the compared classifiers are illustrated in Fig. 6 and Table 1. As we can see, the SVM classification method, which only consider spectral information, can obtain just 83.28% classification accuracy, whereas spectral-spatial-based classification algorithms(EMP, EPF, LMLL, ATVF) tend to obtain much higher than 93%. This phenomenon indicates that the spectral-spatial classification can make full use of spatial information when they are possessed with similar spectral features. In addition, in terms of OA, AA and Kappa coefficient, when compared with the EMP, EPF, JSRC and LMLL spatial-spectral methods, the ATVF also outperforms others. Meanwhile, Table 1 also shows the effects of different methods

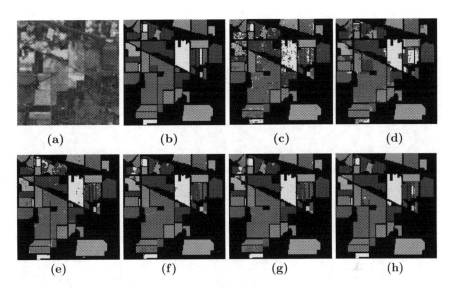

(a)   (b)   (c)   (d)

(e)   (f)   (g)   (h)

Fig. 6. Classification results (Indian Pines image) obtained by (**a**) The false-color composite of the Indian Pines image; (**b**) The original image; (**c**) The SVM method; (**d**) The JSRC method; (**e**) The EMP method; (**f**) The EPF method; (**g**) The LMLL method; (**h**) The ATVF method.

Table 2. Classification accuracies of the SVM, JSRC, EMP, EPF, LMLL, ATVF methods for the Salinas data set with 2% training samples.

Name	Train	Test	SVM	JSRC	EMP	EPF	LMLL	ATVF
Weeds_1	40	1969	99.80	97.21	99.72	98.57	90.10	100.0
	75	3651	99.45	99.76	99.69	99.89	100.0	100.0
Fallow	40	1936	94.73	99.31	98.89	94.76	99.79	99.96
Fallow_P	28	1366	96.87	86.58	99.08	97.50	98.90	97.55
Fallow_S	54	2624	98.87	85.67	96.89	100.0	99.03	99.11
Stubble	79	3880	100.0	97.81	99.41	99.95	99.78	99.84
Celery	72	3507	99.83	95.72	99.45	100.0	99.67	99.89
Graps	225	11046	79.31	99.04	96.12	86.34	88.58	99.58
Soil	124	6079	99.44	100.0	99.59	99.59	99.97	99.78
Corn	21	1047	89.52	96.87	96.27	95.45	96.99	99.13
Lettuce_4wk	21	1047	92.42	91.91	97.28	98.44	94.26	99.85
Lettuce_5wk	39	1888	97.40	89.18	99.8	99.32	100	99.46
Lettuce_6wk	18	898	95.55	67.97	98.83	95.60	97.58	97.97
Lettuce_7wk	21	1049	94.81	70.86	93.87	99.29	98.36	98.57
Vinyard_U	145	7123	77.89	98.03	94.61	93.01	71.29	99.18
Vinyard_T	36	1771	99.01	99.55	99.44	100.0	98.76	100.0
OA			90.68	96.13	97.65	95.35	93.20	99.51
AA			94.66	92.39	98.06	97.47	96.38	99.37
Kappa			89.60	95.69	97.38	94.81	92.42	99.45

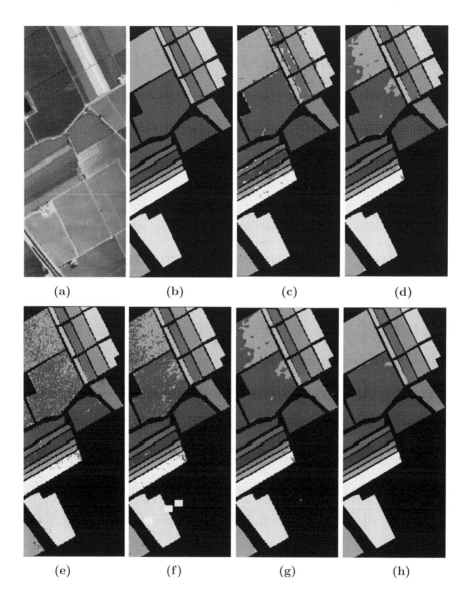

(a)   (b)   (c)   (d)

(e)   (f)   (g)   (h)

Fig. 7. Classification results (Salinas image) obtained by (**a**) The false-color composite of the Salinas image; (**b**) The original image; (**c**) The SVM method; (**d**) The JSRC method; (**e**) The EMP method; (**f**) The EPF method; (**g**) The LMLL method; (**h**) The ATVF method.

on CA, OA, AA, and Kappa coefficients of different categories in the scene. The proposed method is superior to SVM in term of classification accuracy of some spectral surface features. For example, the classification accuracy of Alfalfa increases to 90.53% from 76.94%. The corn classification accuracy increases to 95.96% from 67.2%, and the buildings classification accuracy increases to 99.57% from 72.96%.

The second experiment is performed on the Salinas data set, which uses the data set of the Salinas HSI. Here, the training samples are randomly selected to account for 2% of the reference data, and the rest of the data selected for test. Table 2 and Fig. 7 indicate the classification accuracies of different algorithms, which includes single CA, AA, OA, and Kappa coefficients. As can be seen from Table 2, the overall classification accuracy of the method proposed in the paper is superior to EMP, EPF and LMLL under the default parameter settings. The proposed method is able to significantly improve the classification accuracy compared with the traditional SVM. For example, the Vinyard_U classification accuracy increases to 99.18% from 77.89%. Especially, some classes, such as Weed_1, Weed_2 and Vinyard_T, can obtain 100% classification accuracies. This is because the proposed method reserve edge details to fully use of more information, and OA, AA, Kappa are much higher than other methods.

5 Conclusion

The paper proposes a spectral-spatial classification method for the HSI based on feature extraction, which can improve the classification accuracy for surface features at a low computational cost. Compared with other spectral-spatial classification methods, the major advantage of the proposed ATVF algorithm is that it could reduce the sensitivity of noise to denoise smoothly. Furthermore, the proposed method is able to reserve edge details by further extract the feature filtered image into a transform domain, which is useful for feature extraction with full use of details. Moreover, the results show that the improved algorithm effectively improves the classification accuracy of hyperspectral image. The proposed algorithm has better performance than the traditional ATV with PCA algorithm, and can mine the spatial information of hyperspectral images to a greater extent. Experimental results on four hyperspectral images show that the proposed algorithm yields highly accurate classification results.

References

1. Pal, M., Mather, P.M.: Support vector machines for classification in remote sensing. Int. J. Remote Sens. **26**, 1007–1011 (2005)
2. Benediktsson, J.A., Palmason, J.A., Sveinsson, J.R.: Classification of hyperspectral data from urban areas based on extended morphological profiles. IEEE Trans. Geosci. Remote Sens. **43**, 480–491 (2005)
3. Mura, M.D., Villa, A., Benediktsson, J.A., Chanussot, J.: Classification of hyperspectral images by using extended morphological attribute profiles and independent component analysis. IEEE Geosci. Remote Sens. Lett. **8**, 542–546 (2001)

4. Ghamisi, P., Benediktsson, J.A., Cavallaro, G., Plaza, A.: Automatic framework for spectral-spatial classification based on supervised feature extraction and morphological attribute profiles. IEEE J. Sel. Top. Appl. Earth Obs. Remote Sens. **7**, 2147–2160 (2014)

5. Huang, N.E., Shen, Z., Long, S.R.: The empirical mode decomposition and the Hilbert spectrum for nonlinear and non-stationary time series analysis. Proc. R. Soc. A Math. Phys. Eng. Sci. **454**, 903–995 (1998)

6. Wu, Z., Huang, N.E.: Ensemble empirical mode decomposition: a noise-assisted data analysis method. Adv. Adapt. Data Anal. **1**, 1–41 (2009)

7. Zhang, H., Peng, Q.: Adaptiveim agedenoisingm odelbasedontotalvariation. Opto-Electron. Eng. **60**, 50–53 (2006)

8. Qin, Z., Goldfarb, D., Ma, S.: An alternating direction method for total variation denoising. Optim. Methods Softw. **30**, 594–615 (2015)

9. Liu, W., Wu, C., Xu, T.: Adaptive total variation model for image denoising with fast solving algrithm. Appl. Res. Comput. **28**, 4797–4800 (2011)

10. Chen, Y., Nasrabadi, N.M., Tran, T.D.: Hyperspectral image classification using dictionary-based sparse representation. IEEE Trans. Geosci. Remote Sens. **49**, 973–3985 (2011)

11. Kang, X., Li, S., Benediktsson, J.A.: Spectral spatial hyperspectral image classification with edge-preserving filtering. IEEE Trans. Geosci. Remote Sens. **52**, 2666–2677 (2014)

12. Li, J., Bioucas-Dias, J.M., Plaza, A.: Hyperspectral image segmentation using a new Bayesian approach with active learning. Trans. Geosci. Remote Sens. **49**, 3947–3960 (2011)

Convolutional Neural Network
Based Inter-Frame Enhancement
for 360-Degree Video Streaming

Yaru Li[1], Li Yu[2], Chunyu Lin[1(✉)], Yao Zhao[1], and Moncef Gabbouj[2]

[1] Institute of Information Science, Beijing Jiaotong University, Beijing Key
Laboratory of Advanced Information Science and Network, Beijing, China
cylin@bjtu.edu.cn
[2] Laboratory of Signal Processing, Tampere University of Technology,
Tampere, Finland

Abstract. 360-degree video has attracted more and more attention in recent years. However, it is a highly challenging task to transmit the high-resolution video within the limited bandwidth. In this paper, we first propose to unequally compress the cubemaps in each frame of the 360-degree video to reduce the total bitrate of the transmitted data. Specifically, a Group of Pictures (GOP) is used as a unit to alternately transmit different versions of the video. Each version consists of 3 high-quality cubemaps and 3 low-quality cubemaps. Then, the convolutional neural network (CNN) is introduced to enhance the low-quality cubemaps with the high-quality cubemaps by exploring the inter-frame similarities. It is shown in the experiment that a single CNN model can be used for various videos. The experimental results also show that the proposed method has an excellent quality enhancement compared with the benchmark in terms of PSNR, especially for videos with slow motion.

Keywords: Convolutional neural network · Inter-frame enhancement
360-degree video streaming

1 Introduction

In recent years, VR technology has attracted much attentions with the emergence of HMD products, which has brought dramatic benefits to the entertainment market [1]. The mainstream products, including Oculus Rift [2], Samsung Gear VR [3] and Google Cardboard [4] are driving the development of the consumer market, but the consumers complain that the content deficiencies bring about an unpleasant viewing experience [1]. The reason for this phenomenon is mainly due to the fact that the development of VR technology is immature, such as low-quality, lack of a quasi-targeted content, and video delay.

In general, the video resolution requirement for HMD is 4K × 2K or higher, which is a challenge for existing network bandwidth capabilities. With regard to the projection of 360-degree video, the existing representation methods are Equirectangular [5], Cubemap [6], Pyramidal mapping [7], RD map [8], etc. Some solutions try to expand the deformation on this basis to adapt the modern coding standard by mapping the

© Springer Nature Switzerland AG 2018
R. Hong et al. (Eds.): PCM 2018, LNCS 11165, pp. 57–66, 2018.
https://doi.org/10.1007/978-3-030-00767-6_6

spherical video into a planar video. Among them, the Multi-resolution Equirectangular [5] and Cubemap [6] are the main application methods, and have better performance [9]. The recent research team responsible for future video compression and encoding and decoding technologies, the Joint Video Exploration Team (JVET), has developed the Panorama Projection Test Platform (360Lib) that can combine H.265/High Efficiency Video Coding (HEVC) [10] for research the coding of the 360-degree video. However, existing video coding standards, H.264/Advanced Video Coding [11] and H.265 [10], have not established a unified standard for the 360-degree video. One of the 360-degree video complete mapping transmission methods is to design the encoded transmission video content according to the layout of the rectangular plane. It usually maps the sphere into different shapes and then expands the slicing plane, such as the RD map[8]. The other one is based on viewpoint-adaptive transmission, taking into account the user's subjective evaluation. For example, at the expense of storage and encoding, the server side forms a video transmission of multiple different viewport versions [12] and tile-based encoding transmission based on the current viewport [13].

In this paper, we propose the inter-frame enhancement for unequally compressed the 360-degree video to improve the rate distortion performance. A 360-degree video is created by a fixed camera with multiple lens and contains omnidirectional content in one scene. The relative position for each viewport of the video does not change, and there is no major change shortly, especially indoor space. With this in mind, we utilize the similarity between frames by CNN to enhance the quality of the video. The experimental results show that our method can provide a high rate distortion performance. Our current work is focused on the entire the 360-degree video, and we will use the impact of different areas of the video on the user's viewing experience as a consideration to optimize our solution in future work.

The rest of paper is organized as follows. Section 2 presents popular encoding methods for the 360-degree video and typical convolutional neural networks in improving video quality applications. In Sect. 3, the overall coding scheme is elaborated. The simulation results are evaluated in Sect. 4. Finally, the paper is concluded in Sect. 5.

2   Related Work

2.1   360-Degree Video Transmission Technology

Most of the advanced the 360-degree video transmission technologies use viewpoint adaptive transmission. The proposed viewport-adaptive navigable the 360-degree video delivery [12] transmits a high-quality version from the user's point of view. Each high-quality version comes with a low-quality version of the whole 360-degree view in case the users get video delays when the users switch around. There are some other ways to use block transfer encoding video based tile technology. HEVC-compliant tile-based streaming [13] encodes a 360-degree video into multiple tiles, where the user's viewport is within the blocks that transmitted in high-quality, other viewports are of low-quality, and the results show a bitrate saving of 30% to 40%. Viewport-aware adaptive the 360-degree video streaming [14] produces a good viewing experience by assigning bit-rate of the surrounding area according to the user's interested viewport. In addition, Ultra Wide View presented in [15] divides the 360-degree video into network video, which flexibly reconstructs any view and meets the real-time interaction requirement for the users. [16] calculated tiled BD rate penalties based on spatio-temporal activity metrics. In contrast to the above methods, we introduce a convolutional neural network to enhance video quality based on the inter-frame similarity of the 360-degree video streaming.

2.2 Image Reconstruction Using CNN

In recent years, the convolutional neural network has been widely used in image processing. It successfully handled image classification problems such as [17, 18]. In addition, it also shows a significant effect in the image reconstruction problem. The SRCNN [19], which was originally used for super-resolution enhancement, used a three-layer network output high-resolution images through the three steps of feature extraction and feature representation, feature nonlinear mapping and image reconstruction. So far, the EDSR [20] scheme proposed on the basis of GAN [21] had extracted more features at each layer, resulting in better performance and the best results. When reconstructing compressed images, ARCNN [22] is inspired by SRCNN to enhance the quality of JPEG2000 compressed images using a four-layer network that has functions of extracting features, enhancing features, finding mappings between distortion and sharpness, and reconstruction.

In addition, some schemes used other features as input to guide image reconstruction. For example, Yu et al. [23] proposed a central viewpoint-assisted lateral viewpoint reconstruction in a multi-view video transmission system, and Dong et al. [24] proposed to use edge guidance for convolutional neural network structure sampled on a depth map. We aim to improve the compressed the 360-degree video, so we choose the high quality frames of neighbors as the input of the network so as to output the higher quality frames based on the ARCNN.

3 Proposed Scheme

Motivated by the fact that high-quality central view can provide valuable guidance for the enhancement of low-quality lateral view in [23] and inter-frame images have a great similarity like images between multiple views, we propose to encode the 360-degree video into different quality versions, and then use convolutional neural networks to reconstruct high-quality cubemaps from low-quality cubemaps with inter-frame similarity. Figure 1 outlines the entire framework.

(1) First of all, we unequally compress the 360-degree video into different versions. Compressing a portion of the video into low quality causes the reduction in the size of the video data.
(2) Then we transmit different versions of the 360-degree video at an interval of GOP.
(3) Finally, we use a convolutional neural network to reconstruct the current low-quality frame with the nearest high-quality frame in the previous GOP.

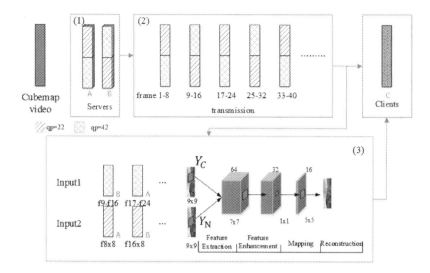

Fig. 1. Framework of the proposed the 360-degree video streaming solution, (1) is a video with unequally compressing, (2) is the 360-degree video transmitted by encoding every GOP into different quality versions, and (3) is an inter-frame enhancement convolutional neural network.

3.1 Unequal Compression

Cubemap is a popular way of representation because the mapping is much simpler for the current game scene production based on OpenGL technology. We choose a layout of the cubes, arranged vertically, as shown in the Fig. 2. Only one boundary of the six faces in this arrangement is not continuous. We assume that each view of the 360-degree video is accessed with the same frequency, and then the video is encoded into two versions (A and B) using the H.264/H.265. The front, left and right sides of the version A are high quality, and the top, back and bottom sides are lower quality. The version B is the opposite of A. In this case, the version A and B have similar-sized video streams, which guarantees a stable bitrate. Two versions of the video streaming are transmitted alternatively. For example, the 1th to 8th frames of version A are transmitted first, then the 9th to 16th frames of version B are transmitted, and so on.

3.2 Convolutional Neural Network for the 360-Degree Video

After receiving two different versions of the video, we extract the Y component of the image as the object of processing due to that is the brightness information and is more sensitive to change by the naked eye. Part (3) of Fig. 1 illustrates that the convolutional neural network consists of four layers, which is similar to [23]. The function of the first layer is feature extraction. The second and third layers are for feature enhancement and non-linear mapping. The last layer has the function of reconstructing the image, which mainly synthesize the final output image. The loss function is the Mean Squared Error (MSE).

We extract one-to-one training data sets from the adjacent GOPs as the input objects. As shown in Fig. 3, the Input1 consists of the low quality frames to be improve in GOPs, which are continuous and different frames. The input2 consists of the nearest frames of the previous GOPs. During the training, the low-quality frames Y_C is the primary reference image as the first dimension. And the nearest neighbor frame Y_N, in the previous GOP of the high-quality frames, is processed as the second dimension. The ground truth is used as a label to constrain the network parameters. The convolutional neural network extracts the features of the two inputs Y_C and Y_N, and then combines the two features together to generate feature vectors. Until the loss function converges to a fixed value, the output model parameters are considered to be the relationship between the input and the label.

4 Experiment Results

In this section, we first introduce the coding parameters and the detailed parameters of the convolutional neural network. Then we evaluate and analyze our solution.

The experiments employ four 360-degree video sequences. As shown in Fig. 4, the resolution of AerialCity, DrivingInCity and DrivingInCountry is 3840 × 1920. The scenes of each video depict the city on the edge of a port with slow movement, the city shooting by a moving car on the expressway, and the country with river and mountain scenery, respectively. Among them, the DrivingInCity and the DrivingInCountry move faster and the speeds of the two are similar, but the DrivingInCountry changes a lot due to the fact that the scene is closer to the camera. The Broadway resolution of the sequence is 6144 × 3072, which is a video captured on a city road. We implement the transformation process from Equirectangular to Cubemap with IEEE 1857.9 subgroup reference test model VRM [25]. In addition, we use JM (H.264) and HM (H.265) to encode video. The QP of the high-quality part low-quality part is 22 and 42, respectively. In a four-layer convolutional neural network, the corresponding dimensions of each layer are 2 × 9×9 × 64, 7 × 7×32, 1 × 1 × 16, and 5 × 5 × 1, respectively. Before going

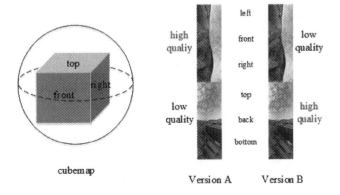

Fig. 2. A projection of the Cubemap mapping and an arrangement of the expansion map (1 × 6).

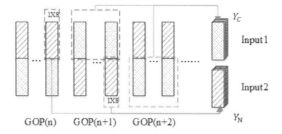

Fig. 3. Extracting two input objects from a 360-degree video streaming.

into the convolutional neural network, we need to perform the data extraction operation on the data. That is, the two input images are cropped into 33×33 pixels. The label image is 21×21. And we set the batch-size size to 128 for batch learning. The learning rate is set as 0.0001. The setting of these parameters promotes the network to quickly learn the mapping relationship between the label image and the input image during the training process.

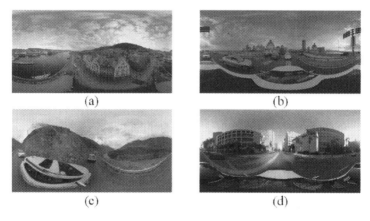

Fig. 4. The images of four 360-degree videos. (a–d) are AerialCity, DrivingInCity, DrivingInCountry and Broadway.

When using JM and HM to encode video, the sequential structure of GOP is set to IPPPPPPP. During the training phase, we use the first 60 frames of video. We first test the sequence of the AerialCity, DrivingInCity and DrivingInCountry which are encoded using H.264. After obtaining optimistic results, we use a training set of three video sequences mixed into the network. The model of mixed training is performed to reconstruct the three sequences using the same model that is generated by mixed training and the PSNR of the test is comparable or even better than a single training test. Figure 5 respectively depicts the PSNR curves of the three 360-degree videos from the 80th frames to the 112th frames. Comparing with the previously coded video, it is not difficult to see that the image quality is improved by the convolutional neural

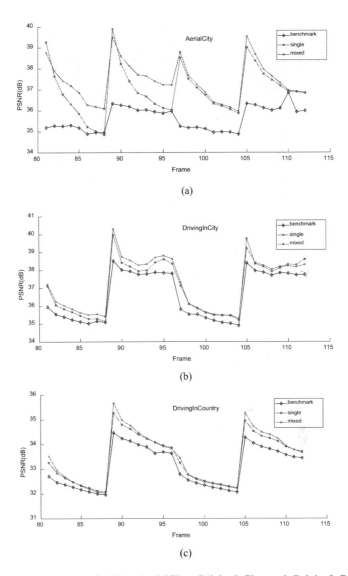

Fig. 5. The PSNR curves of video AerialCity, DrivingInCity and DrivingInCountry use separately trained models and mixed trained models.

network. The inter-frame similarity between the first frame and the auxiliary frame is greatest in each GOP, so they get significant quality enhancement, and the increased PSNR gradually decreases from the most adjacent first frame to the eighth frame. In Fig. 6, a comparison chart of the most neighboring frame and the most distant frame after reconstruction is listed. We can find that the frames which are far away cannot be recovered well. This is due to the relative change in the relative positions of the objects

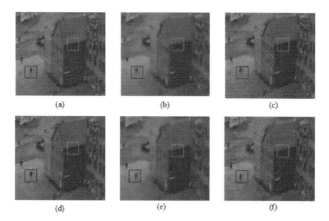

Fig. 6. The details of the nearest frame and the farthest frame from the video AerialCity in a GOP. (a, d) are the nearest and farthest original frames; (b, e) are coded graphs; (c, f) are reconstructions through convolutional neural networks.

in the frames, and the images closer to the auxiliary frames can acquire more similar information.

The PSNR gain plot (Fig. 7) shows that the mixed model possesses similar enhancement effects on the two video sequences, which proves that our scheme is also suitable for different sequences using the same model. The intensity of video inter-frame changes is a factor that affects the enhancement effect, and the higher gain of the AerialCity video sequence indicates that the solution has a better quality improvement for slower motion video. Table 1 lists some parameters of the video and the average PSNR. We only test the PSNR gain of the video Broadway encoded by HM, due to the video resolution limitation of the JM encoder.

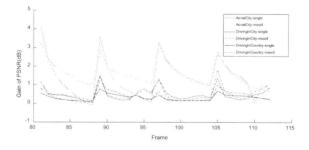

Fig. 7. PSNR gain comparison of Video AerialCity, DrivingInCity and DrivingInCountry that encoded with H.264

Table 1. The average PSNR of the videos encoded with H.264 and H.265

Video	Size	*fps*	$\overline{PSNR}$ (dB)					
			H.264			H.265		
			Benchmark	Single	Mixed	Benchmark	Single	Mixed
AerialCity	4K	30	35.62	36.97	37.42	39.00	39.38	39.36
DrivingInCity	4K	30	36.61	37.12	37.23	40.15	40.34	41.36
DrivingInCountry	4K	30	33.08	33.42	33.48	36.70	36.89	36.84
Broadway	6K	60				40.88	41.16	

5   Conclusion

This paper proposes an inter-frame enhancement scheme for the 360-degree video which uses the similarities among frames to improve the quality of the video. In addition, since the encoded video files have constant bitrate, the network can maintain a stable transmission under a certain bandwidth. The results demonstrate that our solution is effective for the currently widely used H.264 and H.265 encoding technologies. The downside is that we are assuming that the user has the same viewing probability for each viewport in the experiment, so we only design two versions of the videos. The user's experience is particularly important for measuring the quality of a 360-degree video. In the future work, we will design the coding quality of different regions from the user's point of view and the unequal transmission of video for each viewport in inter-frame enhancement transmission.

Acknowledgments. This work was supported in part by the National Natural Science Foundation of China under Grant (61772066), by the Fundamental Research Funds for the Central Universities (2018JBM011).

References

1. Virtual reality augmented reality 2017 report. http://cdn.instantmagazine.com/upload/4666/bom_vrar_2017reportpdf.68ec9bc00f1c.pdf. Accessed 25 May 2018
2. Oculus Rift: http://www.oculus.com/rift/. Accessed 25 May 2018
3. Samsung Gear VR: http://www.samsung.com/global/microsite/gearvr. Accessed 25 May 2018
4. Google Cardboard. https://www.google.com/get/cardboard/. Accessed 25 May 2018
5. Equirectangular. https://en.wikipedia.org/wiki/Equirectangular_projection. Accessed 25 May 2018
6. Ng, K.-T., Chan, S.-C., Shum, H.-Y.: Data compression and transmission aspects of panoramic videos. IEEE Trans. Circuits Syst. Video Technol. **15**(1), 82–95 (2005)
7. Kuzyakov, E., Pio, D.: Next-generation video encoding techniques for 360 video and vr. 2016-07-15) [2017-11-07] (2016). https://code.facebook.com/posts/1126354007399553/next-generationvideo-encoding-techniques-for-360-video-and-vr/. 15 July 2016

8. Fu, C.-W., Wan, L., Wong, T.-T., Leung, C.-S.: The rhombic dodecahedron map: an efficient scheme for encoding panoramic video. IEEE Trans. Multimedia **11**(4), 634–644 (2009)
9. Sreedhar, K.K., Aminlou, A., Hannuksela, M.M., Gabbouj, M.: Viewport-adaptive encoding and streaming of 360-degree video for virtual reality applications. In: 2016 IEEE International Symposium on Multimedia (ISM), pp. 583–586. IEEE (2016)
10. Sullivan, G.J., Ohm, J.-R., Han, W.-J., Wiegand, T., et al.: Overview of the high efficiency video coding (HEVC) standard. IEEE Trans. Circuits Syst. Video Technol. **22**(12), 1649–1668 (2012)
11. Wiegand, T., Sullivan, G.J., Bjntegaard, G., Luthra, A.: Overview of the H.264/AVC video coding standard. IEEE Trans. Circuits Syst. Video Technol. **13**(7), 560–576 (2003)
12. Corbillon, X., Simon, G., Devlic, A., Chakareski, J.: Viewport-adaptive navigable 360-degree video delivery. In: 2017 IEEE International Conference on Communications (ICC), pp. 1–7. IEEE (2017)
13. Zare, A., Aminlou, A., Hannuksela, M.M., Gabbouj, M.: Hevc-compliant tilebased streaming of panoramic video for virtual reality applications. In: Proceedings of the 2016 ACM on Multimedia Conference, pp. 601–605. ACM (2016)
14. Ozcinar, C., De Abreu, A., Smolic, A.: Viewport-aware adaptive 360° video streaming using tiles for virtual reality. In: 2017 IEEE International Conference on Image Processing (ICIP), pp. 2174–2178. IEEE (2017)
15. Ju, R., He, J., Sun, F., Li, J., Li, F., Zhu, J., Han, L.: Ultra wide view based panoramic vr streaming. In: Proceedings of the Workshop on Virtual Reality and Augmented Reality Network, pp. 19–23. ACM (2017)
16. Sanchez, Y., Skupin, R., Hellge, C., Schierl, T.: Spatio-temporal activity based tiling for panorama streaming. In: Proceedings of the 27th Workshop on Network and Operating Systems Support for Digital Audio and Video, pp. 61–66. ACM (2017)
17. Gu, B., Sheng, V.S., Tay, K.Y., Romano, W., Li, S.: Incremental support vector learning for ordinal regression. IEEE Trans. Neural Netw. Learn. Syst. **26**(7), 1403–1416 (2015)
18. Wen, X., Shao, L., Xue, Y., Fang, W.: A rapid learning algorithm for vehicle classification. Inf. Sci. **295**(C), 395–406 (2015)
19. Dong, C., Loy, C.C., He, K., Tang, X.: Learning a deep convolutional network for image super-resolution. In: Fleet, D., Pajdla, T., Schiele, B., Tuytelaars, T. (eds.) ECCV 2014. LNCS, vol. 8692, pp. 184–199. Springer, Cham (2014). https://doi.org/10.1007/978-3-319-10593-2_13
20. Lim, B., Son, S., Kim, H., Nah, S., Lee, K.M.: Enhanced deep residual networks for single image super-resolution. IEEE Conf. Comput. Vis. Pattern Recogn. Workshops **1**, 4 (2017)
21. Ledig, C., Theis, L., Huszar, F., Caballero, J., Cunningham, A., Acosta, A., Aitken, A., Tejani, A., Totz, J., Wang, Z., et al.: Photo-realistic single image super-resolution using a generative adversarial network. In: CVPR, pp. 4681–4690 (2017)
22. Dong, C., Deng, Y., Chen, C.L., Tang, X.: Compression artifacts reduction by a deep convolutional network. In: Proceedings of the IEEE International Conference on Computer Vision, pp. 576–584 (2015)
23. Yu, L., Tillo, T., Xiao, J., Grangetto, M.: Convolutional neural network for intermediate view enhancement in multiview streaming. IEEE Trans. Multimed. **20**(1), 15–28 (2018)
24. Dong, Y., Lin, C., Zhao, Y., Yao, C.: Depth map upsampling using joint edgeguided convolutional neural network for virtual view synthesizing. J. Electron. Imaging **26**(4), 043004 (2017)
25. Wang, Y., Wang, R.: VRM, reference test model of immersive video content coding IEEE, 11857.9 subgroup (2017)

Robust and Index-Compatible Deep Hashing for Accurate and Fast Image Retrieval

Jing Liu[1,2], Dayan Wu[1,2], Wanqian Zhang[1,2], Bo Li[2(✉)], and Weiping Wang[2]

[1] Institute of Information Engineering, Chinese Academy of Sciences,
Beijing 100093, China
{liujing,wudayan,zhangwanqian}@iie.ac.cn
[2] School of Cyber Security, University of Chinese Academy of Sciences, Beijing
100049, China
{libo,weipingwang}@iie.ac.cn

Abstract. Hashing methods have been widely used in large-scale image retrieval. However, the constraints on the hash codes of similar images learned by the previous hashing methods are too strong, which may lead to overfitting and difficult convergence. Besides, the binary codes output by the previous hashing methods are not optimally compatible with the multi-index approach, which is the most effective method for Hamming distance query acceleration. In this paper, we propose a novel Robust and Index-Compatible Deep Hashing (RICH) method to learn compact similarity-preserving binary codes, which focuses on improving the retrieval accuracy and time efficiency simultaneously. With the learned binary codes, we can achieve better results compared with the state-of-the-arts in retrieval accuracy. Meanwhile, remarkable promotions of the retrieval time efficiency have been made in the Hamming distance query process.

Keywords: Deep hashing · Image retrieval
Hamming distance query · The multi-index approach

1 Introduction

With the explosive growth of images on the web, much attention has been devoted to the nearest neighbor search via hashing methods. Mapping image data onto compact and similarity-preserving binary codes is important for large-scale image retrieval. In the literature, existing hashing methods can be grouped into two categories: data-independent methods and data-dependent methods.

In data-independent methods, the hash function is randomly generated which is independent of any training data. LSH [2] and KLSH [8] are representative data-independent methods. LSH uses random linear projections to produce binary codes, and KLSH is a kenelized method for dealing with high-dimensional and non-linear data. There are also some other variants of LSH [6, 13] that have

© Springer Nature Switzerland AG 2018
R. Hong et al. (Eds.): PCM 2018, LNCS 11165, pp. 67–77, 2018.
https://doi.org/10.1007/978-3-030-00767-6_7

been proposed. Due to the limitations of making no use of training data, these methods are usually difficult to achieve satisfactory performance.

Data-dependent methods try to learn the hash function from training data. They can be further divided into unsupervised and supervised methods. Unsupervised methods only use the feature information of data points without using any label information during the learning procedure. ITQ [4] is one of the representative unsupervised methods, which iteratively optimizes the projection matrix of images to minimize the quantization error. In order to deal with the label information, supervised methods are proposed. SDH [15], KSH [1], MLH [14] are well known supervised methods. All of them use hand-crafted features of the training images, which have limitations in capturing the deep semantic information of images thus limit the retrieval accuracy of the learned codes.

Recently, many deep learning based hashing methods have been proposed. CNNH [18] and NINH [9] are early approaches adopting neural networks to learn hash codes. [9,11,12,17,19,20] are also well known methods of this kind. Most of these deep hashing methods are supervised and the supervised information is usually based on pair-wise or triplet labels. Due to the outstanding learning ability of the deep neural networks, they have shown much better performance than the traditional hashing methods.

However, most of the existing deep hashing methods try to make the hash codes of similar images exactly the same, which is unreasonable in some cases. As shown in Fig. 1, two similar images in the semantic space may be very different in visual, which is a common case in the existing real-world image data sets. Under this circumstance, forcing the hash codes of similar images to be exactly the same may lead to overfitting and difficult convergence.

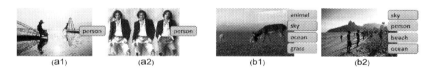

Fig. 1. Two exemplary image pairs from NUS-WIDE. The images in the group (a) are annotated with a single label. The images in the group (b) are annotated with multiple labels. Following previous works, both of them are considered as similar pair, even though they are very different in visual.

Besides, most current deep hashing methods are not compatible with the multi-index approach [3,5], which is an inverted index based method. The method splits each code into disjoint but consecutive blocks and creates a separate index for each block. In the searching phase, the images whose codes have no matching blocks with the query code will be filtered out. Consequently, the candidate images will be checked and ranked. However, the codes generated by previous deep hashing methods are not distributed uniformly, which will slow down the Hamming distance query process with multi-index.

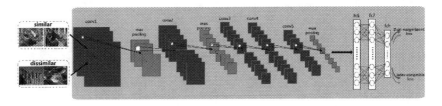

Fig. 2. Overview of the proposed framework. The network consists of 5 convolution layers and 3 fully connected layers. The last layer is hashing layer (fch), which has k output nodes.

In this paper, we introduce a novel framework based on deep learning model, named Robust and Index-Compatible Deep Hashing (RICH). An overview of the proposed framework is illustrated in Fig. 2. Through the proposed architecture, images are first encoded into real-valued feature vectors. Then each vector is converted to a hash code by a hash layer (fch). After that, these hash codes are used in a dual-margin based hashing loss that aims to preserve similarity between images and an index-compatible loss that aims to minimize the number of the matching blocks between the binary codes of dissimilar images. The contributions of this study can be summarized as follows: (1) We present a novel CNN based framework for learning hash functions to improve the retrieval accuracy and time efficiency simultaneously. (2) A loss function is elaborately designed to guide the learning of the neural network. With the proposed loss function, the learned hash functions are more scalable and robust to different data sets. Besides, the hash codes output by RICH are more compatible with the multi-index approach. (3) Extensive experiments demonstrate that the proposed method gains better retrieval accuracy and more robust performance than previous methods. Meanwhile, the retrieval time is also significantly reduced when adopting the multi-index approach.

2   The Proposed Approach

Suppose that we have a training set of N images $\{x_i\}_{i=1}^{N}$. Our goal is to learn the nonlinear hashing function $H : x \rightarrow h \in \{-1,1\}^k$ mapping each image to k-bit binary code. Accordingly, our deep hash function is defined as:

$$h(x) = sign(f(x)) \tag{1}$$

where $f(x)$ indicates the output of layer fch. Therefore, k-bit binary codes can be obtained through k such hash functions $h(x) = [h_1(x), h_2(x), ..., h_k(x)]$. To encourage the fch layer representation to be optimal for hash coding and can be better applied to the multi-index approach, the loss function is elaborately designed.

2.1   Dual-Margin Based Loss Term

Our first goal is to make the codes of similar images close in the Hamming space, while the codes of dissimilar images far away from each other. Meanwhile, different from previous methods, we no longer make the codes of similar images exactly same. To achieve the goal, we propose a dual-margin based loss term as:

$$
L_{hashing} = \frac{1}{2} \sum_{s_{ij} \in S} \{ s_{ij} \max \left(D_h(h(x_i) \cdot h(x_j)) - m_0, 0 \right)
$$
$$
+ (1 - s_{ij}) \max \left(m_1 - D_h(h(x_i) \cdot h(x_j)), 0 \right) \} \tag{2}
$$

where $s_{ij} = 1$ indicates x_i and x_j are similar and $s_{ij} = 0$ implies x_i and x_j are dissimilar. $D_h(h(x_i) \cdot h(x_j))$ denotes the Hamming distance between two codes, and $m_0 > 0, m_1 > 0$ are margin threshold parameters. In the case of $s_{ij} = 1$, which means x_i, x_j are similar images, the first term will punish them only when their Hamming distance exceeds the threshold m_0. In the case of $s_{ij} = 0$, only the second term takes effects, and it will punish dissimilar images mapped to close binary codes when their Hamming distance falls below the margin threshold m_1.

However, it is intractable to directly optimize the Eq. (2), due to the binary constraints on $h(x)$ and discrete Hamming distance (i.e. $D_h(\cdot)$) computing. As implied in previous methods [11], we replace the Hamming distance by Euclidean distance. Meanwhile, in order to reduce the error of quantization, we also impose a regularizer on the real-valued network outputs to approach the desired discrete values $(-1/+1)$. Thus, the complete dual-margin based loss term is rewritten as:

$$
\mathcal{L}_{hashing} = \frac{1}{2} \sum_{s_{ij} \in S} \{ s_{ij} \max \left(||f(x_i) - f(x_j)||_2^2 - m_0, 0 \right)
$$
$$
+ (1 - s_{ij}) \max \left(m_1 - ||f(x_i) - f(x_j)||_2^2, 0 \right) \} \tag{3}
$$
$$
+ \lambda \sum_{i=1}^{N} \left(||| f(x_i) | - \mathbf{1} ||_1 \right)
$$

where $f(x)$ is the continuous output vector of the hash layer. $|| \cdot ||_1$ is the L1-norm of vector, and $|| \cdot ||_2^2$ is the Euclidean-norm. $| \cdot |$ is the element-wise absolute value operation, and $\mathbf{1}$ is a vector of all ones. λ is a trade-off parameter that controls the strength of the regularizer. Note that the higher-order norms of the regularizer are also applicable. We choose the L1-norm on account of its less computational cost, which is beneficial for accelerating the training process.

2.2   Index-Compatible Loss Term

As we mentioned before, making the generated codes better applied to the multi-index approach is our second goal. However, there is a drawback of the previous deep hashing methods that the codes they generate is not uniformly distributed. Therefore, the majority of the codes have exactly the same value in some blocks. As shown in Fig. 3, all the codes have the same value in *block2*, which will make

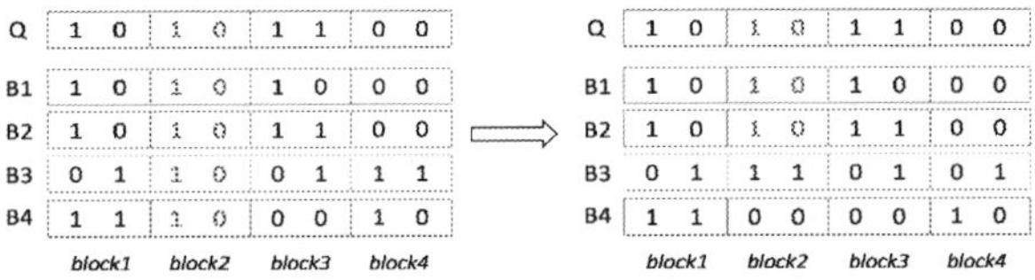

Q	1	0	1	0	1	1	0	0
B1	1	0	1	0	1	0	0	0
B2	1	0	1	0	1	1	0	0
B3	0	1	1	0	0	1	1	1
B4	1	1	1	0	0	0	1	0

block1 block2 block3 block4

$\Longrightarrow$

Q	1	0	1	0	1	1	0	0
B1	1	0	1	0	1	0	0	0
B2	1	0	1	0	1	1	0	0
B3	0	1	1	1	0	1	0	1
B4	1	1	0	0	0	0	1	0

block1 block2 block3 block4

Fig. 3. Q is the hash code of the query image. B1 and B2 are the codes of images similar to the query image, while B3 and B4 are the ones dissimilar to the query image.

all the dissimilar images contained in the candidate sets when adopting the multi-index approach.

To address this problem, for corresponding blocks of the codes between dissimilar image pairs, we push them a certain distance away from each other. The index-compatible loss term is designed as follows:

$$\mathcal{L}_{index} = \frac{1}{2} \sum_{s_{ij} \in S} (1 - s_{ij}) \sum_{t=1}^{b} \max \left(m_2 - ||f^t(x_i) - f^t(x_j)||_2^2, 0 \right) \tag{4}$$

where $f^t(x)$ denotes the t-th block of the hash layer (fch) outputs of image x, and m_2 is a margin threshold parameter.

Integrating the index-compatible loss with the dual-margin based loss, we can get the objective loss function of RICH:

$$\mathcal{L}_{RICH} = \mathcal{L}_{hashing} + \alpha \mathcal{L}_{index} \tag{5}$$

where $\alpha > 0$ is a trade-off parameter between $\mathcal{L}_{hashing}$ and $\mathcal{L}_{index}$.

2.3 Optimization

By substituting Eqs. (3) and (4) into Eq. (5), we rewrite the overall loss function of RICH as follows:

$$
\begin{aligned}
\mathcal{L}_{RICH} = \frac{1}{2} \sum_{s_{ij} \in S} &\left\{ s_{ij} \max \left(||f(x_i) - f(x_j)||_2^2 - m_0, 0 \right) \right. \\
&\left. + (1 - s_{ij}) \max \left(m_1 - ||f(x_i) - f(x_j)||_2^2, 0 \right) \right\} \\
&+ \frac{\alpha}{2} \sum_{s_{ij} \in S} (1 - s_{ij}) \sum_{t=1}^{b} \max \left(m_2 - ||f^t(x_i) - f^t(x_j)||_2^2, 0 \right) \\
&+ \lambda \sum_{i=1}^{N} (|||f(x_i)| - \mathbf{1}||_1)
\end{aligned} \tag{6}
$$

Next, we can employ back-propagation algorithm with mini-batch gradient descent method to train the network. So the gradients of Eq. (6) w.r.t $f(x_i), f(x_j) \forall i, j$ need to be computed. Since the max operation and the absolute value operation is non-differentiable at some certain points, we use subgradients instead, and define subgradients to be 1 at these points. In order to express clearly, we have a separate calculation on the gradients of the regularizer.

$$\frac{\partial \mathcal{L}_{hashing}}{\partial f(x_i)} = (f(x_i) - f(x_j))(s_{ij}I_{dis>m_0} - (1 - s_{ij})I_{dis<m_1}) + \delta(f(x_i))$$

$$\frac{\partial \mathcal{L}_{hashing}}{\partial f(x_j)} = (f(x_j) - f(x_i))(s_{ij}I_{dis>m_0} - (1 - s_{ij})I_{dis<m_1}) + \delta(f(x_j))$$

$$\frac{\partial \mathcal{L}_{index}}{\partial f^t(x_i)} = (1 - s_{ij})(f^t(x_i) - f^t(x_j))I_{||f^t(x_i)-f^t(x_j)||_2^2<m_2}$$

$$\frac{\partial \mathcal{L}_{index}}{\partial f^t(x_j)} = (1 - s_{ij})(f^t(x_j) - f^t(x_i))I_{||f^t(x_i)-f^t(x_j)||_2^2<m_2}$$

(7)

where

$$\delta(x) = \begin{cases} 1, & -1 \le x \le 0 \ or \ x \ge 1 \\ -1, & otherwise \end{cases} \tag{8}$$

where dis denotes the Euclidean distance $||f(x_i) - f(x_j)||_2^2$. And we use the function $I_{condition} = 1$ to indicate $condition$ is true, and $I_{contidion} = 0$ when $condition$ is false. With the computed subgradients over mini-batches, the rest of the back-propagation can be done in standard manner.

3   Experiments

3.1   Evaluation Setup

We conduct extensive experiments on two widely-used benchmark datasets, **CIFAR-10** and **NUS-WIDE**.

CIFAR-10 is a public image dataset, which consists of 60,000 32×32 images belonging to 10 categories. We randomly select 100 images per class (1,000 images in total) as the test query set, 5000 images per class (50,000 images in total) as the training set.

NUS-WIDE contains 269,648 multi-label images collected from Flickr. The association between images and 81 concepts are manually annotated. Following [11,20], we use the images associated with the 21 most frequent concepts, where each of these concepts associates with at least 5,000 images, resulting in a total of 195,834 images. Generally, if two images share at least one same label, they are considered similar, and dissimilar otherwise. We randomly select 2,100 images (100 images per class) for testing queries and the rest is used for training.

We use both deep learning based hashing methods and traditional hashing methods for thorough comparison. Five deep hashing methods: **DSH** [11], **DHN** [20], **DPSH** [10], **NINH** [9] and **CNNH** [18]. For fair comparison, we implement the **DSH** and **DHN** with the same AlexNet network structures. Most of other results are directly reported from previous works. Traditional hashing methods consists of **SH** [16] and **ITQ** [4], both of them use the CNN feature.

We implement RICH based on the open source Caffe[1] framework, and employ the AlexNet [7] neural network architecture, finetune convolutional layers and fully-connected layers that were copied from the pre-trained model. We use the mini-batch stochastic gradient descent with 0.9 momentum. The quantization penalty parameter λ and trade-off parameter α are chosen by cross-validation from 10^{-5} to 10^2 with a multiplicative step-size 10. For the dual margin parameter m_0, we empirically set $m_0 = 4$ for CIFAR-10 and $m_0 = 8$ for NUS-WIDE, which means we allow that there exists $1/2$ different bits between the codes of similar image pairs in CIFAR-10/NUS-WIDE. We set another margin parameter $m_1 = 2k$, which is the same as [11]. That means we expect at least half of the binary codes between dissimilar images to be different. For the parameters in the index-compatible loss term, we split codes to $\frac{k}{2}$ blocks, then we set $m_2 = 4$, which means we try to make each block of dissimilar codes share at least one distinct bit. The multi-index approach is implemented with Lucene-6.4.2[2], which is a high-performance and full-featured text search engine library.

We mainly evaluate the retrieval accuracy and retrieval time efficiency of RICH and other methods. Following [10,11,20], we use the mean Average Precision (mAP) to measure the retrieval accuracy. For NUS-WIDE, following the previous work [10], we calculate the mAP values within the top 5000 returned neighbors. To evaluate the time efficiency, we calculate the overall time of the Hamming distance query process.

3.2 Comparison of Retrieval Accuracy

The mAP results are reported in Table 1, which shows that the proposed RICH method substantially outperforms all the comparison methods.

As shown in Table 1, most of the deep hashing methods perform better than the traditional hashing methods, validating the advantage of learning image representations over using hand-crafted features. Furthermore, DSH achieves a better performance on CIFAR-10 than DPSH, but on NUS-WIDE, DSH is inferior to DPSH. It indicates that they are not robust enough on different data sets. However, RICH performance better on both CIFAR-10 and NUS-WIDE, validating its robustness. We attribute this approvement to the proposed dual-margin based loss term which can effectively prevent overfitting.

To further verify the effectiveness of the proposed dual-margin based loss term, we compare RICH with RICH*. As listed in Table 1, the mAP results of RICH* is obviously lower than RICH, which proves that the dual-margin based loss term can improve the retrieval accuracy. Furthermore, the promotion on NUS-WIDE is more significant, which means the dual-margin based loss is more effective on multi-label images data.

---

[1] http://caffe.berkeleyvision.org/.

[2] https://lucene.apache.org/.

Table 1. Comparison of mAP w.r.t. different number of bits on NUS-WIDE and CIFAR-10. RICH* is a variant of RICH when $m_0 = 0$.

Methods	NUS-WIDE				CIFAR-10			
	16-bits	32-bits	48-bits	64-bits	16-bits	32-bits	48-bits	64-bits
RICH	0.7867	0.7950	0.8177	0.8291	0.8801	0.8892	0.9062	0.9080
RICH*	0.7598	0.7680	0.7694	0.7740	0.8796	0.8839	0.9047	0.9082
DSH	0.7499	0.7602	0.7604	0.7631	0.8538	0.8831	0.9042	0.9065
DHN	0.8153	0.8178	0.8237	0.8215	0.8652	0.8792	0.8921	0.8890
DPSH	0.7752	0.7940	0.8120	0.8253	0.7206	0.7440	0.7570	0.7621
NINH	0.6866	0.7130	0.7155	0.7241	0.5662	0.5580	0.5810	0.5986
CNNH	0.6154	0.6255	0.6080	0.6098	0.4839	0.5090	0.5220	0.5534
ITQ+CNN	0.4235	0.4334	0.4607	0.4303	0.2436	0.2550	0.2610	0.2630
SH+CNN	0.3662	0.3560	0.3834	0.3405	0.1612	0.1610	0.1610	0.1620

3.3   Comparison of Time Efficiency

In general, the multi-index approach consists of two phases: *searching* phase and *checking* phase. In the *searching* phase, the binary code of the query image is first splitted into blocks, then we search the multi-index to identify all the binary codes that contain at least one matching block. In the *checking* phase, we rank the candidate images according to the Hamming distance between the binary codes of the candidate images and query image. We calculate the overall time of these two phases. Note that all the binary codes are stored in memory.

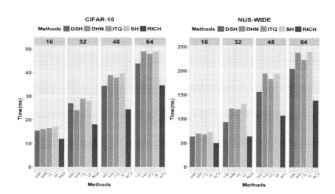

Fig. 4. Comparison of the query time w.r.t. different number of bits on NUS-WIDE and CIFAR-10.

The comparison result is presented in Fig. 4. With the binary codes learned by RICH, the overall query time is significantly reduced comparing to other methods. To be specific, there is a **29.5%–37.7%** reduction of the total query time on CIFAR-10. For NUS-WIDE, there is a **22.3%–47.5%** reduction. We attribute this improvement to the reduction of the candidate images. We calculate the size of the candidate images of different method, and the results are shown in Table 2. We can see that the size of the candidate images are significantly reduced with the codes generated by RICH. Specifically, there is a **43.35%** reduction on NUS-WIDE and **21.80%** reduction on CIFAR-10.

Table 2. Comparison of Recall and the size of candidate images on NUS-WIDE and CIFAR-10 with 32 bits.

Methods	NUS-WIDE					CIFAR-10	
	≥ 1	≥ 2	≥ 3	≥ 4	#candidates	–	#candidates
DSH	0.9985	0.9994	0.9998	0.9999	188741	0.9999	49762
DHN	0.9978	0.9993	0.9997	0.9999	188447	0.9999	49951
RICH	0.9679	0.9888	0.9909	0.9999	106841	0.9999	38942

In Table 2, we further record the recall rate. On CIFAR-10, both the comparison methods and RICH achieve a recall rate of almost 100%, which means that almost no similar images are filtered out in the searching phase. On NUS-WIDE, "$\geq n$" indicates the recall rate of similar images sharing no less than n same labels. Note that with the increasing number of the sharing labels between the database images and the query image, the recall rate increases, which means that the images with higher similarity to the query images are less likely to be filtered out.

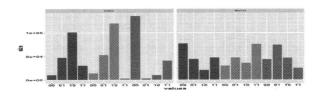

Fig. 5. Distribution of the codes generated by DSH (left) and RICH (right), with 32 bits on NUS-WIDE. The three blocks are randomly selected. The x-axis shows all the possible values of each block, and Y-axis shows the number of images. For the third block, most of the codes generated by DSH are valued "00" which is a extremely non-uniform distribution.

We further visualize the distribution of the codes generated by DSH an RICH. As shown in Fig. 5, the codes output by RICH are more uniform than the ones learned by DSH, which verifies the effectiveness of the index-compatible loss term.

4 Conclusion

In this paper, we propose a robust and index-compatible deep hashing method for accurate and fast image retrieval, named RICH. A loss function is elaborately designed, which consists of a dual-margin based loss term and an index-compatible loss term. With the learned codes, both the retrieval accuracy and the time efficiency are significantly improved. We attribute the improvement of retrieval accuracy to the relaxation of constraints on similar image pairs, and that of the time efficiency to the more uniform distribution of the codes. Extensive experiments validate the effectiveness of the proposed method.

References

1. Chang, S.F.: Supervised hashing with kernels. In: IEEE Conference on Computer Vision and Pattern Recognition, pp. 2074–2081 (2012)
2. Gionis, A., Indyk, P., Motwani, R., et al.: Similarity search in high dimensions via hashing. VLDB **99**, 518–529 (1999)
3. Gog, S., Venturini, R.: Fast and compact hamming distance index. In: SIGIR, pp. 285–294. ACM (2016)
4. Gong, Y., Lazebnik, S., Gordo, A., Perronnin, F.: Iterative quantization: a procrustean approach to learning binary codes for large-scale image retrieval. TPAMI **35**(12), 2916–2929 (2013)
5. Greene, D., Parnas, M., Yao, F.: Multi-index hashing for information retrieval. In: Foundations of Computer Science, pp. 722–731. IEEE (1994)
6. Ji, J., Li, J., Yan, S., Zhang, B., Tian, Q.: Super-bit locality-sensitive hashing. In: International Conference on Neural Information Processing Systems, pp. 108–116 (2012)
7. Krizhevsky, A., Sutskever, I., Hinton, G.E.: Imagenet classification with deep convolutional neural networks. In: NIPS, pp. 1097–1105 (2012)
8. Kulis, B., Grauman, K.: Kernelized locality-sensitive hashing for scalable image search **30**(2), 1895–1900 (2009)
9. Lai, H., Pan, Y., Liu, Y., Yan, S.: Simultaneous feature learning and hash coding with deep neural networks. In: CVPR, pp. 3270–3278 (2015)
10. Li, W.J., Wang, S., Kang, W.C.: Feature learning based deep supervised hashing with pairwise labels. arXiv preprint arXiv:1511.03855 (2015)
11. Liu, H., Wang, R., Shan, S., Chen, X.: Deep supervised hashing for fast image retrieval. In: CVPR, pp. 2064–2072 (2016)
12. Liu, L., Shao, L., Shen, F., Yu, M.: Discretely coding semantic rank orders for supervised image hashing. In: CVPR, pp. 5140–5149 (2017)
13. Mu, Y., Yan, S.: Non-metric locality-sensitive hashing. In: Twenty-Fourth AAAI Conference on Artificial Intelligence, pp. 539–544 (2010)
14. Norouzi, M., Fleet, D.J.: Minimal loss hashing for compact binary codes. In: ICML, pp. 353–360 (2011)
15. Shen, F., Shen, C., Liu, W., Tao Shen, H.: Supervised discrete hashing. In: CVPR, pp. 37–45 (2015)
16. Weiss, Y., Torralba, A., Fergus, R.: Spectral hashing. In: NIPS, pp. 1753–1760 (2009)
17. Wu, D., Lin, Z., Li, B., Ye, M., Wang, W.: Deep supervised hashing for multi-label and large-scale image retrieval. In: ICMR, pp. 150–158. ACM (2017)

18. Xia, R., Pan, Y., Lai, H., Liu, C., Yan, S.: Supervised hashing for image retrieval via image representation learning. In: AAAI Conference on Artificial Intelligence (2014)
19. Yao, T., Long, F., Mei, T., Rui, Y.: Deep semantic-preserving and ranking-based hashing for image retrieval. In: IJCAI, pp. 3931–3937 (2016)
20. Zhu, H., Long, M., Wang, J., Cao, Y.: Deep hashing network for efficient similarity retrieval. In: AAAI, pp. 2415–2421 (2016)

Self-supervised GAN for Image Generation by Correlating Image Channels

Sheng Qian[1], Wen-ming Cao[1], Rui Li[1], Si Wu[2], and Hau-san Wong[1(✉)]

[1] Department of Computer Science, City University of Hong Kong,
Hong Kong, China
cshswong@cityu.edu.hk
[2] Department of Computer Science, South China University of Technology,
Guangzhou, China

Abstract. Current most GAN-based methods directly generate all channels of a color image as a whole, while digging self-supervised information from the correlation between image channels for improving image generation has not been investigated. In this paper, we consider that a color image could be split into multiple sets of channels in terms of channels' semantic, and these sets of channels are closely related rather than completely independent. By leveraging this characteristic of color images, we introduce self-supervised learning into the GAN framework, and propose a generative model called *Self-supervised GAN*. Specifically, we explicitly decompose the generation process as follows: (1) generate image channels, (2) correlate image channels, (3) concatenate image channels into the whole image. Based on these operations, we not only perform a basic adversarial learning task for generating images, but also construct an auxiliary self-supervised learning task for further regularizing generation procedures. Experimental results demonstrate that the proposed method can improve image generation compared with representative methods and possess capabilities of image colorization and image texturization.

Keywords: Image generation · GAN · Self-supervised learning

1 Introduction

Recently increasing attention has been paid to building unsupervised learning models for image generation and representation learning. In general, there are two types of unsupervised learning approaches: (1) a discriminative framework with self-supervised proxy tasks for learning representations; (2) a generative framework for generating data and learning representations [26].

Considering expensive human annotation and plenty of free unlabeled data, self-supervised learning methods directly dig supervised information from the raw data. Based on data characteristics, all of these methods will construct various

ⓒ Springer Nature Switzerland AG 2018
R. Hong et al. (Eds.): PCM 2018, LNCS 11165, pp. 78–88, 2018.
https://doi.org/10.1007/978-3-030-00767-6_8

proxy tasks to learn meaningful representations. In computer vision domains both temporal and spatial clues have been proven to be informative signals for constructing proxy tasks, such as egomotion [1], unsupervised object tracking [23], spatial arrangement [7,18], transformations [8], and context-based reconstruction [20]. Besides, the correlation between image channels is also another important clue, such as colorization [3,4,6,13,14,27] and cross-channel prediction [28].

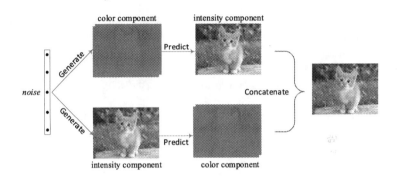

Fig. 1. Image generation by Self-supervised GAN.

Since images are high dimensional with complex patterns, various generative methods have been proposed for achieving better performance of image generation based on the GAN [9] framework. Among them, some methods try to leverage the inherent attributes of images, and focus on improving the architectural design of GAN. For example, [21] exploits the advantages of CNN in image applications, and [5,25,26] design more elaborate network architectures by exploiting structure/style formation [26], multiscale representation [5], and background/foreground composition [25], respectively.

In this paper we expect to incorporate adversarial learning and self-supervised learning into a generative model, and leverage their advantages for improving the performance of image generation. For this purpose, we propose a generative model called *Self-supervised GAN (denoted as SSGAN)*. Specifically, we exploit one of the most basic characteristics of color images as follows: (1) a color image is composed of multiple channels which can be grouped into specific sets based on channels' semantic; and (2) these sets of channels have a close relationship. To simplify the following discussion, we focus on the case where a color image is generally split into the following two components: intensity and color. Considering the above characteristic of color images, as illustrated in Fig. 1, the generation process can be decomposed into the following procedures to generate the whole image: (a) generate two sets of channels; (b) transform from one set to the other set; (c) concatenate these two sets to form the whole data.

Based on these operations, we could combine adversarial learning and self-supervised learning together. Except for performing the adversarial learning task for image generation, we also construct the self-supervised learning task where different sets of channels predict each other using true data to further improve generation. Viewed from another perspective, most of the existing methods directly generate all channels of color images as a whole, and only exploit self-supervised information from true/fake data. Compared with these methods, our proposed method could further dig more self-supervised information from the correlation between image channels. Overall, the main contributions of this work are as follows:

– By leveraging the relationship between color image channels, we propose a generative model which can well incorporate adversarial learning and self-supervised learning and improve the performance of image generation.
– Except for performing image generation, the proposed model also possesses capabilities of image colorization and image texturization.

In the experiments we conduct both qualitative and quantitative evaluation on the benchmark dataset, and compare the proposed method with several representative methods. The experimental results verify the effectiveness of our method.

2   Related Work

2.1   Adversarial Learning

Generally GAN-based methods focus on improving two factors of GAN: the architectural design and the train criteria, since these factors have a great influence on the performance of image generation. For the architectural design, [21] propose to stabilize GAN by applying architecture guidelines of CNN. By further exploiting the inherent attributes of images, [5,26] cascade multiple GANs and adopt a multi-scale strategy, and [24,25] analyze the image formation and decompose image generation into cascaded procedures. Besides, [11,15] design symmetrical architectures to model the cross-domain relationship of two image domains by coupling two GANs in parallel and in cross-linked respectively. For the train criteria, [16] adopts the least squares loss instead of the cross entropy loss used by GAN, and [19] further extends GAN in the f-divergences estimation framework. Differently, [29] rephrases the adversarial learning of GAN from the perspective of an energy-based model. Besides, [2,10] propose to measure the distribution discrepancy using Earth-Mover distance. Instead of weight clipping using by [2,10,17] penalizes the norm of the discriminator's gradient for enforcing a Lipschitz constraint. Overall these GAN-based methods can improve the training stability of models and the performance of image generation.

2.2   Self-supervised Learning

All of the self-supervised methods will leverage discriminative proxy tasks to learn representations well transferred to downstream tasks. By learning representations invariant to transformations, [1] predicts the transformation between a pair of adjacent frames, [23] considers a pair of identically tracked patches from successive frames to make their distance in the latent representation space more closer, and [8] generically forms a set of surrogate classes by applying vast image transformations to images. Considering the spatial arrangement of image patches, [7] predicts the relative position of two image patches, [18] solves the jigsaw puzzle composed of a set of object's patches, and [20] proposes the context-encoder to reconstruct the image region from its contextual region with an adversarial regularization. Some works focus on the problem of image colorization based on the regression model [4,6] or the classification model [13,14,27]. Furthermore, [3] improves the image diversity of colorization via leveraging conditional adversarial learning, and [28] proposes a split-brain auto-encoder by splitting the whole image into multiple channels and performing cross-channel prediction tasks.

3   Preliminary for Adversarial Learning

The GAN framework is an approach for estimating generative models via an adversarial learning process. Specifically, its network architecture is composed of a generator G and a discriminator D. Its objective is to make D to correctly differentiate between the true data and the generated data, and propel G to well capture the data distribution. Considering the training difficulty of the original GAN, we use SNGAN [17] as the baseline model since it shows better generation performance and training stabilization. Formally, the value function and the spectral normalization term adopted by SNGAN are as follows:

$$L_{gan} = \mathbb{E}_{x \sim p_x(x)}[log(D(x))] + \mathbb{E}_{z \sim p_z(z)}[log(1 - D(G(z)))],$$
$$SN(W^l) := W^l/\sigma(W^l) \ \ where \ \ W^l \in \theta, \tag{1}$$

where $p_x(x)$ and $p_z(z)$ are the true data distribution and the prior noise distribution, respectively. $\theta := \{W^1, ..., W^n\}$ is the parameter set of the discriminator's layers, n is the number of layers, and $\sigma(\cdot)$ is the spectral norm of a matrix. More details about the spectral normalization can refer to [17].

4   Self-supervised GAN

In this section we introduce the proposed generative model in detail, and focus on the following aspects: network architecture, adversarial learning for image generation, self-supervised learning for generation regularization, and model training.

4.1   Network Architecture

To perform the basic adversarial learning task and the auxiliary self-supervised learning task, we design an elaborate network architecture as shown in Fig. 2. Specifically, this architecture consists of two types of components for generation and discrimination, and all components are parameterized by deep neural networks. Among them, $S_1 \circ G$ and $S_2 \circ G$ are generators for two sets of channels, where G is the shared part for both sets, and S_1 and S_2 are the splitting parts for each set. Since there are two types of cross-channel prediction: (1) predicting the color component from the intensity component; (2) predicting the intensity component from the color component, we design two transformers T_{12} and T_{21} for predicting one set from the other set. C is a concatenator for combining two sets to form the whole data. D_1, D_2 and D_x are discriminators for the first set of channels, the second set of channels and the whole data, respectively.

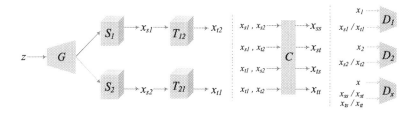

Fig. 2. The network architecture of SSGAN.

4.2   Adversarial Learning for Image Generation

As shown in Fig. 2, given a noise sample $z \sim p_z(z)$ we can generate two splitting channels (x_{s1} and x_{s2}) and two transformed channels (x_{t2} and x_{t1}), and concatenate these channels into four types of the whole data (x_{ss}, x_{st}, x_{ts} and x_{tt}). Overall, they are given by

$$x_{s1} = S_1 \circ G(z), \ x_{s2} = S_2 \circ G(z), \ x_{t2} = T_{12}(x_{s1}), \ x_{t1} = T_{21}(x_{s2});$$
$$x_{ss} = C(x_{s1}, x_{s2}), \ x_{st} = C(x_{s1}, x_{t2}), \ x_{ts} = C(x_{t1}, x_{s2}), \ x_{tt} = C(x_{t1}, x_{t2}). \quad (2)$$

By generating and concatenating image channels, we can build three types of generative models — GM_1, GM_2 and GM_x, as shown in Table 1. These models are responsible for the following adversarial learning tasks respectively: learning the distributions of (1) the first set of channels, (2) the second set of channels, and (3) the whole data. Following SNGAN, the corresponding value functions of these models are as follows:

$$L_1 = \mathbb{E}[log(D_1(x_1))] + \mathbb{E}[log(1 - D_1(x_{*1}))],$$
$$L_2 = \mathbb{E}[log(D_2(x_2))] + \mathbb{E}[log(1 - D_2(x_{*2}))], \quad (3)$$
$$L_x = \mathbb{E}[log(D_x(x))] + \mathbb{E}[log(1 - D_x(x_{**}))],$$

where x_{*1} and x_{*2} denote the generated channels; x_{**} denotes the concatenated whole data; x_1 and x_2 are two sets of channels from the true whole data x. For simplicity, the spectral normalization term of each model is ignored here.

Table 1. Three types of generative models.

Generative model	Components		Discrimination
	Generation		
GM_1	$S_1 \circ G;\ T_{21} \circ S_2 \circ G$		D_1
GM_2	$S_2 \circ G;\ T_{12} \circ S_1 \circ G$		D_2
GM_x	$S_1 \circ G;\ S_2 \circ G;\ T_{12} \circ S_1 \circ G;\ T_{21} \circ S_2 \circ G$		D_x

4.3 Self-supervised Learning for Generation Regularization

Except for adversarial learning for image generation, we further introduce a self-supervised learning task to improve image generation. This task performs a cross-channel prediction by only exploiting true data. Specifically, we split the true data x into x_1 and x_2, reuse transformers T_{12} and T_{21} as cross-channel predictors, and generate two predicting sets of channels — $T_{12}(x_1)$ and $T_{21}(x_2)$. The corresponding loss functions of cross-channel predictors are as follows:

$$L_{T_{12}} = \mathbb{E}[\ell(T_{12}(x_1), x_2)] \quad and \quad L_{T_{21}} = \mathbb{E}[\ell(T_{21}(x_2), x_1)], \tag{4}$$

where $\ell(m, n) = \|m - n\|_p$ measures the reconstruction error of two image channels based on the $\mathbf{L}^p$ norm, and we set $\mathbf{L}^1$ in this paper.

4.4 Model Training

Considering the proposed network architecture and two types of learning tasks, we can train the proposed model in two stages: (1) train these components ($S_1 \circ G$, $S_2 \circ G$ for generation; D_1, D_2, D_x for discrimination) and transformers (T_{12}, T_{21}) independently; and (2) train all components jointly. When jointly training all components, it should be noted that some components are affected by multiple value functions. Hence, we should balance the above value functions.

5 Experiments

We evaluate the proposed SSGAN on the benchmark dataset CIFAR [12], and provide both quantitative and qualitative evaluation. Specifically, we focus on the following aspects: image generation, inspecting the effect of self-supervised learning and channel prediction. For quantitative evaluation of the generation performance, we adopt the inception score (denoted as *IS*) [22]. We choose the *RGB* and *Lab* color spaces, where the *RGB* color space is used for the baselines and the *Lab* color space is used for the SSGAN. Briefly speaking, a whole *Lab* image could be divided into the intensity channel L and the color channels ab in the SSGAN.

Besides, some key configurations of experimental implementation are listed as follows. (1) *Network architecture:* we follow the CNN architectures [17].

(2) *Optimizer:* we use Adam optimizer for optimization with learning rate ($\alpha = 0.0001$) and the first and second order momentum parameters ($\beta_1 = 0.5$ and $\beta_2 = 0.999$) [17]. (3) *Model Training:* to balance the above value functions, we set the coefficient of L_{T_*} as 10 by experience, so that both the adversarial learning task and the self-supervised learning task can contribute to model learning.

5.1   Image Generation

In the SSGAN we can generate four types of the whole image — x_{ss}, x_{st}, x_{ts} and x_{tt}. To compare their performance of image generation, we show four types of generated image samples and list their ISs. In Fig. 3 we can observe that there is not obvious difference between image samples of x_{ss} and x_{st} in terms of visual perception, but image samples of x_{ts} and x_{tt} are inferior than those of x_{ss} and x_{st} in terms of texture and detail (*for a better view by zooming in*). Further, from Table 2 we can see that the IS of x_{st} is the highest, and the ISs of x_{ss} and x_{st} are higher than those of x_{ts} and x_{tt}. Both results indicate that the first type of cross-channel prediction is beneficial to image generation, however the second type of cross-channel prediction does not have a positive effect on image generation.

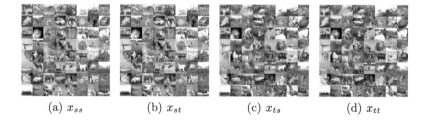

(a) x_{ss} (b) x_{st} (c) x_{ts} (d) x_{tt}

Fig. 3. Four types of image samples generated on CIFAR.

Table 2. Inception scores of four types of the whole image.

Type	Concatenation of (* intensity, * color)	Inception score
x_{ss}	(*splitting, splitting*)	7.44 ± 0.09
x_{st}	(*splitting, transformed*)	**7.70 ± 0.09**
x_{ts}	(*transformed, splitting*)	6.86 ± 0.07
x_{tt}	(*transformed, transformed*)	6.41 ± 0.08

To compare SSGAN with other methods, we also show image samples generated by these methods and list their ISs. In Fig. 4 images generated by SNGAN and SSGAN are clearer than those by other methods, while there is not obvious difference between SNGAN and SSGAN in terms of visual perception. However, from Table 3 we can see that the IS of SSGAN improves almost 0.28 compared with the baseline SNGAN. Besides, SSGAN performs better than other methods which directly generate RGB images as a whole.

(a) DCGAN   (b) WGAN   (c) WGAN-GP   (d) SNGAN   (e) **SSGAN**

Fig. 4. Image samples generated by contrast methods and SSGAN on CIFAR.

Table 3. Inception scores of several representative methods and SSGAN.

Method	Inception score
Real Images	11.24 ± 0.12
DCGAN	6.16 ± 0.17
WGAN	6.41 ± 0.11
WGAN-GP	6.68 ± 0.06
SNGAN	7.42 ± 0.08
SSGAN	$\mathbf{7.70 \pm 0.09}$

5.2   Effect of Self-supervised Learning

In order to evaluate the effectiveness of introducing self-supervised learning, we perform the experiment in which the self-supervised learning for transformer regularization is ignored. In other words, $L_{T_{12}}$ and $L_{T_{21}}$ will be not used for model updating. Here we mainly consider the generated whole image x_{st} and the first type of cross-channel prediction as described in Sect. 4.1. We present the ISs of x_{st} with/without self-supervised learning, and show image samples which consist of the original images and their reconstructed images based on cross-channel prediction. From Table 4 we can see that the IS of x_{st} with self-supervised learning is higher than that of x_{st} without self-supervised learning. As shown in Fig. 5 reconstructed images without self-supervised learning (*the left pair*) fail to infer the color component from the intensity component, while reconstructed images with self-supervised learning (*the right pair*) can better predict the color component. These again indicate that the first type of cross-channel prediction is beneficial to image generation.

Table 4. The effect of self-supervised learning.

Self-supervised	Inception score of x_{st}
Without	7.41 ± 0.07
With	**7.70 ± 0.09**

Fig. 5. Reconstructions based on predicting the color component from the intensity component. Each pair consists of the original image and its reconstruction.

5.3 Cross-Channel Prediction

Since we introduce a self-supervised learning task which performs cross-channel prediction, we could reconstruct a color image if only its intensity component or its color component is provided. In other words, the transformers T_{12} and T_{21} of SSGAN also can be used for image colorization and image texturization, respectively.

We illustrate some examples of image colorization and image texturization in Fig. 6. Specifically, the left subfigure includes original images, the middle subfigure includes reconstructed images based on predicting the color component from the given intensity component, and the right subfigure includes reconstructed images based on predicting the intensity component from the given color component. So the middle subfigure and the right subfigure correspond to image colorization and image texturization, respectively. By comparing original images with two types of reconstructed images, we can see that the transformer T_{12} can infer realistic colors, while T_{21} can not predict very fine texture. Viewed from another perspective, it indicates that when performing cross-channel prediction task, the second type is more difficult to the first type. This maybe explain the inferior generation performance of x_{ts} and x_{tt}.

Fig. 6. Reconstructions based on cross-channel prediction.

6   Conclusion

In this work we propose a generative model called *Self-supervised GAN* for improving image generation by introducing self-supervised learning into the GAN framework. Considering that channels of a color image are tightly correlated, we leverage this inherent attribute of color images and explicitly decompose image generation into multiple procedures. Based on the decomposition of image generation, the correlation between image channels as the self-supervised signal is dug for improving image generation. Hence, except for performing the basic image generation task in the adversarial learning framework, we also build an auxiliary cross-channel prediction task to regularize generation procedures in the self-supervised learning framework. Experimental results demonstrate that the proposed method can improve image generation compared with representative methods, and show capabilities of image colorization and image texturization.

References

1. Agrawal, P., Carreira, J., Malik, J.: Learning to see by moving. In: ICCV, pp. 37–45 (2015)
2. Arjovsky, M., Chintala, S., Bottou, L.: Wasserstein GAN. CoRR arXiv:1701.07875 (2017)
3. Cao, Y., Zhou, Z., Zhang, W., Yu, Y.: Unsupervised diverse colorization via generative adversarial networks. In: Ceci, M., Hollmén, J., Todorovski, L., Vens, C., Džeroski, S. (eds.) ECML PKDD 2017. LNCS (LNAI), vol. 10534, pp. 151–166. Springer, Cham (2017). https://doi.org/10.1007/978-3-319-71249-9_10
4. Cheng, Z., Yang, Q., Sheng, B.: Deep colorization. In: ICCV, pp. 415–423 (2015)
5. Denton, E.L., Chintala, S., Szlam, A., Fergus, R.: Deep generative image models using a laplacian pyramid of adversarial networks. In: NIPS, pp. 1486–1494 (2015)
6. Deshpande, A., Rock, J., Forsyth, D.A.: Learning large-scale automatic image colorization. In: ICCV, pp. 567–575 (2015)
7. Doersch, C., Gupta, A., Efros, A.A.: Unsupervised visual representation learning by context prediction. In: ICCV. pp. 1422–1430 (2015)
8. Dosovitskiy, A., Fischer, P., Springenberg, J.T., Riedmiller, M.A., Brox, T.: Discriminative unsupervised feature learning with exemplar convolutional neural networks. PAMI **38**(9), 1734–1747 (2016)
9. Goodfellow, I.J., et al.: Generative adversarial nets. In: NIPS, pp. 2672–2680 (2014)
10. Gulrajani, I., Ahmed, F., Arjovsky, M., Dumoulin, V., Courville, A.C.: Improved training of wasserstein gans. In: NIPS, pp. 5769–5779 (2017)
11. Kim, T., Cha, M., Kim, H., Lee, J.K., Kim, J.: Learning to discover cross-domain relations with generative adversarial networks. In: ICML, pp. 1857–1865 (2017)
12. Krizhevsky, A., Hinton, G.: Learning multiple layers of features from tiny images (2009)
13. Larsson, G., Maire, M., Shakhnarovich, G.: Learning representations for automatic colorization. In: Leibe, B., Matas, J., Sebe, N., Welling, M. (eds.) ECCV 2016. LNCS, vol. 9908, pp. 577–593. Springer, Cham (2016). https://doi.org/10.1007/978-3-319-46493-0_35

14. Larsson, G., Maire, M., Shakhnarovich, G.: Colorization as a proxy task for visual understanding. In: CVPR, pp. 840–849 (2017)
15. Liu, M., Tuzel, O.: Coupled generative adversarial networks. In: NIPS, pp. 469–477 (2016)
16. Mao, X., Li, Q., Xie, H., Lau, R.Y.K., Wang, Z.: Multi-class generative adversarial networks with the L2 loss function. CoRR arXiv:1611.04076 (2016)
17. Miyato, T., Kataoka, T., Koyama, M., Yoshida, Y.: Spectral normalization for generative adversarial networks. CoRR arXiv:1802.05957 (2018)
18. Noroozi, M., Favaro, P.: Unsupervised learning of visual representations by solving jigsaw puzzles. In: Leibe, B., Matas, J., Sebe, N., Welling, M. (eds.) ECCV 2016. LNCS, vol. 9910, pp. 69–84. Springer, Cham (2016). https://doi.org/10.1007/978-3-319-46466-4_5
19. Nowozin, S., Cseke, B., Tomioka, R.: F-gan: Training generative neural samplers using variational divergence minimization. In: NIPS, pp. 271–279 (2016)
20. Pathak, D., Krähenbühl, P., Donahue, J., Darrell, T., Efros, A.A.: Context encoders: feature learning by inpainting. In: CVPR, pp. 2536–2544 (2016)
21. Radford, A., Metz, L., Chintala, S.: Unsupervised representation learning with deep convolutional generative adversarial networks. CoRR arXiv:1511.06434 (2015)
22. Salimans, T., Goodfellow, I.J., Zaremba, W., Cheung, V., Radford, A., Chen, X.: Improved techniques for training gans. In: NIPS, pp. 2226–2234 (2016)
23. Wang, X., Gupta, A.: Unsupervised learning of visual representations using videos. In: ICCV, pp. 2794–2802 (2015)
24. Wang, X., Gupta, A.: Generative image modeling using style and structure adversarial networks. In: Leibe, B., Matas, J., Sebe, N., Welling, M. (eds.) ECCV 2016. LNCS, vol. 9908, pp. 318–335. Springer, Cham (2016). https://doi.org/10.1007/978-3-319-46493-0_20
25. Yang, J., Kannan, A., Batra, D., Parikh, D.: LR-GAN: layered recursive generative adversarial networks for image generation. CoRR arXiv:1703.01560 (2017)
26. Zhang, H., et al.: Stackgan: text to photo-realistic image synthesis with stacked generative adversarial networks. CoRR arXiv:1612.03242 (2016)
27. Zhang, R., Isola, P., Efros, A.A.: Colorful image colorization. In: ECCV, pp. 649–666 (2016)
28. Zhang, R., Isola, P., Efros, A.A.: Split-brain autoencoders: Unsupervised learning by cross-channel prediction. In: CVPR, pp. 645–654 (2017)
29. Zhao, J.J., Mathieu, M., LeCun, Y.: Energy-based generative adversarial network. CoRR arXiv:1609.03126 (2016)

Deformable Point Cloud Recognition Using Intrinsic Function and Deep Learning

Zhenzhong Kuang[1], Jun Yu[1(✉)], Suguo Zhu[1], Zongmin Li[2],
and Jianping Fan[1,3]

[1] Key Laboratory of Complex Systems Modeling and Simulation,
School of Computer Science and Technology, Hangzhou Dianzi University,
Hangzhou, China
{zzkuang,yujun}@hdu.edu.cn
[2] College of Computer and Communcation Engineering,
China University of Petroleum (Huadong), Qingdao, China
[3] Department of Computer Science, UNC-Charlotte, Charlotte, USA

Abstract. Recognizing 3D point cloud shapes under isometric deformation is an interesting and challenging issue in the geometry field. This paper proposes a novel feature learning approach by using both the model-based intrinsic descriptor and the deep learning technique. Instead of directly applying deep convolutional neural networks (CNN) on point clouds, we first represent the isometric deformation by using a set of local intrinsic functions to grasp the invariant properties of the shape. Then, an effective point CNN network is developed to learn the parameters and perform semantic feature learning in an end-to-end fashion to link the local and global information together for discriminative shape representation and classification. To reduce the computational costs of our CNN network, some simple operations, like downsampling and fusion, are applied to decrease the number of points and the intrinsic dimensions based on our average heat function. The experimental results on multiple standard benchmarks have demonstrated that our proposed algorithm can achieve very competitive results on both the accuracy rates and the computational efficiency.

Keywords: Point cloud recognition · Intrinsic function · Deep learning

1 Introduction

With the exponential growth of the 3D digital objects on the Internet, the representation of geometric shapes has been a central problem in 3D modeling. In particular, point clouds are one of the most primitive and fundamental representations of these objects and working directly with such representation is more critical and challenging [11,12,14]. In recent years, point cloud receives more attention in various fields because of its simplicity and it can also avoid the combinatorial irregularities and complexities of building meshes [12,14].

© Springer Nature Switzerland AG 2018
R. Hong et al. (Eds.): PCM 2018, LNCS 11165, pp. 89–101, 2018.
https://doi.org/10.1007/978-3-030-00767-6_9

This paper focuses on recognizing point cloud shapes under the isometric deformation (e.g. hands bending in Fig. 1) which is one of the most commonly seen shape variations. By using the intrinsic descriptors, traditional methods can invariantly model the isometric shapes, but they cannot always achieve robust representationb because the hand-craft descriptors can only extract limited information compared with the complex geometrical variations [1,8,21].

In recent years, the deep learning technique has boosted the performances in recognition tasks [2,6,23,25,26]. Compared with the traditional methods, deep learning exhibits more flexible properties in model selection with large numbers of data-driven parameters. Many new approaches were proposed by working directly on 3D shapes with new CNN architectures, such as 3D volumetric CNN [2,23], 3D graph CNN [15] and PointNet [17]. Another promising research trend is to convert the 3D shapes to low-dimensional vectors and then apply the existing deep learning frameworks. In [20], the 2D multi-view images are used to train CNN network for shape recognition. In [24], the heat diffusion descriptors were fed to deep neural networks for feature learning and classification.

In this paper, we focus on recognizing isometric shapes in the format of point clouds by addressing two issues: how to learn discriminative features for isometric shape representation, and how to accelerate the learning process on point cloud. We propose a novel 3D feature learning approach to integrate both the low-level and high-level information, so that our systems can automatically interpret rich and diverse semantics of massive shapes. By a series of experimental comparison with state-of-the-art, our algorithm has superior performance for point cloud recognition. The main contributions of this paper are summarized as follows:

- We present an integrated framework for isometric 3D point cloud recognition that extracts discriminative shape features with intrinsic functions.
- We leverage a set of intrinsic functions as low-level descriptors to describe the deformable shape and a subspace sampling strategy is employed to reduce the computational costs and accelerate the deep learning process.
- We describe a novel deep CNN network to integrate the low-level point descriptors for more discriminative high-level semantic feature representation;
- We finally carry out thorough experimental analysis of the proposed approach for shape recognition.

2   Related Work

Deformable Shape Descriptors. A large variety of local shape descriptors have been developed to deal with non-rigid deformations, such as visual similarity approach [8] and diffusion geometry approach [1,21]. For holistic shape representationb of the local features, the bags-of-word (BOW) model [1,5,8] is widely used by extracting the frequency histogram of geometry words, which is further used for shape recognition. Among these descriptors, the heat diffusion theory have demonstrated promising performances on robust feature representation, which has already been used for intrinsic invariant shape analysis, such as global point signature (GPS) [19] and heat kernel signature (HKS) [21].

Point Cloud Descriptors. Compared with 3D meshes, point cloud data is simple without explicit structure information, which poses a more difficult task for feature extraction. Due to the remarkable performance of heat diffusion, the diffusion distance [3] was employed for 3D point cloud recognition [14]. In [12], a retrieval track was organized to recognize the point cloud toys, where diverse feature extraction and learning approaches were compared [8,21]. Although some methods are available in the literature for point cloud representation and recognition, they lack of flexibility in feature modeling and are limited by hand-craft design or empirical assumption.

3D Deep Learning. With the popularity of deep learning, 3D shape recognition also finds new progresses with lots of new algorithms for 3D data representation [2,20,24]. Although prior works have some achievements, they suffer from different drawbacks, such as the low resolution problem of the volumetric CNNs [23] and most of the prior deep learning algorithms focus on processing 3D meshes instead of point clouds. Recently, Su et al. [17] designed a point cloud based CNN network that receives an unordered set of 3D positions and labels. But this approach neglects the latent spatial relations between pairwise points and it cannot address non-rigid shape deformations intelligently.

By investigating the related works, we observe that it is a very interesting task to develop new algorithms by leveraging both the traditional intrinsic approaches and the recent deep learning approaches for point cloud recognition. We need to point out that our proposed approach is different from existing approaches on point cloud recognition. On the one hand, our approach outperforms the traditional methods [8,12,14] by finding a flexible parameterized model to overcome the incomplete BOW model for discriminative shape representation and it can also be transferred to other domains as well. On the other, it is different from most of the deep learning methods [20,23,24] by employing distinct CNN networks on point cloud.

3   Proposed Method

3.1   Overview

Given a 3D shape $S = (V, E)$ with vertex set V and edge set E (which is unknown for point cloud shape), and suppose $F(\cdot)$ is a function defined on S, the goal of isometric shape recognition is to find the best feature $F(\cdot)$ that can: (1) invariantly describe S regardless of the deformations that it suffers, and (2) distinguish it from shapes belong to another category.

According to Fig. 1, our proposed approach consists of two stages: (1) intrinsic function descriptor (IFD) extraction for low-level shape description; (2) discriminative deep IFD (DIFD) feature learning for high-level shape representation. Finally, the learned feature is used for point cloud recognition.

3.2   Low-Level Descriptor Extraction

Intrinsic Embedding. Due to the superior performance of the Laplace-Beltrami (LB) operator related spectral methods for intrinsic representation [1,14,19,21], the following function is recommended for intrinsic embedding

$$p_t(x) = (K_t(\lambda_1)\phi_1(x), K_t(\lambda_2)\phi_2(x), \cdots, K_t(\lambda_\theta)\phi_\theta(x)), \qquad (1)$$

where $x \in V$, λ_i and $\phi_i (1 \leq i \leq \theta)$ are the corresponding eigenvalue and eigenfunction of LB operator, θ represents the number of eigenfunctions, and $K_t(\lambda_i) = e^{-\lambda_i t}$ corresponds to the kernel function. One of the most important properties of the LB operator is that it is invariant to isometric deformations. Because that the eigenfunctions and eigenvalues carry global character, the above equation would lead to robust manifold representation by adopting the top few noiseless items and the eigen-decomposition problem of the LB operator could be solved with linear complexity [1,21]. To alleviate scale selection, Eq. (1) is redefined by accumulating all the possible scales $p(x) = \lim_{T \to \infty} \sum_{t=0}^{T} p_t(x)$ and all the points are moved to the center of the coordinates. An example of the embedding results is shown in Fig. 1(a) a1.

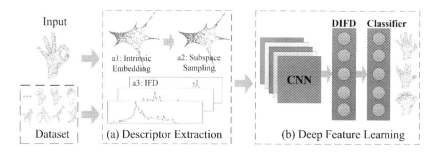

Fig. 1. A brief overview of the proposed approach for isometric 3D point cloud recognition.

Large Margin Subspace Sampling (LMSS). Suppose $\forall S, F(S \to \mathbb{R}^m)$ and $\hat{S}$ be a subspace of S, it is easy to prove that $\exists S_L, \gamma, s.t. S_L + \gamma \subseteq \hat{S} \subseteq S$, $F(\hat{S}) = F(S)$. This theorem reveals that the majority properties of a shape are determined by a set of critical points [1,8,17] which consist of the lower bound S_L of shape S. In general, S_L is unknown and it is approximated by using the subspace sampling result $\hat{S}$ which forms a cover of S_L with a large margin γ.

We propose to sample a subspace $\hat{S} = \{x \in \hat{S}_1(k_1) \cup \hat{S}_2(k_2)\}$ by using our average heat function $h(x) = p^T(x)p(x)$ at each point, where $\hat{S}_1(k_1)$ denotes the most significant k_1 points that have the largest heat values $h(\cdot)$, and $\hat{S}_2(k_2)$ denotes the most non-salient k_2 points that have the smallest heat values $h(\cdot)$. The reason of defining a subspace $\hat{S}$ lies in that the majority properties of a shape are covered by two kinds of points and the remaining points can be seen as the "filler" (or noise for sometimes). Intuitively, an example of the subsampling results is shown in Fig. 1(a) a2, where the red points belong to $\hat{S}_1$, the blue points belong to $\hat{S}_2$ and the others are the "filler" points.

Intrinsic Function Descriptor (IFD). Given a set $\hat{S}$ of critical points and the intrinsic coordinates $\{p(x)\}$, we leverage a concise intrinsic function

$$f(x,l) = \frac{|d(x,v_j) \in bin(x,l)|}{\max_{i,j,\vartheta} |d(v_i,v_j) \in bin(v_i,\vartheta)|}, \tag{2}$$

to describe the local information captured by each point, where $d(x,v_i) = \|\hat{p}(x) - \hat{p}(v_i)\|$ is a Euclidean distance metric, and $|d(v_i,v_j) \in bin(v_i,\vartheta)|$ denotes the number of distances in the ϑ-th bin of vertex v_i. Then, each point cloud shape is represented as a set of intrinsic functions $f = \{f(x) = (f(x,l))_{l=1}^{\vartheta}, \forall x \in \hat{S}\}$.

Our inspiration comes from the traditional shape distributions [3,14,16] to define global shape feature. Instead, we use pairwise distances to define the local descriptors starting from each vertex $x \in \hat{S}$. Traditional methods usually use the BOW model to aggregate the local descriptors for global shape representation, which is usually constrained by the manually defined (coarse) visual dictionary model [1,7,8]. Here, we prefer to explicitly introduce a parameterized function $\hat{f}(\hat{S}|W_0) = \{\hat{f}(x|w_{kl}, b_k), x = v_k\}$ for flexible descriptor aggregation

$$\hat{f}(x|w_{kl}, b_k) = w_{kl}f(\hat{S}(x), l) + b_k \tag{3}$$

which allows our algorithm to learn the parameter set $W_0 = \{w_{kl}, b_k\}$ wisely according to the real data so that we can learn more discriminative features and train more effective classifiers for shape recognition. Besides, it is easy to verify that our descriptor also has multiple desired properties [1,19,21]: (a) *Isometric invariant*, (b) *Scale-invariant*, (c) *Robustness* and (d) *Informative*.

3.3   Deep Feature Learning

Given a set of points $\hat{S}$ and the local feature set $\hat{f}(\hat{S}|W_0)$, the goal of this part is to learn discriminative features for shape recognition

$$F(S) = CNN\{\hat{f}(\hat{S}|W_0), W_l, 1 \leq l \leq \mathcal{L}\} \tag{4}$$

by using the deep CNN networks. Our approach takes effective local shape descriptors as input because the intrinsic coordinate $p(x)$ does not directly support discriminative shape description due to the switch of the eigenfunctions [19,21]. By subspace sampling, we can only obtain a coarse and redundant point set $\hat{S}$ because there are no prior information available to perform accurate sampling, which may produce negative effects for shape representation. We address the problem by using our deep convolutional neural network.

The basic architecture of our network is visualized in Fig. 2, which consists of three components. In the DEM module, the data points are first downsampled by using $r \times 1$ convolutional kernel, and then they are fed to a series of transformation layers with 1×1 kernels to learn discriminative local descriptors, where

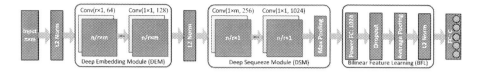

Fig. 2. The basic structure of our network. It takes m-dim descriptors of $n = |\hat{S}|$ points as input, goes through the DEM and DSM modules, outputs a 1024 d vector via maxpooling, and then fed to the BFL module to obtain discriminative features (256 d) for classifier training of C categories. The numbers in brackets, like $(r \times 1, 64)$, denotes the kernel size and layer output.

the parameters in Eq. (3) are determined in this stage. In the DSM module, all the local descriptors are first squeezed by their length m, and then they are fed to a series of transformation layer with 1×1 kernels to enrich the aggregation space for local feature fusion in the following maxpooling layer which works on unordered point clouds to produce a global vector. Although the DSM output can be directly fed to a fully connected (FC) layer for global feature learning, we prefer to enhance the feature discrepancy by using a bilinear feature learning process which outputs a more discriminative global feature for classifier training.

Bilinear Feature Learning (BFL). In recent years, bilinear pooling has been successfully used for fine-grained image recognition [4], but related works are rare for point cloud recognition. In this part, we generalize the method to enrich point cloud features. Given feature $\hat{f}$, we use the following bilinear model to capture the pairwise interactions between feature dimensions

$$z_i = \frac{1}{\delta} \sum_{j=1}^{\delta} (u_j^T \hat{f})^T (u_j^T \hat{f}) = \frac{1}{\delta} \mathbb{1}^T \rho(\xi(\hat{f})) \tag{5}$$

where $z = \{z_i\} \in \mathbb{R}^o$ denotes a set of outputs, $\xi(\hat{f}) = U_i^T \hat{f}$, and $\rho(\xi(\hat{f}))$ is the element-wise square of $\xi(\hat{f})$. By reformulating the matrix $U = [U_i] \in \mathbb{R}^{m \times \delta o}$, $U_i = (u_1, u_2, \cdots, u_\delta) \in \mathbb{R}^{m \times \delta}$, we rewrite z as

$$z = AvgPooling(\rho(\xi(\hat{f})), \delta). \tag{6}$$

As visualized in the BFL module of Fig. 2, we use "Power FC" layer to realize $\rho(\xi(f))$ and a normalized "AvgPooling" layer to produce the final shape feature.

Parallel Network Training (PNT). The sampling operation in our basic net may affect the classification performance because that the size of deep modeling space is significantly reduced when $r > 1$. To alleviate this problem, we design a parallel network in Fig. 3 to improve the performance of our basic network ($r > 1$) in the training process, where L_2 norm layer is the same as the basic network, and batch normalization is used for all layers with ReLU. Then, we minimize the following multi-task learning model to determine shape feature $F(S)$ and the multi-class classifiers

$$L = \alpha_1 L_1 + \alpha_2 L_2 + \alpha_3 L_3 + \alpha_4 L_4 + \alpha_5 L_{reg} \tag{7}$$

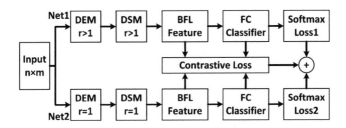

Fig. 3. The parallel network training process.

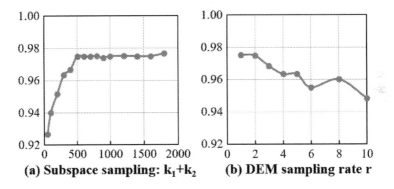

(a) Subspace sampling: k_1+k_2 (b) DEM sampling rate r

Fig. 4. The influencial curves of sampling by (a) large margin subspace sampling and (b) DEM(r) towards classification (the vertical axis).

where $L_1 = -\sum_i y_i p(S|r >1)$ and $L_2 = -\sum_i y_i p(S|r = 1)$ denote the softmax loss function of Net1 and Net2, $L_3 = \|F_{r>1}(S) - F_{r=1}(S)\|$ and $L_4 = \|p_{r>1}(S) - p_{r=1}(S)\|$ denote the contrastive loss for both the feature layer and the classifier layer, $L_{reg} = \|W_{r>1}\| + \|W_{r=1}\|$ is the regularization item to avoid overfitting, and α_i is the importance weight. With a well trained network, our basic network shown in Fig. 2 is leveraged for feature extraction and classification.

Different from prior works [14,17,20,23,24], our approach depends on two stages for point cloud recognition so that we do not need to design complex networks, where the IFD descriptor has helped the deep network to address the deformable invariance problem. Therefore, the training of our deep network would be easier and the process of feature extraction would be faster than the networks without sampling operation. The DEM module could help to further downsample the input data by convolution with $r \times 1$ kernel, which uses data-driven strategy to assist sampling or compression by parameterized fusion. Besides, the DSM module takes the learned DEM local descriptors as input and encodes them to generate the global aggregations and the BFL helps DSM to enrich the discriminative ability for global feature representation.

4   Experimental Results

To validate our approach, we employ three commonly used datasets (SHREC15, SHREC11 and SHREC10) [8–10,22] that have different extent of difficulties for isometric deformable shape recognition. Following prior works [2,24], each dataset is split into train and test sets. The effectiveness of our approach is validated by performing comprehensive comparison with several shape classification and retrieval methods. For performance evaluation, the following popular criteria are employed [8,10,12,20,22]: the top 1 classification accuracy, nearest neighbor (NN), first tier (FT), second tier (ST), discounted cumulated gain (DCG), average precision (AP) and precision-recall (PR) curves.

Table 1. Top 1 classification accuracy (%) of our approach.

Feature	SHREC15	SHREC11	SHREC10
IFD	85.0	85.0	74.3
DIFD	95.1	94.7	85.2
DIFD+BFL	96.3	95.7	86.7
DIFD+BFL+PNT	97.5	96.3	88.5

Table 2. Comparison of the top 1 accuracy rates (%).

Method	Dataset	Shape	Accuracy
ShapeDNA	SHREC15	Mesh	61.2
GPS		Mesh	63.4
PointNet		Point Cloud	48.0
DIFD+BFL+PNT		Point Cloud	97.5
ShapeDNA	SHREC11	Mesh	85.4
GPS		Mesh	83.2
3D MMF		Mesh	95.4
PointNet		Point Cloud	40.0
DIFD+BFL+PNT		Point Cloud	96.3
ShapeDNA	SHREC10	Mesh	82.7
GPS		Mesh	87.2
PointNet		Point Cloud	69.0
DIFD+BFL+PNT		Point Cloud	88.5

Implementation details. All experiments were conducted using Tensorflow, where the parameters of our system is determined by experiment. We uniformly sample 2000 points on each shape as input and we choose $k_1 = 4k_2 = 400$ as the lower bound for large margin subspace sampling. We set the length of IFD descriptor as $\vartheta = 8$ and $r = 2$ for DEM and DSM in Net1, where the number of convolutional layers with 1×1 kernel is set as 1 for both DEM and DSM modules. Figure 4 plots the performance of our approach on different parameters, which is used to specify the best k_1, k_2 and r. We set $\delta = 4$ for BFL, and $\alpha_1 = \alpha_2 = 1$, $\alpha_3 = \alpha_4 = 0.001$, $\alpha_5 = 0.00001$ for Eq. (7).

4.1 Shape Classification Performance

Accuracy of our approach. Table 1 presents the classification results of our approach under different settings. We see that, with the help of deep CNN network, our DIFD descriptor has improved the hand-craft IFD descriptor around 10%. By using BFL instead of FC for global feature learning, DIFD+BFL has improved DIFD more than 1.0% on accuracy. And, the parallel training process can further improve the classification performance about 1.0%.

Table 3. The retrieval performance (%) on the adopted datasets.

Method	Dataset	NN	FT	ST	DCG
DIFD+BFL+PNT	SHREC15	98.5	95.2	97.6	98.4
PointNet		63.3	28.5	37.3	58.3
HKS		91.3	68.6	78.6	87.0
DIFD+BFL+PNT	SHREC11	97.3	93.9	96.6	97.3
PointNet		50.3	25.8	35.5	54.7
3D MMF		98.0	86.9	46.8	97.4
HKS		91.7	75.0	85.8	90.1
DIFD+BFL+PNT	SHREC10	91.0	80.7	91.1	92.5
PointNet		64.0	54.8	67.0	74.7
HKS		85.0	50.3	74.0	80.5

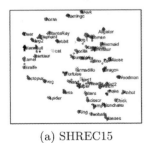

(a) SHREC15	(b) SHREC11	(c) SHREC10

Fig. 5. The t-SNE visualization results by using our DIFD+BFL+PNT feature, where the texts denote the category names.

Comparison with state-of-the-art. In Table 2, we compare our approach DIFD+BFL+PNT with both the traditional hand-craft methods and the deep learning methods: ShapeDNA [18], GPS [19], HKS [1,21], PointNet [17] and 3D MMF [2]. One can observe that our results have achieved comparable performance with the contrast methods across different datasets, which also indicates that our method can achieve very competitive performances with the sate-of-the-art approaches.

4.2 Shape Retrieval Performance

In Fig. 5, we present the t-SNE embedding results [13] of our DIFD+BFL+PNT feature. The results show that, in spite of the limited outliers, most of the shapes from the same category are grouped together and the shapes from different categories are separated.

Retrieval result. For easy comparison with the other methods, we perform a leave-one-out kNN retrieval experiment that each deformed shape is queried against the remaining intra-class shapes [8]. Table 3 presents the comparison results of our approach with other state-of-the-art methods. From this table, we have two main observations: (i) our DIFD+BFL+PNT method has achieved the best overall retrieval performance; (ii) DIFD+BFL+PNT works well on datasets under different strengths of non-rigid deformations. In Fig. 6 and Fig. 7, we plot the AP results and PR curves on the adopted three datasets, respectively.

Complexity and timing. The complexity of constructing the intrinsic embedding and extracting the IFD descriptors is linear to the number of points $O(\xi n)$ and the complexity of our deep network is also linear to the number of points $O(\vartheta n)$, which indicates the scalability of our approach. On the SHREC15 dataset, the average training and testing time of our approach is shown in Table 4, where "Train/Test($r = 1$) without LMSS is the baseline. One can find that, with the help of the DEM downsampling operation by $r = 2$, our algorithm "Train($r = 2$) without LMSS" can speed up "Train($r = 1$) without LMSS" by 43.9%. With the help of LMSS, our algorithm "Train($r = 2$) with LMSS" can speed up "Train($r = 1$) with LMSS" by 37.9%. The overall speed up by using $r = 2$ with LMSS is 82.5% and 64.0% for train and test compared with the baseline, respectively, which suggests the efficiency of our approach.

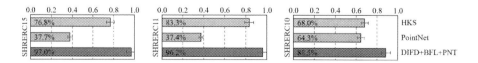

Fig. 6. The barchart results of the average precision.

Table 4. Timings of this paper (in milliseconds) for each epoch with or without large margin subspace sampling (LMSS).

	without LMSS	with LMSS
Train (r = 1)	8487	2390
Train (r = 2)	4759	**1484**
Test (r = 1)	2780	1233
Test (r = 2)	1758	**1001**

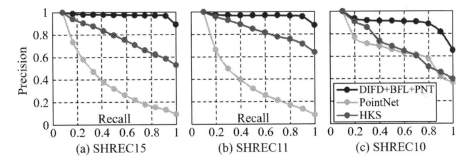

Fig. 7. The precision-recall curves of our proposed approach.

5   Conclusions

In this paper, we have proposed a novel approach for deformable 3D point cloud recognition. With the help of the low-level descriptor, our algorithm can well model the isometric deformation. With the help of the deep feature learning, our algorithm has further improved the feature discriminative ability. By using the LMSS and downsampling operation in DEM and DSM modules, our algorithm has achieved a significant speed up for model learning and feature extraction. By comparing our approach with the state-of-the-art methods on standard benchmarks, all the evaluation results have demonstrated the superior performance of our approach on shape representation, classification and retrieval.

Acknowledgments. This work is partly supported by National Natural Science Foundation of China under Grants 61806063, 61772161, 61622205, 61472110, 61602136 and the Zhejiang Provincial Natural Science Foundation of China under Grant LR15F020002. The authors would like to thank the reviewers and editors for their insightful comments and valuable suggestions which have led to substantial improvements of the paper.

References

1. Bronstein, A.M., Bronstein, M.M., Guibas, L.J., et al.: Shape google: Geometric words and expressions for invariant shape retrieval. ACM Trans. Graph. **30**(1), 1 (2011)
2. Bu, S., Wang, L., Han, P., et al.: 3d shape recognition and retrieval based on multi-modality deep learning. Neurocomputing (2017)
3. Coifman, R.R., Lafon, S.: Diffusion maps. Appl. Comput. Harmonic Anal. **21**(1), 5–30 (2006)
4. Fukui, A., Park, D.H., Yang, D., et al.: Multimodal compact bilinear pooling for visual question answering and visual grounding. arXiv preprint arXiv:1606.01847 (2016)
5. Kuang, Z., Li, Z., Jiang, X.: Retrieval of non-rigid 3D shapes from multiple aspects. Comput.-Aided Des. **58**(C), 13–23 (2015)
6. Kuang, Z., Yu, J., Li, Z., Zhang, B., Fan, J.: Integrating multi-level deep learning and concept ontology for large-scale visual recognition. Pattern Recognit. **78**, 198–214 (2018)
7. Laga, H., Schreck, T., Ferreira, A., et al.: Bag of words and local spectral descriptor for 3D partial shape retrieval. In: Eurographics Workshop on 3D Object Retrieval, pp. 41–48. Citeseer (2011)
8. Lian, Z., Godil, A., Bustos, B.: A comparison of methods for non-rigid 3D shape retrieval. Pattern Recognit. **46**(1), 449–461 (2013)
9. Lian, Z., Godil, A., Fabry, T., et al.: Shrec'10 track: Non-rigid 3D shape retrieval. In: Eurographics Workshop on 3D Object Retrieval, vol. 10, pp. 101–108. The Eurographics Association (2010)
10. Lian, Z., Zhang, J., Choi, S., et al.: Shrec'15 track: non-rigid 3D shape retrieval. In: Eurographics Workshop on 3D Object Retrieval. The Eurographics Association (2015)
11. Liang, J., Lai, R., Wong, T.W., et al.: Geometric understanding of point clouds using laplace-beltrami operator. In: IEEE Conference on Computer Vision and Pattern Recognition, pp. 214–221. IEEE (2012)
12. Limberger, F., Wilson, R., Aono, M., et al.: Point-cloud shape retrieval of non-rigid toys. In: Eurographics Workshop on 3D Object Retrieval. The Eurographics Association (2017)
13. van der Maaten, L., Hinton, G.: Visualizing non-metric similarities in multiple maps. Mach. Learn. **1**(87), 33–55 (2012)
14. Mahmoudi, M., Sapiro, G.: Three-dimensional point cloud recognition via distributions of geometric distances. Graph. Models **71**(1), 22–31 (2009)
15. Maron, H., Galun, M., Aigerman, N., et al.: Convolutional neural networks on surfaces via seamless toric covers. ACM Trans. Graph. **36**(71) (2017)
16. Osada, R., Funkhouser, T., Chazelle, B.: Shape distributions. ACM Trans. Graph. **21**(4), 807–832 (2002)
17. Qi, C.R., Su, H., Mo, K., Guibas, L.J.: Pointnet: Deep learning on point sets for 3D classification and segmentation. In: IEEE Conference on Computer Vision and Pattern Recognition (2017)
18. Reuter, M., Wolter, F.E., Peinecke, N.: Laplace-beltrami spectra as Shape-DNA of surfaces and solids. Comput.-Aided Des. **38**(4), 342–366 (2006)
19. Rustamov, R.M.: Laplace-beltrami eigenfunctions for deformation invariant shape representation. In: Eurographics Symposium on Geometry Processing, pp. 225–233. Eurographics Association (2007)

20. Su, H., Maji, S., Kalogerakis, E., et al.: Multi-view convolutional neural networks for 3D shape recognition. In: IEEE Conference on Computer Vision and Pattern Recognition, pp. 945–953. IEEE (2015)
21. Sun, J., Ovsjanikov, M., Guibas, L.: A concise and provably informative multi-scale signature based on heat diffusion. Comput. Graph. Forum **28**(5), 1383–1392 (2009)
22. Tabia, H., Picard, D., Laga, H., et al.: Compact vectors of locally aggregated tensors for 3D shape retrieval. In: Eurographics Workshop on 3D Object Retrieval, pp. 17–24. The Eurographics Association (2013)
23. Wu, Z., Song, S., Khosla, A., et al.: 3d shapenets: a deep representation for volumetric shapes. In: IEEE Conference on Computer Vision and Pattern Recognition, pp. 1912–1920 (2015)
24. Xie, J., Dai, G., Zhu, F.: Deepshape: deep-learned shape descriptor for 3D shape retrieval. IEEE Trans. Pattern Anal. Mach. Intell. **39**(7), 1335–1345 (2017)
25. Yu, J., Tao, D., Wang, M.: Learning to rank using user clicks and visual features for image retrieval. IEEE Trans. Cybern. **45**(4), 767–779 (2015)
26. Yu, J., Yang, X., Gao, F.: Deep multimodal distance metric learning using click constraints for image ranking. IEEE Trans. Cybern. **47**(12), 4014–4024 (2017)

Feature Synthesization for Real-Time Pedestrian Detection in Urban Environment

Wenhua Fang[1($\boxtimes$)], Jun Chen[1], Tao Lu[2], and Ruimin Hu[1]

[1] National Engineering Research Center for Multimedia Software,
Computer School of Wuhan University, Wuhan 430072, Hubei Province, China
{fangwh,chenj,hrm}@whu.edu.cn
[2] Computer School of Wuhan Institute of Technology, Wuhan 430205,
Hubei Province, China
lutxyl@gmail.com

Abstract. Real-time pedestrian detection is very essential for auto assisted driving system. For improving the accuracy, more and more complicate features are proposed. However, most of them are impracticable for the real-world application because of high computation complexity and memory consumption, especially for onboard embedding system in the unmanned vehicle. In this paper, a novel framework that utilizes reconstruction sparsity to synthesize the feature map online is proposed for real-time pedestrian detection for the early warning system of the unmanned vehicle in real world. In this framework, the feature map is computed by sparse line combination of the representative coefficient and the feature response of trained basis which is learned offline. The efficiency of our method only depends on the dictionary decomposition no matter how complicated the feature is. Moreover, our method is suitable for most of the known complicate features. Experiments on four challenging datasets: Caltech, INRIA, ETH and TUD-Brussels, demonstrate that our proposed method is much efficient (more than 10 times acceleration) than the state-of-the-art approaches with comparable accuracy.

Keywords: Pedestrian detection · Feature synthesization
Sparse representation

1 Introduction

Recently, the unmanned vehicle, as a new transportation which has the merits of energy saving and environmental protection, is getting more and more attentions. Meanwhile, its security is the focus of the debate. As we all know, obstacles identification is one of the core functions of the early warning system of unmanned vehicle, and how to detect pedestrian as soon as possible in real world is one of the key problems in obstacle recognition.

© Springer Nature Switzerland AG 2018
R. Hong et al. (Eds.): PCM 2018, LNCS 11165, pp. 102–112, 2018.
https://doi.org/10.1007/978-3-030-00767-6_10

Pedestrian detection is a very important task in computer vision and has great potential to apply in many fields, such as automatic assisted driving, intelligent traffic management, etc. It is very challenging because of the multiple views, different illuminations, multiple scales and partial occlusion. To overcome these problems, many researchers have proposed a lot of complicated features [1–7] to improve the accuracy of this task, but ignored the efficiency. Take the popular object detector [4] for example. It will take more than 4 s per image with the resolution of 352 × 288 and take even more than 30 s per image with the resolution of 1280 × 720 on the 4-core desktop computer. Nowadays with the wide application of high resolution cameras, this problem is more and more serious and has become a bottleneck for real-time application.

From the common framework of pedestrian detection, we can see that the time cost is proportional to the product of two parts. One is the time complexity of the detector, and the other is the time for probing the object candidates. For the exhaustive search, such as sliding window technique, there are usually tens of thousands of probing candidates for pedestrian classification and location. So many researchers have carried out the work in the above two aspects to improve the efficiency of the pedestrian detection [8–12]. Felzenszwalb et al. [9] proposed a cascade part pruning strategy to speed up the deformable part model by more than ten times. In [8], Yan etal leveraged the low rank constraint on root filter to get a 2D correlation between root filter and feature map, and used the lookup table to speed up the HOG extraction. And it was 4 times faster than the current fastest DPM method with similar accuracy. Besides, many other efforts, called region proposal, have be done to get the object location candidate prior to object detection. In [10], Sande et al proposed a hierarchical framework, named *selective search*, to generate approximate 2000 regions per image by color segmentation and the recall was up to 97%. Compared with the exhaustive search, it was much effective. However, due to the computation complexity of selective search, it is not very fit for the real-time application. Zitnick [11] proposed a method, named *Edgebox*, to gain the region candidates at a lightweight computational cost, but its recall rate is relatively low. Other approaches are also facing similar problems. Recently, deep neural network, named deep learning, has become the state-of-the-art approach in object detection [13,14]. But this kind of methods are too complex to need auxiliary computation equipment, such as GPU, to complete the long-term training and testing.

From the above analysis, we can see that the time consumption of feature extraction is the key for efficient pedestrian detection. Is there an approach that the time consumption is approximately fixed for most of features? In this paper, we proposed a framework based on sparse coding to conduct the feature extraction online by linear combination of features of dictionary atoms extracted offline. This is based on the assumption that the natural image can be linearly combined by the patches sparsely. If the feature satisfies the linear superposition principle, the feature extraction can be synthesized online and the time consumption is just decided by the image decomposition.

2   Related Work

Our work is aspired by [15,16]. The core idea in [15] is the shared representations. And an intermediate representation, called *sparselet*, for deformable part models was proposed for multi-class object detection. In this model, sparse coding of part filters was used to represent each filter as a sparse linear combination of shared dictionary elements, which are the parameters of the part filter. Reconstruction of the original part filter responses via sparse matrix-vector product reduces computation relative to conventional part filter convolutions. The main defect is the sacrifice of the performance. In [16], *sparselet* was reformulated in a general structured output prediction framework leading to larger speedup factors with no decrease in task performance. We think more deeply about the problem. Compared with them, our model has smaller granularity and is more general. Our main contribution is that we first consider the feature response synthesization for feature extraction for any pedestrian detection framework by sparse coding. And we demonstrate that the synthesized feature has comparable performance with fast feature extraction for online pedestrian detection.

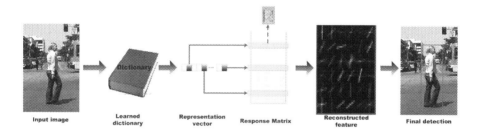

Fig. 1. The framework of feature response synthesization for pedestrian detection.

3   Our Framework

In this section, we will introduce the framework in details. Our model is based on the fact that natural image can be represented by linear combination of redundant bases. There are two stages (offline training and online detecting) in our framework. According to standard sparse representation, we first learn a representative dictionary in the training dataset under the minimum reconstruction error and sparsity constraint. And then the response matrix is created by conducting the feature extraction operation on the items of the dictionary (the item is regarded as patches). On the detecting stage, an input image is represented by the sparse representative vector on the learned dictionary. In the vector, only a few items are non-zero. And then the feature response synthesization is regarded as the linear combination of the representative vector and the row items of the response matrix (shown as Fig. 1).

What conditions should be satisfied if the feature is fit for our framework? We think that the linear superposition principle should be satisfied. The principle is as follows:

$$f(ax + by) = af(x) + bf(y) \tag{1}$$

where $f(*)$ denotes the feature extraction operation. Under certain conditions, we can relax the restriction of linear superposition. For example, if $f(ax) = a^n f(x)$ is satisfied, the feature is still fit for our method.

How many features are fit for our methods? According to the above constraints, most of popular features used in pedestrian detection are all suitable, such as HOG [1], LBP [6], ICF [17], ACF [18]. Because the HOG feature is the basis of some complex approaches, such as DPM [4], our method can accelerate many other complex pedestrian detection frameworks which include the above features.

3.1 Region Proposal

In object detection task, for locating the object, traditional methods scan the image using multiple win-dows with different scales in the zigzag manner, named sliding window strategy, and then discriminate whether it include an object or not in each window. Usually, it will probe more than one million times. So such exhaustive search strategy is not fit for our real-time applications because it is very time-consuming. After analysis of such method, we can observe that a large proportion of probing is in the background. So if the background regions before scanning can be excluded, the efficiency of the detection will be boosted in a large margin. Recently, many efforts are made to generate the object candidates (bounding boxes) for object detection, called region proposal [10,24,25]. Because the decomposition of the whole image based on the trained dictionary is much time-consuming, the region proposal is critical for real-time object detection. After in-depth investigation [24], we take the Edgebox [25] as our region proposal approach, because it is most efficient under the highest recall rate.

3.2 Dictionary Learning

Given a set of image patches $Y = [y_1, \ldots, y_n]$, the standard unsupervised dictionary learning algorithm aims to jointly find a dictionary $D = [d_1, \ldots, d_m]$ and an associated sparse code matrix $X = [x_1, \ldots, x_n]$ by minimizing the reconstruction error as follows.

$$\min_{D,X} ||Y - DX||_F^2 \qquad s.t. \forall i, ||x_i||_0 \le K \tag{2}$$

where x_i are columns of X, the zero-norm $|| \cdot ||_0$ counts the non-zero entries in the sparse code x_i and K is a predefined sparsity level.

Although the above optimization is NP-hard, greedy algorithms such as orthogonal matching pursuit algorithm (OMP) [19,26,27] can be used to efficiently compute an approximate solution. In our experiment, we use K-SVD algorithm [19] to train the discriminative dictionary. In addition, we consider three sparsity inducing regularizers:

(1) Lasso Penalty [28]

$$R_{Lasso}(a) = \lambda_1 \|a\|_1$$

(2) Elastic net penalty [29]

$$R_{EN}(a) = \lambda_1 \|a\|_1 + \lambda_2 \|a\|_2^2$$

These regularizers lead to convex optimization problems, and employ a two step process to get the solution. In the first step, a subset of the activation coefficients is selected to satisfy the constraint $\|a\|_0 \leq \lambda_0$. In the second step, the selection of nonzero variables is fixed (thus satisfying the sparsity constraint) and the resulting convex optimization problem is solved.

3.3   Feature Response Synthesization

Feature response synthesization can be regarded as the linear combination of the representative coefficients and the response of the items of the learned dictionary. Denoting the feature pyramid of an image I as Φ, and $I = [P_1, \cdots, P_N]$, and D_j in $D = [D_1, \cdots, D_K]$ is the atom of D (Dictionary), we have $\Psi * P_i \approx \Psi * (\sum_j \alpha_{ij} D_j) = \sum_j \alpha_{ij}(\Psi * D_j)$, where $*$ denotes the convolution operator. Concretely, we can recover individual part filter responses via sparse matrix multiplication (or lookups) with the activation vector replacing the heavy convolution operation as shown in Eq. 3.

$$
\begin{bmatrix} \Psi * P_1 \\ \Psi * P_2 \\ \vdots \\ \vdots \\ \vdots \\ \vdots \\ \Psi * P_N \end{bmatrix}
\approx
\begin{bmatrix} \alpha_1 \\ \alpha_2 \\ \vdots \\ \vdots \\ \vdots \\ \vdots \\ \alpha_N \end{bmatrix}
\begin{bmatrix} \Psi * D_1 \\ \Psi * D_2 \\ \vdots \\ \Psi * D_K \end{bmatrix}
= AM
\tag{3}
$$

For efficient pedestrian detection, the extraction of some features should be made appropriate adjustments. Take HOG feature for example. It is composed of concatenated blocks. Each block includes 2×2 cells, and each cell is the 8×8 pixels of the image. So the block is 16×16 pixels. The concatenation of histograms of the blocks has two strategies: overlap and non-overlap. In the overlap manner, the sliding step width is usually the width of the cell. In the non-overlap manner, the sliding step width is the width of the block. So the dimension of the feature of the non-overlap is smaller than that of the overlap. But the performance of the feature will be lost by nearly 1% [1]. So the standard HOG feature chooses the overlap manner for better performance. For high acceleration, in this paper, we choose the non-overlap manner.

4   Experiments

For evaluating our method, we conduct the experiments on four challenging pedestrian datasets: Caltech [20], INRIA [1], ETH [21] and TUD-Brussels [22]. The state-of-the-art and classic pedestrian detectors are chosen to test our framework: HOG [1], ChnFtrs [5], ACF [18], HOGLBP [7], LatSvmV2 [4] and Very-Fast [23]. In the experiments, the training and testing data setting is as same as in [18]. We first discuss the relation of the performance versus the sparsity degree, the size of the dictionary, the size of atom. And then we evaluate our method.

4.1   Dictionary Learning vs Performance

Because our method is based on sparse coding, how to select the parameters of dictionary learning directly affects the performance of feature reconstruction. For choosing the best parameters, we conduct some experiments on INRIA Person Dataset and the type of synthesized feature is HOG. INRIA Person Dataset consists of 1208 positive training images (and their reflections) of standing people, cropped and normalized to 64×128, as well as 1218 negative images and 741 test images. This dataset is an ideal setting, as it is what HOG was designed and optimized for, and training is straightforward.

Sparsity Level and Dictionary Size. Figure 3 shows the average precision on INRIA when we change the sparsity level along with the dictionary size using 5×5 patches. We observe that when the dictionary size is small, a patch cannot be well represented with a single codeword. However, when the dictionary size grows and includes more structures in its codes, the $K = 1$ curve catches up, and performs very well. Therefore we use $K = 1$ in all the following experiments.

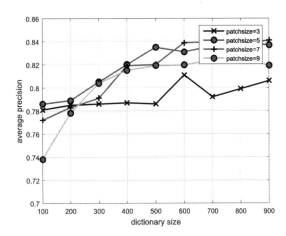

Fig. 2. The patch size vs detection performance on Caltech pedestrian dataset.

Patch Size and Dictionary Size. Next we investigate whether our synthesized features can capture richer structures using larger patches. Figure 2 shows the average precision as we change both the patch size and the dictionary size. While 3×3 codes barely show an edge, 7×7 codes work much better. However, 9×9 patches, may be too large for our setting and do not perform well.

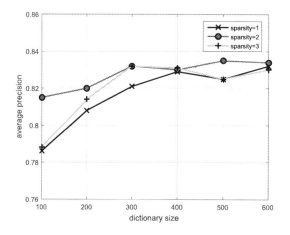

Fig. 3. The sparsity vs detection performance on Caltech pedestrian dataset.

Regularizer. With $K = 1$, one can also use different regularizers to learn a dictionary. Figure 4 compares the detection accuracy with Lasso penalty vs Elastic net penalty on 7×7 patches. The Elastic net penalty is better because it include more constraints to learn discriminative representation.

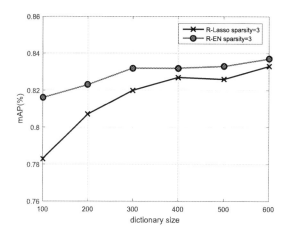

Fig. 4. The regularizer vs detection performance on Caltech pedestrian dataset.

In the following experiments, we set the size of the dictionary, the sparsity degree to be 600, 1 respectively. We set the size of the atom of the dictionary to be 7×7 for better performance.

4.2   Performance Comparison

We just pay attention to whether the performance is lost and the degree of performance loss. As shown in Table 1, we can see the performance comparison between the original detector and the corresponding synthesized detector. As can be seen from the table, the performance degradation is very small, about one percent. Why is the synthesis method a little worse than the original method? We think there are at least two reasons. One is that our method is based on reconstruction error minimum and sparsity constraints, which cause the loss of the discriminative information for pedestrian detection. The other is that we slightly modified the original feature extraction, such as HOG in the non-overlap manner. Compared to high speedup, we think this slight performance degradation is worth.

Table 1. Missing rates of pedestrian detectors (origin vs synthesizer) on four challenging datasets. "origin" denotes the original detector, and "synthesizer" stands for the synthesized feature by our framework.

Detectors	INRIA [1]		Caltech [20]		TUD-Brussels [22]		ETH [21]	
	Origin	Synthesizer	Origin	Synthesizer	Origin	Synthesizer	Origin	Synthesizer
HOG [1]	46	48	68	69	78	78	64	65
ChnFtrs [5]	22	23	56	57	60	61	57	58
ACF [18]	17	18	43	45	50	51	50	52
HOGLBP [7]	39	40	68	68	82	83	55	56
LatSvmV2 [4]	20	23	63	64	70	71	51	52
VeryFast [23]	16	17	53	54	61	62	55	56

4.3   Speed Comparison

The speed of the detector is more important than performance in the real-world applications. In this section, we will show the speed comparison of the above origin detectors and the corresponding synthesized detectors. Because the speed of the detector depends on the resolution.

We just do the statistics and analysis on the INRIA dataset because the results on the other datasets are the same as that on this dataset. The resolution of the image is 640×480 in INRIA testing set. As shown in Table 2, acceleration of the synthesized detector is very obvious. Take the detector HOGLBP [7] for example. The speedup ratio is up to 2000. The speed of original veryfast detector [23] is 50 fps because it is accelerated by GPU. But the speed of our synthesized detector is 110 fps. From the table, experiment results confirm our conjecture that the runtime of our synthesized detector depends on the decomposition of the image based on the dictionary.

Table 2. Speed comparison of pedestrian detectors (origin vs synthesizer) on INRIA person dataset. The unit of speed is the frame of per second.

Detectors	INRIA [1]	
	Origin	Synthesizer
HOG [1]	0.23	**96.5**
ChnFtrs [5]	16.4	**121.2**
ACF [18]	31.9	**125.4**
HOGLBP [7]	0.06	**120.4**
LatSvmV2 [4]	0.6	**108.5**
VeryFast [23]	50	**110.2**

5   Conclusion

In this paper, we proposed a novel framework of feature extraction based on sparse representation. And we give the constraint condition that the feature should satisfy in our framework. At last, we conduct enough experiments on four challenging datasets to evaluate our method. Experiment results demonstrate our method is efficient for pedestrian detection task. In the future, we will seek the efficient dictionary learning method and consider to add the classification error into dictionary learning to add the distinctive information.

Aknowledgment. This research is based upon work supported by National Nature Science Founda- tion of China (No. U1736206), National Nature Science Foundation of China (61671336), National Nature Science Foundation of China (61671332), Technology Research 10 F. Author et al. Program of Ministry of Public Security (No. 2016JSYJA12), Hubei Province Technological Innovation Major Project (No. 2016AAA015), Hubei Province Tech- nological Innovation Major Project (2017AAA123), The National Key Research and Development Program of China (No.2016YFB0100901), Nature Science Foun- dation of Jiangsu Province (No. BK20160386) and National Nature Science Foundation of China (61502354).

References

1. Navneet, D., Bill, T.: Histograms of oriented gradients for human detection. In: Proceedings of the 22nd IEEE Conference on Computer Vision and Pattern Recognition, pp. 886–893, June 2005
2. Sabzmeydani, P., Greg, M.: Detecting pedestrians by learning shapelet features. In: Proceedings of the 24th IEEE Conference on Computer Vision and Pattern Recognition, pp. 1–8, June 2007
3. Bo, W., Ram, N.: Detection of multiple, partially occluded humans in a single image by bayesian combination of edgelet part detectors. In: Proceedings of the Tenth IEEE International Conference on Computer Vision, pp. 90–97, June 2005
4. Pedro, F., David, M., et al.: A discriminatively trained, multiscale, deformable part model. In: Proceedings of the 25th IEEE Conference on Computer Vision and Pattern Recognition, pp. 1–8, June 2008

5. Piotr, D., Zhuowen, T., et al.: Integral channel features. In: Proceedings of the 20th British Machine Vision Conference, pp. 250–258, September 2009
6. Ahonen, T., Hadid, A., Pietikäinen, M.: Face recognition with local binary patterns. In: Pajdla, T., Matas, J. (eds.) ECCV 2004. LNCS, vol. 3021, pp. 469–481. Springer, Heidelberg (2004). https://doi.org/10.1007/978-3-540-24670-1_36
7. Xiaoyu, W., Tony, X., et al.: An HOG-LBP human detector with partial occlusion handling. In: Proceedings of the 26th IEEE Conference on Computer Vision and Pattern Recognition, pp. 32–39, June 2009
8. Junjie, Y., Zhen, L., et al.: The fastest deformable part model for object detection. In: Proceedings of the 32nd IEEE Conference on Computer Vision and Pattern Recognition, pp. 2497–2504, June 2014
9. Pedro, F., Ross, B., et al.: Cascade object detection with deformable part models. In: Proceedings of the 28th IEEE Conference on Computer Vision and Pattern Recognition, pp. 2241–2248, June 2010
10. Uijlings, J. R., Van De Sande, K. E., et al.: Selective search for object recognition. Int. J. Comput. Vis. **104**(2), 154–171 (2013)
11. Zitnick, C.L., Dollár, P.: Edge boxes: locating object proposals from edges. In: Proceedings of the 18th European Conference on Computer Vision, pp. 391–405, September 2014
12. Cheng, M.M., Zhang, Z., et al.: Bing: binarized normed gradients for objectness estimation at 300 fps. In: Proceedings of the 32nd IEEE Conference on Computer Vision and Pattern Recognition, pp. 3286–3293, June 2014
13. Ren, S., He, K., et al.: Faster R-CNN: towards real-time object detection with region proposal networks. In: Proceedings of the 33rd IEEE Transactions on Pattern Analysis and Machine Intelligence, pp. 1–8, June 2015
14. Li, J., Liang, X., et al.: Scale-aware fast R-CNN for pedestrian detection. Comput. Sci. 25–32 (2015)
15. Song, H.O., et al.: Sparselet models for efficient multiclass object detection. In: Fitzgibbon, A., Lazebnik, S., Perona, P., Sato, Y., Schmid, C. (eds.) ECCV 2012. LNCS, pp. 802–815. Springer, Heidelberg (2012). https://doi.org/10.1007/978-3-642-33709-3_57
16. Girshick, R., Song, H.O., et al.: Discriminatively activated sparselets. In: Proceedings of the 30th International Conference on Machine Learning, pp. 196–204, June 2013
17. Dollr, P., Tu, Z., et al.: Integral channel features. In: Proceedings of the 20th British Machine Vision Conference, pp. 7–10, September 2009
18. Dollar, P., Appel, R., et al.: Fast feature pyramids for object detection. IEEE Trans. Pattern Anal. Mach. Intell. **36**(8), 1532–1545, May 2014
19. Aharon, M., Elad, M., et al.: K-SVD: an algorithm for designing overcomplete dictionaries for sparse representation. IEEE Trans. Signal Process. **54**(11), 4311–4322, October 2006
20. Dollar, P., Wojek, C., et al.: Pedestrian detection: an evaluation of the state of the art. IEEE Trans. Pattern Anal. Mach. Intell. **43**(4), 743–761, March 2011
21. Andreas, E., Bastian, L., et al.: Depth and appearance for mobile scene analysis. In: Proceedings of the 25th IEEE International Conference on Computer Vision, pp. 1–8, June 2007
22. Wojek, C., Walk, S., et al.: Multi-cue onboard pedestrian detection. In: Proceedings of the 27th IEEE Conference on Computer Vision and Pattern Recognition, pp. 794–801, June 2009

23. Benenson, R., Mathias, M., et al.: Pedestrian detection at 100 frames per second. In: Proceedings of 30th IEEE Conference on Computer Vision and Pattern Recognition, pp. 2903–2910, June 2012
24. Hosang, J., Benenson, R., Dollar, P., et al.: What makes for effec-tive detection proposals? IEEE Trans. Pattern Anal. Mach. Intell. **38**(4), 814–830, March 2015
25. Zitnick, C.L., Dollr, P.: Edge boxes: locating object proposals from edges. In: Proceedings of the 18th European Conference on Computer Vision, pp. 391–405, September 2014
26. Cotter, S.F., Rao, B.D., et al.: Forward sequential algorithms for best basis selection. IEEE Vis. Image Signal Process. **146**(5), 235 (1999)
27. Mallat, S.G., Zhang, Z.: Matching pursuits with time-frequency dictionaries. IEEE Trans. Signal Process. **41**(12), 3397–3415 (1993)
28. Tibshirani, R.: Regression shrinkage and selection via the lasso. J. Roy. Stat. Soc. **58**(1), 267–288 (1996)
29. Zou, H., Hastie, T.: Regularization and variable selection via the elastic net. J. Roy. Stat. Soc. **67**(2), 301–320 (2005)

Enhancing Person Retrieval with Joint Person Detection, Attribute Learning, and Identification

Jianwen Wu[✉], Ye Zhao, and Xueliang Liu

Hefei University of Technology, Hefei 230601, China
2016110924@mail.hfut.edu.cn

Abstract. Person re-identification receives increasing attention in recent years. However, most works assume the persons have been well cropped from the whole scene images, and only focus on learning features and metrics. This paper considers the person re-identification problem in a real-world scenario, which should consider detection and identification simultaneously. This paper proposes a multi-task learning framework for person retrieval in the wild. Person attribute learning is exploited in our framework to enhance person retrieval. Our work consists of two main contributions: (1) we present a 11 image-level attribute annotations for each image in the large-scale PRW [27] dataset, and (2) we develop an end-to-end person retrieval framework which jointly learns person detector, attribute detectors, and visual embeddings in a multi-task learning manner. We evaluate the effectiveness of the proposed approach on two tasks, i.e. person attribute recognition and person re-identification. Experimental results have demonstrated the effectiveness of the proposed approach.

Keywords: Person retrieval · Object detection
Object re-identification · Deep learning

1 Introduction

Person re-identification (re-ID), has attracted increasing attention for its critical role in video surveillance applications. It aims to retrieval person-of-interest across multiple non-overlapping camera views. Given a probe pedestrian[1] image (query), the task is to rank all the pedestrian images in the gallery set by the similarity/distance between the query and candidate images, and return the most relevant images as retrieval results. It mainly consists of two parts: feature extraction and metric learning. The first part focuses on designing robust hand-crafted features [11,17]. The second part aims to learn a suitable distance/similarity function [10,12,16,24,25]. In recent years, researchers also consider to explore deep learning techniques for person re-identification which can jointly learn features and metrics [2,9,22,23,26,27].

---

[1] In this work, we focus on searching the person-of-interest in a surveillance scene. We will use *person* and *pedestrian* interchangeably in the work.

© Springer Nature Switzerland AG 2018
R. Hong et al. (Eds.): PCM 2018, LNCS 11165, pp. 113–124, 2018.
https://doi.org/10.1007/978-3-030-00767-6_11

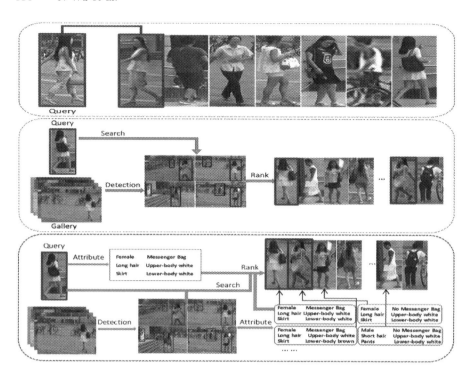

Fig. 1. Comparison between person re-identification, person search and person retrieval improved by attributes. In the 1st row, person re-id aims at matching with manually cropped pedestrians. The 2nd row shows the person search process of previous methods. The query person should find out from the whole gallery images and the red boxes indicate the truly matched person. In the 3rd row, the pedestrian attributes can offer some complementary and effective local information during the person retrieval process.

Although lots of person re-ID methods have been presented, it is still far from being applied in real-world scenarios. Most existing methods all assume that the persons have already been well cropped from the whole scene images, that is to say that the gallery set only contains well-cropped person images. For a real-world application, pedestrian detection and person re-identification should be jointly considered in a unified person retrieval framework. More recently, Xiao et al. [23] designed a person search framework which jointly considers pedestrian detection and person identification in an end-to-end manner. Our work is similar to [23], while we extend [23] to a new multi-task framework which jointly considers pedestrian detection, person re-identification, and also person attribute prediction in an end-to-end learning manner, which is motivated by [13] (Fig. 1). In [13], Lin et al. demonstrated that the performance of person re-identification can be significantly improved by learning person visual embedding and predicting pedestrian attributes simultaneously. Such a multi-task learning framework helps to alleviate overfitting and learn a semantic-aware person visual embedding. In this paper, we propose to incorporate [23] and [13] into a new multi-task framework which simultaneously processes the three sub-tasks. Our main contributions are summarized as follows:

Attribute \ Person ID	1	2	3	4
gender	male	female	female	female
hair-length	short	long	long	long
pants/skirt	pants	skirt	skirt	skirt
wearing Hat	no	no	no	no
carrying backpack	no	no	no	no
carrying messenger bag	no	yes	no	no
carrying handbag	no	no	no	no
bicycling	yes	no	no	no
pose	side	back	front	front
top-clothing yellow	red	white	white	black
bottom-clothing yellow	gray	white	white	gray

Fig. 2. Examples for person attribute annotation. In the left pictures, the different color of bounding boxes represent the different Person Id in the table(Yellow:1, Red:2, Black:3, White:4). Although Person 2 and Person 3 are the same pedestrian, they have the different messenger bag attribute because these people are labeled with regard to visual content. We can't see the female carrying the messenger bag just like the white person actually wear a backpack but we annotate it with "no" in the lower picture (Color figure online).

(1) We have annotated 11 image-level pedestrian attributes for each image in the large-scale PRW [27] dataset: gender (male, female), hair-length (long, short), (skirt/dress, pants), wearing hat (yes, no), carrying backpack (yes, no), carrying handbag (yes, no), carrying messenger bag (yes, no), bicycling (yes, no), poses (front, side, back), top-clothing colors (white, black, brown, gray, red, yellow, green, blue, purple, others), bottom-clothing colors (white, black, brown, gray, red, yellow, green, blue, purple, others). Noted that our labeled pedestrian attributes are image-level, which are different from the identity-level attributes labeled by Lin et al. [13]. In [13], all the images of a person have the same attributes. If a person carries a backpack in an image, the label of the person is always "carrying backpack", although we may cannot see the backpack in some other images of this person or the person may take off the backpack in some time. In our work, an attribute is labeled according by whether we can observe this attribute in this image. The specific examples and details of the pedestrian attributes labeling are shown in Fig. 2.

(2) We present an end-to-end person retrieval framework which jointly considers pedestrian detection, person re-identification, and pedestrian attributes detection in a multi-task deep learning manner. To the best of our knowledge, it is the first work which simultaneously considers the three sub-tasks for person retrieval. Experimental evaluation on the large-scale dataset has clearly showed that such a multi-task learning manner can significantly improve the performance of person matching.

2   Related Work

This paper focuses on tackling person retrieval with the proposed end-to-end multi-task learning framework, which is a combination of person re-identification, person detection, person attribute prediction. In this section, we only briefly introduce some representative works.

(1) **Person re-identification.** Generally, person re-ID consists of two sub-problems, i.e., feature learning and metric learning, separately. Some works focus on designing hand-crafted features [11,17] that are expected to be robust to complex variations in human appearances from different camera views, e.g., Local Maximal Occurrence (LOMO) feature [11] and Gaussian of Gaussian (GOG) [17]. Some works focus on learning an optimal distance/similarity function [10,12,16,24,25] using hand-crafted features to better characterize the similarity relationship between a similar/dissimilar pair of samples. Recently, more researchers in re-ID [2,9,22,23,26,27] prefer to jointly learn feature and metric in an end-to-end manner by training deep neural networks. For the *deep* re-ID methods, identification loss is the first choice, which has been exploited in multiple pioneering works [23,27,28]. For example, Zheng et al. [27] and Xiao et al. [23] proposed to jointly handle both pedestrian detection and identification in an end-to-end framework. Zheng et al. [28] proposed to regularize supervised models in re-ID with unlabeled person samples generated by generative adversarial network, which has effectively improved the baseline. More recently, triplet-based loss functions have attracted more attention in re-ID [1,2,15].

(2) **Person Detection.** Deformable Part Model (DPM) [5], Aggregated Channel Features (ACF) [4] and Locally Decorrelated Channel Features (LDCF) [18], are three representative methods relying on hand-crafted features and linear classifiers for pedestrian detection. Three state-of-the-arts works [14,23,27] have tried to fuse person detection and person re-identification for person retrieval, which is more practical in real-world scenarios.

(3) **Person Attribute Prediction.** Person attributes have been investigated in many works [3,13,20,21]. Su et al. [20] proposed a novel Multi-Task Learning with Low Rank Attribute Embedding framework for person re-identification. Deng et al. [3] released a large-scale pedestrian attribute datasets PETA.

In the PETA dataset, one person only has one image. Recently, Lin et al. [13] proposed a multi-task learning framework for person re-identification, which jointly unifies person attribute predictor and person re-ID. Noted that Lin et al. contributed two ID-level person attribute datasets.

Compared with above-mentioned works, we are the first to present a end-to-end multi-task learning framework for person retrieval which simultaneously learns person detector, person attribute predictor, and person re-identification model. We also contribute an image-level person attribute dataset which can facilitate the research of others. Experiments have demonstrated the effectiveness of our propose approach.

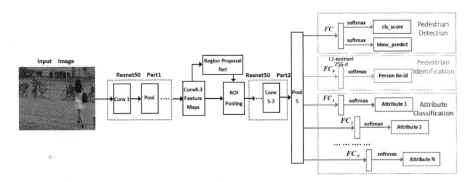

Fig. 3. The architecture of our proposed framework. We use Faster R-CNN [19] as the basic network. During training, it not just predict the location of pedestrians and the ID labels, but also will predict pedestrian attributes. ResNet50 Part1 corresponds to the conv1 layer to the conv4-3 layer of ResNet-50 [7], and ResNet50 Part2 is the part of ResNet-50 from conv4-4 to conv5-3

3   The Proposed Approach

In this work, we aim to address the problem of person retrieval by developing a multi-task deep learning framework that jointly considers pedestrian detection, person re-identification, and pedestrian attributes detection in a single convolutional neural network, as shown in Fig. 3.

Given a whole scene image which contains at least one pedestrian as input, we first use a CNN module to transform the whole scene image from raw-pixel to convolutional feature maps. Based on the feature maps, we employ a region proposal net to predict the region of pedestrians, which will be further passed through a ROI pooling layer, two convolutional layers followed by a global average pooling layer and a fully-connected layer to extract a low-dimensional normalized feature. At training stage, we use several loss functions on the top layer to supervise the network training including pedestrian identification, pedestrian detection, and pedestrian attribute classification.

3.1   Person Detection

When we detect person from the pedestrian images, Faster-RCNN is used as our detector [19]. Faster-RCNN is a very popular object detection method currently, which is unified, deep-learning-based and has a higher accuracy. It merges Region Proposal Network (RPN) and Fast R-CNN into a single network by sharing their convolutional features. And as a fully convolutional network, RPN can simultaneously predict object bounds and scores for each location. When we detect pedestrians, firstly we generate high-quality region proposals by RPN network and then use the Fast R-CNN net for detection.

According to these definitions, Faster R-CNN has an objective function following the multi-task loss. The specific loss function for an image is defined as:

$$\mathcal{L}_{det} = \frac{1}{N_{cls}} \sum_i L_{cls}(p_i, p_i^*) + \lambda \frac{1}{N_{reg}} \sum_i p_i^* L_{reg}(t_i, t_i^*) \tag{1}$$

where L_{cls} stands for the classification loss which is log loss over two classes(person vs. background). For the regression loss, we use $L_{reg}(t_i, t_i^*) = R(t_i - t_i^*)$ where R is the robust loss function (smooth $L1$). And p_i denotes the predicted probability of an anchor being an object while t_i is a 4d vector describing the coordinates of the predicted bounding box. The detailed formula and parameter settings can be referred to [15].

3.2   Pedestrian Attribute Annotation and Prediction

We manually annotate the large scale PRW dataset [27] with 11 pedestrian attributes. We mainly consider the following factors for attribute annotation: gender, hair, clothing types, clothing colors, and poses. In details, 11 attributes are labeled: gender (male, female), hair-length (long, short), (skirt/dress, pants), wearing hat (yes, no), carrying backpack (yes, no), carrying handbag (yes, no), carrying messenger bag (yes, no), bicycling (yes, no), poses (front, side, back), top-clothing colors (white, black, brown, gray, red, yellow, green, blue, purple, others), bottom-clothing colors (white, black, brown, gray, red, yellow, green, blue, purple, others). The statistical distribution of attributes on the PRW.dataset is shown in Fig. 4.

Each person has N attributes($N = 11$). So we use N fully-connected layers together with N softmax loss functions for training person attribute predictors. The probability of assigning pedestrian sample x to the attribute class $k \in 1, ..., n$(where n is the number of classes for a certain attribute) can be written as $p(k|x) = \frac{exp(z_k)}{\sum_{i=1}^n exp(z_i)}$. Similarly, the attribute classification loss can be written as below:

$$\mathcal{L}_{att} = - \sum_{k=1}^n log(p(k))q(k) \tag{2}$$

Let y be the ground-truth attribute label, so that $q(y) = 1$ and $q(k) = 0$ for all $k \neq y$. And the other symbols are the same as the basic softmax loss function.

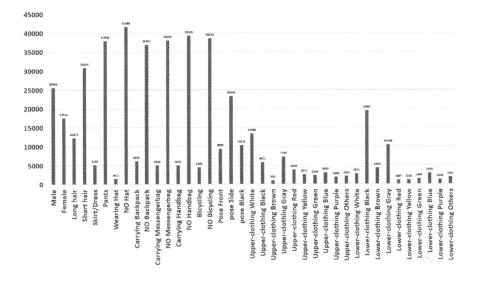

Fig. 4. The distribution of attributes on the PRW dataset.

3.3 Person Re-Identification

In this work, we utilize the semi-supervised online instance matching (OIM) loss [23] for person identification, which can jointly consider labeled identities and unlabeled identities effectively in the training stage. The goal is to maximize the features similarity among the instances of the same person, and minimize the feature similarity between different persons.

In the training stage, the feature vector of an labeled identity inside a mini-batch is denoted as $x \in \mathcal{R}^d$, where d is the feature dimension. A lookup table $\mathbf{V} \in \mathcal{R}^{d \times L}$ is used to store the features of all the labeled identities. For the unlabeled identities, a circular queue is used to store the features of unlabeled identities, $\mathbf{U} \in \mathcal{R}^{d \times Q}$, appearing in recent mini-batches, where Q is the queue size. During forward propagation, we compute the cosine similarity between the mini-batch sample x and all the labeled identities by $\mathbf{V}^T x$. During backward, if the target class-id is t, then the t-th column of the lookup table is updated by $v_t \leftarrow \gamma v_t + (1 - \gamma)x$, where $\gamma \in [0, 1]$, then the t-th column of the lookup table should be ℓ_2-normalized. We can also compute the cosine similarities between the mini-batch sample x and all the unlabeled identities by $\mathbf{U}^T x$. After each iteration, the new feature vectors are pushed into the queue and the out-of-date ones are popped out. The probability of x being recognized as the identity with id i is computed by the following softmax function

$$p_i = \frac{\exp\left(v_i^T x / \tau\right)}{\sum_{j=1}^{L} \exp\left(v_j^T x / \tau\right) + \sum_{k=1}^{Q} \exp\left(u_k^T x / \tau\right)} \tag{3}$$

where the parameter τ is a scale parameter leading to softer probability distribution. The probability of x being recognized as the i-th unlabeled identity in the circular queue can be computed in a similar way. The objective of OIM loss is to maximize the expected log-likelihood $\mathcal{L}_{id} = E_x[\log p_t]$. The gradient of $\mathcal{L}_{id}$ with respect to x can be referred to [23].

The main advantage of the OIM loss [23] is that it can effectively compare the mini-batch sample with all the labeled and unlabeled identities.

By using a person detection loss function, an attribute classification loss function and an identity classification loss function, our framework is trained to predict the location of a person in the pedestrian image with his attribute and identity labels. Here the final loss function is defined as:

$$\mathcal{L} = \mathcal{L}_{det} + \mathcal{L}_{id} + \frac{1}{N} \sum_{i=1}^{N} \mathcal{L}_{att} \qquad (4)$$

where $\mathcal{L}_{det}$, $\mathcal{L}_{id}$ and $\mathcal{L}_{att}$ denote the cross entropy loss of person detection, identity classification and attribute classification, respectively.

4   Experimental Evaluation

4.1   Dataset

The PRW dataset [27] contains 11,816 video frames which was captured by 6 cameras in a university campus. In these 11816 video frames, the PRW dataset provides 43110 manually-marked pedestrian bounding boxes. Among them, the dataset annotates 34304 pedestrians with an ID by 1 to 932 and the overage are assigned of -2. We split the dataset into two subsets for training and testing. In training, we have 482 person with different IDs and 5134 training images. Also the dataset provides 2057 query persons for evaluation with the gallery size is 6112.

4.2   Evaluation Protocol

We use mean average precision (mAP) and the cumulative matching characteristics top-k matching rate (CMC top-K) as performance metrics. The mAP metric reflects the correctness of selecting the query person from the gallery images, inspired from object detection criterion. Its general process is to calculate the average precision (AP) for each query at first, and then average the AP for all queries to get the final mean average precision (mAP). The second evaluation metric is top-K matching rate, which is also used in OIM [23].

4.3   Implementation Details

We use Faster-RCNN [19] as our basic framework for person detection, and then on this basis we implement our work on Caffe [8] deep learning framework. ResNet-50 [7] is utilized as the base network for Faster RCNN detector. We initialize network parameters by using ImageNet pre-trained ResNet50 [7]. When training, we set the iteration number to 50000 and the initial learning rate to 0.001 firstly with the SGD solver. Then the learning rate is changed to 0.0001 after 40 K iterations. Each mini-batch contains two pedestrians scene images. The training lasted 15 hours on a machine with a GTX TITAN X GPU and a Intel i7-6950XCPU, and it will take round 15 min when we conduct evaluation and testing.

4.4 Person Attribute Prediction Accuracy

We test person attribute recognition on the galleries of the PRW dataset. The result is shown in Table 1. We can see that the overall performance is satisfying. The prediction accuracy of gender, hair-length, skirt-dress/pants, hat, backpack, handbag, and bicycling is over 90%, which indicates that such a multi-task learning framework is helpful to learn a discriminative attribute prediction model. The above mentioned attributes are binary attributes, which is easy to predict. Most of the rest attributes, pose, upper-clothing color, and lower-clothing color are multi-dimensional attributes, which are harder than binary attributes.

Table 1. The accuracy (%) of person attribute prediction on the PRW dataset.

Attribute	Accuracy	Attribute	Accuracy
Gender	0.91	Handbag	0.90
Hair-length	0.92	Bicycling	0.98
Skirt-dress/Pants	0.90	Pose	0.83
Hat	0.98	Upper-clothing color	0.82
Backpack	0.93	Lower-clothing color	0.81
Messenger bag	0.87		

4.5 Person Re-Identification Accuracy

On the PRW dataset, we compare the performance of our proposed method with some state-of-the-art methods which combine different three detectors (respective R-CNN [6] detectors of ACF [4], LDCF [18], DPM [5]) and three recognizers (LOMO + XQDA [11], IDE$_{det}$ [27], CWS [27]). These methods break down the person search problem into re-identification and detection two tasks. We just cite their reported results in [27] as a comparison. As shown in Table 2, we can see that different detectors and recognizers can affect the person search performance obviously. The OIM [23] method is our baseline method, which jointly considers person detection and re-identification. Baseline result is obtained using the source code provided in OIM [23] in almost the same setting with our method. Compared with the above-mentioned methods which conduct detection and re-identification as two tasks, OIM achieves better performance because of the joint training and optimization of the identification and detection components.

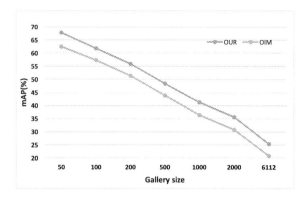

Fig. 5. The mean Average Precision of our method with different gallery sizes on the PRW dataset.

Table 2. Performance (mAP (%) and top-1 (%)) comparison on the PRW dataset.

Method	mAP (%)	top-1 (%)
ACF-Alex + LOMO + XQDA	10.3	30.6
ACF-Alex + IDE$_{det}$	17.5	43.6
ACF-Alex + IDE$_{det}$ + CWS	17.8	45.2
LDCF + LOMO + XQDA	11.0	31.1
LDCF + IDE$_{det}$	18.3	44.6
LDCF + IDE$_{det}$ + CWS	18.3	45.5
DPM-Alex + LOMO + XQDA	13.0	34.1
DPM-Alex + IDE$_{det}$	20.3	47.4
DPM-Alex + IDE$_{det}$ + CWS	20.5	48.3
OIM(Baseline)	20.3	58.9
Ours	24.8	65.5

It should be noted that benefiting from jointly considering person detection, person attribute detection, and person re-identification in a unified framework, our proposed method obtains the best performance 24.8% mAP, and 65.5% top-1. Compared to with our baseline OIM, we improve the mAP by over 4% and top-1 by nearly 7%. The main reason is that such a multi-task deep learning fashion can effectively prevent the overfitting, thus improving the person re-identification accuracy.

As shown in Fig. 5, we also evaluate the performance (mAP, %) of our proposed method with OIM under different settings of gallery size: {50, 100, 200,500, 1000, 2000}. We can observe that although the mAP score gradually decreases with the increase of the gallery size, but our proposed method still can improve the mAP score of OIM by around 4.5%, which further demonstrates the effectiveness of our proposed method.

5   Conclusion

In this work, we present a multi-task deep learning framework for person retrieval. It casts person detection, person attribute prediction, and person re-identification in an end-to-end learning framework, which is different from most existing works that treat the three tasks separately. The main purpose is to leverage such a multi-task learning manner to prevent overfitting. Remarkable improvements on the challenging dataset PRW [27] have demonstrated the effectiveness of our proposed approach. Another attribution of this work is that we manually annotate the large scale PRW dataset [27] with 11 pedestrian attributes, which can be believed to facilitate the further research of this field.

Acknowledgments. This work was supported in part by the National Natural Science Foundation of China (NSFC) under grants 61632007 and 61502139, in part by Natural Science Foundation of Anhui Province under grants 1608085MF128 and in part by Anhui Higher Education Natural Science Research Key Project under grants KJ2018A0545.

References

1. Chen, W., Chen, X., Zhang, J., Huang, K.: Beyond triplet loss: a deep quadruplet network for person re-identification. In: IEEE Conference on Computer Vision and Pattern Recognition (2017)
2. Cheng, D., Gong, Y., Zhou, S., Wang, J., Zheng, N.: Person re-identification by multi-channel parts-based CNN with improved triplet loss function. In: IEEE Conference on Computer Vision and Pattern Recognition, pp. 1335–1344 (2016)
3. Deng, Y., Luo, P., Loy, C.C., Tang, X.: Pedestrian attribute recognition at far distance. In: ACM International Conference on Multimedia, pp. 789–792. ACM (2014)
4. Dollár, P., Appel, R., Belongie, S., Perona, P.: Fast feature pyramids for object detection. IEEE Trans. Pattern Anal. Mach. Intell. **36**(8), 1532–1545 (2014)
5. Felzenszwalb, P.F., Girshick, R.B., McAllester, D., Ramanan, D.: Object detection with discriminatively trained part-based models. IEEE Trans. Pattern Anal. Mach. Intell. **32**(9), 1627–1645 (2010)
6. Girshick, R., Donahue, J., Darrell, T., Malik, J.: Region-based convolutional networks for accurate object detection and segmentation. IEEE Trans. Pattern Anal. Mach. Intell. **38**(1), 142–158 (2016)
7. He, K., Zhang, X., Ren, S., Sun, J.: Deep residual learning for image recognition. In: Proceedings of the IEEE Conference on Computer Vision and Pattern Recognition, pp. 770–778 (2016)
8. Jia, Y., et al.: Caffe: convolutional architecture for fast feature embedding. In: Proceedings of the 22nd ACM International Conference on Multimedia, pp. 675–678. ACM (2014)
9. Li, W., Zhao, R., Xiao, T., Wang, X.: Deepreid: deep filter pairing neural network for person re-identification. In: IEEE Conference on Computer Vision and Pattern Recognition, pp. 152–159 (2014)
10. Li, Z., Chang, S., Liang, F., Huang, T.S., Cao, L., Smith, J.R.: Learning locally-adaptive decision functions for person verification. In: IEEE Conference on Computer Vision and Pattern Recognition, pp. 3610–3617 (2013)

11. Liao, S., Hu, Y., Zhu, X., Li, S.Z.: Person re-identification by local maximal occurrence representation and metric learning. In: IEEE Conference on Computer Vision and Pattern Recognition, pp. 2197–2206 (2015)

12. Liao, S., Li, S.Z.: Efficient PSD constrained asymmetric metric learning for person re-identification. In: IEEE International Conference on Computer Vision, pp. 3685–3693 (2015)

13. Lin, Y., Zheng, L., Zheng, Z., Wu, Y., Yang, Y.: Improving person re-identification by attribute and identity learning. arXiv preprint arXiv:1703.07220 (2017)

14. Liu, H., et al.: Neural person search machines. In: IEEE International Conference on Computer Vision (2017)

15. Liu, J., et al.: Multi-scale triplet CNN for person re-identification. In: ACM on Multimedia Conference, pp. 192–196. ACM (2016)

16. Ma, L., Yang, X., Tao, D.: Person re-identification over camera networks using multi-task distance metric learning. IEEE Trans. Image Process. **23**(8), 3656–3670 (2014)

17. Matsukawa, T., Okabe, T., Suzuki, E., Sato, Y.: Hierarchical gaussian descriptor for person re-identification. In: IEEE Conference on Computer Vision and Pattern Recognition, pp. 1363–1372 (2016)

18. Nam, W., Dollár, P., Han, J.H.: Local decorrelation for improved pedestrian detection. In: Advances in Neural Information Processing Systems, pp. 424–432 (2014)

19. Ren, S., He, K., Girshick, R., Sun, J.: Faster R-CNN: Towards real-time object detection with region proposal networks. In: Advances in Neural Information Processing Systems, pp. 91–99 (2015)

20. Su, C., Yang, F., Zhang, S., Tian, Q., Davis, L.S., Gao, W.: Multi-task learning with low rank attribute embedding for person re-identification. In: IEEE International Conference on Computer Vision, pp. 3739–3747 (2015)

21. Su, C., Zhang, S., Xing, J., Gao, W., Tian, Q.: Deep attributes driven multi-camera person re-identification. In: Leibe, B., Matas, J., Sebe, N., Welling, M. (eds.) ECCV 2016. LNCS, vol. 9906, pp. 475–491. Springer, Cham (2016). https://doi.org/10.1007/978-3-319-46475-6_30

22. Xiao, T., Li, H., Ouyang, W., Wang, X.: Learning deep feature representations with domain guided dropout for person re-identification. In: IEEE Conference on Computer Vision and Pattern Recognition, pp. 1249–1258 (2016)

23. Xiao, T., Li, S., Wang, B., Lin, L., Wang, X.: Joint detection and identification feature learning for person search. In: Proceedings of the IEEE Conference on Computer Vision and Pattern Recognition (2017)

24. Yang, X., Wang, M., Hong, R., Tian, Q., Rui, Y.: Enhancing person re-identification in a self-trained subspace. ACM Trans. Multimedia Comput. Commun. Appl. **13**(3), 27 (2017)

25. Yang, X., Wang, M., Tao, D.: Person re-identification with metric learning using privileged information. IEEE Trans. Image Process. **27**(2), 791–805 (2018)

26. Yi, D., Lei, Z., Liao, S., Li, S.Z.: Deep metric learning for person re-identification. In: International Conference on Pattern Recognition, pp. 34–39. IEEE (2014)

27. Zheng, L., Zhang, H., Sun, S., Chandraker, M., Yang, Y., Tian, Q.: Person re-identification in the wild. In: IEEE Conference on Computer Vision and Pattern Recognition (2017)

28. Zheng, Z., Zheng, L., Yang, Y.: Unlabeled samples generated by gan improve the person re-identification baseline in vitro. In: International Conference on Computer Vision (2017)

End-To-End Learning for Action Quality Assessment

Yongjun Li[1,3], Xiujuan Chai[1,2], and Xilin Chen[1,3(✉)]

[1] Key Lab of Intelligent Information Processing of Chinese Academy of Sciences
(CAS), Institute of Computing Technology, CAS, Beijing 100190, China
{yongjun.li,xiujuan.chai}@vipl.ict.ac.cn, xlchen@ict.ac.cn
[2] Agricultural Information Institute, Chinese Academy of Agricultural Sciences,
Beijing 100081, China
[3] University of Chinese Academy of Sciences, Beijing 100049, China

Abstract. Nowadays, action quality assessment has attracted more and
more attention of the researchers in computer vision. In this paper, an
end-to-end framework is proposed based on fragment-based 3D convolutional neural network to realize the action quality assessment in videos.
Furthermore, the ranking loss integrated with the MSE forms the loss
function to make the optimization more reasonable in terms of both the
score value and the ranking aspects. Through the deep learning, we narrow the gap between the predictions and ground-truth scores as well
as making the predictions satisfy the ranking constraint. The proposed
network can indeed learn the evaluation criteria of actions and works
well with limited training data. Widely experiments conducted on three
public datasets convincingly show that our method achieves the state-
of-the-art results.

Keywords: Action quality assessment
3D convolutional neural network · Deep learning · Ranking loss

1 Introduction

Nowadays, action quality assessment has received more attention in the community of computer vision. It aims at measuring the difference between the standard
actions and the actions we perform. In practice, the action quality assessment
has many potential applications, such as scoring athletes' performance, medical rehabilitation tests, dancing teaching and so on. This paper takes scoring
athletes'performance as the instance to explore the action quality assessment
algorithm. In sports competitions, the inefficiency and subjectivity of manual
judgement diminish the interest seriously. However an automatic scoring system which is efficient and objective can enormously improve this situation [2].

This work was partially supported by 973 Program under contract No2015CB351802,
Natural Science Foundation of China under contracts Nos. 61390511, 61472398,
61532018.

Although the automatic scoring system can bring a lot of benefits, some challenges prevent it from applications, which mainly lie in the following aspects:

1. Each kind of action has its own characteristics and designing a general framework for various kinds of actions is difficult.
2. Some criteria for action quality assessment are not quantifiable and hard to be represented by hand-crafted features.
3. The automatic scoring system should make predictions close to the ground-truth scores and satisfy the ranking constraint. In the training, accomplishing the two goals synchronously is difficult.

To overcome these problem, we propose an end-to-end framework to realize an automatic scoring system. Inspired by the breakthroughs of feature learning with deep learning in video domain [10,16,17], we leverage the 3D convolutional neural network to extract features and these features are the better representation of actions than man-crafted features. However extracting discriminative features from a long sequence is difficult, we segment each video into fragments and extract fragment-features as [7]. Meanwhile, extracting features by fragments ensures that our framework has the capacity to learn the evaluation criteria of actions. Synchronously, we consider the score value constraint and the ranking constraint in the optimization objective. To this end, we integrate the ranking loss with MSE to form the final loss function. Additionally, our framework is common for various kinds of actions.

To summarize, our main contributions are as follows:

1. We propose an end-to-end framework which can indeed learn the evaluation criteria of actions.
2. The ranking loss is integrated with the traditional MSE to form the loss function making the predictions meet the score value constraint and the ranking constraint synchronously.

2   Related Works

There are only a few prior researches in the action quality assessment. The methods can be classified into regression-based method [4,5,7,11] and classification-based method [6,8,9].

The regression-based method directly predicts continuous score to evaluate the quality of an action. In [4], Pirsiavash *et al.* propose a learning-based framework that takes steps towards assessing how well people perform actions in videos. Their approach works by training a regression model from spatiotemporal pose features to scores obtained from expert judges. In [7], they employ the 3D Convolutional Network (C3D) [10] to extract the body-pose features and the C3D+SVR achieves the state-of-art results.

While for the classification-based method, the quality of an action will be classified into different grades. In [6], Parmar *et al.* explore the problem of exercise quality measurement. They collect negative exercise data by asking subjects to deliberately make subtle errors during the exercise. Following that, they use adaboost to classify exercise into "good" or "bad". Literature [8] presents an automatic framework for surgical skill assessment. They introduce video analysis techniques to obtain features.

There are also some researches combining two methods. In [13], Chai *et al.* develop a system on sign evaluation with both classification and regression models. The system first determines whether a sign is the appointed one by the classification model. For the sign which passes the verification, the system will give a score by the regression model.

In summary, most of previous researches tackle this problem by traditional machine learning techniques. They treat the problem as a two-phase task instead of an end-to-end process and only emphasize the score value constraint with ignoring the ranking constraint.

3   End-To-End Score Prediction

Our end-to-end framework is composed of feature extraction and score prediction, as shown in Fig. 1. In order to get discriminative features which can represent complex movement information, we employ the 3D convolutional neural network extractor (C3D extractor), which consists of the first 13 layers of C3D [10], in our framework. Since an action video sequence is too long, it is difficult for C3D extractor to generate a good representation over the whole action. Hence we divide the video into fragments evenly along the temporal dimension. Specifically, one extractor is applied for one fragment and the weights are not

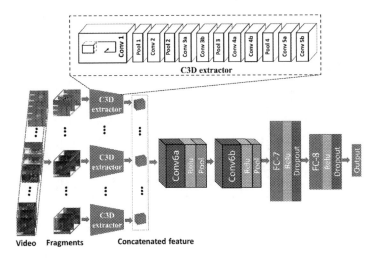

Fig. 1. Overview of our proposed end-to-end learning framework.

shared among C3D extractors, making each C3D extractor be sensitive to just one special movement which is a part of an action, such as the take-off in diving.

Once the fragment-based features are extracted, they will form a concatenated feature immediately. Then the concatenated feature goes through two convolution layers (followed by Relu and $3 \times 3 \times 3$ Max-pooling) to get higher level feature maps. The receptive fields of both convolution layers are $3 \times 3 \times 3$. Finally, two fully-connected layers (followed by Relu and Dropout $= 0.8$) and an output layer are used to regress scores.

4   Loss Function

In our loss function, there are two complementary terms. One is MSE (Eq. 1) which is general applied to constrain score value and the other is ranking loss (Eq. 2) severing to ensure the right order of the predictions. When the predictions violate the ranking constraint, the ranking loss will generate a punishment term. On the contrary, the value of the ranking loss is zero.

$$L_1 = \frac{1}{2n} \sum_{i=1}^{n} (s_i - g_i)^2 \tag{1}$$

$$L_2 = \sum_{i=1}^{n} \sum_{j=1,j>i}^{n} RELU(-(s_j - s_i)sign(g_j - g_i) + \delta) \tag{2}$$

where g_i and s_i are the ground-truth score and prediction of the ith sample in a batch of data respectively. n is batch size. $RELU(\cdot)$ is a rectified linear unit activation and $\delta = 2$ works as the margin.

The final loss function can be written by Eq. 3,

$$L = L_1 + \alpha L_2 + \beta \|w\|^2, \alpha > 0, \beta > 0 \tag{3}$$

where $\|w\|^2$ is L2-regularization item, α and β are used to balance these three items.

For the better optimization, we should balance L_1 and L_2. If one adjusts weights too fast, the other will be weakened. So we make them update weights with similar speeds. The gradients of L_1 and L_2 with respect to s_i can be denoted as Eqs. 4 and 5:

$$\nabla_{s_i} L_1 = \frac{1}{n} (s_i - g_i) \tag{4}$$

$$\nabla_{s_i} L_2 = \alpha \sum_{j=1,j>i}^{n} H(-(s_j - s_i)sign(g_j - g_i) + \delta)$$

$$-\alpha \sum_{j=1,j<i}^{n} H(-(s_i - s_j)sign(g_i - g_j) + \delta) \tag{5}$$

$$H(x) = \begin{cases} 1, x > 0 \\ 0, x \le 0 \end{cases} \tag{6}$$

Let $|\nabla_{s_i} L_1| = |\nabla_{s_i} L_2|$. Thus we can get α by Eq. 7 roughly.

$$\alpha = \frac{|\nabla_{s_i} L_1|}{|\nabla_{s_i} L_2|} \tag{7}$$

5   Experimental Results

In this section, we evaluate our method on three kinds of actions (diving, vault and figure skating) from three public datasets, i.e. UNLV-Diving Dataset [7], UNLV-Vault Dataset [7] and Mit-Skating Dataset [4]. First is the simple introduction of these three datasets. Secondly, the experiment configuration and the evaluation metric are given. Then, we explore the learning capacity of our framework. After that we verify the effectiveness of our loss function. Finally, we compare our method with others.

5.1   Dataset

UNLV-Diving Dataset: This dataset contains 370 Olympic men's 10-meter platform videos, each has roughly 150 frames. The scores vary between 0 (the worst) and 100 (the best). A diving score is determined by the product of execution score (judging quality of a diving) multiplied by the diving difficulty score (fixing agreed-upon value based on diving type).

UNLV-Vault Dataset: This dataset includes 176 videos. These videos are short relatively with an average length of about 75 frames. The scores range from 0 (the worst) to 20 (the Best). A vault score is determined by the sum of the execution score and the difficulty score.

Mit-Skating Dataset: 150 Olympic Figure Skating videos are included in this dataset. Each one has an average of 4200 frames. The scores range from 0 (the worst) to 100 (the Best). In [7], Parmar *et al.* augment the dataset by adding another 21 videos. The figure skating score is obtained as the same way of vault.

In experiments on diving and vault, the training/test data splits are 300/70 and 120/56 respectively. The two data splits are the same as that in [7]. In diving, all videos are padded with zero frames to length 151 and each video is divided into 9 fragments (16 frames per fragment, the rest of frames are used for data augmentation). In vault, all videos are padded to length 100 and the number of fragments is 6 (16 frames per fragment). Each fragment in the two datasets includes 16 frames. The redundant frames are used to augment data by shifting the start frame.

For the Mit-Skating Dataset, 100 videos are used for training and the remaining 71 videos for testing and all videos are padded to length 4500. We drop four out of every five frames because of limited computing resources and divide each video into 9 fragments. Each fragment includes 100 frames.

Because of the limited computing resources and high rate of down-sample, we perform the detailed evaluation on the UNLV-Diving Dataset and the UNLV-Vault Dataset while only give the comparison results on the Mit-Skating Dataset.

5.2   Experiments Setting and Evaluation Metric

In our task, only several hundreds of data are available. In order to address the problem of limited training data, we pre-train the C3D extractor with UCF-101 [12] and augment data by shifting the start frame with a random number within [0,5]. During the training, we adopt different learning rate for the feature extraction (lr = 10e–4) and the score prediction (lr = 10e–3). Learning rate decay is set to 0.45 per 600 iterations. The optimization algorithm is Adam [15].

To evaluate the performance, commonly used spearman rank correlation (SRC) [7] is adopted as our measurement. SRC is a nonparametric measure of rank correlation (statistical dependence between the ranking of two groups of variables). The larger the SRC, the higher the rank correlation between the ground-truth scores and the predictions. Also the high SRC means that the predictions meet the ranking constraint well.

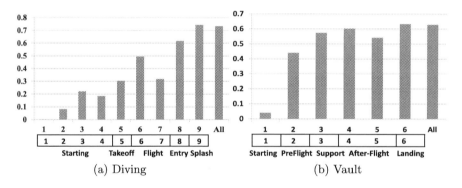

(a) Diving   (b) Vault

Fig. 2. Exploration on the learning capacity of our framework on the UNLV-Diving Dataset and the UNLV-Vault Dataset. The vertical axis indicates SRC. The horizontal axis indicates index of the fragment. In the bottom of each graph, there are 5 stages and the corresponding fragments of each action respectively.

5.3   Learning Capacity of Our Framework

In this subsection, we explore the learning capacity of our framework by predicting the score with different fragments and all fragments. The loss function is MSE.

Every time, we only employ one fragment to predict scores and the other fragments are set to zero. At last, we employ all fragments for score prediction. The results are shown in Fig. 2. According to diving rules [3], a diving can be divided into 5 stages which are Starting, Take-Off, Flight, Entry and Splash. Among these stages, Starting has no contribution to the score and Take-Off determines the score slightly. Flight and Entry are important to the score. Although Splash is irrelevant to the judgement, the small size of splash is a sign of a successful diving. Because, in the end, the accumulation of more minor deviations in preceding movements will result in a larger splash. For vault, it also has 5 stages which are Starting, Pre-Flight, Support, After-Flight and Landing [1] respectively. Similarly, Starting doesn't determine the score and Pre-Flight has small contribution to the score. The other three stages mainly determine the score and the stable Landing can also be regarded as the sign of a perfect vault. Meanwhile we match fragments to different stages roughly at the bottom of Figs. 2(a) and (b). We observe that the fragments corresponding to the stages determining the score have higher SRC than that of other fragments. Specifically, the fragments corresponding to Splash and Landing have the best performance, indicating their important role in judgement. The Splash and Landing can be used to evaluate the quality instead of all fragments. So we only use the last fragment in the following experiments.

The results suggest that our framework can actually learn the evaluation criteria of actions. Figures 3 and 4 show the example frames in the last fragments of two actions and also the corresponding predictions and ground-truth scores.

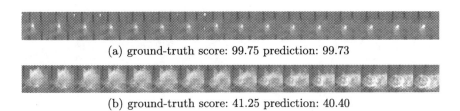

(a) ground-truth score: 99.75 prediction: 99.73

(b) ground-truth score: 41.25 prediction: 40.40

Fig. 3. Example frames in the last fragments of two diving videos and the corresponding predictions and ground-truth scores. The less the splash, the higher the score.

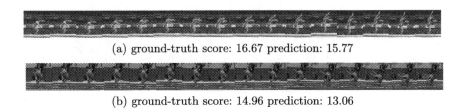

(a) ground-truth score: 16.67 prediction: 15.77

(b) ground-truth score: 14.96 prediction: 13.06

Fig. 4. Example frames in the last fragments of two vault videos and the corresponding predictions and ground-truth scores. The steadier the landing, the higher the score.

5.4   Effectiveness of Loss Function

In this subsection, we verify the effectiveness of our loss function. The results in Table 1 show that the SRC of the ranking loss is higher than that of MSE significantly, indicating the ranking loss is more effective for the ranking constraint than MSE. Further, the combined loss function obtains an extra improvement compared with each single loss term, suggesting that the ranking loss can help MSE to make the prediction meet the ranking constraint.

In Table 1, we also give the mean euclidean distance (MED) between the predictions and the ground-truth scores. It is observed that MSE is able to constrain the score value while the ranking loss doesn't work at all. Only the ranking loss are combined with MSE, the MED get to be small relatively.

So the two loss terms are complementary and their combination contributes to our optimization objective (high SRC and relatively small MED). Finally, we also show the predictions and the ground-truth scores of all test samples, which are sorted from low to high by the ground-truth score, from the two datasets in Fig. 5. As Fig. 5 shows, the predictions and the ground-truth scores have the similar trend.

Table 1. Experimental results of different loss functions on the UNLV-Diving Dataset and the UNLV-Vault Dataset.

Loss function	SRC (Diving)	MED (Diving)	SRC (Vault)	MED (Vault)
MSE	0.7337	9.01	0.6325	1.33
Ranking loss	0.7870	75.62	0.6648	14.63
MSE+Ranking loss	0.8009	7.78	0.7028	2.60

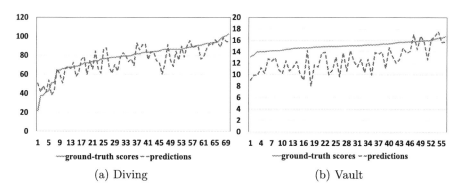

(a) Diving   (b) Vault

Fig. 5. The predictions and ground-truth scores of all test samples in the UNLV-Diving Dataset and the UNLV-Vault Dataset. The Y-axis represents the score value. The X-axis represents index of sample (sorting from low to high by the ground-truth score).

5.5   Comparison with Other Methods

We also compare our method with other state-of-the-art methods [4,7,14] on three datasets, as shown in Table 2.

For diving and vault, literature [7] uses a fixed data split and also implement Pose+DCT by this data split. The results of Pose+DCT here are from [7]. We adopt the same data split as [7].

For figure skating, literature [4] repeats the experiment 200 times with different random data splits and average the results. We just do one random split for the heavy computing cost.

In Table 2, we achieve the state-of-the-art results on the three datasets by using MSE+Ranking loss. Although our method outperforms [4, 7, 14], the results on the UNLV-Vault Dataset and the Mit-Skating Dataset are not breathtaking. The reason may come down to the view variation among the vault videos and down-sample on the skating videos.

Table 2. Comparison of our method with other state-of-the-art methods on the three datasets.

	SRC (Diving)	SRC (Vault)	SRC (Skating)
Pose+DCT [4]	0.53	0.10	0.35
ConvISA [14]	-	-	0.45
C3D+LSTM [7]	0.27	0.05	-
C3D+SVR [7]	0.78	0.66	0.53
Ours (MSE)	0.7435	0.6325	0.4167
Ours (Ranking loss)	0.7870	0.6648	0.2129
Ours (MSE+Ranking loss)	**0.8009**	**0.7028**	**0.5753**

6   Conclusion

This paper proposes an end-to-end learning-based framework to realize action quality assessment. In this framework, we divide videos into fragments and employ C3D extractor to obtain fragment-based features. The fragment-based features ensures our framework has the capacity to learn the evaluation criteria of actions. Specifically, in our loss function, both the score value and the ranking are considered. The combination of MSE and the ranking loss contributes to the better performance. Finally, wide experiments show that our method outperforms other state-of-the-art methods on three public datasets.

References

1. Vault. https://en.wikipedia.org/wiki/Vault-(gymnastics). 2.1.2. Accessed 2018
2. List of Olympic Games Scandals and Controversies. https://en.wikipedia.org/wiki/List-of-Olympic-Games-boycotts
3. FINA-DIVING RULES. http://www.fina.org/content/diving-rules. D8.1.3
4. Pirsiavash, H., Vondrick, C., Torralba, A.: Assessing the quality of actions. In: Fleet, D., Pajdla, T., Schiele, B., Tuytelaars, T. (eds.) ECCV 2014. LNCS, vol. 8694, pp. 556–571. Springer, Cham (2014). https://doi.org/10.1007/978-3-319-10599-4_36
5. Tao, L., et al.: A comparative study of pose representation and dynamics modelling for online motion quality assessment. Comput. Vis. Image Underst. **148**, 136–152 (2016)

6. Parmar, P., Morris, B.: Measuring the quality of exercises. In: 38th Annual International Conference of the IEEE Engineering in Medicine and Biology Society (EMBS), pp. 2241–2244 (2016)
7. Parmar, P., Morris, B.: Learning to score olympic events. In: 30th IEEE Conference on Computer Vision and Pattern Recognition, pp. 76–84 (2017)
8. Zia, A., Sharma, Y., Bettadapura, V., Sarin, E.L., Clements, M.A., Essa, I.: Automated assessment of surgical skills using frequency analysis. In: Navab, N., Hornegger, J., Wells, W.M., Frangi, A.F. (eds.) MICCAI 2015. LNCS, vol. 9349, pp. 430–438. Springer, Cham (2015). https://doi.org/10.1007/978-3-319-24553-9_53
9. Carvajal, J., Wiliem, A., Sanderson, C., Lovell, B.: Towards miss universe automatic prediction: the evening gown competition. In: 23rd International Conference on Pattern Recognition, pp. 1089–1094 (2016)
10. Du, T., Bourdev, L., Fergus, R., Torresani, L., Paluri, M.: Learning spatiotemporal features with 3D convolutional networks. In: International Conference on Computer Vision, pp. 4489–4497 (2015)
11. Venkataraman, V., Vlachos, I., Turaga, P.: Dynamical regularity for action analysis. In: 26th British Machine Vision Conference, pp. 67.1–67.12 (2015)
12. Soomro, K., Zamir, A., Shah, M.: UCF101: A Dataset of 101 Human Actions Classes from Videos in The Wild. arXiv preprint arXiv:1212.0402 (2012)
13. Chai, X., Liu, Z., Li, Y., Yin, F., Chen, X.: SignInstructor: an effective tool for sign language vocabulary learning. In: 4th Asian Conference on Pattern Recognition (2017)
14. Le, Q., Zou, W., Yeung, S., Ng, A.: Learning hierarchical invariant spatio-temporal features for action recognition with independent subspace analysis. In: 24th IEEE Conference on Computer Vision and Pattern Recognition, pp. 3361–3368 (2011)
15. Kingma, D., Ba, J.: Adam: A Method for Stochastic Optimization arXiv preprint arXiv:1412.6980 (2014)
16. Carreira, J., Zisserman, A.: Quo Vadis, action recognition? a new model and the kinetics dataset. In: 30th IEEE Conference on Computer Vision and Pattern Recognition, pp. 4724–4733 (2017)
17. Qiu, Z., Yao, T., Mei, T.: Learning spatio-temporal representation with pseudo-3D residual networks. In: International Conference on Computer Vision, pp. 5534–5542 (2017)

CGANs Based User Preferred Photorealistic Re-stylization of Social Image

Zhen Li, Meng Yuan, Jie Nie$^{(\boxtimes)}$, Lei Huang, and Zhiqiang Wei

Ocean University of China, Qingdao, China
{lizhen,niejie,huangl,weizhiqiang}@ouc.edu.cn,
yuanmeng@stu.ouc.edu.cn

Abstract. In social networks, it is important to re-stylize a randomly taken photo to exhibit a unique individual character. Previous stylization methods either respect to a motivation of improving perceptual quality or artistic style transfer, are neither personalized nor photorealistic. Besides, a strong constraint on scene consistency of reference image is always required, which is not easy to meet for a customized application. In this paper, we propose a customized photorealistic re-stylization method referred to a group of user favorite images with loose scene consistency. To better express user preferred style, reference images are selected from the perspective of photographer where image content and composition are jointly considered and weighed by user preference of light and color. To achieve high perceptual quality, we map image pixels and styles based on Conditional Generative Adversarial Networks. Comprehensive experiments verify our method could improve user preferred photo re-stylization and bring in less artificiality.

Keywords: Social media · User preference · Image stylization
Conditional generative adversarial networks · Photorealistic

1 Introduction

Recently, social networks offer an open and shared platform to display user's daily life especially by the way of photo sharing. Social image is the most common media that exhibits human taste and preference. However, not every photo is taken by a good photographer who understands how to express user's preference. Therefore, a randomly taken photo is required to be stylized before it is released. Usually, filter is employed to re-stylize photos, however, it usually brings in artificiality. Besides, there are very limited number of filters, which could not be personalized. The other way to re-stylize photo is to employ image editor tools such as Photoshop, but performance is highly relied on user's manipulation of software and aesthetic background, so it is hardly extended to the whole group of user. Thus, it is important to realize re-stylization of social image to be photorealistic and user preferred automatically.

At the same time, image re-stylization method could be categorized into two groups if considering the reference model. For models referred to improve the human visual quality, there are image enhancement [1] and image colorization [2] methods aiming at advancing visual perception by adjusting image contrast, brightness, hues, and saturation.

R. Hong et al. (Eds.): PCM 2018, LNCS 11165, pp. 135–146, 2018.
https://doi.org/10.1007/978-3-030-00767-6_13

These methods basically relied on principles of human vision or public knowledge. The other way is to transfer image style from one [3] or a group of images [4]. If the result is required to be photorealistic, these methods always set strict constraint on the selection of reference images. That is, reference image should be with consistent content of input image, which can't always be satisfied.

Thus, this paper tries to realize user preferred photorealistic re-stylization by using a group of user favorite images with loose constraint of scene consistency. Challenges exist in how to select reference images that could better transfer user preference as well as to reduce artificiality. Considering if only one reference image is adopted, since without a strict constraint on scene consistency, semantic in input image won't be properly paired with that in reference image, which will lead to a wrong style transfer result. Therefore, more images could help to provide sufficient semantics. However, more images will bring in diversity of styles, which also will lead to style confusion and increase artificiality. Thus, we investigate following issues to reach our goal:

Firstly, reference images which can best represent the style of user's expectation for input image should be selected. Since our motivation is to achieve photorealistic style, that means, the result could be pretended as was taken by the same photographer. From the perspective of photographer, photo style is mainly relied on content and composition. Therefore, in our cases, image content and composition are two elements taken into account in reference image selection.

Secondly, Inspired by [5], where images are classified into color-based and light-based by considering hue variety, it is observed that users could be classified into two groups according to the style of their favorite photos: light-preferred and color-preferred. For light preferred user, where their favorite images are usually monochromatically, thus photo composition should be higher weighted than content in deciding style. Oppositely, for color preferred user, content is more important. Thus, user style indicates which factor, content or composition, should be more weighted in reference image selection.

Thirdly, high visual quality is required after image re-stylization. However, most existing methods adopt style mapping functions either assuming fit explicit parameter models, which is too limit to ensure a good transfer, or maximizing the likelihood by approximating expectation, which resulted in desaturated and blur results [6].

By sufficient consideration of above issues, this paper proposes a customized photo re-stylization method. This method could stylize an input image referred to an elaborated selection of user preferred photos by jointly considering photo content and composition. Moreover, user preference of light and color is seriously considered to weight impacts of two mentioned factors. Finally, to achieve a vivid and realistic result, we train a Conditional Generative Adversarial Networks (cGANs) by a group of reference images to transfer one given image. Generative Adversarial Networks (GANs) is a recent hot researched model that could produce satisfied samples by achieving Nash equilibrium between the generator and discriminator instead of minimize error between an estimate and the ground truth [7]. Experiments show the effectiveness on how proposed method could re-stylize photo to meet user's preference as well as bring in less artificiality.

2   Related Works

User preferred photorealistic style transfer is to stylize one photo as it was taken by the same photographer. Where the contrast, brightness, hue and saturation, even texture will be revised following this photographer's principle. From the perspective that the result and original image are with content consistency but different styles, image enhancement and image colorization are special cases of image re-stylization respect to different motivations. Thus, various previous methods have been discussed how to re-stylize images. Methods could be categorized from various perspectives, such as global or local, parametric or non-parametric [5], geometry-based or statistical-based [3], etc. Here, we will discuss previous works from the view point of how many reference images were employed, which is clearly for us to compare difference and to conclude contribution of proposed method.

Style Transfer Referred to One Image. Plenty much of works discussed how to transfer color, contrast or texture from one reference image. Reinhard et al. [8] applied a global transfer on Lab color space, which assumed that the distributions of input and reference image could be well fitted by the multivariate Gaussian distribution. However, this assumption turned out to be too restrictive to ensure a good transfer. Many works investigated transfer in locally correspondence regions, for example semantic similar regions [9], or statistical based clusters, [5], but these methods set strong constraint on scene similarity to achieve artificiality free result. With the development of deep neural networks, images could be represented separately by content and style, Gatys et al. [10] introduced a neural algorithm to recombine an artistic style with a natural image content. The limitation of this method is it highlighted the texture style to achieve an artistic result, thus is not suitable for photorealistic style transfer. Therefore, referred to only one reference image, nor enough style correspondence between input image and reference image could be learned unless with strict content consistency.

Style Transfer Referred to a Group of Images. The main issue of style transfer is to learn a mapping function of image representation and style. Thus, more images are required to learn more statistical dependencies between style and image representation. Kang et al. [11] learned a mapping function on image contrast and color by means of distance learning to explicit represent the style transfer model from a group of user preferred images. But a predefined mapping function could enable only limited modifications to the image. Deep neural networks are applied to learn an implicit mapping function by minimizing error between an estimate and the ground truth. Like Cheng et al. [6] trained a multimodal CNN model to re-color image by a large image dataset. For example, 1.3 M ImageNet samples [2], and achieved a photorealistic result. But because it adopted models maximizing the likelihood by approximating expectation, thus resulted in desaturated and blur results. Recently, GANs is hot researched since it could generate real samples instead of expectation. Thus, this paper will investigate how to operate style transfer by a GANs in our situation.

Inspired by Yao et al. [12] who discussed the importance of composition in photo aesthetic evaluation, our work tries to select images by jointly considering content and composition. Moreover, we also observe that user could be classified into two groups according to their preference in light and color. This is an important indicator deciding which factor: content or composition, should be more weighted in reference image selection.

3   Method

Figure 1 illustrates an overview framework of proposed method. Given an input image, our work tries to re-stylize it referring to a selection of user preferred images. Firstly, we propose an image selection method by jointly considering two important factors of stylization: image content and composition. Then, we observe that to weigh these two factors is distinguished according to individual preference. Thus, a user preference assessment module is applied to obtain a user preference indicator to be as a weight to rebalance the influence of each factor. After that, selected images are treated as training images to learn a Conditional Generative Adversarial Networks (cGANs). Finally, we obtain the re-stylized image by inputting the original image through the well trained CGANs.

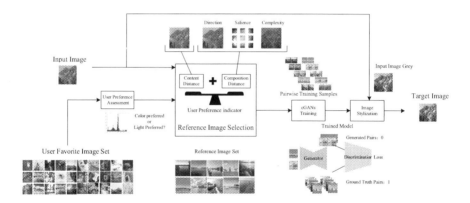

Fig. 1. System workflows.

3.1   Factors Influence Photo Style

From the perspective of photographer, there are two main factors that matter photo style:

Image Content. Usually, user preferred photo style of similar content is always fixed. So choosing images with similar contents could probably represent user's preference. Since content descriptor is not our contribution, here, we adopt the classic Bag of Words (BoW) model to describe image content and measure the content distance by using a normalized norm-1 distance of word histogram. Here, for a pair of image I_i and I_j, the content distance is denoted as $D_{\text{content}}(I_i, I_j)$

Image Composition. Image composition is the art of putting things together with conscious thoughts [15]. Composition is closely related to the aesthetic preference of user. For one certain photographer, he/she may employ similar camera parameters to capture photos with similar spatial layouts, and this leads to a user preferred representation of brightness, contrast, color and texture (e.g. depth of filed). Inspired by Machajdik et al.'s [13] and Yao et al.'s [12] work, we adopt three features that could describe image composition and introduce as following:

Salience Arrangement. Represents the layout of salient object in a photo. The salience map is obtained as described in [14]. To achieve the position of salient object, an image is divided into thirds following the "Rule of Thirds", which is a rule of thumb for good composition photos. For each region, we sum up the salience pixels. Then, we identify whether a region contains a salient object by setting a threshold of the percentage of salient pixels in one region. Finally, we achieve a 9 dimensions' vector, where element may be 1, when the corresponding region is identified containing salient object, or 0, if not. Here, for a pair of image I_i and I_j, we employ the Hamming Distance of salience vectors as the salience arrangement distance denoted as $D_{\text{salience}}(I_i, I_j)$.

Line Directions. Line direction is a significant feature deciding photo composition. The longer, thicker and more dominate the line, the stronger the expressiveness of image composition. Here, we use the Hough transform to detect significant line slopes and length. We divide the directions into 5 categories a $0° \sim 15°$, $15° \sim 75°$, $75° \sim 105°$, $105° \sim 165°$, $165° \sim 180°$. We sum up and normalize the length of lines if their tilt angle distributed in the same direction category. For one image, we achieve a 5 dimensions' vector. And for a pair of image I_i and I_j, we define a normalized distance of line histogram denote as $D_{\text{direction}}(I_i, I_j)$.

Visual Complexity. Measures the detail level of an image. We define the visual complexity of one photo as described in [13] by applying the waterfall segmentation with a predefined Alternating Sequential Filter size (filter size 3 and level 2 of the waterfall hierarchy). We count the result regions as the visual complexity descriptor. Here, for a pair of image Ii and Ij, the complexity distance is denoted as a normalized difference of segment region numbers, which is denoted as $D_{\text{complexity}}(I_i, I_j) D_{\text{complexity}}(I_i, I_j)$.

Finally, the composition distance of image I_i and I_j is a liner combination of above distance with empirical weights w_S, w, w_c respectively.

$$D_{\text{composition}}(I_i, I_j) = w_s D_{\text{salience}}(I_i, I_j) + w_d D_{\text{direction}}(I_i, I_j)$$
$$+ w_c D_{\text{complexity}}(I_i, I_j) \tag{1}$$

3.2   User Preference Indicator

After we achieve the content distance and composition distance, it is necessary to jointly consider the contribution of each factor in selecting reference images. Inspired by [5], it is also observed that user could be divided into two categories according to the style of their preferred images, defined as color-preferred user and light-preferred user. Figure 2 shows two different user styles.

(a) Favorite photo samples of color-preferred User   (b) Favorite photo samples of light-preferred User

Fig. 2. Favorite image samples of color preferred user and light preferred user.

We adopt the algorithm described in [5] to classify images into two categories: color-based image and light-based image. User preference indicator λ is defined as the proportion of light-based images in whole favorite set of one individual. The indicator denotes the possibility of a user to be light-preferred.

3.3   Objective

To sum up, given an input image, reference images should be selected from the whole set of user favorite images. The selection should jointly consider image content distance and composition distance. And user preference plays an important role in balancing the weight between these two factors. Thus, we define our objective as:

$$arg \min_T \sum\nolimits_{I \in T} \left((1 - \lambda)D_{\text{content}}(I, input) + \lambda D_{\text{composition}}(I, input) \right) \qquad (2)$$

where T is a subset of user favorite images containing images most similar as input image under our metric. And λ is the user preference indicator. To obtain T, we calculate image distance between every image in user favorite set with input image and then rank them following a descend order. We choose Top N ranked images as our reference images. Every unknown parameter will be assigned a value in experiment section.

3.4   Pairwise Training Samples

Here, we adopt the cGANs network architecture as proposed in [15] to transfer style from reference images to input image. To train this networks requires pairwise training samples: one is original image, and the other is re-stylized image whose content is consistent with the original image. Since reference images are already user preferred images, thus should be set as results. For the original images, we just transferred them into grey level. Finally, we obtained N pair of training samples to train the cGANs. Training details will be discussed in experiment.

4 Experiments

4.1 Dataset and Implementation Details

Experiments are carried out on our own collected dataset. 60 users posted and liked 61,200 images are downloaded from Flickr. For Each user, we randomly select 20 posted photos as test images, and 2000 as reference candidates. The implementation of our algorithm begins with user preference indicator λ calculation as mentioned in 3.2. For each user, λ is calculated using his/her favorite image collection. Then, distance between test image and every reference candidate image of one user are calculated. The composition distance in Eq. (1) depends on three parameters whose default values are set as follows: $w_s = 0.3$, $w_d = 0.3$ and $w_c = 0.4$. Finally, we choose top $N = 5$ reference images to train the cGANs. For the training, we adopted the same network architecture as described in [18]. Training procedure is implemented on a desktop with an NVIDIA Tesla K40 m and DDR5 12G. All images are with unified size as 256×256. For one test image, 5 reference images are adopted to train the model with 200 epochs following instruction mentioned in 3.4. Every test image is transferred into gray level image firstly and then is input into the trained cGANs to produce a new stylized image as result. We set the original test image as ground truth. We evaluate our method performance by comparing the result and the ground truth.

4.2 Baselines

We compare our method with following baselines:

Reinhard et al.'s [8] method: a classic image color transfer parametric model based on Multivariate Gaussian Distributions. This method transfers style from only one reference image. Here, we set the reference image as the top 1 image in our selection method.

Gatys et al.'s method [10]: a multilayer image representation based method uses a deep CNNs to recombine style representation of reference image and content representation of input image. Here, we set the reference image as top 1 selected by our algorithm.

Zhang et al.'s method [2]: a deep CNNs model trained by 1.3M ImageNet images. It employs a fine grid multimodal color system and achieves more vibrant and perceptually realistic colorization.

Zhang et al.'s method (User retuned): Since above model is not user oriented, we retune it by user's favorite images.

Pix2pix [15]: a cGANs model with same network architecture as we adopt, but is trained by whole user's favorite sets without selection. 2000 user favorite images are trained with 200 epochs.

4.3 Metrics

Our motivation is to stylize image according to user's preference. Thus, a good stylization method has to ensure the stylized image contains properly style preferred by user. We assume user post images are all preferred by this user. Thus, we just measure how similar between the result image and the ground truth. Meanwhile, the result image should be satisfied with visual quality. These two requirements should be satisfied equally. Therefore, we set two objective metrics to measure the performance of above two problems respectively.

The Bhattacharya Coefficient [16] is used to measure the similarity between two histograms. To evaluate how result meet user's preference, we apply it in the Hue, Saturation and Value histograms of the result and the ground truth image. Here we choose HSV color space, which is closely align with the way human vision percepts.

The SSIM [17] can be used to measure the artificiality of re-stylized images by measuring a structural error visibility between result image and the ground truth.

4.4 Results and Analysis

The optimal value for both SSIM and Bhattacharya Coefficient is 1. It refers to result with same visual quality as well as with exactly the same light and color distribution as the ground truth. In the other word, it is artificiality free as well as meets user's

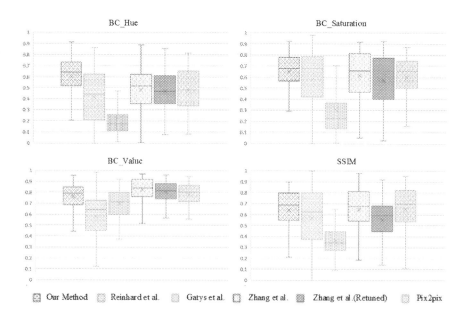

Fig. 3. Box plots of different methods, displaying the distribution of Bhattacharya Coefficients of Hue, Saturation and Value histograms and the distribution of SSIM value.

preference. Figure 3 illustrates the box plot of experiment results. Moreover, Fig. 4 exhibits several samples of result images from our method compared with baselines.

SSIM value indicates the visual quality of re-stylized image compared with ground truth image. It is obviously showed that our method achieves similar performance with best performed methods. However, our model is trained by only 5 reference images, while Pix2pix method is trained over 2000 user favorite images, and Zhang et al.'s method is trained based on 1.3M ImageNet dataset. Also, it is observed that the retuned Zhang et al.'s method fails in obtaining good visual quality. We explained this reason as that there are relative small amount of user preferred images (about 2000 per user) compared with 1.3M ImageNet images, thus connections between pixels and semantics could not be well trained. The reason that Reinhard et al.'s method and Gatys et al.'s method performed poorly on our data set is because these methods require high constraint on scene consistency to achieve photorealistic result, and it could not be satisfied in this application. Moreover, Gatys et al.'s method is designed for artistic style transfer. It brings in obvious artificiality. To sum up, our method achieved satisfied visual perception with lest training cost.

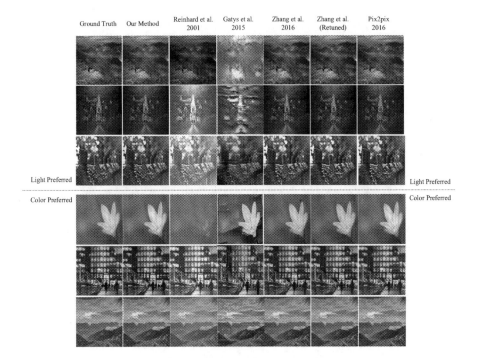

Fig. 4. Results of our method and baseline methods. Ground truth and Results of different methods are showed by column. The first three rows demonstrated the result of a light preferred user (User preference indicator ≥ 0.5) and the last three rows demonstrated that of a color preferred user (User preference indicator <0.5).

The Bhattacharya Coefficient indicates whether the result could meet an individual preference. Our method achieves best performance on Hue and Saturation similarity, and satisfied performance on Value similarity, therefore, it overpasses baseline methods in meeting user's preference. Specially, hue similarity between result image and ground truth image is the most significant indicator of whether user's preference is well learned or not. Our method over performs baseline methods by 44.23% in average, which sufficiently verifies the effectiveness of proposed method. Additionally, compared to Pix2pix, we achieve improvement by 25.00%, and it is directly increased by applying our image selection algorithm. Figure 3 also illustrates results distribution compared with baselines, suggesting our method could perform more stable work than baselines.

Figure 4 demonstrates several samples selected from our dataset. We classify these samples into color-based image and light-based image as in [5]. By exploring our dataset, it is observed that, for Zhang et al.'s method, performance on color based image always achieves better result than that of light based image. That is because it was trained by naturally taken images with higher probability to be a color image. But for our method, by considering the composition aspect in choosing reference images, we select images that could better represent expecting style from the perspective of light preferred user.

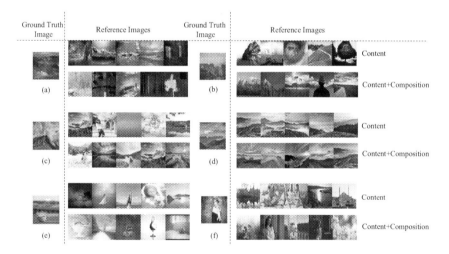

Fig. 5. Reference image selection result of several image samples selected in our database. (a, b) are two samples from light preferred user ($\lambda \geq 0.5$); (c, d, e, f) are samples from color preferred user. For each sample, the first row shows top 5 reference image selected considering only content. The second row shows that with both content and composition consideration.

To intuitively explain the reason why proposed methods achieved better customized performance, Fig. 5 illustrates several samples of reference selection results of our method, to compare the effectiveness of jointly considered image composition and content, we operate the selection by using only content factors. It is intuitively observed that our method achieves better alignment in spacious arrangement with ground truth image, besides, achieves similar level of consistency in content. Comprehensively, we achieve better scene correspondence, and thus leading to a better stylization result.

We also compare the results of training different reference image sets. Table 1 lists the mean value of two adopts objective measurements. The result of jointly considering content and composition ameliorates that of only applying content. It verified the effectiveness of introducing composition consideration in reference image selection.

Table 1. Mean value of Bhattacharya Coefficients in Hue, Saturation, and Value histogram obtained by cGANs trained by different reference image sets.

Content	BC_Hue	BC_Saturation	BC_Value	SSIM
w/o Composition	0.51	0.63	**0.79**	0.60
w/Composition	**0.60**	**0.65**	0.76	**0.64**

5   Conclusion

In this paper, an image re-stylization method is proposed to achieve photorealistic performance and user preference. Given an input image, the expecting style is learned from a group of user favorite images with loose content consistency. In order to select correct reference images, inspired by photography theories, image content and composition are jointly considered. Meanwhile, user preference of light and color is assessed and applied to weigh the importance of above two factors. After that, selected reference images are trained to learn a cGANs based style mapping model. Experiments firstly verify the effectiveness of proposed re-stylization method on meeting user's preference as well as bring in less artificiality. It also confirms that by introducing composition consideration and user preference of light and color, our method could ameliorate the style transfer performance. Future work will concentrate on investigating more user styles to improve customized performance.

Acknowledgments. This work is supported by the National Nature Science Foundation of China (No. 61602430, No. 61702471, No. 61402428), and The Aoshan Innovation Project in Science and Technology of Qingdao National Laboratory for Marine Science and Technology (No. 2016ASKJ07).

References

1. Lee, J.-S.: Digital image enhancement and noise filtering by use of local statistics. IEEE Trans. Pattern Anal. Mach. Intell. **PAMI-2**(2), 165–168 (1980)
2. Zhang, R., Isola, P., Efros, A.A.: Colorful Image Colorization (2016). arXiv:160308511 Cs
3. Faridul, H.S., Pouli, T., Chamaret, C., Stauder, J., Tremeau, A., Reinhard, E.: A Survey of Color Mapping and its Applications (2014)
4. Cheng, W., Jiang, R., Chen, C.W.: Color photo makeover via crowd sourcing and recoloring. In: Proceedings of the 23rd ACM International Conference on Multimedia, New York, NY, USA, pp. 943–946 (2015)

5. Hristina, H., Le Meur, O., Cozot, R., Bouatouch, K.: Style-aware robust color transfer. In: Computational Aesthetics in Graphics, Visualization, and Imaging, Istambul, Turkey (2015)
6. Cheng, Z., Yang, Q., Sheng, B.: Deep colorization. In. Presented at the Proceedings of the IEEE International Conference on Computer Vision, pp. 415–423 (2015)
7. Goodfellow, I: NIPS 2016 Tutorial: Generative Adversarial Networks (2016). arXiv: 170100160 Cs
8. Reinhard, E., Adhikhmin, M., Gooch, B., Shirley, P.: Color transfer between images. IEEE Comput. Graph. Appl. **21**(5), 34–41 (2001)
9. HaCohen, Y., Shechtman, E., Goldman, D.B., Lischinski, D.: Non-rigid dense correspondence with applications for image enhancement. In: ACM SIGGRAPH 2011 Papers, New York, NY, USA, pp. 70:1–70:10 (2011)
10. Gatys, L.A., Ecker, A.S., Bethge, M.: Image style transfer using convolutional neural networks. In: 2016 IEEE Conference on Computer Vision and Pattern Recognition (CVPR), pp. 2414–2423 (2016)
11. Kang, S.B., Kapoor, A., Lischinski, D.: Personalization of image enhancement. In: 2010 IEEE Computer Society Conference on Computer Vision and Pattern Recognition, pp. 1799–1806 (2010)
12. Yao, L., Suryanarayan, P., Qiao, M., Wang, J.Z., Li, J.: OSCAR: on-site composition and aesthetics feedback through exemplars for photographers. Int. J. Comput. Vis. **96**(3), 353–383 (2012)
13. Machajdik, J., Hanbury, A.: Affective image classification using features inspired by psychology and art theory. In: Proceedings of the 18th ACM International Conference on Multimedia, New York, NY, USA, pp. 83–92 (2010)
14. Harel, J., Koch, C., Perona, P.: Graph-based visual saliency. Adv. Neural. Inf. Process. Syst. **19**, 545–552 (2007)
15. Isola, P., Zhu, J.-Y., Zhou, T., Efros, A.A.: Image-to-Image Translation with Conditional Adversarial Networks. arXiv:161107004 Cs (2016)
16. Aherne, F.J., Thacker, N.A., Rockett, P.I.: The Bhattacharyya metric as an absolute similarity measure for frequency coded data. Kybernetika **34**(4), 363–368 (1998)
17. Wang, Z., Bovik, A.C., Sheikh, H.R., Simoncelli, E.P.: Image quality assessment: from error visibility to structural similarity. IEEE Trans. Image Process. **13**(4), 600–612 (2004)
18. Bonneel, N., Sunkavalli, K., Paris, S., Pfister, H.: Example-based video color grading. ACM Trans. Graph **32**(4), 39:1–39:12 (2013)
19. Pitié, F., Kokaram, A.C., Dahyot, R.: Automated colour grading using colour distribution transfer. Comput. Vis. Image Underst. **107**(1), 123–137 (2007)
20. Kapoor, A., Caicedo, J.C., Lischinski, D., Kang, S.B.: Collaborative personalization of image enhancement. Int. J. Comput. Vis. **108**(1–2), 148–164 (2014)

Cross-Modal Event Retrieval: A Dataset and a Baseline Using Deep Semantic Learning

Runwei Situ[1], Zhenguo Yang[1(✉)], Jianming Lv[2], Qing Li[3], and Wenyin Liu[1(✉)]

[1] School of Computer Science and Technology,
Guangdong University of Technology, Guangzhou, China
siturunwei@163.com, zhengyang5-c@my.cityu.edu.hk,
liuwy@gdut.edu.cn
[2] School of Computer Science and Engineering,
South China University of Technology, Guangzhou, China
jmlv@scut.edu.cn
[3] Department of Computer Science, City University of Hong Kong,
Hong Kong, China
itqli@cityu.edu.hk

Abstract. In this paper, we propose to learn Deep Semantic Space (DSS) for cross-modal event retrieval, which is achieved by exploiting deep learning models to extract semantic features from images and textual articles jointly. More specifically, a VGG network is used to transfer deep semantic knowledge from a large-scale image dataset to the target image dataset. Simultaneously, a fully-connected network is designed to model semantic representation from textual features (e.g., TF-IDF, LDA). Furthermore, the obtained deep semantic representations for image and text can be mapped into a high-level semantic space, in which the distance between data samples can be measured straightforwardly for cross-model event retrieval. In particular, we collect a dataset called Wiki-Flickr event dataset for cross-modal event retrieval, where the data are weakly aligned unlike image-text pairs in the existing cross-modal retrieval datasets. Extensive experiments conducted on both the Pascal Sentence dataset and our Wiki-Flickr event dataset show that our DSS outperforms the state-of-the-art approaches.

Keywords: Cross-modal event retrieval · Deep learning · Common space

1 Introduction

The development of the Internet and the emerging social media are changing the way people interact with each other. Meantime, multi-modal online data (e.g. images, texts, audios and videos) is growing rapidly. In reality, data in different modalities can be used for describing the same real-world events [1, 2] (e.g., protests, elections, festivals, natural disasters). For instance, a news website may contain textual descriptions, images and audios to report an event. Cross-modal retrieval aims to retrieve the data in different modalities that are relevant to the same event. Multi-modal data can span

© Springer Nature Switzerland AG 2018
R. Hong et al. (Eds.): PCM 2018, LNCS 11165, pp. 147–157, 2018.
https://doi.org/10.1007/978-3-030-00767-6_14

different feature spaces, known as the problem of "semantic gap", making the measurement on the content similarity among the data more challenging.

Researchers have built up quite a few multimodal datasets for cross-modal retrieval, such as Wikipedia image-text pairs [3], Pascal sentence [4], Pascal VOC [5]. However, these datasets focus on strongly aligned data pairs, e.g., an image of cat and its exact textual descriptions as shown in Fig. 1a. In reality, there may exist more complicated cases which cannot be expressed by one-to-one data pairs. For instance, given a photo of news event (e.g., a protest), users may expect to acquire the relevant textual materials, which are usually not the exact description of the photo but they share the same label of event. Therefore, we call such data weakly aligned data as shown in Fig. 1b.

A ginger kitten asleep on a brown leather sofa.
An orange kitten lay on a leather couch.
A sleepy orange kitten on a brown leather couch.
A small orange kitten is curled up sleeping on a brown leather couch.
Sweet little yellow kitten curled up asleep.

A gray cat laying on a brown table.
A grey cat laying on a dining table.
A grey cat lying on a wooden table.
Black domestic cat lying on brown table.
Grey cat on a dining room table looking at the camera.

(a)   Pascal Sentence dataset

Flickr   Wikipedia

(Reuters) - French President Francois Hollande said on Saturday the attacks in Paris that killed 127 people were "an act of war" organized from abroad by Islamic State with internal help. Some witnesses in the hall said they heard the gunmen shout Islamic chants and slogans condemning France's role in Syria...

To date 70 people have been formally identified, having died following the fire at Grenfell Tower. Also, baby Logan Gomes who was stillborn in hospital on 14 June, has been recorded by police as a victim of the fire. The following victims have been formally identified in agreement with the Coroner...

(b) Wiki-Flickr event dataset

Fig. 1. Examples of strongly aligned image-text pairs from Pascal sentence dataset, and weakly aligned examples from our Wiki-Flickr event dataset. In particular, the corresponding text is the exact description of an image in the strongly aligned data pairs. In contrast, the weakly aligned textual content does not describe an image exactly, but they share the same event label.

In the context of cross-modal event retrieval, using the user-generated content is very challenging to obtain a joint representation for multimodal data. The performance of the traditional techniques, such as CCA [6], CFA [7], is still far from satisfactory. Recently, due to the development of deep learning, significant progress has been made in the fields of speech recognition, image recognition, sentiment classification, and image caption generation. Inspired by these works, we employ deep learning models for cross-modal event retrieval, especially for the weakly aligned data.

In this paper, we utilize deep learning models to learn a common semantic space in order to measure the content similarity between data in different modalities. More specially, we employ a VGG [8] network to transfer semantic knowledge from ImageNet dataset to our Wiki-Flickr event dataset. At the same time, we devise a fully-connected network to extract deep features from raw textual features, e.g., TF-IDF, LDA. Furthermore, we map the images and texts to a common semantic space with high-level semantics, in which the cross-modal data samples can be matched directly by using similarity measurement. The main contributions of this work are the following:

(1) We propose a deep semantic space (DSS) framework based on the VGG network and fully-connected network for cross-modal retrieval. DSS has the advantage of high discriminative power, and hence can be used to deal with weakly aligned data.

(2) We collect a Wiki-Flickr event dataset, where the data are weakly aligned unlike the usual image-text pairs in the existing datasets. We plan to release the dataset for public use later on.

(3) Extensive experiments conducted on the Pascal Sentence dataset and our Wiki-Flickr event dataset show that the proposed DSS outperforms the state-of-the-art approaches.

The rest of this paper is organized as follows. Section 2 reviews the related work on cross-modal retrieval. Section 3 shows the details of our proposed method. Extensive experiments and conclusions are given in Sects. 4 and 5, respectively.

2 Related Work

Various approaches have been proposed to deal with cross-modal retrieval, which can be roughly divided into four categories: subspace learning, hashing-based methods, rank-based methods, and DNN-based methods. We introduce the basic ideas and a few representative approaches in these categories below.

(1) **Subspace learning**. Canonical correlation analysis (CCA) and Kernel-CCA are representative subspace learning approaches, which aim to learn a common subspace shared among different modalities of data by maximizing their correlations.

(2) **Hashing-based methods.** Considering high-dimensional cross-modal data, Bronstein et al. [9] proposed a cross modal similarity sensitive hashing (CMSSH) for efficient cross-modal tasks. Song et al. [10] proposed a novel inter-media hashing (IMH) model to transform multimodal data into a common Hamming space.

(3) **Rank-based methods.** Rank-based methods usually use the strategy of learning to rank. Bai et al. [11] presented supervised semantic indexing (SSI) for cross-lingual retrieval. Grangier et al. [12] proposed a discriminative kernel-based method to solve the problem of cross-modal ranking by adapting the passive-aggressive algorithm.

(4) **DNN-based methods.** The recent DNN-based methods can utilize the advantage of large-scale data, achieving better performance than the traditional approaches. Srivastava et al. [13] proposed to learn a shared representation between different modalities based on restricted Boltzmann machine. Wang et al. [14] proposed a regularized deep neural network (RE-DNN), which is a 5-layer neural network for mapping visual and textual features into a common semantic space. Furthermore, similarity between different modalities can be measured seamlessly. Wei et al. [15] presented a semantic matching method to address the cross-modal retrieval problem. However, shallow networks usually perform well on the small-scale dataset, which may suffer from underfitting when dealing with large-scale datasets.

3   The Proposed Method

This section elaborates our proposed method for cross-modal event retrieval, which uses the VGG network and a fully-connected network to learn the common semantic space. Figure 2 illustrates the overview of our proposed framework.

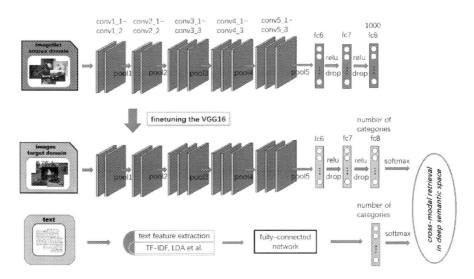

Fig. 2. An overview of our proposed DSS. For images, we use a VGG network to transfer semantic knowledge from ImageNet (in Sect. 3.1). For text, we design a fully-connected network to obtain text semantics (in Sect. 3.2). Finally, the multimodal data is embedded into deep semantic space for cross-modal retrieval (in Sect. 3.3).

3.1   Learning Image Semantics with Knowledge Transfer

Considering the degradation problem faced by the deep learning models on dealing with large-scale dataset, we propose to learn image semantics by VGG network.

More specifically, we fine-tune the VGG network by initializing the parameters with a network pre-trained on the ImageNet dataset. Furthermore, we feed the output of the last fully-connected layer o_I into a softmax, which generates image semantic embedding $S_I \in R^K$ over a number of K event categories. Intuitively, the softmax function maps a K-dimensional vector z to a K-dimensional vector $\sigma(z)$ of real values in the range (0, 1) that add up to 1. The image semantic embedding S_I is defined below:

$$\sigma : R^K \rightarrow \left\{ Z \in R^K | z_i \geq 0, \sum_{i=1}^{K} z_i = 1 \right\} \tag{1}$$

$$z_j = (o_I)_j \quad \text{for} \quad j = 1, \ldots, K. \tag{2}$$

$$(S_I)_j = P(y = j | I) = \sigma(z)_j = \frac{e^{z_j}}{\sum_{i=1}^{K} e^{z_k}} \quad \text{for} \quad j = 1, \ldots, K. \tag{3}$$

where $P(y = j | I)$ represents the predicted probability for the j-th class given a data sample I. $S_I \in R^K$ is the image semantic embedding vector. $(S_I)_j$ represents the j-th element in the vector.

3.2   Learning Text Semantics by Fully-Connected Network

We design a 4-layer fully-connected network based on raw textual features to obtain the text semantics as show in Fig. 3. More specifically, we take term frequency–inverse document frequency (TF-IDF) as an example to illustrate the semantic learning process for the textual content. Stop words have been removed before obtaining vectors consisting of TF-IDF values for the textual documents. The dimension of the vectors is equal to the number of tokens in the corpus. Furthermore, we design a 4-layer fully-connected network to learn the hidden semantics underlying the documents, which is defined below:

$$f(x) = \max(0, x) \tag{4}$$

$$h_t^{(2)} = f^{(2)}\left(W_t^{(1)} \cdot T + b_t^{(1)} \right) \tag{5}$$

$$h_t^{(3)} = f^{(3)}\left(W_t^{(2)} \cdot h_t^{(2)} + b_t^{(2)} \right) \tag{6}$$

$$o_T = f^{(4)}\left(W_t^{(3)} \cdot h_t^{(3)} + b_t^{(3)} \right) \tag{7}$$

where T represents the input TF-IDF features for each document, $f(x)$ represents the rectified linear unit (ReLU) function, i.e., the activation function, and o_T represents the output of the last fully-connected layer.

Finally, o_T is fed into a K-way softmax, which generates text semantic embedding S_T $\in R^K$ over a number of K categories. The text semantic embedding S_T is defined below:

$$\sigma : R^K \rightarrow \{z \in R^K | z_i \geq 0, \quad \sum_{i=1}^{K} z_i = 1\} \tag{8}$$

$$z_j = (o_T)_j \quad \text{for} \quad j = 1, \ldots, K. \tag{9}$$

$$(S_T)_j = P(y = j \mid T) = \sigma(z)_j = \frac{e^{z_j}}{\sum_{i=1}^{K} e^{z_k}} \quad \text{for} \quad j = 1, \ldots, K. \tag{10}$$

where $P(y = j \mid T)$ represents the predicted probability for the j-th class given a data sample T. $S_T \in R^K$ is the text semantic embedding. $(S_T)_j$ represents the j-th element in the vector.

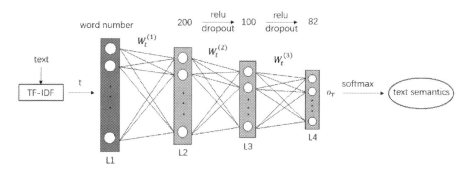

Fig. 3. Learning text semantics by the fully-connected network.

3.3 Semantic Matching in the Deep Semantic Space

As mentioned previously, we obtain unified vectors for both image and text, which is called the deep semantic space (DSS). Therefore, cross-model data samples can be measured directly in DSS by using distance metrics, e.g., Euclidean distance, cosine distance, Kullback-Leibler (KL) divergence, Normalized Correlation (NC). In the experiments, we will investigate the influence of various distance metrics.

4 Experiments

4.1 Dataset and Data Partitions

(1) **Pascal Sentence dataset**: It is a subset of Pascal VOC, which contains 1000 pairs of image and text descriptions (several sentences) from 20 categories. We randomly select 30% pairs from each category for training and the rest for testing. The text-image pairs are strongly aligned data, i.e., text is the exact description of an image, as shown in Fig. 1.

(2) **Wiki-Flickr Event dataset**: We collect 28,825 images and 11,960 text articles, belonging to 82 categories of events. The images are shared by users on Flickr social media, while the text articles are collected from different news media sites, e.g., BBC News, The New York Times, Yahoo News, Google News. In particular, the data is weakly aligned, as a text article is not the exact description of a certain photo, but they share the same event label. Some examples are shown in Fig. 1, and the statistics on the dataset is shown in Fig. 4. For data partitions, 75% of the data samples are used for training, and the rest are used for testing.

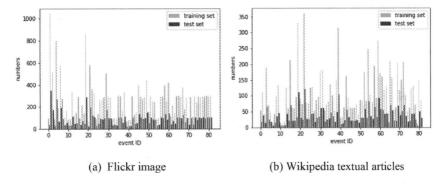

(a) Flickr image   (b) Wikipedia textual articles

Fig. 4. Data partitions of our Wiki-Flickr event dataset.

4.2 Experimental Results

(1) **Implementation Details.** In terms of the images, we crop the images and horizontally flip the images randomly with a given probability of 0.5 for data augmentation. The images are resized to 224 × 224. Then, we normalize the images with mean and standard deviation. The loss function of the network adopts Cross Entropy, which is optimized by using stochastic gradient descent with the momentum of 0.9. The learning rate is 0.001, and the batch size is 16. In terms of the 4-layer fully-connected network for text, we use the same loss function and optimization strategy, the learning rate is 0.01, and the dropout probability is 0.5.

(2) **Evaluation Metrics.** In the experiments, two retrieval tasks are conducted: retrieving text by image queries (denoted as Image → Text) and retrieving images by text queries (denoted as Text → Image). We evaluate the ranking list by mean average precision (MAP). MAP is computed as the mean of average precision (AP) for all the queries.

(3) **Compared with Baselines.** In terms of the performance of our DSS on strongly aligned data pairs, we compare with eight popular approaches on the Pascal Sentence dataset, including CCA [6], CFA [7], JRL [16], LGCFL [17], Multi-modal DBN [18], Corr-AE [19], Bimodal-AE [20], and Deep-SM [15]. The performance of the baselines on the Pascal Sentence dataset is reported in their papers, respectively. As shown in Table 1, DNN-based methods [15, 18–20] tend to achieve better performance than the traditional ones [6, 7, 16, 17]. Overall, our

Table 1. MAP performance of the approaches on Pascal Sentence dataset.

Method	Image → Text	Text → Image	Average
CCA	0.110	0.116	0.113
CFA	0.341	0.308	0.325
JRL	0.416	0.377	0.397
LGCFL	0.381	0.435	0.408
Bimodal-AE	0.404	0.447	0.426
Multimodal DBN	0.438	0.363	0.401
Corr-AE	0.411	0.475	0.443
Deep-SM	0.440	0.414	0.427
Our DSS	**0.472**	**0.495**	**0.484**

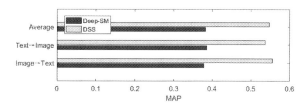

Fig. 5. Comparing DSS with Deep-SM on Wiki-Flickr event dataset.

DSS achieves the best performance, giving significant improvement. Furthermore, we compare the proposed DSS with Deep-SM on the Wiki-Flickr event dataset as shown in Fig. 5, which demonstrates the effectiveness of our DSS for weakly align data.

(4) **Evaluation of Network Structures and Distance Metrics.** We evaluate the influence of adopting different network architectures for pre-training, and using different distance metrics in DSS. The evaluations are conducted on the Wiki-Flickr event dataset as shown in Table 2, from which we make two observations: (1) In terms of the network architectures adopted for pre-training, we can observe that deep models, such as VGG [8], and ResNet [21], achieve better performance than the shallow models, such as AlexNet [22], and SqueezeNet [23]. (2) In terms of the distance metrics, normalized correlation and cosine distance perform better than Euclidean distance and KL-divergence in the context of cross-modal event retrieval.

(5) **Examples of the Retrieval Results.** Intuitively, we take text retrieving images as an example to show the performance of DSS on Wiki-Flickr event dataset in Fig. 6. The top-five images are given in the figure, where the event labels are marked at the lower right corner. Red boxes indicate the mismatched retrieval results, while green boxes indicate the correct results. Our DSS returns three mismatched images in the right example. It is probably due to the fact that the categories of 'Baltimore protests', 'Shooting of Michael Brown' and 'Death of Freddie Gray' share quite a few similar words and images, making them hard to be distinguished.

Table 2. MAP performance of adopting different network architectures and distance metrics in our DSS on Wiki-Flickr event dataset

Architecture	Distance metric	Image → Text	Text → Image	Average
AlexNet	KL-divergence	0.451	0.380	0.416
SqueezeNet		0.447	0.409	0.428
ResNet		0.438	0.456	0.447
VGG		**0.494**	**0.447**	**0.471**
AlexNet	Euclidean distance	0.462	0.410	0.436
SqueezeNet		0.466	0.427	0.447
ResNet		0.461	0.464	0.463
VGG		**0.503**	**0.474**	**0.489**
AlexNet	Cosine distance	0.530	0.495	0.513
SqueezeNet		0.539	0.510	0.525
ResNet		0.556	0.537	0.547
VGG		**0.576**	**0.566**	**0.571**
AlexNet	Normalized correlation	0.532	0.498	0.515
SqueezeNet		0.541	0.510	0.526
ResNet		0.560	0.538	0.549
VGG		**0.578**	**0.570**	**0.574**

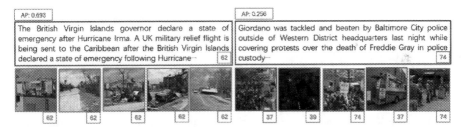

Fig. 6. Two examples of cross-modal retrieval results obtained by DSS on Wiki-Flickr event dataset. Note the numbers refer to the event labels (i.e., 62: Hurricane Irma, 37: Death of Freddie Gray, 39: Shooting of Michael Brown, 74: Baltimore protests, etc.).

5 Conclusion

In this paper, we have proposed a deep semantic space (DSS) learning framework for cross-modal retrieval. DSS embeds multimodal data into a common semantic space with high-level semantics in an end-to-end manner, which casts cross-modal retrieval problem as a homogeneous retrieval task. In particular, we collect a Wiki-Flickr event dataset to advocate the problem of cross-modal retrieval for weakly aligned data. Extensive experiments conducted on a public dataset and Wiki-Flickr event dataset show that our DSS outperforms the state-of-the-art approaches.

Acknowledgments. The authors would like to thank Zehang Lin and Feitao Huang for data collection. This work is supported by the National Natural Science Foundation of China (No. 61703109, No. 91748107, No. U1611461), the Guangdong Innovative Research Team Program (No. 2014ZT05G157), Science and Technology Program of Guangdong Province, China (No. 2016A010101012), and CAS Key Lab of Network Data Science and Technology, Institute of Computing Technology, Chinese Academy of Sciences, 100190, Beijing, China (No. CASNDST201703), and an internal grant from City University of Hong Kong (Project No. 9610367).

References

1. Yang, Z., Li, Q., Lu, Z., Ma, Y., Gong, Z., Liu, W.: Dual structure constrained multimodal feature coding for social event detection from Flickr data. ACM Trans. Internet Technol. **17**(2), 19 (2017)
2. Yang, Z., Li, Q., Liu, W., Ma, Y., Cheng, M.: Dual graph regularized NMF model for social event detection from Flickr data. World Wide Web **20**(5), 995–1015 (2017)
3. Rasiwasia, N., Costa Pereira, J., Coviello, E., Doyle, G., Lanckriet, G. R., Levy, R., Vasconcelos, N.: A new approach to cross-modal multimedia retrieval. In: 18th ACM International Conference on Multimedia, pp. 251–260. ACM (2010)
4. Rashtchian, C., Young, P., Hodosh, M., Hockenmaier, J.: Collecting image annotations using Amazon's Mechanical Turk. In: NAACL HLT 2010 Workshop on Creating Speech and Language Data with Amazon's Mechanical Turk, pp. 139–147. Association for Computational Linguistics (2010)
5. Hwang, S.J., Grauman, K.: Reading between the lines: object localization using implicit cues from image tags. IEEE Trans. Pattern Anal. Mach. Intell. **34**(6), 1145–1158 (2012)
6. Thompson, B: Canonical correlation analysis. In: Encyclopedia of Statistics in Behavioral Science (2000)
7. Li, D., Dimitrova, N., Li, M., Sethi, I. K.: Multimedia content processing through cross-modal association. In: 11th ACM International Conference on Multimedia, pp. 604–611. ACM (2003)
8. Simonyan, K., Zisserman, A.: Very Deep Convolutional Networks for Large-Scale Image Recognition (2014). arXiv preprint arXiv:1409.1556
9. Bronstein, M. M., Bronstein, A. M., Michel, F., Paragios, N.: Data fusion through cross-modality metric learning using similarity-sensitive hashing. In: Computer Vision and Pattern Recognition, pp. 3594–3601 (2010)
10. Song, J., Yang, Y., Yang, Y., Huang, Z., Shen, H.T.: Inter-media hashing for large-scale retrieval from heterogeneous data sources. In: 2013 ACM SIGMOD International Conference on Management of Data, pp. 785–796. ACM (2013)
11. Bai, B., Weston, J., Grangier, D., Collobert, R., Sadamasa, K., Qi, Y., Weinberger, K.: Learning to rank with (a lot of) word features. Inf. Retr **13**(3), 291–314 (2010)
12. Grangier, D., Bengio, S.: A discriminative kernel-based approach to rank images from text queries. IEEE Trans. Pattern Anal. Mach. Intell. **30**(8), 1371–1384 (2008)
13. Srivastava, N., Salakhutdinov, R.: Multimodal learning with deep boltzmann machines. Adv. Neural Inf. Process. Syst. **5**, 2222–2230 (2012)
14. Wang, C., Yang, H., Meinel, C.: Deep semantic mapping for cross-modal retrieval. In: Tools with Artificial Intelligence, pp. 234–241. IEEE (2015)
15. Wei, Y., Zhao, Y., Lu, C., Wei, S., Liu, L., Zhu, Z., Yan, S.: Cross-modal retrieval with cnn visual features: A new baseline. IEEE Trans. Cybern. **47**(2), 449–460 (2017)

16. Zhai, X., Peng, Y., Xiao, J.: Learning cross-media joint representation with sparse and semisupervised regularization. IEEE Trans. Circuits Syst. Video Technol. **24**(6), 965–978 (2014)
17. Kang, C., Xiang, S., Liao, S., Xu, C., Pan, C.: Learning consistent feature representation for cross-modal multimedia retrieval. IEEE Trans. Multimedia **17**(3), 370–381 (2015)
18. Srivastava, N., Salakhutdinov, R.: Learning representations for multimodal data with deep belief nets. In: International Conference on Machine Learning Workshop, vol. 79 (2012)
19. Feng, F., Wang, X., Li, R.: Cross-modal retrieval with correspondence autoencoder. In: 22nd ACM International Conference on Multimedia, pp. 7–16. ACM (2014)
20. Ngiam, J., Khosla, A., Kim, M., Nam, J., Lee, H., Ng, A.Y.: Multimodal deep learning. In: 28th International Conference on Machine Learning, pp. 689–696 (2011)
21. He, K., Zhang, X., Ren, S., Sun, J.: Deep residual learning for image recognition. In: IEEE Conference on Computer Vision and Pattern Recognition, pp. 770–778 (2016)
22. Krizhevsky, A.: One Weird Trick for Parallelizing Convolutional Neural Networks (2014). arXiv preprint arXiv:1404.5997
23. Iandola, F.N., Han, S., Moskewicz, M.W., Ashraf, K., Dally, W.J., Keutzer, K.: SqueezeNet: AlexNet-level accuracy with 50x fewer parameters and <0.5 MB model size (2016). arXiv preprint arXiv:1602.07360

Retinal Vessel Segmentation
via Multiscaled Deep-Guidance

Rui Xu[1,2], Guiliang Jiang[1,2], Xinchen Ye[1,2(✉)], and Yen-Wei Chen[3]

[1] DUT-RU International School of Information Science and Engineering,
Dalian University of Technology, Dalian, China
yexch@dlut.edu.cn
[2] Key Laboratory for Ubiquitous Network and Service Software of Liaoning Province,
Dalian, China
[3] College of Science and Engineering, Ritsumeikan University, Kusatsu, Shiga, Japan

Abstract. Retinal vessel segmentation is a fundamental and crucial step
to develop a computer-aided diagnosis (CAD) system for retinal images.
Retinal vessels appear as multiscaled tubular structures that are variant
in size, length, and intensity. Due to these vascular properties, it is difficult for prior works to extract tiny vessels, especially when ophthalmic
diseases exist. In this paper, we propose a multiscaled deeply-guided neural network, which can fully exploit the underlying multiscaled property
of retinal vessels to address this problem. Our network is based on an
encoder-decoder architecture which performs deep supervision to guide
the training of features in layers of different scales, meanwhile it fuses
feature maps in consecutive scaled layer via skip-connections. Besides, a
residual-based boundary refinement module is adopted to refine vessel
boundaries. We evaluate our method on two public databases for retinal vessel segmentation. Experimental results show that our method can
achieve better performance than the other five methods, including three
state-of-the-art deep-learning based methods.

Keywords: Retinal vessel segmentation · Multiscaled deep-guidance
Deep convolutional neural network

1 Introduction

Retinal vessel is a significant gist for diabetes, glaucoma and arteriosclerosis in
clinical diagnosis. Segmentation of retinal vessels is a fundamental and crucial
step for a CAD system of retinal fundus images. However, segmentation of the
whole vascular trees is not easy. The retinal vasculature is composed of arteries
and veins appearing as elongated features, with their tributaries visible within
the retinal image [1]. Although the intensity profile of the vessel cross-section
could be approximated by a Gaussian or a mixed Gaussian function, local grey
level of blood vessels can vary hugely due to the effect of lighting condition and
retinal pathology. Besides, blood vessel trees are dominated in the multiscaled

© Springer Nature Switzerland AG 2018
R. Hong et al. (Eds.): PCM 2018, LNCS 11165, pp. 158–168, 2018.
https://doi.org/10.1007/978-3-030-00767-6_15

property. Vessel widths are variant in a wide range, from one pixel to several tens of pixels. The terminal parts of vessels are so tiny that they look similar as the surrounding background.

Previous methods of retinal vessel segmentation can be divided into two types. The first type, which we refer to as the model-based method, constructs a mathematical model for retinal vessels according to the properties including intensity, shape, gradient and contrast. Examples of this type are the Gaussian kernel-based filter-banks [2,3], the divergence of normalized gradient vector field [5], kernels based on locally adaptive derivative frames [4] and the active contour-based method [6]. The second type, which we refer to as the learning-based method, resolve the retinal vessel segmentation by using a binary classification framework, where each pixel is classified to be vessels or not. Prior works obey a classical routine, where the designation of hand-crafted features is followed by the training of a discriminative classifier. There are researches that design representative features by using ridges of vessels [7], line operators [8] and 2-D Gabor wavelet filter-banks [9]. Classical techniques, such as K-nearest-neighbor classifier [7], support vector machines [8] and Bayesian classifier [9], are utilized to train classifiers for vessel segmentation. The above-mentioned methods are highly depended on some pre-defined assumptions, and they are easily failed when noise, abnormal lighting conditions and retinal pathology exist due to the mismatching of the assumptions.

In the past several years, deep learning techniques have been extensively evolved in the field of computer vision. After the deep convolutional neural networks (DCNN) exhibit extraordinary performance on the task of image classification [10–13], they are used to deal with other vision-related tasks, such as semantic segmentation [14]. The most-widely used DCNN for segmentation is based on an encoder-decoder architecture [15,16], where feature maps in different scaled layers are learnt. Taking advantage of the inherent multiscaled feature maps can relieve the problem of spatial information loss in DCNN-based methods. For example, skip-connection is adopted to fuse high-level and low-level features in previous works [14,16]. Besides, deep-guidance, which supervises the training of features in different scales, is another technique to exploit multiscaled information in DCNN-based methods, and it is demonstrated to be very efficient to extract detailed edges of objects in natural images [19].

In this paper, we leverage the latest progress on deep learning for the segmentation of retinal vessels on fundus images. We propose a DCNN-based architecture that can fully exploit the multiscaled property of retinal vessels to ensure accurate segmentation especially for tiny vessels. Deep-guidance is adopted in our encoder-decoder based network to guarantee the training of features in specific scales, meanwhile skip-connections integrate feature maps between consecutively high- and low-scaled layers to exhaustively utilize multiscaled information. Besides, a residual-based boundary refinement module [17] is utilized to ensure clear vessel boundaries. We evaluate the propose DCNN-based method on two public datasets, Digital Retinal Images for Vessel Extraction(DRIVE) [7] and

High-Resolution Fundus (HRF) [18], and compare it with other retinal vessel segmentation methods.

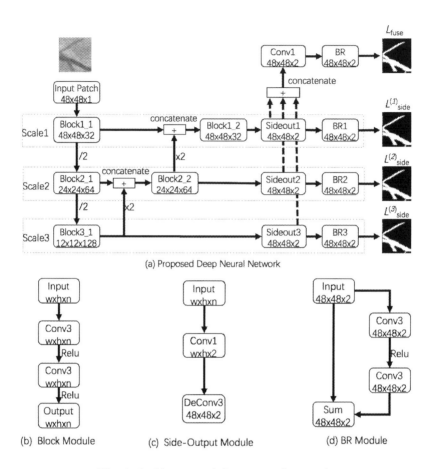

Fig. 1. Architecture of the proposed network

2 Related Works

Deep learning has been applied for the segmentation of retinal vessels in several previous works. Liskowski et al. firstly utilize deep learning techniques to segment vessels on retinal fundus images in [20], where several convolutional layers are followed by fully-connected layers to classify each pixel as vessels or not. Deep-guided convolutional neural networks are adopted in the works [21,22], and a conditional random filed reformulated by a recurrent neural network is added to hold interactions between pixels [22]. Recently, Maninis et al. propose a DCNN-based method to segment vessels and optic discs simultaneously [23].

Compared with these methods, the proposed network is mainly different in two aspects. First, we integrate both deep-guidance and skip connections into our network architecture in order to exhaustively exploit multiscaled information of vessels. Second, a residual-based network module is adopted to learn context information for refinement of vessel boundaries. Such carefully designed network architecture ensures our method to be efficient to extract retinal vessels, even for tiny vessels on pathological images.

3   Approach

Compared to databases in computer vision, medical databases usually have much fewer images. For example, there are only several tens of fundus images in the two widely-used public databases for the evaluation of retinal vessel segmentation. Considering the few amount of data, we choose to develop a patch-based DCNN method instead of an image-based DCNN method that are widely-used in computer vision and usually requires huge numbers of images for training.

As we notice that multiscale is one of the dominate features for retinal vessels on fundus images, we propose a novel network architecture that can learn and integrate multiscaled information to improve the accuracy of retinal segmentation. Inspired from the ideas in the HED network that is originally proposed for edge extraction [19], we utilize the technique of deep-guidance in our network to ensure the training of more representative feature maps on specific scales. To comprehensively take advantage of information embedded in different scales, we adopt skip connections, which is introduced in U-Net [16], to mix feature maps in consecutively high- and low-level scales. Additionally, a boundary refinement module is integrated into our network to sharpen boundaries of vessels for more accurate segmentation results. The details of the proposed patch-based DCNN method are described in this section.

3.1   Network Architecture

The network architecture is illustrated in Fig. 1. The backbone of our network is based on an encoder-decoder architecture. The encoder part gradually generates feature maps in three different stages by using convolution blocks (details are shown by Fig. 1(b)) that is followed by downsampling. Each convolution block is composed by two convolutional layers, and downsampling halves spatial resolution. Feature maps in each stage have a certain spatial resolution, which correspond to a specific scale.

Skip Connections are adopted in the decoder part that gradually recover spatial resolution for accurate segmentation. High-level feature maps are expanded twice in spatial resolution and concatenated with consecutively low-level feature maps. Then, the concatenated features are passed into a convolution block that is composed by two convolutional layers to generate the final feature maps. By using such a kind of network structure, feature maps in two consecutive scales are integrated together.

Deep Guidance is adopted to supervise the training of feature maps for a specific spatial scale. Feature maps on each scale are connected outside to a classifier through a side-output and boundary refinement module. The side-output module is composed by a convolutional and deconvolutional layer (Fig. 1(c)) to calculate score maps that have the same spatial resolution with the input. Though such a network structure, feature maps on each scale are trained with the guidance from a specific loss function.

Boundary Refinement is a residual model based structure whose details are shown by Fig. 1(d). In our network, this module is connected to score maps in side-output paths to ensure vessel boundaries on each scale to be sharper. Besides, it is also utilized in the fusion path that integrate score maps of all scales to achieve optimal vessel segmentation results.

3.2   Network Training

Our network are directly trained from a set of patch-pairs, which are denoted by $S = \{(X_n, Y_n), n = 1, \ldots, N\}$, where $X_n = \{x_j^{(n)}, j = 1, \ldots, |X_n|\}$ is an image patch and $Y_n = \{y_j^{(n)}, j = 1, \ldots, |X_n|\}$, $y_j^{(n)} \in \{0, 1\}$ is the corresponding ground truth of segmentation. For simplicity, we denote all parameters of convolution blocks in the backbone as $\mathbf{W}$. In this paper, our network has three side-output paths, each of which is associated with a side-output module, boundary refinement module and a classifier. If parameters of all modules are denoted as $\mathbf{w} = (\mathbf{w}^{(1)}, \mathbf{w}^{(2)}, \mathbf{w}^{(3)})$, the objective function of side-output paths can be represented by Eq. 1.

$$L_{side}(\mathbf{W}, \mathbf{w}) = \sum_{i=1}^{3} \alpha_i L_{side}^{(i)}(\mathbf{W}, \mathbf{w}^{(i)}) \tag{1}$$

where L_{side} denotes the loss function for side-output paths. As the distributions of vessel or non-vessel pixels in patches are highly unbalanced, we adopt the following class-balanced cross-entropy loss function to calculate Eq. 1.

$$
\begin{aligned}
L_{side}^{(i)}(\mathbf{W}, \mathbf{w}^{(i)}) = &- \beta \sum_{j \in Y_+} \log Pr(y_j = 1 | X; \mathbf{W}, \mathbf{w}^{(i)}) \\
&- (1 - \beta) \sum_{j \in Y_-} \log Pr(y_j = 0 | X; \mathbf{W}, \mathbf{w}^{(i)})
\end{aligned}
\tag{2}
$$

where $\beta = |Y_-|/|Y_+|$ and $1 - \beta = |Y_+|/|Y|$. $|Y_+|$ and $|Y_-|$ denote the vessel and non-vessel ground truth label sets, respectively.

The predicted score maps on each scale is fused together by a fusion-path that includes a convolutional layer, boundary refinement module and a classifier. We denote parameters of all these modules as $\mathbf{h}$, and calculate fusion-path loss function $L_{fuse}(\mathbf{W}, \mathbf{w}, \mathbf{h})$ by using class-balanced cross-entropy. Putting everything together, we minimize the following objective function via standard stochastic gradient descent.

$$L = L_{side}(\mathbf{W}, \mathbf{w}) + L_{fuse}(\mathbf{W}, \mathbf{w}, \mathbf{h}) \tag{3}$$

3.3   Implementation

We implement the proposed network by using public Keras library with the back-end of Tensor-Flow. The network is initialized randomly and directly trained on image-patches that are cropped from retinal images. Hyper-parameters include learning-rate (0.01), mini-batch size (32), drop-out rate (0.2) and side-output loss parameter α_i (1).

Our network takes 48×48 image-patches as input and predict segmentation probability maps of the same resolution. When the network is used to segment a retinal image, a sliding window manner is operated on the image with a stride of 8 pixels, and average probability is calculated in overlapped regions.

4   Experiments

4.1   Data, Preprocessing and Evaluation Protocol

We evaluate our proposed method on two publicly available datasets, which are DRIVE [7] and HRF [18]. The DRIVE database includes 40 fundus images, 33 of which have nearly no-sign of diabetic retinopathy. The rest 7 images only show mild diabetic retinopathy. The HRF database contains 45 images including 15 for healthy patients, 15 for diabetic retinopathy and 15 for glaucomatous patients.

In experiments, all fundus images are processed by the following prepro-cessing: (1) color images are converted to grey-scale images. (2) a histogram equalization routine called CLAHE [24] is operated on all gray-scale images. (3) gamma correction is performed.

We choose 20 images in DRIVE and 30 images in HRF as the training set, and the rest images in two databases as testing set. In order to get image patch pairs for training, 48×48 image patches are randomly cropped from the training set to train deep neural networks. Finally, we obtain 200,000 patches in DRIVE and 240,000 patches in HRF for the training. We evaluate different methods on the testing set by calculating the area under curve (AUC) for receiver operating characteristic (ROC) and precision and recall curve (PR).

4.2   Network Ablation

In this subsection, we evaluate the proposed method by gradually adding the network modules mentioned in Sect. 2. Here, we compared 4 different kinds of network architecture by evaluating the AUC of ROC and PR on DRIVE and HRF databases. The comparison results are given in Table 1. The four kinds of networks are denoted as (1) SCN, which only includes the skip-connections but no deep guidance or boundary refinement, (2) DSN, which only has the deep guidance without the other two network modules (3) MDGN, which exploits multiscaled information by skip-connections and deep guidance but boundary refinement is not used (4) MDGN-BR, which is the proposed network architec-ture using all techniques mentioned in Sect. 2.

Table 1. Experiments of network ablation.

Methods	DRIVE		HRF	
	ROC	PR	ROC	PR
SCN	0.9717	0.8895	0.9727	0.8577
DSN	0.9747	0.8999	0.9707	0.8609
MDGN	0.9786	0.9090	0.9766	0.8802
MDGN-BR	**0.9793**	**0.9104**	**0.9770**	**0.8805**

From Table 1, it can be seen that the AUC of ROC and PR for both SCN and DSN are relatively low. SCN includes skip-connections for integrating feature maps in different scales, however there is no network structure to ensure that feature maps in each scale can be learnt optimally. Thus, it is not able to fully exploit multiscaled information by only using skip-connections. A similar situation happens when only deep guidance is adopted in DSN, which only has network structures to supervise the learning of feature maps in different scales but no structures to fuse them. By utilizing both skip-connections and deep

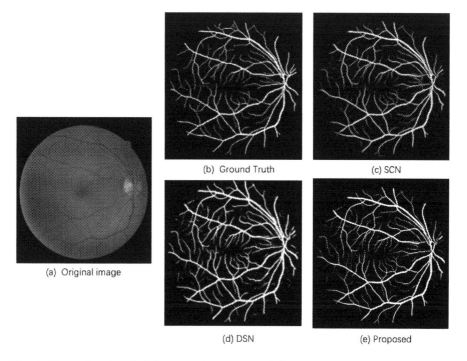

(a) Original image

(b) Ground Truth

(c) SCN

(d) DSN

(e) Proposed

Fig. 2. Examples of probability map (or score map) for vessel segmentation on DRIVE database

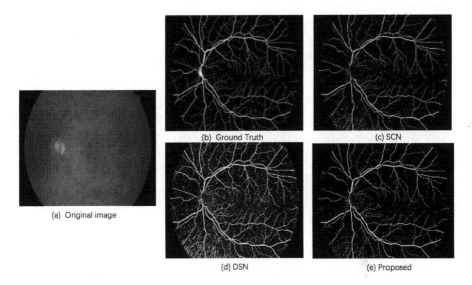

Fig. 3. Examples of probability map (score map) for vessel segmentation on HRF database

guidance, network performance can be highly improved, as shown by MDGN in Table 1. AUC values of both ROC and PR can be further improved when boundary refinement module is adopted, which demonstrates the efficiency of such a network module utilized in our proposed method.

Figures 2 and 3 give an example of vessel-like probability map for the databases of DRIVE and HRF respectively. It can be seen that the proposed method can give higher probability scores for tiny vessels, which visually demonstrates that our network can exhaustively exploit multiscaled information to improve the performance of vessel segmentation. Besides, we also notice that our method is robust to retinal pathology. Compared with DRIVE, HRF includes more cases with severe retinal pathology, which usually degrades performance of vessel segmentation methods. Due to pathological effect, severe short-tubular artifacts exist in the results of SCN and DCN, as shown by the left bottom part in Fig. 3(c) and (d). However, our method is not much affected by retinal pathology and less artifacts exist in our result (Fig. 3(e)).

4.3   Comparison with Other Methods

We compare the proposed network with other methods in this subsection. Table 2 gives the comparison results. Here, we implement two state-of-the-art deep learning-based methods, which are widely used in computer vision and denoted as HED [19] and FCN [14] respectively. We adopt the pre-trained VGG network [11] as the backbones for both of them, and then finely turn network parameters on retinal images. From Table 2, it can be seen that the two deep-learning based method achieve relatively low performances. Both of them take the whole retinal images as the input, however, DRIVE and HRF include only several tens

Table 2. Comparison with different kinds of methods.

Methods	DRIVE		HRF	
	ROC	PR	ROC	PR
HED [19]	0.9662	0.8728	0.9511	0.8022
FCN [14]	0.9558	0.8494	0.9464	0.7815
LADF [4]	0.9636	–	0.9608	–
DCNN-FC [20]	0.9720	–	–	–
Line [8]	0.9633	–	–	–
DRIU [23]	–	0.822	–	–
Proposed	**0.9793**	**0.9104**	**0.9770**	**0.8805**

of retinal images, which are not enough for training. Although fine-turning of a pre-trained network on few images can give relative good segmentation, that can not relieve the full power of the network. This is also the reason why we choose patch-based DCNN method in this paper.

The evaluation quantities for the other 4 methods in Table 2 are directly copied from the original papers. DCNN-FC is the first work that utilizes deep convolutional neural network for retinal segmentation [20], which adopts stacked convolutional and fully-connected layers to classify pixels. LADF [4] takes advantage of locally adaptive derivative frames to design an optimal filter to extract retinal vessels. Line [8] extracts features by using line operator and trains K-NN classifier for pixel classification. DRIU [23] uses a base network architecture on which two set of specialized layers are trained to solve both the retinal vessel and optic disc segmentation. Compared with these methods, the proposed method achieves the highest AUC values for both ROC and PR in the two public databases. These results demonstrate the efficiency of the proposed method for retinal vessel segmentation.

5   Conclusion

In this paper, we propose a novel DCNN-based network to segment retinal vessels from fundus images. In order to exhaustively exploit multiscaled information on retinal images, skip connections and deep guidance are utilized in our network to ensure better learning of features in different scales and fusion of them for vessel segmentation. Besides, boundary refinement module is adopted to make sharper vessel boundaries. By using these techniques, the proposed method outperformed other retinal segmentation methods on the public DRIVE and HRF databases.

Acknowledgement. This work was supported by National Natural Science Foundation of China (NSFC) under Grant 61772106 and Grant 61702078, and by the Fundamental Research Funds for the Central Universities.

References

1. Fraz, M.M., Remagnino, P., et al.: Blood vessel segmentation methodologies in retinal images-a survey. Comput. Methods Programs Biomed. **108**(1), 407–433 (2012)
2. Chaudhuri, S., Chatterjee, S., et al.: Detection of blood vessels in retinal images using two-dimensional matched filters. IEEE Trans. Med. Imaging **8**(3), 263–269 (1989)
3. Hoover, A., Kouznetsova, V., Goldbaum, M.: Locating blood vessels in retinal images by piecewise threshold probing of a matched filter response. IEEE Trans. Med. Imaging **19**(3), 203–210 (2000)
4. Zhang, J., Dashtbozorg, B., et al.: Robust retinal vessel segmentation via locally adaptive derivative frames in orientation scores. IEEE Trans. Med. Imaging **35**(12), 2631–2644 (2016)
5. Lam, B.Y., Yan, H.: A novel vessel segmentation algorithm for pathological retina images based on the divergence of vector fields. IEEE Trans. Med. Imaging **27**(2), 237–246 (2008)
6. Al-Diri, B., Hunter, A., Steel, D.: An active contour model for segmenting and measuring retinal vessels. IEEE Trans. Med. Imaging **28**(9), 1488–1497 (2009)
7. Staal, J., Abramoff, M.D., et al.: Ridge-based vessel segmentation in color images of the retina. IEEE Trans. Med. Imaging **23**(4), 501–509 (2004)
8. Ricci, E., Perfetti, R.: Retinal blood vessel segmentation using line operators and support vector classification. IEEE Trans. Med. Imaging **26**(10), 1357–1365 (2007)
9. Soares, J.V., Leandro, J.J., et al.: Retinal vessel segmentation using the 2-D gabor wavelet and supervised classification. IEEE Trans. Med. Imaging **25**(9), 1214–1222 (2006)
10. Krizhevsky, A., Sutskever, I., Hinton, G.E.: ImageNet classification with deep convolutional neural networks. In: NIPS (2012)
11. Simonyan, K., Zisserman, A.: Very deep convolutional networks for large-scale image recognition. In: ICLR (2015)
12. He, K., Zhang, X., Ren, S., Sun, J.: Deep residual learning for image recognition. In: CVPR (2015)
13. Szegedy, C., Liu, W., et al.: Going deeper with convolutions. In: CVPR (2015)
14. Shelhamer, E., Long, J., Darrell, T.: Fully convolutional networks for semantic segmentation. IEEE Trans. Pattern Anal. Mach. Intell. **39**(4), 640–651 (2017)
15. Badrinarayanan, V., Kendall, A., Cipolla, R.: Segnet: a deep convolutional encoder-decoder architecture for image segmentation. IEEE Trans. Pattern Anal. Mach. Intell. **39**(12), 2481–2495 (2017)
16. Ronneberger, O., Fischer, P., Brox, T.: U-net: convolutional networks for biomedical image segmentation. In: Navab, N., Hornegger, J., Wells, W.M., Frangi, A.F. (eds.) MICCAI 2015. LNCS, vol. 9351, pp. 234–241. Springer, Cham (2015). https://doi.org/10.1007/978-3-319-24574-4_28
17. Peng, C., Zhang, X., Yu, G., Luo, G., Sun, J.: Large kernel matters - improve semantic segmentation by global convolutional network. In: CVPR (2017)
18. Budai, A., Bock, R., Maier, A., Hornegger, J., Michelson, G.: Robust vessel segmentation in fundus images. Int. J. Biomed. Imaging **2013**, 11 (2013)
19. Xie, S., Tu, Z.: Holistically-nested edge detection. In: ICCV (2015)
20. Liskowski, P., Krawiec, K.: Segmenting retinal blood vessels with deep neural networks. IEEE Trans. Med. Imaging **35**(11), 2369–2380 (2016)

21. Mo, J., Zhang, L.: Multi-level deep supervised networks for retinal vessel segmentation. Int. J. Comput. Assist. Radiol. Surg. **12**(9), 1–13 (2017)
22. Fu, H., Xu, Y., Lin, S., Kee Wong, D.W., Liu, J.: DeepVessel: retinal vessel segmentation via deep learning and conditional random field. In: Ourselin, S., Joskowicz, L., Sabuncu, M.R., Unal, G., Wells, W. (eds.) MICCAI 2016. LNCS, vol. 9901, pp. 132–139. Springer, Cham (2016). https://doi.org/10.1007/978-3-319-46723-8_16
23. Maninis, K.-K., Pont-Tuset, J., Arbeláez, P., Van Gool, L.: Deep retinal image understanding. In: Ourselin, S., Joskowicz, L., Sabuncu, M.R., Unal, G., Wells, W. (eds.) MICCAI 2016. LNCS, vol. 9901, pp. 140–148. Springer, Cham (2016). https://doi.org/10.1007/978-3-319-46723-8_17
24. Pizer, S.M., Amburn, E.P.: Adaptive histogram equalization and its variations. Comput. Vis. Graph. Image Process. **39**(3), 355–368 (1987)

Image Synthesis with Aesthetics-Aware Generative Adversarial Network

Rongjie Zhang, Xueliang Liu, Yanrong Guo, and Shijie Hao$^{(\boxtimes)}$

Hefei University of Technology, Hefei 230009, Anhui, China
hfut.hsj@gmail.com

Abstract. With the advance of Generative Adversarial Networks (GANs), image generation has achieved rapid development. Nevertheless, the synthetic images produced by the existing GANs are still not visually plausible in terms of semantics and aesthetics. To address this issue, we propose a novel GAN model that is both aware of visual aesthetics and content semantics. Specifically, we add two types of loss functions. The first one is the aesthetics loss function, which tries to maximize the visual aesthetics of an image. The second one is the visual content loss function, which minimizes the similarity between the generated images and real images in terms of high-level visual contents. In experiments, we validate our method on two standard benchmark datasets. Qualitative and quantitative results demonstrate the effectiveness of the two loss functions.

Keywords: Image synthesis · Generative Adversarial Network
Image aesthetics

1 Introduction

The rapid development of Generative Adversarial Network family sheds light on the task of natural image generation. As the basic idea of GAN [7], the generator tries to produce images as real as possible to confuse the discriminator. Various GAN-based models [1, 2, 8, 8, 24, 28] have been proposed to optimize the instability problems in generating images from different aspects. They have made solid progress in synthesizing natural images by using the standard datasets with legible backgrounds/foregrounds [33], e.g. MNIST [20], CIFAR-10 [18], CUB-200 [36] and so on. In many real-world applications, generating images with good visual aesthetics is highly desirable. Most of the existing GAN models are limited to achieve this goal, as they do not consider the image aesthetics in the learning process.

To address the above issue, in this paper, we propose a novel adversarial network namely AestheticGAN, and synthesize images with better visual aesthetics and plausible visual contents. Our consideration is two-fold. First, people always prefer to images with pleasant appearances, such as vivid color and appropriate

R. Hong et al. (Eds.): PCM 2018, LNCS 11165, pp. 169–179, 2018.
https://doi.org/10.1007/978-3-030-00767-6_16

composition. Therefore, the image generator is expected to be trained with aesthetics awareness. Second, apart from visually appealing, the generated images should also have reasonable visual contents. For example, based on our method, the image scene is quickly recognizable, and the content details are real. So the image generator is also expected to be aware of image semantics. To this end, we design and add two types of loss functions for the DCGAN architecture [29]. The first one is the aesthetics loss, which uses a quantitative score to evaluate the visual aesthetics of an image. The second one is the semantic loss, which measures the high-level semantic similarity between generated and real images [13,21].

The main contributions of this work are listed as follows:

- We attempt to create images with visually appealing images based on adversarial learning. Two types of loss functions are designed and added into the state-of-the-art GAN architecture.
- Extensive experiments are conducted on the AVA and cifar10 datasets. Comparisons in terms of visual appearance, quantitative scores, and user studies all demonstrate the effectiveness of our method.

The remain parts of this paper are organized as follows. We briefly review the related work in Sect. 2, and describe the proposed method in Sect. 3. In Sect. 4, we evaluate our method with qualitative and quantitative experiments. Section 5 finally concludes the paper.

2   Related Works

Since our research is closely related with the fields of GANs and image aesthetics, we briefly introduce their related research in this section.

GAN is a generation model inspired by two-person zero-sum game in Game Theory. Based on the seminal research by Goodfellow et al. [7], many GAN-based variants [1,2,8,24,28] have been proposed, which focus on the model structure extension, in-depth theoretical analysis, and efficient optimization techniques, as well as their extensive applications. For example, in order to solve the problem of disappearance of training gradient, Arjovsky et al. [1] proposes Wasserstein-GAN (W-GAN) and then improves it by adding the gradient penalty [8]. In order to limit the modeling ability of the model, Qi [28] proposes Loss-sensitiveGAN (LS-GAN), which limits the loss function obtained by minimizing the objective function to satisfy the Lipschitz continuity function class, and the authors also give the results of quantitative analysis of gradient disappearance. Further, ConditionalGAN (CGAN) [24] adds additional information(y) to the G and D, where y can be labels or other auxiliary information. InfoGAN [2] is another important extension of GAN, which can obtain the mutual information between hidden layer variables of the input and the specific semantics. Odena et al. [27] proposes that Auxiliary Classifier GAN (AC-GAN) can achieve multiple classification problems, and its discriminator outputs the corresponding tag probability.

Despite of the rapid development of GANs, there are few works that specifically designed for the task of aesthetic image generation.

The computational aesthetics has attracted attentions in recent years [6,14]. The purpose of the research on computational aesthetics is to endow machine with the ability to perceive the attractiveness of an image qualitatively or quantitatively. The extraction of aesthetics-aware features plays a key role in this direction before the deep learning era. Previous research efforts [4,15,23,26,35] have shown some success in extracting aesthetic features. For 3D objects, [10] proposes to employ multi-scale topic models to fit the relationship of features from the multiple views of objects. However, most of them are handcrafted and task-specific. With the continuous development of deep learning, extracting the deep features of aesthetics images becomes the best way to solve the above problems. A lot of CNN-based models such as [22,25] have been proposed to improve the results. The applications are mainly targeted on the task of image aesthetic evaluation [17,22,34]. What's more, Hong et al. [12] propose a multi-view regularized topic model to discover Flickr users's aesthetic tendency and then construct a graph to group users into different aesthetic circles. Based on it, a probabilistic model is used to enhance the aesthetic attractiveness of photos from corresponding circles [11]. Although existing GAN models have achieved great success, they are still limited in producing "beautiful and real" images. Based on adversarial learning, Deng et al. [37] enhance image aesthetics in terms of scene composition and color distribution. This work is different from the theme of our research, as the enhancement model of [37] tries to optimize the parameters of cropping and re-coloring for an existing natural image. For our method, we directly synthesize an image without any prior information on the input side, e.g. a meaningless noise image.

3 Proposed Method

We formulate the problem of automatic aesthetic image generation as an adversarial learning model. We first introduce the overall architecture of our proposed framework shown in Fig. 1. Then we present the details of the newly-added loss functions.

3.1 Overall Framework

Basically, GAN is a pair of neural networks (G;D): the generator G and the discriminator D. G maps a vector z from a noise space N^z with a known distribution p_z into an image space N^x. The goal of G is to generate p_g (the distribution of the samples $G(z)$) to deceive the network D. And goal of D is to try to distinguish p_g (the distribution of a generated image) from p_{data} (the distribution of a real image). These two networks are iteratively optimized against each other in a minimax game (hence namely "adversarial") until the convergence. In this context, the GAN model is typically formulated as a minimax optimization of

$$\min_G \max_D V(D,G) = E_{x \sim p_{data}}[logD(x)] + E_{z \sim p_z(z)}[log(1 - D(G(z)))] \quad (1)$$

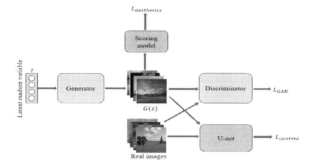

Fig. 1. The overall architecture of the proposed system.

Specifically, as for the structure of G and D, we choose fully convolutional networks as in DCGAN [29]. As shown in Fig. 2, there are a series of fractionally-stride convolutions in G and a series of convolution layers in D.

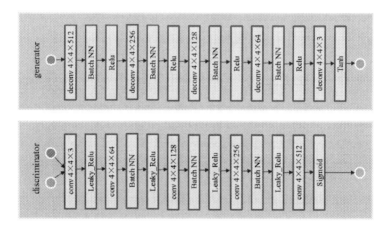

Fig. 2. The network of the G and D.

We can see that the above target function only seeks for the consistency between p_{data} and p_g in a broadly statistical sense. It has no explicit control over the visual appealingness and the content realness. So we extend the total loss function with two additional losses:

$$L_{total} = \alpha_1 L_{GAN} + \alpha_2 L_{aesthetics} + \alpha_3 L_{content} \tag{2}$$

In the formulation, L_{GAN} is the original GAN loss, $L_{aesthetics}$ is the aesthetics loss, and $L_{content}$ is the content loss. α_1, α_2, and α_3 denote their weights. In the following, we introduce the details of the two added losses.

3.2   Loss Function

Aesthetics-Aware Loss. In order to generate a visually appealing image, we propose to apply the aesthetics scoring model [17] to boost the image aesthetics, i.e. maximizing the obtained score, or minimizing $(1 - score)$. The key point is to learn a deep convolutional neural network that is able to accurately rank and rate visual aesthetics. In the network, the scoring ability is subtly encoded in its network architecture in the following aspects. First, the Alexnet [19] is fine-tuned based on a regression loss that predicts continuous numerical value as aesthetic ratings. Second, a Siamese network [3] is used by taking image pairs as inputs, which ensures images with different aesthetic levels have different ranks. The whole network is trained with a joint Euclidean and ranking loss. Moreover, they add attribute and content category classification layers and make the model be aware of fine-grained visual attributes. As demonstrated in [17], the overall aesthetic evaluation model is able to provide aesthetic scores which are well consistent with human rating. Therefore, we use the obtained scores as the aesthetic-aware loss:

$$L_{aesthetics} = \| (1 - S(\tilde{x})) \| \tag{3}$$

where $S(\tilde{x})$ is the aesthetic score of the generated image $\tilde{x}$.

Content-Aware Loss. Our synthesized images are also expected to have meaningful visual semantics. So we design the content-aware loss. In many image processing tasks [13,21], the content loss is considered. It is usually based on the activation maps produced by the ReLU layers of the pre-trained VGG network. Different from measuring pixel-wise distance between images, this loss emphasizes similar feature representation in terms of high-level content and perceptual quality. Since we aim to generate images with both good aesthetics and reasonable details, we need a network that is more suitable to our task. So we replace VGG with a more advanced U-net network [30], as its structure is able to preserve more image details by combining the concept features ("what it is") and the locality features ("where it is"). We denote $\psi_i()$ as the feature map extracted after the $i-th$ convolutional layer of the U-net. Then our content loss is defined as:

$$L_{content} = \frac{1}{C_i H_i W_i} \| \psi_i(\tilde{x}) - \psi_i(x) \| \tag{4}$$

where C_i, H_i, and W_i are the number, height and width of the feature maps, xs are real images and $\tilde{x}$s are generated ones.

3.3   Training Details

In training the proposed GAN model, input images are resized to 96×96 and then randomly cropped to 64×64, which reduces the potential over-fitting problem. The horizontal flipping of cropped images is also applied for random data augmentation. We use the ADAM technique [16] for optimization. As for the learning rates lr_G and lr_D, we set them as 0.002 for both the generator network and the discriminator network. β_1 and β_2 are set as 0.5 and 0.999. We trained the proposed model in the experiments for 10000 epochs with minibatch size of 256. In implementation, we found out that reducing the learning rate during the training process helps to improve the image quality. Therefore, the learning rates are reduced by a factor of 2 in every 1000 epoch. We empirically set $\alpha_1 = 1, \alpha_2 = 0.15, \alpha_3 = 0.1$ in our experiments.

4   Experiments

In this section, we evaluate the proposed AestheticGAN on public benchmark datasets. Apart from the direct visual comparison, we also use quantitative measures and user study to validate its effectiveness.

4.1   Datasets for Training

The Aesthetic Visual Analysis (AVA) dataset is by far the largest benchmark for image aesthetic assessment. Each of the 255,530 images is labeled with aesthetic scores ranging from 1 to 10. In this study we select a subset of them, i.e., 25,000 images, based on the semantic tags provided in the AVA data for analysis. What's more, to further illustrate the applicability of our method, we also compare our model and its competitors on the cifar10 dataset.

4.2   Visual and Quantitative Comparison

We conduct visual comparison between the results of DCGAN and our model in Figs. 3 and 4. First, from Fig. 3, both DCGAN and our model generate images with good appearances at the first glance. However, the DCGAN results have less appropriate image composition, and less realistic image contents. In contrary, we can easily recognize the scene category and image contents of our results. Second, similar trends can be observed from Fig. 4, although they are not as clear as in Fig. 3. Furthermore, by comparing all the resultant images of Figs. 3 and 4, we can see in general that the model trained on AVA has superior performance than the one trained on cifar10 in terms of visual aesthetics, which indicates the data-driven property of GAN-based models.

We adopt the four different metrics for quantitative assessment. The first two are inception score [31] and Freéchet inception distance (FID) [9] that are commonly used in evaluating the performance of GAN-based image synthesis. Since our goal is to make the model aesthetics-aware, we also use two state-of-the-art evaluation models namely NIMA [32] and ACQUINE [5]. Among them, the NIMA estimates aesthetic qualities in aspects of photographing skills and visual appealingness. ACQUINE achieves more than 80% consistency with the human rating. Of note, larger values of inception score/NIMA/ACQUINE, and smaller FID values denote better quality, respectively. From Tables 1 and 2, we can see that our DCGAN+aesthetic+content achieves much better performances than the baseline DCGAN model. Additionally, from Table 2, the aesthetic performance of AVA dataset is consistently better than that of cifar10 dataset, which echoes the above visual results.

We also perform an ablation study. Apart from the baseline DCGAN and our DCGAN+aesthetic+content, we build an intermediate version DCGAN+aesthetic. Figure 5(b) has better lightness, vivid color and composition than Fig. 5(a). Furthermore, the contents in Fig. 5(c) are more realistic than those in Fig. 5(b). The results in Tables 1 and 2 are also consistent with the above observations. This experiment empirically validate the two losses, respectively.

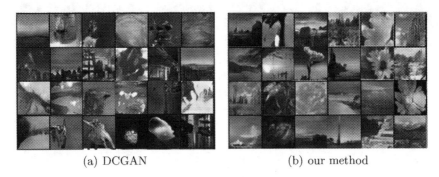

(a) DCGAN   (b) our method

Fig. 3. Comparison of the experiments on the AVA between DCGAN (left) and our method (right)

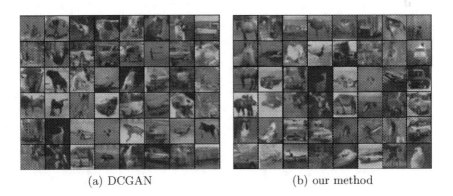

(a) DCGAN   (b) our method

Fig. 4. Comparison of the experiments on the CIFAR10 between DCGAN (left) and our method (right)

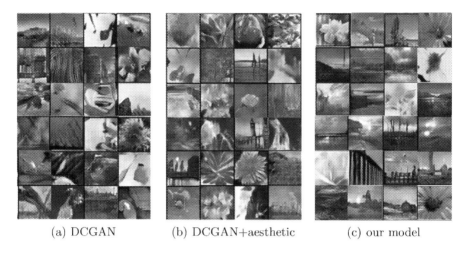

(a) DCGAN   (b) DCGAN+aesthetic   (c) our model

Fig. 5. Result images for 3 different loss functions (Color figure online)

Table 1. Inception scores and FIDs for different methods on CIFAR10 and AVA datasets

Method	Inception score		FID	
	CIFAR10	AVA	CIFAR10	AVA
Real data	$11.24 \pm .12$	$14.37 \pm .68$	$7.82 \pm .11$	$8.15 \pm .09$
DCGAN	$6.64 \pm .14$	$7.45 \pm .29$	$37.71 \pm .24$	$69.81 \pm .15$
DCGAN+aesthetic	$6.92 \pm .17$	$7.69 \pm .26$	$36.64 \pm .32$	$64.93 \pm .23$
DCGAN+aesthetic+content	$7.13 \pm .12$	$8.05 \pm .22$	$34.33 \pm .27$	$62.54 \pm .18$

Table 2. The aesthetic scores of NIMA and ACQUINE for different methods on CIFAR10 and AVA datasets

Method	NIMA		ACQUINE	
	CIFAR10	AVA	CIFAR10	AVA
Real data	$6.54 \pm .25$	$7.98 \pm .69$	8.58 ± 1.29	$9.23 \pm .75$
DCGAN	$4.59 \pm .20$	$5.56 \pm .23$	$5.29 \pm .89$	$6.48 \pm .44$
DCGAN+aesthetic	$4.85 \pm .20$	$5.74 \pm .23$	$5.76 \pm .76$	$6.86 \pm .42$
DCGAN+aesthetic+content	$5.02 \pm .19$	$5.96 \pm .24$	$6.19 \pm .87$	$7.15 \pm .53$

4.3   User Study

We also conduct an experiment of user study. We built a ranking system and distributed it to a total of 30 participants. All participants were shown three sets of 330 images, where each image set were generated by three different loss configurations. We asked all participants to rank the images in range of 1–5, where 1 means the lowest aesthetic quality and 5 is the highest one. In order to avoid the random and systematic errors, the images generated by different loss configurations are listed randomly. Also, we randomly repeatedly provide some images, and ignore the scores when a participant ranked differently on the repeated images. The statistics are shown in Fig. 6, which again demonstrates the effectiveness of the added losses.

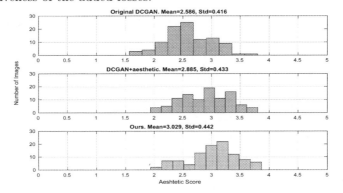

Fig. 6. User study on the AVA dataset

5   Conclusion

In the paper, we proposes a novel AestheticGAN to synthesize more challenging and complex aesthetic images. We enrich the loss function by designing two types of loss functions to train G. The aesthetics-aware loss helps to enhance aesthetic quality of the generated images, while the content-aware loss enforces them to be semantically meaningful. Various experimental results validate the effectiveness of our model. Of note, from Tables 1 and 2, we can see that the overall quality of GAN-generated images is still far from real-world natural images. We plan to narrow this gap by considering fine-grained aesthetic attributes as our future research.

Acknowledgments. This work was supported in part by the National Natural Science Foundation of China (NSFC) under grants 61632007, 61502139, 61772171, and 61702156, in part by Natural Science Foundation of Anhui Province under grants 1608085MF128 and 1808085QF188 and in part by Anhui Higher Education Natural Science Research Key Project under grants KJ2018A0545.

References

1. Arjovsky, M., Chintala, S., Bottou, L.: Wasserstein GAN. arXiv preprint arXiv:1701.07875 (2017)
2. Chen, X., Duan, Y., Houthooft, R., Schulman, J., Sutskever, I., Abbeel, P.: Info-GAN: interpretable representation learning by information maximizing generative adversarial nets. In: NIPS (2016)

3. Chopra, S., Hadsell, R., LeCun, Y.: Learning a similarity metric discriminatively, with application to face verification. In: CVPR. IEEE (2005)

4. Datta, R., Joshi, D., Li, J., Wang, J.Z.: Studying aesthetics in photographic images using a computational approach. In: Leonardis, A., Bischof, H., Pinz, A. (eds.) ECCV 2006. LNCS, vol. 3953, pp. 288–301. Springer, Heidelberg (2006). https://doi.org/10.1007/11744078_23

5. Datta, R., Wang, J.Z.: ACQUINE: aesthetic quality inference engine-real-time automatic rating of photo aesthetics. In: ICMR. ACM (2010)

6. Deng, Y., Loy, C.C., Tang, X.: Image aesthetic assessment: an experimental survey. IEEE Signal Process. Mag. **34**(4), 80–106 (2017)

7. Goodfellow, I., et al.: Generative adversarial nets. In: NIPS (2014)

8. Gulrajani, I., Ahmed, F., Arjovsky, M., Dumoulin, V., Courville, A.C.: Improved training of wasserstein GANs. In: NIPS (2017)

9. Heusel, M., Ramsauer, H., Unterthiner, T., Nessler, B., Klambauer, G., Hochreiter, S.: GANs trained by a two time-scale update rule converge to a nash equilibrium. arXiv preprint arXiv:1706.08500 (2017)

10. Hong, R., Hu, Z., Wang, R., Wang, M., Tao, D.: Multi-view object retrieval via multi-scale topic models. IEEE Trans. Image Process. **25**, 5814 (2016)

11. Hong, R., Zhang, L., Tao, D.: Unified photo enhancement by discovering aesthetic communities from flickr. IEEE Trans. Image Process. **25**, 1124 (2016)

12. Hong, R., Zhang, L., Zhang, C., Zimmermann, R.: Flickr circles: aesthetic tendency discovery by multi-view regularized topic modeling. IEEE Trans. Multimed. **18**, 1555 (2016)

13. Johnson, J., Alahi, A., Fei-Fei, L.: Perceptual losses for real-time style transfer and super-resolution. In: Leibe, B., Matas, J., Sebe, N., Welling, M. (eds.) ECCV 2016. LNCS, vol. 9906, pp. 694–711. Springer, Cham (2016). https://doi.org/10.1007/978-3-319-46475-6_43

14. Joshi, D., et al.: Aesthetics and emotions in images. IEEE Signal Process. Mag. **28**, 94 (2011)

15. Ke, Y., Tang, X., Jing, F.: The design of high-level features for photo quality assessment. In: CVPR. IEEE (2006)

16. Kingma, D.P., Ba, J.: Adam: a method for stochastic optimization. arXiv preprint arXiv:1412.6980 (2014)

17. Kong, S., Shen, X., Lin, Z., Mech, R., Fowlkes, C.: Photo aesthetics ranking network with attributes and content adaptation. In: Leibe, B., Matas, J., Sebe, N., Welling, M. (eds.) ECCV 2016. LNCS, vol. 9905, pp. 662–679. Springer, Cham (2016). https://doi.org/10.1007/978-3-319-46448-0_40

18. Krizhevsky, A., Hinton, G.: Learning multiple layers of features from tiny images. University of Toronto (2009)

19. Krizhevsky, A., Sutskever, I., Hinton, G.E.: Imagenet classification with deep convolutional neural networks. In: NIPS (2012)

20. LeCun, Y.: The MNIST database of handwritten digits (1998). http://yann.lecun.com/exdb/mnist/

21. Ledig, C., et al.: Photo-realistic single image super-resolution using a generative adversarial network. arXiv preprint (2016)

22. Lu, X., Lin, Z., Jin, H., Yang, J., Wang, J.Z.: Rapid: rating pictorial aesthetics using deep learning. In: MM. ACM (2014)

23. Luo, Y., Tang, X.: Photo and video quality evaluation: focusing on the subject. In: Forsyth, D., Torr, P., Zisserman, A. (eds.) ECCV 2008. LNCS, vol. 5304, pp. 386–399. Springer, Heidelberg (2008). https://doi.org/10.1007/978-3-540-88690-7_29

24. Mirza, M., Osindero, S.: Conditional generative adversarial nets. arXiv preprint arXiv:1411.1784 (2014)
25. Murray, N., Marchesotti, L., Perronnin, F.: AVA: a large-scale database for aesthetic visual analysis. In: CVPR. IEEE (2012)
26. Nishiyama, M., Okabe, T., Sato, I., Sato, Y.: Aesthetic quality classification of photographs based on color harmony. In: CVPR. IEEE (2011)
27. Odena, A., Olah, C., Shlens, J.: Conditional image synthesis with auxiliary classifier GANs. arXiv preprint arXiv:1610.09585 (2016)
28. Qi, G.J.: Loss-sensitive generative adversarial networks on lipschitz densities. arXiv preprint arXiv:1701.06264 (2017)
29. Radford, A., Metz, L., Chintala, S.: Unsupervised representation learning with deep convolutional generative adversarial networks. arXiv preprint arXiv:1511.06434 (2015)
30. Ronneberger, O., Fischer, P., Brox, T.: U-net: convolutional networks for biomedical image segmentation. In: Navab, N., Hornegger, J., Wells, W.M., Frangi, A.F. (eds.) MICCAI 2015. LNCS, vol. 9351, pp. 234–241. Springer, Cham (2015). https://doi.org/10.1007/978-3-319-24574-4_28
31. Salimans, T., Goodfellow, I., Zaremba, W., Cheung, V., Radford, A., Chen, X.: Improved techniques for training GANs. In: NIPS (2016)
32. Talebi, H., Milanfar, P.: NIMA: neural image assessment. IEEE Trans. Image Process. **27**, 3998 (2018)
33. Tan, W.R., Chan, C.S., Aguirre, H., Tanaka, K.: ArtGAN: artwork synthesis with conditional categorial GANs. arXiv preprint arXiv:1702.03410 (2017)
34. Tian, X., Dong, Z., Yang, K., Mei, T.: Query-dependent aesthetic model with deep learning for photo quality assessment. IEEE Trans. Multimed. **17**, 2035 (2015)
35. Tong, H., Li, M., Zhang, H.-J., He, J., Zhang, C.: Classification of digital photos taken by photographers or home users. In: Aizawa, K., Nakamura, Y., Satoh, S. (eds.) PCM 2004. LNCS, vol. 3331, pp. 198–205. Springer, Heidelberg (2004). https://doi.org/10.1007/978-3-540-30541-5_25
36. Welinder, P., et al.: Caltech-UCSD birds 200. California Institute of Technology (2010)
37. Yubin Deng, C.C.L., Tang, X.: Aesthetic-driven image enhancement by adversarial learning. arXiv preprint arXiv: 1707.05251 (2017)

Hyperspectral Image Classification Using Nonnegative Sparse Spectral Representation and Spatial Regularization

Xian-Hua Han[1]([⊠]) [iD], Jian Wang[2], Jian De Sun[2], and Yen-Wei Chen[3]

[1] Graduate School of Science and Technology for Innovation, Yamaguchi University, 1677-1 Yoshida, Yamaguchi City, Yamaguchi 753-8511, Japan
`hanxhua@yamaguchi-u.ac.jp`
[2] Shandong Normal University, Jiannan, Shandong, China
`wlockon@163.com,jiandesun@hotmail.com`
[3] Ritsumeikan University, Kusatsu, Shiga, Japan
`chen@is.ritsumei.ac.jp,`
`http://www.mlp.sci.yamaguchi-u.ac.jp/index_EN.html`

Abstract. Hyperspectral image classification is an important technique for variety of applications and a challenge task because of high dimensional features and low SNR. This study proposes a robust hyperspectral image classification approach using nonnegative sparse spectral representation and spatial regularization. Due to the nonnegative composition of material spectra, we investigate a nonnegative sparse representation for spectral analysis, which can decompose the hyper-spectrum into the composite weight, as robust spectral feature, of the possible existed spectral prototypes (called as spectral dictionary) learned from the observed scene, and then conduct the pixel-wise recognition with the decomposed weights instead of raw spectrum. Furthermore, a spatial regularization method for category probability map propagation is explored for integrating the relationship among nearby pixels, and thus can provide more robust classification map. Experimental results on several hyperspectral image datasets validate that the proposed method can achieve much more accurate classification performance than the state-of-the-art approaches.

Keywords: Hyperspectral image classification
Nonnegative sparse representation · Spatial regularization · ADMM

1 Introduction

Hyperspectral imagery consists of hundreds of narrow contiguous wavelength bands with rich spectral information, and thus makes the potential for the pixel-wise classification. Classification using hyperspectral data has received increasing attentions in a variety of applications [1–5], such as land-use and land-cover

© Springer Nature Switzerland AG 2018
R. Hong et al. (Eds.): PCM 2018, LNCS 11165, pp. 180–189, 2018.
https://doi.org/10.1007/978-3-030-00767-6_17

mapping, mineral exploration etc. The intuitive way for pixel-wise classification with hyperspectral data is to directly treat the hundreds of band spectra as pixel feature, and then recognizes the pixel into multiple categories using a classifier such as k-nearest neighbor (k-NN), support vector machine [6,7]. However, because of both the limited number of labeled samples and high dimensionality of the raw spectra, the classification performances by these methods is still needed to be improved for real applications. Thus, some works on hyperspectral image classification concentrated on either feature dimension reduction or semi-supervised classification. Feature reduction consisting of feature extraction using statistical analysis methods [8] such as independent component analysis, principle component analysis, kernel-based approaches [9], and feature selection for taking a group of bands only [10] does not always provide discriminant features for classification. Taking advantage of the availability of large amount of unlabeled samples, some works [11–13] promoted semi-supervised learning by integrating the data structure and the neighbor pixel correlation for improving classification performance. Some other machine learning techniques such as multiple kernel learning [14], relevance vector machine [15] and extreme learning machine (ELM) [16], have also been explored to achieve impressive performances for hyperspectral image classification. Recently, deep learning-based methods have drawn increasing interests in remote sensing image analysis and hyperspectral image classification [17,18], which provides the state-of-the-art performance. In general, the deep learning-based methods requires large number of training samples, which are needed to be labeled manually and is time-consuming. Therefore, Li et al. [19] proposed to form pixel pairs to increase samples for training deep network, and has proved more accurate performance than the one without combined pair samples.

Since neighboring pixel in remote sensing images likely share the same contextual properties, spectral-spatial techniques instead of spectral analysis only have been exploited and manifested impressive performance. Several works [20–23] investigated spatial filters such as mean-, morphological- and anisotropic preprocessing to enhance hyperspectral images and then combine them with spectral analysis framework for classification. On the other hand, integration of the spatial contextual information into the obtained classification probability results by spectral analysis framework has been proved to be an effective tool, which are broadly referred as continuous relaxation or probabilistic relaxation [24,25]. This research line generally conducts smoothing operators on the resulted category probability for imposing the continuity of neighboring labels, which is widely based on the exploring of Markov random Fields (MRFs) [26–28], and significantly improves the classification performance in smooth image regions. However, this kind of method often leads to over-smoothed results across the class boundary regions and thus degrades the classification accuracies for boundary pixels.

In this study, we proposed a robust hyperspectral image classification approach based on nonnegative sparse spectral representation and spatial regularization for improving classification performance. Motivated by the nonnegative property of the spectral composition for any pixel, we explore a nonnegative sparse representation for robust spectral feature extraction, which can decompose the hyper-spectrum into the composite weights of the possible existed spectral prototypes (called as spectral dictionary) learned from the observed scene, and then conduct the pixel-wise classification with the robust spectral feature instead of raw spectrum. In addition, we explore a boundary-aware smoothing method for regularizing the spatial contextual of neighboring pixels on both original spectral and the resulted classification probability spaces, which is robust even for boundary pixels. Experimental results on several widely used hyperspectral images validate that the proposed approach leads to very impressive performance compared with the state-of-the-art methods.

This paper is organized as follows. Section 2 describes the proposed nonnegative sparse representation for robust spectral feature extraction and Sect. 3 introduces the boundary-aware spatial regularization method for enhancing the hyperspectral image and refining the resulted classification probability. Experimental results are presented in Sect. 4, and concluded remarks are summarized in Sect. 5.

2   Nonnegative Sparse Spectral Representation

As the popularly researched topic: unmixing model, in satellite image analysis [29–31] suggests that the hyperspectral pixels in the observed scene can be decomposed into several material composition fraction weights and the corresponding set of pure material spectra, both of which should be nonnegative, we also assume the pixel hyperspectra as a nonnegative sparse representation on a set of learned spectral prototypes formulated as follows:

$$\mathbf{x}_t = \mathbf{D}\alpha, \text{s.t. } \mathbf{D} >= 0, \alpha >= 0, \|\alpha\|_0 < T \tag{1}$$

where $\mathbf{x}_t \in \mathbb{R}^d$ (d: band number in the observed hyperspectral image) denotes the raw spectrum of any pixel, $\mathbf{D} \in \mathbb{R}^{d \times K}$ is the spectral dictionary learned from the labeled samples, and $\alpha \in \mathbb{R}^K$ is the nonnegative sparse representation for $\mathbf{x}_t$. The three constraints in Eq. (1) regularize the nonnegativity of $\mathbf{D}$ and α, sparsity of α, respectively. Then the sparse representation α is prospected to be more robust and discriminated for pixel-wise classification.

Dictionary learning: The dictionary $\mathbf{D}$ can be learned from the labeled samples. Let's denote the labeled samples of cth class as $\mathbf{X}^c = [\mathbf{x}_1^c, \mathbf{x}_2^c, \cdots, \mathbf{x}_{N_c}^c]$ with C classes in the observed scene and the total number of labeled samples is $N = \sum_{c=1}^C = N_c$. In order to give the distinguish spectral atoms, we learn class-specific dictionary using the corresponding labeled samples. Since the defined class such as tree category (different types of trees) may not always have one distinct spectrum only, we learn several spectral prototypes for taking consideration of the spectral variety while not retain all labeled samples as spectral atoms for reducing outliers. Replacing l_0 norm with l_1 norm for sparse constraint, the cth class's dictionary $\mathbf{D}^c \in \mathbb{R}^{d \times K_1}$ can be generated by optimizing the following objective function:

$$(\mathbf{D}^c, \mathbf{A}) = \arg \min_{\mathbf{D}_c, \mathbf{A}} \|\mathbf{X}_c - \mathbf{D}_c \mathbf{A}\|_F^2 + \lambda \|\mathbf{A}\|_1$$
$$s.t. \ \mathbf{A} \geq 0, \mathbf{D}^c \geq 0, \tag{2}$$

Because of the non-negative constraint on both sparse representation $\mathbf{A}$ and dictionary $\mathbf{D}^c$, existing dictionary learning (DL) algorithms, such as K-SVD algorithm [32] cannot be directly used. This study proposes a computationally efficient non-negative DL algorithm, which alternatively updates each atom per iteration via a closed-form solution and calculate the sparse representation with the fixed $\mathbf{D}^c$. For sparse representation calculation with a fixed $\mathbf{D}^c$, the subproblem is formulated as:

$$\mathbf{A} = \arg \min_{\mathbf{A}} \|\mathbf{X}^c - \mathbf{D}^c \mathbf{A}\|_F^2 + \lambda \|\mathbf{A}\|_1, s.t. \ \mathbf{A} \geq 0, \tag{3}$$

For fast convergence, we apply ADMM technique [33] to solve Eq. (3), which can be reformulated as:

$$\mathbf{A} = \arg \min_{\mathbf{A}} \|\mathbf{X}^c - \mathbf{D}^c \mathbf{B}\|_F^2 + \lambda \|\mathbf{A}\|_1, s.t. \ \mathbf{B} = \mathbf{A}, \mathbf{A} \geq 0, \tag{4}$$

After extending Eq. (4) with an augmented Lagrangian function, the variables $\mathbf{B}$ and $\mathbf{A}$ can be alternatively updated; please refer to [33] for the detail information of ADMM optimization. After obtaining $\mathbf{A}$ with the fixed $\mathbf{D}^c$, we fix $\mathbf{A}$ and then update $\mathbf{D}^c$ again by solving:

$$\mathbf{D}^c = \arg \min \|\mathbf{X}^c - \mathbf{D}^c \mathbf{A}\|_F^2, s.t. \ \mathbf{D}^c \geq 0, \tag{5}$$

We exploit a block coordinate descent method [34] for optimizing $\mathbf{D}^c$ by updating one column of $\mathbf{D}^c$ while keeping the others fixed under the nonnegative constraint in each iteration. For each class, the above same procedure is conducted to learn the class-specific dictionary $\mathbf{D}^c$, and then form the global common dictionary $\mathbf{D} = [\mathbf{D}^1, \mathbf{D}^2, \ldots, \mathbf{D}^C]$ for coding any raw spectrum.

Nonnegative sparse representation: By fixing the common dictionary $\mathbf{D}$, the nonnegative sparse representation of any raw spectral (labeled or non-labeled) can be calculated with Eq. (3) using ADMM method as robust feature.

3   Spatial Regularization Method

With the extracted sparse representation of the raw spectral, we conduct a pixel-wise classification using SVM, and the class probabilities can be obtained for each pixel. Let $\mathbf{P} = [\mathbf{p}_1, \mathbf{p}_2, \cdots, \mathbf{p}_N] \in \mathbb{R}^{C \times N}$, where $\mathbf{p}_n = [p_i(1), \cdots, p_n(C)]^T$ for the nth pixel are the C-dimensional multivariate vector of probabilities defined only with spectral information, we aim to refine the probability map with spatial regularization. Denoting $\mathbf{Q} = [\mathbf{q}_1, \mathbf{q}_2, \cdots, \mathbf{q}_N] \in \mathbb{R}^{C \times N}$ be the final probability vectors of the defined categories, the spatial regularization method can be implemented as the following optimization problem:

$$\mathbf{Q} = \arg\min_{\mathbf{Q}} \gamma \|\mathbf{Q} - \mathbf{P}\|_F^2 + (1 - \gamma) \sum_n \sum_{m \in NN_n}$$

$$w_m \|\mathbf{q}_m - \mathbf{q}_n\|, s.t. \ \mathbf{q}_n > 0, \mathbf{1}^T \mathbf{q}_n = 1$$

(6)

where $m \in NN_n$ denotes the mth pixel is the spatial near neighbor of the nth pixel, w_m is the penalty weight of the inconsistency between $\mathbf{q}_m$ and $\mathbf{q}_n$, and γ gives the trade-off between the deviation of the refined probability $\mathbf{Q}$ from $\mathbf{P}$ and the inconsistent degree in local structure. The constraints in Eq. (6) impose the refined vector $\mathbf{q}_n$ representing probabilities should be nonnegative and sum-to-one, and $\mathbf{1}$ is a column vector of C 1s. The penalty weight w_m is based on boundary-aware strategy of the raw hyperspectral image, and is calculated to be

Fig. 1. The used hyperspectral images.

inversely proportional to the threshold gradients of all band image as follows:

$$w_m = \exp(-\sum_{b=1}^{d} f(grad(\mathbf{X}^{(b)}))_m \qquad (7)$$

where $\mathbf{X}^{(b)}$ is the bth band image and $f(grad())$ represents the threshold function of the gradient image for any band by shrinking small gradient magnitude to zero while retaining large values. $\sum_{b=1}^{d} f(grad(\mathbf{X}^{(b)}))$ results in large magnitude on edge discontinuities in an image and 0 in smooth regions. By conducting exponential operator on this, where σ is the decay rate, the weight w_m would have large value in smooth regions for regularizing the nearby pixel and small value in discontinuous regions for relaxing the smoothing on the nearby pixels. For solving Eq. (6), we exploit a projected iterative scheme consisting of iteratively updating $q_n(k)$ with current state, and project the probability vector of each pixel onto the probability simplex (satisfy the constrains of Eq. (6)). The updating for $q_n^{(t)}(k)$ at the tth step is formulated as:

$$q_n^{(t)}(k) = \frac{\gamma p_n(k) + (1-\gamma)\sum_{m \in NN_n} w_m q_n^{(t-1)}(k)}{\gamma + (1-\gamma)\sum_{m \in NN_n} w_m} \qquad (8)$$

and then project it onto probability simplex with the following equation:

$$q_n^{(t)}(k) = \frac{q_n^{(t)}(k)}{\sum_{k=1}^{K} q_n^{(t)}(k)} \qquad (9)$$

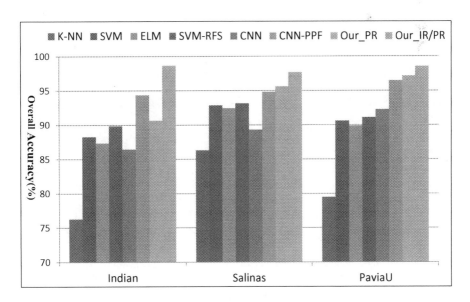

Fig. 2. Compared overall accuracy of our methods (Our_IR and Our_IR/PR) and the state-of-the-art approaches including k-NN, SVM, SVM-RFS, ELM, CNN and CNN-PPF [19] on the three hyperspectral datasets.

We conduct the above iteration procedure for updating $\mathbf{Q}$ until the change between the two continuous steps are smaller than a predefined threshold, or arriving a predefined maximum iteration number. Finally, the pixel is decided to the category label with the largest probability. In addition, the proposed spatial regularization method can be also applied to the original hyperspectral cube for reducing noise, which is prospected for further improving classification performance.

4   Experimental Results

We validate the proposed method with three benchmark hyperspectral datasets including Indian Pines, Salinas, and University of Pavia scenes (denoted as PaviaU). The Indian Pines and Salinas data were collected by Airborne Visible/Infrared Imaging Spectrometer (AVIRIS) sensor in northwestern Indiana and Salinas Valley, California, respectively. Indian Pines image comprising 145×145 pixels has 220 spectral channels in the 0.4 to 2.45 μm region of the visible and infrared spectrum with a spatial resolution of 20 m, and 16 categories with different pixel samples. Salinas data with a spatial resolution of 3.7 m, 512×217 pixels, and same channel number as Indian Pines data, also has 16 different classes such as vegetables, bares oils, and vineyard fields. While PaviaU data with 103 spectral bands was gathered by the Reflective Optics System Imaging Spectrometer (ROSIS) sensor, and has spatial coverage of 610×340 pixels covering the city of Pavia, Italy.

Following the experimental setting in [20], we randomly select 200 labeled pixels per category for training and the remainder pixels in the ground truth map as test. Since the labeled pixel numbers of some categories in Indian data are less than 200, we only use 8 classes with more than 200 labeled pixels in our experiments. The used parameters in our experiments are set as: (1) γ: 0.1, (2) the spectral atom number in each class: $K_k = 10$, (3) the maximum iteration number in spatial regularization method: 10, (4) the sparse parameter $\lambda = 0.001$. The classification performance is measured as Overall Accuracy (OA) of all classes. Figure 1 provides the compared performances with several state-of-the-art methods including k-NN, SVM, SVM-RFS, ELM, CNN and CNN-PPF [19] on the three hyperspectral datasets. We conducted the experiments using spatial regularization on probability map (denoted as Our_PR), and also both hyperspectral image/probability space (denoted as Our_IR/PR). It can be seen from Fig. 2 that our proposed method manifests the best overall accuracy for all three datasets. In addition, we also change the atom number of each class from 5 to 30 (denoted $K = 5\times, 10\times, \cdots$, since the total atom number is multiplication of each class's atom and class number, and show the overall accuracy and *kappa* coefficient in Fig. 3, which manifests the impressive performance of our proposed method even with small number of dictionary.

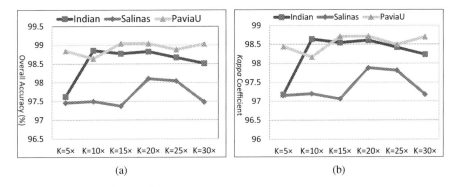

Fig. 3. Experimental results with different atom numbers. (a) Overall accuracy; (b) *kappa* coefficient.

5   Conclusion

This study proposed a novel hyperspectral image classification method for robust feature extraction using nonnegative sparse representation and enhancement/refinement of the original hyperspectral cube and the category probability map using spatial regularization. The proposed nonnegative sparse method can not only learn the adaptive hyperspectral prototypes for decomposing the raw image into meaningful representation but also consider the variety of the spectral even for same defined category to provide accurate and robust feature in the target scene. In addition, a spatial regularization method with boundary-aware constraint was proposed for integrating spatial relationship of nearby pixels, which can be applied to both the original hyperspectral cube and the category probability map resulted by a pixel-wise classifier. Experiments on three benchmark hyperspectral datasets validated that the proposed method achieved much more accurate performance than state-of-the-art approaches.

References

1. Bioucas-Dias, J., Plaza, A., Camps-Valls, G., Scheunders, P., Nasrabadi, N., Chanussot, J.: Hyperspectral remote sensing data analysis and future challenges. IEEE Geosci. Remote Sens. Mag. **1**(2), 6–36 (2013)
2. Plaza, A., et al.: Recent advances in techniques for hyperspectral image processing. Remote Sens. Environ. **113**, 110–122 (2009)
3. Li, W., Du, Q.: Gabor-filtering-based nearest regularized subspace for hyperspectral image classification. IEEE J. Sel. Topics Appl. Earth Observ. Remote Sens. **7**(4), 1012–1022 (2014)
4. Li, J., Marpu, P.R., Plaza, A., Bioucas-Dias, J.M., Benediktsson, J.A.: Generalized composite kernel framework for hyperspectral image classification. IEEE Trans. Geosci. Remote Sens. **51**(9), 4816–4829 (2013)
5. Huang, X., Zhang, L.: An SVM ensemble approach combining spectral, structural, and semantic features for the classification of highresolution remotely sensed imagery. IEEE Trans. Geosci. Remote Sens. **51**(1), 257–272 (2013)

6. Waske, B., Benediktsson, J.A.: Fusion of support vector machines for classification of multisensor data. IEEE Trans. Geosci. Remote Sens. **45**(12–1), 3858–3866 (2007)

7. Guo, B., Gunn, S., Damper, R., Nelson, J.: Customizing kernel functions for svm-based hyperspectral image classification. IEEE Trans. Image Process. **17**(4), 622–629 (2008)

8. Wang, J., Chang, C.-I.: Independent component analysis-bases dimensionality reduction with applications in hyperspectral image analysis. IEEE Trans. Geosci. Remote Sens. **47**(4), 1586–1600 (2006)

9. Kuo, B.-C., Li, C.-H., Yang, J.-M.: Kernel nonparametric weighted feature extraction for hyperspectral image classification. IEEE Trans. Geosci. Remote Sens. **44**(6), 1139–1155 (2009)

10. Guo, B., Damper, S.G.R., Nelson, J.: Band selection for hyperspectral image classification using mutual information. IEEE Geosci. Remote Sens. Lett. **3**(4), 522–526 (2006)

11. Camps-Valls, G., Marsheva, T.B., Zhou, D.: Semi-supervised Graph-based hyperspectral image classification. IEEE Geosci. Remote Sens. Lett. **45**(10), 3044–3054 (2007)

12. Dopido, I., Li, J., Marpu, P.R., Plaza, A., Bioucas-Dias, J.M., Benediktsson, J.A.: Semisupervised self-learning for hyperspectral image classification. IEEE Trans. Geosci. Remote Sens. **51**(7), 4032–4044 (2013)

13. Persello, C., Bruzzone, L.: Active and semisupervised learning for the classification of remote sensing images. IEEE Trans. Geosci. Remote Sens. **52**(11), 6937–6956 (2014)

14. Gu, Y.F., Wang, C., You, D., Zhang, Y.H., Wang, S.Z., Zhang, Y.: Representative multiple kernel learning for classification in hyperspectral imagery. IEEE Trans. Geosci. Remote Sens. **55**(7), 2852–2865 (2012)

15. Mianji, F.A., Zhang, Y.: Robust hyperspectral classification using relevance vector machine. IEEE Trans. Geosci. Remote Sens. **49**(6), 2100–2112 (2011)

16. Samat, A., Du, P., Liu, S., Li, J., Cheng, L.: E2LMs: ensemble extreme learning machines for hyperspectral image classification. IEEE J. Sel. Topics Appl. Earth Observ. Remote Sens. **7**(4), 1060–1069 (2014)

17. Hu, W., Huang, Y., Wei, L., Zhang, F., Li, H.: Deep convolutional neural networks for hyperspectral image classification. J. Sens. **2015**, Article no. 258619 (2015)

18. Zhao, W., Du, S.: Spectral-spatial feature extraction for hyperspectral image classification: a dimension reduction and deep learning approach. IEEE Trans. Geosci. Remote Sens. **54**(8), 4544–4554 (2016)

19. Li, W., Wu, G.D., Zhang, F., Du, Q.: Hyperspectral image classification using deep pixel-pair features. IEEE Trans. Geosci. Remote Sens. **55**(2), 844–853 (2016)

20. Velasco-Forero, S., Manian, V.: Improving hyperspectral image classification using spatial preprocessing. IEEE Geosci. Remote Sens. Lett. **6**(2), 297–301 (2009)

21. Yildirim, I., Ersoy, O.K., Yazgan, B.: Improvement of classification accuracy in remote sensing using morphological filter. Adv. Space Res. **36**(5), 1003–1006 (2005)

22. Wang, Y., Niu, R., Yu, X.: Anisotropic diffusion for hyperspectral imagery enhancement. IEEE Sens. J. **10**(3), 469–477 (2010)

23. Ghamisi, P., Benediktsson, J., Sveinsson, J.: Automatic spectral-spatial classification framework based on attribute profiles and supervised feature extraction. IEEE Trans. Geosci. Remote Sens. **52**(9), 5771–5782 (2014)

24. Huang, X., Lu, Q., Zhang, L., Plaza, A.: New postprocessing methods for remote sensing image classification: a systematic study. IEEE Trans. Geosci. Remote Sens. **52**(11), 7140–7159 (2014)

25. Tarabalka, Y., Fauvel, M., Chanussot, J., Benediktsson, J.A.: SVM and MRF-based method for accurate classification of hyperspectral images. IEEE Geosci. Remote Sens. Lett. **7**(4), 736–740 (2010)

26. Adams, J.B., Smith, M.O., Johnson, P.E.: Combining support vector machines and Markov random fields in an integrated framework for contextual image classification. IEEE Trans. Geosci. Remote Sens. **51**(5), 2734–2752 (2013)

27. Zhang, B., Li, S., Jia, X., Gao, L., Peng, M.: Adaptive Markov random field approach for classification of hyperspectral imagery. IEEE Geosci. Remote Sens. Lett. **8**(5), 973–977 (2011)

28. Sun, L., Wu, Z., Liu, J., Xiao, L., Wei, Z.: Supervised spectral? Spatial hyperspectral image classification with weighted Markov random fields. IEEE Trans. Geosci. Remote Sens. **53**(3), 1490–1503 (2015)

29. Moser, G., Serpico, S.: Spectral mixture modeling: a new analysis of rock and soil types at the viking lander 1 site. J. Geophys. Res. **91**, 8098–8112 (1986)

30. Settle, J., Drake, N.: Linear mixing and the estimation of ground cover proportions. Int. J. Remote Sens. **14**(6), 1159–1177 (1993)

31. Chang, C.-I., Wu, C.-C., Liu, W., Ouyang, Y.-C.: A new growing method for simplex-based end member extraction algorithm. IEEE Trans. Geosci. Remote Sens. **44**(10), 2804–2819 (2006)

32. Aharon, M., Elad, M., Bruckstein, A.: K-SVD: an algorithm for designing overcomplete dictionaries for sparse representation. IEEE Trans. Signal Process. **54**(11), 4311–4322 (2006)

33. Boyd, S., Parikh, N., Chu, E., Peleato, B., Eckstein, J.: Distributed optimization and statistical learning via the alternating direction method of multipliers. Found. Trends Mach. Learn. **3**(1), 1–122 (2011)

34. Friedman, J., Hastie, T., Hing, H., Tibshirani, R.: Pathwise coordinate optimization. Ann. Appl. Statist. **1**(2), 302–332 (2007)

Adaptive Aggregation Network for Face Hallucination

Jin Guo, Jun Chen[✉], Zhen Han, Han Liu, Zhongyuan Wang, and Ruimin Hu

National Engineering Research Center for Multimedia Software,
School of Computer Science, Wuhan University, Wuhan 430072, China
gjjggjjg@163.com, hanzhen_1980@163.com, liuhanooo@163.com,
wzy_home@163.com, chenj@whu.edu.cn, hrm1964@gmail.com

Abstract. Face hallucination refers to obtaining a clean face image from a degraded ones. The degraded face is assumed to be related to the clean face through the forward imaging model that account for blurring, sampling and noise. In recent years, many methods have been proposed and improved well progress. These methods usually learn a regression function to reconstruct the entire picture. However, there are huge differences among the optimal learned regression functions in different regions. In other words, the learned regression function needs to process all regions, which makes it difficult to reconstruct a satisfactory picture. As a result, the reconstructed images in some regions are relatively smooth. In order to address the problem, we present a novel face hallucination framework, called Adaptive Aggregation Network (AAN), which uses the aggregation network to guide face hallucination. Our network contains two branches: aggregation branch and generator branch. Specifically, our aggregation branch can explore regression function from low-resolution (LR) to high-resolution (HR) images in different regions, and aggregate the regions by the similarity of the regression function. Then generator module can be used to make a specific hallucination on the selected regions to get a better reconstruction result. After evaluating on datasets, our model was proved to be above the state-of-the-art methods in terms of effectiveness and accuracy.

Keywords: Face hallucination · Adaptive aggregation
Regression function

1 Introduction

Face hallucination, as a representative of low-level vision tasks, is the process of reconstructing a clean face image from the degraded observation. It is not only

Research supported by National Key R&D Program of China (No. 2017YFC0803700), National Nature Science Foundation of China (U1736206, U1611461, 61671332), Natural Science Foundation of Hubei Province (2016CFB573), Hubei Province Technological Innovation Major Project (2016AAA015, 2017AAA123).

R. Hong et al. (Eds.): PCM 2018, LNCS 11165, pp. 190–199, 2018.
https://doi.org/10.1007/978-3-030-00767-6_18

a fundamental problem in face analysis, but can be used as a preprocessor for tasks such as face recognition [1], face alignment [2]. In practical applications, however, the face images captured by surveillance cameras are generally of poor quality and difficult to use directly.

Over the past few decades, many conventional methods [3–6] have been proposed to solve the problem. They assume the correlation between the degraded face image with clean ones, and focus on learning a mapping from degraded images to clean images. However, most methods involve a large number of optimization problems during the reconstruction phase, making it difficult to implement high-performance applications. Due to the complex environment of the degenerative process, the consistency of the hypothesis of degraded face images with clean images is not well. Therefore, the results produced are often unsatisfactory.

Recently, with the development of convolutional neural networks, many methods [7–10] based on deep learning have been used for image reconstruction. Face prior knowledge and spatial structure information are often used as additional information for face hallucination. Despite their high reconstruction quality, most of the methods typically suffer from two major drawbacks. First, there are huge differences in the structure of different regions, making it difficult to generate a mapping that satisfies all regions. Second, in the noisy environment, the prior information and spatial structure of face will be destroyed which makes it difficult to generate satisfactory results.

Therefore, how to reconstruct an HR face image in a noisy environment becomes a difficult problem. Inspired by the recent success of aggregation network in computer vision tasks [11,17–19], we propose the Adaptive Aggregation Network to deal with noise face hallucination. Our network contains two branches: aggregation branch and generator branch. The aggregation branch can cluster face images into two robust regions in a data-driven manner through the similarity of the regression function. Then the generator branch can be used to make a specific face hallucination of the selected regions.

The main contribution of this paper is that we propose an effective model to deal with the face hallucination in noise environments. The noise face hallucination is often difficult to reconstruct a satisfactory result due to the complex degradation process and the destruction of the face prior structure. Compared with other methods, our method not only provides robust face structure information under noise conditions, but turns a complex face hallucination problem into two relatively simple sub-problems. The empirical results show that our designed network surpasses the state-of-the-art methods in terms of effectiveness and efficiency.

2   Related Work

2.1   Face Hallucination and Image Super-Resolution

Face hallucination is a special case of image super resolution, which introduces face prior structure information to reconstruct face images. Early techniques assumed that the face was in a controlled setting with small variations. Ma et al. [3] utilize face priors information to reconstruct HR face by solving a constrained least squares problem (known as Least Square Representation (LSR)). Yang et al. [4] thought that low-resolution and high-resolution faces have similar sparse priors and reconstruct HR faces through the low-dimensional projections. Later, on the basis of locality and sparseness, Jiang et al. [5,6] proposed a Local Constraint Representation (LcR) method to obtain a better reconstruction face. However, these methods require the face to be landmark detection beforehand which often cannot achieve good results when the images are seriously degraded.

In recent years, many deep learning methods have been used for image hallucination and have achieved great progress. In particular, Dong et al. [7] first proposed a super-resolution convolutional network (SRCNN) for image reconstruction through equally performing sparse coding. Kim et al. [8] proposed a deep convolutional network to achieve better reconstruction performance by skipping connections and learning residuals between HR with LR. Zhou et al [10] proposed a bi-channel convolutional neural network for facial hallucination. They point out the importance of input image, and use full-connected layers to restore HR face. Tuzel et al. [9] added global face information to the network and reconstructed the face image by considering global and local constraints.

2.2   Adaptive Aggregation Network

Adaptive aggregation network developed in recent work and has benefited various tasks [11–13], such as object classification [11] and human pose estimation [12,13]. Since contextual information is important for computer vision problems, most of these works attempt to acquire dense features adaptively by focusing on the top information. Recent proposed Residual Attention Network [11] achieves state-of-the-art results on image classification task. A deep network module capturing top information is used to adaptive aggregation module. The aggregation module is applied to the input image to get important regions and then feed to another deep network module for classification. Chen et al. [12] used a stacked hourglass network structure to fuse information from multiple-context to predict human pose, and benefits from global and local information.

3   Proposed Method

3.1   Problem Formulation

We denote the input noise face image and corresponding clean image as X and Y, respectively. The process of getting the noise LR image from HR image can be modeled as

$$X = DBY + N \tag{1}$$

where D, B and N respectively denote downsampling operator, blur operator and additive noise operator.

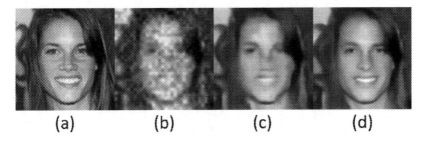

(a)   **(b)**   **(c)**   **(d)**

Fig. 1. Examples of experimental results. (a) Target image. (b) LR face image with noise. (c) Result of directly training [8]. (d) Result of our method.

For a given LR image, the face hallucination network F is expected to predict a hallucinated face as similar as the ground truth HR image by minimizing the mean square error (MSE).

$$L = \frac{1}{N_I}||F(X) - Y||_F^2 \tag{2}$$

However, we found that the result obtained by directly training [8] on the image domain (direct network) is not satisfactory. In Fig. 1, we show an example of a hallucinated face image which is used in the training process. In Fig. 1(c) we see that the hallucinated image by directly training has severe smooth in some details. In general, we observe that the learned regression function is performed on the entire picture, which means it need take into account various situations. However, the optimal learned regression function in different regions is different. In other words, the regression function need deal with all regions, which makes it hard to learn well. As a result, the reconstruction results in some areas are relatively smooth.

3.2 The Network Architecture

In order to solve the problem, we propose an effective model for noise face hallucination. The detailed structure of the network is shown in Fig. 2. It is divided into two branches: aggregation branch and generator branch. According to the similarity of the regression parameters, the aggregation branch can adaptively aggregate face regions into two categories. Then the generator branch can be targeted to recover HR images for selected regions.

We denote the networks input as X. The network can be summarized as

$$L = \frac{1}{N_I}||(G_1(X, \xi_1) - Y)M(X, \Phi) + (G_2(X, \xi_2) - Y)(1 - M(X, \Phi))||_F^2 \tag{3}$$

where M, G represents the output of aggregation branch and generator branch and ξ, Φ denotes the parameters to be learned. The aggregation branch aggregates the face regions into two categories as $M(X, \Phi)$ and $1 - M(X, \Phi)$. Then each generator branch can be targeted to recover HR images for selected regions. Finally, the reconstructed faces of different generator branches are added to generate the final output.

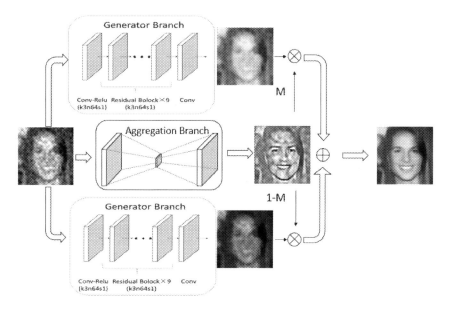

Fig. 2. The detailed structure of network

In aggregation branch, the aggregation network can not only serve as a mask selector during forward inference, but also can guide generator branch gradient update during backward propagation. In the generator branch, the gradient for input image is:

$$\frac{\mathrm{d}(G_1(X,\xi_1)-Y)M(X,\varPhi)}{\mathrm{d}\xi_1} = M(X,\varPhi)\frac{\mathrm{d}G_1(X,\xi_1)}{\mathrm{d}\xi_1} \tag{4}$$

This property allows the generator module to better reconstruct the selected regions. The aggregation branch can prevent unrelated data to update the parameters of the generator branch.

In addition, the excellent recovery of the generator branches in turn causes the aggregation branches to cluster more similar structures into the region. Assuming that the parameters of our generator branch are fixed, the loss function of our network is:

$$\arg\min_{\varPhi}||G_1(X,\xi_1)M(X,\varPhi)+G_2(X,\xi_2)(1-M(X,\varPhi))-Y||_F^2 \tag{5}$$

In order to minimize the loss, our aggregation branch can get greater weight for better reconstruction regions, which means the aggregation branch cluster more similar structures into categories. Through the similar process of alternating minimization, our network is constantly optimized to generate better results.

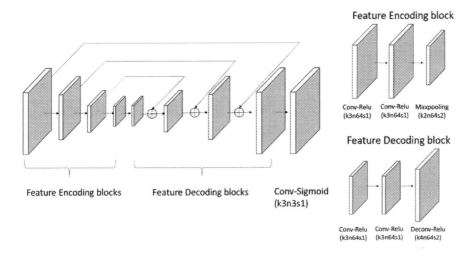

Fig. 3. The detailed structure of aggregation branch

Aggregation Branch. Our aggregation branch adopts a similar Hour-Glass structure, which used to human pose estimation [12,13], to cluster face regions. The detailed structure of the aggregation network is shown in Fig. 3. The network consists of multiple Feature Encoding blocks and Feature Decoding blocks. Each pair of feature encoding block and feature decoding block brings the feature representation into a new spatial scale, so that the whole network can process information on different scales. To effectively consolidate and preserve spatial information in different scales, the hourglass block uses a skip connection mechanism between symmetrical layers. Specifically, the feature information of input is quickly collected through multiple feature encoding blocks, and the feature decoding blocks amplifies the feature information to the same scale as the input. Finally a sigmoid layer normalizes the output range to [0, 1] to get a mask. Compared with CNN, the Hour-Glass structure can obtain a wider range of input information with less computational cost.

Generator Branch. For the generator branch, it be served to generate face images and can be adapted to any state-of-the-art network structures. Considering the success of residual network [14] in computer vision tasks, we choose the residual block as our network basic unit to hallucination face images. The generator branch consists of a cascade of multiple residual blocks. Each residual block contains two convolutional layers, then the input data is passed through a skip connection for element-wise sum with the output of the last convolutional layer.

4   Experiments

4.1   Dataset

We evaluated extensive experiments in Celebrity Face Attributes (CelebA) dataset [1]. The CalebA dataset contains 202,599 face images with 10,177 celebrity identities which is a very common dataset for face-related training. In our experiment, we first aligned the images with Mtcnn method [15] and crop the center image patches with size of 128×128 as the HR face images to be processed. Then we generate LR images by applied blur operation, down-sampling operation, and noise adding operation on the HR image. We set fixed Gaussian blur kernel b = 1.0, down-sampling factor 4 and we consider three noise levels $\sigma = 5$, 15 and 25. We select 22k faces from the dataset, of which 20k face images are trained and the rest are used for testing.

4.2   Parameter Settings

In our aggregation network, the number of feature encoding blocks and feature decoding blocks is 4, and each residual block consists of 2 convolution layers with kernels size of 3×3. In addition to the feature map of input layer and each branch's output layer is 3, the feature map of other layer is 64. For implementation, we train our model with the Tensorflow platform. The model is trained using the Adam optimization algorithm with an initial learning rate of 1e−3. We total train 50 epochs, and the later 20 epochs with learning rate of 1e−4. Training our network on celebA dataset takes about 6 h on 1 Titan X GPU.

Table 1. Quantitative comparison under Gaussian noise.

Method	$\sigma = 5$			$\sigma = 15$			$\sigma = 25$		
	PSNR	SSIM	FSIM	PSNR	SSIM	FSIM	PSNR	SSIM	FSIM
Bicubic	25.0573	0.7538	0.8421	22.9231	0.6607	0.8192	20.5045	0.5423	0.7781
BCCNN	25.5556	0.7640	0.8520	23.1615	0.6672	0.8209	21.7181	0.5946	0.7952
GLN	25.9653	0.7990	0.8785	25.2131	0.7548	0.8532	23.6237	0.7108	0.8353
SRCNN	27.4359	0.8144	0.8857	25.4672	0.7534	0.8532	23.9494	0.6902	0.8265
VDSR	28.1647	0.8389	0.8994	26.0401	0.7767	0.8659	24.4137	0.7061	0.8341
Ours	**28.6401**	**0.8533**	**0.9056**	**26.4691**	**0.7991**	**0.8762**	**24.9496**	**0.7394**	**0.8476**

4.3   Comparisons

We compare our approach with two types of methods: general image super-resolution methods and face hallucination approaches. For general image SR methods, we compare with SRCNN [7] and VDSR [8]. For face hallucination methods, we choose GLN [9] and BCCNN [10] as the contrast methods. Then we use the widely used PSNR (peak signal to noise ratio), SSIM (structural similarity) and FSIM (feature similarity) [16] to evaluate the reconstructed face.

Target   Bicubic   BCCNN   GLN   SRCNN   VDSR   Ours

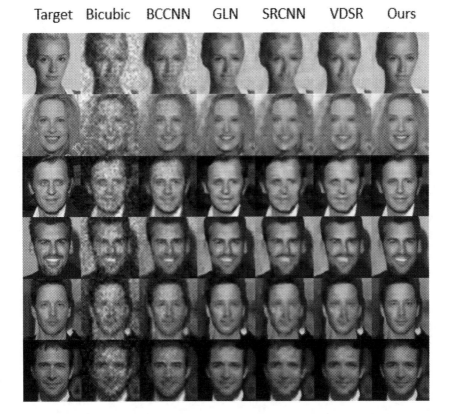

Fig. 4. Comparison of the hallucinated HR images with noise level 25.

4.4   Results

We use test sets to generate several types of LR surfaces with different noise level. Figure 4 shows the performance of our model and comparisons with other methods. It has been observed that Zhou's BCCNN [10] cannot eliminate noise. The final result of BCCNN is partly from the input noise face image, so the network does not converge well to obtain better results. Tuzel's GLN [9], like BCCNN, introduced the structural information of the face. However, in the noisy environment, the prior structural information of the face is not stable enough to converge well to obtain satisfactory results. Dong's SRCNN [7] also cannot remove the noise because there are only 3 convolutional layers and the parameters of the network are too small to produce satisfactory results. Kim's VDSR [8] makes the face clean and has more facial detail compare with SRCNN. However, this method uses the same regression function for all regions, resulting in poor reconstruction of some facial regions. Obviously, our method produces better results which not only removes the noise but preserves more face features information.

Table 1 shows the results of comparing our model with other state-of-the-art methods at different noise level. In terms of PSNR, SSIM and FSIM indicators, our model is much better than all comparison methods.

5   Conclusion

In this paper, we present a novel face hallucination framework which uses adaptive aggregation network to guide noise face hallucination. Our network contains two branches: aggregation branch and generator branch. Specifically, our aggregation branch can explore mapping relationships from LR to HR images in different regions, and aggregate the regions by the similarity of the mapping. Then generator branch can be used to make a specific face hallucination of the selected regions to get a better reconstruction result. The experimental results show that our model achieves state-of-the-art performance in noise face hallucination.

References

1. Liu, Z., Luo, P., Wang, X., Tang, X.: Deep learning face attributes in the wild. In: IEEE International Conference on Computer Vision, pp. 3730–3738 (2015)
2. Zhang, Z., Luo, P., Loy, C.C., Tang, X.: Learning deep representation for face alignment with auxiliary attributes. IEEE Trans. Pattern Anal. Mach. Intell. **38**(5), 918–930 (2016)
3. Yang, J., Wright, J., Huang, T.S., Ma, Y.: Image super-resolution via sparse representation. IEEE Trans. Image Process. **19**(11), 2861–2873 (2010)
4. Yang, C.Y., Liu, S., Yang, M.H.: Structured face hallucination. In: Computer Vision and Pattern Recognition, pp. 1099–1106(2013)
5. Jiang, J., Hu, R., Han, Z., Lu, T., Huang, K.: Position-patch based face hallucination via locality-constrained representation. In: IEEE International Conference, pp. 212–217 (2012)
6. Jiang, J., Hu, R., Wang, Z., Han, Z.: Noise robust face hallucination via locality-constrained representation. IEEE Trans. Multimed. **16**(5), 1268–1281 (2014)
7. Dong, C., Loy, C.C., He, K., Tang, X.: Learning a deep convolutional network for image super-resolution. In: Fleet, D., Pajdla, T., Schiele, B., Tuytelaars, T. (eds.) ECCV 2014. LNCS, vol. 8692, pp. 184–199. Springer, Cham (2014). https://doi.org/10.1007/978-3-319-10593-2_13
8. Kim, J., Lee, J.K., Lee, K.M.: Accurate image super-resolution using very deep convolutional networks. In: IEEE Conference on Computer Vision and Pattern Recognition, pp. 1646–1654 (2016)
9. Tuzel, O., Taguchi, Y., Hershey, J.R.: Global-local face upsampling network. arXiv preprint arXiv:1603.07235 (2016)
10. Zhou, E., Fan, H., Cao, Z., Jiang, Y., Yin, Q.: Learning face hallucination in the wild. In: AAAI, pp. 3871–3877 (2015)
11. Wang, F., et al.: Residual attention network for image classification. arXiv preprint arXiv:1704.06904 (2017)
12. Chen, Y., Shen, C., Wei, X.S., Liu, L., Yang, J.: Adversarial posenet: a structure-aware convolutional network for human pose estimation. CoRR, abs/1705.00389, 2 (2017)

13. Newell, A., Yang, K., Deng, J.: Stacked hourglass networks for human pose estimation. In: Leibe, B., Matas, J., Sebe, N., Welling, M. (eds.) ECCV 2016. LNCS, vol. 9912, pp. 483–499. Springer, Cham (2016). https://doi.org/10.1007/978-3-319-46484-8_29

14. He, K., Zhang, X., Ren, S., Sun, J.: Deep residual learning for image recognition. In: IEEE Conference on Computer Vision and Pattern Recognition, pp. 770–778 (2016)

15. Zhang, K., Zhang, Z., Li, Z., Qiao, Y.: Joint face detection and alignment using multitask cascaded convolutional networks. IEEE Signal Process. Lett. **23**(10), 1499–1503 (2016)

16. Zhang, L., Zhang, L., Mou, X., Zhang, D.: FSIM: a feature similarity index for image quality assessment. IEEE Trans. Image Process. **20**(8), 2378–2386 (2011)

17. Sun, Y., Liang, D., Wang, X., Tang, X.: Deepid3: face recognition with very deep neural networks. arXiv preprint arXiv:1502.00873 (2015)

18. Caicedo, J.C., Lazebnik, S.: Active object localization with deep reinforcement learning. In: IEEE International Conference on Computer Vision, pp. 2488–2496 (2015)

19. Gregor, K., Danihelka, I., Graves, A., Rezende, D.J., Wierstra, D.: DRAW: a recurrent neural network for image generation. arXiv preprint arXiv:1502.04623 (2015)

Multi-decoder Based Co-attention for Image Captioning

Zhen Sun, Xin Lin, Zhaohui Wang, Yi Ji[(⊠)], and Chunping Liu[(⊠)]

Soochow University, Suzhou, China
{jiyi, cpliu}@suda.edu.cn

Abstract. Recently image caption has gained increasing attention in artificial intelligence. Existing image captioning models typically adopt visual mechanism only once to capture the related region maps, which is difficult to attend the regions relevant to each generated word effectively. In this paper, we propose a novel multi-decoder based co-attention framework for image captioning, which is composed of multiple decoders that integrate the detection-based mechanism and free-form region based attention mechanism. Our proposed approach effectively produce more precise caption by co-attending the free-form regions and detections. Particularly, given the "Teacher-Forcing", which leads to a mismatch between training and testing, and exposure bias, we use a reinforcement learning approach to optimize. The proposed method is evaluated on the benchmark MSCOCO dataset, and achieves state-of-the-art performance.

Keywords: Co-attention · Image captioning · Multi-decoder

1 Introduction

Nowadays automatically generating captions of images has gained much attention in both academia and industry. The main challenge is to generate the caption in line with human's understanding. When human look at an image, he/she cannot get all information at first glance, but observes a small part of whole scene at a time. The similar attention mechanism, which can attend to salient parts of an image while generating its caption, is introduced. Inspired by recent research of machine translation, the mainstream image captioning [3, 20] approaches use an encoder-decoder framework [21], which is composed of a Convolution Neural Network (CNN) [11] for image encoding, and a Long Short-Term Memory (LSTM) [5] for decoding to generate caption.

Most visual attention mechanisms in image caption can be classified into free-form region based attention mechanisms [2, 12, 15, 21] and detection-based attention mechanisms [1, 9]. Xu et al. [21] proposed the first visual attention model for image captioning which incorporates the spatial attention on convolutional features of images into the encoder-decoder framework through the soft and hard attention mechanisms. Chen et al. [2] proposed a unified SCA-CNN framework for effectively integrating spatial, channel-wise, and multi-layer visual attention in CNN features for image captioning. Both of them utilize the free-form region based attention mechanism. Anderson et al. [1] proposed a combined bottom-up and top-down attention mechanism (detection-based attention mechanism) for image captioning and VQA. Our attention

R. Hong et al. (Eds.): PCM 2018, LNCS 11165, pp. 200–210, 2018.
https://doi.org/10.1007/978-3-030-00767-6_19

mechanisms is similar to Lu et al. [13] while our work make use of the two attention mechanisms by multi-decoders for image captioning. For the free-form region based methods, attention mechanism can attend both global visual context and specific foreground objects for inferring words caused by no restriction on location. However, this mechanism may result in frequent notice of some irrelevant areas and context. For the detection-based attention methods, the attention mechanism is designed to relate the previous words to pre-specified detection boxes. By calculating the attention weights of all detected boxes, the attention mechanism is more effective for generating words related to foreground objects. However, the above models for image captioning just focus on one type of image regions (i.e., free-from image regions or detection boxes). In contrast, our proposed approach effectively integrates both attention mechanisms with multi-decoder to make full use of the complementary information.

While most of the existing methods train their model by maximizing the likelihood estimation, some latest attempts have been made to use reinforcement learning (RL) to deal with the exposure bias. Ren et al. [17] used a Policy Network and a Value Network to collaborate to produce corresponding image caption. The Policy network evaluates the current state to produce the next word distribution, and the Value network evaluates the global possible expansion results in the current state. Similarly, Zhang et al. [22] investigated training image captioning by using actor-critic reinforcement learning. Rennie et al. [18] presents the SCST (Self-Critical Sequence Training) approach to directly optimize the CIDEr [19] metric with greedy decoding at test-time. In this paper, our RL-based approach is related to SCST but ours is designed for our co-attention model. We summarize the main contributions as:

1. We propose a novel multi-decoder using co-attention framework for image captioning. It can effectively integrate the free-form region based and detection-based attention mechanisms for generating accurate caption.
2. We improve our model by using a reinforcement learning with sentence reward supervision.
3. Our proposed model achieves the state-of-the-art performance on the benchmark MSCOCO dataset.

2   Proposed Method

In this paper, we adopt the encoder-decoder framework for image caption generation, which is composed by a CNN and a Faster-RCNN [16] for encoding an input image to a vector and 3 LSTMs for decoding the vector into a sequence of words. As illustrated in Fig. 1, our model takes full advantage of the two attention mechanisms to generate more detailed and accurate captions.

We first describe the encoder for image caption in Sect. 3.1, then introduce decoders in Sect. 3.2, fusion layer in Sect. 3.3, and RL learning for image captioning in Sect. 3.4.

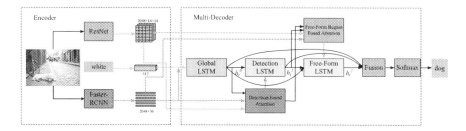

Fig. 1. Overview of the proposed multi-decoder based co-attention framework for image captioning. When predicting *dog* from the captioning *a white dog laying on the sidewalk next to a bike*, we use the previous word *white* to guide to attend the detections and free-form regions for generating *dog*

2.1   Image Encoding

We first encode the given image *I* to the spatial image features with pre-trained ResNet-101 on ImageNet [8]. The input image are not resized and the output feature map is the last convolution layer feature map denoted as V_F, which has dims of $2048 \times 14 \times 14$. Then we adopt the Faster-RCNN framework to obtain objection boxes in the input image. In this paper, we select the top-ranked 36 detection boxes for our detection-box features and adopt the ResNet-101 to get the visual features of each detection boxes donated as $V_D = [v_1, \ldots, v_{36}]$, $v_i \in R^{2048}$.

2.2   Decoding

We characterize the first LSTM layer as a Global LSMT for obtaining the rough image feature related to the previous words, and the second LSTM layer is denoted as Detection LSTM with a detection-based attention mechanism for attending detections. The last LSTM layer is described as Free-Form LSTM with a free-form region based attention mechanism for attending free-form regions.

Global LSTM. This decoder can get the global context from the global image feature related to the previous word for our Detection LSTM. The input vector to the Global LSTM at each time step consists of the mean-pooled detection feature $\bar{v} = \frac{1}{36} \sum_i v_i$ and an encoding of the previously generated word, given by:

$$x_t^0 = [\bar{v}, w_e \pi_t] \tag{1}$$

where $w_e \in R^{E \times D}$ is a word embedding matrix for a vocabulary of size D and π_t is one-hot encoding of the input word at time-step *t*. Our word embedding is randomly initialized without pre-training.

Detection LSTM. Detection LSTM can decode the detection-based feature with detection-based attention mechanism for better generating the words related to foreground objects. At each time step *t*, given output h_t^0 of the Global LSTM, we generate a normalized attention weight $\alpha_{i,t}$ for each of the 36 image detection features, and a convex combination of all input features of all input features will be input to Free-Form LSTM as follows:

$$V_d = f_d\left(V_G, V_D, h_t^0\right) \tag{2}$$

where f_d is the detection-based attention function that will be detailed in below, V_G is the mean-pooled feature of the free-form region feature V_F, h_t^0 is the hidden state of the Global LSTM. The input to detection-based attention LSTM consists of the detection-based attention image feature V_d, concatenated with the output h_t^0 of the Global LSTM, given by:

$$x_t^1 = \left[V_d, h_t^0\right] \tag{3}$$

Detection-Based Attention. Given the hidden state $h_t^0 \in R^M$, the whole image representation $V_G \in R^{2048}$ and the detection representation $V_D \in R^{2048 \times 36}$, we transform each representation to a common semantic space as h_{att1}, V_{ga} and V_{da}, and transform V_D as V_{dm}, then obtain the detection-based attention visual feature representation V_d as

$$h_{att1} = \tanh\left(W_{att1} h_t^0 + b_{att1}\right) \tag{4}$$

$$V_{ga} = \tanh\left(W_{ga} V_G + b_{ga}\right) \tag{5}$$

$$V_{da} = \tanh(W_{da} V_D + b_{da}) \tag{6}$$

$$V_{dm} = \tanh(W_{dm} V_D + b_{dm}) \tag{7}$$

$$C_1 = Norm_2\left(h_{att1} \circ V_{ga} \circ V_{da}\right) \tag{8}$$

$$\alpha_d = softmax(W_{c1} C_1 + b_{c1}) \tag{9}$$

$$V_d = \sum_i^{36} \alpha_d(i) V_{dm}(i) \tag{10}$$

where $W_{att1} \in R^{H \times M}$, $W_{ga} \in R^{H \times 2048}$, $W_{da} \in R^{H \times 2048}$, $W_{dm} \in R^{M \times 2048}$, b_{att1}, b_{ga}, $b_{da} \in R^H$, $b_{dm} \in R^M$ are the learnable parameters for linear transformation, $\circ$ indicates element-wise multiplication, $W_{c1} \in R^{36 \times H}$ and $b_{c1} \in R^{36}$ are learnable parameters of learning the attention weights α_d for each detection box feature. The transformed hidden state h_{att1} and whole-image feature V_{ga} are replicated to the same dimension of V_{da} for calculating the joint representation.

Free-Form LSTM. Free-Form LSTM can effectively get the global visual context and specific foreground object for inferring words caused by no restriction on location. Given output h_t^1 of the Detection LSTM, at each time step t, we generate a normalized attention weight $\beta_{i,t}$ for each of the 14×14 image grid features, and a convex combination of all input features of all input features will be input to Free-Form LSTM as follows:

$$V_f = f_g\left(V_F, V_M, h_t^1, V_d\right) \tag{11}$$

where f_g is the free-form region based attention function that will be detailed in below, V_M is the mean-pooled feature of the detection feature V_D, h_t^1 is the hidden state of the Detection LSTM. The input to Free-Form LSTM consists of the free-form region based attention image feature, concatenated with the linear transformation of previous hidden states h_t^0 and h_t^1, given by:

$$x_t^2 = \left[V_f, \tanh(w_m[h_t^0, h_t^1] + b_m)\right] \tag{12}$$

where $w_m \in R^{M \times 2M}$ and $b_m \in R^M$ are learned weights and biases. We use the output of the Global LSTM to prevent the disappearance of gradients and loss of information transfer.

Free-Form Region Based Attention. Given the whole-image representation $V_M \in R^{2048}$, the attended visual feature $V_d \in R^M$, Detection LSTM hidden state h_t^1, free-form representation $V_F \in R^{2048 \times 14 \times 14}$, we embed them into a common space as h_{att2}, V_{ma} and V_{fa}, and embed V_F as V_{fm}. Then attended free-form representation V_f can be defined as

$$h_{att2} = \tanh\left(W_{att2}\left(h_t^1 + V_d\right) + b_{att2}\right) \tag{13}$$

$$V_{ma} = \tanh(W_{ma}V_M + b_{ma}) \tag{14}$$

$$V_{fa} = \tanh\left(W_{fa}V_F + b_{fa}\right) \tag{15}$$

$$V_{fm} = \tanh\left(W_{fm}V_F + b_{fm}\right) \tag{16}$$

$$C_2 = Norm_2\left(h_{att2} \circ V_{ma} \circ V_{fa}\right) \tag{17}$$

$$\alpha_f = \text{softmax}\left(W_{c2}C_2 + b_{c2}\right) \tag{18}$$

$$v_f = \sum_i^{196} \alpha_f(i)V_{fm}(i) \tag{19}$$

where $W_{att2} \in R^{H \times M}, W_{ma} \in R^{H \times 2048}, W_{fa} \in R^{H \times 2048}, W_{fa} \in R^{M \times 2048}$ are learned parameters, $b_{att2}, b_{ma}, b_{fa}, b_{fm}$ are the bias parameters. We reshape the transformed free-form representation to H $\times$ 196. h_{att2}, V_{ma} are spatially replicated to 196 grid for match the spatial size of the free-form feature. $W_{c2} \in R^{196 \times H}$ and $b_{c2} \in R^{196}$ are the learnable parameters of learning the attention weights.

2.3 Fusion Layer

We use a fully connected layer to fusion the output of the three decoders for calculating the probability over a vocabulary D of possible words at time t:

$$h_t^e = \tanh\left(w_e\left[h_t^0, h_t^1, h_t^2\right] + b_e\right) \tag{20}$$

$$p_t = \text{softmax}\left(w_p h_t^e + b_p\right) \tag{21}$$

where $w_e \in R^{M \times 3M}, w_p \in R^{D \times M}$ and $b_e \in R^D$ are learned parameters.

2.4 RL Learning for Image Captioning

In Fig. 2, we show the RL learning process. Firstly, we train the network by defining a loss function that minimizes the cross-entropy loss,

$$L_{XE}(\theta) = -\sum_{t=0}^{T-1} \log(p_\theta(y_t|y_{0:T-1}, I)) \tag{22}$$

where the y_t is the ground-truth word, and the θ is the parameters of the captioning model. For fair comparison with recent works, we directly optimize the CIDEr metric by using a reinforcement learning approach following the Rennie et al. [18]. The objective of RL-based training is minimize the negative expected reward:

$$\nabla_\theta L_{RL}(\theta) = -r(s_{0:T-1})\nabla_\theta \log p_\theta(s_{0:T-1}) \tag{23}$$

where $s_{0:T-1} = (s_0, \ldots, s_{T-1})$ and s_t is the word sampled from the model at the time t. We follow the SCST method and adapt a baseline to reduce the variance of the gradient estimate. We finally approximate the expected gradient:

$$\nabla_\theta L_{RL}(\theta) = -(r(s_{0:T-1}) - r(\check{s}_{0:T-1}))\nabla_\theta p_\theta(s_{0:T-1}) \tag{24}$$

where $r_{(\check{s}_{0:T-1})} = (\check{s}_0, \ldots, \check{s}_{T-1})$ and $\check{s}_t$ is the word obtained by greedily decoding the current model. This gradient tends to increase the probability of the sampled captions that return higher reward than current model and suppress the samples that have worse reward than greedy decoding results.

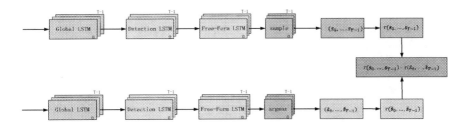

Fig. 2. Illustration of the proposed model with RL learning. The top row is the model under training that get the reward from the sampled sentence. The bottom row shows the reward obtained by its inference procedure

3 Experiments

3.1 Datasets and Evaluation Metrics

We evaluate our model on the public MSCOCO 2014 captions dataset [3] that contains 82,783 images in training set, 40,504 images in validation set and 40775 images in test set. Because the test set does not have the ground truth for offline testing, we follow the setting of Karpathy splits [10]. This split utilizes 113,287 training images with five captions each, and selects 5,000 images for validation and 5000 images for test from original validation set.

We use the BLEU (B@1, B@2, B@3, B@4) [14], METEOR (MT) [4], ROUGE-L (R-L) [6] and CIDEr (CD) [19] scores to evaluate our proposed model. What's more, we further evaluate our model performance in MSCOCO test set, and submit to the online MSCOCO evaluation server for online comparison with the other state-of-the-art methods.

3.2 Implementation Details

In this paper, we set the number of hidden units of each LSTM M to 1000, the number of hidden units H in the attention layer to 512, the size of input word embedding E to 512, and the vocabulary size of the word embedding D to 9487. Our Faster R-CNN implementation uses an IoU threshold of 0.7 for region proposal suppression and 0.3 for objects class suppression. In our experiments, we first train our model by the cross-entropy cost and use Adam optimizer with an initial learning rate of 5×10^{-4} and a momentum parameter of 0.9. After that, we train our RL-based approach with the pre-trained checkpoints as initializations and use Adam with a learning rate 5×10^{-5}. In test time, we use a beam size of 5 to decode captions for increasing the performance of greedy decoding.

3.3 Quantitative Analysis

In Table 1, we show the comparative experiments to prove the effectiveness of our proposed method. We first train those models with the standard cross-entropy loss. After that, we use reinforcement-learning method to optimize the model. For fair comparison, we adopted the SCST method to optimize CIDEr score. It is worth noting that all results is without fine-tuning of the ResNet-101. As can be seen from Table 1, our GDF

Table 1. The optimized model performance of GF, GD, G2F, G2D, GFD and GDF on the MSCOCO Karpathy test split.

Approach (CIDEr)	B@1	B@2	B@3	B@4	MT	R-L	CD
GF	77.9	61.5	46.3	35.2	26.9	56.1	114.3
GD	79.3	61.8	46.9	36.4	27.5	56.6	117.2
G2F	78.2	61.2	45.8	35.1	27.1	56.2	113.5
G2D	78.3	61.2	46.8	36.1	27.4	56.3	116.4
GFD	79.5	63.6	49.2	37.4	28.1	57.9	119.0
GDF	**80.4**	**64.6**	**50.2**	**38.2**	**28.6**	**58.5**	**122.8**

The highest entry for each evaluation metric highlight in boldface

($LSTM_{Global}$ + $LSTM_{Detection}$ + $LSTM_{Free-Form}$) model gets the best results. When we use the two-layer LSTM, the GD ($LSTM_{Global}$ + $LSTM_{Detection}$) model is better than GF ($LSTM_{Global}$ + $LSTM_{Free-Form}$). When we used the three-layer LSTM, we first try the G2F ($LSTM_{Global}$ + $2LSTM_{Detection}$) and G2D ($LSTM_{Global}$ + $2LSTM_{Free-Form}$) models, and find that the results of the model are worse than the models of the two layers. As a single attention mechanism can lead to over-fitting problems, we use the fusion of two attention mechanisms. However, considering the effect of different order on the result, we test the GFD and GDF ($LSTM_{Global}$ + $LSTM_{Free-Form}$ + $LSTM_{Detection}$) model respectively, and finally find that GDF achieves better results.

Table 2 compares our GDF results with the state-of-the-art results on Karpathy splits of the MSCOCO dataset. We can see our model achieves the best performance in all metrics.

Table 2. Comparisons of the image captioning performance of the state-of-art on the MSCOCO Karpathy test split.

Approach	B@1	B@2	B@3	B@4	MT	R-L	CD
Google NIC [20]	–	–	–	27.7	–	23.7	85.5
Hard-Attention [21]	70.7	49.2	34.4	24.3	–	–	–
Soft-Attention [21]	71.8	50.4	35.7	25.0	–	–	–
Adaptive [12]	74.2	58.0	43.9	33.2	26.6	–	108.5
SCA-CNN [2]	71.9	54.8	41.1	31.1	25.0	–	–
SCST:Att2in [18]	–	–	–	34.8	26.9	56.3	115.2
Up-Down [1]	79.8	–	–	36.3	27.7	56.9	120.1
Stack-Cap [7]	78.6	62.5	47.9	36.1	27.4	56.9	120.4
GDF	**80.4**	**64.6**	**50.2**	**38.2**	**28.6**	**58.5**	**122.8**

Our GDF model achieves significant gains across all metrics

Table 3 reports the performance of our GDF model trained with CIDEr optimization on the official MSCOCO evaluation server. We can see that our approach

Table 3. Leaderboard of published image captioning models on the online MSCOCO test server.

Approach	B@1		B@4		MT		R-L		CD	
	c5	c40	c5	c40	c5	c40	c5	c40	c5	c40
Google NIC† [20]	71.3	89.5	30.9	58.7	25.4	34.6	53.0	68.2	94.3	94.6
Hard-Attention [21]	70.5	88.1	27.7	53.7	24.1	32.2	51.6	65.4	86.5	89.3
Adaptive† [12]	74.8	92.0	33.6	63.7	26.4	35.9	55.0	70.5	104.2	105.9
SCA-CNN [2]	71.2	89.4	30.2	57.9	24.4	33.1	52.4	67.4	91.2	92.1
SCST:Att2in† [18]	78.1	93.7	35.2	64.5	27.0	35.5	56.3	70.7	114.7	116.7
Up-Down† [1]	80.2	**95.2**	36.9	68.5	27.6	36.7	57.1	72.4	117.9	120.5
Stack-Cap [7]	77.8	93.2	34.9	64.6	27.0	35.6	56.2	70.6	114.8	118.3
GDF	**80.2**	94.6	**37.5**	**68.7**	**28.3**	**37.5**	**58.0**	**73.2**	**118.3**	**120.9**

Our GDF model achieves comparable performance with the state-of-the-art across all metrics. † indicates ensemble model outperforms other models. Though some metrics is slightly lower than the up-down model because Up-Down model is an ensemble of 4 models optimized for CIDEr with different initializations.

3.4   Qualitative Analysis

To prove the validity of our proposed model, we visualize some our co-attention maps for different words generated by our captioning model in Fig. 3. For the generated word *grass*, Figs. 3 shows that our free-form region based attention can attend the correct image regions and detection-based attention attends incorrect regions. For the generated word *giraffes*, detection-based attention perform better.

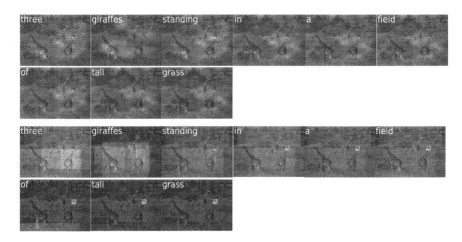

Fig. 3. For each generated word, we visualize the attended image regions for free-form region based attention (top row) and detection-based attention (bottom row)

4   Conclusion

In this paper, we propose a novel multi-decoder using co-attention framework for image captioning. We use multiple LSTM networks to integrate free-form region based attention and detection-based attention for making full use of the complementary information. By contrasting with the current state-of-the-art methods, we have found that our model achieves better performance on the online MSCOCO test server. In the future research, we will introduce external reasoning information for better understanding of the relationship between objects.

Acknowledgments. This work was partially supported by National Natural Science Foundation of China (NSFC Grant Nos. 61773272, 61272258, 61301299, 61572085, 61170124, 61272005), Provincial Natural Science Foundation of Jiangsu (Grant Nos. BK20151254, BK20151260), Science and Education Innovation based Cloud Data fusion Foundation of Science and Technology Development Center of Education Ministry (2017B03112), Six talent peaks Project in

Jiangsu Province (DZXX-027), Key Laboratory of Symbolic Computation and Knowledge Engineering of Ministry of Education, Jilin University (Grant No. 93K172016K08), and Provincial Key Laboratory for Computer Information Processing Technology, Soochow University.

References

1. Anderson, P., et al.: Bottom-up and top-down attention for image captioning and VQA. arXiv:1707.07998 (2017)
2. Chen, L., et al.: SCA-CNN: spatial and channel-wise attention in convolutional networks for image captioning, pp. 6298–6306 (2016)
3. Chen, X., et al.: Microsoft coco captions: data collection and evaluation server. arXiv:1504. 00325 (2015)
4. Denkowski, M., Lavie, A.: Meteor universal: language specific translation evaluation for any target language. In: Proceedings of the Ninth Workshop on Statistical Machine Translation, pp. 376–380 (2014)
5. Donahue, J., Anne Hendricks, L., Guadarrama, S., Rohrbach, M., Venugopalan, S., Saenko, K., Darrell, T.: Long-term recurrent convolutional networks for visual recognition and description. In: Proceedings of the IEEE Conference on Computer Vision and Pattern Recognition, pp. 2625–2634 (2015)
6. Flick, C.: Rouge: a package for automatic evaluation of summaries. In: The Workshop on Text Summarization Branches Out, p. 10 (2004)
7. Gu, J., Cai, J., Wang, G., Chen, T.: Stack-captioning: coarse-to-fine learning for image captioning. arXiv:1709.03376 (2017)
8. He, K., Zhang, X., Ren, S., Sun, J.: Deep residual learning for image recognition. In: Proceedings of the IEEE Conference on Computer Vision and Pattern Recognition, pp. 770–778 (2016)
9. Johnson, J., Karpathy, A., Fei-Fei, L.: Densecap: fully convolutional localization networks for dense captioning. In: Proceedings of the IEEE Conference on Computer Vision and Pattern Recognition, pp. 4565–4574 (2016)
10. Karpathy, A., Fei-Fei, L.: Deep visual-semantic alignments for generating image descriptions. In: Proceedings of the IEEE Conference on Computer Vision and Pattern Recognition, pp. 3128–3137 (2015)
11. Krizhevsky, A., Sutskever, I., Hinton, G.E.: Imagenet classification with deep convolutional neural networks. In: Advances in Neural Information Processing Systems, pp. 1097–1105 (2012)
12. Lu, J., Xiong, C., Parikh, D., Socher, R.: Knowing when to look: adaptive attention via a visual sentinel for image captioning. In: Proceedings of the IEEE Conference on Computer Vision and Pattern Recognition (CVPR), vol. 6 (2017)
13. Lu, P., Li, H., Zhang, W., Wang, J., Wang, X.: Co-attending free-form regions and detections with multi-modal multiplicative feature embedding for visual question answering. arXiv:1711.06794 (2017)
14. Papineni, K., Roukos, S., Ward, T., Zhu, W.J.: Bleu: a method for automatic evaluation of machine translation. In: Proceedings of the 40th Annual Meeting on Association for Computational Linguistics, pp. 311–318. Association for Computational Linguistics (2002)
15. Pedersoli, M., Lucas, T., Schmid, C., Verbeek, J.: Areas of attention for image captioning. In: ICCV-International Conference on Computer Vision (2017)

16. Ren, S., He, K., Girshick, R., Sun, J.: Faster R-CNN: towards real-time object detection with region proposal networks. In: Advances in Neural Information Processing Systems, pp. 91–99 (2015)
17. Ren, Z., Wang, X., Zhang, N., Lv, X., Li, L.J.: Deep reinforcement learning-based image captioning with embedding reward. arXiv:1704.03899 (2017)
18. Rennie, S.J., Marcheret, E., Mroueh, Y., Ross, J., Goel, V.: Self-critical sequence training for image captioning. In: CVPR, vol. 1, p. 3 (2017)
19. Vedantam, R., Lawrence Zitnick, C., Parikh, D.: Cider: consensus-based image description evaluation. In: Proceedings of the IEEE Conference on Computer Vision and Pattern Recognition, pp. 4566–4575 (2015)
20. Vinyals, O., Toshev, A., Bengio, S., Erhan, D.: Show and tell: a neural image caption generator. In: 2015 IEEE Conference on Computer Vision and Pattern Recognition (CVPR), pp. 3156–3164 (2015)
21. Xu, K., et al.: Show, attend and tell: neural image caption generation with visual attention. In: International Conference on Machine Learning, pp. 2048–2057 (2015)
22. Zhang, L., et al.: Actor-critic sequence training for image captioning. arXiv:1706.09601 (2017)

Simultaneous Occlusion Handling and Optical Flow Estimation

Song Wang[1,2] and Zengfu Wang[1,2(✉)]

[1] Institute of Intelligent Machines, Chinese Academy of Sciences,
Hefei 230031, Anhui, China
[2] University of Science and Technology of China, Hefei 230026, Anhui, China
`zfwang@ustc.edu.cn`

Abstract. Occlusion handling and optical flow estimation is a chicken-and-egg problem. In this paper, we propose our method which can handle occlusion and estimate optical flow simultaneously. First of all, we use the backward interpolation strategy to gain the warped image, then we obtain the occlusion relationship by comparing the pixels before the movement. After that we use the occlusion relation to correct the warped image and get the occlusion coefficient. Later, using the occlusion coefficient to modify the energy function. Finally, the corrected energy function and warped image are used to estimate the final optical flow results. We evaluate our method on some popular datasets such as Flying Chairs and MPI-Sintel. Experimental results demonstrate that the proposed method improves the accuracy of current optical flow estimation methods significantly.

Keywords: Optical flow estimation · Occlusion handling
Image warping

1 Introduction

The concept of optical flow, which describes the apparent motion between images or video frames, arises from the studies of biological visual systems. The methods of optical flow estimation can be roughly divided into three categories. The most predominant type is based on variational methods, where a local; color-based matching of the pixel is combined with a global smoothness assumption. And then the coarse-to-fine warping schemes are proposed to optimize large displacement optical flow problems, which were previously used by Lucas and Kanade [8] in local techniques. The second type is based on sparse matching methods, converting the optical flow estimation into the matching problem while in the early days, using the pyramid based structure to optimize the search space has been proposed by Bergen et al. [3]. The third type is based on the approximate nearest neighbor fields (ANNF) methods, it can efficiently perform an optical flow estimation on the full image resolution [2]. In addition to the above three methods and the mutual integration [5], the Convolutional Neural Network (CNN)

© Springer Nature Switzerland AG 2018
R. Hong et al. (Eds.): PCM 2018, LNCS 11165, pp. 211–220, 2018.
https://doi.org/10.1007/978-3-030-00767-6_20

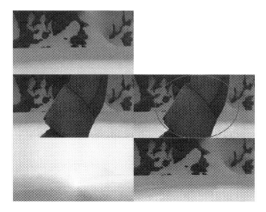

Fig. 1. Illustration of the occlusion artifacts, the two sets of images are from Flying Chairs and MPI-Sintel datasets respectively. In each set, the left column is image pair and ground truth, and the one on the right is warped image with occlusion and our warped image.

based methods are getting considerable attention in recent years. Flownet [7], Flownet2.0 and SPN [10] are all the end-to-end convolutional architectures. DM-CNN [12], MC-CNN [15], PatchBatch and many others CNN-based methods are of matching based architectures.

The end-to-end convolutional architecture is at times hard to train, and the results are unreliable from time to time. The matching-based convolutional architecture has the same drawback as the sparse matching methods, it can only achieve sparse optical flow estimation. Besides, some of the results are confusing caused by occlusions involved missing of their counterpart in the next frame. And many descriptors, whether it's CNN-based or not, are not well localized, thus the precision of the motion estimates are lower than the variational techniques. The most serious problem of the ANNF-based methods is it usually contained many outliers which are difficult to identify. Most of outliers are result from the initial matching errors and the absence of the regularization terms in ANNF. The variational methods can acquire the sub-pixel level optical flow estimation at case, as a result, they are widely used as a post-processing, e.g., EpicFlow, for a variety methods. Due to the limitation of the optimization function, the coarse-to-fine warping schemes are chosen. The uniqueness of warped images makes it lose the ability of error correction but for the pyramid-based methods it is of great importance and we will propose more details in Sect. 2.

To overcome the problems mentioned above, in this paper, we proposed a novel pyramid-based method which can joint optical flow estimation and occlusion handling. Unlike previous methods using the traditional image warping strategies, such as bi-liner interpolation or spline-based interpolation, in order to obtain the warped images, our method which can warp images and handle occlusion based on the color constraint simultaneously. Meanwhile, we found that the occlusion coefficients from the occlusion detection and our affine-based filtering improve the precision of the motion estimates significantly. Figure 1 shows the comparison of two different warping strategies, what we need to point out is that the occlusion artifacts in the warped images are usually the fundamental cause of the error in optical flow estimation. We found that on the right set, even if the warped image calculated by ground truth is used, there will be arms left in the image. Using this warped image and the previous frame to estimate the optical flow, it will fall into infinite iteration.

The contributions of this study could be summarized as follows:

Firstly, we provide an improved coarse-to-fine warping scheme which can joint optical flow estimation and occlusion handling.

Secondly, an effective occlusion estimation strategy is proposed which can correct the warped image and the energy function simultaneously.

Thirdly, we present an efficient affine-based filtering strategy which can correct the warped image and further improvement of the results.

The experimental results on Flying Chairs and MPI-Sintel dataset show that our method significantly improves the accuracy of optical flow estimation.

The remainder of this paper is organized as follows. The coarse-to-fine warping schemes, our methods, including the warping strategy, the occlusion coefficients, and our affine-based filtering strategy are given in Sect. 2. Experimental results and analyses are presented in Sect. 3. Our conclusion and future work are summarized in Sect. 4.

2   Method

Whether it is CNN-based methods or not, the coarse-to-fine warping schemes are the widely recognized methods for estimating optical flow field. In this section, we briefly introduce the coarse-to-fine optical flow estimation schemes in Sect. 2.1. In the following Sect. 2.2, we propose our warping method and the occlusion coefficients. And our affine-based filtering are presented in Sect. 2.3.

2.1   Coarse-To-Fine Optical Flow Estimation Schemes

Optical flow is the pattern of apparent motion of objects in adjacent frames, I_1 and I_2. In general, the first step is to construct the image spatial pyramids for both images, respectively. The overview of the coarse-to-fine optical flow estimation algorithm is presented in Fig. 2. More specifically, we construct the spatial pyramids with n levels for both images firstly. We denote the k-th levels of pyramid of I_i as I_i^k, $i \in \{1, 2\}$, $k \in \{1, 2, ..., n\}$. It should be point out that the bottom levels of pyramids, which labeled I_1^1 and I_2^1, are the images in original size. Then, we estimate the optical flow F_n at the coarsest resolution level using I_1^n and I_2^n. We optimize the following model for optical flow. For all the parameters and E_* in this equation, we suggest [5] for more details.

$$E(\omega) = E_{color}(\omega) + \gamma E_{gradient}(\omega) + \alpha E_{smooth}(\omega) + \beta E_{match}(\omega) + E_{desc}(\omega) \tag{1}$$

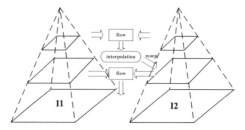

Fig. 2. Illustration of the coarse-to-fine spatial pyramidal optical flow estimation scheme.

Whereafter, the warped image $w(I_2^{n-1}, F_n)$ is obtained based on the optical flow F_n. Specifically, the result F_n needs to be interpolated before image warping. Next, with I_1^{n-1} and $w(I_2^{n-1}, F_n)$, the optical flow F_{n-1} is estimated successively. Finally, the optical flow of every pixel in I_1 against I_2 can be effectively estimated after finite times of iteration.

2.2   Our Image Warping Strategy

The concept of optical flow describes the apparent motion between images or video frames. There are two possibilities for each pixel in the I_1 image after motion, one is to remain on the I_2 image and the other is virtually invisible in the I_2 image. There are two reasons for the disappearance, one is that the displacement is too large and out of the field, and the other is blocked after the movement. The warped images express the scene before motion, and the traditional method only uses the optical flow information and the I_2 images. However, from the above analysis, we find that some information in the I_2 images is missing since occlusion or other reasons. Without taking into account this problem, the pixels of the occluded regions are replaced by those of the occlusion regions in the warped images wrongly. More specifically, the warped images record the moment before the motions, at which the occlusion regions and the occluded regions are separate. When the two regions coincide in I_2, only the occlusion regions are retained. Based on above description and discussion, we propose our novel image warping method which can discriminate and eliminate the ghosting artifacts in the warped images effectively.

If the optical flow fields are reliable, the correct warped image should be consistent with the I_1 image. Based on this hypothesis, we propose a method which is combining the I_1 images to obtain the more accurate warped images. First, we establish the mapping relations between the warped image and the I_2 image. By comparing pixels in the corresponding positions in the I_1 image and the I_2 image, we retain the most similar mapping relations, while the other relations are abandoned. We compensate the pixel using the pixel from the I_1 image if there is no pixel in the I_2 image corresponding to the current position. For the other case, we compare the pixels of the corresponding positions in the I_2 image and the I_1 image. If the difference between them is more than a certain threshold, which place will be identified as the occlusion area, then the current position in the warped image needs to be filled with the pixel from the I_1 image.

Considering that the optical flow is sub-pixel level, that is, the corresponding pixel is obtained by the bilinear interpolation, so we use the maximum value of the pixel difference in the 4-neighborhood as the threshold. We consider two strategies, the local threshold means each position has its threshold, and the global threshold can be obtained by multiplying the maximum of the local threshold by a coefficient. We find that the latter is superior in terms of the computational efficiency and the final robustness.

The optical flow may have some error at the coarse layer, if we use pixels in the I_1 image to compensate the warped image may cause this regions loss the ability to correction its estimated. Therefore, we estimate the occlusion coefficient which can be integrated into the variational model to reduce the weight of the occlusion regions in the energy function. We tried a variety of methods, and obtained the best performance which is the weight of the filling area was set to 0, the other areas were set to 1. After normalization, the occlusion coefficient is the parameter θ for the E_{color} in the variational model (Eq. 1).

2.3   Affine-Based Filtering

The classical optical flow estimation implements the median filtering of intermediate flow fields during optimization which can improve the robustness of the results. Sun et al. [11] proved that it actually leads to higher energy function, meaning that these methods are not optimizing the original objective function. In fact, the pixels in the neighborhood are not consistent, so we propose an edge based image filtering strategy. Figure 3 shows that even the two positions without edges in the same object could have different optical flow because of their depth. And we found that the difference of the optical flow caused by the depth is well fitted by affine transformation.

Based on the above analysis, we propose a segmentation based affine filtering method. First, we use SLIC [1] to get over-segmentation results. Then, for each

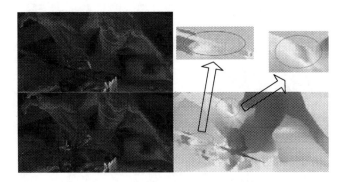

Fig. 3. Illustration of the different optical flow within the same body.

superpixel, partial least squares (PLS) method is used to fit an affine formula. The optical flow field $F(p)$ is using a affine estimator at a pixel $p \in superpixel_i$ as $F(p) = A * p + t$, where A and t are the parameters of an affine formula estimated for each superpixel i. In practice, we use our affine-based filtering after several iterations. Because we find that the optical flow calculated in the coarse layer is not satisfied with the affine formula, and it reduces the robustness of our algorithm.

3 Experiments

In this Section, we evaluate our proposed method on some popular datasets such as Flying Chairs [7] and MPI-Sintel [6] datasets. And the MPI provide two versions: the Final version (MPIfinal) contains motion blur and atmospheric effects, while the Clean version (MPIclean) does not include these effects. The endpoint error (EPE) and the average of endpoint error (AEE) are normally worked as the measures of optical flow estimation. We compare a variety of methods that have been proposed in recent years, including the traditional methods and the CNN-based methods.

Brox et al. [4] reported the image pyramidal downsampling factor of 0.95 produces much better results than 0.5. While Sun et al. [11] demonstrated that when using a convex penalty, the downsampling factor become unimportant and a standard factor of 0.5 can be chosen. A large number of experiments verify that excessive downsampling factor not only bring huge computational cost, but also cause massive errors in optical flow. Through the experiment, we achieved a good

Table 1. Average end-point error (AEE) on different datasets using different downsampling strategies

Methods\AEE	Flying chairs	MPIclean	MPIfinal
LDOF_0.98[a]	4.4122	4.3154	6.2055
OURS_0.98[b]	**3.6965**	3.8558	5.5077
LDOF_0.80	4.1697	3.7201	5.7437
OURS_0.80	3.7331	3.5523	5.3904
LDOF_0.80+16[c]	4.1152	3.7077	5.7484
OURS_0.80+16	3.7197	**3.5387**	**5.3641**
LDOF_0.50	4.9211	4.2852	5.8581
LDOF_0.50+16	4.2997	3.8682	5.7101

[a] 0.98 is the downsampling factor, code from https://www.cs.cmu.edu/~katef/LDOF.html.
[b] OURS is using our information transfer strategy method.
[c] We adjust the minimum downsampling scale from 40 to 16 in order to increase the number of lower sampling layers.

result with the downsampling factor is set to 0.8, and the minimum sampling size is set to 16. What's more, we found that our proposed information transfer strategy is effective on all the different parameter tests. Due to space limitations, we summery our numbers of experiments concern downsampling levels and factor then the result are summerized in Table 1.

3.1   Occlusion Handling

The L1-norm potentials have been widely used since its robustness, it is considered to have a certain ability to handle occlusion. In this sub-section, we tried first to extract the occlusion information from the matching result. For each match ($I_1^{pixelA}, I_2^{pixelB}$), we check whether ($I_2^{pixelB}, I_1^{pixelA}$) is the best match in the backward direction. This consistency check way removes many false matches, particularly those caused by occlusion. Based on the double-check, we can get the occlusion area mask.

We contrasted a large number of methods for extracting occlusion information, which is summarized in Table 2. The first item in Table 2 is the original LDOF method, it does not extract any occlusion information. If an area went through double check and the counterpart cannot be found then we define this area as the occlusion area while this kind of way of obtaining occlusion information can be defined as the "matching" item. The "div" term using divergence information to obtain occlusion results. The remaining three items are using modified warped image, the occlusion coefficient or both of them at the same time. Through the ablation experiments, we found that the occlusion coefficient has a greater impact than the modified warped image on the quality of the optical flow estimation. The optical flow information of the coarse layer is not accurate enough, so sometimes it could result in undesirable errors to the warped image, and the occlusion coefficient directly excludes the occlusion area off the variational model, so the result is better than the previous strategy.

Table 2. Average end-point error (AEE) on different occlusion handling strategies

Methods\AEE	Flying chairs	MPIclean	MPIfinal
LDOF	4.1152	3.7077	5.7484
matching	3.9206	3.6915	5.7417
div	3.7832	3.5391	5.3961
OURS1_I2warped	4.0003	3.6536	5.5237
OURS2_coef	3.7866	3.6256	5.4802
OURS3_coef+I2warped	**3.7197**	**3.5387**	**5.3641**

Table 3. Average end-point error (AEE) on methods with/without affine-based filtering (AF)

Methods\AEE	Flying chairs	MPIclean	MPIfinal
LDOF	4.1152	3.7077	5.7484
LDOF+AF	3.9692	3.6745	5.6638
OUR	3.7197	3.5387	5.3716
OUR+AF	**3.6282**	**3.5000**	**5.3596**

3.2 Affine-Based Filtering

In this section, we demonstrate the effect of our affine-based filtering method. Here we will show the experimental results in two different cases to demonstrate the effectiveness of our method.

The first two rows use the traditional image warp method, and the last two rows use our information transfer strategy. We found that in both cases our affine-based filtering can improve the results, which are summarized in Table 3.

3.3 Final Results

To integrate the smoothing weight [9] and the methods mentioned above, the final results can be achieved, after detailed comparison with the current optical flow estimation methods in Table 4, our methods applied to several training sequences are given in Fig. 4.

As shown in Fig. 4, the original LDOF method almost fails to estimate the optical flow of image pairs in the fourth column, while our method achieves the acceptable results based on our information transfer strategy and iterative fine tune in the last column. In the first row, the LDOF almost lost the details of the hollowed back of the chair, but we manage to kept it well. We do better than the LDOF in the detail reservation of the wheels of the chair in the second row.

Table 4. Average end-point error (AEE) on different methods

Methods\AEE	Flying chairs	MPIclean	MPIfinal
DeepFlow [13]	3.53	3.19	4.4
PCA-Flow [14]	\	4.04	5.18
classic+NLP [11]	3.93	4.13	5.9
FlownetS [7]	2.71	4.5	5.45
FlownetC [7]	**2.19**	4.31	5.87
SPN [10]	2.63	4.12	5.57
LDOF (baseline) [5]	4.41	4.32	6.21
Our method	2.96	**3.01**	**5.09**

Fig. 4. Illustration of optical flow estimation on Flying Chairs dataset. In each row left to right: I_1, I_2, ground truth, optical flow estimation of the LDOF, optical flow estimation of our method. For all methods, the endpoint error (EPE) is printed in the image.

In the next two rows, the LDOF has hardly detected the chairs, but our method achieves the acceptable results. In the last row, the optical flow results in the upper right corner prove that our method achieves good results in the overlap of moving objects, but the chairs back in the left side is missing, probably due to over-smoothing and further research is needed. Since we can greatly reduce the mismatching on the borders of motion, it has achieved clearer contours of the chairs in our optical flow images.

The results show that our method is far superior to the baseline method and has a significant advantage over the CNN-based method proposed in recent years.

4   Conclusion

This paper presents a novel pyramid-based method which can joint optical flow estimation and occlusion handling. We can correct the energy function and the warped image simultaneously. In addition, we present an efficient affine-based filtering strategy which can further improve the results. Experiments on the popular datasets, such as Flying Chairs and MPI-Sintel, demonstrate that our approach significantly improves the accuracy of optical flow on overall images, and our approach achieves an impressive result that surpasses the current leading learning-based method. Furthermore, our proposed method can be applied to various CNN-based methods. In the future work, it is a worth direction with deep learning involvement with our model.

References

1. Achanta, R., Shaji, A., Smith, K., Lucchi, A., Fua, P., Süsstrunk, S.: Slic superpixels compared to state-of-the-art superpixel methods. IEEE Trans. Patt. Anal. Mach. Intell. **34**(11), 2274–2282 (2012)
2. Bailer, C., Taetz, B., Stricker, D.: Flow fields: dense correspondence fields for highly accurate large displacement optical flow estimation. In: Proceedings of the IEEE International Conference on Computer Vision, pp. 4015–4023 (2015)
3. Bergen, J.R., Anandan, P., Hanna, K.J., Hingorani, R.: Hierarchical model-based motion estimation. In: Sandini, G. (ed.) ECCV 1992. LNCS, vol. 588, pp. 237–252. Springer, Heidelberg (1992). https://doi.org/10.1007/3-540-55426-2_27
4. Brox, T., Bruhn, A., Papenberg, N., Weickert, J.: High accuracy optical flow estimation based on a theory for warping. In: Pajdla, T., Matas, J. (eds.) ECCV 2004. LNCS, vol. 3024, pp. 25–36. Springer, Heidelberg (2004). https://doi.org/10.1007/978-3-540-24673-2_3
5. Brox, T., Malik, J.: Large displacement optical flow: descriptor matching in variational motion estimation. IEEE Trans. Patt. Anal. Mach. Intell. **33**(3), 500–513 (2011)
6. Butler, D.J., Wulff, J., Stanley, G.B., Black, M.J.: A naturalistic open source movie for optical flow evaluation. In: Fitzgibbon, A., Lazebnik, S., Perona, P., Sato, Y., Schmid, C. (eds.) ECCV 2012. LNCS, vol. 7577, pp. 611–625. Springer, Heidelberg (2012). https://doi.org/10.1007/978-3-642-33783-3_44
7. Dosovitskiy, A., et al.: FlowNet: learning optical flow with convolutional networks. In: Proceedings of the IEEE International Conference on Computer Vision, pp. 2758–2766 (2015)
8. Lucas, B.D., Kanade, T., et al.: An iterative image registration technique with an application to stereo vision (1981)
9. Monzón, N., Salgado, A., Sánchez, J.: Regularization strategies for discontinuity-preserving optical flow methods. IEEE Trans. Image Process. **25**(4), 1580–1591 (2016)
10. Ranjan, A., Black, M.J.: Optical flow estimation using a spatial pyramid network. arXiv preprint arXiv:1611.00850 (2016)
11. Sun, D., Roth, S., Black, M.J.: A quantitative analysis of current practices in optical flow estimation and the principles behind them. Int. J. Comput. Vis. **106**(2), 115–137 (2014)
12. Thewlis, J., Zheng, S., Torr, P.H., Vedaldi, A.: Fully-trainable deep matching. arXiv preprint arXiv:1609.03532 (2016)
13. Weinzaepfel, P., Revaud, J., Harchaoui, Z., Schmid, C.: DeepFlow: large displacement optical flow with deep matching. In: Proceedings of the IEEE International Conference on Computer Vision, pp. 1385–1392 (2013)
14. Wulff, J., Black, M.J.: Efficient sparse-to-dense optical flow estimation using a learned basis and layers. In: Proceedings of the IEEE Conference on Computer Vision and Pattern Recognition, pp. 120–130 (2015)
15. Zbontar, J., LeCun, Y.: Computing the stereo matching cost with a convolutional neural network. In: Proceedings of the IEEE Conference on Computer Vision and Pattern Recognition, pp. 1592–1599 (2015)

Handwritten Chinese Character Recognition Based on Domain-Specific Knowledge

Qian Liu[1], Danqing Wang[1], Hong Lu[2(✉)], and Chaopeng Li[1]

[1] Shanghai Key Laboratory of Intelligent Information Processing,
School of Computer Science, Fudan University, Shanghai, China
[2] Shanghai Engineering Research Center for Video Technology and System,
School of Computer Science, Fudan University, Shanghai, China
honglu@fudan.edu.cn

Abstract. Although some encouraging progress has been achieved in handwritten Chinese character recognition (HCCR), handwritten Chinese address recognition (HCAR) remains an ongoing challenge. Few methods achieve satisfying performance on it due to more irregular distortion and overlapping between characters. In this paper, we first extract keywords from the address by a specially designed key character classifier. Then we use a single character network to recognize the place names. In order to take advantage of hierarchical relationships among place names, we construct an address database from the Chinese administrative divisions and design an error-correction method to improve the recognition of place names. Experiments on handwritten Chinese address datasets demonstrate the effectiveness of the proposed method.

Keywords: Handwritten address · Chinese recognition
Deep learning · Domain-specific knowledge

1 Introduction

With the development of digitization, automatic handwritten text recognition has become ever more important for information storage, management and retrieval. Due to the complicated structure and large number, handwritten Chinese character recognition (HCCR) is much difficult compared with digit and English character recognition. On the other hand, due to the rapid development of deep learning, the performance of single Chinese character recognition has been significantly improved. For common handwritten Chinese character set, the recognition accuracy is above 90% [3,4,11,14]. Even so, the recognition of Chinese texts in real scene still fails to achieve satisfactory results.

In digits and English texts, the overlap between characters is small and leads to high segmentation accuracy. Then the text's recognition is performed on a single character. However, the situation is different for Chinese text recognition.

© Springer Nature Switzerland AG 2018
R. Hong et al. (Eds.): PCM 2018, LNCS 11165, pp. 221–231, 2018.
https://doi.org/10.1007/978-3-030-00767-6_21

On one hand, the structure of a single Chinese character is much more complex than a digit and an English character. A Chinese character is composed of many smaller strokes and radicals. On the other hand, strokes and radicals between adjacent Chinese characters prone to be overlapping with each other. Then the Chinese characters are difficult to be segmented. At present, most research on handwritten Chinese character recognition focuses on single character recognition. And the Chinese text recognition largely depends on segmentation. Thus the detection performance of texts is not satisfactory.

In this paper, we propose to combine domain knowledge to improve the recognition performance of handwritten Chinese texts. We focus on Chinese address recognition. We segment texts with hierarchical information and use the address database to build an error correction mechanism. Specifically, we first summarize 15 key characters from hierarchical information in Chinese address. Then we train a 16-class classifier with a class 'X' of characters not belonging to one of the above 15 key characters. Based on this classifier, the texts can be split into hierarchies. Each level of the hierarchy consists of a place. Second, to recognize the place names, we first segment the texts into single characters and apply a single character classifier. We form a candidate list for this character. Finally, we utilize candidate list to correct the possible recognition errors. If the recognition result is not in the hierarchy of the address database, the next candidate on the list is selected to replace it. We compare the performance of our proposed method by incorporating domain-specific knowledge with that just on the single character recognition method. And our method can obtain promising recognition result.

The rest of this paper is organized as follows. Section 2 provides a brief introduction on handwritten Chinese address recognition methods. Section 3 describes our proposed method, which includes keyword recognition and place name recognition. Section 4 presents our constructed handwritten Chinese address dataset and the recognition performance of our proposed method. Conclusion and future work are outlined in Sect. 5.

2   Related Work

In this section, we review the work on character recognition and handwritten Chinese character recognition.

2.1   Single Character Recognition

With respect to single character recognition, there are three categories, i.e. radical-based [5,8], stroke-based [7,10,12] and holistic methods [6]. Although radical-based and stroke-based methods need to improve the performance of segmentation and radical or stroke recognition, the features extracted can better represent the structures of Chinese characters. The development of deep learning makes it possible to deal with more complicated characters with less time and higher performance [14]. Among these three categories of methods, holistic methods are commonly used.

Holistic recognition methods ignore the internal structure of Chinese character's. It takes the single character as a binarization image and uses image processing methods on the image. Most of the methods extract features, such as directional or gradient features, and then apply traditional machine learning or deep learning algorithms. With further development of deep learning methods, much research builds deep network to achieve end-to-end recognition. Many methods obtain promising performance in Chinese Handwritten Recognition Competition on CASIA Online and Offline Chinese Handwriting Databases [3,4,11]. And deeper networks show much more improvements [2,14].

Some methods combine machine learning with deep learning. They use features extracted by traditional methods as input. [14] uses 8-direction directMap to represent characters, and passes it through a stack of 11 layers network. An adaptation layer is optional to adapt the network to different writers' styles. The method is tested on the dataset of ICDAR-2013 offline and online competition and the accuracy of recognition is larger than 96.9%.

2.2 Handwritten Chinese Address Recognition

Our study focuses on handwritten Chinese address recognition (HCAR). [9] uses the hierarchical information of addresses to segment the handwritten address characters and obtains promising performance on handwritten Japanese address recognition. The method summarizes some key characters from large scale of different addresses and their combination. Strokes are split into lines and combined into key characters. The combinations which are created according to the rule of the Chinese five-level administration are used to verify the recognition of key characters' sequences. Afterwards, key characters split the lines into words, which are recognized by dynamic dictionary of geographical names built during training. Because of complex hierarchies and difficulty of Chinese character recognition, the recognition performance of Chinese addresses is inferior to the Japanese address. On the other hand, incorporating domain-specific knowledge can benefit the recognition.

[1] takes advantage of address information to segment and merge the input address character string. The method first over-segments the input string into radical series and then selects 100 candidate merging paths from all possible paths according to radicals' geometric features. In order to find optimal merging path, the candidate paths are evaluated on a hidden Markov model (HMM) and Viterbi algorithm. The address information is used to evaluate the prior probability of each character and the transition probability between characters on HMM.

3 Method

In this section, we introduce our proposed method on handwritten Chinese address recognition. There are two separate networks in our proposed method,

Table 1. The administrative divisions of the People's Republic of China and the keywords. There are some Chinese keywords with different administrative meanings. However we do not distinguish them in our extraction.

Level	Chinese keyword
First-level	(Province) (Municipality City) (Region)
Second-level	(Prefecture-level City) (Prefecture) (Prefecture)
Third-level	(County-level City) (County) (District)
Fourth-level	(Subdistrict) (Town) (Township)
Others	(Village) (Community) (Group)
	(Team) (Street) (Alley) (Road)

Fig. 1. The simplified handwritten Chinese character recognition network. (The dot parts has similar network structure)

keyword recognition network and single character recognition network. The first network only divides characters into 16 classes. And the other network is a standard Chinese character recognition network with 3755 characters.

3.1 Keyword Recognition Network

In order to take advantage of hierarchical relationship of address, we first create an address database from the Chinese administrative divisions. We summarize 15 keywords from this address database according to the division rule (i.e. the last character of the place name), which are described in detail in Table 1. These keywords represent the hierarchical relationship of administrative place names in China. Then we train a special classifier with 16 classes (including a category 'X' for characters not belonging to the 15 classes). This network is designed to extract keywords. Based on the extracted keywords, the original address texts are split into a series of place names. The division itself is much more important than the key character recognition results. If we misclassify one keyword to another one, we will also get the right division result. And the wrong keyword will be corrected using the following error-correction algorithm. This division between keywords and place names is more accurate than the common segmentation used for text recognition.

3.2   Single Character Recognition Network

Since deep learning has obtained promising performance in handwritten Chinese character recognition, we use deep neural network to build our classifier. Specifically, we first resize the original image of the single character to 32×32. Then we build the network with 11 layers. Eight of these layers are convolutional layers and two of them are fully connected layers. We pad the input feature map planes with zeros in all convolutional layers to get the output that is the same size as the input. This avoids reducing the dimension in intermediate layers to 1×1. Every two convolution layers are followed by a max-pooling layer. After eight convolutional layers and four max-pooling layers, we get 400 feature maps with 2×2 size. We flatten these feature maps and concatenate them to a vector with the dimensionality of 1600. This vector is passed to next two fully connected layers. The last layer is a softmax layer for classification.

This network shares the same idea with that in [14], but there are some differences on the network setting. Firstly, we remove the adaptation layer. The adaptation layer assumes that the training set shares the same class distribution of the test set. However, our training set has uniform class distribution while our test set is handwritten address samples, where the distribution of different characters varies from each other. Thus we drop the adaptation layer. Secondly, we use pixels instead of 8-direction directMap as input. During the training, we find that the original setting is costly especially in the preprocessing of 8-direction directMap feature and the fully connected layers. Therefore we adjust our network for less training time and GPU memory. The 8-direction directMap doesnot show much advantage in the keyword recognition network, so we replace it by pixels. We also change the unit number of fully connected layers to reduce the parameters. This simplification saves considerable memory for our experiment. The structure of our network is shown in Fig. 1. The keyword recognition network and single character recognition network share a same network architecture except the last softmax layer.

For the single character recognition network, we classify all characters into 3755 categories. It is the total number of the first-level Chinese characters according to GB 2312-80. In Chinese address database, we find that among all 73 million entries, there are 55 characters in the first-level, 401 in the second, 1056 in the third, 3007 in the fourth, and 4837 in the fifth. This means that there are some characters which are not in GB 2312-80. Taking this into consideration, we only choose addresses in GB 2312-80 as the first level Chinese characters.

3.3   Address Recognition Mechanism

To obtain a better recognition result, we take advantage of domain-specific knowledge, i.e. the hierarchical information of Chinese address. To be more specific, the key character recognition network is first used to classify the Chinese address into a mixture of keywords and unknown class 'X'. Take the string as an example, the key character classifier will get a string of XXXXX. Keywords in this string show the division line of the original input address. The original

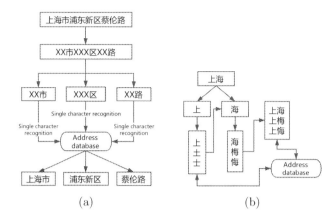

Fig. 2. Address recognition mechanism. (a) The process of address recognition. (b) The error correction mechanism.

input image will be divided into several place name words according to division lines. In this example, the image section of, and will be obtained. This process is shown in Fig. 2(a).

However, place names can also contain keywords. For example, "Distric"() is a keyword in the hierarchical structure of Chinese address. On the other hand, is a place name and should not be split into and. If the division only depends on the position of keywords, we will get a wrong result in this example.

Therefore, it is necessary to combine the address information with keyword segmentation. In the tree of address database, we add two extra attributes. These two attributes indicate the shortest and longest length of next level place name respectively. For each split place name, the algorithm compares its length with the range of the corresponding parent node. If the length of the place name is shorter, then we merge it with the next split place name and get a new place name word. If the length is longer, the entire address will be divided further. In the example above, since the minimum length of all child nodes under the same parent node (the parent of) is 5, the length of is not within the scope and will be merged with to get the correct place name. In fact, has a particular region suffix in Chinese address and the actual name of the place is. Except for this special case, the keywords are seldom included in the place name.

After division, these images of place names will be fed to the single character classifier according to the error-correction algorithm as shown in Fig. 2(b). The error-correction algorithm maintains a candidate list for each character and check the correctness of recognition result by the address database. The method will choose the most similar text by matching algorithm. To be more specific, we take as an example. First, the character will be fed to the single character classifier. During the recognition, several similar characters will be obtained as candidates for this character, including, and so on. After that, the most possible one will be taken and searched in the first level of the address tree (the root at the zero level is China). The search result is (Shanghai). Then we get the correct answer. However, if the most possible candidate obtained is, then there is no node in the first level starting with it. In this case, the algorithm will take the next candidate of the input character, possibly, and repeat the above process. Until finally the current candidate in the address database can be found. This process is repeated for each candidate. If all candidates of the character fail, it will go back to the previous character and check its next candidate. In the worst case that all the candidates of the first character in the place name fail, the closest result in the corresponding level of address database will be selected according to the text similarity.

4   Experiment

In this section, we introduce the dataset we construct and the experimental results.

4.1   Dataset

Single handwritten Chinese character set is selected from HCL2000 offline handwritten Chinese character dataset, belonging to Beijing University of Post and Telecommunications [13]. This dataset is currently the largest offline handwritten Chinese character library, including 3755 first level Chinese characters according to GB 2312-80, written by 1000 people. A single character is a 64×64 binary matrix.

In the training of key character recognition network, we label the samples of HCL2000 offline handwritten Chinese character dataset into 15 keywords and one 'X' category. For the single character recognition network, we use the original 3755 characters.

In the collection of handwritten address samples, we use synthetic address data to verify the performance of the algorithm. The main process of creating synthetic address data is as follows. First, we select the same person's handwritten samples using writers' information from HCL2000. Then we link these samples into a legitimate address text based on address hierarchical rules and add some artificial overlap and distortion to simulate the real situation. Totally, we create 1000 handwritten Chinese addresses for our experiment. Examples are given in Fig. 3.

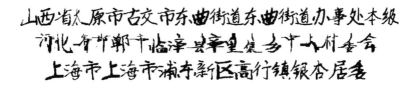

Fig. 3. Examples of our dataset.

4.2   Chinese Address Database

The address database is primarily used for rule-based validation and error correction. We collect 73,000 five-level addresses on a national scale through the administrative divisions of the People's Republic of China in 2013 (up to December 31, 2013) to construct it.

The address database is organized as a tree structure. The root of the tree is China and each node is a place name. The nodes are arranged according to the address' hierarchical rules. Each path from root to the leaves is a standard address. In the algorithm, we use the hierarchical information to assist the recognition of the entire address. For example, "Province"-"Prefecture-level City"-"District"-"Township" and "Province"-"Prefecture-level City"-"District"-"Subdistrict" are both reasonable hierarchical relationships in Chinese administrative divisions. Place names divided by these keywords are in different ranks of the address. The place name before "Province" is the first level of the address tree and the place name after Township is the lowest level. We check the recognition result of the place name in its level. This method reduces the recognition difficulty and improves the performance.

4.3   Results

The keyword recognition network classifies all characters into 16 categories, and achieves an accuracy of 98.9% in our key character dataset. Our key character dataset is selected from ICDAR-2013 offline and online competition dataset. We modify the categories of all characters with 15 keywords and one class 'X'.

We first test our single character classifier on ICDAR-2013 offline competition database. The results are tabulated in Table 2. The performance of our simplified network is higher than other methods and close to the original network [14]. The reason lies that we use pixels instead of 8-direction directMap to describe the character image and remove the adaptation layer. Take the keyword network and the error correction mechanism into consideration, the simplification will reduce time and memory with marginal reduction on performance. Then it demonstrates the effectiveness of our single Chinese character recognition.

Table 2. Recognition results on ICDAR-2013 offline HCCR competition database [14].

No	Method	Accuracy
1	Liu [3]	92.39%
2	Yin [3]	95.77%
3	Liu [11]	95.31%
4	Yin [4]	97.39%
5	Kim [2]	97.51%
6	Zhang [14]	**97.91%**
7	Our Method	97.71%

Table 3. Accuracy on 1000 handwritten Chinese address recognition. (Recognition means we only use single character classifier to recognize the handwritten texts. Mechanism means we combine the classifier with our address mechanism.)

	Character	Place name	Address
Recognition	92.49%	74.60%	7.10%
Mechanism	95.64%	94.68%	88.89%
Improvement	**3.15%**	**20.08%**	**81.79%**

Then, we test our method on 1000 standard addresses in our address datasets, which have been described above. The results are tabulated in Table 3. It can be observed that, after training on 3755 characters, the recognition accuracy of character in addresses has reached higher than 92%. This accuracy is lower than single character recognition because there are overlapping between characters. With the combination of address database, this accuracy has increased to 95%.

Most of the mistakes in single character recognition come from the inproper recognition of the first character of the place name. The method tends to depend much on address matching than recognition result. So if the first character is misclassified and is not corrected afterwards, it will affect all the following steps. In word recognition, the performance is not good. And with hierarchical information, the performance can be enhanced. In address recognition, the gap is much larger. Accuracy of simply applying single character recognition to texts is lower than 10%. This means that few of these addresses can be correctly recognized. However, with the address database information, the recognition performance can be improved much.

In our experiment, we use key character classifier to split the original address. Division accuracy is 96.70%, which is marginally lower than keyword recognition. This happens because that some place name words contain keywords. This division error can be rectified in our error correction part. We have carried out our experiments with a remote computing platform SuperVessel provided by IBM with a multicore cpu POWER8E (@2.061 GHz), a 2 GB RAM and GPU Tesla K40m (@12 GB). The average recognition time is approximately 0.5s for one address image.

Figure 4 shows the comparison of the recognition results of the two typical addresses. In the first example, the characters are wrongly recognized as. And after the retrieval and check on the address database, the result is corrected. In the second example, the keyword classifier obtains (Alley) instead of (County-level City) by mistake. And the result still belongs to keywords. Therefore, this error does not affects the division of the address. The keyword is rectified in the place name retrieval. On the other hand, there are already too many errors in the place name. It leads to no appropriate match in the address database. In this case, the algorithm will choose the most possible result by text similarity in the corresponding level of the address database and has a high possibility to get the wrong result.

Sample	山西省太原市古交市东曲街道办事处本级
Direct	山西省太原市古交市东曲街道办章沁本级
Result	山西省太原市古交市东曲街道办事处本级

Sample	河北省邯郸市临漳县章里集乡中上村委会
Direct	河北省邯郸市临漳县辛里虞乡牛上村委会
Result	河北省邯郸市临漳县章里集乡大章村委会

Fig. 4. Typical errors in address recognition (Sample: correct address, Direct: single character recognition, Result: incorporating hierarchical information).

5   Conclusions and Future Work

After applying deep learning to recognize handwritten characters, the accuracy has been improved. However, the recognition performance of handwritten Chinese texts is not satisfactory. Thus, we just try to use prior knowledge of the context through their usage scenario. In this paper, we propose a handwritten Chinese address recognition method by incorporating domain-specific knowledge. We obtain promising results by combining hierarchical address information and single character recognition. Domain-specific knowledge and simple natural language processing can be combined with single character classifier to deal with handwritten address characters recognition. Experimental results demonstrate that our proposed method is effective.

References

1. Fu, Q., Ding, X., Liu, C., Jiang, Y.: A hidden Markov model based segmentation and recognition algorithm for chinese handwritten address character strings. In: IEEE 8th International Conference on Document Analysis and Recognition, pp. 590–594 (2005)
2. Kim, I.J., Xie, X.: Handwritten hangul recognition using deep convolutional neural networks. Int. J. Doc. Anal. Recogn. **18**(1), 1–13 (2015)
3. Liu, C.L., Yin, F., Wang, D.H., Wang, Q.F.: Chinese handwriting recognition contest 2010. In: Chinese Conference on Pattern Recognition, pp. 1–5 (2010)
4. Liu, C.L., Yin, F., Wang, D.H., Wang, Q.F.: Online and offline handwritten chinese character recognition: benchmarking on new databases. Patt. Recogn. **46**(1), 155–162 (2013)
5. Shi, D., Gunn, S., Damper, R.: Handwritten chinese radical recognition using nonlinear active shape models. IEEE Trans. Patt. Anal. Mach. Intell. **25**(2), 277–280 (2003)
6. Srihari, S.N., Yang, X., Ball, G.R.: Offline chinese handwriting recognition: an assessment of current technology. Front. Comput. Sci. China **1**(2), 137–155 (2007)
7. Su, Y.M., Wang, J.F.: Decomposing chinese characters into stroke segments using SOGD filters and orientation normalization. In: International Conference on Pattern Recognition, vol. 2, pp. 351–354 (2004)

8. Wang, A.B., Fan, K.C.: Optical recognition of handwritten chinese characters by hierarchical radical matching method. Patt. Recogn. **34**(1), 15–35 (2001)

9. Wang, C., Hotta, Y., Suwa, M., Naoi, S.: Handwritten chinese address recognition. In: International Workshop on Frontiers in Handwriting Recognition, pp. 539–544 (2004)

10. Wang, Q., Chi, Z., Feng, D.D., Zhao, R.: Hidden Markov random field based approach for off-line handwritten chinese character recognition. In: International Conference on Pattern Recognition, vol. 2, pp. 347–350 (2000)

11. Yin, F., Wang, Q.F., Zhang, X.Y., Liu, C.L.: ICDAR 2013 chinese handwriting recognition competition. In: IEEE 12th International Conference on Document Analysis and Recognition, pp. 1464–1470 (2013)

12. Zeng, J., Liu, Z.: Markov random fields for handwritten chinese character recognition. In: IEEE 8th International Conference on Document Analysis and Recognition, vol. 1, pp. 101–105 (2005)

13. Zhang, H., Guo, J., Chen, G., Li, C.: HCL2000 - a large-scale handwritten chinese character database for handwritten character recognition. In: IEEE 10th International Conference on Document Analysis and Recognition, pp. 286–290 (2009)

14. Zhang, X.Y., Bengio, Y., Liu, C.L.: Online and offline handwritten chinese character recognition: a comprehensive study and new benchmark. Patt. Recogn. **61**, 348–360 (2017)

Attention to Refine Through Multi Scales for Semantic Segmentation

Shiqi Yang and Gang Peng$^{(\boxtimes)}$

Key Laboratory of Ministry of Education for Image Processing and Intelligence Control, School of Automation, Huazhong University of Science and Technology, Wuhan, China
{albert_yang,penggang}@hust.edu.cn

Abstract. This paper proposes a novel attention model for semantic segmentation, which aggregates multi-scale and context features to refine prediction. Specifically, the skeleton convolutional neural network framework takes in multiple different scales inputs, by which means the CNN can get representations in different scales. The proposed attention model will handle the features from different scale streams respectively and integrate them. Then location attention branch of the model learns to softly weight the multi-scale features at each pixel location. Moreover, we add an recalibrating branch, parallel to where location attention comes out, to recalibrate the score map per class. We achieve quite competitive results on PASCAL VOC 2012 and ADE20K datasets, which surpass baseline and related works.

Keywords: Semantic segmentation · Attention model · Multi-scale
Context

1 Introduction

With the booming of deep learning, many visual tasks have made significant progress. For instance, semantic segmentation, also known as image labeling or scene parsing which aims at giving label for each pixel, has made great breakthroughs in recent years. Efficient semantic segmentation can facilitate plenty of other missions such as image editing.

Recent approaches for semantic segmentation are all almost based on Fully Convolutional Network (FCN) [13], which outperforms the traditional methods by replacing the fully connected layers with convolutional layers in classification network. The follow-up works have extended the FCN from several points of view. Some works [2,14] have introduced the coarse-to-fine structure with upsample modules like deconvolution to give the final mask prediction. And due to the usage of pooling layer, spatial size has decreased largely, for which dilated (or atrous) convolution [6,20] has been employed to increase the resolution of intermediate features and hold the same receptive field simultaneously.

© Springer Nature Switzerland AG 2018
R. Hong et al. (Eds.): PCM 2018, LNCS 11165, pp. 232–241, 2018.
https://doi.org/10.1007/978-3-030-00767-6_22

Other works mainly focus on two directions. One is to post-process the prediction from the CNN through Conditional Random Field (CRF) to get smooth output. These works [1,6,22] are actually ameliorating the localizing ability of the framework. Another direction is to ensemble multi-scale features. Because features from lower layers in CNN have more spatial information and ones from deeper layers have more semantic meaning and less location information, it is rational to integrate representations from various positions since location information is important for semantic segmentation. The first type method for multi-scale combines features from different stages with skip connection to get fused features for mask prediction, such as [6,13,18]. And another type is to resize input to several scales and pass each one with a shared network, it will produce final prediction using the fusion of multi-stream resulting features. There are also methods trying to exploit the capability of global context information, like ParseNet [12] which adds a global pooling branch to extract contextual features. And PSPNet [21] adopts a pyramid pooling module to embed global context information to achieve accurate scene perception.

Attention model has been all the rage in natural language processing area, such as [3], and it has also shown its effectiveness in computer vision and multimedia community recently [4,16,17,19]. It allows model to focus on specific relevant features. Attention-to-scale [7] is the first approach to introduce attention model into semantic segmentation for multi-scale. It takes in different scale inputs. For each scale, the attention model produces a weight map to weight features at each location, and the weighted sum of score maps across all scales is then used for mask prediction. But it only utilizes the feature from specific layer to generate attention, which may omit many contextual details, and this can not ensure that the attention model can guide network to get precise results.

Referring to attention-to-scale, we propose a new attention model in this paper, which also takes in multi-scale inputs but integrates features from different layers, similar to hypercolumns [10]. The attention model has two branch outputs, *i.e.*, one for location attention through which it drives network to focus on large objects or regions for small scale input and pay attention to small targets for large scale just like attention-to-scale, another branch is to recalibrate the score map per class since resulting features from several stages carry contextual information. The outputs from attention model will be applied to multi-scale stream predictions, and final mask prediction is a weighted sum of all these streams.

Our contributions are two aspects as follows:

(1) We introduce a novel attention model into multi-scale streams semantic segmentation framework, the final mask prediction is produced by merging the predictions from multiple streams.
(2) The attention model utilizes fused features from different positions of CNN, which carry more contextual information, and has two branch outputs, where one is for location attention and another is for recalibrating.

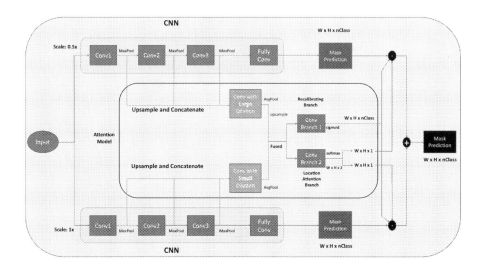

Fig. 1. Architecture of semantic segmentation framework with the proposed attention model. The attention model takes in features from different stages in CNN just like hypercolumns [10], and then it adopts convolutional layer with different dilation to process features for each scale respectively. Attention model produces two kinds of weight maps which are applied to multiple streams predictions. The final mask prediction is a sum of all streams.

2   Proposed Methods

2.1   Attention Model with Multi-scales

Like we mention before higher-layer features contain more semantic information and lower ones carry more location information. Fusion of information from several spatial scales will improve the accuracy of prediction in semantic segmentation. In addition, multi-scale aggregation also catch more contextual representations since some operations like pooling will dispose of the global context information, leading to local ambiguities which will be discussed later. It is the reason why multi-scale fusion gained a lot of popularity.

Since our work is extended from attention-to-scale [7], here we give a brief review on it. In attention-to-scale, the images are resized to several scales which will be fed to a weight-shared CNN, and the attention model takes as input the directly concatenating features from penultimate layer in each scale stream. The attention model consists of two convolutional layers and will produce n channels scores map, where n means the number of input scales. The attention model is expected to adaptively find the best weights on scales. But there exists some problems. The features from penultimate layer surely contain semantic representations, but they lack essential localization and global information fed to the attention model to achieve precise prediction. And we also posit that simply concatenating features from certain position is not conducive to lead the attention model to learn soft weight across scales. Seeing that the attention model is to put large weights on the large object or region in small-scale stream and gives large weights to the small targets in large-scale stream, we think it is rational to handle features from different scales respectively before integrating them.

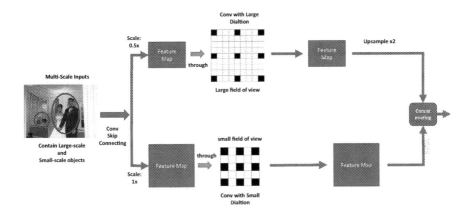

Fig. 2. Convolution with different dilation for different scale. Convolution with large dilation has large field of view while convolution with small dilation has small field of view.

Inspired by hypercolumns, we adopt the philosophy of it. Like depicted in Fig. 1, features from different stages in CNN get upsampled to same size and then we concatenate them all. To keep computation cost at bay, we choose the size of features after two pooling operation as the appointed resolution to do upsampling by bilinear interpolation. Through this way, the acquired features carry more localization and context information.

It is well-known that the structure of network has an impact on the range of pixels of the input image which correspond to a pixel of the feature map. In other words, filters will implicitly learn to detect features at specific scales due to the fixed receptive field. To accomplish our motivation of attention model which is to adaptively put weights on corresponding scale, we add a unique convolutional layer with unequal dilation for each scale. This process is demonstrated in Fig. 2. Convolution with large dilation has large field of view (FOV) and is expected to catch the long-span interlink of pixels for large scale object or region in small scale stream, and small dilation convolution is deployed to encode target of small scale in large scale stream. After the dilated convolution, the features will be concatenated, resulting one contains much more abundant and context information.

By the way, the two-stream CNNs in Fig. 1 are actually the same one when implemented in practice, just like Siamese Network.

2.2   Two Branch Outputs of Attention Model

The concatenated features will go through two parallel convolutional branches: location attention branch and recalibrating branch.

In common with attention-to-scale, the attention model will produce soft weights for multiple scales (we refer to it as location attention). Assuming the number of input scale is n, and the size of mask prediction, which is denoted as P^s for scale s, is W × H, $nClass$ means the class number of the objects. The location attention output by the model is shared across all channels. After the refinement of local attention, the mask predictions, denoted as M_i^s, are described as:

$$M_{i,c}^s = \sum_{s=1}^n l_i^s \cdot P_{i,c}^s \qquad (1)$$

The l_i^s is computed by:

$$l_i^s = \frac{\exp\left(wl_i^s\right)}{\sum_{j=1}^n \exp\left(wl_i^j\right)} \qquad (2)$$

where wl_i^s is the score map produced by the location attention branch at position $i \in [0, W*H-1]$ for scale s, before the softmax layer of course.

And since the fused features fed to the attention model contain context information, we want to make full use of them to eliminate some degrees of class ambiguity, $i.e.$, to utilize contextual relationship to enhance the ability of classification. The lack of ability to collect contextual information may increase the chance of misclassification in certain circumstances. To take an example, neural network sometimes tends to take apart a large-scale object into several regions of different classes [11], or maybe classify a boat on the river as a car and so on in scene parsing [21] (these can be observed among visualization results in Sect. 3.1). To deal with these issues, we add a recalibrating branch parallel to location attention. It has the same architecture as location attention branch which means containing two convolutional layers, except that output channel changes to $nClass$ and $sigmoid$ activation is deployed instead of softmax. This branch aims to find the interdependencies between adjacent objects or regions using the integrating features, and its output is used for recalibrating the score maps before the location attention refinement. Because the contextual relationship stay the same in different scale, the recalibrating outputs are shared across all scales. So the final mask prediction for each stream can be described as:

$$M_{i,c}^s = \sum_{s=1}^n l_i^s \cdot [P_{i,c}^s \otimes wr_{i,c}] \qquad (3)$$

where the $\otimes$ means element-wise multiplication and $wr_{i,c}$ means output in position i in channel $c \in [0, n-1]$ produced by recalibrating branch. Another choice for recalibrating branch is to predict bias per position in each channel instead of multiplication. But it will bring around 1% performance decrease according to our experiment.

And the ultimate mask prediction is as below, where M^s is the mask prediction of scale s:

$$M_{final} = \sum_{s=1}^{n} M^s \qquad (4)$$

As for the loss function, we follow the setting of attention-to-scale, *i.e.*, the total loss function is sum of $1+S$ cross entropy loss functions for segmentation, where S symbolizes number of scales and one for final prediction.

3   Experimental Results

We experiment our method on two benchmark datasets: PASCAL VOC 2012 [8] and ImageNet scene parsing challenge 2016 dataset [23] (it is from ADE20K [24], hereinafter we refer to it as ADE20k).

For all training, we only train the network with 2 scales, *i.e.*, 1x upsample and 0.5x upsample. As for the different dilation, we set it to 2 for small scale and 12 for large scale. And we use the poly learning rate policy [12], meaning current learning rate is computed by multiplying $(1 - \frac{iter}{max\_iter})^{power}$ to base learning rate, where the power is set to 0.9. We refer to the layers in the last stage where gives mask prediction as decoder, layers previous to decoder are encoder. Learning rate of decoder is 10 times that of encoder. All experiments are implemented using PyTorch on a NVIDIA TITAN Xp GPU.

3.1   PASCAL VOC 2012

The PASCAL VOC 2012 [8] segmentation dataset consists of 20 foreground object classes and a background class. The PASCAL VOC 2012 dataset we use is augmented with extra annotation by Hariharan *et al.* [9], resulting in 10582 training images. In experiment we report performance results on original PASCAL VOC 2012 validation set.

DeepLab-LargeFOV [5] is chosen as base model. Since our work is extended from attention-to-scale, in order to compare fairly, we reproduce the DeepLab-LargeFOV and attention-to-scale based on it by ourselves, following the set of

Table 1. Results on PASCAL VOC 2012 validation set. There exists 2 scale streams: 1x and 0.5x. The mIoU means mean intersection of union [13].

Method	mIoU
Baseline (DeepLab-LargeFOV)	61.40%
Merged with MaxPooling	63.88%
Merged with AvgPooling	64.07%
Attention-to-Scale	64.74%
Our method	**67.98%**

Table 2. Ablation study for proposed method on PASCAL VOC 2012. The multistage means hypercolumns-like feature integration from different positions. Diverse dilation means utilizing different dilated convolution for multi-scale features. Extra branch means adding recalibrating branch. *-The base model is actually attention-to-scale. †-No diverse dilations means using standard convolution instead.

Method	Multi-stage	Diverse dilations†	Location attention	Extra branch	mIoU
Base model*			√		64.74%
Base model+	√		√		65.80%
Base model++	√	√	√		66.83%
Our method	√	√	√	√	**67.98%**

attention-to-scale [7]. All these experiments use VGG16 [15] as skeleton CNN, which is pretrained on ImageNet. Our reproduction of them yields performance of 61.40% and 64.74% on the validation set respectively. The performance of attention-to-scale is lower than original paper, but the follow-up experiments still can verify effectiveness of our proposed method since ours is directly built on attention-to-scale. Noted that both of attention-to-scale and our work adopt extra supervision, meaning adding softmax loss function for each scale stream. The results of experiment are demonstrated in Table 1.

Merged with Pooling in Table 1 means adopting pooling operation as fusion approach for multi-scale stream instead of attention model. It can be seen that our method surpasses baseline and attention-to-scale by 6.58% and 3.24% respec-

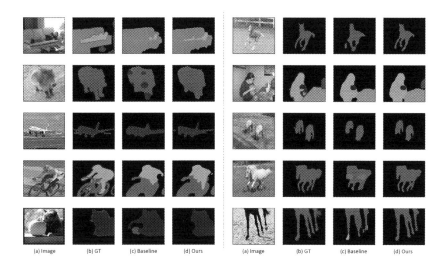

Fig. 3. Representative visual segmentation results on PASCAL VOC 2012 dataset. Images are from train and val set. GT means ground truth, and baseline means attention-to-scale approach. Our proposed method produces more accurate and detailed results.

tively. Furthermore, we conduct additional experiments for ablation study of each module in our method. We cut off certain modules from our proposed method, re-train and report the performance of remainder, which is shown in Table 2. Please noted that base model without all these modules is actually attention-to-scale approach. As you can see, the modules we design indeed take effect on segmentation task.

Since the attention-to-scale has verified the motivation which we share with by visualizing weight maps produced by the attention model, we don't replicate this experiment on our proposed model. Turning to qualitative results, some representative visual comparisons are provided between attention-to-scale and our method in Fig. 3. We observe that unlike attention-to-scale, our method can get finer contour in some cases and probability of breaking down a large-scale object into several pieces decreases. Our results contain much more detailed structure and more accurate pixel-level categorization, which we posit it comes from the utilization of multi-scale and context information as well as the extra branch.

3.2   ADE20K

ADE20K dataset first shows up in ImageNet scene parsing challenge 2016. It is much more challenging since it has 150 labeled classes for both objects and background scene parsing. It contains around 20K and 2K images in the training and validation sets respectively.

We deploy ResNet34-dilated8 [20] (not resnet50 because of limited GPU memory) as base CNN to investigate several different methods. Besides applying attention-to-scale and our proposed attention model, we also experiment on Pyramid Scene Parsing (PSP) [21] module as a comparison, which is a state-of-the-art approach on ADE20K dataset to the best of our knowledge. The experiment results are presented in Table 3. The PSP here doesn't contain auxiliary loss in original paper. We can see that our proposed attention model outperforms other methods, and achieves 4.40% improvement on mIoU over baseline. Besides, we also embed both the PSP module and proposed attention module in baseline and it obtains further performance improvement.

Table 3. Results on ADE20K validation set. *- Two multi-scale attention methods take as input two scale streams: 1x and 0.5x.

Method	mIoU	Pixel accuracy
ResNet34-dilated8 (Baseline)	32.67%	76.41%
Baseline + attention-to-scale*	35.11%	76.82%
Baseline + PSP	36.43%	78.01%
Baseline + our attention model*	**37.07%**	78.57%
Baseline + our attention model* + PSP	**38.21%**	79.29%

4   Conclusion

In this paper, we propose a novel attention model for semantic segmentation. The whole CNN framework takes in multi-scale streams as input. Features from different stage of CNN are fused, then resulting one in each scale goes through convolutional layers with different dilation, which are expected to catch distinctive context relationship for different scales. After that, all these features get concatenated and resulting one is fed into two parallel convolution output branches of the attention model. One of the branches is location attention, aiming to pay soft attention to each location across channels. Another one is designed to fully utilize contextual information to deal with class ambiguity by recalibrating the prediction per location for each class. Experiments on PASCAL VOC 2012 and ADE20K show that proposed method make a significant improvement.

References

1. Arnab, A., Jayasumana, S., Zheng, S., Torr, P.H.S.: Higher order conditional random fields in deep neural networks. In: Leibe, B., Matas, J., Sebe, N., Welling, M. (eds.) ECCV 2016. LNCS, vol. 9906, pp. 524–540. Springer, Cham (2016). https://doi.org/10.1007/978-3-319-46475-6_33
2. Badrinarayanan, V., Kendall, A., Cipolla, R.: SegNet: a deep convolutional encoder-decoder architecture for image segmentation. IEEE Trans. Patt. Anal. Mach. Intell. **39**(12), 2481–2495 (2017)
3. Bahdanau, D., Cho, K., Bengio, Y.: Neural machine translation by jointly learning to align and translate. arXiv preprint arXiv:1409.0473 (2014)
4. Chen, J., Zhang, H., He, X., Nie, L., Liu, W., Chua, T.S.: Attentive collaborative filtering: multimedia recommendation with item-and component-level attention. In: Proceedings of the 40th International ACM SIGIR Conference on Research and Development in Information Retrieval, pp. 335–344. ACM (2017)
5. Chen, L.C., Papandreou, G., Kokkinos, I., Murphy, K., Yuille, A.L.: Semantic image segmentation with deep convolutional nets and fully connected CRFs. arXiv preprint arXiv:1412.7062 (2014)
6. Chen, L.C., Papandreou, G., Kokkinos, I., Murphy, K., Yuille, A.L.: DeepLab: semantic image segmentation with deep convolutional nets, atrous convolution, and fully connected CRFs. IEEE Trans. Patt. Anal. Mach. Intell. **40**(4), 834–848 (2018)
7. Chen, L.C., Yang, Y., Wang, J., Xu, W., Yuille, A.L.: Attention to scale: scale-aware semantic image segmentation. In: Proceedings of the IEEE Conference on Computer Vision and Pattern Recognition, pp. 3640–3649 (2016)
8. Everingham, M., Van Gool, L., Williams, C.K.I., Winn, J., Zisserman, A.: The PASCAL Visual Object Classes Challenge 2012 (VOC2012) Results. http://www.pascal-network.org/challenges/VOC/voc2012/workshop/index.html
9. Hariharan, B., Arbeláez, P., Bourdev, L., Maji, S., Malik, J.: Semantic contours from inverse detectors. In: 2011 IEEE International Conference on Computer Vision (ICCV), pp. 991–998. IEEE (2011)
10. Hariharan, B., Arbeláez, P., Girshick, R., Malik, J.: Hypercolumns for object segmentation and fine-grained localization. In: Proceedings of the IEEE Conference on Computer Vision and Pattern Recognition, pp. 447–456 (2015)

11. Li, X., et al.: FoveaNet: perspective-aware urban scene parsing. In: Proceedings of the IEEE International Conference on Computer Vision, pp. 784–792 (2017)
12. Liu, W., Rabinovich, A., Berg, A.C.: ParseNet: Looking wider to see better. arXiv preprint arXiv:1506.04579 (2015)
13. Long, J., Shelhamer, E., Darrell, T.: Fully convolutional networks for semantic segmentation. In: Proceedings of the IEEE Conference on Computer Vision and Pattern Recognition, pp. 3431–3440 (2015)
14. Noh, H., Hong, S., Han, B.: Learning deconvolution network for semantic segmentation. In: Proceedings of the IEEE International Conference on Computer Vision, pp. 1520–1528 (2015)
15. Simonyan, K., Zisserman, A.: Very deep convolutional networks for large-scale image recognition. arXiv preprint arXiv:1409.1556 (2014)
16. Song, X., Feng, F., Han, X., Yang, X., Liu, W., Nie, L.: Neural compatibility modeling with attentive knowledge distillation. arXiv preprint arXiv:1805.00313 (2018)
17. Wang, F., et al.: Residual attention network for image classification. In: Proceedings of the IEEE Conference on Computer Vision and Pattern Recognition, pp. 3156–3164 (2017)
18. Xia, F., Wang, P., Chen, L.-C., Yuille, A.L.: Zoom better to see clearer: human and object parsing with hierarchical auto-zoom net. In: Leibe, B., Matas, J., Sebe, N., Welling, M. (eds.) ECCV 2016. LNCS, vol. 9909, pp. 648–663. Springer, Cham (2016). https://doi.org/10.1007/978-3-319-46454-1_39
19. Xu, K., et al.: Show, attend and tell: Neural image caption generation with visual attention. In: International Conference on Machine Learning, pp. 2048–2057 (2015)
20. Yu, F., Koltun, V.: Multi-scale context aggregation by dilated convolutions. arXiv preprint arXiv:1511.07122 (2015)
21. Zhao, H., Shi, J., Qi, X., Wang, X., Jia, J.: Pyramid scene parsing network. In: Proceedings of the IEEE Conference on Computer Vision and Pattern Recognition, pp. 2881–2890 (2017)
22. Zheng, S., et al.: Conditional random fields as recurrent neural networks. In: Proceedings of the IEEE International Conference on Computer Vision, pp. 1529–1537 (2015)
23. Zhou, B., Zhao, H., Puig, X., Fidler, S., Barriuso, A., Torralba, A.: Semantic understanding of scenes through the ADE20K dataset. arXiv preprint arXiv:1608.05442 (2016)
24. Zhou, B., Zhao, H., Puig, X., Fidler, S., Barriuso, A., Torralba, A.: Scene parsing through ADE20K dataset. In: Proceedings of the IEEE Conference on Computer Vision and Pattern Recognition (2017)

Frame Segmentation Networks
for Temporal Action Localization

Ke Yang[✉], Peng Qiao, Qiang Wang, Shijie Li, Xin Niu, Dongsheng Li,
and Yong Dou

National Laboratory for Parallel and Distributed Processing College of Computer,
National University of Defense Technology, Changsha, China
yangke13@nudt.edu.cn

Abstract. Temporal action localization is an important task of computer vision. Though many methods have been proposed, it still remains an open question how to predict the temporal location of action segments precisely. Most state-of-the-art works train action classifiers on video segments pre-determined by action proposal. However, recent work found that a desirable model should move beyond segment-level and make dense predictions at a fine granularity in time to determine precise temporal boundaries. In this paper, we propose a Frame Segmentation Network (FSN) that places a temporal CNN on top of the 2D spatial CNNs. Spatial CNNs are responsible for abstracting semantics in spatial dimension while temporal CNN is responsible for introducing temporal context information and performing dense predictions. The proposed FSN can make dense predictions at frame-level for a video clip using both spatial and temporal context information. FSN is trained in an end-to-end manner, so the model can be optimized in spatial and temporal domain jointly. Experiment results on public dataset show that FSN achieves superior performance in both frame-level action localization and temporal action localization.

Keywords: Temporal action localization
Convolutional neural network · Frame-level action prediction

1 Introduction

In recent years, temporal action localization has been extensively studied by researchers in computer vision. A lot of works have been tried to solve this problem [6,9,10,14,15,25], but how to perform temporal action localization precisely is still an open question. Temporal action localization aims to detect action instances in the untrimmed videos, including their temporal boundaries and categories. Most works adopt the detection by classification framework which

This work was supported by the National Basic Research Program of China (973) under Grant No.2014CB340303 and the National Natural Science Foundation of China under Grants U1435219, 61402507 and 61572515.

R. Hong et al. (Eds.): PCM 2018, LNCS 11165, pp. 242–252, 2018.
https://doi.org/10.1007/978-3-030-00767-6_23

is widely used in object detection task [7]. First, action segment proposals are generated by action proposal methods or sliding windows. Then various features are extracted on action segment proposals and action classifiers are trained on these extracted features.

A recent work claimed that action prediction at a finer temporal granularity contributes to more precise temporal action localization results [14]. This finding encourages us to perform action prediction at a fine granularity rather than at segment-level. To achieve this goal, some techniques can be adapted: (1) Recurrent Neural Network (RNN); (2) 3D CNN; (3) 2D CNN; In [24], a Long Short Term Memory (LSTM) network is proposed to model these temporal relations via multiple input and output connections. However, it is claimed that RNN will introduce temporal smoothing that is harmful to precise temporal localization task in [14,25]. In [14], 3D CNN is reformed to accomplish this goal. 3D CNN is designed to classify a whole video clip. To perform frame-level predictions, a Convolutional-De-Convolutional (CDC) layer is developed to upsample the temporal resolution. However, 3D CNN's model parameters increase significantly relative to 2D CNN and 3D CNN is much more data hungry than 2D CNN [2]. The increase in the number of parameters makes the computational resource and training time consumption increase significantly. Meanwhile, there are very few pre-trained 3D CNN models available [19].

In order to reduce the number of parameters while modeling spatio-temporal information, a valid method is to decompose the spatial and temporal dimensions of 3D CNN. *We propose to combine 2D spatial CNNs and 1D temporal CNN instead of 3D CNN to model spatio-temporal information.* The 1D temporal CNN is placed on top of the 2D CNNs. We have also considered separable 3D convolution for space and time, but this operation changes the internal structure of the network and might invalidate the application of pre-training model weights which is important for relatively small dataset.

In recent years, due to the rapid development of image recognition, 2D CNN has been developed by leaps and bounds, deeper and deeper networks with stronger capacity are being proposed [8,17]. These state-of-the-art 2D CNN models pre-trained on ImageNet [4] can be transferred to action recognition with a small computational and time cost [16,22]. 2D CNN has already been able to model spatial information successfully. However, 2D CNNs classify each frame using 2D CNN without consideration of temporal information which is important for video understanding. Therefore, we consider stacking a 1D temporal CNN on top of a 2D CNN to model temporal information. As shown in Fig. 1, 2D CNNs take single images as input and model the spatial information. *All these 2D CNNs share the weights.* A 1D temporal CNN is placed on top of the 2D CNN. *The temporal CNN takes a sequence of feature vectors from 2D CNN and outputs dense predictions of each input frame through a single pass.* This forms our Frame Segmentation Network (FSN). FSN allows the model to take multiple video frames as input and output predictions for every input frame. And we can easily control the temporal receptive field size by setting different kernel size and step size for the temporal CNN. FSN can be trained in an end-to-end manner.

Our contributions can be concluded as follows: (1) Our FSN can make dense predictions with temporal context information and can be trained in an end-to-end manner. (2) FSN achieves competitive results in both per-frame action localization and segment-level action localization. It is also worth noting that FSN can be easily updated by simply changing the 2D CNN to a more powerful one, allowing FSN to benefit from the progress of image classification network.

2 Related Work

Action Recognition: For a quite long period of time, the conventional features such as Improved Dense Trajectory Feature (iDTF) [20] were in a dominant position in the field of action recognition. In recent years, thanks to ImageNet dataset [4], Convolutional Neural Networks (CNN) such VGG [17], ResNet [8] have gradually been proposed and adapted to perform action recognition, but the performance is still poor since they can only capture appearance information. To model motion information in videos, various two-stream CNNs which take both RGB images and optical flow as input have significantly improve the action recognition performance and surpass the conventional features [16,22]. To explicitly model spatio-temporal feature directly from raw videos, a 3D CNN architecture called C3D is proposed in [19].

Temporal Action Localization: A typical framework used in many state-of-the-art temporal action localization systems [12,15] is detection by classification framework, which is borrowed from object detection task. First, various features are extracted on the action segments pre-determined by action proposals. Then action classifiers are trained on these features to classify these action segments. In order to design a model specific to temporal localization, in [13], statistical length and language modeling are used to represent temporal and contextual structures. In [1] a sparse learning framework is proposed to retrieve action segment proposals of high recall.

In recent years, deep neural networks are used widely to improve performance of temporal localization. In [25], a Long Short Term Memory (LSTM)-based agent is trained using REINFORCE to learn both which frame to look in the next step and when to emit an action segment output. In [6], a LSTM based framework is designed to take pre-extracted CNN features to output temporal action proposal. In [24], a MultiLSTM network is proposed to model these temporal relations via multiple input and output connections. In [27], a Pyramid of Score Distribution Feature (PSDF) is proposed and PSDF is taken as input into the RNN to improve temporal consistency. A novel Single Shot Action Detector (SSAD) network is proposed to skip the proposal generation step via directly detecting action instances in untrimmed video [10]. In [19], an end-to-end framework named Segment-CNN (S-CNN) is proposed to perform action localization via multi-stages 3D CNNs. Convolutional-De-Convolutional (CDC) is proposed to perform action predictions in every frame in [14] and then the frame-level action predictions are used to refine the action segment boundaries from S-CNN to generate more precise segment-level predictions.

3   Frame Segmentation Networks

3.1   Motivation of Frame Segmentation Networks

The 3D CNN is naturally suitable for encoding video data, but the amount of parameters of 3D CNN is too large, making the training process consume a lot of computing resources and time. For example, a 11-layer 3D CNN [19] has about 80M parameters while a 50-layer ResNet [8] only has about 25M parameters. To reduce the model parameter number, we turn to model spatio-temporal information using a combination of 2D spatial CNNs and 1D temporal CNN rather than 3D CNN. At the same time, state-of-the-art 2D CNNs, such as VGG [17] and ResNet [8], have been adapted for action recognition task and get the state-of-the-art results [22], we can use any one of them to initialize our 2D CNN. And with the 1D temporal CNN, we can predict action class scores at the original temporal resolution rather than output a single video-level label. The training pipeline of FSN is shown in Fig. 1.

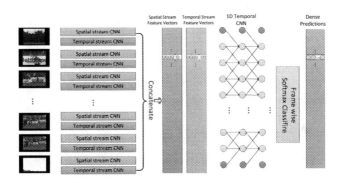

Fig. 1. Overall training pipeline of FSN. Our FSN takes a sequence of video frames as input. Then a sequence of feature vectors are extracted for every input video frame. Then the feature vectors sequence are passed to a 1D temporal CNN and output a sequence of scores of the same time length. Next, these scores are passed through a frame wise softmax classifier to output the dense action predictions for every input frame. This figure is for illustrative purpose only, we omit the upsampling layer and the specific structure of temporal CNN. Please zoom in for better viewing.

3.2   Feature Extraction

First of all, video clip frames are fed into a deep 2D CNN for feature extraction. A video clip V contains T frames can be divided into N consecutive snippets. We choose the center frame of each snippet $C = \{c_i\}_{i=1}^{N}$ to represent the snippet. When snippet's length is 1, the center frame is the snippet itself.

Each center frame is processed by a 2D CNN to extract the representation as $f_{c_i} = \phi(V; c_i) \in \mathbb{R}^D$. Then all $\phi(V; C)$ of the center frames are concatenated together to $f_{con} \in \mathbb{R}^{N \times D}$. Then f_{con} is fed to a 1D temporal CNN.

Our FSN does not depend on the choice of 2D CNN feature extraction model. In our experiments, a two stream network like network for action recognition: Temporal Segment Network (TSN) [22] is investigated due to its superior performance on action recognition task. Feature vectors after global pooling layer are used.

3.3   Temporal CNN

Representation of the input video clip f_{con} is a simple concatenation of feature vectors of multiple frames. A desirable model is that can take temporal context information to predict each frame of input video through a single pass. The temporal CNN enables the entire model to see the temporal context when predict the current frame. The temporal CNN is a 1D CNN with L layers. Following network architecture design of classic CNN model VGG [17], all the kernel sizes are set to 3. All kernels' steps are set to 1, since dense predictions needs the model to preserve the time length. Convolutional kernels with step size of 1 raise a problem that the temporal receptive field size will be too small, but too small temporal receptive field is harmful to the prediction precision. To solve this problem, we adapt spatial dilated convolution [26] to *temporal dilated convolution* to enlarge the temporal receptive field. Temporal dilated convolution is used in all but the first layers of temporal CNN. Temporal CNN has three convolution layers except the classification layer. Dilated rate is set to 1,2 and 4 respectively from first to last convolution layers.

Bilinear Upsampling and Classifier. The output of temporal CNN is $X = \{x_i\}_{i=1}^{N}$. $x_i \in \mathbb{R}^{K+1}$ is the score vector of the ith snippet. N is snippet number. K is the action category number. $(K + 1)$ dimension score vector corresponds to K action categories and 1 background category. FSN needs to perform frame-level predictions, thus when the snippet length is larger than 1 frame, we need to upsample the number of score vector to the frame number of input video clip. We choose bilinear upsampling which has no parameter following [11]. $X \in \mathbb{R}^{N \times (K+1)}$ is upsampled to $X_{up} \in \mathbb{R}^{T \times (K+1)}$ by $X_{up} = BilinearUpsampling1D(X)$. T is the frame number of input video clip as described in Sect. 3.2. Then the upsampled feature vectors X_{up} are passed through a frame wise softmax layer as follows:

$$\overline{x}^j_{up\ i} = \frac{exp\left(x^j_{up\ i}\right)}{\sum_{k=1}^{K+1} exp\left(x^k_{up\ i}\right)} \tag{1}$$

where $x^j_{up\ i}$ denotes the jth dimension of the ith input frame's score vector $x_{up\ i}$. $\overline{x}^j_{up\ i}$ denotes the jth dimension of the ith input frame's softmax score vector $\overline{x}_{up\ i}$. We have presented the pipeline of FSN. It is clearly that FSN can be trained in an end-to-end manner.

3.4   Model Training and Prediction

Training Data Construction of FSN. Training data of FSN consists of video clips with temporal length T frames. T can be an arbitrary value since temporal CNN is a 1D fully convolutional network. Following [14], and considering that the temporal stream network of Temporal Segment Network [22] takes 5 adjacent optical flows as input, we set T to 35 frames which is a multiple of 5. Therefore, the snippet number N for FSN is $35/5 = 7$. We slide temporal window of length T on the videos and only keep the segments include at least 5 frames belongs to action instances. We also re-sample the segments to get a balance training dataset.

Model Training. We implement FSN based on Keras [3] and Temporal Segment Network (TSN) [22]. For experiments on THUMOS'14, we first finetune TSN on UCF101. For FSN, the *frame wise* cross-entropy loss $\mathcal{L}$ is as follows:

$$\mathcal{L} = \frac{1}{B} \sum_{b=1}^{B} \sum_{t=1}^{L} \sum_{k=1}^{K+1} \left(-y_n^{(k)}[t] \log \left(\frac{\exp\left(O_n^{(k)}[t] \right)}{\sum_{j=1}^{K+1} \exp\left(O_n^{(j)}[t] \right)} \right) \right) \tag{2}$$

where B denotes batch size, L denotes video length, K denotes the action category. We use Stochastic Gradient Descent (SGD) to train FSN network. We train all layers of FSN with learning rate 0.0001 and mini-batch size 12. We set momentum to 0.9 and weight decay to 0.0005. Training iteration is about 60000.

FSN Model Prediction. First, we introduce frame-level action predictions. During test, we slide the FSN on the whole untrimmed videos without temporal overlapping. We get action predictions of every frame in the test set.

Then, we introduce segment-level action localization predictions. We follow [23] to get the segment-level results. After frame-level action predictions, we have the predictions of all frames. Then we can generate segment-level action predictions by grouping frame-level action scores. First, we take threshold processing on classification scores and we get a string of "0" and "1" (0 stands for background frame, and 1 inversely). Then we group adjacent frames of "1" to get segment-level results. Thresholds are uniformly selected from 0 to 1 with an interval of 0.1.

4   Experiments

4.1   Dataset

We evaluate FSN network on the challenging dataset THUMOS'14 [9].

THUMOS'14 Dataset. THUMOS'14 has 101 action classes. Training set is directly taken from UCF101 dataset [18]. Validation set consists of 1010 untrimmed videos. Test set consists of 1574 untrimmed videos. Temporal action detection task in THUMOS'14 challenge is dedicated to localize the action instances in untrimmed video and involves 20 action classes. And there are 200 videos in validation set and 213 videos in test set that contain the action instances of these 20 classes. We train FSN on the 200 validation videos and test on the 213 test videos. On THUMOS'14, We evaluate FSN network on both frame-level action localization and segment-level action localization tasks.

4.2 Frame-Level Action Localization

First, we evaluate FSN network in predicting action labels for every frame in the whole video. This task can take multiple frames as input to take into account temporal information. Following [24], we evaluate frame-level prediction as a retrieval problem. For each action class, we rank all the frames in the test set by their confidence scores and compute Average Precision (AP) for this class. And mean AP (mAP) is computed by average the AP of 20 action classes.

In Table 1, we compare our FSN network with state-of-the-art methods. All the results are quoted from [14,24]. Single-frame CNN stands for frame-level VGG-16 2D CNN model proposed by [17]. Two-stream CNN [16] is the frame-level 2D CNN model which takes both optical flow and RGB images as input. LSTM stands for the 2D CNN + LSTM model [5]. MultiLSTM represents the LSTM with temporal attention mechanism [24]. CDC denotes the convolutional-de-convolutional network proposed in [14]. We denote our FSN network as **FSN**. Single-frame CNN only takes into account appearance information in a single frame. Two-stream CNN takes both appearance information in a single frame and motion information from six adjacent frames as input. LSTM and MultiL-STM can utilize temporal information to make frame-level predictions. CDC is based on 3D CNN, can model spatio-temporal information and make dense predictions by upsampling. Our FSN use 2D CNN to abstract spatial semantics, and use a temporal CNN to pursue temporal context information for dense predictions for each input frame. FSN achieves significant performance improvement relative to other methods. We also report performance of each single stream network in Table 1.

Table 1. Frame-level action localization mAP on THUMOS'14.

Method	mAP
Single-frame CNN [17]	34.7
Two-stream CNN [16]	36.2
LSTM [5]	39.3
MultiLSTM [24]	41.3
CDC [14]	44.4
FSN RGB	47.5
FSN Flow	41.4
FSN	**53.5**

4.3 Temporal Action Localization

Given frame-level action predictions, we can get segment-level action localization results using various strategies. As described in Sect. 3.4, we use multiple threshold frame grouping method to obtain the segment-level localization results. Finally, we perform post-processing steps such as non-maximum suppression (NMS). NMS IoU threshold is 0.1 lower than the IoU threshold used during the evaluation. We evaluate our model on THUMOS'14 dataset.

We perform evaluation using mAP as frame-level action localization evaluation. For each action class, we rank all the predicted segments by their confidence results and calculate the AP using official evaluation code. One prediction is correct when its temporal overlap intersection-over-union (IoU) with a ground truth action segment is higher than the threshold, so evaluation under various IoU threshold is necessary. We evaluate our model under IoU threshold from 0.3 to 0.7 following most works [14,15,25]. Results are shown in Table 2, our model denoted as **FSN** achieves competitive results compared to current state-of-the-art results.

Table 2. Segment-level action localization mAP on THUMOS'14. IoU threshold values are ranged from 0.3 to 0.7. '-' in the table indicates that results of that IoU value are not available in the corresponding papers. Some of the results were not reported in the published papers, we contacted the authors for these results.

IoU threshold	0.3	0.4	0.5	0.6	0.7
Wang et al. [21]	14.6	12.1	8.5	4.7	1.5
Heilbron et al. [1]	-	-	13.5	-	-
Escorcia et al. [6]	-	-	13.9		-
Oneata et al. [12]	28.8	21.8	15.0	8.5	3.2
Richard and Gall [13]	30.0	23.2	15.2	-	-
Yeung et al. [25]	36.0	26.4	17.1	-	-
Yuan et al. [27]	33.6	26.1	18.8	-	-
S-CNN [15]	36.3	28.7	19.0	10.3	5.3
CDC [14] + S-CNN [15]	40.1	29.4	23.3	13.1	7.9
SSAD [10]	43.0	35.0	24.6	15.4	7.7
TPC [23]	44.1	37.1	28.2	20.6	12.7
FSN RGB	40.7	33.6	24.9	17.5	10.6
FSN Flow	36.4	28.0	20.0	11.9	6.2
FSN	**51.8**	**41.5**	**32.1**	**22.9**	**14.7**

4.4 Discussions

In this subsection, for purpose of saving time, we only use RGB stream network of TSN.

Table 3. Segment-level action localization mAP on THUMOS'14. IoU threshold values are ranged from 0.3 to 0.7.

IoU threshold	0.3	0.4	0.5	0.6	0.7
RGB FSN w/o temporal CNN	38.2	27.6	19.7	12.2	6.9
RGB FSN	**40.7**	**33.6**	**24.9**	**17.5**	**10.6**

The Necessity of Modeling Temporal Information. *TSN can also perform dense predictions via proper adaptation, do we really need the temporal CNN to model temporal information?* To answer this question, we design an ablation experiment. The temporal CNN is replaced by a 1D convolution layer of which both the kernel size and step size are 1. The number of the convolutional layer's output nodes is $K + 1$, K is the number of action categories. The convolutional layer is followed by a upsampling layer and frame wise softmax layer that are the same as FSN. We denote this model as **FSN w/o Temporal CNN**. FSN w/o Temporal CNN can only see a single frame when it predict a frame, since the kernel size is 1. We train FSN w/o Temporal CNN and our FSN with the same training set and evaluate them using the same metric. RGB FSN w/o Temporal CNN's frame-level performance is 42.5 (mAP), while our RGB FSN's is 47.5 as shown in Table 1. The segment-level experiment results are shown in Table 3. These results are reasonable since in temporal action localization task, only looking at current frame is very likely to make false predictions. A bigger

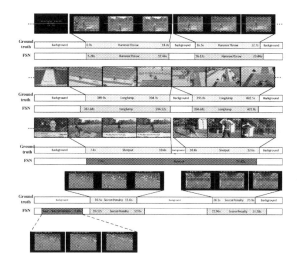

Fig. 2. Visualization of temporal localization results FSN. The above first two examples show a few successful cases. The last two examples show a few failed cases. In order to have a good display of results, we did not draw the timeline strictly according to the ratio. Please zoom in for better viewing.

temporal receptive field size helps improve performance. For example, for the *CleanAndJerk* action category, you might make false positive predictions when the jerk failed.

4.5 Qualitative Results

We show qualitative results of FSN on segment-level localization results in Fig. 2. The first two examples are two successful cases. The last two examples includes a few failure case. In the third example, the prediction of FSN merges two instances. In this case, these two instances are too close in time and the time interval is too small. In the fourth example, the predictions of FSN have false positive instances. The common false positive instance is that soccer penalty shooter is walking to the penalty position. This might be because the receptive field size is still not big enough. This encourage us to exploit other techniques such as multiple resolution in future.

5   Conclusion

In this paper, we proposed a Frame Segmentation Network (FSN) that model spatio-temporal information using a combination of 2D spatial CNNs and a 1D temporal CNN for precise temporal action localization. Our FSN achieves competitive performance on both frame-level and segment-level action localization.

References

1. Caba Heilbron, F., Carlos Niebles, J., Ghanem, B.: Fast temporal activity proposals for efficient detection of human actions in untrimmed videos. In: CVPR (2016)
2. Carreira, J., Zisserman, A.: Quo vadis, action recognition? a new model and the kinetics dataset. In: CVPR (2017)
3. Chollet, F., et al.: Keras (2015). https://github.com/fchollet/keras
4. Deng, J., Dong, W., Socher, R., Li, L.J., Li, K., Fei-Fei, L.: ImageNet: a large-scale hierarchical image database. In: CVPR (2009)
5. Donahue, J., et al.: Long-term recurrent convolutional networks for visual recognition and description. In: CVPR (2015)
6. Escorcia, V., Caba Heilbron, F., Niebles, J.C., Ghanem, B.: DAPs: deep action proposals for action understanding. In: Leibe, B., Matas, J., Sebe, N., Welling, M. (eds.) ECCV 2016. LNCS, vol. 9907, pp. 768–784. Springer, Cham (2016). https://doi.org/10.1007/978-3-319-46487-9_47
7. Girshick, R., Donahue, J., Darrell, T., Malik, J.: Rich feature hierarchies for accurate object detection and semantic segmentation. In: CVPR, pp. 580–587 (2014)
8. He, K., Zhang, X., Ren, S., Sun, J.: Deep residual learning for image recognition. In: CVPR (2016)
9. Idrees, H., et al.: The THUMOS challenge on action recognition for videos "in the wild". In: CVIU (2017)
10. Lin, T., Zhao, X., Shou, Z.: Single shot temporal action detection. In: ACM Multimedia (2017)

11. Long, J., Shelhamer, E., Darrell, T.: Fully convolutional networks for semantic segmentation. In: CVPR, pp. 3431–3440 (2015)
12. Oneata, D., Verbeek, J., Schmid, C.: The LEAR submission at THUMOS (2014)
13. Richard, A., Gall, J.: Temporal action detection using a statistical language model. In: CVPR (2016)
14. Shou, Z., Chan, J., Zareian, A., Miyazawa, K., Chang, S.F.: CDC: convolutional-de-convolutional networks for precise temporal action localization in untrimmed videos. In: CVPR (2017)
15. Shou, Z., Wang, D., Chang, S.F.: Temporal action localization in untrimmed videos via multi-stage CNNs. In: CVPR (2016)
16. Simonyan, K., Zisserman, A.: Two-stream convolutional networks for action recognition in videos. In: NIPS (2014)
17. Simonyan, K., Zisserman, A.: Very deep convolutional networks for large-scale image recognition. In: ICLR (2015)
18. Soomro, K., Zamir, A.R., Shah, M.: UCF101: a dataset of 101 human actions classes from videos in the wild. arXiv preprint arXiv:1212.0402 (2012)
19. Tran, D., Bourdev, L., Fergus, R., Torresani, L., Paluri, M.: Learning spatiotemporal features with 3D convolutional networks. In: ICCV (2015)
20. Wang, H., Schmid, C.: Action recognition with improved trajectories. In: ICCV (2013)
21. Wang, L., Qiao, Y., Tang, X.: Action recognition and detection by combining motion and appearance features. In: THUMOS14 Action Recognition Challenge (2014)
22. Wang, L., et al.: Temporal segment networks: towards good practices for deep action recognition. In: Leibe, B., Matas, J., Sebe, N., Welling, M. (eds.) ECCV 2016. LNCS, vol. 9912, pp. 20–36. Springer, Cham (2016). https://doi.org/10.1007/978-3-319-46484-8_2
23. Yang, K., Qiao, P., Li, D., Lv, S., Dou, Y.: Exploring temporal preservation networks for precise temporal action localization. In: AAAI (2018)
24. Yeung, S., Russakovsky, O., Jin, N., Andriluka, M., Mori, G., Fei-Fei, L.: Every moment counts: dense detailed labeling of actions in complex videos. arXiv preprint arXiv:1507.05738 (2015)
25. Yeung, S., Russakovsky, O., Mori, G., Fei-Fei, L.: End-to-end learning of action detection from frame glimpses in videos. In: CVPR (2016)
26. Yu, F., Koltun, V.: Multi-scale context aggregation by dilated convolutions. arXiv preprint arXiv:1511.07122 (2015)
27. Yuan, J., Ni, B., Yang, X., Kassim, A.A.: Temporal action localization with pyramid of score distribution features. In: CVPR (2016)

Stereo Matching Based on Density Segmentation and Non-Local Cost Aggregation

Jianning Du[1,2], Yanbing Xue[1,2(✉)], Hua Zhang[1,2,3], and Zan Gao[1,2]

[1] Key Laboratory of Computer Vision and System (Ministry of Education),
Tianjin University of Technology, Tianjin 300384, China
Jianningdu0217@163.com,
{Xueyb0718,hzhang}@tjut.edu.cn
[2] Tianjin Key Laboratory of Intelligence Computing and Novel Software
Technology, Tianjin University of Technology, Tianjin 300384, China
[3] Tianjin Sino-German University of Applied Sciences, Tianjin 300350, China

Abstract. Recently, segment-tree based Non-Local cost aggregation algorithm, which can provide extremely low computational complexity and outstanding performance, has been proposed for stereo matching. The segment-tree (ST) based method integrated the segmentation information with non-local cost aggregation. However, the segmentation method used in the ST method results in under-segmentation so that some of the edges crossing the boundary will be preserved. On the other hand, pixel-level color information can not represent different patterns (smooth regions, texture and boundaries) well. So, only using the color information to establish the weight function is not enough. We proposed a density information based ST for non-local cost aggregation method. The core idea of the algorithm includes: (1) SLIC based method via density information is used to segment image. This clustering feature (density feature) and over-segmentation method are more suitable for stereo matching. (2) In generating sub-MST and linking all the sub-MSTs, we use density information to establish the weight function. We not only consider the color information but also the density information when establishing the weight formula. Performance evaluations on 31 Middlebury stereo pairs show the proposed algorithm outperforms better than other state-of-the-art aggregated based algorithms.

Keywords: Stereo matching · Non-local · Density · Segmentation
Superpixel

1 Introduction

Stereo matching's main objective is to find the corresponding matching point in different viewpoint images to calculate the disparity and obtain the depth information of the target point by using a geometric method. Stereo matching is an important branch of computer vision. Applications for stereo matching include 3D environment perception and modeling, robot navigation, object tracking and detection, and so on. A stereo algorithm usually consists of four steps [1]: matching cost computation, cost

© Springer Nature Switzerland AG 2018
R. Hong et al. (Eds.): PCM 2018, LNCS 11165, pp. 253–263, 2018.
https://doi.org/10.1007/978-3-030-00767-6_24

aggregation, disparity computation and disparity refinement. In general, the stereo matching algorithms are broadly classified into global and local approaches.

Global algorithms usually make explicit smoothness assumption, and minimize a predefined energy function to obtain optimal results. The problem can be solved by using graph cut [2], dynamic programming [3] and loopy belief propagation [4, 5], among others. Despite the reliable matching results obtained, global algorithms are often time-consuming. Local algorithms compute the matching cost firstly and then perform the cost aggregation to get a locally optimized cost volume. The problem of the cost aggregation step is how to choose optimal local support regions for each pixel. The approach [6] that utilized guided filtering [7] achieved good results very efficiently. However, the selected support regions of these methods are often limited in a pre-defined window.

Recently, Yang proposed a non-local cost aggregation (NLCA) algorithm, which was implemented on a MST structure and based on it to perform a tree-based filtering [8, 9]. This method discarded the local algorithm thinking of the support window and proposed the idea of cost aggregation based on the global MST. However the MST construction of the NLCA algorithm is not perfect because only the color cue is considered in weight function, which leads to some mistakes at the boundary of two objects with similar color distribution.

Mei [10] proposed a segment-tree based algorithm (ST) which introduced segmentation information into non-local cost aggregation framework. Yao [11] improved the structure of ST by building a novel segmentation strategy. The advantage of ST method is introduced a judgment condition that allowed it to take into account the area information of the image when established the structure of MST. However, the two methods [10, 11] do not consider the image region information very rigorously. The segmentation method [12] used in ST often leads to under-segmentation, which makes some wrong edges (edges that crossed the boundary) preserved. In additional, the color difference used to calculate the weight of edge in generating MST is just using the pixel-level information, which may affect the structure of the MST and thus affect the stereo matching results.

Rui [13] proposed an image segmentation method using density information as feature to segment image. He pointed out that in contrast to the pixel values, pixel' density can reflect the pixel regional information in the local region. Moreover, density channel can be utilized as the metric to identify image components (such as smooth regions, texture and boundaries). Therefore, the SLIC based method to segment the image via the proposed density channel is used in this paper. This can overcome some of the weaknesses of under-segmentation and use a more robust feature (density feature) for image segmentation. On other hand, traditional weight function just uses the pixel-level information (color difference) and did not consider regional information. Fortunately, density information can reflect the pixel regional information in the local region, thus density and color information are used to establish the weight function.

The rest of this paper is organized as follows: Sect. 2 is an overview of the workflow of the non-local framework. We introduce the new structure of MST with density segmentation method, new weight function to generate MST in Sect. 3. A discussion of the strategies for constructing different tree structures is also provided in Sect. 3. Experimental results and performance evaluations are shown in Sect. 4. Finally, we draw the conclusions and discuss the future work in Sect. 5.

2 Non-Local Cost Aggregation

Let $C_d^A(p)$ denote the matching cost for pixel p at disparity level d, the non-local aggregated cost $C_d^A(p)$ is computed as a weighted sum of $C_d(q)$:

$$C_d^A(p) = \sum_{q \in I} S(p, q) * C_d(q) \tag{1}$$

where q covers every pixel in image I. This is different from traditional aggregation methods, where q is limited in a local region around p. $S(p, q)$ is a weighting function, which denotes the contribution of pixel q to pixel p.

$$S(p, q) = \exp\left(-\frac{D(p, q)}{\sigma}\right) \tag{2}$$

$S(p, q)$ denotes the supported weight from p to any pixel on a path $P(p, q)$ of the tree; $D(p, q)$ denotes the weights summation on $P(p, q)$, λ is a parameter of user-specified. Yang [8, 9] showed that the aggregated costs for all the pixels can be efficiently computed by traversing the tree structure in two sequential passes. In the first pass, the tree is traced from the leaf nodes to the root node. For a pixel p (root node), its cost values are not updated until all its children have been visited:

$$C_d^{A\uparrow}(p) = C_d(p) + \sum_{q \in Ch(p)} S(p, q) \cdot C_d^{A\uparrow}(q) \tag{3}$$

where $C_d^{A\uparrow}(p)$ denotes the intermediate aggregated costs, and the set $Ch(p)$ contains the children of pixel p. In the second pass, the tree is traversed from top to bottom. Starting from the root node, the aggregated costs are passed to the sub-trees. For a pixel p, its final aggregated costs are determined with its parent as follows:

$$
\begin{aligned}
C_d^A(p) = {}& S(\Pr(p), p) \cdot C_d^A(\Pr(p)) \\
& + \left(1 - S^2(\Pr(p), p)\right) \cdot C_d^{A\uparrow}(p)
\end{aligned} \tag{4}
$$

3 Approach

As mentioned above we can infer that the critical problem is the ST structure, which directly determines the final result of non-local cost aggregation. So the segmentation algorithm plays an important role for constructing the ST structure. Once the tree is constructed, the support weight between a pair of pixels is decided by the distance between them on the tree. It is also essential that the weight function of edges is established with good information. Our work exactly follows the ST framework, and the significant difference is that we employ a different segmentation method: SLIC based image segmentation with density feature. Moreover, the Prim algorithm is used to generate sub-tree for each local region and Kruskal algorithm is used to link these sub-MST to a MST. The weight functions of the edges are also redefined according to the pixel similarity composed by color information and density information.

3.1   SLIC Based Image Segmentation with Density Feature

As Rui [13] says, traditional simple pixel-level features can't effectively distinguish these components since nature images are complicated and generally have many patterns for different types of regions. Therefore, in this section, the SLIC [15] based method is used to segment the image via the proposed density channel [13]. Formulate the pixel density mathematically as:

$$d(p) = \sum_{q \in \aleph_p} d_s(p,q) \cdot d_c(p,q) \tag{5}$$

The similarity of two pixels is formulated by mapping the Euclidean distance to the Gaussian space. The spatial weighted term d_s and the color weighted term d_c are:

$$ds = \exp\left(-\frac{Ds(p,q)^2}{2\delta s^2}\right) \tag{6}$$

$$d_c = \exp\left(-\frac{Dc(p,q)^2}{2\delta c^2}\right) \tag{7}$$

where D_s and D_c denote the Euclidean distance on the spatial space and Lab color space respectively. δs and δc are the variance of Gaussian kernel and control the influence of the spatial and color weight. $\aleph_p$ is the neighboring pixel around pixel p. For computational efficiency, we set $\aleph_p$ to be a size (e.g., (7, 7)). δs and δc are equal to 150, 90 in the experiment. Aggregate pixels based on the similarity between cluster centers and neighboring pixels is formulated as:

$$d(p,q) = |d(p) - d(q)| \tag{8}$$

Figure 1 shows maps produced by the GS04 [12] method and SLIC based image segmentation with density feature (writing as D-SLIC), respectively. We can see that GS04's, in Fig. 1(a), under-segmentation error (the yellow area showed in Fig. 1(a)) is high and only a small part of the color edges belongs to the true depth boundaries. This may cause an object to be split into different parts and different objects to be split into one part, which cause ST structure error. By taking advantage of the over-segmentation results, the superpixels will not cover the tiny elements' boundaries. Moreover, when the disparity of one pixel is wrong, it will have a great influence on the pixels in one segment. When using the over-segmentation method, there are only a few pixels in one segment. Therefore, only a few pixels will be significant affected, which will reduce the error rate of disparity. Fortunately, SLIC based image segmentation with density feature method not only is a kind of over-segmentation method, but also uses the density feature that reflect the pixel regional information. We have tried the SLIC method proposed in [15], but SLIC method could not work well. The SLIC method has a relatively fast speed, so based on this segmentation method, the real-time performance of stereo matching can be better achieved.

(a) GS04   (b)D-SLIC

Fig. 1. (a) Result of GS04. (b) Results of D-SLIC based method. $\delta s = 150$; $\delta c = 60$; Number of superpixel = 900 (Color figure online)

3.2   MST (Minimum Spanning Tree) Generation and Weight Function

In the prior of segmentation, we now need to generate a MST for the entire image for cost aggregation. We use the Prim and Kruskal algorithms to generate the MST. And in different spanning tree processes, color information and density information is combined to establish the weight formula of edge.

MST (Minimum Spanning Tree) Generation. We use the Prim algorithm to generate the sub-MST for each segment. The sub-MST of segment is set up from the cluster center, which can be described as high similarity to local others and as having a relatively large distance from other cluster center. We choose the pixel with the highest density (the point with the highest degree of similarity to the surrounding pixels) in each local region as the center point. Then gradually spread out from the center point to generate sub-MST. Why we set the pixel with the highest density as a starting point to generate sub-MST? The pixel with the highest density represents high similarity with the surrounding pixels. The way of using the Prim algorithm to establish the sub-MST start from the pixel with the highest density is more robust. Then, all sub-MSTs need to link into the final MST for cost aggregation. In this paper, the Kruskal method is used to link all the sub-trees into the final MST.

This algorithm combined with Prim and Kruskal has several advantages: (1) we can generate sub-MST in each local region in parallel. (2) This method decrease time complexity. (3) This step-by-step establishment of a MST can reduce errors that are preserved the edges crossing the depth boundaries.

Weight Function. The two types of edges used in this article are shown in Fig. 2. Fortunately, the density channel is a powerful means to identify different patterns (smooth regions, texture and boundaries), and in contrast to the pixel values, pixel' density can reflect the pixel regional information in the local region. So the density information and color information are combined to build weight function. Because edge2 crosses the segment, the densities of p and q are very small. That means the density difference of p and q is closed to zero that there is no need to consider. In this paper, different weight calculation formulas are used in generating sub-MSTs on each local region and linking these sub-MSTs to a MST. In generating the sub-MST for each local region process, the weights of edges are dependent on color and density difference. For an edge connecting pixel s and r, its weight is rewritten as follows:

$$\omega_{edge1} = \lambda \frac{|\mathrm{I}(s) - \mathrm{I}(r)|}{\Delta \mathrm{I}} + (1 - \lambda) \frac{|d(s) - d(r)|}{\Delta d} \tag{9}$$

For a color image I, $|\mathrm{I}(s) - \mathrm{I}(r)|$ is the maximum value estimated from three RGB channels. And $|d(s) - d(r)|$ is absolute value estimated from density channels. λ is user-defined parameter to adjust the effect of color and density on weight formulas. $\Delta \mathrm{I}$ and Δd are normalization parameter. After the sub-tree is generated, sub-MSTs should be linked to a MST. It means that the edge between s and r crosses the segment. In this process, the weight function is decided as follows:

$$\omega_{edge2} = \frac{|\mathrm{I}(s) - \mathrm{I}(r)|}{\Delta \mathrm{I}} \tag{10}$$

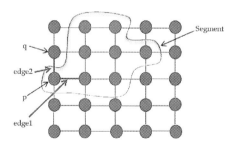

Fig. 2. The two types of edges and the segment (Color figure online)

4   Experimental Results

In this part, we have three sections to show the experimental results. Because there is no improvement for post-processing in this article, so there is no disparity refinement step introduced in all sections. We also set the same error threshold for each disparity map (1.0 pixel). For all the algorithms we have tested the error rates in non-occluded regions. All the algorithms are implemented on a same PC platform with a 3.40 GHz Intel Core i7 CPU, 16 GB RAM and 64-bits OS; furthermore, all the implementations are estimated by using the C++ code. These methods are evaluated on the same 31 stereo pairs. We use 31 Middlebury data sets [14], including the four commonly used data sets, to give a more reliable evaluation of the performance. For our method we set $\{\sigma, k, \lambda\} = \{0.08, 600, 0.55\}$ on Tsukuba, Venus, Teddy and Cones; and other 27 Middlebury stereo pairs $\{\sigma, k, \lambda\} = \{0.04, 1000, 0.55\}$.

4.1 Evaluations on SLIC Based Image Segmentation via Density Feature

This part shows the performance evaluations of the proposed image segmentation method in ST-2 [10], IST-2 [11], and D-S-ST (with SLIC via density feature, the second to last column in Table 1). Except for the image segmentation method, all the steps are same. We present the WTA disparity results of Aloe and Baby1 data sets in Fig. 3. For the reference images (Aloe and Baby1), GS04 [12] segmentation method used in ST-2 [10] and IST-2 [11] produces not well enough (Fig. 3(b, c, f, g)). Figure 3 (d, h) show our segmentation used in non-local cost aggregation improved disparity estimation. We can see that in some of boundaries, our algorithm has been improved (marked by yellow area). The penultimate column of Table 1 shows that this method (D-S-ST) which segmentation method belongs to over-segmentation has a good experimental effect.

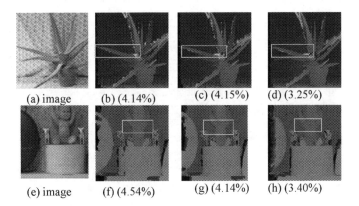

Fig. 3. Comparisons of different segmentation based stereo matching algorithms. (a), (e) reference images. (b), (f) Results of ST-2. (c), (g) Results of IST-2. (d), (h) Results of proposed D-S-ST. All the bad pixels are marked with red dots and the error rates in non-occluded regions are indicated below. Error threshold is 1.0 pixel. (Color figure online)

4.2   Evaluations on Different Weight Function

This part shows the performance evaluations of the different weight function in D-S-ST framework. The benefits of the different edge weight are showed in Fig. 4. For the reference images (bowling2 and books), weight with only color information (D-S-ST) produces not well enough (Fig. 4(b, e)). By including density information (D-S-DST) leads to much improved disparity estimation (Fig. 4(c, f)). When calculating the weights, not only the pixel level information but also the pixel area information are added, so that the error rate of pixel disparity in the un-textured area is reduced. And in some texture-less areas (marked by purple area), good matching is achieved after adding density information. The last column of Table 1 shows that this method (D-S-DST) has a good experimental effect than D-S-ST.

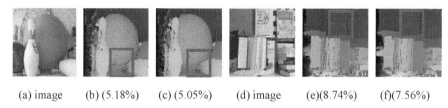

(a) image   (b) (5.18%)   (c) (5.05%)   (d) image   (e)(8.74%)   (f)(7.56%)

Fig. 4. Disparity maps of Bowling2. (a) (d) reference image. (b) (e) D-S-ST with no density. (c) (f) D-S-ST with density. All the error pixels in non-occluded regions are marked with red color and the error rates are indicated below. Error threshold is 1.0 pixel. (Color figure online)

4.3   Evaluations on Other Methods and the Proposed Method

In this section, seven non-local based cost aggregation methods are evaluated with various stereo data sets: D-S-DST (with new segmentation and new weight function), D-S-ST (with new segmentation), Segment-Tree (ST-2) [10], improved Segment-Tree (IST-2) [11] and MST-CD2 [16] and Cross-Scale: CS-ST and CS-Guided Filter (CS-GF) [17]. Because this paper is based on non-local stereo matching, the experiments we want to compare are also based on non-local stereo matching methods. These methods are proposed based on non-local stereo matching methods in recent years.

Table 1 shows a comparison of the quantitative evaluation results produced by stereo matching algorithms with the near-real time computation performance. Among the 31 Middlebury stereo pairs, D-S-DST ranks 1 of 17/31 stereo pairs (marked with fonts in corresponding row). We can draw that the method proposed provides better results than any of the other methods. The best numerical results are marked in bold.

For visual comparison, we present the WTA disparity results (without post-processing) of Teddy, Venus, Flowerpots, Dolls data sets in Fig. 5. Compared to other methods, our algorithm improved the results with the density segmentation information and weights of edges composed by color information and density information, especially in large texture-less regions and around depth boundaries. Although the running time increases, our algorithm still performs shorter time than CS-ST and CS-GF algorithms. Especially for CS-GF, it performs the longest implementing time.

Table 1. Performance evaluations (error rate: %) without disparity refinement on 31 Middlebury stereo pairs by seven algorithms

Stereo Pairs	CS-GF	ST-2	CS-ST	MST-CD2	IST-2	D-S-ST	D-S-DST
Tsukuba	2.35_5	1.65_2	1.90_3	2.36_6	$\mathbf{1.49_1}$	2.47_7	2.06_4
Venus	1.26_5	0.92_4	2.63_6	0.64_2	0.73_3	0.64_2	$\mathbf{0.55_1}$
Teddy	6.87_2	7.48_6	$\mathbf{5.53_1}$	7.31_5	7.58_7	6.97_3	7.11_4
Cones	$\mathbf{3.19_1}$	3.50_4	4.55_7	3.56_5	3.62_6	3.24_2	3.24_2
Aloe	5.61_7	4.51_5	4.90_6	3.79_3	4.15_4	$\mathbf{3.25_1}$	3.26_2
Art	9.20_3	10.99_7	10.46_6	9.70_5	9.35_4	8.79_2	$\mathbf{8.22_1}$
Baby1	3.98_2	4.79_6	4.43_7	5.83_6	4.06_3	4.21_4	$\mathbf{3.23_1}$
Baby2	$\mathbf{4.17_1}$	16.32_7	15.15_6	11.20_4	15.05_5	6.50_3	4.72_2
Baby3	5.68_6	5.27_5	6.27_7	5.22_4	4.23_3	3.40_2	$\mathbf{3.27_1}$
Books	8.74_2	9.57_6	9.84_7	9.08_5	9.06_4	8.74_2	$\mathbf{7.71_1}$
Bowling1	13.03_3	16.48_6	22.11_7	15.46_5	13.91_4	8.86_2	$\mathbf{6.75_1}$
Bowling2	6.62_3	10.29_6	10.91_7	8.52_5	6.68_4	5.18_2	$\mathbf{5.05_1}$
Cloth1	1.31_7	0.41_5	0.77_6	0.37_4	0.32_3	$\mathbf{0.16_1}$	0.18_2
Cloth2	3.52_5	3.55_6	4.06_7	2.59_4	1.99_3	1.93_2	$\mathbf{1.81_1}$
Cloth3	2.26_6	1.69_5	2.60_7	1.54_4	1.51_3	0.98_2	$\mathbf{0.88_1}$
Cloth4	1.75_7	1.13_3	1.69_6	1.14_4	1.19_5	0.85_2	$\mathbf{0.66_1}$
Dolls	4.95_5	5.92_7	5.65_6	4.88_3	4.94_4	$\mathbf{4.46_1}$	4.58_2
Flowerpots	12.22_3	16.02_6	15.77_5	16.20_7	13.21_4	5.73_2	$\mathbf{5.25_1}$
Lampshade1	9.47_2	10.79_7	10.63_6	10.40_4	10.42_5	9.78_3	$\mathbf{8.19_1}$
Lampshade2	16.53_4	21.17_7	15.35_3	20.11_6	18.46_5	9.75_2	$\mathbf{9.45_1}$
Laundry	13.12_4	13.04_2	14.69_5	13.11_3	14.94_7	13.31_6	$\mathbf{12.76_1}$
Midd1	36.92_7	32.97_6	26.47_3	27.87_4	29.10_5	$\mathbf{23.89_1}$	24.68_2
Midd2	33.33_5	32.44_4	$\mathbf{24.56_1}$	34.71_6	40.92_7	27.07_3	26.15_2
Moebius	9.56_7	8.36_5	8.67_6	7.71_3	7.83_4	$\mathbf{7.16_1}$	7.59_2
Monopoly	25.11_7	24.57_6	22.96_3	24.64_5	23.17_4	15.65_2	$\mathbf{14.85_1}$
Plastic	$\mathbf{29.55_1}$	35.79_4	42.32_7	41.50_6	36.72_5	35.34_3	31.39_2
Reindeer	7.23_5	7.01_4	8.33_6	8.47_7	6.49_3	6.30_2	$\mathbf{5.92_1}$
Rock1	2.43_6	2.12_4	2.61_7	2.18_5	1.89_3	1.66_1	$\mathbf{1.66_1}$
Rock2	1.59_6	1.54_5	2.07_7	1.52_4	1.17_2	$\mathbf{1.14_1}$	1.19_3
Wood1	4.30_4	5.17_5	5.69_6	8.45_7	3.43_3	2.93_2	$\mathbf{2.40_1}$
Wood2	2.92_6	2.75_5	6.10_7	$\mathbf{0.94_1}$	1.96_4	1.20_3	0.94_2
Avg. Error	9.15_3	10.25_5	10.27_7	10.00_5	9.58_4	7.46_2	$\mathbf{6.95_1}$
Avg.Rank	4.41_4	5.16_6	5.38_7	4.58_5	4.09_3	2.38_2	$\mathbf{1.58_1}$
Avg.Time	7.68 s	0.87 s	1.70 s	1.64 s	1.33 s	1.48 s	1.56 s

Error threshold is 1.0 pixel and unit of time is second. Only non-occluded regions are evaluated

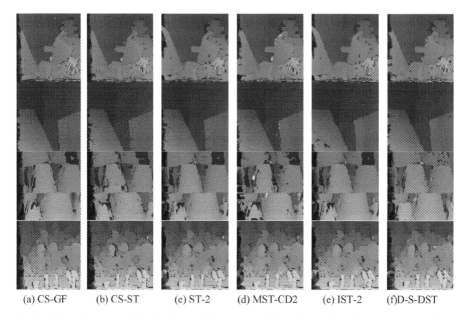

(a) CS-GF   (b) CS-ST   (c) ST-2   (d) MST-CD2   (e) IST-2   (f)D-S-DST

Fig. 5. Disparity maps of Teddy, Venus, Flowerpots, Dolls without disparity refinement ordered from top to bottom rows. The error pixels are marked with red color dots and the error threshold is 1.0 pixel. (Color figure online)

5 Conclusions

In this paper, we improved the ST based stereo algorithm by using a density segmentation algorithm, and we propose new weight function (color information and density information) to generate MST and perform cost aggregation. Performance evaluations show that the proposed algorithm outperforms than other five aggregated based algorithms on 31 Middlebury stereo pairs and time consuming does not increase too much. According to the experimental results, the proposed algorithm provides a better performance in terms of both matching accuracy and computation efficiency than previous methods. In the future, we plan to employ various segmentation algorithms into the ST based stereo algorithmic framework for obtaining superior results. In addition, in this paper, segmenting image and generating the minimum spanning tree is a step-by-step implementation, so in the future we will accomplish the segmenting image and generating the minimum spanning tree at the same time.

Acknowledgment. This research has been supported by National Natural Science Foundation of China (U1509207, 6147227, 61572357 and 61872270) .Tianjin Education Committee science and technology development Foundation (No. 2017KJ254).

References

1. Scharstein, D., Szeliski, R.: A taxonomy and evaluation of dense two-frame stereo correspondence algorithms. Int. J. Comput. Vis. **47**(1–3), 7–42 (2002)
2. Boykov, Y., Veksler, O., Zabih, R.: Fast approximate energy minimization via graph cuts. In: The Proceedings of the Seventh IEEE International Conference on Computer Vision 2002, vol. 1, pp. 377–384 (2002)
3. Veksler, O.: Stereo correspondence by dynamic programming on a tree. In: IEEE Computer Society Conference on Computer Vision and Pattern Recognition, CVPR 2005. vol. 2, pp. 384–390 (2005)
4. Sun, J., Zheng, N.N., Shum, H.Y.: Stereo matching using belief propagation. IEEE Trans. Pattern Anal. Mach. Intell. **25**(7), 787–800 (2003)
5. Yang, Q., Wang, L., Ahuja, N.: A constant-space belief propagation algorithm for stereo matching. In: IEEE Conference on Computer Vision and Pattern Recognition 2010, pp. 1458–1465 (2010)
6. Hosni, A., Rhemann, C., Bleyer, M., Rother, C., Gelautz, M.: Fast cost-volume filtering for visual correspondence and beyond. IEEE Trans. Pattern Anal. Mach. Intell. **35**(2), 504–511 (2001)
7. Tang, X.: Guided image filtering. IEEE Trans. Pattern Anal. Mach. Intell. **35**(6), 1397–1409 (2010)
8. Yang, Q.: A non-local cost aggregation method for stereo matching. In: Computer Vision and Pattern Recognition 2012, vol. 157, pp. 1402–1409 (2012)
9. Yang, Q.: Stereo matching using tree filtering. IEEE Trans. Pattern Anal. Mach. Intell. **37**(4), 834–846 (2015)
10. Mei, X., Sun, X., Dong, W., Wang, H., Zhang, X.: Segment-tree based cost aggregation for stereo matching. In: IEEE International Conference on Acoustics, Speech and Signal Processing 2013, vol. 9, pp. 313–320. IEEE (2013)
11. Yao, P., et al.: Segment-tree based cost aggregation for stereo matching with enhanced segmentation advantage. In: International Conference on Acoustics, Speech and Signal Processing, vol. 10, p. 1109 (2017)
12. Felzenszwalb, P., Huttenlocher, D.: Efficient graph-based image segmentation. Int. J. Comput. Vis. **59**(2), 167–181 (2004)
13. Rui, L., Fang, L.: Cluster sensing superpixel and grouping. In: Computer Vision and Pattern Recognition Workshops 2016, pp. 1350–1358 (2016)
14. Scharstein, D., Szeliski, R.: Middlebury stereo evaluation. http://vision.middlebury.edu/stereo/eval/
15. Achanta, R., Shaji, A., Smith, K., Lucchi, A., Fua, P., Süsstrunk, S.: SLIC superpixels compared to state-of-the-art superpixel methods. IEEE Trans. Pattern Anal. Mach. Intell. **34**(11), 2274–2282 (2012)
16. Yao, P., Zhang, H., Xue, Y., Zhou, M., Xu, G., Gao, Z.: Iterative color-depth MST cost aggregation for stereo matching. In: IEEE International Conference on Multimedia and Expo 2016, pp. 1–6 (2016)
17. Zhang, K., Fang, Y., Min, D., Sun, L., Yang, S., Yan, S., Tian, Q.: Cross-scale cost aggregation for stereo matching. In: Computer Vision and Pattern Recognition 2014, pp. 1590–1597 (2014)

Single Image Super Resolution
Using Local and Non-local Priors

Tianyi Li[1], Kan Chang[1(✉)], Caiwang Mo[1], Xueyu Zhang[1], and Tuanfa Qin[1,2]

[1] School of Computer and Electronic Information, Guangxi University,
Nanning 530004, China
changkan0@gmail.com

[2] Guangxi Key Laboratory of Multimedia Communications and Network
Technologies, Guangxi University, Nanning 530004, China

Abstract. The task of single image super resolution (SR) is to generate a plausible high-resolution (HR) image from a given low-resolution (LR) measurement. This paper presents a reconstruction-based single image SR method. Firstly, an adaptive-shape non-local means (AS-NLM) model is proposed by taking the local structures around pixels into consideration. Afterwards, AS-NLM is utilized to further improve the existing non-local steering kernel regression (NLSKR) model, achieving a new model called I-NLSKR. To obtain superior performance, AS-NLM and I-NLSKR are combined, leading to a new SR algorithm named SRLNP (SR using local and non-local priors). Experimental results demonstrate that SRLNP outperforms many existing methods in both objective and subjective evaluations.

Keywords: Super-resolution · Non-local means
Steering kernel regression · Regularization prior

1 Introduction

Single image super resolution (SR) is used to produce a high resolution (HR) image from a low resolution (LR) input. This task is challenging since we need to recover the missing high frequency information in the HR image. Typically, there are three basic types of methods, including the interpolation-based methods, the learning-based methods and the reconstruction-based methods.

The interpolation-based methods include the bilinear interpolation, the bicubic interpolation, etc. This kind of methods has low complexity yet produces inferior results. For images full of details and edges, obvious artifacts such as ringing and aliasing might exist in the SR output.

Usually, the learning-based methods learn the basic models for the SR task from external datasets. For example, in [13,14], the LR and HR dictionaries are

This work was supported in part by Natural Science Foundation (NSF) of China under Grants 61761005 and 61761007, and in part by the NSF of Guangxi under Grants 2016GXNSFAA380154 and 2016GXNSFAA380216.

R. Hong et al. (Eds.): PCM 2018, LNCS 11165, pp. 264–273, 2018.
https://doi.org/10.1007/978-3-030-00767-6_25

learned offline, and the sparse representations in the LR space and the HR space are assumed to be the same. Due to the fact that solving the l_1 minimization problem is time-consuming, the collaborative representation is applied in [9,10]. Recently, the methods based on deep learning have also been proposed [4,7,12]. However, the deep learning-based methods require parallel techniques such as GPU to perform a long-time training.

The reconstruction-based methods have also attracted much attention recently. Different kinds of prior information are utilized during the SR reconstruction, and the representative priors includes the sparsity-based priors [5,6], the local priors [5,8] and the non-local priors [1,15,16], etc. Note that by combining different types of image priors, some methods [2,3,5,6,17] are able to produce high-quality SR results. For instance, to take advantage of both the local and the non-local image priors, Zhang et al. [17] directly assemble the steering kernel regression (SKR) model [8] and the non-local means (NLM) model [1]. Nevertheless, as the SKR-based image prior and the NLM-based image prior are independently incorporated to the reconstruction framework in [17], the performance of this method is limited.

In this paper, we propose a reconstruction-based single image SR method, the contributions are threefold: Firstly, to enhance the performance of the NLM-based prior, we propose an adaptive shape NLM (AS-NLM) model, where the shape of the neighbourhood of each pixel is no longer fixed. Secondly, the proposed AS-NLM is further used to improve the existing non-local steering kernel regression (NLSKR) [15,16]. Finally, both the AS-NLM and the improved NLSKR models are integrated into a reconstruction-based SR framework, and an algorithm is designed for solving the optimization problem.

The remainder of this paper is organized as follows. The related background is briefly reviewed in Sect. 2. The proposed method is detailed in Sect. 3. Experimental results are presented in Sect. 4 to demonstrate the effectiveness of the proposed method. In Sect. 5, we conclude this paper.

2 Related Work

2.1 Non-Local Means (NLM)

The NLM model, which was first proposed by Buades et al. [1], is a powerful tool for characterizing the non-local structure of nature images. In this model, it is supposed that each pixel in an image can be expressed by the weight sum of its non-local neighbours. Taking the ith pixel as an example, the weight of its lth non-local neighbour can be calculated as

$$w_{i,l}^{N} = \left[\exp \left(-\frac{\|\mathbf{N}_i - \mathbf{N}_l\|_2^2}{h_1} \right) \right] / c_i \tag{1}$$

where h_1 is a parameter which affects the decay of the exponential expression; $\mathbf{N}_i$ and $\mathbf{N}_l$ respectively denote the square image patches centered at the ith and the lth pixels; c_i is a normalization factor for the ith pixel.

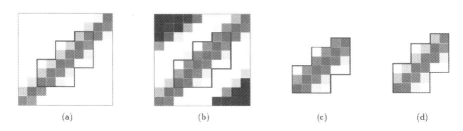

(a)   (b)   (c)   (d)

Fig. 1. Illustration of the benefit of using adaptive-shape patches. The patches (a) and (b) are square. Consequently, the Euclid distance between these two patches is far. The patches (c) and (d) have shapes adapted to the content, so their Euclid distance becomes close.

2.2 Steering Kernel Regression (SKR)

Different from the NLM model, the kernel regression model exploits the relationship between each pixel and its nearby neighbours. The regression function is the solution to the following problem

$$\hat{z}(\mathbf{x}_i) = \arg\min_{z_i} \sum_{j \in Q(\mathbf{x}_i)} [y_j - z(\mathbf{x}_i)]^2 w_{i,j}^{\mathrm{K}} \tag{2}$$

where $Q(\mathbf{x}_i)$ represents the index set of the local neighbours of the pixel at $\mathbf{x}_i$, and $\mathbf{x}_i \in \mathbb{R}^{2 \times 1}$ is a vector containing the horizontal and vertical positions of the ith pixel; y_j stands for the jth measured data; $w_{i,j}^{\mathrm{K}}$ is the weight of the jth local neighbour, and it penalizes the pixels which are less similar to the one at $\mathbf{x}_i$. According to the steering kernel (SK) proposed in [8], $w_{i,j}^{\mathrm{K}}$ is computed as follows

$$w_{i,j}^{\mathrm{K}} = \frac{\sqrt{\det(\mathbf{C}_{i,j})}}{2\pi h_2} \exp\left(\frac{(\mathbf{x}_j - \mathbf{x}_i)^T \mathbf{C}_{i,j} (\mathbf{x}_j - \mathbf{x}_i)}{2h_2}\right) \tag{3}$$

where $\mathbf{C}_{i,j} \in \mathbb{R}^{2 \times 2}$ denotes a gradient covariance matrix at the jth local neighbour of the ith target pixel, and h_2 is a smoothing parameter.

3 The Proposed Method

3.1 Adaptive-Shape NLM (AS-NLM)

In the classical NLM model, the shape of image patches $\mathbf{N}_i$ and $\mathbf{N}_j$ in Eq. (1) is square. However, the square patch can not well capture the local structure information nearby the target pixel. As a result, the weight calculated by (1) is inaccurate, especially in the images containing fine details.

In order to address this problem, we propose an adaptive-shape NLM (AS-NLM) model. The benefit of using the adaptive-shape (AS) patches is illustrated in Fig. 1. It is obvious that if the shape of image patch can be adapted to the

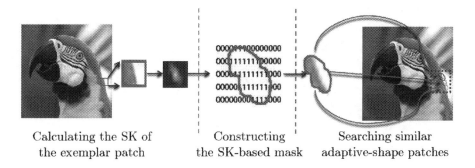

Calculating the SK of   Constructing   Searching similar
the exemplar patch   the SK-based mask   adaptive-shape patches

Fig. 2. The procedure of AS searching

image content, the local structure around the target pixel (i.e., the center pixel in an image patch) can be well characterized. By doing so, the accuracy of the weights in NLM model will be enhanced.

Figure 2 shows how the AS searching works. Firstly, we extract the square patch centered at the ith target pixel. To well describe the surrounding structures of the target pixel, the SK [8] of the square patch is computed according to Eq. (3). Secondly, a SK-based mask is generated by

$$
M_{i,j} = \begin{cases} 0 & w_{i,j}^{\mathrm{K}} < T_{\mathrm{s}} \\ 1 & w_{i,j}^{\mathrm{K}} \geq T_{\mathrm{s}} \end{cases} \tag{4}
$$

where $M_{i,j}$ denotes the mask value at position $\mathbf{x}_j$ in the square patch, and T_{s} is a threshold. If $w_{i,j}^{\mathrm{K}} \geq T_{\mathrm{s}}$, it means that the jth pixel in the patch is very similar to the ith target pixel. Therefore, the jth pixel should be included for the further patch searching. Afterwards, we multiple the SK-based mask of the ith target pixel with all the non-local square patches in the search window, so as to change the shape of the image patches. Finally, the Euclid distance between each non-local AS patch and the AS exemplar patch is computed, and the N_{p} nearest similar patches are located.

For the lth non-local neighbour, its corresponding AS-NLM weight is calculated as

$$
w_{i,l}^{\mathrm{AN}} = \left[\exp\left(-\frac{\|\mathbf{M}_i \circ (\mathbf{N}_i - \mathbf{N}_l)\|_2^2}{h_1} \right) \right] / c_i \tag{5}
$$

where $\mathbf{M}_i$ stands for the mask vector of the ith target pixel, the jth element of which is $M_{i,j}$; $\circ$ represents Hadamard product.

3.2 Improved NLSKR (I-NLSKR)

The NLSKR model was first introduced in [15]. Later on, it was employed in the reconstruction-based SR framework [16]. The basic idea of NLSKR is to

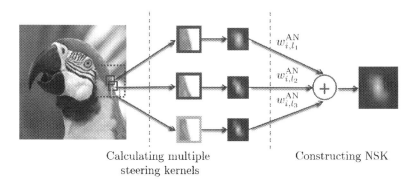

Calculating multiple   Constructing NSK
steering kernels

Fig. 3. The procedure of establishing the NSK

utilize the NLM model to improve the performance of the SKR model. As a new AS-NLM model has been presented in Sect. 3.1, we can use AS-NLM to further improve the NLSKR model.

Figure 3 shows how to build the non-local steering kernel (NSK) in the improved NLSKR (I-NLSKR) model. For the ith patch, its N_p nearest similar patches, which are found by the AS-NLM, are directly chosen as the non-local neighbours. Afterwards, the steering kernels of all the N_p non-local patches are calculated. Then we compute the weight sum of the obtained kernels, leading to a NSK as follows

$$\mathbf{W}_i^{\mathrm{NSK}} = \sum_l w_{i,l}^{\mathrm{AN}} \mathbf{W}_l^{\mathrm{K}} \tag{6}$$

where $\mathbf{W}_l^{\mathrm{K}}$ is a diagonal matrix, and its elements on principal diagonal are calculated by Eq. (3). Similar to [16], the estimation of the ith target pixel can be computed as

$$\hat{X}_i = \mathbf{e}^T \left(\mathbf{\Phi}^T \mathbf{W}_i^{\mathrm{NSK}} \mathbf{\Phi} \right)^{-1} \mathbf{\Phi}^T \mathbf{W}_i^{\mathrm{NSK}} \mathbf{F}_i \mathbf{X} \tag{7}$$

where $\mathbf{\Phi}$ denotes the polynomial basis from Taylor expansion [15]; $\mathbf{F}_i$ stands for a matrix to extract the ith patch from image $\mathbf{X}$; $\mathbf{e}$ is a column vector, where the first element is one and the rest are zeros.

3.3 The SR Algorithm Using Local and Non-local Priors (SRLNP)

AS-NLM is a non-local model, while I-NLSKR is a local model. Both of them are incorporated into our proposed SR reconstruction framework, leading to

$$\hat{\mathbf{X}} = \operatorname*{argmin}_{\mathbf{X}} \left\{ \begin{array}{l} \|\mathbf{Y} - \mathbf{DHX}\|_2^2 + \lambda_1 \sum_{i \,\in\, \mathbf{X}} \left\| X_i - \mathbf{w}_i^{\mathrm{AN}} \mathbf{R}_i \mathbf{X} \right\|_2^2 \\[2mm] + \lambda_2 \sum_{i \,\in\, \mathbf{X}} \left\| X_i - \mathbf{w}_i^{\mathrm{IN}} \mathbf{F}_i \mathbf{X} \right\|_2^2 \end{array} \right\} \tag{8}$$

Algorithm 1. SRLNP

Input: Y, D, H
Use the bicubic interpolation to get an initial HR estimation $\mathbf{X}^0$
For $t = 1, \cdots, T_m$
  If $\mathrm{mod}(t, P) == 1$
    Perform AS searching on $\mathbf{X}^{t-1}$ and calculate $\{M_{i,j}\}$ according to (4).
    Compute $\{w_{i,l}^{\mathrm{AN}}\}$ according to (5), and update matrix $\mathbf{A}$.
    Obtain $\mathbf{W}_i^{\mathrm{NSK}}$ via (6), and update matrix $\mathbf{B}$.
  End if
  Update the SR result $\mathbf{X}^t$ through Eq. (10).
End for
return $\mathbf{X}^{T_m}$

---

where $\mathbf{Y}$ stands for the LR image, $\mathbf{D}$ and $\mathbf{H}$ represent the operations of downsampling and blurring, respectively; X_i is the ith pixel in the HR image; matrix $\mathbf{R}_i$ is responsible for extracting the non-local similar pixels of X_i; λ_1 and λ_2 are two trade-off parameters; $\mathbf{w}_i^{\mathrm{AN}}$ is a row vector, with each element being equal to $w_{i,l}^{\mathrm{AN}}$; $\mathbf{w}_i^{\mathrm{IN}} = \mathbf{e}^T \left(\boldsymbol{\Phi}^T \mathbf{W}_i^{\mathrm{NSK}} \boldsymbol{\Phi} \right)^{-1} \boldsymbol{\Phi}^T \mathbf{W}_i^{\mathrm{NSK}}$.

In problem (8), the first term is the data fidelity term. The second and the third terms respectively require that the reconstructed pixels in the HR image should be close to the weight sums of their non-local neighbours and local neighbours. As the local and non-local models are complementary, jointly considering them will lead to superior performance.

To solve problem (8), a concise form is reformulated as

$$\hat{\mathbf{X}} = \underset{\mathbf{X}}{\mathrm{argmin}} \, \|\mathbf{Y} - \mathbf{DHX}\|_2^2 + \lambda_1 \|(\mathbf{I} - \mathbf{A}) \mathbf{X}\|_2^2 + \lambda_2 \|(\mathbf{I} - \mathbf{B}) \mathbf{X}\|_2^2 \qquad (9)$$

where $\mathbf{I}$ denotes the identity matrix. The ith row of matrix $\mathbf{A}$ is $\mathbf{w}_i^{\mathrm{AN}}\mathbf{R}_i$, while the ith row of matrix $\mathbf{B}$ is $\mathbf{w}_i^{\mathrm{IN}}\mathbf{F}_i$.

Since the whole objective function of problem (9) is quadratic, the gradient descent method can be used to solve the problem, i.e.,

$$\mathbf{X}^{t+1} = \mathbf{X}^t + \tau \left\{ \begin{array}{c} \mathbf{H}^T \mathbf{D}^T \left(\mathbf{Y} - \mathbf{DHX}^t \right) - \lambda_1 \left(\mathbf{I} - \mathbf{A} \right)^T \left(\mathbf{I} - \mathbf{A} \right) \mathbf{X}^t \\ - \lambda_2 \left(\mathbf{I} - \mathbf{B} \right)^T \left(\mathbf{I} - \mathbf{B} \right) \mathbf{X}^t \end{array} \right\} \qquad (10)$$

where t is the number of iterations, and τ is the step of the gradient descent method.

So far, all the details of our algorithm have been presented. We name the algorithm as SRLNP (super resolution using local and non-local priors), and summarize it as Algorithm 1. In Algorithm 1, T_m is the maximum number of iterations. Since the quality of the SR result becomes better with the growth of the number of iterations, it is necessary to update the matrices $\mathbf{A}$ and $\mathbf{B}$ multiple times. However, to balance between the performance and the complexity, they are updated once every P iterations.

Table 1. The PSNR (dB) and SSIM results obtained by different methods (×3).

Images	Zeyde [14]	ANR [9]	A+ [10]	NLSKR [16]	SKR-NLM [17]	ASDS [6]	SRLNP
butterfly	24.89	24.67	26.76	26.25	26.32	27.10	**27.94**
	0.8304	0.8204	0.8843	0.8992	0.9084	0.9100	**0.9220**
parrots	28.27	28.41	29.55	29.54	29.50	29.77	**30.14**
	0.8600	0.8707	0.8834	0.9098	0.9126	0.9105	**0.9143**
parthenon	26.21	26.24	**26.84**	26.31	26.61	26.56	26.67
	0.6913	0.7130	0.7228	0.728	0.7369	0.7336	**0.7373**
bike	23.05	23.29	24.20	23.95	24.19	24.29	**24.51**
	0.7030	0.7416	0.7663	0.7867	0.7995	0.7937	**0.8009**
flower	27.70	27.82	28.81	28.7	28.95	28.94	**28.95**
	0.7801	0.8058	0.8228	0.8447	**0.8520**	0.8469	0.8484
girl	32.72	32.85	33.41	33.44	33.46	33.43	**33.50**
	0.7873	0.8088	0.8100	0.8238	0.8233	0.8225	**0.8241**
hat	29.75	29.90	30.90	30.37	30.77	30.85	**30.98**
	0.8215	0.8306	0.8487	0.8689	0.8750	0.8750	**0.8757**
leaves	24.20	23.93	25.89	25.41	26.26	26.48	**26.89**
	0.8131	0.8142	0.8746	0.8917	0.9119	0.9084	**0.9158**
plants	31.59	31.61	33.06	33.13	**33.75**	33.34	33.45
	0.8598	0.8742	0.8929	0.9124	**0.9196**	0.9106	0.9163
raccoon	28.24	28.53	28.88	28.92	29.10	29.13	**29.15**
	0.7020	0.7436	0.7378	0.7597	0.7611	**0.7667**	0.7616
Average	27.66	27.73	28.83	28.60	28.89	28.99	**29.22**
	0.7849	0.8023	0.8244	0.8424	0.8500	0.8478	**0.8516**

4   Experimental Results

In this section, the effectiveness of the proposed SRLNP algorithm is evaluated. Six methods are selected for comparison, including Zeyde's method [14], ANR [9], A+ [10], SKR-NLM [17], NLSKR [16], and ASDS [6]. The ten test images used in [6] are chosen for testing. To generate the degraded LR images, the original images are first blurred by a Gaussian kernel with a size of 7×7 and a standard deviation of 1.0, and then downsampled by a factor of 3. Note that in the original implementations of Zeyde's method [14], ANR [9] and A+ [10], the simple bicubic downsampling is applied to generate the LR images. Therefore, for a fair comparison, we re-train the models of these three methods according to the blur kernel used in our experiments.

The basic settings of SRLNP are: the size of the square patches for both AS-NLM and I-NLSKR is 7×7. h_1 in Eq. (5) and h_2 in Eq. (3) are respectively set as 65 and 6.25. The threshold in Eq. (4) is selected as 0.03. To balance between complexity and quality, the size of the search window is chosen as 17×17, and the number of similar patches N_p is 17. The parameters τ, λ_1, and λ_2 are respectively selected as 9.3, 0.01 and 0.001. The maximum number of iterations T_m is 1000, and the matrices $\mathbf{A}$ and $\mathbf{B}$ are updated once every $P = 100$ iterations.

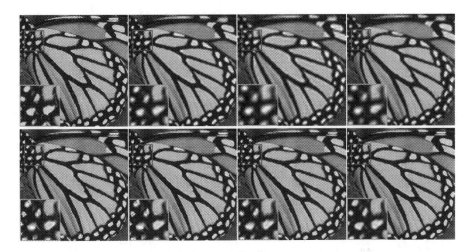

Fig. 4. Comparison of the SR results of image *butterfly* (×3). From left to right and from top to bottom: original image, Zeyde's method [14], ANR [9], A+ [10], NLSKR [16], SKR-NLM [17], ASDS [6], SRLNP. Please zoom in for better viewing.

Fig. 5. Comparison of the SR results of image *leaves* (×3). From left to right and from top to bottom: original image, Zeyde's method [14], ANR [9], A+ [10], NLSKR [16], SKR-NLM [17], ASDS [6], SRLNP. Please zoom in for better viewing.

The PSNR and SSIM [11] of the 10 test images are listed in Table 1. We can see that in most cases, SRLNP achieves the highest PSNR/SSIM values. The average PSNR/SSIM gains of SRLNP over ASDS [6] are 0.23dB/0.0038, while these gains become 0.33dB/0.0016 when compared with SKR-NLM [17]. The subjective SR results of images *butterfly* and *leaves* are shown in Figs. 4 and 5, respectively. It can be observed that the results of Zeyde [14], ANR [9] and A+

Table 2. The average PSNR (dB) and SSIM results on 10 test images obtained by different regularization terms (×3).

Methods	SKR [8]	NLM [1]	NLSKR [16]	SKR-NLM [17]	AS-NLM	SRLNP
PSNR(dB)	28.46	28.78	28.60	28.89	29.03	29.22
SSIM	0.8406	0.8488	0.8424	0.8500	0.8507	0.8516

Table 3. The average running time on 256×256 images.

Methods	Zeyde [14]	ANR [9]	A+ [10]	NLSKR [16]	SKR-NLM [17]	ASDS [6]	SRLNP
Time(s)	0.8	1.4	1.8	123.5	145.6	206.7	165.4

[10] are blurry. Compared with the other methods, the HR images reconstructed by SRLNP contain sharper edges and less artifacts.

To better evaluate the effectiveness of SRLNP, we additionally list the average PSNR/SSIM results obtained by different regularization terms in Table 2. We can conclude that: Firstly, SKR-NLM [17] is superior to independently utilizing the SKR [8] or NLM [1] regularization term. Secondly, as non-local structures are considered in NLSKR [16], it outperforms SKR [8]. Thirdly, the proposed AS-NLM produces results better than the conventional NLM [1] since the local information is well described. Finally, by jointly considering the complementary models AS-NLM and I-NLSKR, the proposed SRLNP algorithm is able to deliver the best quality of results among all the sophisticated models.

The average running time on 256×256 images required by different methods is listed in Table 3. The time is measured under MATLAB 2014b, and on a computer with Intel(R) Core(TM) i5-4460 CPU and 8G RAM. We can see that: the three learning-based methods [9,10,14] are much faster than the reconstruction-based methods; because two sophisticated models are incorporated in SRLNP, it is slower than SKR-NLM [17], but still faster than ASDS [6]. In SRLNP, the complexity of AS searching and updating the matrix $\mathbf{A}$ is $O(NP_s^2W_s^2)$, where N is the number of pixels in an image, $P_s \times P_s$ is the size of non-local patches, and $W_s \times W_s$ is the size of the search window; updating the matrix $\mathbf{B}$ requires $O(NP_s^6)$; computing the $\mathbf{X}$ in the tth iteration needs $O(N^2)$.

5   Conclusion

In this paper, we propose two sophisticated models AS-NLM and I-NLSKR. By jointly considering both models, a single image SR algorithm SRLNP is presented. Experiments show that SRLNP can achieve the best performance among the tested methods. However, the two models incorporated in SRLNP result in a relative high computational burden. In the future work, we will try to accelerate the SRLNP algorithm by using parallel techniques such as GPU.

References

1. Buades, A., Coll, B., Morel, J.M.: A review of image denoising algorithms, with a new one. SIAM Multiscale Model. Simul. **4**(2), 490–530 (2005)
2. Chang, K., Ding, P.L.K., Li, B.: Single image super-resolution using collaborative representation and non-local self-similarity. Sig. Process. **149**, 49–61 (2018)
3. Chang, K., Ding, P.L.K., Li, B.: Single image super resolution using joint regularization. IEEE Sig. Process. Lett. **25**(4), 596–600 (2018)
4. Dong, C., Loy, C.C., He, K., Tang, X.: Learning a deep convolutional network for image super-resolution. In: Fleet, D., Pajdla, T., Schiele, B., Tuytelaars, T. (eds.) ECCV 2014. LNCS, vol. 8692, pp. 184–199. Springer, Cham (2014). https://doi.org/10.1007/978-3-319-10593-2_13
5. Dong, W., Zhang, L., Shi, G., Li, X.: Nonlocally centralized sparse representation for image restoration. IEEE Trans. Image Process. **22**(4), 1620–1630 (2013)
6. Dong, W., Zhang, L., Shi, G., Wu, X.: Image deblurring and super-resolution by adaptive sparse domain selection and adaptive regularization. IEEE Trans. Image Process. **20**(7), 1838–1857 (2011)
7. Kim, J., Lee, J., Lee, K.M.: Accurate image super-resolution using very deep convolutional networks. In: IEEE Conference on Computer Vision and Pattern Recognition (CVPR), pp. 1646–1654. IEEE, Las Vegas (2016)
8. Takeda, H., Farsiu, S., Milanfar, P.: Kernel regression for image processing and reconstruction. IEEE Trans. Image Process. **16**(2), 349–366 (2007)
9. Timofte, R., Smet, V.D., Gool, L.V.: Anchored neighborhood regression for fast example-based super resolution. In: IEEE International Conference on Computer Vision (ICCV), pp. 1920–1927. IEEE, Sydney (2013)
10. Timofte, R., De Smet, V., Van Gool, L.: A+: adjusted anchored neighborhood regression for fast super-resolution. In: Cremers, D., Reid, I., Saito, H., Yang, M.-H. (eds.) ACCV 2014. LNCS, vol. 9006, pp. 111–126. Springer, Cham (2015). https://doi.org/10.1007/978-3-319-16817-3_8
11. Wang, Z., Bovik, A.C., Sheikh, H.R., Simoncelli, E.P.: Image quality assessment: from error visibility to structural similarity. IEEE Trans. Image Process. **13**(4), 600–612 (2004)
12. Wang, Z., Liu, D., Yang, J., Han, W., Huang, T.: Deep networks for image super-resolution with sparse prior. In: IEEE International Conference on Computer Vision (ICCV), pp. 370–378. IEEE, Santiago (2015)
13. Yang, J., Wright, J., Huang, T.S., Ma, Y.: Image super-resolution via sparse representation. IEEE Trans. Image Process. **19**(11), 2861–2873 (2010)
14. Zeyde, R., Elad, M., Protter, M.: On single image scale-up using sparse-representations. In: Boissonnat, J.D., et al. (eds.) Curves and Surfaces 2010. LNCS, vol. 6920, pp. 711–730. Springer, Heidelberg (2012). https://doi.org/10.1007/978-3-642-27413-8_47
15. Zhang, H., Yang, J., Zhang, Y., Huang, T.S.: Non-local kernel regression for image and video restoration. In: Daniilidis, K., Maragos, P., Paragios, N. (eds.) ECCV 2010. LNCS, vol. 6313, pp. 566–579. Springer, Heidelberg (2010). https://doi.org/10.1007/978-3-642-15558-1_41
16. Zhang, K., Gao, X., Li, J., Xia, H.: Single image super-resolution using regularization of non-local steering kernel regression. Sig. Process. **123**, 53–63 (2016)
17. Zhang, K., Gao, X., Tao, D., Li, X.: Single image super-resolution with non-local means and steering kernel regression. IEEE Trans. Image Process. **21**(11), 4544–4556 (2012)

Plenoptic Image Compression via Simplified Subaperture Projection

Haixu Han[(⊠)], Jin Xin[(⊠)], and Qionghai Dai[(⊠)]

Graduate School at Shenzhen, Tsinghua University, Beijing, China
Haixu.han@qq.com, jin.xin@sz.tsinghua.edu.cn

Abstract. In this paper, a generalized encoding architecture is proposed for compressing plenoptic images to reduce the redundancy among the subaperture images. Based on the homography analysis, simplification of subaperture projection is performed due to high correlations among adjacent views. Then, a pseudo-video sequence consisting of central subaperture images and a sequence consisting of residual images between the central image and the adjacent images are generated by the proposed partitioning and reordering methods. They are compressed by temporal coding tools in HEVC. The experimental results demonstrate that the proposed method outperforms state-of-the-art pseudo-video compression methods by an average of 25.3%/27.4%/21.7%/36.7%/5.3% bitrate saving with comparable/lower computational complexity.

Keywords: Plenoptic image coding · Subaperture image reordering
Subaperture projection · HEVC

1 Introduction

Portable light field photography, which can capture not only the spatial intensity information but also directional information of light rays in a three-dimensional (3D) scene by handheld devices, has attracted broad interests both from academy and industry. Plenoptic cameras like Lytro [1], which insert a microlens array into the light path between the main lens and the sensor (as shown in Fig. 1), are typical devices. Due to the distinct intensity distribution in the captured plenoptic images in super-high-resolution, efficient compression methods are fundamentally desired for transmission and storage.

The existing approaches for plenoptic image coding can be classified into two categories: approaches that compress the plenoptic image directly and approaches that compress the pseudo-video generated from the plenoptic image. Approaches in the first category compress the plenoptic images directly using an image encoder like JPEG [2] or spatial coding tools in a video encoder, like high efficiency video coding (HEVC) [3]. Some predictive coding methods like displacement intra prediction mode [4, 5], disparity-based compensation mode [6, 7], self-similarity compensated prediction [8, 9] are proposed, in which the current block is predicted by a spatially matched block. However, the approaches in this group cannot fully exploit the optical imaging correlations among micro-images. For the pseudo-video based compression methods in the second category, a video sequence is generated from subaperture images and then

© Springer Nature Switzerland AG 2018
R. Hong et al. (Eds.): PCM 2018, LNCS 11165, pp. 274–284, 2018.
https://doi.org/10.1007/978-3-030-00767-6_26

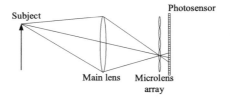

Fig. 1. Architecture of a standard plenoptic camera.

compressed using temporal coding tools in a video encoder. Various scanning topologies like line-mapping, rotational scanning, raster scanning, and U-shape scanning were proposed to reorder the images in the pseudo video [10, 11]. Additionally, Perra et al. [12–14] partitioned the plenoptic image into tiles and treated each tile as a frame of a pseudo-temporal sequence for encoding. Also, some compression methods [15–17] utilize multiview video coding (MVC) to explore the relations among the adjacent subaperture images. And some predictive methods using selected compressed views by linear approximation [18] or sparse representation [19] are proposed to reduce the overhead bits. Compared with the approaches in the first category, the approaches in the second category can achieve much higher compression efficiency with an overhead of high computational complexity.

In this paper, a pseudo-video coding method is proposed for plenoptic image compression. The correlations among the subaperture images extracted from the plenoptic image, corresponding to the images from different viewpoints, are analyzed utilizing view projection. According to the analysis of view relations, simplification of subaperture projection is performed and an optimal grouping method for subaperture images is proposed, in which subaperture images are categorized into two sets: the central images and the residual images, i.e. the difference between the central image and the adjacent images. Then the images in each set are reordered individually according to the proposed reordering methods to generate a pseudo-video sequence. They are compressed by temporal coding tools in HEVC. The experimental results demonstrate that the proposed approach outperforms state-of-the-art pseudo-video coding methods by significant bitrate reductions with comparable/lower computational complexity.

The rest of this paper is organized as follows. Section 2 introduces the subaperture projection and simplification. The proposed system architecture is presented in Sect. 3. Section 4 presents the experimental results. Section 5 concludes the paper.

2 Simplified Subaperture Projection

2.1 Subaperture Image Extraction

According to the architecture of a standard plenoptic camera, as shown in Fig. 1, the light rays from a point of the object are diverged by microlens array and imaged on the sensor as a group of pixels [20], called a macropixel or an elemental image (EI), which can record both spatial and angular light information. Based on Ng's optical analysis

[21] and light field decoding [22], a subaperture image can be easily extracted by picking out the co-located pixels from all the macropixels, like that depicted in Fig. 2 using light field decoding toolbox [22, 23]. Different subaperture images correspond to pixels at different positions in the macropixel, which represent the different perspectives of the scene. Therefore, the geometric relation between two subaperture images can be revealed by subaperture projection.

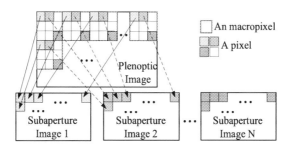

Fig. 2. Subaperture image extraction from a plenoptic image.

2.2   Subaperture Projection and Simplification

Subaperture projection can be described by a homography [24], which is a projective mapping between two images of a planar surface in the 3D space and describes the translation and rotation created by the variation of pixel position. Considering the subaperture images are equivalent to views captured by a 2D planar camera array, the projection between a pair of matching points on the image plane of the two subaperture images, denoted by $\mathbf{x} = (x, y, w)^T$ and $\mathbf{x'} = (x', y', w')^T$, in the Euclidean 2D-space can be formulated by a linear transformation as [24]:

$$x'/w' = \frac{r_{00}x + r_{10}y + t_x}{p_x x + p_y y + 1}, y'/w' = \frac{r_{01}x + r_{11}y + t_y}{p_x x + p_y y + 1},\tag{1}$$

where r_{ij} is a rotation; t_x and t_y represent a horizontal and vertical translation, respectively; p_x and p_y are scaling factors. Normalizing w and w' to be 1 and considering the image planes of two subaperture images are planar with only translations, Eq. (1) can be written as:

$$\mathbf{x'} = \mathbf{Hx}, \mathbf{H} = \begin{bmatrix} 1 & 0 & t_x \\ 0 & 1 & t_y \\ 0 & 0 & 1 \end{bmatrix},\tag{2}$$

where $\mathbf{H}$ is a homography, which can be derived by SIFT [25, 26] and the direct linear transform (DLT) [24] algorithms.

In order to analyze the correlation among different views utilizing subaperture projection, we divide the extracted subaperture images into several groups with different grouping size including 3×3 and 5×5 (as shown in Fig. 3(a) and (b), the grid in black in Fig. 3 represents the group boundary). In different categories, the **H**s are derived between a central image (a red block in Fig. 3) and eight images at different positions (green blocks in Fig. 3 with index 1–8). Table 1 summarizes the retrieved average value of t_x and t_y for the eight images at different positions using the images (c)–(e) [28] in Fig. 6 as instances, which have different well-focused object distances. It can be found that in the first group including 3×3 subaperture images, both t_x and t_y are less than half-pixel, because of the extremely small disparity among subapertures, which are possible to be quantized to zero. However, for the projection among subaperture images with intervals in the second category, many values of t_x and t_y are larger than half-pixel that cannot be ignored. Thus, a simplification of subaperture projection is considered to be performed due to high correlation among adjacent subaperture images. Simplifying homography to be an identity matrix, high correlation and similarity among adjacent views can be employed directly in compression to reduce the overhead bits.

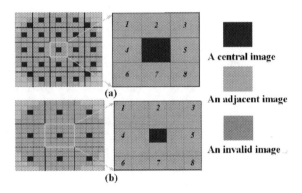

Fig. 3. Grouping methods: (a) 3×3; (b) 5×5.

Table 1. Average retrieved average value of t_x and t_y.

	POS	1	2	3	4	5	6	7	8
3×3	t_x	−0.06	0.16	0.10	−0.01	0.09	−0.10	−0.12	0.00
	t_y	0.11	−0.08	−0.11	0.25	−0.01	0.33	0.16	0.30
5×5	t_x	−0.61	0.01	0.95	−0.60	2.18	1.58	−0.11	−0.17
	t_y	1.12	−0.07	0.73	1.15	0.08	1.06	0.24	0.28

3 Proposed Compression Methods

3.1 Proposed Compression Architecture

The encoding architecture is proposed in Fig. 4. As shown in the figure, after the analysis and simplification of subaperture projection, the extracted subaperture images are divided into 3×3 groups as shown in Fig. 3(a), in which the images can be categorized into two sets: central images and adjacent images. Subtracting the central image from the adjacent ones, residual images that have a large amount of zero coefficients are generated. Then, *Reordering* method proposed by us generates two pseudo-video sequences: one for central images and one for the residual images. They are fed to HEVC encoder for compression.

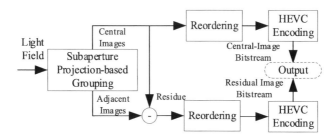

Fig. 4. The proposed generalized encoding architecture.

3.2   Proposed Reordering Methods

Since the subaperture images represent different perspectives of the scene, reordering the central images and the residual images is to generate two pseudo video sequences with higher temporal correlations to further improve the compression efficiency. Figure 5 depicts the proposed reordering methods, which consist of two methods for the central images: *Zigzag* and *S-Shape* scan in Fig. 5(a) and (b), respectively, and three methods for the residual images: *Zigzag*, *S-Shape* and *Circle* scan in Fig. 5(c)–(e), respectively. The arrows represent the temporal order in the pseudo video sequence, which determines the reference frames available during encoding. Notice that, for the images at the four corners, the position of the central images and residual images are

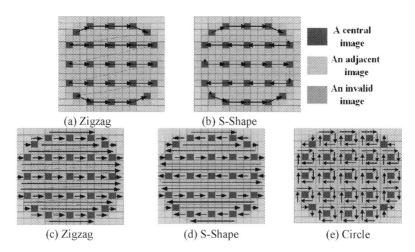

Fig. 5. Reordering methods for: (a), (b): central images; and (c), (e): residual images.

adjusted according to that in Fig. 3(a) due to the existence of those images with poor quality (grey blocks in Fig. 5). The compression efficiency is compared among the combinations of the reordering methods for the central images and the residual images in the next section.

4 Experimental Results

To demonstrate the efficiency of the proposed compression method, two sets of experiments have been conducted in this section. First, the efficiency of the combinations of reordering methods are compared. Then, the efficiency of the proposed method is compared with the state-of-the-art pseudo video based compression methods.

The sequence of the central images and that of the residual images are encoded by HEVC independently using reference software HM-16.9 SCM8.0 [27]. During encoding the sequence of the residual images, bit depth is set to be 10-bit since the residual value varies from -255 to 255. Five plenoptic images downloaded from the light-field image dataset [28] and five captured by our *Lytro Illum* camera with spatial resolution of 7728×5368 and angular resolution of 15×15, as shown in Fig. 6, are tested. The end-to-end processing workflow of plenoptic images including pre-processing, encoding, decoding and rendering adopts the method recommended by JPEG Pleno [29, 30], which is initiated by JPEG for standardization of plenoptic image coding. Before feeding into the encoder, the sequences are converted from RGB colour space to YCbCr4:4:4 colour space. Compression is performed using the configurations of Lowdelay P Main Profile [31] at QP 16,22,28,34. BD-bitrate defined in [32] is used to measure the compression efficiency among the approaches, in which the bitrate is the total bits consumed by the central images and the residual images, and the PSNR is calculated between the reconstructed subaperture images and the subaperture images rendered from the original plenoptic image. The rendering method was proposed by [22] and also recommended by light field compression evaluation method in [33]. For the proposed method, the reconstructed subaperture images consist of the reconstructed central images and those generated by adding the reconstructed residues to the reconstructed central images.

First, the compression efficiency is compared among the combinations of the proposed reordering methods to find the one with the highest efficiency. A representative pseudo video coding method: line scan mapping (LSM) method proposed in [10] is selected as the anchor for performance comparison. LSM reorders the subaperture images using raster scan and encodes the sequence directly by HEVC in this paper.

Since there are two scanning methods for the central images and 3 methods for the residual images, as shown in Fig. 5, altogether six combinations are available. Table 2 summarizes the results of compression efficiency for the combinations. As shown in the table, if compare the performance of reordering for the central images individually, shown in the column of *Central Images*, *Zigzag* presents an average of 2.17% bitrate reduction relative to *S-Shape*, which benefits the compression of central images. Thus, combining it with the reordering methods of residual images, the three combinations outperform the other three obviously, although all the proposed combinations can

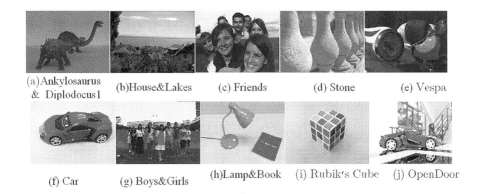

(a) Ankylosaurus   (b)House&Lakes   (c) Friends   (d) Stone   (e) Vespa
& Diplodocus1

(f) Car   (g) Boys&Girls   (h)Lamp&Book   (i) Rubik's Cube   (j) OpenDoor

Fig. 6. Tested plenoptic images: (a)–(e): images from EPFL dataset [28]; (f)–(j): images captured by our own *Lytro Illum* camera.

outperform LSM with significant bitrate reductions. Among the three (*Zigzag + Zigzag, Zigzag + S-Shape* and *Zigzag + Circle*), *Zigzag + Circle* works the best with an average bitrate reduction of 25.29% over the anchor. Hence, *Zigzag + Circle* is selected as the best mode for the proposed method.

Secondly, the effectiveness of the proposed method is demonstrated by comparing with the five state-of-the-art compression algorithms: LSM in [10], rotation scan mapping (RSM) in [10], horizontal zigzag scan (HZS) in [11], pseudo-video based on tiling and arrangement (TPVA) in [12], and pseudo-video compression using linear approximation prior (LAP) in [18], which got the first place of light field compression competitions in ICIP2017. Similar to LSM, RSM and HZS encode the reordered subaperture images directly. TPVA partitions the plenoptic image into 19×19 tiles, each of which has 464×320 pixels, and treats a tile as a frame in the sequence to perform encoding [12]. LAP selects several central views to generate a pseudo-video for compression, while other views are reconstructed by linear approximation [18]. In order to achieve an accurate evaluation and comparison, the bitrates of these compared approaches are aligned with the proposed methods by adjusting the QP.

Using *Zigzag + Circle* as the best reordering method, denoted by *Proposed*, the comparison results are shown in Table 3. It is obvious that the proposed compression method outperforms LSM/RSM/HZS/TPVA/LAP by 25.29%/27.38%/21.71%/36.67%/5.27% bitrate reduction on average. Randomly taking "House&Lakes" and "Friends_01" as instances, the rate-distortion curves of the five approaches are depicted in Fig. 7. It is obvious that the proposed method provides robust compression efficiency improvement especially at low bitrates. Although the efficiency improvement provided by the proposed method varies due to the textural correlations among the subaperture images and the residual images, it still outperforms the other approaches obviously for the most tested images because of greatly reducing the redundancy among the central images and the adjacent images.

Table 2. Comparison among the reordering methods.

Image Index	Central Images	Total performance vs. LSM [10]						
	Zigzag vs. S-Shape	Zigzag + Zigzag	Zigzag + S-Shape	Zigzag + Circle	S-Shape + S-Shape	S-Shape + Zigzag	S-shape + Circle
(a)	1.21%	−27.86%	−28.63%	−25.16%	−28.18%	−28.95%	−26.03%
(b)	−2.06%	−22.38%	−21.60%	−22.42%	−21.33%	−20.62%	−21.26%
(c)	−2.47%	−36.28%	−36.25%	−36.05%	−34.81%	−34.78%	−34.61%
(d)	−2.41%	−40.79%	−40.82%	−36.37%	−39.23%	−39.26%	−34.88%
(e)	−3.91%	−36.24%	−36.46%	−36.75%	−33.71%	−33.95%	−34.24%
(f)	−2.11%	−16.82%	−16.22%	−19.30%	−15.12%	−14.56%	−17.68%
(g)	−3.05%	−37.03%	−37.25%	−38.35%	−34.99%	−35.26%	−36.35%
(h)	−2.70%	−9.84%	−9.36%	−11.85%	−7.90%	−7.47%	−9.99%
(i)	−1.28%	−15.29%	−14.59%	−18.34%	−14.04%	−13.39%	−17.21%
(j)	−2.97%	−7.08%	−7.98%	−8.32%	−4.36%	−5.24%	−5.55%
Average	−2.17%	−24.96%	−24.92%	−25.29%	−23.37%	−23.35%	−23.78%

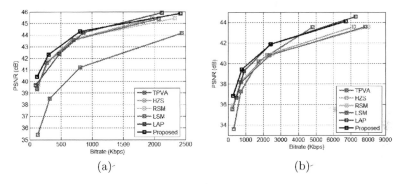

Fig. 7. Rate-distortion curves for: (a) House&Lake; and (b) Friends_01.

The section is concluded by comparing the computational complexity among the approaches. The execution time is retrieved for each approach using a PC with Intel® Core™ E5-2620 CPU @ 2.4 GHz with 64 GB RAM and 64-bit Windows Server operating system. The relative ratios between the proposed approach and the other five approaches are listed in Table 4. As shown in the table, the computational complexity of the proposed approach is comparable with that of LSM, RSM and HZS, lower than that of TPVA and a bit higher than LAP. Combined with compression efficiency performance shown in Table 3, it can be concluded that the proposed method outperforms the approaches LSM/RSM/HZS/TPVA with much higher rate-distortion-complexity performance and is comparable with LAP.

Table 3. Compression efficiency comparison with other methods.

Image Index	Proposed vs.				
	LSM [10]	RSM [10]	HZS [11]	TPVA [12]	LAP [18]
(a)	−25.16%	−12.54%	−18.49%	−40.28%	−13.31%
(b)	−22.42%	−24.24%	−19.02%	−73.32%	−21.31%
(c)	−36.05%	−35.41%	−31.32%	−29.72%	−13.26%
(d)	−36.37%	−32.66%	−29.25%	−38.95%	6.43%
(e)	−36.75%	−34.51%	−31.28%	−66.47%	−10.33%
(f)	−19.30%	−32.97%	−18.52%	−31.19%	4.97%
(g)	−38.35%	−37.98%	−33.03%	−22.61%	−13.73%
(h)	−11.85%	−17.46%	−11.24%	−29.24%	7.37%
(i)	−18.34%	−25.62%	−20.82%	−43.85%	−13.43%
(j)	−8.32%	−20.46%	−4.09%	8.94%	13.89%
Average	−25.29%	−27.38%	−21.71%	−36.67%	−5.27%

Table 4. Computational complexity comparison

Proposed/methods	Methods				
	LSM	RSM	HZS	TPVA	LAP
	1.069	1.071	1.077	0.820	2.174

5 Conclusions

This paper proposes a novel compression architecture for plenoptic images, which can efficiently exploit the correlations among the subaperture images. According to simplification of subaperture projection, two pseudo video sequences are generated and compressed: one consists of several central images selected from the subaperture images, the other consists of residual images generated by subtracting the central image from the adjacent images. Experimental results demonstrate its superior improvement in compression efficiency relative to state-of-the-art methods with comparable or even lower computational complexity. The future work will focus on exact subaperture projection methods to further improve the compression performance.

Acknowledgement. This work was supported in part by Shenzhen project JCYJ20170307 153135771 and Foundation of Science and Technology Department of Sichuan Province 2017JZ0032c, China.

References

1. Lytro. https://www.lytro.com/
2. Higa, R.S., Chavez, R.F.L., Leite, R.B.: Plenoptic image compression comparison between JPEG, JPEG2000 and SPITH. Cyber J. JSAT **3**(6), 1–6 (2013)

3. Sullivan, G.J., Ohm, J., Han, W.J., et al.: Overview of the high efficiency video coding (HEVC) standard. IEEE Trans. Circuits Syst. Video Technol. **22**(12), 1649–1668 (2012)
4. Li, Y., Sjostrom, M., Olsson, R., Jennehag, U.: Coding of focused plenoptic contents by displacement intra prediction. IEEE Trans. Circuits Syst. Video Technol. **26**(7), 1308–1319 (2016)
5. Li, Y., Olsson, R., Sjostrom, M.: Compression of unfocused plenoptic images using a displacement intra prediction. In: IEEE International Conference on Multimedia Expo Workshops (ICMEW), pp. 1–4, July 2016
6. Liu, D., An, P., Ma, R., et al.: Disparity compensation based 3D holoscopic image coding using HEVC. In: 2015 IEEE China Summit and International Conference on Signal and Information Processing (ChinaSIP), pp. 201–205 (2015)
7. Conti, C., Kovács, P.T., Balogh, T., et al.: Light-field video coding using geometry-based disparity compensation. In: 2014: The True Vision-Capture, Transmission and Display of 3D Video, pp. 1–4. IEEE (2014)
8. Conti, C., Soares, L.D., Nunes, P.: HEVC-based 3D holoscopic video coding using self-similarity compensated prediction. Sig. Process. Image Commun. **42**, 59–78 (2016)
9. Conti, C., Nunes, P., Soares, L.D.: HEVC-based light field image coding with bi-predicted self-similarity compensation. In: IEEE International Conference on Multimedia Expo Workshop (ICMEW), pp. 1–4, July 2016
10. Dai, F., Zhang, J., Ma, Y., Zhang, Y.: Lenselet image compression scheme based on subaperture images streaming. In: IEEE International Conference on Image Processing, pp. 4733–4737. IEEE (2015)
11. Zhao, S., Chen, Z., Yang, K., Huang, H.: Light field image coding with hybrid scan order. In: 2016 Visual Communications and Image Processing (VCIP), Chengdu, pp. 1–4 (2016)
12. Perra, C., Assuncao, P.: High efficiency coding of light field images based on tiling and pseudo-temporal data arrangement. In: IEEE International Conference on Multimedia and Expo Workshops (ICMEW), pp. 1–4 (2016)
13. Perra, C., Giusto, D.: JPEG 2000 compression of unfocused light field images based on lenslet array slicing. In: 2017 IEEE International Conference on Consumer Electronics (ICCE), Las Vegas, NV, pp. 27–28 (2017)
14. Perra, C., Giusto, D.: Raw light field image compression of sliced lenslet array. In: 2017 IEEE International Symposium on Broadband Multimedia Systems and Broadcasting (BMSB), Cagliari, pp. 1–5 (2017)
15. Liu, D., Wang, L., Li, L., et al.: Pseudo-sequence-based light field image compression. In: IEEE International Conference on Multimedia and Expo Workshops (ICMEW), pp. 1–4 (2016)
16. Shi, S., Gioia, P., Madec, G.: Efficient compression method for integral images using multi-view video coding. In: 18th IEEE International Conference on Image Processing, pp. 137–140 (2011)
17. Ahmad, W., Olsson, R., Sjostrom, M.: Interpreting plenoptic images as multi-view sequences for improved compression. In: 2017 IEEE International Conference on Image Process (ICIP), pp. 1–4, September 2017
18. Zhao, S., Chen, Z.: Light field image coding via linear approximation prior. In: 2017 IEEE International Conference on Image Process (ICIP), pp. 1–4, September 2017
19. Tabus, I., Helin, P., Astola, P.: Lossy compression of lenslet images from plenoptic cameras combing sparse predictive coding and JPEG2000. In: 2017 IEEE International Conference on Image Process (ICIP), pp. 1–4, September 2017
20. Lam, E.Y.: Computational photography with plenoptic camera and light field capture: tutorial. J. Opt. Soc. Am. A **32**(11), 2021–2032 (2015)
21. Ng, R.: Digital light field photography. Ph.D. thesis, Stanford University, 2006

22. Dansereau, D.G., Pizarro, O., Williams, S.B.: Decoding, calibration and rectification for lenselet-based plenoptic cameras. In: 2013 IEEE Conference on Computer Vision and Pattern Recognition (CVPR), Portland, OR, , pp. 1027–1034 (2013)

23. Light Field toolbox. http://www.mathworks.com/matlabcentral/fileexchange/49683-light-fieldtoolbox-v0-4

24. Hartley, R., Zisserman, A.: Multiple View Geometry in Computer Vision. Cambridge University Press, Cambridge (2000)

25. Lowe, D.G.: Distinctive Image Features from Scale-Invariant Keypoints. Kluwer, Hingham (2004)

26. Fischler, M.A., Bolles, R.C.: Random sample consensus: a paradigm for model fitting with applications to image analysis and automated cartography. Commun. ACM **24**(6), 381–395 (1981)

27. Downloaded from: https://hevc.hhi.fraunhofer.de/svn/svn_HEVCSoftware/tags/HM-16.9 +SCM-8.0/

28. Light-field image dataset. http://mmspg.epfl.ch/EPFL-light-field-image-dataset

29. Ebrahimi, T., Foessel, S., Pereira, F., Schelkens, P.: JPEG Pleno: toward an efficient representation of visual reality. IEEE Multimed. **23**(4), 14–20 (2016)

30. ISO/IEC JTC 1/SC29/WG1 JPEG: JPEG Pleno Call for Proposals on Light Field Coding. Doc. N74014, Geneva, Switzerland, January 2017

31. Au, O.C., Zhang, X., Pang, C., Wen, X.: Suggested common test conditions and software reference configurations for screen content coding. In: Joint Collaborative Team on Video Coding (JCT-VC), Torino, JCTVC-F696, July 2011

32. Bjontegaard, G.: Calculation of average PSNR difference between RD-curves. ITU-T VCEG-M33 (2001)

33. Light field compression evaluation. http://mmspg.epfl.ch/files/content/sites/mmspl/files/shared/LF-GC/CFP.pdf

Gaze Aware Deep Learning Model for Video Summarization

Jiaxin Wu[1], Sheng-hua Zhong[1], Zheng Ma[2(✉)], Stephen J. Heinen[2],
and Jianmin Jiang[1]

[1] College of Computer Science and Software Engineering,
Shenzhen University, Shenzhen, China
jiaxin.wu@email.szu.edu.cn, {csshzhong,jianmin.jiang}@szu.edu.cn
[2] The Smith-Kettlewell Eye Research Institute, San Francisco, CA, USA
{zma,heinen}@ski.org

Abstract. Video summarization is an ideal tool for skimming videos. Previous computational models extract explicit information from the input video, such as visual appearance, motion or audio information, in order to generate informative summaries. Eye gaze information, which is an implicit clue, has proved useful for indicating important content and the viewer's interest. In this paper, we propose a novel gaze-aware deep learning model for video summarization. In our model, the position and velocity of the observers' raw eye movements are processed by the deep neural network to indicate the users' preferences. Experiments on two widely used video summarization datasets show that our model is more proficient than state-of-the-art methods in summarizing video for characterizing general preferences as well as for personal preferences. The results provide an innovative and improved algorithm for using gaze information in video summarization.

Keywords: Video summarization · Gaze information
Convolutional neural networks

1 Introduction

With the development of modern technologies, video data is experiencing an explosive growth. Recent statistics show that about 300 h of videos are uploaded to YouTube every minute.[1] It is time-consuming for users to browse videos to find the ones they are interested in. Video summarization is an effective tool to give frame- or shot-based summaries for videos [20]. A good summary often requires a well-designed computer vision model, and will significantly shorten the length of the original video but still keep the most interesting parts. This often leads to a more efficient viewing experience and lets users browse relevant information within a short period of time.

---

[1] https://fortunelords.com/youtube-statistics/.

© Springer Nature Switzerland AG 2018
R. Hong et al. (Eds.): PCM 2018, LNCS 11165, pp. 285–295, 2018.
https://doi.org/10.1007/978-3-030-00767-6_27

Many previous studies describe how to build a better computer vision model for video summarization [13,21,22,24,27]. However, video summarization is still a challenging problem for several reasons. First, there is no clear definition of "content of interest" in a video. Second, users' expectations of a good summary can vary from person to person.

Most video summarization methods tried to find attractive content from explicit information in the video. For example, Song et al. calculated three kinds of traditional visual descriptors for each video frame [19]. Gygli et al. tried to detect faces or persons from the visual appearances of frames [6]. Other methods discovered interesting video content from temporal information [24] or audio track [10].

Implicit information can also provide useful information for video summarization. For example, heart rate data and facial expressions of observers have been used to generate video summaries [1]. Here, we propose that users' gaze information during video watching is also an important type of implicit information. When incorporated into video summarization models, it should significantly improve model performance.

Gaze information has been demonstrated to be effective in revealing user's interest in many tasks. For example, fixation points have been identified as helpful markers for indicating regions of interest in segmentation tasks [14] and face recognition task [3]. Also, gaze data has been used to determine children's points of interest in advertisements [8]. However, only a few studies on video summarization used eye gaze information. Xu et al. applied fixation counts as attention scores, and extracted features of gaze regions to generate video summaries [23]. Salehin el al. used smooth pursuit to detect important frames for video summarization [16]. Both methods determined the pattern of eye movement before using specific patterns for video summarization. In the current paper, we integrate raw eye movement data into our proposed model to generate a good summary.

Deep learning networks have achieved great successes in many video-related tasks [12,17,27–29]. Thus, our video summarization model is based on a novel three-stream convolutional neural networks. The overview of our proposed method is shown in Fig. 1. We use both explicit and implicit information in the model. The spatial stream uses RGB images as input to represent the visual appearance of the video. The temporal stream uses multi-frame motion vectors to convey the motion information across frames. The gaze stream uses observers' eye movements to indicate human's preferences. To the best of our knowledge, we are the first to integrate the position and velocity of the raw eye movement data into a deep learning network for video summarization. We evaluate our proposed method on two video summarization benchmark datasets: SumMe [6] and TVSum [19]. Experimental results show that our proposed model is more effective in generating attractive video summaries for individuals, as well as multiple users, than other methods, demonstrating the importance of eye gaze data.

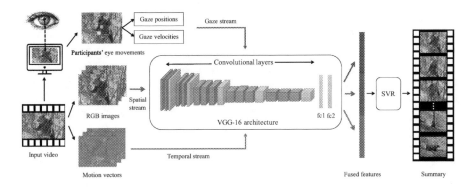

Fig. 1. Overview of the proposed method.

2   Gaze Aware Deep Learning Model

The pipeline of our proposed gaze aware deep learning model for video summarization (GVS) is displayed in Fig. 1. There are three streams: (1) gaze stream; (2) spatial stream; and (3) temporal stream. The gaze stream uses participants' eye movements (gaze positions or gaze velocities) to indicate viewers' preferences. The spatial stream utilizes the RGB images of videos to represent the visual appearance. The temporal stream applies multi-frame motion vectors to represent motion information across frames. These streams are implemented with the convolutional neural networks. We combine the output of each stream as features for each video frame. Then, the fused features are inputted to the support vector regression algorithm (SVR) [5] to predict highlight scores. Finally, the summary is comprised of those frames which have higher predicted scores.

2.1   The Input of Gaze Stream

The gaze stream takes the position and velocity of participants' raw eye movements as input. Firstly, we build up a representation S based on gaze position (GP representation) for the gaze stream. Due to the internal noise of the eye tracker, there is an approximate 1 degree error in the gaze data. Thus, we build up the GP representation by constructing a Gaussian distribution centered at collected fixation point. If the participant's fixation point is (x^t, y^t) in frame t, and the width and height of the input video are denoted as w and h, then the GP representation $S(x, y) \in \mathbb{R}^{w \times h}$ is calculated by:

$$S(x,y) = \frac{1}{2\pi\sigma^2} e^{-\frac{(x-x^t)^2+(y-y^t)^2}{2\sigma^2}} \tag{1}$$

where σ equals to 1 degree of visual angle. For multiple fixation points, we construct Gaussian distributions for each fixation point. When the distance of a pair of fixation points is smaller than σ, the overlap part takes the sum of the values in the overlapped distributions.

We also build up a representation based on gaze velocity (GV representation). We assume that GV representation of an arbitrary frame t is denoted as V_t. It records the velocity of the GP representation $S(x, y)$ from frame t to frame $t+1$. V_t^x and V_t^y are the horizontal and vertical components of V_t. We stack multiple GV representations to represent the movement information of fixation points across frames. A $2L$ input based on gaze velocities (GV input) for gaze stream is formed by stacking the GV representations from frame t to next $L-1$ frame. The GV input $\Upsilon_t \in \mathbb{R}^{w \times h \times 2L}$ for frame t is defined as:

$$\begin{cases} \Upsilon_t(2i-1) = V_{t+i-1}^x \\ \Upsilon_t(2i) \quad = V_{t+i-1}^y \end{cases}, 1 \le i \le L. \tag{2}$$

2.2   The Input of Temporal Stream

We construct multi-frame motion vectors to convey the motion information of the video to the model. Motion vectors have been proved to contain useful movement information for action recognition [25] and can be directly extracted as they are already encoded in compressed videos. The motion vectors of an arbitrary frame t are denoted as M_t. It contains the displacement from frame t to frame $t+1$. M_t^x and M_t^y represent the horizontal and vertical components of M_t. A $2L$ input (multi-frame motions vector) is stacked to convey the motion information between frame t and the next $L-1$ frames. L represents the stacking length. The multi-frame motion vectors $F_t \in \mathbb{R}^{w \times h \times 2L}$ for an arbitrary frame t is defined as:

$$\begin{cases} F_t(2i-1) = M_{t+i-1}^x \\ F_t(2i) \quad = M_{t+i-1}^y \end{cases}, 1 \le i \le L. \tag{3}$$

2.3   Gaze Aware Deep Model for Video Summarization

The overall flowchart of our proposed method is shown in Fig. 1. Given an input video, we first construct multi-frame motion vectors and prepare the input of gaze stream using the collected human eye movement data. Then, the gaze representations, RGB images and multi-frame motion vectors are delivered to the gaze stream, spatial stream and temporal stream respectively. Each stream is implemented with a VGG-16 deep convolutional neural networks [18]. The architecture is $C64 - C64 - C128 - C128 - C256 - C256 - C256 - C512 - C512 - C512 - C512 - C512 - C512 - F4096 - F4096 - F2$ with 13 convolution layers (represented by C with the number of neurons) and 3 fully-connected layers (represented by F with the number of neurons).

In the training stage, the input sources with their corresponding labels included in the datasets are utilized to train each stream. In the test stage, we combine the output of the second fully-connected layer (a 4096-dimensional vector) of each stream to be the features for each video frame. Then, the fused features are inputted to the subsequent SVR algorithm [5] to predict an interesting score for each frame. Finally, we construct the final video summary according to their predicted interesting scores. The final summary is comprised of those video frames with highest Q percentage of the predicted highlight scores. We use the standard toolbox LIBSVM [2] to perform SVR. Radial Basis Function (RBF) is applied as the kernel function. In addition, a grid search is run to find the optimal parameter settings.

3   Experimental Study

3.1   Datasets

All experiments were conducted on two video summarization benchmark datasets: SumMe [6] and TVSum [19]. The SumMe dataset contains 25 videos depicting various events such as cooking, plane landing, and base jumping. The video length varies from 1 to 7 min. Each video contains at least 15 user-annotated summaries and frame-level important scores. The important score indicates the possibility of a frame to be included in the final summary. The TVSum dataset contains 50 videos from 10 categories (e.g., parade, animal grooming, and bee keeping). The length of the video varies from 1 to 11 min. It provides 20 user annotations for each video, which represent as shot-level impor-tant scores of 1 to 5 (indicating not important to very important). Each shot lasts for 2 seconds.

3.2   Human Gaze Data Collection

We collected gaze data from six participants on the SumMe dataset and two participants on the TVSum dataset. Participants were asked to freely watch the entire video. The audio was muted to ensure only visual stimuli were used to direct the eye movements. Participants' eye movements were recorded with an SR Research EyeLink 1000 video-based eye tracker at 1000 Hz. In this paper, we refer to the persons who provided their personal summaries for the SumMe and TVSum datasets as users, and the persons we recorded gaze data as participants, and use their individual ID numbers to represent them.

3.3   Gaze Consistency Analysis

We first analyze the consistency of participants' eye movements to test the valid-ity of using the data we collected as an indicator of different observers' points of interest. If eye movements of participants are consistent, then gaze information may reflex some common interests from different observers, which can be helpful for generating satisfying summary results. We conducted down-sampling on the collected gaze data according to the frame rate of each video. NaN values were used for blinks.

In the analysis, two types of measurements are performed on the collected data: the similarity of fixation locations and the correlation of gaze trajectories. The analyses are conducted over six collected participants' eye movement data on SumMe dataset. Pairwise comparison is used, leading to 15 participant pairs.

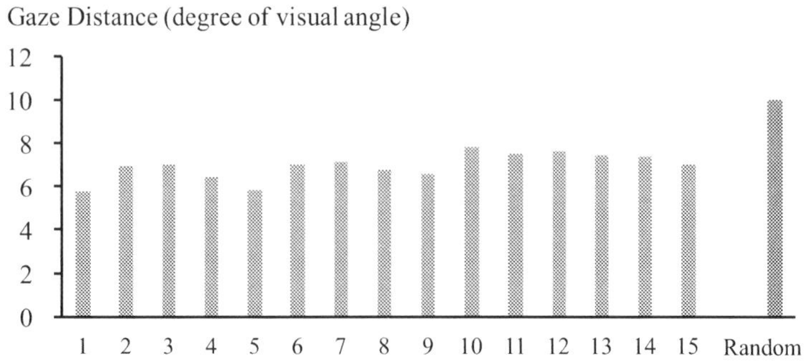

Gaze Distance (degree of visual angle)

Fig. 2. The average gaze distances of each participant pair on SumMe.

We use index $(1 \sim 15)$ to specify participant pairs. We first shuffle the temporal order of participants' eye movement sequences in each video to generate random sequences of eye movement. Then, we calculate the corresponding statistics based on the generated random data. These are repeated for 1000 times, and the average result is regarded as the random baseline. Blinks are excluded from the consistency analysis.

The Similarity of Fixation Locations. This analysis uses fixation distance to investigate gaze consistency in a single frame. The distance $d^t_{i,j}$ between participant i's and participant j's fixation points in frame t is calculated as follows:

$$d^t_{i,j} = \sqrt{(x^t_i - x^t_j)^2 + (y^t_i - y^t_j)^2} \tag{4}$$

where (x^t_i, y^t_i) and (x^t_j, y^t_j) represent the fixation points of participant i and participant j in frame t.

Figure 2 shows that the gaze distance of each pair is significantly smaller than the random baseline.

The Correlation of Gaze Trajectories. In this part, we try to find out if fixation locations are consistent in a whole video by calculating the correlation between two participants' gaze locations across a whole video. We suppose that vectors v_i and v_j denote the gaze trajectories of participant i and participant j in a video. We compute the correlation coefficients $r_{i,j}$ between these two vectors:

$$r_{i,j} = \frac{Cov(v_i, v_j)}{\sqrt{Cov(v_i, v_i)Cov(v_j, v_j)}} \tag{5}$$

where $Cov(x, y)$ calculates the covariance of the two variables x and y.

Figure 3 shows correlation coefficient values of gaze movement on X-axis and Y-axis by different lines. We can see that the values of two random baselines (yellow and green dashed lines) are around zero, which means that on average two

Correlation Coefficients

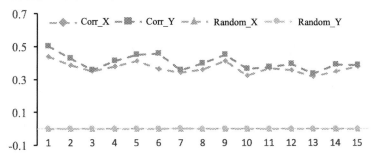

Fig. 3. The comparison of the average correlation coefficients of gaze trajectories with the random baselines on SumMe.

random fixation points are not correlated. However, the correlation coefficient values of participants' pairs are substantially higher than the random baselines.

From the above two types of measurements, we can see that the participants' eye movements are quite similar. Their gaze locations synonym are consistent in a single frame as well as a whole video.

3.4 Evaluation of the Gaze Aware Deep Learning Model for Video Summarization

In this part, we apply our proposed gaze aware deep learning models to video summarization tasks on two benchmark datasets and evaluate their performances.

Experimental Settings. We compare the automatic summary (A) with the human-annotated summarizations (H) and report the F-score to evaluate the performance of compared methods [6,19]. F-score is defined as:

$$F = \frac{2 \times p \times r}{p + r}, \tag{6}$$

$$r = \frac{\#matched \quad pairs}{\#frames \quad in \quad H} \times 100\%, p = \frac{\#matched \quad pairs}{\#frames \quad in \quad A} \times 100\% \tag{7}$$

where r is the recall and p is the precision. We report the mean F-score and the highest F-score by comparing the predicted summaries with the user summaries. The highest F-score also implies the comparison between the automatic summary and its most similar user summary. For SumMe, we apply the evaluation code provided by Gygli et al. [6]. For TVSum, we utilize the code provided by [27].

In SumMe, we randomly select two-thirds of the videos for training and apply the rest of the videos for testing. In TVSum, we randomly select 80 percent of the videos as training data and the rest as test data [13,27]. For the final summary, we apply the widely-used setting [6, 19] to set the summary length $Q = 15$. The ConvNets are pre-trained on the ImageNet dataset [4] to avoid the over-fitting. We implement our proposed method on Caffe platform [9] with the Tesla K80 GPUs. Motion vectors are extracted by applying the toolbox provided by Zhang et al. [26]. We follow [17] to set the stacking length L of GV representation and multi-frame motion vectors equal to 10.

We implemented different versions of our proposed gaze aware deep learning model for video summarization (GVS). In the spatial stream, they all use RGB images as the input. The temporal stream always takes input using multi-frame motion vectors (MV). In the gaze stream, there could be gaze position (GP) or gaze velocities (GV). Some versions of the models only take two of the three-stream structure (e.g. GVS-RGB&GP).

Comparison with State-of-the-art Methods. To demonstrate the effectiveness of our proposed method, we compare our work with several state-of-the-art methods, including Creating Summaries from User Videos (CSUV) [6], Summarizing Web Videos using Titles (SWVT) [19], Exemplar-based Subset Selection (ESS) [26], Learning Submodular Mixtures of Objectives (LSMO) [7], Video Summarization with Long Short-term Memory (dppLSTM) [27] and Unsupervised Video Summarization with Adversarial LSTM Networks (SUM-GAN$_{sup}$) [13]. Generally, CSUV, SWVT and ESS were based on hand-crafted features while LSMO, dppLSTM and SUM-GAN$_{sup}$ were based on deep learning methods.

The comparison results are shown in Table 1. Among the existing methods, for both datasets, deep learning based methods outperform the traditional models. More specifically, dppLSTM and SUM-GAN$_{sup}$ gain the best performances on average highest F-score. They take advantages of the latest deep learning architectures such as long-short term memory (LSTM) and generative adversarial network (GAN). For the SumMe dataset, our proposed three-stream models (GVS-RGB&MV&GP and GVS-RGB&MV&GV) outperform the two-stream models (GVS-RGB&MV, GVSRGB&GP and GVS-RGB&GV), suggesting that eye movement, when combined with spatial and temporal information, is an effective clue to indicate user's preferences. When compared with the state-of-the-art methods, our proposed methods gain considerable improvement on the SumMe dataset. We achieve higher average mean F-score (AMF) and average highest F-score (AHF). For the TVSum dataset, our proposed methods achieve slight improvement of AMF and significant improvement of AHF. These results suggest that our proposed architecture is competitive or even more proficient in generating similar summary with most of the user annotations and also capable of satisfying individual preferences. In the comparison with other deep learning methods, our proposed models even outperform the dppLSTM and SUM-GAN$_{sup}$, suggesting the importance of including the gaze stream for video summarization.

It is worth noting that although the participants who provided gaze information are not the users who selected video summaries, the proposed method can still improve the summary results by using the eye movement. It implies

Table 1. The performance comparisons of our proposed methods and the state-of-the-art on SumMe and TVSum datasets.

	Method	SumMe		TVSum	
		AMF	AHF	AMF	AHF
Existing methods	CSUV	23.4%	39.4%	----	----
	SWVT	26.6%	----	50.0%	----
	ESS	----	40.9%	----	----
	LSMO	----	39.7%	----	----
	dppLSTM	17.7%	42.9%	58.7%	78.6%
	SUM-GAN$_{sup}$	----	43.6%	61.2%	----
Proposed methods	GVS-RGB&MV	35.4%	56.3%	**62.8%**	83.8%
	GVS-RGB&GP	32.6%	53.9%	62.0%	83.6%
	GVS-RGB&MV&GP	35.8%	57.2%	62.2%	84.2%
	GVS-RGB&GV	32.8%	53.8%	62.2%	**84.4%**
	GVS-RGB&MV&GV	**36.0%**	**57.3%**	62.0%	84.0%

that gaze data is a robust source of information for video summarization tasks. Another interesting result of our experiment is that for the TVSum dataset, our proposed gaze-based model (GVS-RGB&GP and GVS-RGB&GV) gain similar performance with non gaze-based model GVS-RGB&MV. This suggests that for some videos, eye movement information might provide similar clues to the traditional temporal information (e.g. motion vectors).

4   Conclusion and Future Work

In this paper, we try to use participants' eye movements to indicate essential and attractive video content. A novel gaze aware deep model is proposed for video summarization. We first show that the gaze data of different participants are very similar to each other. Later, we apply our gaze aware model on two video summarization benchmark datasets. Results show that our proposed methods outperform several state-of-the-art methods, and also demonstrate that with the help of gaze information, our proposed model can generate better summaries.

Although we utilize the gaze data of the participants, who were not the users who annotated the summary, our model can still obtain better video summarization performance. We also used relatively small sample sizes. These imply that gaze data from a small number of participants could improve video summarization performance with very little additional work. In the future, we will build an end to end architecture for video summarization with eye movement information based on recent advances in webcam-based eye tracking [15] and human eye movements prediction [11].

Acknowledgments. This work was supported by the National Natural Science Foundation of China (No. 61502311, No. 61620106008), the Natural Science Foundation of Guangdong Province (No. 2016A030310053, 2016A030310039, 2017A030310521), the Science and Technology Innovation Commission of Shenzhen under Grant (No. JCYJ2016 0422151736824), Shenzhen Emerging Industries of the Strategic Basic Research Project under Grant (No. JCYJ20160226191842793), the Shenzhen high-level overseas talents program, the Tencent "Rhinoceros Birds"- Scientific Research Foundation for Young Teachers of Shenzhen University (2016), the National Institutes of Health Grant (5T32EY025201-03), and the Smith-Kettlewell Eye Research Institute Grant.

References

1. Chakraborty, P.R., Tjondronegoro, D., Zhang, L., Chandran, V.: Automatic identification of sports video highlights using viewer interest features. In: ICMR, pp. 55–62 (2016)
2. Chang, C.C., Lin, C.J.: Libsvm: a library for support vector machines. ACM TIST **2**(3), 1–27 (2011)
3. Chuk, T., Chan, A., Hsiao, J.: Hidden markov model analysis reveals better eye movement strategies in face recognition. In: CogSci (2015)
4. Deng, J., et al.: Imagenet: a large-scale hierarchical image database. In: CVPR, pp. 248–255 (2009)
5. Drucker, H., Burges, C.J.C., Kaufman, L., Smola, A.J., Vapnik, V.: Support vector regression machines. In: NIPS, pp. 155–161 (1997)
6. Gygli, M., Grabner, H., Riemenschneider, H., Van Gool, L.: Creating summaries from user videos. In: Fleet, D., Pajdla, T., Schiele, B., Tuytelaars, T. (eds.) ECCV 2014. LNCS, vol. 8695, pp. 505–520. Springer, Cham (2014). https://doi.org/10.1007/978-3-319-10584-0_33
7. Gygli, M., Grabner, H., Van Gool, L.: Video summarization by learning submodular mixtures of objectives. In: CVPR (2015)
8. Holmberg, N., Holmqvist, K., Sandberg, H.: Children's attention to online adverts is related to low-level saliency factors and individual level of gaze control. JEMR **8**(2), 1–10 (2015)
9. Jia, Y., et al.: Caffe: convolutional architecture for fast feature embedding. CoRR abs/1408.5093 (2014)
10. Jiang, W., Cotton, C., Loui, A.C.: Automatic consumer video summarization by audio and visual analysis. In: ICMR, pp. 1–6 (2011)
11. Li, Y., Fathi, A., Rehg, J.M.: Learning to predict gaze in egocentric video. In: ICCV, pp. 3216–3223 (2013)
12. Liu, Y., Zhong, S.H., Li, W.: Query-oriented multi-document summarization via unsupervised deep learning. In: AAAI, pp. 1699–1705 (2012)
13. Mahasseni, B., Lam, M., Todorovic, S.: Unsupervised video summarization with adversarial LSTM networks. In: CVPR (2017)
14. Mishra, A.K., Aloimonos, Y., Cheong, L.F., Kassim, A.: Active visual segmentation. TPAMI **34**(4), 639–653 (2012)
15. Papoutsaki, A., Sangkloy, P., Laskey, J., Daskalova, N., Huang, J., Hays, J.: Webgazer: Scalable webcam eye tracking using user interactions. In: IJCAI, pp. 3839–3845 (2016)
16. Salehin, M.M., Paul, M.: A novel framework for video summarization based on smooth pursuit information from eye tracker data. In: ICMR, pp. 692–697 (2017)

17. Simonyan, K., Zisserman, A.: Two-stream convolutional networks for action recognition in videos. In: NIPS, pp. 568–576 (2014)
18. Simonyan, K., Zisserman, A.: Very deep convolutional networks for large-scale image recognition. CoRR abs/1409.1556 (2014)
19. Song, Y., Vallmitjana, J., Stent, A., Jaimes, A.: Tvsum: summarizing web videos using titles. In: CVPR, pp. 5179–5187 (2015)
20. Truong, B.T., Venkatesh, S.: Video abstraction: a systematic review and classification. ACM TOMM **3**(1), 1–37 (2007)
21. Wu, J., Zhong, S.H., Jiang, J., Yang, Y.: A novel clustering method for static video summarization. MTAP **76**(7), 9625–9641 (2017)
22. Wu, J., Zhong, S.H., Ma, Z., Heinen, S.J., Jiang, J.: Foveated convolutional neural networks for video summarization. MTAP (2018)
23. Xu, J., Mukherjee, L., Li, Y., Warner, J., Rehg, J.M., Singh, V.: Gaze-enabled egocentric video summarization via constrained submodular maximization. In: CVPR, pp. 2235–2244 (2015)
24. Yao, T., Mei, T., Rui, Y.: Highlight detection with pairwise deep ranking for first-person video summarization. In: CVPR, pp. 982–990 (2016)
25. Zhang, B., Wang, L., Wang, Z., Qiao, Y., Wang, H.: Real-time action recognition with enhanced motion vector CNNs. In: CVPR, pp. 2718–2726 (2016)
26. Zhang, K., Chao, Wei, L., Sha, F., Grauman, K.: Summary transfer: exemplar-based subset selection for video summarization. In: CVPR (2016)
27. Zhang, K., Chao, W.-L., Sha, F., Grauman, K.: Video summarization with long short-term memory. In: Leibe, B., Matas, J., Sebe, N., Welling, M. (eds.) ECCV 2016. LNCS, vol. 9911, pp. 766–782. Springer, Cham (2016). https://doi.org/10.1007/978-3-319-46478-7_47
28. Zhong, S.H., Liu, Y., Li, B., Long, J.: Query-oriented unsupervised multi-document summarization via deep learning model. ESWA **42**(21), 8146–8155 (2015)
29. Zhong, S.H., Liu, Y., Liu, Y.: Bilinear deep learning for image classification. In: ACM MM, pp. 343–352 (2011)

Three-Stream Action Tubelet Detector for Spatiotemporal Action Detection in Videos

Yutang Wu[1,2], Hanli Wang[1,2(✉)], and Qinyu Li[1,3]

[1] Department of Computer Science and Technology, Tongji University,
Shanghai 201804, People's Republic of China
hanliwang@tongji.edu.cn
[2] Key Laboratory of Embedded System and Service Computing,
Ministry of Education, Tongji University,
Shanghai 200092, People's Republic of China
[3] Department of Computer Science, Lanzhou City University,
Lanzhou 730070, People's Republic of China

Abstract. In recent years, human action detection in videos has gained wide attention. Instead of detection frame by frame, a model named action tubelet (ACT) detector detects human actions sequence by sequence and achieves remarkable performances on both accuracy and speed in the form of two streams. In this work, a three-stream action tubelet detector (three-stream ACT detector) is proposed which adds an extra pose stream to obtain more information about human actions and fuses three streams by weighted average compared to the two-stream architecture. The experimental results on the benchmark UCF-Sports, J-HMDB and UCF-101 datasets demonstrate that the proposed three-stream ACT detector framework is able to boost the performance of human action detection.

Keywords: Human action detection · Three-stream architecture
Action tubelet detector · Pose stream

1 Introduction

With the rapid development of multimedia technologies in recent years, hundreds of millions of videos either long or short are produced every day. In view of this, video understanding has been a hot topic in the area of computer vision. As an important part of video understanding, action detection which not only classifies actions but also locates them in both space and time are now gaining more and

This work was supported in part by National Natural Science Foundation of China under Grants 61622115 and 61472281, Program for Professor of Special Appointment (Eastern Scholar) at Shanghai Institutions of Higher Learning (No. GZ2015005), Shanghai Engineering Research Center of Industrial Vision Perception & Intelligent Computing (17DZ2251600), and IBM Shared University Research Awards Program.

© Springer Nature Switzerland AG 2018
R. Hong et al. (Eds.): PCM 2018, LNCS 11165, pp. 296–306, 2018.
https://doi.org/10.1007/978-3-030-00767-6_28

more attention. There are many challenges to which action detection task faces, including noisy background, moving viewpoint, intra-class diversity, low-quality video data and so on, which makes it still unable to rival other well-studied visual computing tasks.

Hitherto, most state-of-the-art methods for spatiotemporal action detection in videos come down to the procedure as follows. A detector is designed to detect human actions at every time step, and then a linking algorithm is applied to link those eligible detections over time to build final spatiotemporal action tubes. The detectors used in these approaches are based on convolutional neural network (CNN) based object detectors [2,3,8,11–14]. Faster region-based CNN (R-CNN) [14] and single shot detector (SSD) [8] are proved to be the mainstream frameworks utilized in action detection task, the former more accurate, the latter faster. Saha et al. [15] and Peng et al. [10] both extend faster R-CNN to a two-stream variant, one stream for appearance and the other for motion, to better localize actions appearing in videos. Region proposals are processed from the appearance and motion region proposal networks (RPNs) respectively and then detection is produced by fusion [15]. Further in [10], the appearance and motion RPNs are combined before classification and box regression, then a multi-region scheme is applied to improve performance. Although these two methods [10,15] achieve remarkable results in accuracy, the processing speed is limited. To solve this problem, Singh et al. [16] work out a new framework to replace faster R-CNN with SSD and devise a greedy linking algorithm, making action detection to run at a real-time speed.

However, the aforementioned approaches treat videos as a set of separate frames that are processed by detector one at a time, which neglects the temporal trait of videos. More recently, as the extension of [16], the two-stream action tubelet (ACT) detector is introduced in [6] to surpass this limitation. The two-stream ACT detector performs action detection at the sequence level, that's to say, at a time a fixed number of consecutive frames instead of a single one are processed and a set of tubelets are generated accordingly. It yields state-of-the-art results when taking both accuracy and processing speed into consideration.

The success of two-stream architectures mentioned above has verified that the optical flow cue is very useful for deep networks to learn to detect actions in videos. Due to the complexity of action detection tasks, more visual cues are considered to be employed to better understand actions. Human pose cue has been proved to greatly benefit action recognition as shown in [18]. It explicitly captures the position of human body parts, making it easier to recognize actions. Inspired by these, a three-stream ACT detector is proposed in this work, which incorporates an extra stream similar to the appearance and motion ones but instead taking human pose as input to further capture the movements of human body parts. Experiments are carried out on the benchmark UCF-Sports [1], J-HMDB [5] and UCF-101 [17] datasets, which demonstrate that the proposed three-stream ACT detector is able to boost the performance of human action detection, with accuracy improved and processing speed maintained. The rest of this paper is organized as follows. Section 2 describes the details of the proposed three-stream ACT detector. The experimental results are shown and analyzed in Sect. 3. Finally, Sect. 4 concludes this work.

2   Proposed Three-Stream ACT Detector

This section first introduces an overview (Sect. 2.1) of the proposed three-stream ACT detector framework and then reveals its details mainly including system input (Sect. 2.2), the network to generate action tubelets (Sect. 2.3), fusion of three streams (Sect. 2.4) and the linking algorithm for building final action tubes (Sect. 2.5).

2.1   System Overview

An overview of the proposed three-stream ACT detector framework is illustrated in Fig. 1. Given a video, K consecutive frames are processed at a time (*e.g.*, frames $\{f_1, f_2, \cdots, f_K\}$, frames $\{f_2, f_3, \cdots, f_{K+1}\}$, and so on). The K frames, *i.e.*, raw images, together with the corresponding optical flow and human pose are respectively sent into three streams where for each stream an ACT detector is applied to classify and regress anchor cuboids (the extension to anchor boxes). The classification and regression of these three streams are then fused, converting anchor cuboids with fixed spatial extent into regressed tubelets with associated scores. For each frame only a small set of eligible tubelets are left as candidates after selection. Then an efficient linking algorithm is devised to iteratively add the tubelets to a set of existing tubelet links over time, building final action tubes for the entire video.

2.2   System Input

The inputs to the proposed three-stream ACT detector include raw images, optical flow and human pose. In this work, both optical flow and human pose are computed from raw RGB images, and human pose takes the form of human body part segmentation.

Optical Flow. Optical flow is computed for each frame according to [4]. Its horizontal and vertical components, as well as the magnitude are stacked to form the three dimensions of an image, which are then scaled by a constant s and converted into the [0, 255] range.

Human Pose. Considering that Fast-Net [9] achieves a good trade-off between quality and runtime in monocular road segmentation, its framework is used here for human body part segmentation. As shown in Fig. 2, this up-convolutional architecture is composed of a contractive part and a corresponding expansive part. The contractive part is similar to a classification network which provides coarse representation while the expansive part takes in the representation and outputs a high quality segmentation with the same resolution of the input image. In this work, Fast-Net is trained on the J-HMDB action recognition dataset which is fully annotated on human body part masks and joint positions to adapt to the task of body part segmentation. At test time, the high-resolution body part segmentation is converted into a three-dimensional image and each label is mapped into a RGB value that is predefined.

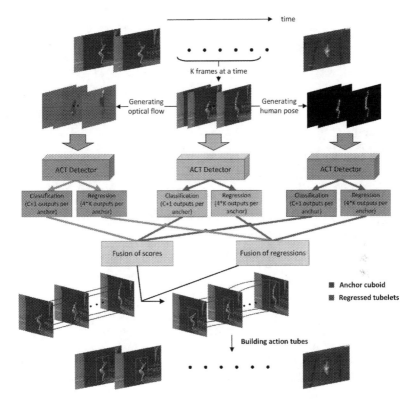

Fig. 1. Overview of the proposed three-stream ACT detector.

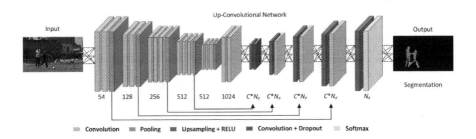

Fig. 2. Architecture of human body part segmentation.

2.3 Action Tubelet Detector

The action tubelet detector is designed mainly based on [6]. In this work, some parameters of the data layer are modified to accept the input of human pose. The architecture is shown in Fig. 3. Given K frames, each frame is sent into a set of SSD convolutional layers to compute features. The features from the same SSD layer are stacked together and then respectively sent into two convolutional layers, one is for classification and the other is for regression. Note that for each anchor cuboid, the classification layer outputs C scores for C classes and one extra score for background class, besides, the regression layer outputs $4 \times K$ coordinates with 4 for each frame.

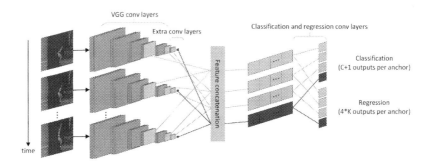

Fig. 3. Architecture of action tubelet detector.

In the training phase, only those sequences of frames in which the ground-truth action occurs in every frame are considered. For all the anchor cuboids, *Pos* is used to denote the set of positive anchor cuboids from which each cuboid has an overlap larger than σ with a ground-truth tubelet, and *Neg* corresponds to the set of the rest. Note that the overlap between tubelets is measured by averaging the intersection over union (IoU) of boxes over the K frames. Let $x_{ij}^l = \{0, 1\}$ be an indicator whose value is 1 if the i-th anchor cuboid is matched to the j-th ground-truth tubelet of label l. The overall loss of ACT detector is composed of localization loss L_{loc} and confidence loss L_{conf} as

$$L(x, s, p, g) = \frac{1}{N}(L_{conf}(x, s) + L_{loc}(x, p, g)), \tag{1}$$

where $N = \sum_{i,j,l} x_{ij}^l$ is the number of matched anchor cuboids. The confidence loss is based on a softmax function defined as

$$L_{conf}(x, s) = - \sum_{i \in Pos}^{N} x_{ij}^l \log(\hat{s}_i^l) - \sum_{i \in Neg} \log(\hat{s}_i^0), \tag{2}$$

where x refers to an anchor cuboid, s means confidence score, $\hat{s}_i^l$ denotes the confidence score of the i-th anchor cuboid for label l, and $\hat{s}_i^0$ stands for the confidence score of the i-th anchor cuboid classified to the background class. On the other hand, the localization loss employs a L1 based Smooth loss function between the predicted regression (p) and the ground-truth tubelet (g). Following the setting of SSD, the regression is for the center (cx, cy) as well as the width (w) and height (h) of each box in the anchor cuboid (a) as

$$L_{loc}(x, p, g) = \frac{\sum_{i \in Pos}^{N} \sum_{m \in \{cx, cy, w, h\}} x_{ij}^l \sum_{k=1}^{K} smooth_{L1}(p_i^{m_k} - \hat{g}_j^{m_k})}{K},$$

$$\hat{g}_j^{cx_k} = (g_j^{cx_k} - a_i^{cx_k})/a_i^{w_k}, \qquad \hat{g}_j^{cy_k} = (g_j^{cy_k} - a_i^{cy_k})/a_i^{h_k}, \qquad (3)$$

$$\hat{g}_j^{w_k} = log(\frac{g_j^{w_k}}{a_i^{w_k}}), \qquad \hat{g}_j^{h_k} = log(\frac{g_j^{h_k}}{a_i^{h_k}}).$$

2.4   Stream Fusion

For each stream, a set of coordinate regressions with associated scores are computed. The initial fusion method combines appearance and motion streams by just averaging the scores from these two streams, which ignores the different contribution that the two streams make to the detection results. In this work, for each anchor cuboid the final score on a certain label is defined as a weighted average of the corresponding confidence scores from the proposed three streams as

$$S_i^l = \alpha R_i^l + \beta F_i^l + \gamma P_i^l, \qquad (4)$$

where S_i^l, R_i^l, F_i^l and P_i^l denote the final score, the scores from the appearance stream, motion stream and pose stream for the i-th anchor cuboid on label l, respectively. Note that the sum of α, β and γ equals 1 for the convenience of comparing the contribution rates among three streams. Considering that appearance is more relevant for regressing bounding box especially for actions with limited motion or rough body part segmentation, the proposed fusion method only keeps the regressions from the RGB stream. After fusion, a set of action tubelets are produced by applying fused coordinate regressions and fused associated scores on the same set of anchor cuboids.

2.5   Linking Algorithm

As for building action tubes, this work just follows the linking method of [6], which is robust to missed detections. Given a video, for each set of K frames only the top N tubelets are kept for each class after applying non-maximum supression. Tubelet linking is done separately for each class and a link's score is the average scores of its component tubelets. At time f, each of the links generated up to time $f - 1$ and sorted by score in the decreasing order adds the highest scored tubelet starting at time f, on the premise that the added tubelet has an overlap larger than τ with the last tubelet of the link. Once a tubelet is added to a link, it cannot be picked by other links. Finally, the completed links are transformed into action tubes using temporal smoothing [6].

3   Experimental Result

3.1   Dataset and Evaluation Metric

In this work, the proposed three-stream ACT detector is evaluated on three benchmark datasets, including UCF-Sports [1], J-HMDB [5] and UCF-101 [17]. The following reviews these three datasets briefly and presents the metrics used for evaluation.

Dataset. For the UCF-Sports dataset [1], 150 videos for 10 sport classes are included. Each video is trimmed to action and provided with bounding box annotation for all frames. The training and test split method introduced in [7] is used. As far as the J-HMDB dataset [5] is concerned, it contains 928 videos for 21 actions. Each video is truncated to the duration of action. There are three train/test splits and the results averaged on these three splits are reported. The UCF-101 dataset [17] contains 3207 videos for 24 labels. Unlike the former two datasets, UCF-101 videos are longer and untrimmed. The results are only reported on the first split.

Evaluation Metric. Four evaluation metrics are employed in the experiments. They are defined in [6] to have a better comparison, including (1) *frame-mAP* which shows the frame-level performance of action detection, (2) *frame-MABO* which addresses the precision on action localization at the frame level, (3) *frame-CLF* which represents the classification accuracy at the frame level, and (4) *video-mAP* that presents the video-level performance of action detection.

3.2   Implementation

The proposed three-stream ACT detector framework is built on the caffe toolbox. As for the computation of optical flow, the constant s used to scale the three values of flow is set to 16. In the human pose generation period, Fast-Net is trained using SGD with the contractive part initialized with the VGG architecture. Each of these three streams in ACT detector is fine-tuned from VGG-Net which is pre-trained on the ILSVRC dataset. Data augmentation including sampling, resizing and photometric distortions is applied to make the model more robust to various input shapes. In order to maintain a balance between positive and negative training samples, hard negative mining is used following [8]. The resolution of input images is re-scaled to 300×300 and K, the number of frames as the input to the system, is set to 6 according to the experiments of [6].

3.3   Evaluation of Three-Stream ACT Detector

At first, the detection results from the proposed three streams are combined equally, in other words, the weight ratio of three groups of confidence scores is set to 1:1:1, which provides a fair and objective environment to compare the proposed three-stream ACT detector with [6], just considering the influence caused by the extra pose element. The frame-level metrics including *frame-mAP*, *frame-MABO* and *frame-CLF* are evaluated with an illustration shown in Table 1, and Table 2 displays the results on the video-level metric, *i.e.*, *video-mAP* at various overlap thresholds. Note that the overlap threshold 0.5:0.95 refers to an average value of *video-mAP* with an overlap thresholds ranging from 0.5 to 0.95 at the interval of 0.05.

Table 1. Comparison of frame-level metrics on UCF-Sports and J-HMDB.

Dataset	Method	*frame-mAP*	*frame-MABO*	*frame-CLF*
UCF-Sports	[6]	87.7	85.0	83.4
	Ours	**90.1**	**85.5**	**84.6**
J-HMDB	[6]	65.7	84.6	62.1
	Ours	**66.4**	**85.2**	**63.1**

Table 2. Comparison of *video-mAP* on UCF-Sports and J-HMDB.

Dataset	Method	Threshold			
		0.2	0.5	0.75	0.5:0.95
UCF-Sports	[6]	92.7	92.7	78.4	58.8
	Ours	**94.3**	92.7	**84.5**	**59.9**
J-HMDB	[6]	74.2	73.7	52.1	44.8
	Ours	74.2	73.7	**53.3**	**45.2**

As shown in Tables 1 and 2, almost all the values of frame-level and video-level metrics achieved by the proposed three-stream ACT detector are higher than those of [6], which verifies that pose cue indeed plays an important role in boosting the performance of action detection. More concretely, by adding human pose, there exists 1%~3% performance improvement on both the UCF-Sports and J-HMDB datasets. The notable gains can be explained by that pose stream provides explicit position of human body parts, which makes it easier and more accurate to locate and classify actions and then enhance the whole detection precision.

Since human pose is proved to work, further experiments are performed by reallocating the associated weights of the proposed three streams to explore the contribution degrees of the three visual cues on enhancing the performance of action detection. In this work, the contribution degrees of the three visual cues are assumed to be different from each other, and the total weight 1 is divided into 6 parts, with each of the three streams allocated one or two or three parts. The weight ratio is configured from 1:1:1 to six different values including 1:2:3, 1:3:2, 2:1:3, 2:3:1, 3:1:2 and 3:2:1. The results of both frame-level and video-level

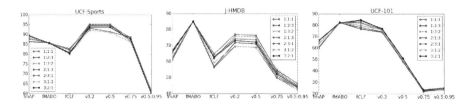

Fig. 4. Comparison on different weight ratios.

metrics are displayed in Fig. 4. It's not hard to see that fusion at the weight ratio 2:3:1 outperforms other fusion ways on the whole. This outcome shows that optical flow is very important for action detection, suggesting that the temporal information contained in videos contributes greatly in classifying and locating video actions. The RGB stream weights second owing to its clear representation for appearance which has been proved to be very helpful to detect actions in previous action detection methods. Although human pose alone doesn't lead the final results, it indeed gives a gratifying improvement when combining with optical flow, for that temporal dynamics of the position of human body parts over time are obtained, which further explores the temporal context of videos. In the future, more weight ratios will be considered.

3.4   Comparison with State-of-The-Arts

In this part, the proposed three-stream ACT detector is compared with several state-of-the-art approaches with the comparative *video-mAP* results shown in Table 3. From the results, it can be seen that the proposed framework outperforms all the other methods. Specifically speaking, the proposed three-stream ACT detector achieves remarkable performances compared to [6] which runs at a real-time speed on UCF-Sports, J-HMDB and UCF-101, with an overall increase of 4% in accuracy on the former two datasets and 24~28 frames per second in running speed. As compared to [10] with the multi-region (MR) strategy enabled or disabled, the proposed three-stream ACT detector also achieves better outcomes especially at high overlap thresholds. In a word, the proposed three-stream ACT detector brings a dramatic improvement on the performance of action detection.

Table 3. Comparison of *video-mAP* to state-of-the-art approaches.

Method	UCF-Sports				J-HMDB				UCF-101			
	0.2	0.5	0.75	0.5:0.95	0.2	0.5	0.75	0.5:0.95	0.2	0.5	0.75	0.5:0.95
[15]	-	-	-	-	72.6	71.5	43.3	40.0	66.7	35.9	7.9	14.4
[10] w/o MR	94.8	94.8	47.3	51.0	71.1	70.6	48.2	42.2	71.8	35.9	1.6	8.8
[10] with MR	94.8	94.7	-	-	74.3	73.1	-	-	72.9	-	-	-
[16]	-	-	-	-	73.8	72.0	44.5	41.6	73.5	46.3	15.0	20.4
[6]	92.7	92.7	78.4	58.8	74.2	73.7	52.1	44.8	77.2	**51.4**	22.7	25.0
Ours	**95.1**	**95.1**	**86.4**	**61.4**	**77.0**	**76.4**	**54.3**	**46.5**	**77.4**	51.2	**23.0**	**25.2**

4   Conclusion

In this work, a three-stream ACT detector framework is proposed for spatiotemporal action detection in videos. The proposed framework adds an extra pose stream which takes human pose as input to obtain the temporal dynamics of the position of human body parts. Unequal weights are respectively allocated to the proposed three streams, including RGB image, optical flow and human pose, on the fusion stage. The experimental results demonstrate that the proposed three-stream ACT detector is able to boost the performance of human action detection.

References

1. Brox, T., Bruhn, A., Papenberg, N., Weickert, J.: High accuracy optical flow estimation based on a theory for warping. In: Pajdla, T., Matas, J. (eds.) ECCV 2004. LNCS, vol. 3024, pp. 25–36. Springer, Heidelberg (2004). https://doi.org/10.1007/978-3-540-24673-2_3
2. Girshick, R.: Fast R-CNN. In: 2015 IEEE International Conference on Computer Vision (ICCV), pp. 1440–1448 (2015)
3. Girshick, R., Donahue, J., Darrell, T., Malik, J.: Rich feature hierarchies for accurate object detection and semantic segmentation. In: 2014 IEEE Conference on Computer Vision and Pattern Recognition (CVPR), pp. 580–587 (2014)
4. Gkioxari, G., Malik, J.: Finding action tubes. In: 2015 IEEE Conference on Computer Vision and Pattern Recognition (CVPR), pp. 759–768 (2015)
5. Jhuang, H., Gall, J., Zuffi, S., Schmid, C., Black, M.J.: Towards understanding action recognition. In: 2013 IEEE International Conference on Computer Vision (ICCV), pp. 3192–3199. IEEE (2013)
6. Kalogeiton, V., Weinzaepfel, P., Ferrari, V., Schmid, C.: Action tubelet detector for spatio-temporal action localization. In: 2017 IEEE International Conference on Computer Vision (ICCV), pp. 4405–4413 (2017)
7. Lan, T., Wang, Y., Mori, G.: Discriminative figure-centric models for joint action localization and recognition. In: 2011 IEEE International Conference on Computer Vision (ICCV), pp. 2003–2010 (2011)
8. Liu, W., et al.: SSD: single shot multibox detector. In: Leibe, B., Matas, J., Sebe, N., Welling, M. (eds.) ECCV 2016. LNCS, vol. 9905, pp. 21–37. Springer, Cham (2016). https://doi.org/10.1007/978-3-319-46448-0_2
9. Oliveira, G.L., Burgard, W., Brox, T.: Efficient deep models for monocular road segmentation. In: 2016 IEEE/RSJ International Conference on Intelligent Robots and Systems (IROS), pp. 4885–4891 (2016)
10. Peng, X., Schmid, C.: Multi-region two-stream R-CNN for action detection. In: Leibe, B., Matas, J., Sebe, N., Welling, M. (eds.) ECCV 2016. LNCS, vol. 9908, pp. 744–759. Springer, Cham (2016). https://doi.org/10.1007/978-3-319-46493-0_45
11. Redmon, J., Divvala, S., Girshick, R., Farhadi, A.: You only look once: Unified, real-time object detection. In: 2016 IEEE Conference on Computer Vision and Pattern Recognition (CVPR), pp. 779–788 (2016)
12. Redmon, J., Farhadi, A.: Yolo9000: better, faster, stronger. In: 2017 IEEE Conference on Computer Vision and Pattern Recognition (CVPR), pp. 6517–6525 (2017)
13. Redmon, J., Farhadi, A.: Yolov3: An incremental improvement. arXiv preprint arXiv:1804.02767 (2018)

14. Ren, S., He, K., Girshick, R., Sun, J.: Fast R-CNN: Towards real-time object detection with region proposal networks. In: 2015 Advances in Neural Information Processing Systems (NIPS), pp. 91–99 (2015)
15. Saha, S., Singh, G., Sapienza, M., Torr, P.H., Cuzzolin, F.: Deep learning for detecting multiple space-time action tubes in videos. arXiv preprint arXiv:1608.01529 (2016)
16. Singh, G., Saha, S., Sapienza, M., Torr, P., Cuzzolin, F.: Online real-time multiple spatiotemporal action localisation and prediction. In: 2017 IEEE Conference on Computer Vision and Pattern Recognition (CVPR), pp. 3637–3646 (2017)
17. Soomro, K., Zamir, A.R., Shah, M.: UCF101: A dataset of 101 human actions classes from videos in the wild. arXiv preprint arXiv:1212.0402 (2012)
18. Zolfaghari, M., Oliveira, G.L., Sedaghat, N., Brox, T.: Chained multi-stream networks exploiting pose, motion, and appearance for action classification and detection. In: 2017 IEEE International Conference on Computer Vision (ICCV), pp. 2923–2932 (2017)

Multi-person/Group Interactive Video Generation

Zhan Wang[✉], Taiping Yao, Huawei Wei, Shanyan Guan, and Bingbing Ni

Shanghai Key Laboratory of Digital Media Processing and Transmission,
Shanghai Jiao Tong University, Shanghai , China
{trillion_power,sndlytp,shyanguan,nibingbing}@sjtu.edu.cn,
weihuawei26@gmail.com

Abstract. Human motion generation from caption is a fast-growing and promising technique. Recent methods employ the latest hidden states of a recurrent neural network (RNN) to encode the skeletons, which can only address Coarse-grained motions generation. In this work, we propose a novel human motion generation framework which can simultaneously consider the temporal coherence of each individual action. Our model consists of two components: *Semantic Extractor, Motion Generator*. The *Semantic Extractor* can map caption into semantical guidance for fine motion generation. The *Motion Generator* can model the long-term tendency of each individual action. In addition, the *Motion Generator* can capture global location and local dynamics of each individual action such that more fine-grained activity generation can be guaranteed. Extensive experiments show that our method achieves a superior performance gain over previous methods on two benchmark datasets.

Keywords: Fine-grained · Temporal · Human motion

1 Introduction

Generating video from caption is a promising and valuable technique for advertising production and filmmaking. In the recent, although there have been extensive efforts on video generation [5, 16, 18, 20, 26], they can just generate videos obeying the distribution learned from the training data. In addition, what will be generated cannot be controlled at runtime, resulting in a coarse and incoherent generation. To address this issue, we propose a semantic driven video generation framework. Moreover, the temporal continuity and consistency of each individual motion, as well as their spatial influence are taken into full consideration. As a result, our proposed framework specializes in generating video with complex movement and frequent interaction. The main goal of our work is generating multi-person interaction activity videos, such as pushing and punching. Considering this prior information, we attempt to extract semantic constraints from the caption for every individual motion, which will guide the subsequent movements generation. For this purpose, we propose a *Semantic Extractor*. We adopt

© Springer Nature Switzerland AG 2018
R. Hong et al. (Eds.): PCM 2018, LNCS 11165, pp. 307–317, 2018.
https://doi.org/10.1007/978-3-030-00767-6_29

attention mechanism to execute the semantic extraction process. Unlike previous attention networks, which just focus on one point [25], our *Semantic Extractor* can attend to different points for each individual motion simultaneously.

The extracted semantic information of each individual motion is fed into the *Motion Generator*, which is responsible for generating motion videos of the group interaction. Recently, Generative Adversarial Networks (GANs) [7] and Variational Auto-Encoders (VAEs) [6,11] have gained great success on image generation. However, they are weak in the generation of videos, which demands good modeling ability for long-term dynamics. To this end, we propose a temporal VAE to automatically generate motion videos. At each timestep, we generate a frame from the current encoded latent space, and the characteristics of the current space are transferred to the next one. So that the generated action is smooth and continuous. Naturally, the generated video will have a better visual perception.

To sum up, we first extract different semantic guidances from caption. Then the guidance are sent to assist the generation of each individual motion and refine the generated action videos. Our proposed video generation framework achieves a superior performance gain over previous methods on two benchmark datasets.

2   Related Work

Recent years have witnessed the emergence of video generation like GAN [7] and VAE [4,11,21], which has been the hot spot in the industrial and academic research area. Existing methods typically employ generative models [15,17,19] to learn the distribution of the training data. VGAN [24] divides a video into foreground and background. And a generator combines them to approach a real video. TGAN [18] uses a temporal generator to generate a set of latent variables and uses a generator to converts them into an image sequence. Though a great success has been achieved by these methods, the process of generation is relatively random as there does not involve any semantic guidance over what will be generated at runtime.

To address the issue, some researchers have developed some techniques which can generate video from captions in recent past [17,23]. Generating images/videos from captions is relatively more difficult because of the complexity of the solution space. There are several attempts on caption guided video generation. Karol *et al.* [8] proposed a recurrent attentive architecture to improve the effect of the image generation. And a famous work Sync-DRAW is proposed by Gaurav *et al.* [14], which uses a VAE and a recurrent attention mechanism to maintain the structural integrity of objects. Tanya *et al.* [13] combines the use of long-term and short-term spatiotemporal contexts to generate videos on unseen captions maintaining a consistency between consecutive frames.

Compared with the above-mentioned methods, our proposed framework adequately considers the consistency and difference of each two adjacent frames. In our temporal VAE, the learned high-level characteristics of the current

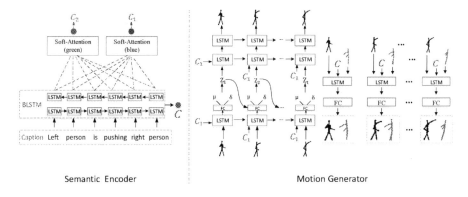

Semantic Encoder   Motion Generator

Fig. 1. Overview of our proposed network, we first encode the caption using attention model and get different latent representation. Then a temporal VAE architecture guides to generate each individual's motion based on different semantic representation and generate reasonable group interaction sequence.

generation are well preserved and transferred to the next generation. After the generation of each individual motion, our *Refiner* can further adjust the part movements of each frame.

3 Method

While previous works [13,14] either extract the semantic representation without fine-grained attention or generate videos directly giving different focuses to different components. To address this issue, on the one hand, we adopt soft attention model to extract fine-grained semantic representation for each individual motion, on the other hand, instead of generating multi-person interaction directly, we first propose a temporal VAE architecture to generate each individual motion respectively assisting by the extracted semantic representation. And we usefully connected layers to refine each individual motion and reconstruct a reasonable group interaction.

As illustrated in Fig. 1, Our proposed network consists of two parts. First, a semantic extractor uses a soft-attention model based on a bi-directional LSTM to extract different latent representations of the fine-grained caption. Then, the temporal coherence in generating motion sequence of each individual modeled by our proposed temporal VAE architecture models. Finally, the refiner based on fully connected layer refines the interaction among individuals to achieve a reasonable human motion sequence.

Supposing a caption contains N words and describes an interaction of M individuals. We define C^k as the extracting semantic representation of the k-th individual ($k = 1, 2, \cdots, M$). Instead of generating videos directly, we first generate skeleton joints. Each individual contains 15 skeleton joints, S_t^k is a 30-dimensional vector, which denotes the xy coordinates of the skeleton joints of the k-th individual at t time step. Obviously, S^k denotes the skeleton joints of the k-th individual in all frames. After generating the skeleton joints for every individual, we generate interaction video given a reference image.

3.1 Semantic Extractor

To generate the fine-grained human motion sequence we expect, we must provide the fine-grained description of the interaction. We generate each individual motion via pay attention to the specific words in the caption. For example, '*the left tall person is punching the right short person*' is a fine-grained caption. When we generate the motion of the left person, considering this prior information, we should focus on these words: '*left*', '*tall*', '*is*' and '*punching*'.

For *semantic extractor*, we use a bi-directional LSTM to model it, which is consists of a forward LSTM and a backward LSTM. Since the input caption contains N words, we denote by $(h_1^f, \cdots, h_N^f)$ as forward hidden states. Similarity, $(h_1^b, \cdots, h_N^b)$ denotes backward hidden states. Then we obtain an annotation for each word by concatenating the forward hidden state h_j^f and the backward one h_j^b, i.e., $h_j = \left[h_j^f; h_j^b \right]$ $(j = 1, 2, \cdots, N)$. Finally, we obtain the latent representation $H = [h_1, h_2, \cdots, h_N]$. We need focus on specific words when generating each individual's motion sequence. We adopt a soft-attention layer to create the long-term context representation C^k which is shown in Fig. 1. The output of soft-attention is then given by:

$$
\begin{aligned}
C^k &= attention(H) \\
&= \alpha_1^k h_1 + \alpha_2^k h_2 + ... + \alpha_N^k h_N
\end{aligned}
\tag{1}
$$

$$
\alpha_i^k = \frac{exp\left((V^k)^T tanh\left(W^k h_i + b^k \right) \right)}{\sum_{j=1}^{N} exp\left((V^k)^T tanh\left(W^k h_j + b^k \right) \right)}
\tag{2}
$$

where V, W, b are the network parameters. The representation C without attention defined as Eq. 3. After encoding, C^k is used to guide the motion of k-th individual, and C will guide the global generation in the following refinement.

$$
C = \frac{1}{N}(h_1 + h_2 + ... + h_N)
\tag{3}
$$

3.2 Motion Generation

For the k-th individual, let $Y^k = \{Y_1, Y_2, \cdots, Y_n\}$ denotes the distribution over his motion in the video, Y_i denotes his motion at i-th time step, where n denotes the n frames of the video. Then we model the conditional distribution of generating random motion sequence Y^k as $P(Y^k|C^k)$ for given C^k. To consider the time correlation in generated motion sequence, we redefine $P(Y^k|C^k)$ as follows:

$$
P(Y^k|C^k) = \prod_{i=1}^{n} P(Y_i|Y_{i-1}, \cdots, C^k)
\tag{4}
$$

Our objective is to maximize the likelihood of generating a reasonable motion sequence based on the input C^k and individual skeleton joints S^k. $P(Y^k|C^k)$ is a complex multi-modal distribution, which models the various possibilities of generating a video given a caption. For example, Given the caption '*the left tall person is punching the right short person*', the framework can generate different locations and different heights of two individuals, even different time slices of the punching interaction. In order to model the $P(Y^k|C^k)$, we design a temporal VAE architecture to capture the temporal coherence and generate motion sequence. Unlike [12,14] and [13], our recurrent model is adapted to capture the temporal coherence between frames, not to iteratively improve the quality of the generation. We first encode the C^k and S^k of the k-th individual. Then, we introduce a latent variable Z to our proposed temporal VAE, and the sample z denotes sampling a particular value of Z. Therefore, we define the $P(Y^k|C^k)$ as follows:

$$P(Y^k|C^k) = \mathcal{N}\left(f\left(z, C^k\right), \sigma^2 * \mathcal{I}\right) \tag{5}$$

In the training stage, for the k-th individual, let C^k denotes the caption representation, $S^k = \{S_1^k, S_2^k, \cdots, S_T^k\}$ denotes the skeleton sequence data. T is the time step. First, a LSTM encoder encodes the skeleton data based on C^k. Then, we model the latent variable $Z = \{Z_1, Z_2, \cdots, Z_T\}$. And we use Z_t to denote the latent variable of t time step. As shown in Fig. 1, all the latent variables Z also depend on the previous latent variable except for Z_1 (i.e., the latent variable at first time step), the equation as follows:

$$h^k = LSTM_{enc}(S^k, C^k) \tag{6}$$

$$Z_t \sim P(Z_t|Z_{1:t-1}) \tag{7}$$

$$Z' = LSTM_{enc-z}(Z) \tag{8}$$

$$O^k = LSTM_{dec}(Z', C^K) \tag{9}$$

Note the h^k denotes the output of the LSTM encoder for the k-th individual, which has a different footnote with caption representation h_j. Similar to [11], we achieve the mean and variance via equation shown as follow:

$$\mu_t = W_\mu(h_t, Z_{1:t-1}) \tag{10}$$

$$\sigma_t = exp(W_\sigma(h_t, Z_{1:t-1})) \tag{11}$$

$$P(Z_t|Z_{1:t-1}) = \mathcal{N}(\mu_t, \sigma_t) \tag{12}$$

After getting the latent variable Z, we use an LSTM to model the temporal coherence of Z. Finally, another LSTM is adopted as a decoder which generates the skeleton sequence data O^k. In test stage, Z_t is sample from standard gaussian distribution $\mathcal{N}(0, I)$. Then the decoder outputs the generated motion sequence.

Supposing there are M individuals in an interaction, the temporal VAE architecture will output motion sequence of all individuals, $O = \{O^1, O^2, \cdots, O^M\}$. However, these motion sequences are generated respectively without considering the interaction between each other. To address this issue, we propose a refiner to refine the interaction among individuals.

Our proposed refiner consists of LSTM and three fully connected layers which are illustrated in Fig. 1. The final output of refiner is reasonable human motion skeleton sequence, which set decoding motion sequence O and caption representation C (shown in Eq. 3) as input. Since we have predicted each person's skeleton, we use the method in [2] to generate videos based on generated skeleton sequences. Given an appearance image and generated skeleton sequences, the video generator outputs a corresponding video.

3.3   Loss Function

Similar to a standard VAE [11], the loss function for our proposed network is composed of two types of losses. The first loss is the reconstruction loss L_X, computed as the L_2 loss between the original motion sequences and the generated motion sequences. The second is the KL-divergence loss, L_Z, defined between a latent prior $P(Z_t)$ and $Q(Z_t|h_t, Z_{1:t-1}) \sim \mathcal{N}(\mu^t, (\sigma^t)^2)$ and summed over all T timesteps and all K individuals. Namely, we assume prior to a standard normal distribution and loss is given as follows:

$$L_X = \sum_1^K \sum_1^T (S_{k,t} - \hat{S_{k,t}})^2 \tag{13}$$

$$L_Z = \sum_1^K \left\{ \frac{1}{2} \left(\sum_1^T \mu_t^2 + \sigma_t^2 - log\sigma_t^2 \right) - \frac{T}{2} \right\} \tag{14}$$

$$L_{all} = L_X + L_Z \tag{15}$$

where K is the number of the individuals involved in the interactive activity, T is the length of frames we want to generate. And $S_{k,t}$ denotes the skeleton joints of the k-th individual at t time step, while $\hat{S_{k,t}}$ is the corresponding ground truth.

3.4   Implementation Details

Our implementation is based on Tensorflow [1]. The hidden state size of semantic bi-direction LSTM and motion generator LSTM are set as 512 and 256. In order to avoid exploding gradient in LSTMs, we apply gradient clipping by 5. We train our framework with stochastic gradient descent. We minimize the KL divergence loss and l2-loss from motion generator to update the parameters of the network. We use the Adam optimizer [10] and fixed learning rate of 0.0002 and momentum term of 0.5. The latent noise code has 100 dimensions, which we sample from a normal Gaussian distribution in the testing stage. Besides, we set batch size as 16. Training typically took 24 h on a GPU (TITAN X).

(a) The result of Sync with caption 'A left tall person is punching a right short person'

(b) Our result with caption 'A left tall person is punching a right short person'

Fig. 2. The comparison (with caption) with Sync [14] and Attentiv [13] on SBU.

4 Experiment

In this section, we will first compare our method with state-of-the-art methods with caption and make an in-depth analysis. Furthermore, we compare our proposed method with several baseline methods without caption guidance. Finally, the quantitative analysis of our proposed method is also performed. The evaluations are performed on the following two human action datasets:

SBU Dataset. SBU dataset [27] is an interaction dataset with two subjects. It about 300 sequences of 8 class interactions, including 'approaching', 'pushing', 'kicking', 'punching', 'exchanging objects', 'hugging' and 'shaking hands'. Due to the lack of training data, we augment the data by flipping. Besides we run 5-fold training and testing as suggested in [27] for each activity.

Choi's New Dataset. Choi's New Dataset [3] is composed of 32 videos clips with 6 collective activities: gathering, talking, dismissal, walking together, chasing and queueing. Similarity, we augment data by flipping. Besides, we run 3-fold training and testing for each activity.

For all the dataset mentioned above, we use pose estimation algorithm [2] to get skeleton joints of each video. And we perform 10 frames video generation experiment on both two datasets. We use a sliding window method to obtain these fixed-length videos since all of these videos are larger than 10 frames. *e.g.*, if we have a 15 frames video, we will extract the frames from 1*st* to 10*th* at the first time. We use this method to get about 600 videos in each collection.

As for caption achieved, we manually created captions such as 'A right short person is pushing a left tall person.'. And for group dataset (Choi's New Dataset), one of the captions is 'A left short person and a middle tall person and a right tall person is gathering.'. All these captions must include fine-grained words like ('left high person') and interactive words ('is punching'). We first use the semantic extractor to get a semantic feature. We then perform our proposed fine-grained motion generator to achieve human motion sequence based on the caption representation. Finally, we achieve the fine-grained motion video based on our video generation module (Fig. 3).

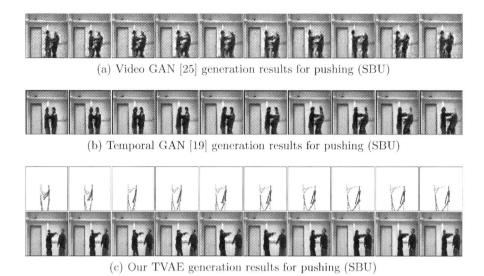

(a) Video GAN [25] generation results for pushing (SBU)

(b) Temporal GAN [19] generation results for pushing (SBU)

(c) Our TVAE generation results for pushing (SBU)

Fig. 3. The results of video generation without caption and comparison with two state-of-the-art methods on SBU dataset. Note our method shows the generated skeleton sequence.

4.1 Comparisons with the State-Of-The-Art Methods

To evaluate the effect of our generated video based on a caption, we provide a qualitative comparison for our proposed temporal VAE with serval state-of-the-art methods. The results on SBU and Choi's dataset [27] are shown in Fig. 2. Meanwhile, we retrain the *Attentive Semantic Draw* [13] and *Sync-Draw* [14] models on SBU dataset with that caption labeled by us. We resize the inputs size to 64×64 and perform *'read'* and *'write'* attention model for each channel. The final RGB results are the concatenation of the three channel. Note that we use the same results for these two methods because they use the same official released code. As illustrated in Fig. 2, we observe that the body (head or arm) of some frames of the *Sync-Draw* [14] and *Attentive Semantic Draw* have disappeared. It demonstrates that our generated video is more in accordance with realistic scene than them. The key reason is that these methods only consider global information rather than capture global location and local dynamics of each individual motion.

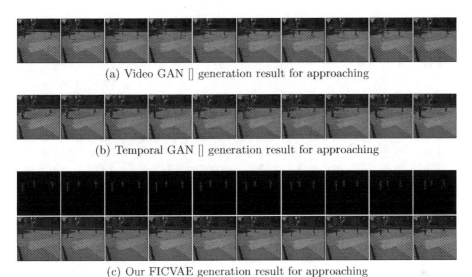

(a) Video GAN [] generation result for approaching

(b) Temporal GAN [] generation result for approaching

(c) Our FICVAE generation result for approaching

Fig. 4. The results of video generation without caption and comparison with two state-of-the-art methods on Choi's Dataset. Note our method shows the generated skeleton sequence.

4.2 Comparisons with the Video Generation Method Without the Caption

We further compared our temporal VAE (TVAE) with several baselines without the caption. For comparison, we use two models, (1) a video GAN proposed by Carl *et al.* [24], *i.e.*, VGAN. (2) A Temporal GAN (TGAN) proposed by Masaki *et al.* [18]. We train VGAN with the normal GAN loss, where we resize the input video to the resolution 64×64. And we train TGAN with WGAN loss on that two datasets mentioned above. Similarly to VGAN, we also resize input to resolution 64×64. The experimental results on SBU dataset [27] and Chios' New dataset [3] are illustrated in Fig. 4, because the spatial-temporal convolution cannot effectively handle the temporal relationship, the temporal coherence of our results are better than that two models and achieve higher quality. As for TGAN, using a temporal generator to generate latent variables cannot consider the coherence of adjacent latent variable. It demonstrates that our framework can effectively improve the quality of human motion video generation.

4.3 Quantitative Analysis

Since our network can generate videos based on a fine-grained caption, it can even be employed to artificially create a labeled dataset. Hence, we quantitatively evaluated performance on classifying actions on SBU dataset [27]. To make a fair analysis, we first use a 3D spatio-temporal convolution network [22] to extract the feature, which is trained on Sports 1M Dataset [9]. Then we trained an

Table 1. Recognition Accuracies (%) on the generated sequence for SBU Dataset and The Choi's New Dataset.

Method	Approaching (SBU)	Kicking (SBU)	Punching (SBU)	Pushing (SBU)	Gathering (Choi's)	Queueing (Choi's)	Walking together (Choi's)
GT	70.45%	73.28%	68.84%	66.37%	65.66%	67.81%	71.36%
Ours	68.46%	70.21%	63.77%	61.65%	61.13%	62.77%	70.96%
Sync	66.28%	67.61%	61.36%	60.03%	59.60%	60.23%	68.94%
VGAN	56.74%	53.48%	50.03%	49.64%	50.73%	52.18%	58.69%
TGAN	67.49%	65.78%	60.03%	59.32%	60.24%	58.37%	66.19%

SVM on real SBU Dataset [27] with action labels corresponding to the captions. We test the classifying accuracy on sync-draw, attentive-draw and our TVAE method. It demonstrates that the generated sequences can be recognized by the off-the-shelf classifier. The results are shown in Table 1.

5   Conclusion

Considering the fact that existing video generation methods are weak in generating videos with complex interactions, we present a novel framework which divides video generation into generations of each individual motion and combines them to an integral video involving the interactions among them. As evident from the results, our approach is highly promising in capturing the various details of object motion.

References

1. Abadi, M., et al.: Tensorflow: a system for large-scale machine learning. In: OSDI, vol. 16, pp. 265–283 (2016)
2. Cao, Z., Simon, T., Wei, S.E., Sheikh, Y.: Realtime multi-person 2D pose estimation using part affinity fields. In: CVPR (2017)
3. Choi, W., Savarese, S.: A unified framework for multi-target tracking and collective activity recognition. In: Fitzgibbon, A., Lazebnik, S., Perona, P., Sato, Y., Schmid, C. (eds.) ECCV 2012. LNCS, vol. 7575, pp. 215–230. Springer, Heidelberg (2012). https://doi.org/10.1007/978-3-642-33765-9_16
4. Chung, J., Kastner, K., Dinh, L., Goel, K., Courville, A.C., Bengio, Y.: A recurrent latent variable model for sequential data. In: NIPS, pp. 2980–2988 (2015)
5. Denton, E.L., et al.: Unsupervised learning of disentangled representations from video. In: NIPS, pp. 4417–4426 (2017)
6. Fabius, O., van Amersfoort, J.R.: Variational recurrent auto-encoders. arXiv:1412.6581 (2014)
7. Goodfellow, I., et al.: Generative adversarial nets. In: NIPS, pp. 2672–2680 (2014)
8. Gregor, K., Danihelka, I., Graves, A., Rezende, D.J., Wierstra, D.: Draw: a recurrent neural network for image generation. arXiv:1502.04623 (2015)

9. Karpathy, A., Toderici, G., Shetty, S., Leung, T., Sukthankar, R., Fei-Fei, L.: Large-scale video classification with convolutional neural networks. In: CVPR, pp. 1725–1732 (2014)

10. Kingma, D.P., Ba, J.: Adam: A method for stochastic optimization. arXiv:1412.6980 (2014)

11. Kingma, D.P., Welling, M.: Auto-encoding variational bayes. arXiv:1312.6114 (2013)

12. Mansimov, E., Parisotto, E., Ba, J.L., Salakhutdinov, R.: Generating images from captions with attention. arXiv:1511.02793 (2015)

13. Marwah, T., Mittal, G., Balasubramanian, V.N.: Attentive semantic video generation using captions. In: 2017 ICCV, pp. 1435–1443. IEEE (2017)

14. Mittal, G., Marwah, T., Balasubramanian, V.N.: Sync-draw: Automatic video generation using deep recurrent attentive architectures. In: ACMMM, pp. 1096–1104. ACM (2017)

15. van den Oord, A., Kalchbrenner, N., Espeholt, L., kavukcuoglu, k., Vinyals, O., Graves, A.: Conditional image generation with pixelcnn decoders. In: Lee, D.D., Sugiyama, M., Luxburg, U.V., Guyon, I., Garnett, R. (eds.) NIPS, pp. 4790–4798. Curran Associates, Inc. (2016)

16. Oord, A.v.d., Kalchbrenner, N., Kavukcuoglu, K.: Pixel recurrent neural networks. arXiv:1601.06759 (2016)

17. Reed, S., Akata, Z., Yan, X., Logeswaran, L., Schiele, B., Lee, H.: Generative adversarial text to image synthesis. arXiv:1605.05396 (2016)

18. Saito, M., Matsumoto, E., Saito, S.: Temporal generative adversarial nets with singular value clipping. In: ICCV, pp. 2830–2839 (2017)

19. Salakhutdinov, R., Larochelle, H.: Efficient learning of deep boltzmann machines. In: ICAISC, pp. 693–700 (2010)

20. Sohn, K., Lee, H., Yan, X.: Learning structured output representation using deep conditional generative models. In: Cortes, C., Lawrence, N.D., Lee, D.D., Sugiyama, M., Garnett, R. (eds.) NIPS, pp. 3483–3491. Curran Associates, Inc. (2015)

21. Sohn, K., Lee, H., Yan, X.: Learning structured output representation using deep conditional generative models. In: NIPS, pp. 3483–3491 (2015)

22. Tran, D., Bourdev, L., Fergus, R., Torresani, L., Paluri, M.: Learning spatiotemporal features with 3d convolutional networks. In: ICCV, pp. 4489–4497. IEEE (2015)

23. Venugopalan, S., Rohrbach, M., Donahue, J., Mooney, R., Darrell, T., Saenko, K.: Sequence to sequence-video to text. In: ICCV, pp. 4534–4542 (2015)

24. Vondrick, C., Pirsiavash, H., Torralba, A.: Generating videos with scene dynamics. In: NIPS, pp. 613–621 (2016)

25. Xu, K., Ba, J., Kiros, R., Cho, K., Courville, A., Salakhudinov, R., Zemel, R., Bengio, Y.: Show, attend and tell: Neural image caption generation with visual attention. In: ICML, pp. 2048–2057 (2015)

26. Yan, Y., Xu, J., Ni, B., Zhang, W., Yang, X.: Skeleton-aided articulated motion generation. In: ACMMM, pp. 199–207. ACM (2017)

27. Yun, K., Honorio, J., Chattopadhyay, D., Berg, T.L., Samaras, D.: Two-person interaction detection using body-pose features and multiple instance learning. In: CVPRW. IEEE (2012)

Image Denoising Based
on Non-parametric ADMM Algorithm

Xinchen Ye[1(✉)], Mingliang Zhang[1,2], Qianyu Yan[1], Xin Fan[1],
and Zhongxuan Luo[1]

[1] DUT-RU International School of Information Science and Engineering,
Key Laboratory for Ubiquitous Network and Service Software of Liaoning Province,
Dalian University of Technology, Dalian, China
`yexch@dlut.edu.cn`
[2] School of Mathematical Sciences, Dalian University of Technology, Dalian, China

Abstract. Image denoising is one of the most important tasks in image processing. In this paper, we propose a new method called Non-ParaMetric Alternating Direction Method of Multiplier (ADMM) algorithm (NPM-ADMM). We utilize the standard ADMM algorithm to solve the noisy image model and update the parameters via back propagation by minimizing the loss function. In contrast to the previous methods which are required to set the parameters carefully to approach better results, the proposed method can automatically learn the related parameters without the need of manually specifying. Furthermore, the filter coefficients and the nonlinear function in the regularization term are also learned together with the parameters, rather than fixed. Experiments on image denoising demonstrate our superior results with fast convergence speed and high restoration quality.

Keywords: Image denoising · Prior learning · Neural network
Alternating direction method of multiplier

1 Introduction

In computer vision, image denoising is a basic image restoration problem. The denoising model can be expressed by

$$y = x + \eta, \tag{1}$$

where $x \in \mathbf{R}^{M \times N}$ is the original image, $y \in \mathbf{R}^{M \times N}$ is the corrupted image and $\eta \in \mathbf{R}^{M \times N}$ is an additive zero-mean white Gaussian noise. In recent years, there are many scholars who focus on image denoising and have proposed many efficient methods. Those methods can be generally divided into

This work was supported by National Natural Science Foundation of China (NSFC) under Grant 61702078, and by the Fundamental Research Funds for the Central Universities.

R. Hong et al. (Eds.): PCM 2018, LNCS 11165, pp. 318–328, 2018.
https://doi.org/10.1007/978-3-030-00767-6_30

three classes: local filtering methods, global optimization methods and learning based methods. Local filtering methods such as mean filtering, median filtering and transform-domain filtering [3] are simple but usually have unsatisfied visual result. The global optimization methods are the predominant approaches in the past decades, which have a sound mathematical framework to guarantee the existence of the solutions. One of these approaches is based on variational method to construct an energy function of image and to minimize it, such as total variation (TV)[26]. Perona et al. [17] proposed a nonlinear diffusion model from partial differential equation's (PDEs) perspective, which can preserve edges and eliminate noise simultaneously. It is worth noting that the global optimization methods are handcrafted, and the model parameters are set carefully to obtain denoised images.

To avoid manually tuning parameters, many scholars attempt to automatically learn related parameters [6,11,16,21]. For example, Chen et al. [6] established a trainable nonlinear reaction diffusion framework to restore an image. Schmidt et al. [21] proposed a cascade of shrinkage fields (CSF) model which replaces regularization term with a flexible shrinkage function - Gaussian radial basis functions (RBFs) kernel [23] to learn parameters. Kim et al. [11] designed a deep networks to implicitly represent the regularization term.

The idea of our proposed method is similar to [21], while the difference is that we employ alternating direction method of multiplier (ADMM) algorithm [2,24] to solve denoising model rather than half-quadratic optimization method which usually has slower convergence rate, especially for non-convex problem and demands the penalty parameters to close infinity to get a optimal solution [22]. Our main contributions can be summarized as follows:

- The proposed method (NPM-ADMM) can automatically learn the related parameters including the filter coefficients and the nonlinear function, and the process of solving the model can be viewed as a optimization network by minimizing the loss function.
- We utilize two kinds of learnable regularization functions to train our network and prove its effectiveness in our experiments.
- The proposed model has good denoising performance and can rapidly converge to the desired solutions with lower training loss.

2   Related Work

We briefly review the denoising methods based on priors, together with the non-parametric image restoration methods presented recently. To approximate the regularization term in denoising, l_0-norm is a natural measurement, but it usually leads to nondeterministic polynomial (NP) hard problems. The total variation (TV) proposed by [26] is one of the most famous techniques which has good performance on edge-preserving. Subsequently, Bredies et al. [4] improved on TV by using total generalized variation (TGV), and Krishnan et al. [12] substituted l_p-norm for TV. Note that the above parameters of models are manually designed to obtain a good denoising result.

In terms of non-parametric image restoration methods, Chen et al. [6] proposed a trainable nonlinear reaction diffusion model from the perspective of Partial differential equation. Kim et al. [11] and Meinhardt et al. [16] proposed to learn proximal mapping using denoising network instead of the handcrafted denoiser. Chan et al. [5] put forward a Plug-and-Play ADMM algorithm and its main idea is to regard the proximal mapping as a denoising step, namely a denoiser which could be any off-the-shelf image denoising algorithms.

In our proposed method, we establish a learning-based denoising model which can automatically learn the corresponding parameters, especially the filter coefficients as well as the nonlinear function in the regularization term, and derive an algorithm using augmented Lagrange multiplier (ALM) method [13] plus ADMM framework.

3   Proposed Method

3.1   Denoising Model and Its ADMM Algorithm

From the perspective of MAP estimation [18], the image restoration problem can be expressed as

$$\hat{x} = \arg\min_x \quad f(x) + \lambda g(x), \tag{2}$$

where $f(x)$ and $g(x)$ are known as data term and regularization term respectively, $f(x) = \frac{1}{2}\|y-x\|^2$ in particular and λ is a regularization parameter which controls the balance between $f(x)$ and $g(x)$.

Based on the observation and analysis [10], the gradient of x (Dx, D: the difference matrix) is subject to a heavy-tailed distribution, so Dx has sparsity in general, or in other words the gradient has less non-zero value. Usually, the heavy-tailed distributions are replaced with Laplacian distributions or Gaussian distributions which can be approximately modeled by using l_1-norm as a regularizer [25], i.e., $\|Dx\|_1$. However, $\|Dx\|_1$ is hypothetical and handcrafted, which can not fully capture the image statistical characteristics. So in this paper, we will adopt learnable nonlinear function $g(\cdot)$ to replace the l_1 penalty, namely

$$\hat{x} = \arg\min_x \quad \frac{1}{2}\|y - x\|^2 + \lambda g(Dx), \tag{3}$$

where D is a learnable filter operator replacing the fixed gradient operator. Note that D could be Discrete Wavelet Transform (DWT), Discrete Cosine Transform (DCT), and so on, and the regularization term $g(\cdot)$ could be a sparse prior, e.g., l_q-norm ($0 \leq q \leq 1$). Moreover, D and $g(\cdot)$ are not chosen or fixed, but learnable.

To solve problem (3), we leverage ALM algorithm [13] and introduce an auxiliary variable z to decouple the data and regularization term, yielding following augmented Lagrangian function

$$L_\rho(x, z, \alpha) = \frac{1}{2}\|y - x\|^2 + \lambda g(z) + <\alpha, Dx - z> + \frac{\rho}{2}(Dx - z)^2, \tag{4}$$

where α is a Lagrangian multiplier and ρ is a penalty parameter.

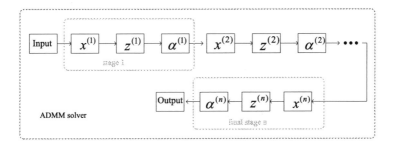

Fig. 1. An iterative process of ADMM algorithm: the parameters are fixed and the variables are updated stage by stage. The iteration process from the **Input** to the **Output** is described as an ADMM solver.

According to the optimality condition, the Eq. (4) is minimized to obtain the corresponding closed form solutions, resulting the following iteration:

$$x^{(k+1)} = \mathcal{F}^{-1}\mathcal{F}\left\{\frac{y - D^T\alpha^{(k)} + \rho D^T z^{(k)}}{I + \rho D^T D}\right\} \tag{5}$$

$$z^{(k+1)} = Dx^{(k+1)} + \frac{\alpha^{(k)}}{\rho} - \frac{\lambda}{\rho}\nabla g(z^{(k)}) \tag{6}$$

$$\alpha^{(k+1)} = \alpha^{(k)} + \rho(Dx^{(k+1)} - z^{(k+1)}), \tag{7}$$

where $\mathcal{F}(\cdot)$ and $\mathcal{F}(\cdot)^{-1}$ denote the discrete Fourier transform (DFT) and its inverse transform respectively. The Eq. (6) is reformulated as a shrinkage function $S(\cdot)$

$$z^{(k+1)} = S(Dx^{(k+1)} + \frac{\alpha^{(k)}}{\rho}; \frac{\lambda}{\rho}), \tag{8}$$

and we select two different functions, i.e., piecewise linear function and Gaussian RBFs [23] to approximate $S(\cdot)$ separately, see Sect. 4 for details.

3.2 NPM-ADMM Algorithm

The i-th iterative process of ADMM algorithm, which covers the updating from x^i to α^i, is called the i-th stage. In Fig. 1, the parameters are fixed and the variables are updated stage by stage in ADMM algorithm, where the iteration process from the **Input** to the **Output** is also described as a normal ADMM solver. It is not trivial to manually select the appropriate fixed parameters for ADMM solver. The proposed NPM-ADMM algorithm can automatically update these parameters to overcome this difficulty. Our NPM-ADMM algorithm is illustrated in Fig. 2, where the variables are updated by ADMM solver with fixed parameters and the loss function with regard to the final stage variables and ground truth images is minimized to update the parameters via back propagation.

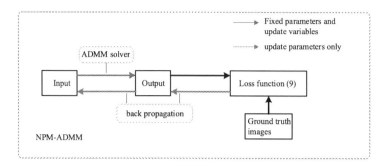

Fig. 2. NPM-ADMM algorithm: We use ADMM solver with fixed parameters to update variables and update the parameters via back propagation by minimizing the loss function.

3.3 Model Training and Parameters Updating

In this section, we minimize loss function to learn the model parameters $\Theta_{1,...,T} = \{\Theta_t\}_{t=1}^T = \{\lambda_t, \rho_t, D_t, S(\cdot)_t\}_{t=1}^T$ which are used to iterate in ADMM algorithm and T is the total number of stages. Given K training image pairs $\left\{y^{(k)}, x_{gt}^{(k)}\right\}_{k=1}^K$, the loss function is defined as

$$\ell(\Theta_{1,...,T}) = \frac{1}{2K} \sum_{k=1}^K \|x_{T_i}^{(k)} - x_{gt}^{(k)}\|_2^2, \tag{9}$$

where $y^{(k)}$ and $x_{gt}^{(k)}$ are the training data and ground truth images respectively, and $x_{T_i}^{(k)}$ represents the k-th output of the T-th stage (or iteration) of ADMM algorithm in the i-th epoch. Note that an epoch is defined as the process in turn from 1 to T stage. The gradient of the loss function with respect to the parameters Θ_t is generally calculated by chain rule. Subsequently, the corresponding parameter can be updated by estimating the descent direction using L-BFGS method [14,20].

Combining ADMM algorithm with the loss function about $\left\{x_T^{(k)}, x_{gt}^{(k)}\right\}_{k=1}^K$, the NPM-ADMM algorithm is reformulated as a bi-level optimization problem

$$\begin{cases} \Theta^* = & \arg\min_{\Theta} \dfrac{1}{2K} \sum_{k=1}^K \|x_{T_i}^{(k)} - x_{gt}^{(k)}\|_2^2 \\ (x_{T_i}^{(k)}, z_{T_i}^{(k)}, \alpha_{T_i}^{(k)}) = & \arg\min_{x^{(k)}, z^{(k)}, \alpha^{(k)}} L(x^{(k)}, z^{(k)}, \alpha^{(k)}; \Theta^*), \end{cases}$$

where $\Theta = \Theta_{1,...,T}$. Now the proposed NPM-ADMM algorithm can be explained as follows: the down level problem is considered as a forward propagation to generate variable $x_{T_i}^{(k)}$ using ADMM solver and the up level problem is considered as a back propagation to update parameters Θ by minimizing loss function of

$\left\{ x_T^{(k)}, x_{gt}^{(k)} \right\}_{k=1}^{K}$. The proposed NPM-ADMM algorithm can be summarized in Algorithm 1.

Algorithm 1. NPM-ADMM algorithm

Input: the noisy images $\left\{ y^{(l)} \right\}_{l=1}^{L}$, the maximum number of epochs t_{max}, and the final
     stage s_{final},
Output: net k
 1: **Initialize:** $k = 0$, $\Theta^0 = \Theta_{1,\ldots,S}^0 = \left\{ \lambda_s^0, \rho_s^0, D_s^0, S(\cdot)_s^0 \right\}_{s=1}^{s_{final}}$, the loss function $l(\Theta^0)$
 2: **for** $epoch = 1$: t_{max} **do**
 3:     **for** $s = 1 : s_{final}$ **do**
 4:         calculate $x_{epoch}^s, z_{epoch}^s, \alpha_{epoch}^s$ by ADMM solver;
 5:     **end for**
 6:     **for** $s = s_{final} : 1$ **do**
 7:         calculate the gradients of the parameters Θ^{epoch} by chain rule;
 8:     **end for**
 9:     update the parameters Θ^{epoch} by L-BFGS ;
10:     **if** $l(\Theta^k) > l(\Theta^{epoch})$ **then**
11:         $k = k + 1$;
12:         **Save net** k;
13:     **end if**
14: **end for**

3.4 Initialization of Training Model

In our model, we choose D as DCT basis which can be obtained by running Matlab code *dctmtx.m* and set the initial filter coefficients as a identity matrix. The penalty parameter ρ and regularization parameter λ are set as 0.2, 0.01 respectively. Two regularization functions, namely piecewise linear function and Gaussian RBFs [23] are utilized to solve z-problem. We aim to learn $S(\cdot)$ using piecewise linear function that is determined by a list of control points $\{p_i, q_i\}_{i=1}^{N}$ where $\{p_i\}_{i=1}^{N}$ are predefined positions according to uniform distribution within [-1,1], and $\{q_i\}_{i=1}^{N}$ are the values at these positions to be learned. For Gaussian RBFs, it is employed to represent the derivative of the regularization term $g(\cdot)$, instead of $S(\cdot)$, leading to the following form

$$g'(x) = \sum_{l=1}^{M} \delta_l \exp(\frac{-(x - \mu_l)^2}{2\gamma_l^2}), \tag{10}$$

where $M = 63$ is the number of Gaussian kernels, δ_l $(l = 1, \ldots, M)$ are weights which need to be learned. The scaling parameters $\gamma_l = 10$ $(l = 1, \ldots, M)$, and the equidistant centers μ_l $(l = 1, \ldots, M)$ are chosen at $[-310 : 10 : 310]$.

Table 1. The comparison between piece-wise linear function and Gaussian RBFs.

Method	NPM-ADMM$_{3\times3}^5$-rbf	NPM-ADMM$_{3\times3}^5$	NPM-ADMM$_{5\times5}^5$-rbf	NPM-ADMM$_{5\times5}^5$
$\sigma = 15$	28.54	30.81	28.78	**31.16**
$\sigma = 25$	27.33	28.31	27.48	**28.61**

Table 2. The PSNR results on 68 images from [15] for image denoising with $\sigma = 15$ and $\sigma = 25$. The results of the KSVD, BM3D, ARF-4 and FoE are quoted from Chen *et al.* [6].

Method	KSVD	BM3D	ARF-4	FoE	BLS-GSM	NPM-ADMM$_{5\times5}^5$
$\sigma = 15$	30.87	31.08	30.70	30.99	30.56	**31.16**
$\sigma = 25$	28.28	28.56	28.20	28.40	28.10	**28.61**

Table 3. Average PSNR on 68 images from [15] for image denoising with $\sigma = 15$ and $\sigma = 25$. The filters size is set as 3×3 and 5×5 respectively.

Method	CSF$_{pw}^5$	CSF$_{3\times3}^5$	CSF$_{5\times5}^5$	NPM-ADMM$_{3\times3}^5$	NPM-ADMM$_{5\times5}^5$
$\sigma = 15$	29.99	30.79	31.14	30.81	**31.16**
$\sigma = 25$	27.47	28.31	28.60	28.31	**28.61**

4   Experiments

To prepare training data, we choose 400 images from the BSD500 dataset [15] and crop a 180×180 pixel region from each of those images. Subsequently, Gaussian noise is added to the cropped images with standard deviation $\sigma = 15$ and $\sigma = 25$ as the training samples. For model test, we evaluate the denoising performance on 68 test images suggested by [15], which is a test set widely used for image denoising. The restoration performance is evaluated based on the Peak Signal to Noise Ratio (PSNR) over the test images. All our experiments are implemented in plain Matlab codes, and are run under a desktop with Intel(R) Xeon(R) 2.10-GHz E5-2620 CPU and 128.0-GB RAM.

Here, our method, i.e., NPM-ADMM$_{n\times n}^t$ denotes that the total stage number is t with $n^2 - 1$ filters of the size $n \times n$. We leverage two different functions: piece-wise linear function and Gaussian RBFs to approximate the shrinkage function $S(\cdot)$ and compare their differences in NPM-ADMM algorithm for image denoising. Although Gaussian RBFs have many desired properties, such as better smoothness and fast convergence speed, piecewise linear function usually has better denoising results in our experiments. Table 1 shows that the PSNR of piece-wise linear function is higher than Gaussian RBFs about 1 dB. Therefore, piece-wise linear function is adopted in all our experiments. In Table 2, we compare the proposed NPM-ADMM with 5 state-of-the-art methods: KSVD [8], BM3D [7], ARF [1], FoE [9] and BLS-GSM [19]. We downloaded these

Fig. 3. Image denoising results for noise level $\sigma = 25$, obtained by different methods: (a) Noisy (b) KSVD [8] (c) FoE [9] (d) BLS-GSM [19] (e) NPM-ADMM$_{5\times5}^{5}$

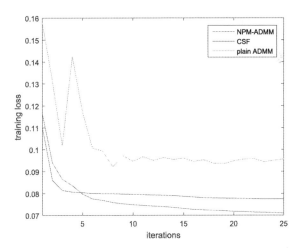

Fig. 4. The training loss for three methods: plain ADMM method, NPM-ADMM and CSF method [21]. For the plain ADMM method, the penalty parameters are updated by setting fixed step size and the filter coefficients are fixed as well as the nonlinear function. For CSF and NPM-ADMM, the final stage is 5 and the filter size is 3×3.

algorithms from the corresponding author's homepage and the parameters are set to recommended values in their papers. Table 2 manifests that our method has achieved stage-of-the-art results and BM3D generates the comparable results to ours (0.08 dB lower than ours for $\sigma = 15$). Other methods, i.e., BLS-GSM, is worse than our method about 0.5 dB. Figure 3 illustrates the denoising results obtained by different methods for noise level $\sigma = 25$ and our NPM-ADMM algorithm acquires the best denoising results.

In Table 3, we compare the proposed NPM-ADMM with the CSF [21] method. The reason why the final stage is chosen as 5 is that the PSNR is increased by 0.05 dB on average when we train the NPM-ADMM model with stage $= 10$, but the training and testing time are increased by a half. For the filters size, if we set 48 filters of size 7×7 compared with NPM-ADMM$_{5\times5}^5$, the denoising performance is slightly increased by 0.06 dB, and the training and testing time are almost doubled. So in Table 3, the filters size is set as 3×3 and 5×5 respectively and the results indicate that NPM-ADMM is better than CSF for both filters size.

We also compared the training loss for three methods: plain ADMM method, NPM-ADMM and CSF method in Fig. 4. It is shown that the proposed NPM-ADMM has lower training loss than CSF after the 5-th iteration.

5   Conclusion

In this paper, we propose a novel framework called NPM-ADMM, which can be used for image denoising. Contrary to the traditional ADMM algorithm which need to select the parameters carefully, NPM-ADMM can automatically update these parameters. Experiments on image denoising demonstrate good performances in contrast with other state-of-the-art methods. There is a problem that our trained model is a specific network. For example, the trained model based on noise level $\sigma = 15$ usually leads to inferior performance for noise level $\sigma = 25$. So one possible direction is to train a universal model that can deal with all the noise levels.

References

1. Barbu, A.: Training an active random field for real-time image denoising. IEEE Trans. Image Process. (TIP) **18**(11), 2451–2462 (2009)
2. Boyd, S., Parikh, N., Chu, E., Peleato, B., Eckstein, J.: Distributed optimization and statistical learning via the alternating direction method of multipliers. Found. Trends Mach. Learn. **3**(1), 1–122 (2011)
3. Boyle, R., Thomas, R.: Computer Vision: A First Course. Blackwell Scientific Publications (1988)
4. Bredies, K., Kunisch, K., Pock, T.: Total generalized variation. SIAM J. Imag. Sci. **3**(3), 492–526 (2010)
5. Chan, S., Wang, X., Elgendy, O.: Plug-and-play admm for image restoration: fixed-point convergence and applications. Proc. IEEE Trans. Comput. Imaging **3**(1), 84–98 (2016)

6. Chen, Y., Yu, W., Pock, T.: On learning optimized reaction diffusion processes for effective image restoration. In: Proceedings of IEEE Conference on Computer Vision and Pattern Recognition (CVPR), pp. 5261–5269 (2015)

7. Dabov, K., Foi, A., Katkovnik, V., Egiazarian, K.: Image denoising by sparse 3-d transform-domain collaborative filtering. IEEE Trans. Image Process. (TIP) **16**(8), 2080–2095 (2007)

8. Elad, M., Matalon, B., Zibulevsky, M.: Image denoising with shrinkage and redundant repersentations. In: Proceedings of the IEEE Conference on Computer Vision and Pattern Recognition (CVPR), pp. 1924–1931 (2006)

9. Gao, Q., Roth, S.: How well do filter-based mrfs model natural images? In: Proceedings of German Association for Pattern Recognition (DAGM), pp. 62–72 (2012)

10. Jia, Y., Darrell, T.: Heavy-tailed sistances for gradient based image descriptors. In: Proceedings of Advances in Neural Information Processing Systems (NIPS), pp. 397–405 (2011)

11. Kim, Y., Jung, H., Min, D., Sohn, K.: Deeply aggregated alternating minimization for image restoration. In: Proceedings of IEEE Conference on Computer Vision and Pattern Recognition (CVPR), pp. 284–292 (2017)

12. Krishnan, D., Fergus, R.: Fast image deconvolution using hyper-laplacian priors. In: Proceedings of Advances in Neural Information Processing Systems (NIPS), pp. 1033–1041 (2009)

13. Lin, Z., Chen, M., Ma, Y.: The augmented lagrange multiplier method for exact recovery of corrupted low-rank matrices. arXiv preprint arXiv:1009.5055 (2010)

14. Liu, D., Nocedal, J.: On the limited memory bfgs method for large scale optimization. Proc. Math. Program. **45**(1–3), 503–528 (1989)

15. Martin, D., Black, M.J.: Fields of experts. Int. J. Comput. Vis. (IJCV) **82**(2), 205–229 (2009)

16. Meinhardt, T., Moeller, M., Hazirbas, C., Cremers, D.: Learning proximal operators: using denoising networks for regularizing inverse imaging problems. In: Proceedings of IEEE International Conference on Computer Vision (ICCV), pp. 1781–1790 (2017)

17. Perona, P., Malik, J.: Scale-space and edge detection using anisotropic diffusion. Proc. IEEE Trans. Pattern Anal. Mach. Intell. **12**(7), 629–639 (1990)

18. Poor, H.V.: An Introduction to Signal Detection and Estimation, 2nd edn. Springer, Heidelberg (1998). https://doi.org/10.1007/978-1-4757-2341-0

19. Portilla, J., Strela, V., Wainwright, M., Simoncelli, E.: Image denoising using scale mixtures of gaussians in the wavelet domain. IEEE Trans. Image Process. (TIP) **12**(11), 1338–1351 (2003)

20. Schmidt, M.: minfunc (2013). http://mloss.org/software/view/529.html

21. Schmidt, U., Roth, S.: Shrinkage fields for effective image restoration. In: Proceedings of IEEE Conference on Computer Vision and Pattern Recognition (CVPR), pp. 3791–3799 (2015)

22. Taylor, G., Burmeister, R., Xu, Z., Singh, B., Patel, A., Goldstein, T.: Training neural networks without gradients: a scalable admm approach. In: Proceedings of International Conference on International Conference on Machine Learning, pp. 2722–2731 (2016)

23. Vert, J.P., Tsuda, K., Scholkopf, B.: A primer on kernel methods. In: Proceedings of Kernel Methods in Computational, pp. 35–70 (2004)

24. Wang, H., Banerjee, A., Luo, Z.: Parallel direction method of multipliers. In: Proceedings of Advances in Neural Information Processing Systems (NIPS), pp. 181–189 (2014)

25. Wang, Y., Li, K., Yang, J., Ye, X.: Intrinsic decomposition from a single RGB-D image with sparse and non-local priors. In: Proceedings of IEEE International Conference on Multimedia & Expo (ICME), pp. 1201–1206 (2017)
26. Wang, Y., Yang, J., Yin, W., Zhang, Y.: A ner alternating minimization algorithm for total variation image reconstruction. SIAM J. Imag. Sci. **1**(3), 248–272 (2008)

Satellite Image Scene Classification via ConvNet With Context Aggregation

Zhao Zhou[1], Yingbin Zheng[1(✉)], Hao Ye[1], Jian Pu[2], and Gufei Sun[3]

[1] Shanghai Advanced Research Institute, Chinese Academy of Sciences, Shanghai, China
zhengyb@sari.ac.cn
[2] East China Normal University, Shanghai, China
[3] ZhongAn Technology, Shanghai, China

Abstract. Scene classification is a fundamental problem to understand the high-resolution remote sensing imagery. Recently, convolutional neural network (ConvNet) has achieved remarkable performance in different tasks, and significant efforts have been made to develop various representations for satellite image scene classification. In this paper, we present a novel representation based on a ConvNet with context aggregation. The proposed two-pathway ResNet (ResNet-TP) architecture adopts the ResNet [1] as backbone, and the two pathways allow the network to model both local details and regional context. The ResNet-TP based representation is generated by global average pooling on the last convolutional layers from both pathways. Experiments on two scene classification datasets, UCM Land Use and NWPU-RESISC45, show that the proposed mechanism achieves promising improvements over state-of-the-art methods.

Keywords: Scene classification · Convolutional neural network
ConvNet · Residual learning · Context aggregation

1 Introduction

With the growing deployment of remote sensing instruments, satellite image scene classification, or scene classification from high-resolution remote sensing imagery, has drawn attention for its potential applications in various problems such as environmental monitoring and agriculture. Multiple challenges exist to produce accurate scene classification results. Large intra-class variation in the same scene class is a common issue. Moreover, the semantic gap between the scene semantic and the image features could further increase the difficulties of robust classification. Thus, the design of suitable representations on satellite images to deal with the challenges is of fundamental importance.

Great progress has been achieved in the recent years with the utility of representations based on the convolutional neural networks (ConvNet), which led to breakthroughs in a number of computer vision problems like image classification. The typical ConvNet including AlexNet [2], SPP-net [3], VGG [4], and

© Springer Nature Switzerland AG 2018
R. Hong et al. (Eds.): PCM 2018, LNCS 11165, pp. 329–339, 2018.
https://doi.org/10.1007/978-3-030-00767-6_31

GoogleNet [5], has also been applied to the task of satellite image scene classification. As the image number of the satellite image datasets are order-of-magnitude smaller than that of the image classification datasets (e.g., ImageNet [6]) and may not sufficient to train the robust deep models, these ConvNet based methods usually employ the off-the-shelf pre-trained deep networks (e.g., in [7–15]). The activations of the layers or their fusion are considered as the visual representation and sent to the scene classifiers. Evaluations on the benchmarks show that the deep learning based features often outperform previous handcrafted features.

The number of stacked layers in most current deep networks for satellite images is relatively small. For example, [11] design classification systems based on the 7-layer architecture of AlexNet [2] or its replication CaffeNet [16], and [12,13] employ the 16-layer VGG architecture [4]. Recent evidence suggests that deeper convolutional networks are more flexible and powerful with high modeling capacity for image classification [1,17]. Some previous works (e.g., [8,14]) employ the Residual Networks (ResNet) [1] as one of the basic models. However, the effectiveness of these deeper models and how their performance depends on the number of layers are still not fully exploited for remote sensing images.

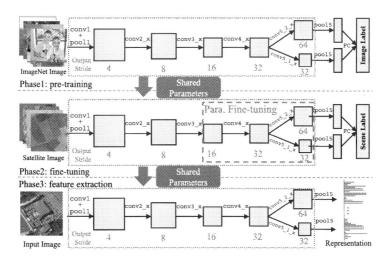

Fig. 1. Pipeline of the proposed framework with two-pathway ResNet (ResNet-TP). The network is pre-trained using ImageNet database (Phase 1). Phase 2 produces the fine-tuned network with the satellite image dataset. A given satellite image goes through the network and the representation is generated from the global average pooling on the last convolutional layers (Phase 3).

In this work, we focus on the problem of deeper ConvNet with context aggregation, and introduce an image representation built upon a novel architecture for satellite images, which adopts the ResNet [1] as backbone. The two-pathway ResNet (or ResNet-TP abbreviatedly) is proposed, and Fig. 1 illustrates the pipeline. The proposed structure aims to aggregate the contextual information to enhance the feature discrimination. The input images go through two paths of convolutional operations after a few layers: one path follows the default building block settings, and another path incorporates the dilation within convolutional layers to expand the receptive field. Training the deeper ConvNet is usually more difficult and may lead to higher risk of overfitting, especially when using the relatively small remote sensing dataset. Therefore, we also employ the transfer learning strategy to reuse the parameters learned from image classification dataset. The idea of constructing contextual representations has been taken in several previous remote sensing works, e.g., [7,11]. These approaches use a single spatial pyramid pooling [3] on the feature maps of last convolutional layer, which is usually tiny after progressively resolution reduction of previous operations. ResNet-TP is designed with contextual pathways *before* last convolutional layers, and is able to alleviate the loss of spatial acuity caused by tiny feature maps. To evaluate our proposed framework, we report the evaluations on the recent NWPU-RESISC45 [12] and the UC Merced (UCM) Land Use dataset [18]. Our representation is compared with several recent approaches and achieves state-of-the-art performance.

2 Methodology

2.1 ResNet Architecture

We begin with a brief review of the ResNet and the residual learning to address the training issue of deeper neural networks, which is the foundation to win the ILSVRC&COCO 2015 competition for the tasks of ImageNet detection, ImageNet localization, COCO detection, and COCO segmentation [1]. First, downsampling is performed directly by one 7×7 convolutional layer and one max-pooling (over 2×2 pixel window) with stride 2, respectively. The main component used to construct the architecture is the stacked convolutional layers with shortcut connections. Such building block is defined as

$$\mathcal{H}(\mathbf{x}) = \mathcal{F}(\mathbf{x}, \{W_i\}) + W_s \mathbf{x}, \tag{1}$$

where $\mathbf{x}$ and $\mathcal{H}(\mathbf{x})$ are the input and output of the building block, $\mathcal{F}(\cdot)$ is the residual mapping function to be learned, W_i is the parameters of the convolutional layers, and W_s is linear projection matrix to ensure the dimension matching of $\mathbf{x}$ and $\mathcal{F}$ (W_s is set as identity matrix when they are with the same dimension). The operation $\mathcal{F}(\cdot) + W_s \mathbf{x}$ is performed by a shortcut connection and element-wise addition. There are usually two or three layers within one building block, and two typical building blocks are shown in Fig. 2, where the basic building block is for 18/34-layer and the bottleneck building block is for 50/101/152-layer in [1]. The convolution is performed with the stride of 2 after a few building blocks to reduce the resolution of feature maps. Unlike previous architectures such as AlexNet [2] and VGG [4], ResNet has no hidden fully-connected (FC) layers; it ends with a global average pooling and then a N-way FC layer with softmax (N is the number of classes). We refer the reader to [1] for more details.

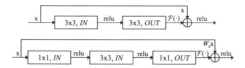

Fig. 2. The building block with the residual function $\mathcal{F}$. IN and OUT denote number of in-plane and out-plane, respectively. Top: the basic building block. Bottom: the bottleneck building block.

Table 1. Configuration of the groups in ResNet-TP Architecture. Suppose the input image is with size 224×224. Basic(IN, OUT) and Bottleneck(IN, OUT) denote the basic and bottleneck building block with number of in-plane IN and out-plane OUT (see Fig. 2). '$\times n_i$' indicates stacking n_i blocks, where $[n_2, n_3, n_4, n_5] = [2, 2, 2, 2]$ for 18 layer, $[3, 4, 6, 3]$ for 34/50 layer, $[3, 4, 23, 3]$ for 101 layer.

Group	Block		Output size, dilation
	18/34 layer	50/101 layer	
conv1+pool1	[7×7, 64]; Max Pooing		56×56, 1
conv2_x	Basic(64,64)$\times n_2$	Bottleneck(64,256)$\times n_2$	56×56, 1
conv3_x	Basic(128,128)$\times n_3$	Bottleneck(128,512)$\times n_3$	28×28, 1
conv4_x	Basic(256,256)$\times n_4$	Bottleneck(256,1024)$\times n_4$	14×14, 1
conv5_2_x	Basic(512,512)$\times n_5$	Bottleneck(512,2048)$\times n_5$	14×14, 2
conv5_1_x			7×7, 1

2.2   Context Aggregation

We now elaborate the construction of ResNet-TP. The architecture of the network is summarized in Table 1. In general, the network contains six groups of layers or building blocks. Group conv1+pool1 consist of the 7×7 convolutional layer and the max-pooling, and conv2_x to conv4_x are with a stack of building blocks. All their configurations follow the generic design presented as in Sect. 2.1, and differ only in the depth of blocks. Consider an input image with 224×224 pixels, group conv4_x is with output stride of 16 and thus its feature map size is 14×14.

We introduce group with dilation convolutional layers, which has been shown to be effective in many tasks such as semantic segmentation [19,20], video analysis [21,22], RGB-D [23], and DNA modeling [24]. The two-pathway architecture is made of two streams: a pathway with normal building blocks (conv5_1_x) and another with larger receptive fields (conv5_2_x). The dilation is operated on the 3×3 convolutional layer in the building block. Let $\mathbf{x}$ be the input feature map and $\mathbf{w}$ be the filter weights associated with the dilation convolutional layer, the output $\mathbf{y}$ for position $\mathbf{p} = (p_1, p_2)$ is defined as:

$$\mathbf{y}(\mathbf{p}) = \sum_{\mathbf{d} \in \mathcal{G}_d} \mathbf{w}(\mathbf{d}) \cdot \mathbf{x}(\mathbf{p} + \mathbf{d}) \tag{2}$$

where $\mathcal{G}_d = \{(-d, -d), (-d, 0), \ldots, (0, d), (d, d)\}$ is the grid for the 3×3 filters and d is the dilation. We set the dilation $d = 2$ for conv5_2_x, and the layers in conv5_1_x can also be considered as a special case with $d = 1$. The motivation for this architectural design is that we would like the prediction to be influenced by two aspects: the visual details of the region around each pixel of the feature map as well as its larger context. In fact, ResNet-TP is degenerated to the standard ResNet when conv5_2_x and its subsequent layers are removed. Finally, we connect the last convolutional hidden layers in both pathways with the global average pooling followed by the FC layer with softmax to perform a prediction of the labels.

2.3   Model Training and Implementation Details

The ResNet-TP architecture is with a large amount of parameters to train. A traditional remote sensing dataset contains thousands of high-resolution satellite images, which is far less than the image classification datasets for training the state-of-the-art deep learning models. Following previous works [7,8], training of ResNet-TP is based on the transfer learning strategy and Fig. 1 illustrates the overall framework of the proposed ResNet-TP based scene representation.

The whole training procedure as well as the feature extraction are carried out via the open source PyTorch library and an Nvidia Titan X (Pascal) GPU. The first phase is to get a pre-trained model using the ImageNet database [6]. During this process, due to the network with only conv5_1_x pathway having the same structure with the original ResNet, we set the weights of conv1_x to conv5_1_x with the existing PyTorch ResNet models[1]. Directly updating model from this initialization lead to performance drop, as the parameters of conv5_2_x are randomly initialized. On the other hand, it is time-consuming if the model is trained from scratch, since ImageNet contains millions of images. Here we make a compromise by learning the weights of conv5_2_x and its subsequent layers from the network with only conv5_2_x pathway and by frozen of conv1_x to conv4_x[2]. We compare this pre-training strategy with the model trained from scratch under ResNet-TP-18, and find that they are with similar performance on ImageNet validation set, while its training is much faster.

For the fine-tuning phase, we only fine-tune the building block groups after conv3_x by using the training satellite images and their labels due to the limitation of the GPU memory. We take random rotated, mirrored, or scaled images for data augmentation during fine-tuning. Finally, the representation is obtained from the global average pooling in both pathways, and the linear SVM classifier with default setting $C = 1$ is carried out for a fair comparison with previous works [7,12–14].

---

[1] The download link can be found from https://github.com/pytorch/vision/blob/master/torchvision/models/resnet.py.

[2] The stochastic gradient descent (SGD) is used with batch size of 64 and momentum of 0.9. The learning rate is initially set to be 0.01 and is divided by 10 every 30 epochs.

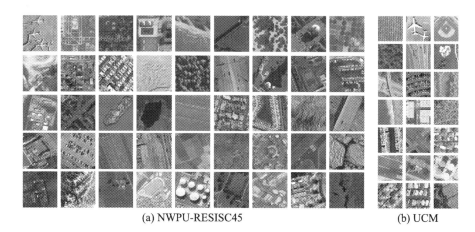

(a) NWPU-RESISC45   (b) UCM

Fig. 3. Scene categories from the datasets.

Table 2. Overall accuracies and standard deviations (%) of the proposed methods and state-of-the-arts under different training ratios on the NWPU-RESISC45 dataset. The results of pre-trained (PT-*) and fine-tuned (FT-*) ConvNets are reported in [12], and results of BoCF are from [13].

Network	Training ratios		Network	Training ratios	
	10%	20%		10%	20%
PT-AlexNet	76.69 ± 0.21	79.85 ± 0.13	BoCF-AlexNet	55.22 ± 0.39	59.22 ± 0.18
PT-GoogleNet	76.19 ± 0.38	78.48 ± 0.26	BoCF-GoogleNet	78.92 ± 0.17	80.97 ± 0.17
PT-VGG-16	76.47 ± 0.18	79.79 ± 0.15	BoCF-VGG-16	82.65 ± 0.31	84.32 ± 0.17
FT-AlexNet	81.22 ± 0.19	85.16 ± 0.18	Triplet networks [15]	–	92.33 ± 0.20
FT-GoogleNet	82.57 ± 0.12	86.02 ± 0.18	ResNet-TP-18	87.79 ± 0.28	91.03 ± 0.26
FT-VGG-16	87.15 ± 0.45	90.36 ± 0.18	ResNet-TP-101	**90.70± 0.18**	**93.47±0.26**

3   Experiments

To evaluate the effectiveness of the proposed method, we compare it with several state-of-the-art approaches on two remote sensing scene classification datasets, including the recent proposed 45-Class NWPU-RESISC45 dataset [12] and the widely used 21-Class UCM Land Use dataset [18].

3.1   NWPU-RESISC45

The NWPU-RESISC45 dataset contains 31500 remote sensing images extracted from Google Earth covering more than 100 countries and regions. Each scene class is composed of 700 images with the spatial resolution varied from about 30 to 0.2 m per pixel. Sample images and the scene categories are shown in Fig. 3(a), and we wrap the images into the size of 224×224. We follow the official train/test split strategy with two training ratios, i.e., 10% (10% for training and

90% for testing) and 20% (20% for training and 80% for testing). We repeat the evaluations ten times under each training ratio by randomly splitting the dataset and also report the mean accuracy and standard deviation.

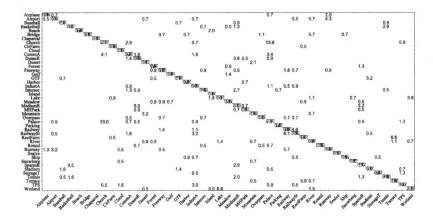

Fig. 4. Confusion matrices under the training ratio of 20% by using ResNet-TP-101 on the NWPU-RESISC45 dataset.

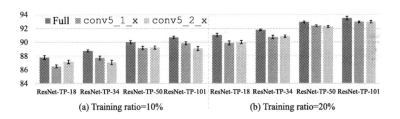

Fig. 5. Evaluation of the ResNet-TP parameters and components with different training ratio on the NWPU-RESISC45 dataset. conv5_1_x and conv5_2_x indicates the network with only one stream of ResNet-TP.

We compare ResNet-TP based representation with several baselines and state-of-the-art approaches. Among them, the first group contains several well-known baseline descriptors, including the pre-trained or fine-tuned AlexNet [2], GoogleNet [5], and VGG-16 [4]. Table 2 shows that the proposed representation outperforms all the baseline descriptors as well as the state-of-the-art approaches shown in the right part of Table 2, including the Bag of Convolutional Features (BoCF) [13] and a very recent triplet networks [15]. In Fig. 4, we report the confusion matrix and detail classification accuracy for each scene label using training ratios of 20%.

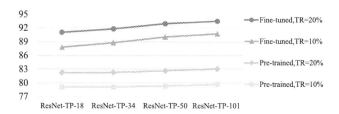

Fig. 6. Comparison of the pre-trained and fine-tuned ResNet-TP models on the NWPU-RESISC45 dataset. TR indicates training ratio.

Network and Training Ratio. We also study the performance of different network settings and training ratios, and results are given in Fig. 5. Adding the layers in ResNet-TP architecture and context aggregation by the two-pathways boost the classification accuracy. We conjecture that the applied single pathway may not be the one at which the network responds with optimal confidence, and context aggregation with multiple pathways increase the robustness. In addition, we observe that increasing training images (70 to 140 per class) lead to significant performance gains, probably due to the scene variation and data diversity in the NWPU-RESISC45 dataset.

Pre-trained vs. *Fine-Tuned.* Our last experiment on NWPU-RESISC45 evaluates the alternative method for ResNet-TP model generation. While the fine-tuned method follows the pipeline of Fig. 1, the pre-trained approach is only composed of phase 1 and 3 in the figure, and the model parameters are directly learned from ImageNet. The comparison between the curves in Fig. 6 verifies that for both training ratios using the fine-tuned network is important toward a more discriminant representation. We also find that the fine-tuned method outperforms the pre-trained method even though the training images is half of which for pre-trained.

3.2   UCM Land Use

The UCM Land Use dataset contains 2100 aerial scene images extracted from United States Geological Survey (USGS) national maps. Each land use class is composed of 100 images with the spatial resolution of 1 ft and the size of 256×256 pixels. The sample images are illustrated in Fig. 3(b). As UCM Land Use dataset is with relatively small and the results on it are already saturated, in this paper we focus on the performance w.r.t. the number of training images. Figure 7 shows the effect of training image number in the representation. We observe significant performance gains when the number of training images increases from 10 to 50, after which the performance tends to be saturated. Another observation is that the result of ResNet-TP-50 is similar to the accuracy of ResNet-TP-101 in most of the comparisons, indicating that the computation could be saved by ResNet-TP-50 with marginal performance drop.

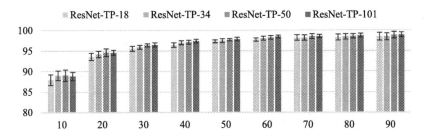

Fig. 7. Evaluation of the ResNet-TP models with different number of training images on the UCM dataset.

Table 3. Overall accuracies and standard deviations (%) of the proposed methods and state-of-the-arts on the UCM dataset. '–' indicates that the results are not available in the corresponding paper.

Number of images	5	50	80
MKL [25]	64.78 ± 1.62	88.68 ± 1.10	91.26 ± 1.17
SPP-net MKL [7]	75.33 ± 1.86	95.72 ± 0.50	96.38 ± 0.92
AlexNet-SPP-SS [9]	–	–	96.67 ± 0.94
VGG-16 [10]	–	94.14 ± 0.69	95.21 ± 1.20
ResNet50 [8]	–	–	98.50 ± 1.40
ResNet-TP-50	$\mathbf{77.07 \pm 1.73}$	$\mathbf{97.68 \pm 0.26}$	$\mathbf{98.56 \pm 0.53}$

We also compare the results of proposed representation with several state-of-the-art approaches. Table 3 summarizes the overall accuracy and standard deviation of all the classes. As can be seen from the table, the ResNet-TP based representation shows very competitive performance with different number of training images, which is significantly better than the other representations when the training images are limited. We also notice previous approach ResNet152_EMR [14] is also a ResNet-152 based representation and reach the accuracy of 98.90% by combining information from multiple layers with larger input image size (320×320). When the input image size is set to 224×224, the classification accuracy is 98.38%, which is inferior to ours with fewer layers. We believe that ResNet-TP based representation is also complementary to these mixed-resolution methods since they focus on different levels of information, which will be examined in the future work.

4 Conclusion

In this work, we have introduced ResNet-TP, a two-pathway convolutional network with context aggregation to generate a discriminant representation for satellite image scene classification. Through empirical scene classification experiments, we have shown that proposed ResNet-TP based representation is more effective than previous deep features, generating very competitive results on the UCM Land Use and NWPU-RESISC45 datasets. For future work, we plan to incorporate multi-scale and multiple layers into the ResNet-TP based representation, and also explore the performance benefits of a combination of this representation with other features.

Acknowledgments. This work was supported in part by grants from National Natural Science Foundation of China (No. 61602459) and Science and Technology Commission of Shanghai Municipality (No. 17511101902 and No. 18511103103).

References

1. He, K., Zhang, X., Ren, S., Sun, J.: Deep residual learning for image recognition. In: IEEE Conference on Computer Vision and Pattern Recognition (CVPR), pp. 770–778 (2016)
2. Krizhevsky, A., Sutskever, I., Hinton, G.E.: Imagenet classification with deep convolutional neural networks. In: Neural Information Processing Systems (NIPS), pp. 1097–1105 (2012)
3. He, K., Zhang, X., Ren, S., Sun, J.: Spatial pyramid pooling in deep convolutional networks for visual recognition. IEEE Trans. Pattern Anal. Mach. Intell. **37**(9), 1904–1916 (2015)
4. Simonyan, K., Zisserman, A.: Very deep convolutional networks for large-scale image recognition. In: International Conference on Learning Representations (ICLR) (2015)
5. Szegedy, C., et al.: Going deeper with convolutions. In: IEEE Conference on Computer Vision and Pattern Recognition (CVPR) (2015)
6. Deng, J., Dong, W., Socher, R., Li, L., Li, K., Fei-Fei, L.: ImageNet: a large-scale hierarchical image database. In: IEEE Conference on Computer Vision and Pattern Recognition (CVPR), pp. 248–255 (2009)
7. Liu, Q., Hang, R., Song, H., Li, Z.: Learning multiscale deep features for high-resolution satellite image scene classification. IEEE Trans. Geosci. Remote Sens. **56**(1), 117–126 (2018)
8. Scott, G.J., England, M.R., Starms, W.A., Marcum, R.A., Davis, C.H.: Training deep convolutional neural networks for land-cover classification of high-resolution imagery. IEEE Geosci. Remote Sens. Lett. **14**(4), 549–553 (2017)
9. Han, X., Zhong, Y., Cao, L., Zhang, L.: Pre-trained alexnet architecture with pyramid pooling and supervision for high spatial resolution remote sensing image scene classification. Remote Sens. **9**(8), 848 (2017)
10. Xia, G.S., et al.: Aid: a benchmark data set for performance evaluation of aerial scene classification. IEEE Trans. Geosci. Remote Sens. **55**(7), 3965–3981 (2017)
11. Han, X., Zhong, Y., Cao, L., Zhang, L.: Pre-trained alexnet architecture with pyramid pooling and supervision for high spatial resolution remote sensing image scene classification. Remote Sens. **9**(8), 848 (2017)

12. Cheng, G., Han, J., Lu, X.: Remote sensing image scene classification: benchmark and state of the art. Proc. IEEE **105**(10), 1865–1883 (2017)
13. Cheng, G., Li, Z., Yao, X., Guo, L., Wei, Z.: Remote sensing image scene classification using bag of convolutional features. IEEE Geosci. Remote Sens. Lett. **14**(10), 1735–1739 (2017)
14. Wang, G., Fan, B., Xiang, S., Pan, C.: Aggregating rich hierarchical features for scene classification in remote sensing imagery. IEEE J. Sel. Top. Appl. Earth Obs. Remote. Sens. **10**(9), 4104–4115 (2017)
15. Liu, Y., Huang, C.: Scene classification via triplet networks. IEEE J. Sel. Top. Appl. Earth Obs. Remote. Sens. **11**(1), 220–237 (2018)
16. Jia, Y., et al.: Caffe: convolutional architecture for fast feature embedding. In: ACM International Conference on Multimedia (MM), pp. 675–678 (2014)
17. Huang, G., Liu, Z., van der Maaten, L., Weinberger, K.Q.: Densely connected convolutional networks. In: IEEE Conference on Computer Vision and Pattern Recognition (CVPR) (2017)
18. Yang, Y., Newsam, S.D.: Bag-of-visual-words and spatial extensions for landuse classification. In: SIGSPATIAL International Conference on Advances in Geographic Information Systems, pp. 270–279 (2010)
19. Yu, F., Koltun, V.: Multi-scale context aggregation by dilated convolutions. In: International Conference on Learning Representations (ICLR) (2016)
20. Chen, L.C., Papandreou, G., Kokkinos, I., Murphy, K., Yuille, A.L.: Deeplab: semantic image segmentation with deep convolutional nets, atrous convolution, and fully connected CRFs. IEEE Trans. Pattern Anal. Mach. Intell. **40**(4), 834–848 (2018)
21. Lea, C., Flynn, M., Vidal, R., Reiter, A., Hager, G.: Temporal convolutional networks for action segmentation and detection. In: IEEE Conference on Computer Vision and Pattern Recognition (CVPR) (2017)
22. Xu, B., Ye, H., Zheng, Y., Wang, H., Luwang, T., Jiang, Y.G.: Dense dilated network for few shot action recognition. In: ACM International Conference on Multimedia Retrieval (ICMR), pp. 379–387 (2018)
23. Zheng, Y., Ye, H., Wang, L., Pu, J.: Learning multiviewpoint context-aware representation for RGB-D scene classification. IEEE Signal Process. Lett. **25**(1), 30–34 (2018)
24. Gupta, A., Rush, A.M.: Dilated convolutions for modeling long-distance genomic dependencies. arXiv preprint arXiv:1710D.01278 (2017)
25. Cusano, C., Napoletano, P., Schettini, R.: Remote sensing image classification exploiting multiple kernel learning. IEEE Geosci. Remote Sens. Lett. **12**(11), 2331–2335 (2015)

VAL: Visual-Attention Action Localizer

Xiaomeng Song and Yahong Han[✉]

School of Computer Science and Technology, Tianjin University, Tianjin, China
{songxiaomeng,yahong}@tju.edu.cn

Abstract. In this paper, we focus on solving a new task called TALL (Temporal Activity Localization via Language Query). The goal of it is to use nature language queries to localize actions in longer, untrimmed videos. We propose a new model called VAL (Visual-attention Action Localizer) to address it. Specifically, it employs voxel-wise attention and channel-wise attention on last conv-layer feature maps. These two visual attention are designed corresponding to the characteristics of feature maps. They can enhance the visual representations and boost the cross-modal correlation extraction process. Experimental results on TaCoS and Charades-STA datasets both show the effectiveness of our model.

Keywords: Temporal Activity Localization via Language Query
Voxel-wise attention · Channel-wise attention

1 Introduction

For the past few years, temporal localization (temporal action detection) task draw highly attention [7,10]. The goal of it is to localize actions in longer, untrimmed videos given a predefined action-label list. However, this traditional task has some limitations to meet the practical needs: First, action is complex. A label with only one or two words cannot accurately describe an action that consists of actors, scene, objects and their relations. Second, action is diverse. Even the predefined action-label list is relatively large [14], it is hard to cover all possible actions in the wild.

To go beyond these limitations, Gao *et al.* [1] recently proposed a new task called TALL (Temporal Activity Localization via Language). As Fig. 1 illustrates, instead of giving a predefined action-label list, TALL is to use natural language sentences (queries) to localize actions, i.e., output the start time and end time (or frame index) of the action that matches the query in an untrimmed video. This new task takes a step closer to the real-world applications, but also brings challenges and difficulties. As a muti-modal task, addressing it needs to design suitable representations for videos and queries respectively, and also needs to capture the correlations between them accurately. Previous work [1] used both one-dimensional vectors to represent visual and language information, and fuse them by element-wise multiplication, element-wise addition and vector concatenation. However, we argue that only using one-dimensional vectors to represent

© Springer Nature Switzerland AG 2018
R. Hong et al. (Eds.): PCM 2018, LNCS 11165, pp. 340–350, 2018.
https://doi.org/10.1007/978-3-030-00767-6_32

videos is not enough. Since the inherent spatial-temporal structures of videos are destroyed during the flat feature extraction process, this way may weaken or ignore the important visual cues. In addition, only employing linear operations such as multiplication or addition to fuse representations of two modal can not fully explore the correlations between them.

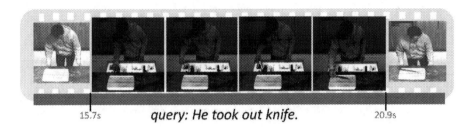

15.7s   *query: He took out knife.*   20.9s

Fig. 1. An example of TALL task.

To address these concern, we propose a new model called VAL (Visual-attention Action Localizer) to solve the TALL task. It utilizes C3D [9] to extract visual representations and Skip-Thought [13] to extract query representations, respectively. Different from previous work that only extract one-dimensional features from *fc6* layer of C3D, we additionally extract feature maps from *conv5_3* layer. Since these feature maps preserve the spatial and temporal dimensions, they possess more complete visual information. More importantly, we employ visual attention on these feature maps to enhance the visual representations and boost the cross-modal correlation extraction process. Visual attention has been shown effectiveness in many prior works [3,6,16] and enhancing visual representations is an important topic which has been discussed in various works [4,12,15]. Specifically, we employ voxel-wise attention (v-att) and channel-wise attention (c-att) corresponding to the characteristics of feature maps. Voxel denotes every point in three-dimensional space, which is the same as the concept of pixel for two-dimensional space. Since feature maps generated from *conv5_3* layer are also three-dimensional, containing height, weight and time dimensions, we apply attention on every voxel to gather visual cues through the whole visual space. Our goal is to force our model to find related points with queries as guidance and assemble them into semantic-relevant regions. Besides, feature maps are multi-channel. As [8] mentioned, different filters in the same convolution layer can be viewed as semantic detectors from different standpoints. Thus the generated multiple channels correspond to different semantic context. Applying attention to channels can be seen as a process to select relevant visual cues on the demand of the queries from another perspective. In addition, as [5] mentioned, low-layer filters can detect low-level visual cues like edges or corners while high-layer filters can detect high-level visual cues like objects or parts. Since feature maps generated from *conv5_3* layer already fused complete and high-level semantic information through multi-layer convolution operations, we apply the channel-wise attention directly on them can capture the complex correlations between visual information and language information.

In summary, we propose a new model called VAL to solve the TALL task. Particularly, it employs voxel-wise attention and channel-wise attention, which are designed corresponding to the characteristics of visual representations and to address limitations of previous work. Experimental results show the effectiveness of our model on two benchmarks: TACoS [11] and Charades-STA [1].

2 Related Work

Many prior works are proposed to address temporal localization task: Shou *et al.* proposed a multi-stage CNNs [7], which consists of proposal, classification and localization network. Zhao *et al.* presented SSN [10] to model the temporal structure of each action instance via a structured temporal pyramids.

Recently, Gao *et al.* extend this task by introducing a new task called TALL [1]. Instead of using a predefined action-label list, it targets to use natural language queries to localize actions, which is close to the real-world requirements. Gao *et al.* also proposed a model named CTRL [1] to solve this task. CTRL achieved superior performances on TACoS [11] and Charades-STA [1] datasets. However, it seems ignore the spatial-temporal structure of videos and lacks of fully exploration of correlations between visual and language information. To address these limitations, we propose a new model named VAL. It employs voxel-wise attention and channel-wise attention which designed corresponding to the characteristics of visual representations.

Voxel-wise attention is designed to capture the inherent spatial-temporal structure of videos and gather complete visual cues in whole visual space. Channel-wise attention was used in some prior works of varied visual tasks, such as SCA-CNN [8] and SE-Net [2]. Inspired by these works, we employ channel-wise attention in our model to solve TALL. Since it also can be seen as a process to select relevant visual cues on the demand of the semantic information, integrating it with voxel-wise attention enhances visual representations and boosts the cross-modal correlation extraction process.

3 Model

Our model VAL contains four modules: a visual encoder module to extract visual representations, a query encoder module to extract query representations, an attention module to employ voxel-wise attention and channel-wise attention, and a decoder module to generate predicted results. We detailed introduce each module below. The overview of our model is shown in Fig. 2.

3.1   Visual Encoder Module

First we use sliding window method to generate candidate clips. We employ various length of sliding windows (32,64,128,256,512) through the whole video with overlap between adjacent clips is 80% for dense sampling. Then we sample 16 frames uniformly in each clip and feed them into C3D to extract clip-level visual representations. For each clip i, we extract feature $V_i(\in R^{2\times7\times7\times512})$ from $conv5\_3$ layer and feature $u_i(\in R^{4096})$ from $fc6$ layer. Since u_i gathers global visual information and V_i preserves the local visual information, using both of them to represent a clip is complete. We also take consideration on context information. Thus for each u_i, we concatenate the feature u_{i-1} from its previous clip and u_{i+1} from its next clip with it. Then we embed this concatenated feature into 4096-dimension to get a global visual representation U_i for each clip i.

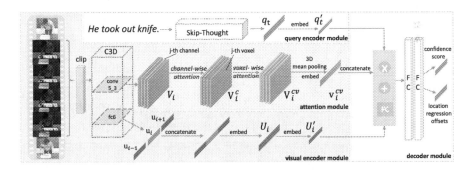

Fig. 2. The overview of our model. Our model consists of four modules: a visual encoder module to extract visual representations, a query encoder module to extract query representations, a decoder module to generate predicted results, and an attention module to employ voxel-wise attention and channel-wise attention. Specifically, we employ four types of attention (each single one and their two combinations) in our model, we illustrate the channel-voxel attention (cv-att) in this figure.

3.2   Query Encoder Module

We use an off-the-shelf tool Skip-Thought [13] to extract the sentence-level query representation q_t for each query t. Skip-Thought is a pretrained sentence encoder, it embeds each sentence into a fixed dimension space ($\in R^{4800}$). The reason we use Skip-Thought instead of LSTM which is widely used in encoder tasks is that the scales of TACoS and Charades-STA datasets are not large enough to train the LSTM, as mentioned in [1].

3.3 Attention Module

In attention module, we apply voxel-wise attention and channel-wise attention on V_i which preserve the spatial and temporal dimensions. More importantly, we not only employ each single attention, but also their various combinative types for detailed investigate. Thus we employ four types of attention in our model in total. We detailed describe each type below.

Voxel-Wise Attention. We employ voxel-wise attention (v-att) to force model to focus on related points in whole visual space with queries as guidance and gather them into semantic-relevant regions. Specifically, we use query representation q_t to calculate voxel-wise attention weight s_{ij} for each voxel j ($1 \leqslant j \leqslant 2 \times 7 \times 7$) in V_i, then assign them to corresponding voxel to get a voxel-attention weighted feature map $V_i^v (\in R^{2 \times 7 \times 7 \times 512})$. The process is illustrated as below:

$$\alpha_{ij} = W^T tanh(W_v V_{ij} + W_q q_t)$$

$$s_{ij} = exp(\alpha_{ij}) / \sum_{j=1}^{2 \times 7 \times 7} exp(\alpha_{ij}) \tag{1}$$

$$V_{ij}^v = s_{ij} V_{ij}$$

where W, W_v and W_q are trainable parameters. W_v and W_q transform representations of two modal to same attention dimension space.

Channel-Wise Attention. As different channels can be seen as semantic responders corresponding to different filters which can be seen as semantic detectors, we employ channel-wise attention (c-att) to select relevant visual cues on the demand of the queries from another perspective. Specifically, we use query representation q_t to calculate a channel-wise attention weight β_{ij} for each channel j ($1 \leqslant j \leqslant 512$) in V_i, then assign them to corresponding channel to get a channel-attention weighted feature map $V_i^c (\in R^{2 \times 7 \times 7 \times 512})$:

$$\beta_{ij} = U^T tanh(U_v V_{ij} + U_q q_t)$$

$$m_{ij} = exp(\beta_{ij}) / \sum_{j=1}^{512} exp(\beta_{ij}) \tag{2}$$

$$V_{ij}^c = m_{ij} V_{ij}$$

where U, U_v and U_q are trainable parameters. U_v and U_q transform representations of two modal to same attention dimension space.

Combination. Since voxel-wise attention and channel-wise attention are both processes to catch semantic-relevant visual cues, we combine them to facilitate the effectiveness. There are two types of combination. The first type is called voxel-channel attention (vc-att) which employs voxel-wise attention before channel-wise attention. Specifically, we first apply voxel-wise attention on initial feature maps V_i, then adopt channel-wise attention on generated feature maps V_i^v for further refine and get final attention weighted feature maps V_i^{vc}.

We use F_v and F_c to represent voxel-wise attention and channel-wise attention operations respectively, thus the whole process is shown as follows:

$$V_i^v = F_v(V_i, q_t)$$
$$V_i^{vc} = F_c(V_i^v, q_t) \tag{3}$$

The second type is called channel-voxel attention (cv-att) which employs channel-wise attention before voxel-wise attention to get final attention weighted feature maps V_i^{cv}. Similar to the vc-att, the formula is shown as:

$$V_i^c = F_c(V_i, q_t)$$
$$V_i^{cv} = F_v(V_i^c, q_t) \tag{4}$$

3.4 Decoder Module

Multi-modal Fusion. First we embed global visual representation U_i and query representation q_t into the same 1024-dimension to get new vectors U_i' and q_t'. Then we use element-wise addition $(+)$, element-wise multiplication $(\times)$ and concatenation $(||)$ followed by a 1024-dimension Fully Connected layer *(FC)* to combine them as [1]. More importantly, we additionally combine the attention weighted feature maps with them. We first employ three-dimensional mean pooling through the spatial and temporal dimensions of feature maps generated from attention module, then flat feature maps and feed generated one-dimensional vector into 1024-dimension. Finally we concatenate it with other outputs from three operations to get a final multi-model representation f_m as formula below, where v_i^* means the 1024-dimensional vector generated from attention weighted feature maps of four types of attention we described above:

$$f_m = (q_t' \times U_i') \,||\, (q_t' + U_i') \,||\, FC(q_t' \,||\, U_i') \,||\, v_i^* \tag{5}$$

Decoding. After obtaining the final multi-model representation, we feed it into two FC layers to generate predicted results. The last FC layer outputs one confidence score and two location regression offsets. Specifically, the confidence score denotes the matching degree between the clip and the query. And location regression offsets denote the differences between ground-truth offsets and predicted offsets. The ground-truth offsets are calculated as below:

$$t_g^s = s - s_c$$
$$t_g^e = e - e_c \tag{6}$$

where t_g^s and t_g^e denote start time offset and end time offset separately, s and e denote start and end time of the ground-truth clip, s_c and e_c denote start and end time of the clip which generated by sliding window method. The location regression offsets can help model refine the temporal boundaries of the input clip, thus localize the matched action more accurately.

Loss Function. We calculate our loss as [1]. The loss contains two part: L_{score} for the confidence score, and L_{offset} for the location regression offsets. The whole loss L is formulated as:

$$L = L_{score} + \omega L_{offset} \tag{7}$$

where ω is a hyper-parameter to balance two losses. L_{score} is calculated as:

$$L_{score} = \frac{1}{N}\sum_{i=1}^{N}[\lambda log(1 + exp(-cs_{i,i})) + \sum_{j=1,j\neq i}^{N}(log(1 + exp(cs_{i,j})))] \tag{8}$$

where N is the batch size, $cs_{i,j}$ is the confidence score between clip i and query j. λ is a hyper-parameter to balance positive (matched) and negative (mismatched) training samples. This loss encourages matched pairs have positive scores while mismatched pairs have negative scores. And L_{offset} is calculated as:

$$L_{offset} = \frac{1}{N}\sum_{i=1}^{N}[R(t_g^s - t_i^s) + R(t_g^e - t_i^e)] \tag{9}$$

where t_g^s and t_g^e represent ground-truth offsets, t_i^s and t_i^e are predicted offsets output by model. R is the L1 distance function. This loss encourages the smallest differences between predicted offsets and ground-truth offsets, i.e., the predicted boundaries get closer to the ground-truth boundaries.

4   Experiment

We conduct experiments on two datasets: TACoS [11] and Charades-STA [1]. For a fair comparison, we use the same way as [1] to collect training samples for two datasets. In total, there are 17344 clip-sentence pairs in TACoS and 18131 clip-sentence pairs in Charades-STA. We evaluate our model VAL with four types of visual attention on these two datasets, and make comparisons with other methods in previous work [1] to suggest the effectiveness of our model VAL. The results are shown in the Tables 1 and 2. We also display prediction examples between our model VAL with different types of attention in Fig. 3.

From the results we can see that our model achieve best performances on both datasets. Since we use the same visual encoder, sentence encoder and loss function as the CTRL(reg-np) which show best performances in [1] for a fair comparison, the improvements suggest the effectiveness of our attention module. We argue that because voxel-wise attention and channel-wise attention enhance the visual representations and boost the correlation extraction process between visual information and natural language information. Besides, performances of our model with combinative attention are better than the model with single attention, demonstrating that integrate two attention can facilitate the function of each one. We argue the reason is that voxel-wise attention and channel-wise attention both can be viewed as visual cues extraction process with queries as guidance. The difference between them is that they gather information from

Table 1. Comparison of different methods on TACoS

Method	R@1	R@1	R@1	R@5	R@5	R@5
	IoU = 0.5	IoU = 0.3	IoU = 0.1	IoU = 0.5	IoU = 0.3	IoU = 0.1
Random	0.83	1.81	3.28	3.57	7.03	15.09
Verb	1.62	2.62	6.71	3.72	6.36	11.87
Verb+Obj	8.25	11.24	14.69	16.46	21.50	26.60
VSA-RNN	4.78	6.91	8.84	9.10	13.90	19.05
VSA-STV	7.56	10.77	15.01	15.50	23.92	32.82
CTRL(aln)	10.67	16.53	22.29	19.44	29.09	41.05
CTRL(loc)	10.70	16.12	22.77	18.83	31.20	45.11
CTRL(reg-p)	11.85	17.59	23.71	23.05	33.19	47.51
CTRL(reg-np)	13.30	18.32	24.32	25.42	36.69	48.73
VAL(c-att)	14.08	19.63	25.08	25.44	36.74	49.25
VAL(v-att)	14.62	19.67	24.69	25.64	36.93	50.06
VAL(vc-att)	14.47	19.57	24.37	**26.57**	37.25	50.72
VAL(cv-att)	**14.74**	**19.76**	**25.74**	26.52	**38.55**	**51.87**

Table 2. Comparison of different methods on Charades-STA (For CTRL, we use the **newest** results updated in their *Github* recently for comparison)

Method	R@1	R@1	R@5	R@5
	IoU = 0.5	IoU = 0.7	IoU = 0.5	IoU = 0.7
Random	8.51	3.03	37.12	14.06
VSA-RNN	10.50	4.32	48.43	20.21
VSA-STV	16.91	5.81	53.89	23.58
CTRL(aln)	17.69	5.91	55.54	23.79
CTRL(loc)	20.19	6.92	55.72	24.41
CTRL(reg-p)	19.22	6.64	57.98	25.22
CTRL(reg-np)	21.42	7.15	59.11	26.91
VAL(c-att)	20.60	6.77	59.09	26.04
VAL(v-att)	21.99	7.69	59.20	27.53
VAL(vc-att)	22.45	8.65	60.05	26.61
VAL(cv-att)	**23.12**	**9.16**	**61.26**	**27.98**

various perspectives: the former is from whole visual space, focus on 'where'; the later is from semantic content, focus on 'what'. Thus combining them can boost the whole semantic-relevant visual cues collection process, i.e., fully explore the correlations between queries and videos. Besides, we also find that our model with voxel-wise attention outperforms the model with channel-wise attention. It is partly because the voxel, i.e., the spatial and temporal structure of the clip plays more crucial role in finding semantic-relevant regions. Moreover, although results of cv-att and vc-att are close, cv-att is sightly better. It is may because cv-att first employs c-att which can capture semantic information from different channels, so it enhances the semantic-level visual representations and facilitate the next v-att process which gather visual cues from whole visual space. We argue that this processing sequence is more favorable to capture the visual cues. We also display some attention visualization examples in Fig. 4.

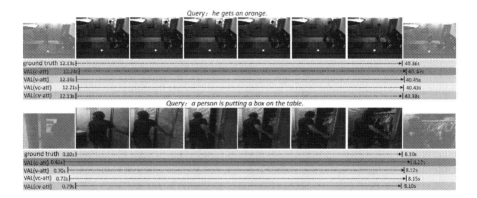

Fig. 3. Comparison between prediction results from VAL with different types of attention on TACoS and Charades-STA.

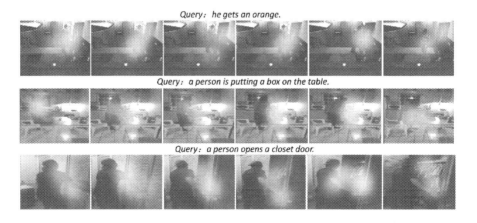

Fig. 4. Attention visualization examples of VAL (cv-att).

5 Conclusion

We propose a model called VAL to address the TALL task, the goal of which is to use natural language queries to localize actions in longer, untrimmed videos. VAL employs voxel-wise attention and channel-wise attention which designed corresponding to characteristics of feature maps. These two visual attention enhance the visual representations and boost the cross-modal correlation extraction process. Experimental results on TACoS and Charades-STA datasets all demonstrate the effectiveness of our model compared with other methods.

Acknowledgments. This work is supported by the NSFC (under Grant U150920 6,61472276) and Tianjin Natural Science Foundation (no. 15JCYBJC15400).

References

1. Gao, J., Sun, C., Yang, Z., Nevatia, R.: TALL: temporal activity localization via language query. In: ICCV, pp. 5277–5285 (2017)
2. Hu, J., Shen, L., Sun, G.: Squeeze-and-excitation networks. In: CVPR (2018)
3. Yang, Z.W., Han, Y.H., Wang, Z.: Catching the temporal regions-of-interest for video captioning. In: ACM MM, pp. 146–153 (2017)
4. Hong, R., Zhang, L., Tao, D.: Unified photo enhancement by discovering aesthetic communities from Flickr. In: IEEE Transactions on Image Processing, pp. 1124–1135 (2016)
5. Zeiler, M.D., Fergus, R.: Visualizing and understanding convolutional networks. In: Fleet, D., Pajdla, T., Schiele, B., Tuytelaars, T. (eds.) ECCV 2014. LNCS, vol. 8689, pp. 818–833. Springer, Cham (2014). https://doi.org/10.1007/978-3-319-10590-1_53
6. Wang, B., Xu, Y.J., Han, Y.H., Hong, R.C.: Movie question answering: remembering the textual cues for layered visual contents. In: AAAI. (2018)
7. Shou, Z., Wang, D., Chang, S.F.: Temporal action localization in untrimmed videos via multi-stage CNNs. In: CVPR, pp. 1049–1058 (2016)
8. Chen, L., et al.: SCA-CNN: spatial and channel-wise attention in convolutional networks for image captioning. In: CVPR, pp. 6298–6306 (2017)
9. Du, T., Bourdev, L., Fergus, R., Torresani, L., Paluri, M.: Learning spatiotemporal features with 3D convolutional networks. In: ICCV, pp. 4489–4497 (2015)
10. Zhao, Y., Xiong, Y., Wang, L., Wu, Z., Tang, X., Lin, D.: Temporal action detection with structured segment networks. In: ICCV, pp. 2933–2942 (2017)
11. Regneri, M., Rohrbach, M., Dominikus, W., Thater, S.: Grounding action descriptions in videos. TACL **1**, 25–36 (2013)
12. Hong, R., Hu, Z., Wang, R., Wang, M., Tao, D.: Multi-view object retrieval via multi-scale topic models. In: IEEE Transactions on Image Processing, pp. 5814–5827 (2016)
13. Kiros, R., et al.: Skip-thought vectors. In: NIPS, pp. 3294–3302 (2015)
14. Heilbron, F.C., Escorcia, V., Ghanem, B., Niebles, J.C.: ActivityNet: a large-scale video benchmark for human activity understanding. In: CVPR, pp. 961–970 (2015)

15. Hong, R., Zhang, L., Zhang, C., Zimmermann, R.: Flickr circles: aesthetic tendency discovery by multi-view regularized topic modeling. IEEE Trans. Multimedia **18**, 1555–1567 (2016)
16. Xu, Y.J., Han, Y.H., Hong, R.C., Tian, Q.: Sequential video VLAD: training the aggregation locally and temporally. IEEE Trans. Image Process. **27**, 4933–4944 (2018)

Incremental Nonnegative Matrix Factorization with Sparseness Constraint for Image Representation

Jing Sun[1], Zhihui Wang[1], Haojie Li[1(✉)], and Fuming Sun[2]

[1] Dalian University of Technology, Dalian 116300, China
hjli@dlut.edu.cn
[2] Liaoning University of Technology, Jinzhou 121001, China

Abstract. Nonnegative matrix factorization (NMF) is a powerful method of data dimension reduction and has been widely used in face recognition. However, existing NMF algorithms have two main drawbacks. One is that the speed is too slow for large matrix factorization. The other is that it must conduct repetitive learning when the training samples or classes are incremental. In order to overcome these two limitations and improve the sparseness of the data after factorization, this paper presents a novel algorithm, which is called incremental nonnegative matrix factorization with sparseness constraint. By using the results of previous factorization involved in iterative computation with sparseness constraint, the cost of computation is reduced and the sparseness of data after factorization is greatly improved. Compared with NMF and INMF, the experimental results on some face databases have shown that the proposed method achieves superior results.

Keywords: Nonnegative matrix factorization · Incremental
Sparseness constraint

1 Introduction

With the increasing development of network technology in the information age, people can access to a variety of large-scale high-dimensional data [1–4] everywhere. How to obtain useful information [5,6] from these data has become a research focus. For the efficient handling of these large-scale data, the crucial step is to reduce the dimension of them. On the other hand, for matrix is the most commonly used expression of these data model, most of the data can be represented and stored in the form of a matrix. As an example, an image with a matrix can correspond exactly and matrix column represents the image, matrix line for pixel. Therefore matrix factorization is one of the effective methods to achieve large-scale data dimension reduction. Generally speaking, matrix factorization is

Supported by the National Natural Science Foundation of China (NSFC) under Grants 61572244, 61472059 and 61772108.

R. Hong et al. (Eds.): PCM 2018, LNCS 11165, pp. 351–360, 2018.
https://doi.org/10.1007/978-3-030-00767-6_33

non-unique and many different methods like this have been proposed by incorporating different constraints with different criteria. The canonical techniques include Principal Component Analysis (PCA), Independent Component Analysis (ICA), Singular Value Decomposition (SVD), Vector quantization (VQ), Non-negative Matrix Factorization (NMF) [7–10], etc. Among them, NMF is a novel matrix decomposition method that emerged in recent years, and it has been widely used in face recognition [11–13], image classification, image retrieval, document clustering, biomedical engineering and other fields.

After the research and development for many years, a plenty of improved methods have been proposed on the basis of NMF. Hoyer [14] applied NMF to the sparse code, and proposed the Nonnegative Sparse Coding (NSC). Furthermore, he also presented the Sparse NMF (SNMF) algorithm that incorporates the sparseness constraint into NMF for increasing the sparseness of NMF method. Liu et al. [15] developed the Constrained Nonnegative Matrix Factorization (CNMF), which takes the label information as additional constraints into the objective function. The single objective function realization of semi-supervised decomposition, and the lack of it is unable to maintain the geometric structure of sample space. Cai et al. [16] proposed Graph Regularized Nonnegative Matrix Factorization (GNMF) algorithm for face recognition. GNMF imposes manifold into NMF, which makes it feasible to preserve geometric structure of data after it maps the original image into low dimension space. Ding et al. [17] introduced the Semi-NMF decomposition algorithm, which is not sensitive to the positive and negative input data. This method forces the encoding matrix of decomposition is nonnegative, but there is no limit on the basic matrix; Jiang et al. [18] presented Graph regularized Nonnegative Matrix Factorization with Sparseness Constraint (GNMFSC) algorithm, which not only considers the geometric information of the data, but also enforces the sparseness constraint on the coefficient matrix. Chen et al. [19] developed block nonnegative matrix factorization based on incremental study, the large matrix per class is divided into blocks, and then each small matrix is decomposed, finally the synthesis of small matrix will be decomposed; Bucak et al. [20] introduced Incremental Nonnegative Matrix Factorization (INMF) algorithm, which imposes incremental study into NMF, so as to avoid repetitive learning of the basic matrix and the coefficient matrix when the training samples or classes are increased. In addition, this algorithm reduces the cost of computation.

Motivated by recent progresses in matrix factorization, we proposed a novel NMF method for image representation by exploiting constraint conditions. The NMF proposed is referred as Incremental Nonnegative Matrix Factorization with Sparseness Constraint (INMFSC) to represent the image in a more reasonable way. In the first place, incremental study is introduced into NMF to reduce the computational time of re-learning phase; Secondly, on the basis of INMF, by enforcing an additional sparseness constraint, the proposed method possesses the merit of INMF and the sparseness of data after factorization is greatly improved. Finally, we carried out the extensive experiments on several common databases to validate the effectiveness and efficiency of the method adopted in this paper. The corresponding iterative formula and algorithm steps for the optimization problem are also given in the next few sections.

2   A Brief Review of NMF

NMF algorithm is described as follows: given a data matrix $X = [x_1, x_2, \cdots, x_n] \in R^{m \times n}$, each column of X is an m-dimensional data point, where $x_i \in R^{m \times n}$ is a sample vector. The goal of NMF is to seek two nonnegative matrices $U \in R^{m \times r}$ and $V \in R^{n \times r}$ $(k \leq mn/(m+n))$. Especially, the elements of two matrices are all negative. In order to guarantee that the similarity between X and UV^T is the highest, it reduces to minimize the following objective function.

$$O_F = \|X - UV^T\|^2 = \sum_{i,j} \left(x_{ij} - \sum_{k=1}^{K} u_{ik} v_{jk} \right)^2 \ s.t. \ U \geq 0, V \geq 0 \qquad (1)$$

where O_F is referred as the Frobenius norm. The objective function of NMF is convex with respect to one variable matrix U or V. Multiplicative update method is a well-known NMF method, which can find a local minimum in (1). The updating rules for the Euclidean distance objective function are as follows:

$$u_{ik} = u_{ik} \frac{(XV)_{ik}}{(UV^T V)_{ik}} \qquad (2)$$

$$v_{jk} = v_{jk} \frac{(X^T V)_{jk}}{(V^T V U^T)_{jk}} \qquad (3)$$

where $U = [u_{ik}]$, $V = [v_{jk}]$.

At the very beginning of the iterative update process, the two nonnegative matrices U_0 and V_0 are initialized at random. The iterative update procedure is executed repeatedly according to the updating rules until the given terminal condition is met. Ultimately, the final U and V can be obtained.

3   Incremental NMF with Sparseness Constraint

3.1   INMF

In this section, we will give a brief review on INMF: Let U_k and V_k respectively represent the basic matrix and the coefficient matrix of the sample set X_k after decomposition, where k indicates the number of sample. If we define the F_k as the objective function with corresponding to the NMF as

$$F_k = \|X_k - U_k V_k^T\|^2 = \sum_{i=1}^{m} \sum_{j=1}^{k} \left(x_{ij} - \left(U_k V_k^T \right)_{ij} \right)^2 s.t. \ U \geq 0, V \geq 0 \quad (4)$$

Every time, when the $(k+1)$th sample x_{k+1} arrives, F_{k+1} can be defined as (5)

$$F_{k+1} = \|X_{k+1} - U_{k+1} V_{k+1}^T\|^2 = \sum_{i=1}^{m} \sum_{j=1}^{k+1} \left(X_{ij} - \left(U_{k+1} V_{k+1}^T \right)_{ij} \right)^2 \qquad (5)$$

INMF assumes that the first k columns of the coefficient matrix V_{k+1} would not be changed when the new sample x_{k+1} comes and $V_{k+1} = [V_k, v_{k+1}]$. Meanwhile, the computation cost will be reduced significantly. F_{k+1} can be further expressed as the sum of the F_k with the residual error f_{k+1} introduced by the new sample x_{k+1}.

$$
F_{k+1} \approx \sum_{i=1}^{m}\sum_{j=1}^{k} \left(X_{ij} - \left(U_k V_k^T\right)_{ij} \right)^2 + \sum_{i=1}^{m} \left((x_{k+1})_i - \left(U_{k+1}v_{k+1}^T\right)_i \right)^2 \tag{6}
$$
$$
= F_k + f_{k+1}
$$

where $f_{k+1} = \sum_{i=1}^{m} \left((x_{k+1})_i - \left(U_{k+1}v_{k+1}^T\right)_i \right)^2$.

In order to save space, here we just list the iterative updating algorithms of INMF in (7) and (8).

$$
(v_{k+1})_a = (v_{k+1})_a \frac{\left(x_{k+1}^T U_{k+1}\right)_a}{\left(v_{k+1}U_{k+1}^T U_{k+1}\right)_a} \tag{7}
$$

$$
(U_{k+1})_{ia} = (U_{k+1})_{ia} \frac{(X_k V_k + x_{k+1}v_{k+1})_{ia}}{\left(U_{k+1}V_k^T V_k + U_{k+1}v_{k+1}^T v_{k+1}\right)_{ia}} \tag{8}
$$

3.2 INMFSC

Combined with the ideas of INMF and sparseness constraint, the INMFSC algorithm is proposed in this section. The algorithm achieves the effect of optimizing data sparseness, because the sparse constraints based on l_2-norm [21,22] regularization are added to the coefficient matrix. INMFSC is obtained:

$$
F_k = \|X_k - U_k V_k^T\|^2 + \beta\|V_k\|_2^2 \quad s.t.\ U \geq 0, V \geq 0 \tag{9}
$$

where β is the sparse coefficient and $\beta \in (0,1)$. When the $(k+1)$th sample x_{k+1} arrives, F_{k+1} can be defined as

$$
F_{k+1} = \|X_{k+1} - U_{k+1}V_{k+1}^T\|^2 + \beta\|V_{k+1}\|_2^2 \tag{10}
$$

So F_{k+1} can be rewritten as

$$
F_{k+1} \approx \sum_{i=1}^{m}\sum_{j=1}^{k} \left(X_{ij} - \left(U_k V_k^T\right)_{ij} \right)^2 + \sum_{i=1}^{m} \left((x_{k+1})_i - \left(U_{k+1}v_{k+1}^T\right)_i \right)^2
$$
$$
+ \beta\sum_{i=1}^{k}\sum_{j=1}^{r} (V_{k+1})_{ij}^2 = F_k + f_{k+1}
$$

where $f_{k+1} = \sum_{i=1}^{m} \left((x_{k+1})_i - \left(U_{k+1}v_{k+1}^T\right)_i \right)^2 + \beta v_{k+1}^2$.

As follows, we can deduce the updating rule of the incremental part v_{k+1} of the coefficient matrix by using the gradient descent method.

$$(v_{k+1})_a = (v_{k+1})_a - \eta_a \frac{\partial F_{k+1}}{\partial (v_{k+1})_a} \tag{11}$$

In (11), $\partial F_{k+1}/\partial (v_{k+1})_a$ is the partial derivative of F_{k+1} with respect to $(v_{k+1})_a$ and is given as

$$\frac{\partial F_{k+1}}{\partial (v_{k+1})_a} \cong \frac{\partial}{\partial (v_{k+1})_a} (F_k + f_{k+1}) = \frac{\partial f_{k+1}}{\partial (v_{k+1})_a}$$

$$= 2 \sum_{i=1}^{m} \left(\left((x_{k+1})_i - \left(U_{k+1} v_{k+1}^T \right)_i \right) \left(- (U_{k+1})_{ia} \right) \right) + 2\beta (v_{k+1})_a \tag{12}$$

$$= -2 \left(x_{k+1}^T U_{k+1} \right)_a + 2 \left(v_{k+1} U_{k+1}^T U_{k+1} \right)_a + 2\beta (v_{k+1})_a$$

η_a is the step size in the negative gradient direction and is calculated by

$$\eta_a = \frac{(v_{k+1})_a}{2 \left(v_{k+1} U_{k+1}^T U_{k+1} \right) + 2\beta (v_{k+1})_a} \tag{13}$$

After substituting (12) and (13) into (11), the update rule equation for $(v_{k+1})_a$ yields as

$$(v_{k+1})_a = (v_{k+1})_a \frac{\left(x_{k+1}^T U_{k+1} \right)_a}{\left(v_{k+1} U_{k+1}^T U_{k+1} \right)_a + \beta (v_{k+1})_a} \tag{14}$$

In the same way, the update rules of the basic matrix can be obtained as below.

$$(U_{k+1})_{ia} = (U_{k+1})_{ia} \frac{(X_k V_k + x_{k+1} v_{k+1})_{ia}}{\left(U_{k+1} V_k^T V_k + U_{k+1} v_{k+1}^T v_{k+1} \right)_{ia}} \tag{15}$$

The specific steps of the INMFSC algorithm are given below:

1. Matrix U and matrix V are initialized randomly;
2. For the sample set X (including k training samples), using the update rules (2) and (3) for iterative computation, until it meets the convergence conditions;
3. Everytime, U and V are updated according to the update rules (14) and (15) when a new training sample v_{k+1} comes, until the convergence conditions are satisfied.

4   Experiment and Result Analysis

4.1   Databases and Parameters Selection

1. **ORL-32.** This database is created by AT&T lab at the University of Cambridge in the United Kingdom. There are 400 pictures with 10 different images for 40 distinct subjects in this dataset. For some subjects, the images were taken at different times, varying the lighting, facial expression (open/closed eyes, smiling/not smiling) and facial details (glasses/no glasses).

2. **COIL20.** This database is collected and produced by Columbia University, which consists of 1440 images with 20 different objects (toy duck, cups, etc.), each object is viewed from varying angles and sampled $5°$ apart while the object is rotating on a turntable, so each object possesses 72 pictures.

Besides, some example images from two databases are displayed as Fig. 1.

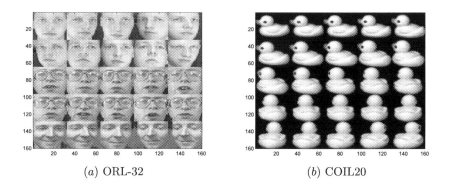

(a) ORL-32   (b) COIL20

Fig. 1. The selected instances from two databases

In the experiment, we selected 5 images from each class as the training samples for ORL-32 [23]. Similarly, we selected 36 images for COIL20 [24], then the residual images are treated as new training samples. And INMFSC is set as follows: $\beta = 0.3$ for ORL-32 and $\beta = 0.2$ for COIL20. The dimension of the basic matrix is randomly chosen from r $(r = 20, 30, \cdots, 100)$, we repeated the experiment 20 times for each r and the average result is recoded as the final result. The initial and incremental iteration number of the gradient descent method is 200.

In our experiments, two metrics, i.e., accuracy (AC) and mean running time [25], are used to evaluate the clustering performance. As for AC, it is used to measure the percentage of correct labels obtained. Here, the mean running time is not described in detail.

4.2   Clustering Results

To prove the effectiveness of the proposed algorithm, the experimental results are compared with NMF and INMF algorithms. Table 1 shows the detailed AC by three methods on ORL-32 and COIL20, respectively.

From the Table 1, we can see that the proposed INMFSC outperforms INMF and NMF under the same reduced dimension in both databases. In ORL-32, the average AC of INMFSC algorithm is 4.13% higher than NMF and 2.88% higher than INMF. In COIL20, the average AC of INMFSC algorithm is 2.57% higher than NMF algorithm and 1.55% higher than INMF algorithm. We also observe that incremental methods can keep the accuracy increasing smoothly with the dimension number of the basic matrix increasing; while the canonical

Table 1. The AC (%) on ORL-32 and COIL20, respectively

r	ORL-32			COIL20		
	NMF	INMF	INMFSC	NMF	INMF	INMFSC
20	73.25	73.88	74.00	66.25	67.30	72.78
30	76.25	76.50	78.98	79.00	79.02	79.67
40	75.00	78.01	81.21	79.85	80.58	82.14
50	77.25	79.10	81.85	80.25	80.69	82.77
60	79.50	79.50	82.17	84.75	83.19	84.30
70	78.63	80.02	83.50	80.50	84.02	84.80
80	81.00	81.01	85.35	83.01	84.69	85.42
90	82.03	82.51	85.66	83.63	84.72	85.83
100	79.96	83.05	87.36	85.00	87.17	87.63
Avg.	78.09	79.34	82.22	80.24	81.26	82.81

method would like the accuracy fluctuating. This is because every time when we rerun the NMF algorithm, the initial values for U and V are randomly assigned; different initial values would lead to different local minimum value, which in turn affect the accuracy significantly.

Figure 2 illustrates the mean running time of the selected algorithms in both databases, and r is set as follows: $r = 36$ for ORL-32 and $r = 20$ for COIL20.

From this figure we can see that the mean running time of INMFSC and INMF have an obvious advantage over the traditional NMF algorithm. The running time of INMFSC algorithm is much less than NMF approach, and close to that of the INMF method, which means that INMFSC achieved better accuracy than NMF while faster than it. Therefore, the computation efficiency is highly improves.

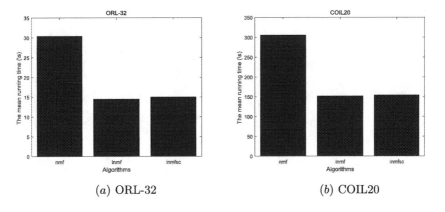

(a) ORL-32   (b) COIL20

Fig. 2. The mean running time of two databases

4.3   Sparseness Study

In this paper, we use a sparseness measure based on the relationship between the L_1 norm and L_2 norm [26]:

$$sparseness\,(x) = \frac{1}{n-1}\left[n - \left(\|x\|_1/\|x\|_2\right)^2\right] \qquad (16)$$

where n is the dimensionality of the vector X, $sparseness\,(x) \in [0,1]$, $\|\cdot\|_1$ denotes L_1 norm and $\|\cdot\|_2$ denotes L_2 norm. The value of $sparseness\,(x)$ is close to 1, and the sparseness of the vector is higher.

Next, we compute the sparseness of the basis vectors learned by three algorithms according to (16) on two databases, respectively. Figures 3 and 4 demonstrate the basis vectors obtained by NMF, INMF and INMFSC.

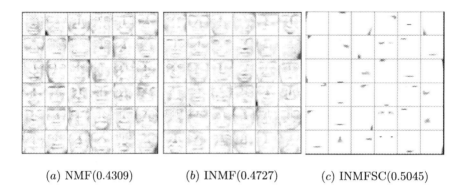

(a) NMF(0.4309) (b) INMF(0.4727) (c) INMFSC(0.5045)

Fig. 3. Basis images in ORL-32

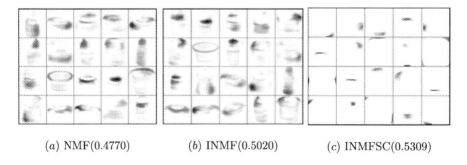

(a) NMF(0.4770) (b) INMF(0.5020) (c) INMFSC(0.5309)

Fig. 4. Basis images in COIL20

From these figures, it can be observed that the basis vectors generated by INMFSC approach are sparser than those generated by other methods. Those results indicate that INMFSC can learn a better parts-based representation than other algorithms.

5   Conclusions

In this paper, a novel incremental study method named INMFSC is proposed for image representation, and the corresponding iterative formula and algorithm steps are given. This paper carried out experiments on the ORL-32 and COIL20 databases, clustering accuracy and running time as the evaluation criteria to measure the effectiveness of the INMFSC algorithms. Experimental results show that the clustering accuracy of INMFSC algorithm can achieve more superior performance than other existing algorithms. At the same time, the efficiency of operating is greatly improved. Finally, we measure the sparseness of INMFSC method, the result shows that the sparseness of INMFSC is the highest, so our approach can obtain the better parts-based representations. However, with the change of r, INMFSC algorithm would make the clustering accuracy fluctuate. So our future work will focus on how to resolve this issue.

References

1. Hong, R.C., Hu, Z., Wang, R., Wang, M., Tao, D.: Multi-view object retrieval via multi-scale topic models. IEEE Trans. Image Process. **12**(25), 5814–5827 (2016)
2. Liu, C.Y., He, L., Li, Z.T., Li, J.: Feature-driven active learning for hyperspectral image classification. IEEE Trans. Geosc. Remote Sens. **1**(56), 341–354 (2018)
3. Hong, R.C., Zhang, L., Zhang, C., Zimmermann, R.: Flickr circles: aesthetic tendency discovery by multi-view regularized topic modeling. IEEE Trans. Multimedia **8**(18), 1555–1567 (2016)
4. Li, Z.T., Jiao, Y.F., Yang, X.S., Zhang, T.Z., Huang, S.C.: 3D attention-based deep ranking model for video highlight detection. IEEE Trans. Multimedia **PP**(99), 1–13 (2018)
5. Liu, J.W., Zha, Z.J., Chen, X.J., Wang, Z.L., Zhang, Y.D.: Dense 3D-convolutional neural network for person re-identification in videos. ACM Trans. Multimedia Comput. Commun. Appl. **4s**(14) (2018)
6. Tang, J.H., Shu, X., Qi, G.J., Wang, M., Yan, S., Jain, R.: Tri-clustered tensor completion for social-aware image tag refinement. IEEE Trans. Pattern Anal. Mach. Intell. **8**(39), 1662–1674 (2017)
7. Lee, D.D., Seung, H.S.: Algorithms for non-negative matrix factorization. In: 14th Neural Information Processing Systems, Vancouver, Canada, pp. 556–562. Advances in Neural Information Processing Systems (2001)
8. Liu, F., Liu, M., Zhou, T., Qiao, Y., Yang, J.: Incremental robust nonnegative matrix factorization for object tracking. In: Hirose, A., Ozawa, S., Doya, K., Ikeda, K., Lee, M., Liu, D. (eds.) ICONIP 2016. LNCS, vol. 9948, pp. 611–619. Springer, Cham (2016). https://doi.org/10.1007/978-3-319-46672-9_68
9. Zhang, X., Chen, D., Wu, K.: Incremental nonnegative matrix factorization based on correlation and graph regularization for matrix completion. Int. J. Mach. Learn. Cybern. **15**(36), 1–10 (2018)
10. Sun, J., Cai, X.B., Sun, F.M., Hong, R.C.: Dual graph-regularized constrained nonnegative matrix factorization for image clustering. KSII Trans. Internet Inf. Syst. **5**(11), 2607–2627 (2017)
11. Jian, M., Lam, K.M., Dong, J.: A novel face-hallucination scheme based on singular value decomposition. Pattern Recogn. **11**(46), 3091–3102 (2013)

12. Jian, M., Lam, K.M.: Simultaneous hallucination and recognition of low-resolution faces based on singular value decomposition. IEEE Trans. Circ. Syst. Video Technol. **11**(25), 1761–1772 (2015)
13. Liu, J.W., et al.: Multi-scale triplet CNN for person re-identification. In: 2016 ACM Conference on Multimedia, Amsterdam, Netherlands, pp. 192–196. ACM Digital Library (2016)
14. Hoyer, P.O.: Non-negative matrix factorization with sparseness constrains. J. Mach. Learn. Res. **5**(9), 1457–1469 (2004)
15. Liu, H.F., Wu, Z.H., Li, X.L.: Constrained non-negative matrix factorization for image representation. IEEE Trans. Pattern Anal. Mach. Intell. **7**(34), 1299–1311 (2012)
16. Cai, D., He, X.F., Han, J.W.: Graph regularized non-negative matrix factorization for data representation. IEEE Trans. Pattern Anal. Mach. Intell. **8**(33), 1548–1560 (2011)
17. Ding, C., Li, T., Jordan, M.: Convex and semi-nonnegative matrix factorizations. IEEE Trans. Syst. Man Cybern. Part B Cybern. **1**(32), 45–55 (2010)
18. Jiang, W., Li, H., Yu, X.G., Yang, B.R.: Graph regularized non-negative matrix factorization with sparseness constraints. Comput. Sci. **1**(40), 218–256 (2013)
19. Pan, B.B., Chen, W.S., Xu, C.: Incremental learning of face recognition based on block non-negative matrix factorization. Appl. Res. Comput. **1**(26), 117–120 (2009)
20. Bucak, S.S., Gunsel, B.: Incremental subspace learning via non-negative matrix factorization. Pattern Recogn. **7**(34), 788–797 (2009)
21. Kong, D.G., Ding, C., Huang, H.: Robust nonnegative matrix factorization using L_{21}-norm. In: 20th ACM Conference on Information and Knowledge Management, Glasgow, UK, pp. 673–682. ACM, New York (2011)
22. Dai, L.Y., Feng, C.M., Liu, J.X., Zheng, C.H.: Robust nonnegative matrix factorization via joint graph Laplacian and discriminative information for identifying differentially expressed genes. Complexity **40**(2017), 1–11 (2017)
23. Chen, W.S., Li, Y.G., Pan, B.B., Chen, B.: Incremental learning based on block sparse kernel nonnegative matrix factorization. In: 10th International Conference on Wavelet Analysis and Pattern Recognition, Jeju, South Korea, pp. 219–224. IEEE (2016)
24. Zheng, J.W., Chen, Y., Jin, Y.T., Wang, W.L.: Incremental locality preserving nonnegative matrix factorization. In: 6th International Conference on Advanced Computational Intelligence, Hangzhou, China, pp. 135–139. IEEE (2013)
25. Yu, Z.Z., Liu, Y.H., Li, B., Pang, S.C., Jia, C.C.: Incremental graph regulated nonnegative matrix factorization for face recognition. J. Appl. Math. **1**(2014), 1–10 (2014)
26. Zeng, S.N., Gou, J.P., Deng, L.M.: An Antinoise Sparse Representation Method for Robust Face Recognition via joint l_1 and l_2 Regularization, 1st edn. Pergamon Press, Oxford (2017)

A Data-Driven No-Reference Image Quality Assessment via Deep Convolutional Neural Networks

Yezhao Fan[1], Yuchen Zhu[1], Guangtao Zhai[1(✉)], Jia Wang[1], and Jing Liu[2]

[1] Institute of Image Communication and Network Engineering,
Shanghai Jiao Tong University, Shanghai, China
{yezhaofan,zyc420,zhaiguangtao,jiawang}@sjtu.edu.cn
[2] School of Electrical and Information Engineering, Tianjin University,
Tianjin, China
jliu_tju@tju.edu.cn

Abstract. Nowadays, image quality assessment (IQA) is important for many applications, such as transmission, compression or reproduction. And no-reference image quality assessment (NRIQA) has received extensive attention because there is no need to use original images. Especially in this artificial intelligence and big data age, it is impossible to get the reference images every time. However, many conventional NR-IQA algorithms extracted features by hand and designed only for one specific distortion type. In this paper, we will propose a NR-IQA method to predict image quality accurately with great generalization ability. We first use gaussian kernel regression to do the data augmentation owing to the limited dataset. Then, a NR-IQA model based on convolutional neural network (CNN) is trained, named Deep No-Reference Image Quality Assessment (DNRIQA), including five convolutional layers and three pooling layers for feature extraction, and three fully connected layers for regression. Finally, the experiments show that this approach achieves the state-of-the-art performance on LIVE dataset, and further cross dataset experiments prove that the model has excellent generalization ability. Experimental results of this paper is very competitive compared with other algorithms.

Keywords: Image quality assessment · No-Reference Image Quality Assessment · Gaussian kernel regression · Deep NoReference Image Quality Assessment

1 Introduction

As we all know, with the development of network technology, it is much easier to communicate or transmit images with each other by Internet services like Fackbook, Instagram and Twitter. The sharing of digital images has occurred tremendous growth in the past few years. However, since the distorted images

© Springer Nature Switzerland AG 2018
R. Hong et al. (Eds.): PCM 2018, LNCS 11165, pp. 361–371, 2018.
https://doi.org/10.1007/978-3-030-00767-6_34

usually loss massive information, it is of great importance to guarantee the quality of the images that the users received. It can evaluate and ensure the images fitted to transmit and receive. So, the task of image quality assessment becomes increasingly important.

IQA is identified as the characteristic of measuring images level of degradation as calculated by human visual system (HVS), and it aims to measure the distortion between the distorted images and the reference images. The distorted images in publicly IQA datasets include LIVE [1], CSIQ [2], TID 2013 [3] and so on. If the reference images are available, known as full-reference image quality assessment (FR-IQA) [3,4], they can be applied to quantify the differences directly with the distorted images. The classical FR measures, such as FSIM [6] and VIF [7], get a high correlation with human opinion.

However, reference images do not always exist, especially in the field of industrial production. Since the NR-IQA measures are not dependent on the reference images, and it can directly calculate image degradations by features exploited. So it is more useful and flexible for practical application. Many NR-IQA methods have been proposed [21–24]. One class including DIIVINE [8], BRISQUE [9], is using hand-crafted features, such as edge, color and so on. Another class learns features automatically, like CORNIA [10] and other algorithms based convolutional neural network (CNN) approaches [11,12], which are more efficient. Many IQA algorithms adopt machine learning for different purposes to improve the performance of experiment results. Support vector regression(SVR) is used to extracted features and regressed the quality score in FR-IQA [13]. Natural scene statistics (NSS) is used for discriminating distorted images from reference images [9,14] in NR-IQA.

Recently, CNN has been exploited in many computer vision tasks, such as object detection [15], object classification [16] and so on. Unlike the conventional algorithms, CNN gets the features automatically. These works prove that in other computer vision tasks, it will also have outstanding performance. There are also many attempts to use deep learning for NRIQA problems [25–27]. Kang et al. first applied CNN for NR-IQA problem with automatically learned features [11]. Sebastian Bosse presents a NR-IQA method based on the deep convolutional neural network [12]. Kim and Lee build a twostage CNN NR-IQA model and generate a FR-IQA methodto calculate local quality scores by proxy patch labels [18]. Li proposed a Deep Convolutional Neural Network to predict the quality scores [19].

In this paper, because of the limited data, it is important to do the data augmentation. We adopt the Gaussian kernel regression, which is a kernel based linear smoother algorithm. By this method, a nonlinear regression function can be fitted to expand data. To the best of our knowledge, this is the first time for machine learning to do the data augmentation for training NR-IQA model based deep convolutional neural network. The data is the core of deep learning and our approach is purely data-driven and does not rely on hand-crafted features or other types of prior domain knowledge about the human visual system (HVS) or image statistics.

We will start by introducing the details of the proposed method in Sects. 2 and 3 will show the experiments results. Section 4 will draw a conclusion.

2   Proposed Method

The proposed method of using deep convolutional neural network (DCNN) for image quality assessment is as follows.

2.1   Data Augmentation

Since the IQA datasets like LIVE [1], CSIQ [2] and TID 2013 [3] have limited data (about 150 images) for one distortion, we have to do data augmentation for training DNRIQA to prevent overfitting. We first use the gaussian kernel regression to fit the nonlinear regression function for the distortion. Gaussian kernel regression is a regression algorithm which does not rely on any iterative learning. It is a technique for non-linear regression. It takes a weighted average of the surrounding points and the equation is given as follows:

$$K(x^*, x_i) = e^{-\frac{(x_i - x^*)^2}{2\sigma^2}} \tag{1}$$

$$y^* = \frac{\sum_{i=1}^{N} K(x^*, x_i) y_i}{\sum_{i=1}^{N} K(x^*, x_i)} \tag{2}$$

here x_i is i_{th} training example input, y_i is i_{th} training example output, K is the kernel function, x^* is the query point and y^* is the predicted output. Figure 1 shows the results of the gaussian blur from LIVE database after using gaussian kernel regression.

Then, we get 20 points from the function fitted and every kind of images the database have. As we can see in Fig. 2. These points represent the relationship between the Peak Signal to Noise Ratio (PSNR) and score. PSNR is the most popular index to measure the quality of distorted images. It is defined via mean squared error (MES) for a m × n distorted image. To get the relation of the PSNR and noise, we add different gaussian blur noise degrees on the original image and calculate the corresponding PSNR. These two relationship can give the correlation of PSNR and noise to create new distorted images to do the data augmentation.

After creating new distorted images, we get overlapping 224 × 224 patches by 20-pixel-stride taken from images for further data augmentation. Assigning each patch quality score as its source original score in that the images in LIVE dataset have the homogeneous distortions. During the test time, we average the predicted patch scores for each image to obtainthe image quality score. Through taking small patches as the input, we have a even much larger train set compared to only using original given dataset. Each distortion has about 200,000 images finally.

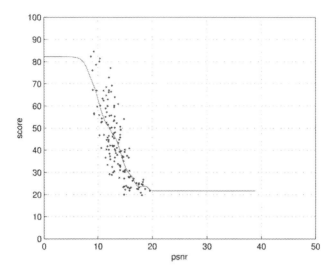

Fig. 1. The function fitted of gaussian blur from LIVE database.

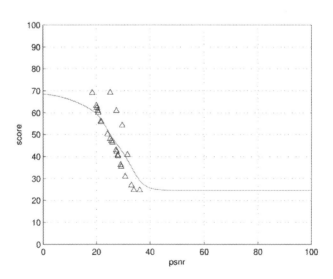

Fig. 2. The relationship of PSNR and score based on the 20 points we created and the original signal image points from gaussian blur in LIVE.

2.2   Network Architecture

AlexNet [16] is a classical convolutional neural network. The architecture of the network is shown in Fig. 3. It consists of eight layers, including five convolutional layers, three pooling layers and three fully-connected layers. The output of each layer is also as shown in Fig. 3. Owing to the goal of regression, the FC3 layer output is designed for 1. It is appropriate to use simple model because of the limited data to prevent overfitting. We input the created image patches from train set for regression training. Firstly, we use pre-trained model for transfer learning, which is BVLC alexnet model. It is trained by 360,000 iteration, and the best performance is 57.258% for validation accuracy and 1.83948 for loss. It obtains the top-1 accuracy and a top-5 accuracy 80.2% on the ILSVRC-2012 validation set. Using the first five convolutional layers weights and training the last three fully-connected layers. Then training our network with images we created and labeled by scores. Finally, predicting quality scores for every input image patches with the size of 224×224 and average the test set image score as the results. In the following section, we will discuss the details of the experiments.

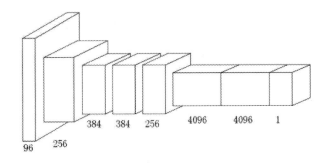

Fig. 3. The architecture of our network.

3   Experiment

3.1   Dataset

The following two datasets are used in the experiments.

(1) LIVE [1]: With five different distortions, the dataset consists of 29 reference images and 779 distorted images. The distortions are Fast Fading (FF), Gaussian blur (BLUR), White Gaussian (WN), JP2k compression (JP2K) and JPEG compression (JPEG) at different degradation levels. Each image is provided with Differential Mean Opinion Scores (DMOS), in the range between 0 and 100. The higher DMOS the image has, the lower quality of it.

(2) TID 2013 [3]: There are 25 reference images and 3000 distorted images affected by 24 different distortions. We consider four distortions that are appeared in LIVE which are BLUR, WN, JP2k and JPEG. Each image is marked with the Mean Opinion Score (MOS) in the range from 0 to 9. Unlike DMOS, higher MOS represents higher quality.

3.2 Evaluation

We are going to use two classical measures to evaluate our proposed method. (1) Pearsonlinear correlation coefficient (PLCC). (2) Spearman Rank Order Correlation Coefficient (SROCC). PLCC measures the dependence of two variables and SROCC represents how well one quantity can be used as the monotonic function of another variable [20]. We randomly select 60% of distorted images as the train set, 20% as the validation set, and the rest 20% as the test set.

3.3 Gaussian Kernel Regression

Because of the limited data, it is probably overfitting without data augmentation. It occurs when the model is excessively complex, for example, having too many parameters related to the train set. A overfitted model has poor predictive performance on test set, in that it overreacts to minor variations in the train set. So, we first use Gaussian kernel regression to fit function for every distortion. As we can see in Fig. 4, they are relationships between PSNR and score of distortions such as (a) Fast Fading, (b) Gaussian Blur, (c) White noise, (d) JPEG compression and (e) JPEG2000 compression.

Because each image has its own attribute, it is essential to create new images separately. And then, to create new images with scores, we select 20 points from each fitted function in that it is not convinced to fit a function with only a few points. As shown in Fig. 2, these 25 points consist of a new function. This gives the relationship between PSNR and score. Based on the database, we also get the relation between noise and score. These two can form the score and noise function to create new images. Based on this function,we can create a lot of new images. In this experiment, we create 20 images for one original image. Some gaussian blur created images with scores are shown in Fig. 5. The new images are based on the degree of gaussian filter. It is obvious that the true scores are greater than predicted, but other original scores may lower. So, it does not affect the results.

Table 1. PLCC and SROCC of different algorithms on live dataset.

SROCC	JP2K	JPEG	WN	BLUR	FF	ALL
PSNR	0.870	0.885	0.942	0.763	0.874	0.866
SSIM	0.939	0.946	0.964	0.907	0.941	0.913
FSIM	0.970	0.981	0.967	0.972	0.949	0.964
DIIVINE	0.913	0.910	**0.984**	0.921	0.863	0.916
BLIINDS-II	0.929	0.942	0.969	0.923	0.889	0.931
BRISQUE	0.914	0.965	0.979	0.951	0.877	0.940
CORNIA	0.943	0.955	0.976	0.969	0.906	0.942
CNN	0.952	**0.977**	0.978	0.962	0.908	0.956
DeepCNN	0.945	0.941	0.964	0.969	0.907	0.935
DNR-IQA	**0.957**	0.961	**0.984**	**0.970**	**0.933**	**0.957**
PLCC	JP2K	JPEG	WN	BLUR	FF	ALL
PSNR	0.873	0.876	0.926	0.779	0.870	0.856
SSIM	0.921	0.955	0.982	0.893	0.939	0.906
FSIM	0.910	0.985	0.976	0.978	0.912	0.960
DIIVINE	0.922	0.921	**0.988**	0.923	0.888	0.917
BLIINDS-II	0.935	0.968	0.980	0.938	0.896	0.930
BRISQUE	0.922	0.973	0.985	0.951	0.903	0.942
CORNIA	0.951	0.965	0.987	0.968	0.917	0.935
CNN	0.953	**0.981**	0.984	0.953	0.933	0.953
DeepCNN	**0.973**	0.955	0.981	**0.984**	**0.955**	0.956
DNR-IQA	0.961	0.974	0.976	0.971	0.942	**0.957**

3.4 Experiment Performance

Table 1 shows the experiment results compared with other NR-IQA state-of-the-art methods on the LIVE database. It is obvious to see our propoesd method is the best NRIQA experiment performance. From the overall evaluation, our DNR-IQA outperformed previous NR-IQA methods.

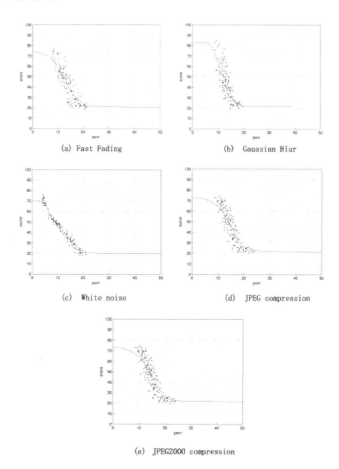

Fig. 4. The relationship between PSNR and score of distortions such as (a) Fast Fading, (b) Gaussian Blur, (c) White noise, (d) JPEG compression and (e) JPEG2000 compression.

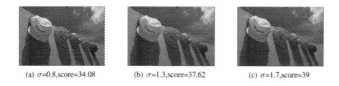

Fig. 5. The created images with scores by data augmentation.

3.5   Cross Data Test

Table 2 shows that our model has great generalization ability. We use our model to test four kinds of distortions that LIVE and TID2008 are in commom. The results are not satisfied, probably because in LIVE dataset each image is provided with Differential Mean Opinion Scores (DMOS), in the range between 0 and 100. But in TID 2013, each image is marked with the Mean Opinion Score (MOS) in the range from 0 to 9.

Table 2. SROCC and PLCC obtained by model trained on live and tested on TID 2013.

	LIVE	TID 2013
SROCC	0.957	0.904
PLCC	0.957	0.885

4   Conclusion

In this paper, we developed a DNR-IQA algorithm. Our method aims to do quality assessment through data augmentation to learn weights and achieves state-of-the-art experiment performance on LIVE dataset. Furthermore we proved that our model has great robustness to estimate quality in TID 2013 dataset. So, it has many potential applications in image reconstruction or enhancement.

Acknowledgement. This work was supported by the National Natural Science Foundation of China under Grant 61521062, and Grant 61527804.

References

1. Sheikh, H.R., Sabir, M.F., Bovik, A.C.: A statistical evaluation of recent full reference image quality assessment algorithms. IEEE Trans. Image Process. **15**(11), 3440–3451 (2006). A Publication of the IEEE Signal Processing Society
2. Larson, E.C., Chandler, D.M.: Most apparent distortion: full-reference image quality assessment and the role of strategy. J. Electron. Imaging **19**(1), 011006 (2010)
3. Ponomarenko, N., Jin, L., Ieremeiev, O.: Image database TID2013: peculiarities, results and perspectives. Signal Process. Image Commun. **30**, 57–77 (2015)
4. Wang, Z., Bovik, A.C., Sheikh, H.R.: Image quality assessment: from error visibility to structural similarity. IEEE Trans. Image Process. **13**(4), 600–612 (2004). A Publication of the IEEE Signal Processing Society
5. Xue, W., Zhang, L., Mou, X.: Gradient magnitude similarity deviation: a highly efficient perceptual image quality index. IEEE Trans. Image Process. **23**(2), 684–695 (2014)
6. Zhang, L., Zhang, L., Mou, X.: FSIM: a feature similarity index for image quality assessment. IEEE Trans. Image Process. **20**(8), 2378 (2011)

7. Sheikh, H.R., Bovik, A.C., Veciana, G.D.: An information fidelity criterion for image quality assessment using natural scene statistics. IEEE Trans. Image Process. **14**(12), 2117–2128 (2005)
8. Moorthy, A.K., Bovik, A.C.: Blind image quality assessment: from natural scene statistics to perceptual quality. IEEE Trans. Image Process. **20**(12), 3350–64 (2011). A Publication of the IEEE Signal Processing Society
9. Mittal, A., Moorthy, A.K., Bovik, A.C.: No-reference image quality assessment in the spatial domain. IEEE Trans. Image Process. **21**(12), 4695 (2012). A Publication of the IEEE Signal Processing Society
10. Doermann, D., Kang, L., Kumar, J., et al.: Unsupervised feature learning framework for no-reference image quality assessment. In: IEEE Conference on Computer Vision and Pattern Recognition, pp. 1098–1105. IEEE Computer Society (2012)
11. Kang, L., Ye, P., Li, Y., et al.: Convolutional neural networks for no-reference image quality assessment. In: Computer Vision and Pattern Recognition, pp. 1733–1740. IEEE (2014)
12. Bosse, S., Maniry, D., Wiegand, T., et al.: A deep neural network for image quality assessment. In: IEEE International Conference on Image Processing, pp. 3773–3777. IEEE (2016)
13. Narwaria, M., Lin, W.: SVD-based quality metric for image and video using machine learning. IEEE Trans. Syst. Man Cybern. Part B Cybern. **42**(2), 347 (2012). A Publication of the IEEE Signal Processing Society
14. Saad, M.A., Bovik, A.C., Charrier, C.: Blind image quality assessment: a natural scene statistics approach in the DCT domain. IEEE Trans. Image Process. **21**(8), 3339–52 (2012)
15. Girshick, R., Donahue, J., Darrell, T., et al.: Rich feature hierarchies for accurate object detection and semantic segmentation. In: IEEE Conference on Computer Vision and Pattern Recognition, pp. 580–587. IEEE Computer Society (2014)
16. Krizhevsky, A., Sutskever, I., Hinton, G E.: ImageNet classification with deep convolutional neural networks. In: International Conference on Neural Information Processing Systems, pp. 1097–1105. Curran Associates Inc. (2012)
17. Liang, Y., Wang, J., Wan, X., Gong, Y., Zheng, N.: Image quality assessment using similar scene as reference. In: Leibe, B., Matas, J., Sebe, N., Welling, M. (eds.) ECCV 2016. LNCS, vol. 9909, pp. 3–18. Springer, Cham (2016). https://doi.org/10.1007/978-3-319-46454-1_1
18. Kim, J., Lee, S.: Fully deep blind image quality predictor. IEEE J. Sel. Top. Signal Process. **11**(1), 206–220 (2017)
19. Li, Y., Po, L.M., Feng L, et al.: No-reference image quality assessment with deep convolutional neural networks. In: IEEE International Conference on Digital Signal Processing, pp. 685–689. IEEE (2017)
20. Video Quality Experts Group. Final report from the Video Quality Experts Group on the validation of objective models of video quality assessment, Phase II (FR TV2) (2003)
21. Zhai, G., Cai, J., Lin, W.: Cross-dimensional perceptual quality assessment for low bit-rate videos. IEEE Trans. Multimed. **10**(7), 1316–1324 (2008)
22. Gu, K., Zhai, G., Yang, X.: Using Free energy principle for blind image quality assessment. IEEE Trans. Multimed. **17**(1), 50–63 (2014)
23. Zhai, G., Cai, J., Lin, W.: Three dimensional scalable video adaptation via user-end perceptual quality assessment. IEEE Trans. Broadcast. **54**(3), 719–727 (2008)
24. Min, X., Ma, K., Ke, G.: Unified blind quality assessment of compressed natural, graphic and screen content images. IEEE Trans. Image Process. **26**, 5462–5474 (2017)

25. Min, X., Gu, K., Zhai, G.: Blind quality assessment based on pseudo reference image. IEEE Trans. Multimed. **20**, 2049–2062 (2017)
26. Min, X., Zhai, G., Gu, K.: Blind image quality estimation via distortion aggravation. IEEE Trans. Broadcast. **PP**(99), 1–10 (2018)
27. Zhai, G., Wu, X., Yang, X.: A psychovisual quality metric in free-energy principle. IEEE Trans. Image Process. **21**(1), 41–52 (2012)

An Asian Face Dataset and How Race Influences Face Recognition

Zhangyang Xiong[1]($\bowtie$), Zhongyuan Wang[1], Changqing Du[2],
Rong Zhu[1], Jing Xiao[1], and Tao Lu[3]

[1] NERCMS, School of Computer, Wuhan University, Wuhan 430072, China
xiong_zhangyang@qq.com
[2] School of Information Engineering, Qujing Normal University,
Qujing 655000, China
[3] Hubei Key Laboratory of Intelligent Robot,
School of Computer Science and Engineering, Wuhan Institute of Technology,
Wuhan 430073, China

Abstract. The face recognition scheme based on deep learning can give the best face recognition performance at present, but this scheme requires a large amount of labeled face data. The currently available large-scale face datasets are mainly Westerners, only containing few Asians. In practice, we have found that models trained using these data sets are lower in accuracy in identifying Asians than Westerners. Therefore, the establishment of a large-scale Asian face dataset is of great value for the development and deployment of face related applications for Asians. In this paper, we propose a simple semi-automatic approach to collect face images from Internet and build a large-scale Asian face dataset (AFD) containing 2019 subjects and 360,000 images. To the best of our knowledge, this is the largest Asian face image dataset proposed so far. To illustrate the quality of AFD, we train 3 different models with the same CNN structure yet by different training datasets (AFD, WebFace, mixed WebFace&AFD) and verify them on one Western and two Asian face testing datasets. Extensive experimental results show that the model by our AFD outperforms counterparts by a large margin for Asian face recognition. We have made the AFD dataset public to facilitate face recognition development for Asians.

Keywords: CNN · Face recognition · Asian face dataset · Race

1 Introduction

In recent years, face recognition has achieved great progress with the help of deep convolutional neural network (CNN). Usually, there are two essential issues in CNN based tasks, such as the structure of CNN and the training dataset. Large face datasets are important for advancing face recognition research, but most public face datasets mainly consist of Western face images, with only a few Asian face images included. For Asian face recognition, the deep learning models trained by these datasets do not provide satisfactory recognition accuracy comparable to that of Westerners.

© Springer Nature Switzerland AG 2018
R. Hong et al. (Eds.): PCM 2018, LNCS 11165, pp. 372–383, 2018.
https://doi.org/10.1007/978-3-030-00767-6_35

The construction of face datasets is tedious because a lot of work must be done to clear large amounts of raw data. To facilitate this task, we have developed a semi-automatic method for constructing a face dataset that detects faces in images returned by public persons retrieved on Internet, and then automatically discards those not belonging to each queried person. We create a collection name indexes and then gather photos from Web based on the collection. Considering that a single photo may contain multiple persons or not only facial images, we detect the individual's face based on the MTCNN method [7]. Finally, because of incorrect labeling or face detection errors, some candidate faces do not match with the actual individuals. We therefore use the Google Face-net CNN model [11] for face matching to eliminate false candidates. Following this way, we offer a face image dataset specific for Asian, including 2019 individuals and total about 360,000 images, called AFD. AFD is currently the largest public Asian face image dataset as we know. This dataset can be applied to training face recognition CNN or other purposes, which is not only used for academic research but also for real applications. AFD can be downloaded at https://github.com/X-zhangyang/AFD-dataset.

To illustrate the quality of AFD, we use different datasets (merely Western faces, merely Asian faces and mixture) to train same model, and then compare model's performance on three different testing datasets (one is Western and the other are Asian). This way, we exploit how training datasets of different races influence experimental results in face recognition. Particularly, we intend to prove that the model trained by Asian faces is able to give better recognition performance on Asian faces than the model trained by Western faces.

The major contributions are highlighted as follows.

(1) Based on the massive Internet photos, we propose a universal semi-automatic method to build new face image datasets with high classification rate.
(2) We create the largest Asian face dataset so far, containing 360,000 face images by 2019 individuals. In contrast, the second largest Asian face dataset CASIA-FaceV5 merely includes 2500 images by 500 individuals.
(3) We experimentally prove that different races influence face recognition performance in terms of the consistency of training and testing datasets. Particularly, our built Asian face dataset largely outperforms Western face datasets for Asian face recognition.

2 Related Work

In face recognition, training data and algorithm are two significant issues. Nowadays, CNN as well as its variants are mainstream algorithms, the structures of which have been becoming deeper and wider. Using deeper and wider structure means we need a larger scale of training dataset. Otherwise, if training data is insufficient, the CNN will become overfit and produces poor recognition performance. Many face data sets have been used in industry and academia which consist of Western face images, have been made public. In this section, we present some popular face image datasets.

Labeled Face in the Wild (LFW) [1]. A dataset of face photographs is designed for studying the problem of unconstrained face recognition. The dataset contains more than 13,000 human facial images collected from Web. Each face has been labeled with the name of the person picture. Among them 1680 faces have two or more distinct photos in the dataset. The only constraint on these faces is that they were detected by the Viola–Jones [13] face detector. More details can be found in website. LFW dataset was very popular in past few years, as the developed algorithm based on it mostly can achieve up to 99% [14]. It also describes a pipeline to build a face image dataset.

FaceScrub [2]. The FaceScrub dataset was created from Internet, followed by manually checking and cleaning the results. It comprises a total of 106,863 face images of male and female 530 celebrities, with about 200 images per person. As such, it is one of the largest public face databases.

CASIA-WebFace [3]. The CASIA-WebFace is one of largest face image datasets, almost from Western race. It includes about 1000 subjects and 494,414 face images. This dataset is widely used in face recognition, especially used to train CNNs. Nevertheless, not all face images are detected and annotated correctly in this dataset. There are no overlapped images between this dataset and LFW. Again, this dataset mainly consists of Western face images.

MegaFace [4]. The MegaFace dataset is the largest publicly available facial recognition dataset with a million faces and their respective bounding boxes. All images were obtained from Flickr (Yahoo's dataset) and licensed under creative commons. MegaFace has become most popular in face recognition.

YouTube Faces [5]. A database of face videos is designed for studying the problem of unconstrained face recognition in videos. The dataset contains 3,425 videos of 1,595 different people. All the videos are downloaded from YouTube. An average of 2.15 video clips are available for each subject, and the average length of a video clip is in 181.3 frames.

FaceDB [6]. FaceDB offers an anonymous "face-to-face" authentication service with high accuracy liveness detection. It supports multimodal biometrics software completely in-house that can be easily integrated in any applications.

CASIA-FaceV5 [15]. This dataset contains 2,500 color facial images of 500 subjects. All images are Asian human face. The number of images is too small to train a CNN because the CNN will become overfitted.

AFD. Our created dataset AFD contains 360,000 color face images from 2019 different individuals. On average, each subject takes about 178 facial images in diversified poses, expressions, deformations, and even lighting conditions. Therefore, this dataset is particularly suitable for training face applications based on deep learning, given that it is not only large-scale, but also rich in poses and lighting. All face images are gathered from Web by Asian names and are also discarded from the wrong labeled ones. To the best of our knowledge, AFD is the largest Asian face dataset proposed so far.

Table 1 tabulates some typical face datasets, where most datasets mainly contain Western human facial images. In order to find out how different races influence face recognition, we manage to build an Asian dataset, termed Asian Face Dataset (AFD). More details will be shown in the next section.

Table 1. Popular facial image datasets.

Datasets	Subjects	Total images	Races
LFW [1]	5,749	13,233	Mainly Western
FaceScrub [2]	530	107,818	Mainly Western
CASIA-WebFace [3]	10,575	494,414	Mainly Western
MegaFace [4]	4,030	4,400,000	Mainly Western
YouTube Faces [5]	1,595	Videos	Mainly Western
FaceDB [6]	23	1521	Mainly Western
CASIA-FaceV5 [15]	500	2500	Only Asian
AFD	2,019	360,000	Only Asian

3   Building AFD Dataset and Training CNN

The previous methods to build a dataset are to collect images from Internet and then tag each image. This paper proposes a new pipeline GDC (Gather-Detect-Classify) to build face image datasets. As most human images are already labeled when uploaded to Internet (usually are tagged with person name, although some images may be mismatched to their labels), we firstly create a name-index set before collecting images and then collect subject images according to the set.

Since the single photo may contain multiple faces and furthermore the faces are not separated from the entire person's image, we need to use face detection method to isolate individual's face. In addition, photos may be incorrectly associated with an individual due to incorrect annotations when people upload photos. For this reason, face recognition techniques need to be used to eliminate false match by classifying matched and mismatched candidates. Moreover, we train face recognition models based on the established dataset to verify its validity.

3.1   GDC Pipeline

A. Gather Raw Images

The main difference of our proposed GDC pipeline from DAR (Detection-Alignment-Recognition) [1] pipeline is that we build a name-index set which is used to collect intended images before gathering raw images. In practice, we create a large name-index set, covering about 2500 individuals. Given that movie star photos are easier to collect from Web, our name-index set contains a large number of Asian stars. According to the constructed name-index set, we search for their photo images one by one, resulting in a raw picture collection about 630,000 images. All images are saved in JEPG format with the largest size being 500 × 500 pixels.

B. Detect Face Images

Although 630,000 pictures were collected from the Internet, what we really care about is facial images. Note that there may appear multiple persons in one picture, or the image may contain entire person's photo rather than face. We therefore use MTCNN [7] to detect faces in each image so as to separate individual's face image. Since MTCNN cannot detect faces with almost 100% accuracy, and some raw images have multiple faces, we end up with more than 910,000 face images totally, taking a variety of mismatched faces with the specific subject. For the total of 2019 individuals, each one takes about 450 faces, including diverse facial poses and expressions as well as mislabeled and morphed faces. After detecting the faces, we crop them horizontally and vertically to 250×250 pixels about the center to exclude the noisy background. The ultimate collection can be treated as the preliminary version of AFD dataset.

C. Classify Labeled Face Images

Specifically, we consider collections of face images obtained by running a face detector on images returned from a search engine by person's name. Within each collection for a specific person, some are false positives (e.g., non-faces) found by the imperfect face detector, and a number of them belong to other people appearing in the same image as the queried person or people in images irrelevant to the query. We refer the all faces acquired for a person as candidates, and those mismatched faces as outliers. Our goal is to remove the outliers among the detected candidates for each queried person, so that we obtain faces belonging just to him/her and a cleaned dataset overall. As CNN-based face recognition has given promising results, we use Google Face-net CNN model [2, 11] to select the right face images for each person.

3.2   Discarding Mismatched Candidates

In order to determine whether the collected and detected faces really belong to a person, we need a face image as a benchmark. We manually select a right labeled and clear frontal facial image as the benchmark. By comparing the candidate faces with the benchmark, we decide whether they belong to the same person. To ensure metric accuracy, this similarity comparison is performed in the feature domain instead of the pixel domain. Specifically, we firstly input the benchmark and all candidate face images of the same individual into a trained CNN model to obtain a feature vector for each image, and then compute the Euclidean distance between the benchmark and each candidate.

We observe that images of the same person usually enjoy smaller Euclidean distances, while images belonging to different individuals have larger Euclidean distances. When a Euclidean distance equals to 0, it means these two images are exactly same. Therefore, we need to set a threshold to determine whether to dismiss the candidate face image or not. Since a larger threshold retains more candidates but meanwhile mistakes a certain one, the appropriate threshold directly affects the availability of the dataset.

We sort the distance vector from small to large and then depict it in Fig. 1. We find that it has a steep front and back end and an approximately linear distribution in the middle. Obviously, the element at the steep front should be accepted, and the element at the steep end should be rejected instead. The key is how to determine the attribution of the middle element.

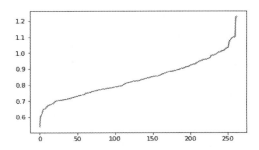

Fig. 1. The distribution of the sorted Euclidean distances by index vs. distance value.

We examine the mean distance of each person, by averaging all the Euclidean distances with

$$D_{mean} = \frac{\sum_{i=0}^{n} D_i}{n} \tag{1}$$

Where D_i represents the Euclidean distance between the i-th candidate face and the benchmark face.

Ranging all D_i from small to large, we find that they are unevenly distributed. For most individuals, their candidate faces are correctly associated. Consequently, most distance values are statistically small and are seldom larger than D_{mean}. In this case, the average distance can be used as a suitable classification criterion. More specifically, the distances less than D_{mean} usually give a correct classification, while those larger than D_{mean} instead give wrong classification.

However, for some individuals, most of the candidates do not match their labels. In this case, D_{mean} is mostly close to the very end D_i. Due to the large number of outliers, the average distance is also very large, so that it is obviously not appropriate to use only the average as the threshold. Even if the distance is less than D_{mean}, the correct classification rate is still very low. Therefore, we need to decrease the threshold against D_{mean} to improve the correct classification rate. In order to eliminate the interference of many outliers, we consider a parameter that is not much affected by high proportions and large values of outliers. Median filtering is a non-linear signal processing technique that can effectively suppress noise and outliers based on sorting statistics theory. We therefore use the median filtering of the distance vector to produce a more robust parameter D_s. In general, for a vector containing many large outliers, its median filtering result is smaller than its average. But for insurance, we still examine the minimum of the two parameters. We choose the minimum one between D_{mean} and D_s as the optimal threshold using

$$D_f = \min(D_{mean}, D_s) \tag{2}$$

For a variety of candidate faces for a particular individual, we exclude candidate faces whose distance values are greater than the threshold, leaving only candidates that are less than the threshold. After dismissing mislabeled candidates, AFD is reduced to 360,000 images from the preliminary 910,000 images, with 178 faces for an individual on average. Subject to the classification accuracy, AFD still takes a very small number of mismatched face images. Actually, the existence of few mismatched images is hopefully to improve the robustness of the model training against data errors and noise. As a concrete example, Fig. 2 shows facial images for 6 individuals under varied poses and expressions, where each row corresponds to the same person.

Fig. 2. Six individuals under varied poses or expressions in AFD dataset.

3.3   Training CNN Models

To validate our built face dataset, we need to perform verification on face recognition by training the same CNN model with different training datasets, including ours and publicly available ones. The widespread CNN structures in face recognition include Alex-net [8], VGG [9] and Inception [10]. Considering the comprehensive performance, we implement our CNN structure based on Inception-Resnet [11], using the triplet loss [12] to guide convergence.

4   Experiments

This section experimentally validates the performance of the established Asian face dataset in face recognition. For fair comparison, we use 3 different datasets to train the same CNN structure, such as WebFace, AFD and mixed WebFace&AFD. Few Asian face images are available in WebFace, but AFD contains only Asian face images. Models' details are tabulated in Table 2.

Table 2. Three models trained with the same structure but different training datasets.

Models	Structures	Training datasets	Total Images
Model_1	Inception + Triplet Loss	WebFace (Western)	490,000
Model_2	Inception + Triplet Loss	AFD (Asian)	360,000
Model_3	Inception + Triplet Loss	WebFace&AFD	850,000

The testing images are retrieved from the face datasets and are taken under uncontrolled real-world situations (unconstrained environments). Three trained models will be verified on three different testing datasets, shown in Table 3. Among them, one is for Westerner and two are for Asian.

Table 3. Three different testing datasets, including 1 Western face dataset and 2 Asian face datasets.

Datasets	Positive pairs	Negative pairs	Testing methods	Races
LFW	3000	3000	Cross-validation & ROC	Western
CASIA-FaceV5	3000	3000	Cross-validation & ROC	Asian
RealPhoto	116	7772	Mean Euclidean distance	Asian

Each testing dataset includes a large number of positive pairs (two facial images are from the same identity) and negative pairs (two facial images are from different identities). We took part of faces from public LFW and CASIA-FaceV5 to form the first and second test data sets. To make the test closer to the actual application environment, we build the third testing dataset RealPhoto by ourselves, which contains 68 individuals. Each of them has one Chinese ID card photo and several photos taken under the actual situation. Note that that all face images in training datasets (Table 2) and training datasets (Table 3) are mutually exclusive.

As for experiments on LFW and CASIA-FaceV5, we use cross validation to verify three above models. For visual classification tasks such as face recognition, excellent models should be able to reduce intra-class differences and increase inter-class differences at the same time. For this reason, in the third test, we mainly examine the ability of the model to distinguish between the same person's face and different people's faces. This will be compared by measuring the Euclidean distances of the feature vectors given by the model.

4.1   Results on LFW

The test results on LFW and their corresponding ROC curves are shown in Figs. 3 and 4, respectively. Green line, red line and blue line represent models trained by WebFace, AFD and WebFace&AFD, respectively. The plot illustrates that model trained by mixture dataset achieves the highest accuracy, and meanwhile model trained by WebFace gives a higher accuracy than model trained by AFD. It seems reasonable since WebFace and LFW are both Western face images while AFD contains Asian faces.

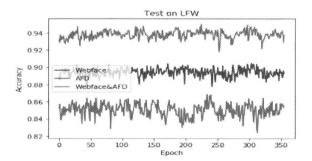

Fig. 3. The results of three models on LFW testing dataset. Blue line by WebFace&AFD tells the best accuracy, about 94%. Green line by WebFace indicates the accuracy of 89%. Red line by AFD shows the lowest mean accuracy, about 85%. (Color figure online)

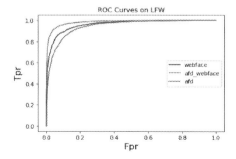

Fig. 4. Three different models' ROC curves on LFW, highly agreeing with the results shown in Fig. 3.

4.2   Results on CASIA-V5

Results on CASIA-V5 and their corresponding ROC curves are shown in Figs. 5 and 6, respectively. The curves illustrate that red line by AFD shows the best performance and green line by WebFace shows the worst instead. Testing dataset CASIA-V5 only includes Asian face images. As we expect, the model trained by Asian face images (AFD) outperforms the models trained by Western faces (WebFace) or mixed races (WebFace&AFD) for Asian face recognition. Once again, it is evident that face recognition performance is best when the face race in the test data set is consistent with the training data set.

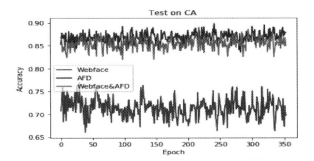

Fig. 5. The results of three models on CASIA-V5 testing dataset. Red line is trained on AFD and green line is trained on WebFace. The mean accuracy of red line is about 87%, and green line seems to be around 72%. (Color figure online)

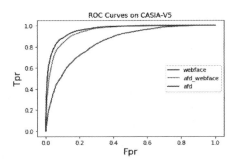

Fig. 6. Three different models' ROC curves on CASIA-V5, highly agreeing with the results shown in Fig. 5.

4.3   Results on RealPhoto

Three models are also verified on a real dataset RealPhoto built by ourselves. This testing dataset includes 68 standard Chinese ID card photos and 116 facial photos taken in real environment. There are the maximum number of pair is 7888 (a pair consists of 1 ID photo and 1 captured photo). Among them, there exist 116 pairs of same identity labels and 7772 pairs of different identity labels. We calculate the mean Euclidean distances with respect to same and different pairs. Same pair means that those two images belong to the same individual, and different pair indicates that they are from different individuals. We examine the difference between the average Euclidean distance of different pair and the average Euclidean distance of the same pair. If the difference is large, there is a large gap between the same pair and different pair, then the corresponding model has a higher accuracy of distinguishing the face image.

Results on the average Euclidean distance of same pair, the average Euclidean distance of different pair and their difference as well are shown in Table 4. Again, in this Asian face testing dataset, the model trained by AFD achieves the largest Euclidean gap 0.58327, while the model trained by WebFace gives the smallest gap 0.24718. This again shows that in Asian face recognition, the model trained with the Asian face dataset works best.

Table 4. Mean Euclidean distances of same and different pairs on testing dataset RealPhoto. The value of the fourth column is equal to the third column minus the second column. The larger the value of the fourth column, the better effect of training with the corresponding data set. The 0.58327 mean AFD dataset achieves best performance.

Training datasets	Mean distances of same pair	Mean distances of different pair	Difference between the two
AFD	0.83953	1.42280	**0.58327**
WebFace	0.69745	0.94463	0.24718
WebFace&AFD	0.72562	1.11575	0.39013

5   Conclusion

In this paper, we propose a semi-automatic way to collect face photos from Internet and build a large scale Asian face dataset containing 2019 subjects and 360,000 images, called AFD. To the best of our knowledge, the size of this dataset rank first in the literature from the perspective of Asian face datasets. Further, we explore how different races of facial image datasets (Western and Asian) affect face recognition performance. In conclusion, the model trained by Asian face images is able to provide better face recognition performance on Asian people than the model trained by Western faces. More generally, when the training dataset agrees with the testing dataset in terms of race, the more promising recognition performance will be achievable. Therefore, for face recognition task under real situations, it is strongly recommended that the agreeable training set be used.

So far, more than 80% facial images in AFD are Chinese. In the future, we would like to collect more face images, such as Japanese and Korean, to build a larger and more comprehensive Asian face image dataset.

Acknowledgements. We'd like to thank Institute of Automation, Chinese Academy of Sciences (CASIA) for offering CASIA-FaceV5 and CASIA-WebFace datasets. The research was supported by National Natural Science Foundation of China (61671332, 61502354, U1736206), Hubei Province Technological Innovation Major Project (2017AAA123), National Key R&D Project (2016YFE0202300), Hubei Provincial Natural Science Fund Key Project (2018CFA024), and Joint Project of Yunnan Provincial Science and Technology Department (2017FH001-060).

References

1. Huang, G.B., Ramesh, M., Berg, T., Learned-Miller, E.: Labeled faces in the wild: a database for studying face recognition in unconstrained environments, vol. 1, no. 2. Technical Report 07-49, University of Massachusetts, Amherst (2007)
2. Ng, H.W., Winkler, S.: A data-driven approach to cleaning large face datasets. In: IEEE International Conference on Image Processing (ICIP), pp. 343–347 (2014)
3. Yi, D., Lei, Z., Liao, S., Li, S.Z.: Learning face representation from scratch. arXiv preprint arXiv:1411.7923 (2014)
4. Miller, D., Brossard, E., Seitz, S., Kemelmacher-Shlizerman, I.: MegaFace: a million faces for recognition at scale. arXiv preprint arXiv:1505.02108 (2015)
5. Wolf, L., Hassner, T., Maoz, I.: Face recognition in unconstrained videos with matched background similarity. In: CVPR, pp. 529–534. IEEE (2011)
6. Jesorsky, O., Kirchberg, K.J., Frischholz, R.W.: Robust face detection using the Hausdorff distance. In: Bigun, J., Smeraldi, F. (eds.) AVBPA 2001. LNCS, vol. 2091, pp. 90–95. Springer, Heidelberg (2001). https://doi.org/10.1007/3-540-45344-X_14
7. Chen, D., Ren, S., Wei, Y., Cao, X., Sun, J.: Joint cascade face detection and alignment. In: Fleet, D., Pajdla, T., Schiele, B., Tuytelaars, T. (eds.) ECCV 2014. LNCS, vol. 8694, pp. 109–122. Springer, Cham (2014). https://doi.org/10.1007/978-3-319-10599-4_8
8. Krizhevsky, A., Sutskever, I., Hinton, G.E.: imagenet classification with deep convolutional neural networks. Commun. ACM **60**, 84–90 (2017)
9. Simonyan, K., Zisserman, A.: Very deep convolutional networks for large-scale image recognition. arXiv preprint arXiv:1409.1556 (2014)
10. Szegedy, C., et al.: Going deeper with convolutions. In: 2015 IEEE Conference on Computer Vision and Pattern Recognition (CVPR), pp. 1–9 (2015)
11. Szegedy, C., Ioffe, S., Vanhoucke, V., Alemi, A.A.: Inception-v4, Inception-ResNet and the impact of residual connections on learning. In: AAAI, vol. 4 (2017)
12. Schroff, F., Kalenichenko, D., Philbin, J.: FaceNet: a unified embedding for face recognition and clustering. In: 2015 IEEE Conference on Computer Vision and Pattern Recognition (CVPR), pp. 815–823 (2015)
13. Viola, P., Jones, M.J.: Robust real-time face detection. Int. J. Comput. Vis. **57**, 137–154 (2004)
14. Learned-Miller, E., Huang, G.B., RoyChowdhury, A., Li, H., Hua, G.: Labeled faces in the wild: a survey. In: Kawulok, M., Celebi, M.E., Smolka, B. (eds.) Advances in Face Detection and Facial Image Analysis, pp. 189–248. Springer, Cham (2016). https://doi.org/10.1007/978-3-319-25958-1_8
15. http://biometrics.idealtest.org/findTotalDbByMode.do?mode=Face

Towards Stereo Matching Algorithm
Based on Multi-matching Primitive Fusion

Renpeng Du[1], Fuming Sun[1(✉)], and Haojie Li[2]

[1] Liaoning University of Technology, Jinzhou, China
sunwenfriend@hotmail.com
[2] Dalian University of Technology, Dalian, China

Abstract. Classical adaptive support weight (ASW) algorithm has poor robustness and high computational complexity for stereo matching in the case of relatively low texture and complex texture regions. To solve this issue, a novel stereo matching algorithm based on the multi-matching primitive is proposed by combining color matching primitive with gradient matching primitive and integrating the correlation. This algorithm consists of three stages: initial matching cost stage, aggregation stage of cost function and parallax post-processing stage. In the first stage, we design a cost function incorporating color primitives and gradient primitives. In the second stage, we develop an adaptive matching window based on the relationship between RGB color and the space distance. In the last stage, we perform parallax post-processing by Left-Right Consistency check and adaptive weight median filtering based on Sub-Pixel. Experimental results showed that the proposed algorithm has good performance in the case of low texture and complex texture regions compared with ASW.

Keywords: Adaptive support weight · Multi-matching primitive
Correlation

1 Introduction

As a core issue in binocular vision, stereo matching algorithms are widely used in multiple areas, such as unmanned navigation and robot navigation. The basic principle of stereo matching algorithm is to obtain a two-dimensional map of a scene through a binocular camera and then get the disparity of a pair of matching points using a certain matching algorithm, further get the depth information of this scene. Because of its superior performance, stereo matching algorithm has been deeply investigated. However, the stereo matching algorithm would not provide good performance under the conditions of the complexity of the texture or the discontinuity of the scene depth, which limits the scope of its use in natural scene. To enhance its generalization ability, some variants of stereo matching algorithms have been proposed recent years [1–9]. According to the calculation methods, they can be divided into two categories: global algorithm and local algorithm [1, 2]. The global stereo matching algorithm features high computational complexity and low efficiency performance while calculating energy function, and hence unable to meet the actual needs. But, the local stereo matching algorithm has the advantage of good Real-Time performance and the disadvantage of

© Springer Nature Switzerland AG 2018
R. Hong et al. (Eds.): PCM 2018, LNCS 11165, pp. 384–395, 2018.
https://doi.org/10.1007/978-3-030-00767-6_36

relatively low accuracy performance since local data items are only operated in the stage of the cost function aggregation. The matching accuracy of stereo matching algorithm has been qualitatively improved since the classical Adaptive Support Weight (ASW) algorithm was proposed [3]. Furthermore, the local stereo matching algorithm can be subdivided into three categories according to matching primitives: region algorithm [4], feature algorithm [5], and phase algorithm [6].

ASW algorithm has the disadvantage of low accuracy performance in complex texture regions due to single matching primitive and fixed window, and poor real time performance because of the high computational complexity of the support weight. In order to further improve the accuracy of ASW algorithm, Zhu et al. [7] exploited gradient matching primitive to calculate the initial cost function, and used the gray information of neighboring pixels to design an adaptive window. However, this algorithm still features very low real time and also weak robustness because the correlation of Multi-matching primitives is not considered and the complicated computation of the support weight is not abandoned. Lin et al. [8] used linear functions to fit the Gaussian weight function in the original algorithm. The Gaussian pyramid sample is obtained by Down-Sampling the original image and the cost aggregation function is calculated by the hierarchical clustering algorithm. As a result, both the real-time and accuracy of the algorithm are improved. As a result, both the real time and accuracy of the algorithm are improved. However, the algorithm has weak robustness for blur sampling image due to the limitation of sampling times. Recently, Men et al. [9] abandoned some obviously wrong disparity by referring to the coincidence degree between the adaptive window of the reference image and the matched image, and deleted the calculation of complex weights in the cost aggregation stage. But the accuracy of this algorithm is not better than the classical algorithm because the correlation between the reference image and the cost function is not taken into account in cost function aggregation stage.

In view of the problems aforementioned, in order to further improve the performance of the ASW algorithm in the practical scene, a novel stereo matching algorithm based on the Multi-matching primitive is proposed by combining color matching primitive with gradient matching primitive and incorporating the correlation between the cost function and the reference image. Specific measures of the proposed method are described as follows. In the initial matching cost stage, the color matching primitive and the gradient matching primitive are firstly combined, and then the proportion coefficient α of the matching element of the color and the gradient is adaptively adjusted by using the Cones image on the Middlebury platform [10]. In the aggregation stage of cost function, in order to solve the problem of low accuracy of the ASW on complex texture conditions sine the fixed window is adopted in traditional method, the adaptive window method is explored by expanding the central pixel to be matched according to the color information and spatial distance information between the pixels. Meanwhile, in order to reduce the time complexity of the algorithm, the task of cost function aggregation is finished by using correlation [11] between the initial matching cost function value and the reference image RGB values instead of the traditional complex weight operations. Finally, the Left-Right Consistency (LRC) detection process and an adaptive sub-pixel based median filtering operation are performed to further improve the accuracy of the algorithm in the disparity map [12, 13]. Compared with other algorithms on the Middlebury platform, experimental results demonstrated that the average error rate of the disparity map obtained by the proposed method has been greatly improved.

2   Adaptive Support Weight (ASW) Algorithm

In the stereo matching algorithm, the two images are assumed to satisfy the epipolar constraint that the matching points corresponding to the left and the right images are in the same row position. The core of ASW is that the support of neighboring pixels is only valid if they are coming from the same depth so that the central pixel has the same disparity with the pixel to be matched when the support weights are used to calculate the similarity between image pixels. Therefore, the support weight w of the pixels is proportional to the disparity probability Pr in the window as follows:

$$w(p, q) \propto \Pr(d_p = d_q) \tag{1}$$

where p is the pixel to be matched, q is other pixel except p in the window, d is the disparity related to the color and spatial distance of the image.

The support weight w is also represented as

$$w(p, q) = k \cdot f(\Delta c_{pq}, \Delta g_{pq}) \tag{2}$$

where Δc_{pq} and Δg_{pq} represent the distance between p and q in LAB color space and geometric space respectively, k is the proportionality coefficient, f is the Laplacian kernel function. The distance Δc_{pq} and Δg_{pq} are independent each other.

$$f(\Delta c_{pq}, \Delta g_{pq}) = f(\Delta c_{pq}) \cdot f(\Delta g_{pq}) \tag{3}$$

where Δc_{pq}, Δg_{pq} are calculated as shown in formulas (4) and (5).

$$\Delta c_{pq} = \sqrt{(L_p - L_q)^2 + (a_p - a_q)^2 + (b_p - b_q)^2} \tag{4}$$

$$\Delta g_{pq} = \sqrt{(p_x - q_x)^2 + (p_y - q_y)^2} \tag{5}$$

where the component L is used to represent the brightness of a pixel in the Lab color space, its range is [0, 100] from pure black to white; a represents the range from red to green, its range is [127, −128]; b represents the range from yellow to blue, its range is [127, −128]; (p_x, p_y) and (q_x, q_y) are coordinate value in geometric space. The Laplacian kernel function is used to define the intensity as follows:

$$f(\Delta c_{pq}) = \exp(-\frac{\Delta c_{pq}}{\gamma_c}) \tag{6}$$

$$f(\Delta g_{pq}) = \exp(-\frac{\Delta g_{pq}}{\gamma_p}) \tag{7}$$

where γ_c, γ_p are obtained by experiment, $\gamma_c = 7$, $\gamma_p = 36$ here (The value is generally related to the window size). Then the cost function is aggregated below.

$$E(P,\overline{P_d}) = \frac{\sum_{q\in N_p,\overline{q}_d\in N_{\overline{p}_d}} w(p,q)\cdot w(\overline{p}_d,\overline{q}_d)\cdot e_0(q,\overline{q}_d)}{\sum_{q\in N_p,\overline{q}_d\in N_{\overline{p}_d}} w(p,q)\cdot w(\overline{p}_d,\overline{q}_d)} \tag{8}$$

where the Initial matching cost $e_0(q,\overline{q}_d)$ is shown as follows:

$$e_0(q,\overline{q}_d) = \sum_{c\in\{r,g,b\}} |I_c(q) - I_c(\overline{q}_d)| \tag{9}$$

where $I_c(q)$ and $I_c(\overline{q}_d)$ are the gray value in the fixed window of the reference image and the image to be matched, the disparity of the two pixels is d. Finally, the final disparity map is determined by the WTA (Winner-Takes-All) method.

3   The Improved Stereo Matching Algorithm Based on ASW

3.1   The Algorithm Flowchart

The flowchart of the improved algorithm based on ASW proposed is shown in Fig. 1, which mainly includes 4 stages: Read a pair of images (the left and right images), Calculate initial matching cost, Aggregate cost function and Perform Parallax Post-Processing (mainly includes LRC check and filter operation).

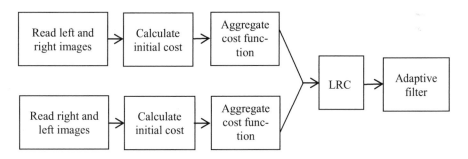

Fig. 1. The improved algorithm flowchart.

3.2   Initial Matching Cost with Colour and Gradient

The traditional ASW algorithm uses gray density of the pixel as the matching primitive. In recent years, many scholars have made improvements on this basis, including using RGB pixel information or gradient information to calculate the initial matching cost. But due to the complex texture in natural scene, it is difficult to characterize the image features by a single matching primitive. And in some particularly complex cases, the initial cost function obtained through matching primitives does not meet the actual characteristics of the scene obviously. Therefore, the truncation thresholds are set by the mean of the gradient primitive and the RGB color primitive in this paper. The information of the RGB color and gradients of the image are fused by the idea of Kalman filtering. Then the coefficient α is adjusted to make improvements adaptively. The specific process is given below.

Set the color and the gradient thresholds t1, t2, calculate the initial matching cost e_g and e_c, shown as follows:

$$e_g(q, \bar{q}_d) = \min(|I_{rg}(q) - I_{lg}(\bar{q}_d)|, t1) \tag{10}$$

$$e_c(q, \bar{q}_d) = \min(\frac{1}{3} \cdot \sum_{c \in (r,g,b)} |I_c(q) - I_c(\bar{q}_d)|, t2) \tag{11}$$

where $I_{rg}(q)$ and $I_{lg}(\bar{q}_d)$ are the gradient values of pixel in the reference image and the image to be matched, $I_c(q)$ and $I_c(\bar{q}_d)$ are the RGB values of the pixel in the reference image and the image to be matched.

Adjust the coefficient α to calculate the final initial matching cost adaptively, shown as follows:

$$e_0(q, \bar{q}_d) = \alpha \cdot e_c(q, \bar{q}_d) + (1 - \alpha) \cdot e_g(q, \bar{q}_d) \tag{12}$$

3.3   Adaptive Window

The ASW algorithm has poor robustness in complex texture areas because the window size is fixed. Here, adaptive window method is adopted according to the color and spatial distance between pixels. As is known the center pixel $p(x, y)$ to be matched, and in the x and y direction of each neighboring pixel are $p(x - 1, y)$, $p(x + 1, y)$ and $p(x, y - 1)$, $p(x, y + 1)$. Different from the traditional adaptive window expanded through the gray value of the pixels, we use RGB color to expand the adaptive window which is expanded when the Three-Channels information of the neighboring pixel with the central pixel satisfy the following formula at the same time.

$$I_{r,g,b}(x, y) - I_{r,g,b}(x - 1, y) < t \tag{13}$$

where t is assumed as the color threshold, let $t \in (0, 1)$. If the calculation of the cost function aggregation is too complex and the window size is too wide due to the existence of repeated texture regions in the scene, the neighboring pixels are expanding with the formula (13). So the cut-off length is set according to the geometric characteristics of the image in this paper. The window length is truncated when the following formula is satisfied.

$$\sqrt{(p(x) - q(x))^2 + (p(y) - q(y))^2} > L_{\max} \tag{14}$$

where $p(x)$ and $p(y)$ are the horizontal and vertical coordinate values of the center pixel, $q(x)$ and $q(y)$ are the horizontal and vertical coordinate values of the neighboring pixels. Adaptive window is shown in Fig. 2. Here set the minimum expanding length $L_{min} = 5$ and the cut-off length $L_{max} = 11$.

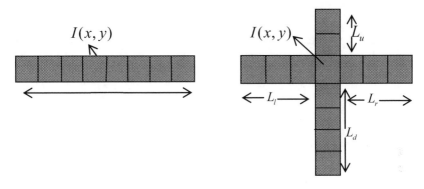

(a) Horizontal expansion   (b) The final renderings by window expansion

Fig. 2. Adaptive window.

First, the central pixel in the window is expanded horizontally in Fig. 2(a). If the final length of the window is less than the minimum arm length L_{min}, the original length is replaced by L_{min}. But if the distance between the neighbouring pixel and the central pixel is longer than the cut-off length, the original length is replaced by the cut-off length. Then, the final rendering by window expansion is shown in Fig. 2(b). In Fig. 2 (b), four different length, including up L_u, down L_d, left L_l, right L_r, are gotten with the central point. Finally, the row and column size of adaptive window are given as follows:

$$Rows = L_u + L_d \tag{15}$$

$$Cols = L_l + L_r \tag{16}$$

3.4   Cost Function Aggregation Algorithm

After calculating the initial matching cost, we perform the cost function aggregation, which is different from the complex weight calculation in the ASW algorithm. Both the correlation between the cost function and the reference image RGB density, and the correlation between the reference image and the image to be matched are to be considered [11]. First, we need to calculate the variance and the covariance between the cost function and the reference image RGB density and obtain the correlation function. Then the correlation function is subtracted by the initial matching cost before performing cost function aggregation. The experimental results show that the performance of the algorithm, including accuracy and the Real-Time, has been greatly improved.

It needs to calculate the mean value of the convolution in the RGB density and the initial matching cost function in the adaptive window, represented below.

$$m_{ce}(p)_{c\in\{r,g,b\}} = \frac{1}{n} \cdot \sum_{q\in N_p} I_c(q) \cdot e(q) \tag{17}$$

where p is the coordinates of the central pixel, q is the coordinate of each pixels including the central one, N_p is the adaptive window mentioned before, and n is the number of pixels. After calculating the mean of convolution function in each channel, we can calculate the covariance function, shown as follows:

$$v_{ce}_{c\in\{r,g,b\}} = m_{ce} - m_c \cdot m_e \tag{18}$$

where m_c, m_e are the mean values of the RGB element and the initial cost function in the adaptive window respectively. Their specific formulas are shown below.

$$m_c(p)_{c\in\{r,g,b\}} = \frac{1}{n} \cdot \sum_{q\in N_p} I_c(q) \tag{19}$$

$$m_e(p)_{c\in\{r,g,b\}} = \frac{1}{n} \cdot \sum_{q\in N_p} e_c(q) \tag{20}$$

The variance matrix θ of the Three-Channels value of the reference image in the adaptive window is calculated, shown below.

$$\theta = \begin{bmatrix} S_{rr} & S_{rg} & S_{rb} \\ S_{rg} & S_{gg} & S_{gb} \\ S_{rb} & S_{gb} & S_{bb} \end{bmatrix} \tag{21}$$

Each value of the matrix θ is calculated as shown below.

$$s_{cc}(p)_{c\in\{r,g,b\}} = \frac{1}{n} \cdot \sum_{q\in N_p} I_c(q) \cdot I_c(q) - \sum_{q\in N_p} I_c(q) \cdot \sum_{q\in N_p} I_c(q) \tag{22}$$

By (17)–(22) we can get the Coefficient matrix, shown below.

$$\gamma_{kn}_{n\in\{1,2,3\}} = \frac{v_{ce}}{\theta} \tag{23}$$

As v_{ce} ($c \in \{r, g, b\}$) is one 1×3 vector. We can know that the value of γ_{kn} is a vector that contains three channels. The convolution of the Three-Channels pixels in the reference image with the correlation coefficient is subtracted by the mean of initial cost function in the adaptive window so that the initial matching cost are more independent. The final initial matching cost is shown as follows:

$$e_0(p, \bar{p}_d) = m_e - \sum_{\substack{q \in N_p, c \in \{r, g, b\} \\ n \in (1, 2, 3)}} I_c(q) \cdot \gamma_{kn}(q) \tag{24}$$

As γ_{kn} is a 1×3 vector, $n \in (1, 2, 3)$. The cost function aggregation is done in the adaptive window after the initial matching cost has been calculated. The specific formula is shown in formula (25).

We only operate the aggregation process for reference images without taking into account the image to be matched in order to improve the real-time performance of the algorithm. Finally, the WTA (Winner-Takes-All) method is used to choose the disparity corresponding to the minimum value of cost function aggregation as the pixel value of the disparity map, shown as follows.

$$E(d) = \frac{1}{n} \cdot \sum_{q \in N_p} e_0(q, \bar{q}_d) + \sum_{\substack{q \in N_p, c \in \{r, g, b\} \\ n \in (1, 2, 3)}} I_c(q) \cdot \gamma_{kn}(q) \tag{25}$$

$$D(p) = \arg\min_d (E(d)) \tag{26}$$

3.5 Parallax Post-processing

LRC Check. In the stereo matching algorithm, the occlusion problem has been inevitable due to the existence of disparity in left and right images. In order to obtain the final disparity map, we perform LRC check at first.

The disparity d_l is calculated when the left image as the reference image, the disparity d_r is calculated when the right image as the reference image. Assume that the following formula is satisfied.

$$|d_l - d_r| > \delta \tag{27}$$

where δ is the threshold, $\delta \in (0, 1)$, here $\delta = 1$ is set. When the absolute value in formula (27) is larger than δ, it is considered as a blocking point. Then the smaller disparity is filled into the blocking point.

Sub-Pixel Adaptive Median Filtering. After running the cost function aggregation, the disparity map often has some salt and pepper noise. It is necessary to make median filter for the image. However, the traditional filtering algorithms often neglect the correlation between pixels. But in this paper, the pixels in the window are given different weights based on the difference of color and spatial distance between pixels, shown below.

$$w(x,y) = \exp(\frac{-k1}{\gamma_c \cdot \gamma_c} + \frac{-k2}{\gamma_d \cdot \gamma_d}) \tag{28}$$

where γ_c, γ_d are the constant and obtained by experiment, set $\gamma_c = 0.1, \gamma_d = 9, k_1$ and k_2 are determined by the difference between the central pixel and the neighboring pixels in the color space and the geometric space, shown below.

$$k1 = \sqrt[3]{\sum_{q \in N_p} \sum_{c \in \{r,g,b\}} I_c(p) - \sum_{c \in \{r,g,b\}} I_c(q)} \tag{29}$$

$$k2 = \sqrt[3]{\sum_{(x,y) \in N_p} (x-x')^2 \cdot (y-y')^2} \tag{30}$$

The window size is 19×19. Adaptive median filtering is performed after the weight of each pixel is obtained in the window. The specific processes are given as follows. The gray value of each pixel besides the central pixel is multiplied by the respective weight to obtain a new gray value in the window, shown below.

$$I'(q) = w \cdot I(q) \tag{31}$$

Sort the new value of each pixel including the original central point in the window, and then take the two pixels' values $I'(q_1)$ and $I'(q_2)$ which are the closest to the median value, so to obtain the mean value of the new sub-pixel gray value to replace the original one of the center pixel. It is calculated by the following formula.

$$I(p) = \frac{I'(q_1) + I'(q_2)}{2} \tag{32}$$

4   Experimental Results

Experimental operating environment include Intel core i7-6700, 3.5G Hz CPU, 12G RAM, Matlab2016a and middlebury platform. As shown in Fig. 3, the first column is the tsukuba, teddy, venus test images; the second column is the standard disparity map; the third column is the disparity map with classic ASW algorithm; the fourth column is the disparity map by the proposed algorithm.

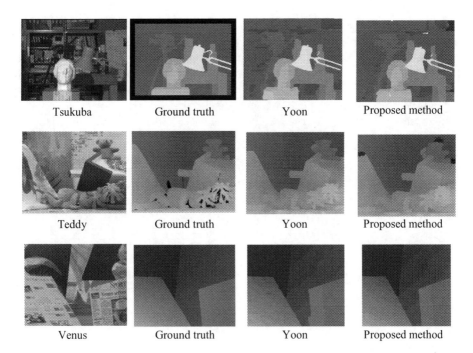

Fig. 3. The original image and the disparity map.

In experiment, the parameters are set as follows: $L_{min} = 5$, $L_{max} = 11$, $t_1 = 2$, $t_2 = 7$, $\alpha = 0.11$, $\gamma_c = 0.1$, $\gamma_d = 0.9$, the window size is 19×19. The disparity map has three test indicators on the middlebury platform: nocc (error rate in non-occlusion areas), all (error rate in all areas), disc (error rate in depth discontinuities areas). Comparing ASW and the improved algorithm based on the gradient with ASW by De-Maeztu, the specific comparison is shown in Table 1 below.

Table 1. Evaluation table of matching algorithm (unit: %).

Algorithm	Tsukuba			Venus			Teddy		
	Nocc	All	Disc	Nocc	All	Disc	Nocc	All	Disc
Proposed	1.65	1.98	6.51	0.26	0.44	2.34	6.73	12.2	15.4
ASW	1.38	1.85	6.90	0.71	1.19	6.13	7.88	13.3	18.6
GradAW	2.26	2.63	8.99	0.99	1.39	4.92	8.00	13.1	18.6

Table 2. Comparison of average error rate.

Algorithm	ASW	GradAW	IterAdaptWgt	HCFilter	Proposed
Average	6.67	6.55	6.08	5.67	5.54

From Table 1, compared with ASW and its variants, stereo matching algorithm based on multi-matching primitive fusion has obvious advantages under Venus and Teddy in all regions, non-occlusion regions and depth discontinuous regions. And in the Tsukuba image, there is also some improvement in non-occlusion areas. Compared with GradAW algorithm based on single gradient primitive, Tsukuba, Venus and Teddy have lower error rate in all regions. The average error rate of the algorithm on the middlebury platform is shown in Table 2, and the considerable results are achieved.

Table 3. Time complexity comparison (unit: ms).

Algorithm	Tsukuba	Venus	Teddy
Proposed method	507	983	2740
ASW	23754	41560	86431

Time complexity comparison between ASW and the proposed method is given in Table 3. From Table 3, we can draw a conclusion that the proposed method has good computation performance.

5   Conclusion

A novel stereo matching algorithm based on multi-matching primitive is proposed by combining color matching primitive and gradient matching primitive, and taking the correlation between initial matching cost and the reference image RGB density into account. In the initial matching cost stage, the matching cost is calculated by adjusting the ratio coefficients of gradient primitive and color primitive based on the Kalman filter idea. In the cost function aggregation stage, the traditional method is limited to the fixed window and does not consider the correlation of neighboring pixels. Therefore, we explore the color and spatial distance relationship of the neighboring pixels with the central pixel for adaptive window expansion. We perform cost function aggregation by exploiting the correlation between initial matching cost and the reference image RGB density so that the time complexity of the algorithm has been greatly reduced but its accuracy is not affected. Finally, the parallax post-processing operations is run by using the LRC check and a Sub-Pixel adaptive median filtering algorithm. The experimental results showed that the proposed algorithm is superior than most of the current algorithms on the middlebury platform. In the future, we will consider the corresponding research work in the field of image enhancement [14] and image search [15].

Acknowledgment. This work was founded by the National Natural Science Foundation of China (Grant No. 61572244, 61272214) and the major science and technology platform funds from the Liaoning Provincial Education Department (No. JP2016015).

References

1. Zhu, S., Yang, L.: Stereo matching algorithm with graph cuts based on adaptive watershed. Acta Optica Sinica **33**, 221–229 (2013)
2. Ge, X., Xing, S., Xia, Q., Wang, D., Hou, X., Jiang, T.: Semi-global stereo matching algorithm based on tree structure. Comput. Eng. **42**, 243–248 (2016)
3. Yoon, K.J., Kweon, I.S.: Locally adaptive support–weight approach for visual correspondence search. In: IEEE Computer Society Conference on Computer Vision and Pattern Recognition, vol. 2, pp. 924–931 (2005)
4. Zhang, K., Fang, Y., Min, D., Sun, L., Yang, S., Yan, S., et al.: Cross-scale cost aggregation for stereo matching. In: IEEE Conference on Computer Vision and Pattern Recognition, pp. 1590–1597 (2014)
5. Tan, X., Sun, C., Sirault, X., Furbank, R., Pham, T.D.: Feature matching in stereo images encouraging uniform spatial distribution. Pattern Recognit. **48**, 2530–2542 (2015)
6. Guo, L.Y., Sun, C.Y., Zhang, G.Y., Wu, J.H.: Variable window stereo matching based on phase congruency. Appl. Mech. Mater. **380–384**, 3998–4001 (2013)
7. Zhu, S., Li, Z.: A stereo matching algorithm using improved gradient and adaptive window. Acta Opt. Sin. **35**, 115–123 (2015)
8. Lin, Y., Lu, N., Lou, X., Zou, F., Yao, Y., Du, Z.: Matching cost filtering for dense stereo correspondence. Math. Probl. Eng. **2013**(4) (2013). (2013-9-30)
9. Men, Y., Zhang, G., et al: Adaptive window stereo matching algorithm based on pixel expansion. J. Harbin Eng. Univ. **39**(3), 547–553 (2018)
10. Hosni, A., Rhemann, C., Bleyer, M., Rother, C., Gelautz, M.: Fast cost-volume filtering for visual correspondence and beyond. IEEE Trans. Pattern Anal. Mach. Intell. **35**(2), 504–511 (2013)
11. He, K., Sun, J., Tang, X.: Guided image filtering. IEEE Trans. Pattern Anal. Mach. Intell. **35**(6), 1397–1409 (2013)
12. De-Maeztu, L., Villanueva, A., Cabeza, R.: Stereo matching using gradient similarity and locally adaptive support-weight. Pattern Recognit. Lett. **32**, 1643–1651 (2011)
13. Psota, E.T., Kowalczuk, J., Carlson, J., Pérez, L.C.: A local iterative refinement method for adaptive support-weight stereo matching. In: International Conference on Image Processing, Computer Vision, and Pattern Recognition (2012)
14. Hong, R., Zhang, L., Tao, D.: Unified photo enhancement by discovering aesthetic communities from flickr. IEEE Trans. Image Process. **25**(3), 1124–1135 (2016)
15. Hong, R., Li, L., Cai, J., Tao, D., Wang, M., Tian, Q.: Coherent semantic-visual indexing for large-scale image retrieval in the cloud. IEEE Trans. Image Process. **26**(9), 4128–4138 (2017)

An Effective Image Detection Algorithm for USM Sharpening Based on Pixel-Pair Histogram

Hang Gao[1], Mengting Hu[1], Tiegang Gao[2(✉)], and Renhong Cheng[1]

[1] College of Computer and Control Engineering,
Nankai University, Tianjin 300381, China
[2] College of Software, Nankai University, Tianjin 300381, China
gaotiegang@nankai.edu.cn

Abstract. USM sharpening is a popular method for enhancement of image quality, detection of image sharpening has attracted much attention in recent years. A novel image sharpening detection algorithm is proposed in this paper. In the scheme, different from some image forensic schemes, which used Cb or Cr channel of YCbCr color model to extract image features for forensics, in this paper, color images are firstly transformed into the YCbCr model, then the luminance channel of YCbCr color model is selected to extract pixel-pair histogram features based on four directional differential matrixes, these features within some threshold scope constitute the final image features. LIBSVM is used to implement classification for real and sharpened image. Widely used UCID database is employed to conduct test with various sharpening strength and range. Experimental results show that the proposed algorithm has superior performance; extensive comparisons with some existing algorithms show that it outperforms state-of-art methods investigated, even if the sharpening intensity is very weak ($\sigma = 0.3$).

Keywords: USM sharpening · Pixel-pair histogram · Image forensic

1 Introduction

Over the past decades, it is easier to generate fabricated digital images using photo editing tools. Thus, it sometimes is very difficult to distinguish fake image from the genuine ones. This phenomenon may lead to serious social problems, as fabricated image can cause much fear in society, and it may also cause misjudgment if the counterfeit image is used for a paragon of learning.

To cope with the problem, researchers have presented many effective image forensic algorithms to detect tampered image. For example, people have presented some forensic method to detect copy-paste operation, image splicing manipulation, contrast enhancement embellishing and filtering operation [1–8].

Recently, as it has remarkable ability of learning the necessary features from a huge amount of images, convolutional neural network (CNN) has been used to detect various image manipulations such as contrast enhancement, median filtering and JPEG compression attacks [9–13]. Among them, Chen et al. firstly use CNN based scheme to

© Springer Nature Switzerland AG 2018
R. Hong et al. (Eds.): PCM 2018, LNCS 11165, pp. 396–407, 2018.
https://doi.org/10.1007/978-3-030-00767-6_37

detect image tampering by median filtering [9]. Wang and Barni et al. presented JPEG compression detection technique based on CNN, Wang and Zhang even achieve forgery localization [10, 11]. Salloum et al. designed a splicing localization algorithm by utilizing a multi-task fully convolutional network (MFCN) [12]. Sun et al. presented a detection algorithm for contrast enhancement of image based on CNN, in the scheme, different from conventional CNN that accept the pixel value of image to be its input, the input of CNN is gray-level co-occurrence matrix (GLCM) [13].

Image sharpening aims at emphasizing texture and drawing viewer focus and enhancing the contour of the image, image clarity and quality can be effectively improved by image sharpening. Unsharp mask (USM) technique is a most widely used sharpening methods, some editing software for image embellishment utilize USM to enhance the visual effect of image. At present, some detection algorithms of image sharpening operation have been given. For example, Cao et al. firstly studied the detection for image sharpening; the ability of detection is got by considering histogram distortion and the intensity of overshoot item [14, 15]. After that, Ding et al. suggested a local binary pattern (LBP) based detection algorithm for image sharpening [16]. Inspired by the above scheme, a method by extracting edge perpendicular binary coding (EPBC) is given to detect USM based image tampering [17]. Gu et al. presented sparse coding based image sharpening detection [18]. Recently, in order to detect image sharpening of weak strength, based on the basis of EPBC method, Ding et al. recently presented a new image sharpening detection method based on edge perpendicular ternary coding (EPTC), and achieved better performance [19].

A novel image sharpening detection algorithm based on pixel-pair histogram is proposed in this letter. Different from some existing algorithms, which utilize RGB space or chrominance channel (Cr or Cb) to extract features [20], luminance component (Y) in YCrCb space is used to extract feature of pixel-pair histogram in this letter. Experiments are implemented on widely used database UCID. Test results and analyses show that the suggested method achieves superior performance for sharpening detection on all databases, it outperforms current methods.

The other contents of the paper are given as follows. The introduction on USM, pixel-pair histogram and differential matrix are described in Sect. 2. The establishment of experimental environment and creation of dataset is described in Sect. 3. Experimental results and analyses are shows in detail in Sect. 4, and conclusions and some remarks are discussed in the last section.

2 Some Preliminaries

In this section, some preliminaries, such as USM sharpening, pixel-pair histogram, color space models and differential matrixes will be presented.

2.1   Brief Review of USM Sharpening

USM sharpening can increases the acutance of an image by boosting contrast between adjacent areas by means of an unsharp mask of linear or nonlinear filter. The following two steps are often used to finish the sharpening operation [14–16]:

(1) Imposed a high-pass filter on original image I to produce an unsharp mask M.

$$M = I \otimes H \tag{1}$$

where, $\otimes$ represents convolution operator, H is a high-pass filter.

(2) Scale the mask with a sharpening strength, then add it to the original image, thus, sharpened image is produced through the following equation

$$I' = I + \lambda M \tag{2}$$

where λ is the scaling ratio, it is also regarded as the sharpening intensity of image. It controls the magnitude of each overshoot, but it does not affect the width of the edge.

Another expression of unsharp mask for the purpose of noise decreasing is given by the following formula

$$M = I - IG_\sigma \tag{3}$$

where G_σ is Gaussian low-pass filter with variance σ, the parameter σ may be used to limit the scope of the sharpening. In general, to achieve apparent visual effect, sharpening strength is set to be $\lambda \geq 0.8$, and a smaller σ may enhance image detail.

2.2   Pixel-Pair Histogram

In essence, the pixel-pair histogram is a matrix, where the value of each location (i, j) represents the number of times that the pixel pairs with the intensities i and j. Traditionally, it is defined in image spatial domain, and it has been used in the steganography and image forensic [21].

The typical example of pixel-pair histogram in spatial domain is depicted in Fig. 1. It is assumed that, an image is given in Fig. 1(a). Then, rearrange the pixels in Zigzag or other sequence into a row vector, such as shown in Fig. 1(b). The corresponding pixel-pair histogram is presented in Fig. 1(c).

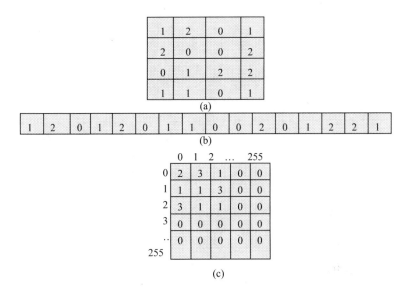

Fig. 1. Example of pixel-pair histogram of a typical image using column ordering form (a) Typical image (b) 1D-vector (c) Pixel-pair histogram

2.3 YCbCr Chrominance Channel

There are several different color models, and different traces of tampering vary in different color models. It has been shown that better performance of image forgery detection has been achieved in grayscale and RGB color systems, and it is found that detection performance also can be enhanced by using chromatic channel in YCbCr model [20].

YCbCr color model consists of luminance (Y) and chrominance (Cb and Cr) components, the relation between RGB and YCbCr is given in formula (4)

$$
\begin{pmatrix} Y \\ Cb \\ Cr \end{pmatrix} = \begin{bmatrix} 0.299 & 0.587 & 0.177 \\ -0.299 & -0.587 & 0.886 \\ 0.701 & -0.587 & -0.114 \end{bmatrix} \begin{bmatrix} R \\ G \\ B \end{bmatrix} + \begin{bmatrix} 16 \\ 128 \\ 128 \end{bmatrix} \tag{4}
$$

In general, Y component can be well perceived by visual system, and chrominance components are often used to detect image forgery.

2.4 Differential Matrix

For some image components or channel, it is donated by matrix D(x, y), where, x and y are the corresponding value as that of the width and height, and D(x, y) is the pixel value of the position (x, y). Then, four directional differential matrixes of $D(x, y)$ are calculated by the following formula

$$D_h(x,y) = |D(x,y)| - |D(x+d,y)|$$
$$D_v(x,y) = |D(x,y)| - |D(x,y+d)|$$
$$D_d(x,y) = |D(x,y)| - |D(x+d,y+d)| \qquad (5)$$
$$D_m(x,y) = |D(x+d,y)| - |D(x,y+d)|$$

where, x and y are index of position information. $D_h(x,y)$ and $D_v(x,y)$ are the differential matrixes in horizontal direction and vertical direction, respectively; $D_d(x,y)$ and $D_m(x,y)$ represent matrixes generated in the diagonal direction and the secondary diagonal direction, respectively. d is the control parameter of distance.

3   The Proposed Scheme

Although USM is a wonderful tool in image sharpening, however, too much sharpening may also introduce "halo artifacts". These halos artifacts will become a problem when they are clearly visible at the intended viewing distance, this phenomenon can be mitigated by using a smaller radius value for the unsharp mask. Thus, in practical application, weak sharpening is the widely used mean to enhance the quality of image, and EPTC based algorithm has been proposed to detect the image operation using weak sharpening, and it has achieved better performance.

To detect image sharpening tampering, the following steps are followed to extract the image features, which can be used to distinguish the sharpened image and the original one. The flowchart of the proposed scheme is depicted in Fig. 2.

(1) Convert RGB space of color image into the YCbCr color model. Firstly, transform the RGB color image into the image expressed by YCbCr color mode using formula (4).
(2) Extract the luminance Y component.
(3) Calculate four differential matrixes on Y component.
(4) Compute pixel-pair histogram for every differential matrix for a given positive threshold T and parameter d. That is to say, rearrange every differential matrix into a row vector using column ordering pattern, and then, for a given positive threshold T, the pixel-pair histograms within interval [−T, T] are calculated, thus four pixel-pair histograms matrixes are got, and the total dimension of matrixes is $4 \times (2 \times T + 1)^2$. As the correlation of pixels in an image is high, when the threshold T becomes larger, it gives no much contribution on the characteristic of the differential matrix.

The contour figure of pixel-pair histogram of differential matrix with $T = 4$ for original image and sharpened image are shown in Fig. 3(a–b). The comparison of feature between original image and sharpened one is given in Fig. 3(c).

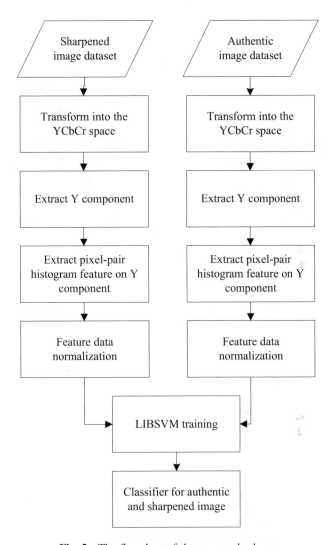

Fig. 2. The flowchart of the proposed scheme

In the proposed scheme, the Y component is selected to implement image sharpening detection, because the sharpening operation emphasizes the edges in the image, or the differences between adjacent light and dark sample points in an image. Some tests about the detection effect of components on the sharpened image have been implemented, the test is conducted on the channel of Y, Cb, Cr and grey for sharpened image by parameters $\sigma = 1$ and $\lambda = 0.5$. Detection accuracy (ACC) for Y component is 98.6547%, ACC of Cb channel is 83.6323%, ACC of Cr is 93.2735%, and ACC for grey channel is 96.1883%.

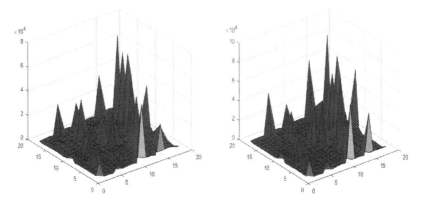

(a) Contour of pixel-pair histogram for real image (b) Contour of pixel-pair histogram for sharpened image

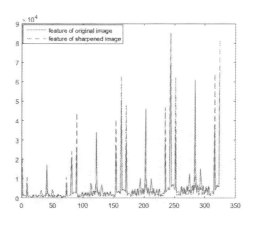

(c) Comparison between pixel-pair histogram vector

Fig. 3. Comparisons of pixel-pair histogram between the original and sharpened image

The comparisons between receiver operating characteristic curves (ROC) are given in Fig. 4. Obviously, the detection performance in Y channel is best compared with other components.

The parameter T is chosen experimentally based on the balance between the accuracy of detection and dimension of the feature. Here, the test is conducted for sharpened image by parameters $\sigma = 1$ and $\lambda = 0.8$. The ACC for different T is summarized in Table 1.

It can be seen from Table 1 that, the higher ACC of detection is achieved when the parameter T = 4. Thus, the threshold T is set 4 for all experiments in the paper.

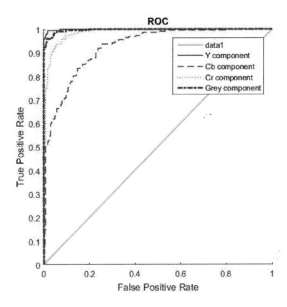

Fig. 4. The performance comparisons of sharpening detection in different components

Table 1. Detection accuracy for different T

T	ACC (%)
2	98.0492
3	98.5426
4	99.2152
5	98.5426
6	98.2063

4   Experimental Results and Analysis

In this section, some experimental condition, datasets and experimental results are discussed, in the meantime, some comparisons on detection performance between the proposed scheme and the existing algorithm are also given.

4.1   Dataset

To conduct detection test, the dataset used here is UCID Database [22], which includes 1338 uncompressed TIFF images on a variety of topics including natural scenes and man-made objects, both indoors and outdoors. It is a widely used database for USM detection test.

Here, in order to test the detection performance of different sharpening strengths and scopes. 7 datasets are produced by different sharpening parameters, they are denoted as follows:

$UCID^{071}$ Sharpened dataset sharpened by parameters $\sigma = 0.7$, $\lambda = 1$.
$UCID^{1-1}$ Sharpened dataset sharpened by parameters $\sigma = 1$, $\lambda = 1$.
$UCID^{131}$ Sharpened dataset sharpened by parameters $\sigma = 1.3$, $\lambda = 1$.
$UCID^{103}$ Sharpened dataset sharpened by parameters $\sigma = 1$, $\lambda = 0.3$.
$UCID^{105}$ Sharpened dataset sharpened by parameters $\sigma = 1$, $\lambda = 0.5$.
$UCID^{108}$ Sharpened dataset sharpened by parameters $\sigma = 1$, $\lambda = 0.8$.
$UCID^{131}$ Sharpened dataset sharpened by parameters $\sigma = 1$, $\lambda = 1.3$.

4.2   Experimental Method

In our experiments, popular LIBSVM classifier is selected to implement classification, and a RBF kernel is used [23]. In the experiments, the five-fold cross-validation is used to get the best kernel parameters for the SVM, all the authentic image are labeled as +1, while the sharpened images are labeled as −1. The software for implementation is Windows platform, and Matlab R2015b is used for conduct test. In test, threshold $T = 4$, 5/6 data is used for training and the others are used for testing.

To evaluate the performance, the True Positive Rate (TPR), True Negative Rate (TNR), Accuracy and Precision are used to verify the effectiveness of the classifier. They are defined as follows.

$$TPR = \frac{TP}{TP + FN} \tag{6}$$

$$TNR = \frac{TN}{TN + FP} \tag{7}$$

$$Accuracy = \frac{TP + TN}{TP + FN + FP + TN} \tag{8}$$

$$precision = \frac{TP}{TP + FP} \tag{9}$$

where, TP (True Positive) stands for the number of correct classified image for all the authentic image; FN (False Negative)is the number of incorrect classified image for all the authentic image; TN (True Negative) is the number of correct classified image for all the sharpened images; FP (False Positive) means the number of incorrect classified image for the entire sharpened images.

4.3   Experimental Results

For different sharpening intensity and scope, the test results are given in Table 2 for datasets.

Table 2. Detection accuracy in UCID database

Dataset	TPR	TNR	ACC
$UCID^{071}$	0.9910	0.9955	0.9933
$UCID^{1-1}$	0.9910	0.9955	0.9933
$UCID^{131}$	0.9955	0.9910	0.9933
$UCID^{103}$	0.9641	0.9776	0.9709
$UCID^{105}$	0.9910	0.9821	0.9865
$UCID^{108}$	0.9955	0.9910	0.9933
$UCID^{113}$	0.9910	0.9955	0.9933

From the experimental results, it can be seen that, the algorithm achieves better performance in UCID database even if the sharpening strength is weak.

4.4   Comparison with Related Methods

In this section, the comparisons of detection performance in UCID dataset between the proposed scheme and the existing algorithms are summarized in Table 3.

Table 3. Comparisons of detection accuracy in UCID database

Sharpened dataset	Our's scheme	EPTC (N = 7) [19]	Sparse coding [18]	EPBC (N = 7) [17]	Cao [15]
$UCID^{071}$	0.9933	0.9507	0.9665	0.9250	0.8464
$UCID^{1-1}$	0.9933	0.9573	0.9796	0.9135	0.9163
$UCID^{131}$	0.9933	0.9632	0.9856	0.9225	0.9355
$UCID^{103}$	0.9709	0.9468	/	0.8644	0.7196
$UCID^{105}$	0.9865	0.9555	0.9250	0.8967	0.7996
$UCID^{108}$	0.9933	0.9600	0.9687	0.9217	0.8752
$UCID^{113}$	0.9933	0.9852	98.90	0.9480	0.9222

It can be observed from the Table 2, the proposed scheme outperforms all the methods summarized in the Table 2. Even if the sharpening strength is very weak ($\sigma = 1$, $\lambda = 0.3$), the detection accuracy is also around 97%, this is a very obvious enhancement in accuracy.

5 Conclusion

A detection algorithm of image USM sharpening with high performance is proposed in this paper. The luminance component of YCbCr color model is used to extract pixel-pair histogram features in the scheme; the experiments are implemented on UCID database. Sharpening databases generated by different sharpening strengths and scopes are tested using LIBSVM classifier. Experimental results show that better performance is achieved even if the sharpening strength is weak. The test on other databases will be researched in the future.

Acknowledgements. The work was supported by the Program of Natural Science Fund of Tianjin, China (Grant No. 16JCYBJC15700).

References

1. Cao, Y., Gao, T., Yang, Q.: A robust detection algorithm for copy-move forgery in digital images. Forensic Sci. Int. **214**(1–3), 33–43 (2012)
2. Fadl, S.M., Semary, N.A.: Robust copy-move forgery revealing in digital images using polar coordinate system. Neurocomputing **265**, 57–65 (2017)
3. Zhao, X., Wang, S., Li, S., Li, J.: Passive image-splicing detection by a 2-D noncausal Markov model. IEEE Trans. Circuits Syst. Video Technol. **25**(2), 185–199 (2015)
4. He, P., Jiang, X., Sun, T., Wang, S.: Detection of double compression in MPEG-4 videos based on block artifact measurement. Neurocomputing **228**, 84–96 (2017)
5. Yang, J., Xie, J., Zhu, G., Kwong, S., Shi, Y.: An effective method for detecting double JPEG compression with the same quantization matrix. IEEE Trans. Inf. Forensics Secur. **9**(11), 1933–1942 (2014)
6. Cao, G., Zhao, Y., Ni, R., Li, X.: Contrast enhancement-based forensics in digital images. IEEE Trans. Inf. Forensics Secur. **9**, 515–525 (2014)
7. Yang, L., Gao, T., Xuan, Y., Gao, H.: Contrast modification forensics algorithm based on merged weight histogram of run length. Int. J. Digit. Crime Forensics **8**(2), 27–35 (2016)
8. Kang, X., Stamm, M.C., Peng, A., Liu, K.J.R.: Robust median filtering forensics using an autoregressive model. IEEE Trans. Inf. Forensics Secur. **8**, 1456–1468 (2013)
9. Chen, J., Kang, X., Liu, Y., Wang, Z.J.: Median filtering forensics based on convolutional neural networks. IEEE Signal Process. Lett. **22**, 1849–1853 (2015)
10. Wang, Q., Zhang, R.: Double JPEG compression forensics based on a convolutional neural network. EURASIP J. Inf. Secur. **2016**(1), 23 (2016)
11. Barni, M., Bondi, L., Bonettini, N., et al.: Aligned and non-aligned double JPEG detection using convolutional neural networks. J. Vis. Commun. Image Represent. **49**, 153–163 (2017)
12. Salloum, R., Ren, Y., Jay Kuo, C.-C.: Image splicing localization using a multi-task fully convolutional network (MFCN). J. Vis. Commun. Image Represent. **51**, 201–209 (2018)
13. Sun, J.-Y., Kim, S.-W., Lee, S.-W., Ko, S.-J.: A novel contrast enhancement forensics based on convolutional neural networks. Signal Process. Image Commun. **63**, 149–160 (2018)
14. Cao, G., Zhao, Y., Ni, R.: Detection of image sharpening based on histogram aberration and ringing artifacts. In: Proceedings of IEEE International Conference on Multimedia and Expo (ICME), pp. 1026–1029 (2009)

15. Cao, G., Zhao, Y., Ni, R., et al.: Unsharp masking sharpening detection via overshoot artifacts analysis. IEEE Signal Process. Lett. **18**(10), 603–606 (2011)
16. Ding, F., Zhu, G., Shi, Y.Q.: A novel method for detecting image sharpening based on local binary pattern. In: Shi, Y.Q., Kim, H.-J., Pérez-González, F. (eds.) IWDW 2013. LNCS, vol. 8389, pp. 180–191. Springer, Heidelberg (2014). https://doi.org/10.1007/978-3-662-43886-2_13
17. Ding, F., Zhu, G., Yang, J., et al.: Edge perpendicular binary coding for USM sharpening detection. IEEE Signal Process. Lett. **22**(3), 327–331 (2015)
18. Gu, Y., Wang, S., Lin, X., Sun, T.: USM sharpening detection based on sparse coding. In: Proceedings of 2016 International Conference on Digital Image Computing: Techniques and Applications (DICTA), pp. 1–5 (2016)
19. Ding, F., Zhu, G., Dong, W., Shi, Y.: An efficient weak sharpening detection method for image forensics. J. Vis. Commun. Image Represent. **50**, 93–99 (2018)
20. Hussain, M., Saleh, S.Q., Bebis, G., Muhammad, G., Aboalsamh, H.: Evaluation of image forgery detection using multi-scale weber local descriptors. Int. J. Artif. Intell. Tools **24**(4), 1–27 (2015)
21. Shabanifard M., Shayesteh M.G., Akhaee M.A.: Forensic detection of image manipulation using the Zernike moments and pixel-pair histogram. IET Image Process. **7**(9), 817–828 (2013)
22. Schaefer, G., Stich, M.: UCID—an uncompressed colour image database. Storage Retr. Methods Appl. Multimed. **5307**, 472–480 (2003)
23. https://www.csie.ntu.edu.tw/~cjlin/libsvm/

Skeletal Bone Age Assessment Based on Deep Convolutional Neural Networks

Pengyi Hao, Yijing Chen, Sharon Chokuwa, Fuli Wu[✉], and Cong Bai

Zhejiang University of Technology, Liuhe Road 288, Hangzhou, China
fuliwu@zjut.edu.cn

Abstract. Bone Age Assessment (BAA) is a pediatric examination performed to determine the difference between children's skeletal bone age and chronological age, the inconsistency between the two will often indicate either hormonal problems or abnormalities in the skeletal system maturity. Previous works to upgrade the tedious traditional techniques had failed to address the human expert inter-observer variability in order to significantly refine BAA evaluations. This paper proposes a deep learning method that detects and segments carpal bones as the region of interests within the left hand and wrist radiographs, and then feed the image data into a deep convolutional neural network. Tests are then made to determine whether it is more efficient to use full hand radiographs or segmented regions of interest, and also made comparisons with some CNN models. Evaluations show that the proposed method can dramatically increase the accuracy.

Keywords: Bone Age Assessment · Carpal bones region of interest Classification

1 Introduction

Bone Age Assessment (BAA) is a medical examination performed by pediatricians and pediatric endocrinologists to determine the difference between children's skeletal bone age and real age (in years). The inconsistency between the two will often indicate either hormonal problems or abnormalities in the skeletal system maturity [9,10]. The results obtained from BAA can enable pediatricians to make the predictions such as an estimation of the required time for a child to grow, the estimated puberty age and a child's ultimate height. The assessment also gives an insight into how to monitor progress and eventually treat conditions that inhibit normal growth. Thus, BAA is a vital technique in pediatric endocrinology, orthodontics and pediatric orthopaedics for assessing childrens skeletal system maturity.

The most common and widely accepted traditional methods of performing BAA is by obtaining a radiograph of the left wrist, hand and fingers. The acquired X-ray image is then compared with samples of labelled images within a bone development standard atlas, comprising of children with the same age

© Springer Nature Switzerland AG 2018
R. Hong et al. (Eds.): PCM 2018, LNCS 11165, pp. 408–417, 2018.
https://doi.org/10.1007/978-3-030-00767-6_38

and gender [2,4,10]. The oldest way for attaining a child's age is based on two versions of standardised main atlases; Greulich and Pyle (GP) atlas [4] and the Tanner-Whitehouse (TW) atlas [10]. While, the radiographs used in the GP method were obtained over eight decades ago, hence, it may be difficult to assess bone age accurately nowadays, While the TW is more objective than the GP [5], it takes relatively longer to perform an assessment using the TW.

BoneXpert was the first most celebrated attempt which utilizes an automated implementation for BAA, yielding a considerable high accuracy [15]. BoneXpert makes use of the active appearance model (AAM), to automatically segment 15 bones in the hand and wrist and then determine either the GP or TW bone age based on shape, intensity, and textural features. Although this approach has successfully managed to yield high accuracy over the traditional methods, it still bears some shortcomings; this system is depended on the tie between the bone age and chronological age and hence cannot give direct predictions. Previous works which used methods like canny edge detection with fuzzy classification made use of morphological features belonging to carpal bones but this method does not generalise to children above 7 years old [13,16]. The other paper which used the method of SVD fully connected NN utilised fixed- sized feature vectors from SIFT descriptions with SVD [11], other papers used automatic carpal bone area extraction and dealing the information of carpal bone with support vector regression [8] or random forest regression [9] to do a prediction. These methods lifted the BAA to a new height of automation, however they faced data limitations for training and validation hence also lacked robustness.

Recently, with the coming of the big data age, deep learning is applied to more and more areas [6]. The medical field has also incorporated these techniques to replace some of the conventional methods, hence easing and accurately performing various applications [1,8]. Deep CNNs have capabilities to discover multiple levels of distributed representations by learning high-level abstractions in medical image data [3]. In the field of bone age assessment, Spampinato et al. [14] tested several deep learning methods to assess skeletal bone age automatically. Lee et al. [7] mainly focused on the preprocess of DICOM images using a CNN model and then bone ages were predicted based on the common classification CNN model.

In this paper, we propose a deep learning approach which initially employs segmentation of the region of interest (ROI) consisting of carpal bones from the left-hand wrist radiographs. Secondly, we propose a detection CNN for detecting the ROI in X-ray images and then we also give a classification CNN to predict the bone age of the patient. The advantages of our method include full automation over the use of manual techniques, the utilisation of the carpal bones over the whole hand, and robustness since this evaluation is extended over the age group from 0 to 18 years. Finaly, we evaluate the method using whole hand and the method using ROIs, we also compare our CNN model with the famous VGGNet-16 [12]. Analysis of bone age is a complicated process even for experts; hence this papers main objective is to utilise an automated bone age evaluation method which could reasonably reduce the cost of assessment of bone age, by decreasing the time that radiologists spend in predicting the bone age and also lessening the bias caused by an inter-observer.

2   Proposed Method

In this section, the details of proposed BAA method will be given. It mainly includes three parts: data preprocessing, ROIs detection and segmentation, and classification. Its flowchart is shown in Fig. 1. The original radiographs are pre-processed at first, then a detection CNN is explored to obtain carpal bones that are called ROIs in this work. After training a classification CNN, we can get predictions of bone ages.

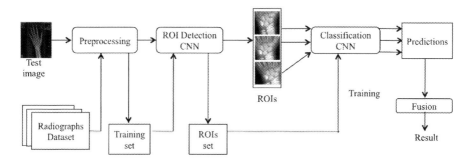

Fig. 1. The flowchart of proposed approach for bone age assessment.

2.1   Preprocessing

The original radiographs have different colour backgrounds with various sizes and the hospital annotations are situated at various positions around the radiograph. The first row in Fig. 2 shows some samples of the input radiographs used in this study. The original radiographs also have a variety of noises. These noises will give a wrong guidance in the way of learning features. So we need to preprocess them, which can not only eliminate the noises in the original images as much as possible but also reduce the image size for efficiently training. As we all know, bones are the most important part of the whole image since our target is bone age assessment. The preprocess should be able to distinguish bones from other noises. We sample pure background, background with line, background with words, bones with background and pure bones from the original radiographs, then train a network using VGG structure, with relu function and categorical crossentropy.

For every X-Ray image, we did sliding window operation with size 32×32. Each sample patch by sliding window operation did a classification using the trained CNN model. Based on the classification result, we can get a hand bone label-map by assigning pixels labeled as pure bones to white and other pixels to black. Since bones with background may have some unneeded information, then extracting the largest contiguous contour and filling the hand bone are needed to correct the output of model. Finally the hand bones are got from the original radiographs, some examples are shown in the second row of Fig. 2. These preprocessed images are then used as input of ROI detection CNN.

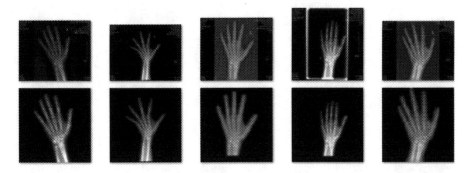

Fig. 2. Samples of radiographs. The first row shows the original radiographs. The second row shows their preprocessed results.

2.2 ROI Detection and Segmentation

Figure 3 presents the process of ROI detection and segmentation. A sliding window with a size of 100×100 is used to move across the whole preprocessed radiograph while returning us a set of 100×100 images for each radiograph. This set gives an intuition of the likely location of the ROI.

Since the upper top of the original radiographs did not contain the ROI, these were ignored, hence one 512×512 image produced 25 sub-images. Most of these slices did not include the ROI; they were either dark patches or included the phalanges, metacarpal and a large portion of the ulna and radius bones. In order to remove the dark image slices, the mean pixel intensity for each image was calculated. The average mean pixel intensity for all images was 130 but this could not be used as the intensity threshold due to the presence of some outliers, because hospitals use different radiography machines producing various intensities for the radiographs. After considering a couple of intensity thresholds, it performed better in eliminating the dark images which did not include ROI when chose threshold to be 85. After this elimination process, the resulting set still had images that did not contain carpal bones. Part of the set was then labelled as positive (containing ROI) and negative (without ROI) in order to pave way for a detection CNN for the ROI. This set was then fed into a detection CNN with the structure as shown on Fig. 4. The model achieved an accuracy of 93% in detecting the ROI. The images predicted correctly from the detection CNN consisted of 3 ROI images of size 100×100 per one original full hand image; they were then fed into the BAA classification CNN.

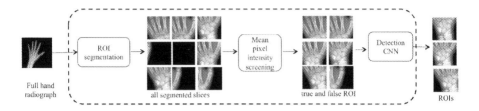

Fig. 3. The flowchart of the ROI segmentation and ROI detection process.

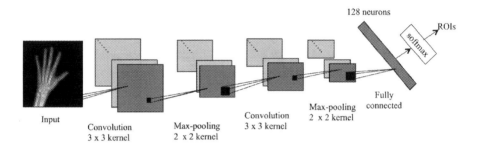

Fig. 4. ROI detection CNN architecture.

2.3   Classification CNN

We give a classification CNN here, whose structure is illustrated in Fig. 5. All the Conv Blocks are similar, with some convolutional 2D layers and ReLU activation layers, end with MaxPooling 2D layers. Fully-connected block includes a Flatten layer, two Dense layers with ReLU activation. Then through a Droupout layer and a softmax Dense layer, we can get the class label for the input. The convolution kernels are 24, 48, 96, 200 respectively, and both of the two Dense layers have 596 filters. We also explore VGGNet-16 [12] in this work. We implemented these two networks under the open source deep learning library Keras. Upon training, an SGD optimizer was used with a learning rate of 0.001. We carried out of 100 epochs with different epochs. As for VGGNet-16, the batch size is 15. For the proposed one, the batch size is 30.

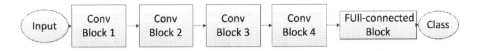

Fig. 5. Structure of the simplified version of VGGNet-13.

2.4 Data Augmentation

Overfitting is a major problem in deep learning algorithms if the dataset is too small therefore data augmentation techniques were used to enlarge the dataset and increase robustness. Rotations ranging from $0°$ to $350°$ with $10°$ increments produced 36 synthetic images. The brightness was altered by adding an integer ranging from 10 to 70, resulting in 7 synthetic images. If there exists one pixel larger than 200 in the original X-ray image, we change for decreasing brightness by adding an integer ranging from -10 to -70.

3 Experiments

3.1 Dataset and Evaluations

The Children's Hospital, Zhejing University School of Medicine of China, provided us 945 radiographs from patients with chronological age of 0–18 years, including the patients chronological age and the bone age. Figure 6 shows the bone age radiographs distribution for both male and female left hand and wrist. Due to the limitation of the dataset size, gender was not considered as this would further reduce the dataset.

For evaluations, we will compare two methods and two classification models. In the first method, noted as M1, the whole hand that encompassed of all the left hand wrist bones (i.e. phalanges, metacarpal, carpal, ulna and radius bones) was preprocessed as shown in Sect. 2.1 and then used as input image to the classification CNN shown in Sect. 2.3. Images of two different sizes 512×512 and 224×224 were then used as the model input to the classification CNN. The second method is denoted as M2 that is described in Sect. 2, which mainly used the automatically cropped ROIs, then ROIs were used as the model input to the classification CNN. Fusion rules (i.e. the average rule and the maximum rule) were employed for calculating the accuracy of the second methos (M2) because 3 ROI images were present for each original full hand radiograph. M2's CNN network gave 3 predictions for the 3 singular ROI, thus when using the average rule the mean of these 3 predictions is the final result; and when using the maximum rule the 3 predictions were evaluated to find the most frequent prediction which would then be the final prediction. At the same time, we will also compare the performance of our proposed classification model given in Sect. 2.3 and the commonly used VGGNet-16 using different sizes of inputs.

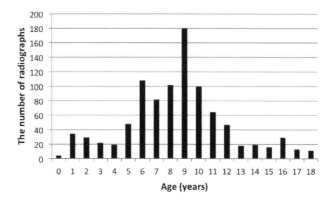

Fig. 6. Bone age radiographs distribution in the used dataset.

3.2   Results

Initially 140 radiographs from the dataset were randomly chosen encompassing of the age from 0 to 18 years. These images were used for testing models. The accuracy results obtained from the methods are shown in Table 1. Firstly, the results show that VGGNet-16 can get a little higher accuracy of totally correct than our model under the M1 method no matter with the input size of 512×512 or 224×224. But our model performed a litter better than VGGNet-16 when radiographs were assigned an age within 1 year of ground truth. By using M2, our model got much higher accuracy than VGGNet-16. Our model can assign 43.3% radiographs to be the correct age, and 74.2% radiographs were assigned an age within 1 year of ground truth and 88.1% were assigned an age within 2 years of ground truth. What is more, no matter VGGNet-16 or our model, using the input with 512×512 performs better than the input size of 224×24; the results also show that utilising automatically segmented ROIs is a viable method for significantly improving the accuracy in comparison to using the full hand; the results attained portray that the average rule outperforms the maximum rule. The accuracy of totally correct for each model and each method is not very high. The reason may be that the age groups from 0 to 4 years have bad preprocessing result which restricts the system from being applicable to all age groups. The other reason is that the imbalance in the age groups distribution in the dataset, for example there are more radiographs for the age groups from 6 years to 11 years in contrast to the age groups from 12 years to 18 years.

Sample predictions achieved by the our classification CNN are shown in Fig. 7 illustrating the ground truth versus the models prediction. Here, the predictions of M2 used average rule. In Fig. 7, the first two rows are correctly predicted, both methods can get the same bone age with ground truth. The last two rows are wrong predictions that may due to the presence of soft tissue reduced the hand-to-background ratio. Using a machine with a GPU of Nvidia GeForce GT 730M 2G, it averagely takes one and half minutes from reading an original radiograph to get the final bone age assessment, which is much more efficient than traditional metholds.

Table 1. Comparison between different methods using different classification models in terms of accuracy.

	Type	Correct	±1 year	±2 years
VGGNet-16	M1 (224 × 224)	0.371	0.571	0.779
	M1 (512 × 512)	0.464	0.641	0.771
	M2 (Maximum)	0.411	0.645	0.834
	M2 (Average)	0.428	0.684	0.862
Proposed CNN	M1 (224 × 224)	0.364	0.593	0.721
	M1 (512 × 512)	0.414	0.673	0.712
	M2 (Maximum)	0.412	0.698	0.844
	M2 (Average)	0.433	0.742	0.881

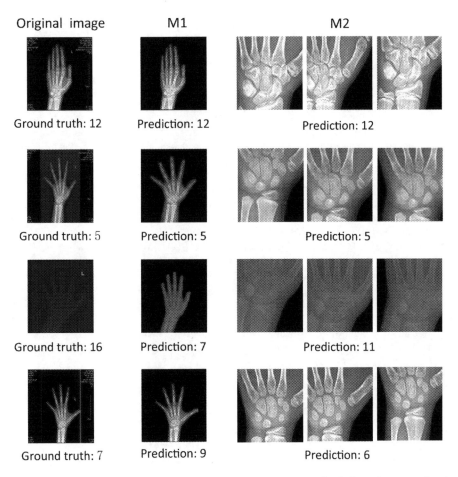

Original image   M1   M2

Ground truth: 12   Prediction: 12   Prediction: 12

Ground truth: 5   Prediction: 5   Prediction: 5

Ground truth: 16   Prediction: 7   Prediction: 11

Ground truth: 7   Prediction: 9   Prediction: 6

Fig. 7. The examples of bone age assessments using two methods based on our classification CNN.

4   Conclusions

In this paper, we gave a deep-learning method to automatically detect and segment carpal bones as a region of interest. We also gave a classification CNN for BAA evaluation. This work made a comparison of the use of full hand and wrist and the carpal bones while utilising a data driven approach with a classification CNN. The technique proposed in this study has great capability to yield a much higher accuracy result however the presence of a number of limiting factors restrained its performance, like the amounts of data. Given a larger training dataset this accuracy can be improved.

Acknowledgments. The work is supported by Zhejiang Provincial Natural Science Foundation of China under grants No. LY18F020034, LY18F020032, and National Natural Science Foundation of China under grants No. 61502424 and No. 61801428 and partially supported by the Ministry of Education of China under grant of No. 2017PT18 and the Zhejiang University Education Foundation under grant of No. K18-511120-004 and No. K17-511120-017. This work is also supported by The Research of Real Doctor AI Research Center.

References

1. Anthimopoulos, M., Christodoulidis, S., Ebner, L., Christe, A., Mougiakakou, S.: Lung pattern classification for interstitial lung diseases using a deep convolutional neural network. IEEE Trans. Med. Imaging **35**(5), 1207–1216 (2016)
2. Gaskin, C., Kahn, S., Bertozzi, J., Bunch, P.: Skeletal Development of the Hand and Wrist: A Radiographic Atlas and Digital Bone Age Companion (2011)
3. Greenspan, H., van Ginneken, B., Summers, R.: Guest editorial deep learning in medical imaging: overview and future promise of an exciting new technique. IEEE Trans. Med. Imaging **35**(5), 1153–1159 (2016)
4. Greulich, W., Pyle, S.: Radiographic Atlas of Skeletal Development of the Hand and Wrist. Stanford University Press, Stanford (1959)
5. Khan, K., Elayappen, A.: Bone Growth Estimation Using Radiology (Greulich-Pyle and Tanner-Whitehouse Methods). In: Preedy, V. (ed.) Handbook of Growth and Growth Monitoring in Health and Disease. Springer, New York (2012). https://doi.org/10.1007/978-1-4419-1795-9_176
6. LeCun, Y., Bengio, Y., Hinton, G.: Deep learning. Nature **521**(7553), 436 (2015)
7. Lee, H., et al.: Fully automated deep learning system for bone age assessment. J. Digit. Imaging **30**(4), 427–441 (2017)
8. Liskowski, P., Krawiec, K.: Segmenting retinal blood vessels with deep neural networks. IEEE Trans. Med. Imaging **35**(11), 2369–2380 (2016)
9. Bonawitz, C.A.: Practical pediatric imaging: diagnostic radiology of infants and children. Acad. Radiol. **6**(10), 607 (1999)
10. Poznanski, A.: Assessment of skeletal maturity and prediction of adult height (TW2 method). Am. J. Dis. Child. **131**(9), 1041–1042 (1977)
11. Seok, J., Hyun, B., Kasa-Vubu, J., Girard, A.: Automated classification system for bone age x-ray images. In: 2012 IEEE International Conference on Systems, Man, and Cybernetics (SMC), pp. 208–213. IEEE (2012)
12. Simonyan, K., Zisserman, A.: Very deep convolutional networks for large-scale image recognition. In: ICML (2015)

13. Somkantha, K., Theera-Umpon, N., Auephanwiriyakul, S.: Bone age assessment in young children using automatic carpal bone feature extraction and support vector regression. J. Digit. Imaging **24**(6), 1044–1058 (2011)
14. Spampinato, C., Palazzo, S., Giordano, D.: Deep learning for automated skeletal bone age assessment in x-ray images. Med. Image Anal. **36**, 41–51 (2016)
15. Thodberg, H.H., Kreiborg, S., Juul, A., Pedersen, K.: The bonexpert method for automated determination of skeletal maturity. IEEE Trans. Med. Imaging **28**(1), 52–66 (2009)
16. Zhang, A., Gertych, A., Liu, B.J.: Automatic bone age assessment for young children from newborn to 7-year-old using carpal bones. Comput. Med. Imaging Graph. **31**(4–5), 299–310 (2007)

Panoramic Image Saliency Detection by Fusing Visual Frequency Feature and Viewing Behavior Pattern

Ying Ding[1,2], Yanwei Liu[1(✉)], Jinxia Liu[3], Kedong Liu[1], Liming Wang[1], and Zhen Xu[1]

[1] SKLOIS, Institute of Information Engineering, Chinese Academy of Sciences,
Beijing, China
liuyanwei@iie.ac.cn
[2] School of Cyber Security, University of Chinese Academy of Sciences,
Beijing, China
[3] Zhejiang Wanli University, Ningbo, China

Abstract. The panoramic images are widely used in many applications. Saliency detection is an important task for panoramic image processing. Traditional saliency detection algorithms that are originally designed for conventional flat-2D images are not efficient for panoramic images due to their particular viewing way. Based on this consideration, we propose a novel saliency detection algorithm for panoramic images by fusing visual frequency feature and viewing behavior pattern. By extracting the spatial frequency information in viewport domain and computing the center-surround contrast of them for the whole panoramic image, the visual frequency feature for saliency detection is accurately obtained. Further more, the context of user's viewing behavior is integrated with visual frequency feature to generate the final saliency map. The experimental results show that the proposed algorithm is superior to the state-of-the-art algorithms when Pearson Correlation Coefficient (CC) is used as the evaluation metric.

Keywords: Panoramic image · Saliency detection · Viewport ·
Virtual reality

1 Introduction

In recent years, the panoramic image has attracted people's attention greatly since it provides a wide field of view that can capture the surrounding scene as much as possible. Via a head-mounted display (HMD), the panoramic image presents the viewer an immersive visual experience. Based on such feature, the

This work was supported in part by National Natural Science Foundation of China under Grant 61771469 and Zhejiang Provincial Natural Science Foundation of China under Grant LY17F010001.

© Springer Nature Switzerland AG 2018
R. Hong et al. (Eds.): PCM 2018, LNCS 11165, pp. 418–429, 2018.
https://doi.org/10.1007/978-3-030-00767-6_39

panoramic images have been widely used in many virtual reality (VR) applications.

Panoramic images are much larger than the traditional images in spatial size. People usually watch only a small part of the whole image. Thus, distinguishing the salient region from the whole image is very important for the panoramic image compression, transmission and VR image/video editing [1]. In the past, there were many classic saliency detection algorithms for traditional flat-2D images. Although these saliency detection algorithms can be applied to panoramic images directly, they usually cannot obtain the perfect results due to several issues that are brought in by the particular features of panoramic image.

First, the saliency detection that is directly applied on the panoramic image cannot effectively capture the users' attentions. The reason is that the traditional flat-2D image and the panoramic image are viewed in different way. For the flat-2D image, the user watches the image directly on the screen. The content that is perceived by the human eye is totally same with the image. However, the panoramic image is usually displayed via a set of viewports in HMD. The viewport is generated from a part of data in the panoramic image using several geometry projections. The viewport content that is perceived by the human eye is a bit different from that in corresponding region of the panoramic image.

Second, unlike that in traditional flat-2D image, the saliency in panoramic image is also affected by users' viewing behavior besides the visual attention features. This is because the viewport position is completely determined by the viewer's eye and head orientation. According to the human observing habitation, it is uncomfortable for people to look up or down for a long time. People prefer to scan the scene in the circular range that is parallel to human eye.

To deal with the above issues, in this paper we propose to detect saliency for panoramic images by fusing the visual frequency feature and viewing behavior pattern. To cope with the first issue, we extract the visual spatial frequency feature in the viewport domain. The receptive field of human ganglion cells is a concentric structure composed of a central excitatory region and a peripheral inhibition region. The Difference of Gaussians (DoG) filtering for an image can model this visual characteristic well [14]. Thus, DoG can be firstly used in the viewport domain to extract the visual spatial frequency difference for each pixel and then the center-surround contrast of them is computed in the panoramic image domain to generate the visual frequency feature saliency map. For the second issue, the general viewing behavior pattern for different panoramic images is summarized based on the practical head movement data. By establishing a mathematical model, the functional relationship between the pixel position (latitude) and the users' viewing probability is exploited to generate the viewing behavior saliency map.

The rest of the paper is organized as follows. In Sect. 2, we overview the related work. Section 3 presents the proposed panoramic image saliency detection framework. Next, Sect. 4 provides the experimental results. Finally, we conclude the paper in Sect. 5.

2   Related Work

For traditional flat-2D images, most existing saliency detection algorithms generally fall into two categories: bottom-up and top-down algorithms. The bottom-up saliency detection is a kind of data-driven algorithm. The typical one of the bottom-up algorithms is the central-neighborhood feature integration algorithm proposed by Itti et al. [2]. Another one is the histogram-based contrast approach that utilizes the sparse histogram representation of the color and global contrast [3]. Unlike the bottom-up algorithm, top-down algorithms are usually task-driven that entails supervised learning with class labels. Liu et al. [4] calculated the saliency of each pixel by learning the saliency of each underlying feature. BY using human eye tracking data set, Kienzle et al. [5] obtained the saliency map from the low-level features of the image through the non-linear mapping. Although the above algorithms performed well for traditional flat-2D images, they usually cannot generate the panoramic saliency map accurately due to the difference between contents in the panoramic image and the user's viewport.

With increasing applications of panoramic images, the saliency detection algorithms for panoramic images gradually appear. Currently, there are two categories of panoramic image saliency detection algorithms: the updated algorithms from those originally used for traditional images and the data analysis/learning-based algorithms.

For the first category, the traditional saliency detection algorithms are directly used for different projection domains of panoramic images. In [6], Bogdanova et al. proposed a sphere-domain saliency detection algorithm. Maugey et al. [7] proposed to aggregate and analyze 2D saliency detectors in the cubemap projection domain of panoramic image. These algorithms detected saliency using the traditional algorithms with considering new features of panoramic image. However, the user's behavioral characteristics were completely ignored. In addition, the way that the user views the panoramic image via extracting the viewport was ignored. Therefore, the obtained saliency map cannot accurately indicate the user's attention.

In recent years, machine learning has been used for saliency detection. Monroy et al. [8] presented an end-to-end CNN (Convolution Neural Network) that is specifically tailored for getting saliency maps for panoramic images. In [9], Abreu et al. proposed a method to transform the gathered viewport center trajectories data into saliency map. In conjunction with the analysis of the angular velocity of the head, Upenik et al. [10] described a simple model to obtain fixation locations and continuous saliency map from head direction trajectories for VR content. Similarly, Sitzmann et al. [11] analyzed gaze and head orientation data of users to extract saliency map based on fixation bias feature. These algorithms used the machine learning or data analysis to capture the general saliency cues. However, they neglected the viewport-domain content feature, and there is still room to improve the saliency detection efficiency for them.

3   The Proposed Saliency Detection Framework

Given consideration to the particular characteristics of panoramic image, we design a saliency detection algorithm by fusing the visual frequency feature in viewport domain and the human viewing behavior context. The whole panorama saliency detection framework is shown in Fig. 1. First, the DoG in viewport domain is performed to obtain the visual frequency feature saliency map. Second, human viewing behavior saliency map is estimated in terms of a viewing behavior model that is trained from the head movement data. Finally, the saliency map is obtained by merging the visual frequency feature saliency map and the viewing behavior saliency map.

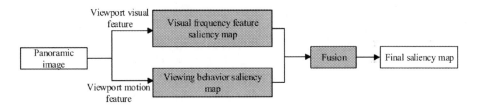

Fig. 1. The proposed saliency detection framework.

3.1   Visual Frequency Feature Saliency Map

To extract the visual frequency feature accurately, we need to render the viewport and then extract the spatial frequency feature in the viewport domain. Figure 2 shows the viewport rending process. The left rectangle represents the panoramic image. The sphere represents a spherical representation of the panoramic image. Viewport is the ABCD tangent plane to the sphere at point Q, which is the center of the viewport. As shown in Fig. 2, the viewport rendering is a pixel coordinate projection process from the panoramic image to the sphere and finally to the viewport plane. For example, E is a point on the panoramic image, its projection on the sphere is e, and E' is its projection on the viewport.

We assume that $S(x, y, z)$ represents Cartesian coordinate of a point on the sphere, $E(u, v)$ denotes the 2D homogeneous coordinate of its projection on the panoramic image, and $V(m, n)$ is the 2D homogeneous coordinate of its projection on the viewport. It is simple to convert the panoramic image to sphere domain. The longitude and latitude coordinates of the sphere are directly linked to a grid of horizontal and vertical coordinates. The projection is written as

$$\begin{pmatrix} x \\ y \\ z \end{pmatrix} = \begin{pmatrix} \cos\left(2\pi\left(u - 0.5\right)\right) & 0 \\ 0 & \sin\left(\pi\left(0.5 - v\right)\right) \\ -\sin\left(2\pi\left(u - 0.5\right)\right) & 0 \end{pmatrix} \begin{pmatrix} \cos\left(\pi\left(0.5 - v\right)\right) \\ 1 \end{pmatrix} \quad (1)$$

The pixel projection from sphere to the viewport is similar to pinhole camera model. By using perspective transformation, pixels on the sphere are projected

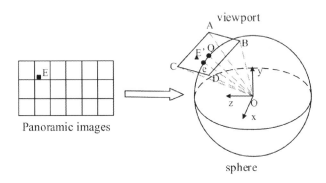

Fig. 2. Viewport rendering.

onto the viewport. The point on the sphere intersects the viewport plane through the light passing through the center O of the sphere. The center of the viewport is tangent to the sphere. For the convenience of computing, the sphere is always rotated until the viewport is perpendicular to the z-axis of the sphere. Denote (θ, φ) the longitude and latitude of the central point of viewport, and the rotation matrix R is defined as

$$
R = \begin{bmatrix}
\cos(\theta + \pi/2) & -\sin(\theta + \pi/2)\sin\varphi & \sin(\theta + \pi/2)\cos\varphi \\
0 & \cos\varphi & \sin\varphi \\
-\sin(\theta + \pi/2) & -\cos(\theta + \pi/2)\sin\varphi & \cos(\theta + \pi/2)\cos\varphi
\end{bmatrix} \tag{2}
$$

Then one pixel at $V(m, n)$ in the viewport can be obtained as

$$
\begin{pmatrix} m \\ n \\ 1 \end{pmatrix} = \frac{1}{z} \begin{pmatrix} f_x & 0 & m_0 \\ 0 & f_y & n_0 \\ 0 & 0 & 1 \end{pmatrix} \begin{pmatrix} f & 0 & 0 & 0 \\ 0 & f & 0 & 0 \\ 0 & 0 & 1 & 0 \end{pmatrix} \begin{pmatrix} R & t \\ 0^T & 1 \end{pmatrix} \begin{pmatrix} x \\ y \\ z \\ 1 \end{pmatrix} \tag{3}
$$

where (m_0, n_0) is the coordinate of the point Q in the viewport domain. f is the radius of the sphere. f_x and f_y are the focal length expressed in pixels. They are related to the size of the viewport and the field of view of each eye in the HMD. If w and h are the width and height of the viewport image respectively, and fov_x and fov_y are the horizontal and vertical fields of view per eye in the HMD, we have $f_x = \frac{w}{2\tan\left(\frac{fov_x}{2}\right)}$ and $f_y = \frac{h}{2\tan\left(\frac{fov_y}{2}\right)}$.

As mentioned above, the viewport is viewed directly by the user. Accordingly, the saliency detection should consider the visual response feature of the human eye on the viewport. It has been found biologically that the receptive field of human ganglion cells is a concentric structure composed of a central excitatory region and a peripheral inhibition region [14]. The excitement of the central area is strong. The inhibitory peripheral mechanism has a weaker effect but has a larger area. The central area and the surrounding area are functionally antagonistic to each other. Similarly, the DoG model consists of two Gaussian filters:

one has a higher intensity and a narrower range, and the other has a weaker intensity and a wider range. The former well describes the central mechanism of strong excitement in the receptive field. The latter well describes the weaker but larger area of inhibitory peripheral mechanisms in the receptive field. The response intensity at each point of the entire receptive field is perfectly described by the difference between these two Gaussian filters. Therefore, the DoG model can successfully simulate the receptive field of concentric antagonism of retinal ganglion cells.

To capture the visual frequency feature accurately, the DoG is utilized to process the viewport image as

$$D_V(m,n) = \frac{1}{2\pi\sigma_1^2} e^{-\frac{m^2+n^2}{2\sigma_1^2}} - \frac{1}{2\pi\sigma_2^2} e^{-\frac{m^2+n^2}{2\sigma_2^2}} \tag{4}$$

where (m,n) is the coordinate of the point in the viewport domain. σ_1 and σ_2 are the standard deviations of the Gaussian filters ($\sigma_1 = 1.6\sigma_2$). In the whole panoramic image, each pixel position corresponds to one position on the spherical image and there is a viewport centered on it. Only the center of the viewport is both on the sphere and on the viewport. To avoid the projection distortion of each pixel in the panoramic image for rendering multiple viewports, only when the pixel is centered on the viewport, the DoG value for this pixel is extracted to characterize the visual saliency feature on the whole viewport. Since each pixel in the panoramic image can obtain one DoG value on the viewport, the DoG value can be inversely projected to the panoramic image to form the DoG value map $D_V(u,v)$, where (u,v) is the pixel position in the panoramic image.

$D_V(u,v)$ depicts the visual frequency characteristic in each viewport. In terms of the perception mechanism of the human eye, the larger the difference between the current viewport and the other adjacent ones, the easier it is to be concerned. Thus, we utilize Euclidean distance between the DoG value of each pixel and the average DoG value over all pixels in the image to characterize the saliency of panoramic image as

$$S_V = \left\| D_V(u,v) - \overline{D_V} \right\| \tag{5}$$

where S_V is the visual frequency feature saliency map. $D_V(u,v)$ is the DoG value at (u,v) position in the panoramic image. $\overline{D_V}$ is the average DoG value over all pixels in the panoramic image. It is computed as $\overline{D_V} = \frac{1}{W \cdot H} \sum_{u=1}^{W} \sum_{v=1}^{H} D_V(u,v)$, where W and H are the width and height of the panoramic image, respectively.

3.2   Viewing Behavior Saliency Map

In the 360-degree space, the viewing popularities of different fields of view are also different. The viewing probability of a viewport is directly linked to the user's viewing behavior. Usually, via a HMD the user's head is located at the center of sphere to view a panoramic image. When viewing the image, users can watch all directions: back and forth, left and right, up and down. Normally, it is very uncomfortable for the user to look up or down for a long time. Undoubtedly, the probability that the user will watch left and right will be much greater than that of watching top and bottom. Thus, the users' viewing behaviors also indicate the saliency distribution over the whole panoramic image. Based on the user's head position data [12, 13], we plot the viewing probability for each latitude over the panoramic image in Fig. 3. In Fig. 3, the blue dots indicate the distribution of users' viewing position data. Around zero radian, the viewing probability achieves the highest value. The area between $-\pi/3$ and $\pi/3$ accounts for most of the viewing probability. It can be seen from Fig. 3 that the relationship between viewing probability and latitude approximately conforms a mathematical function.

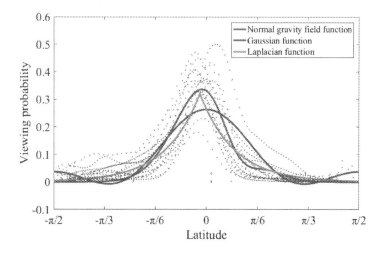

Fig. 3. The relationship between latitude and viewing probability.

The shape of viewing probability distribution over the panoramic image is very similar to those of Gaussian and Laplacian distributions. Thus, Gaussian and Laplacian functions can be used to fit the user's behavioral function. Besides the above two distribution functions, we also fit the viewing probability distribution over latitudes by a normal gravity field function. Physically, the normal gravity field is only related to the latitude, and it keeps unchanged along the longitude direction. It also has a maximum value at the equator and a minimum value at two poles. Hence, the normal gravity field function can be used to model the user's behavioral function. Based on the "Training-dataset" of "Salient360!" data set [12,13], the fitting functions are obtained as

$$P_g = -0.1671e^{\frac{(\varphi-17.59)^2}{15.15^2}} + 0.3774e^{\frac{(\varphi-2.361)^2}{27.15^2}} \tag{6}$$

$$P_l = \frac{12.91}{2 \times 20.13}e^{-\frac{|\varphi+3.769|}{20.13}} \tag{7}$$

$$P_n = 0.2428\left(1 - 0.8561\sin 2\varphi - 0.4927\sin 22\varphi\right) \tag{8}$$

where φ denotes the latitude. P_g, P_l, P_n are the Gaussian, Laplacian, and normal gravitational field functions, respectively.

Three functions's fitting curves are also shown in Fig. 3. In Table 1, their fitting performances are summarized. Obviously, P_g is the best among all the three functions based on the R-square and RMSE (root mean squared error) metrics. Therefore, the Gaussian function P_g is adopted to characterize the user's viewing behavior pattern in our algorithm and the viewing behavior saliency map S_B is obtained from it directly.

Table 1. Fitting performances of the three functions

Function	R-square	RMSE
P_g	0.91	0.030
P_l	0.82	0.042
P_n	0.68	0.062

3.3 Salience Map Fusion

Finally, we combine visual frequency feature saliency map and viewing behavior saliency map to get the final saliency map. We adopt two fusion functions to generate the final saliency map: Normalized and Sum (NS), Normalized and Maximum (NM). NS is the simplest method. It believes that the visual frequency feature saliency map and the viewing behavior saliency map are independent. NS fusion is

$$S_{NS} = \frac{1}{2}N\left(S_V\right) + \frac{1}{2}N\left(S_B\right) \tag{9}$$

where S_V is the visual frequency feature saliency map, S_B is the viewing behavior saliency map, $N\left(\cdot\right)$ is the normalization operation. For the NM algorithm, the final saliency value is determined by the maximum value between the visual frequency feature saliency and the viewing behavior saliency. It is written as

$$S_{NM} = max\left(N\left(S_V\right), N\left(S_B\right)\right) \tag{10}$$

4   Experimental Results

In this section, we evaluate the proposed saliency detection algorithm on "Salient360!" data set [12,13]. The database consists of 85 images, of which 60 images are used for training, and 25 images are used for testing. It includes various types of images, such as daytime and night, indoor and outdoor, simple background and complex background, etc. What's more, the data set provides the ground-truth data for all images.

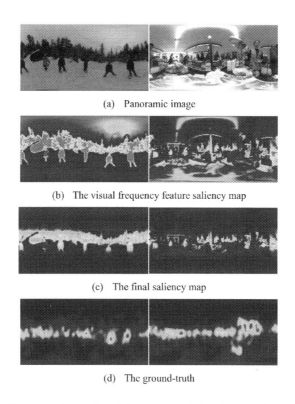

(a)   Panoramic image

(b)   The visual frequency feature saliency map

(c)   The final saliency map

(d)   The ground-truth

Fig. 4. The example results of the proposed algorithm with NM fusion.

Figure 4 shows the example results of the proposed algorithm with NM fusion. It can be seen from Fig. 4 that the saliency detection based on visual frequency features can detect a significant portion of the image saliency. After merging the viewing behavior saliency map, the obtained saliency map (Fig. 4 (c)) shows more similar shape with the ground truth saliency image. It illustrates that the viewing behavior context can improve the saliency detection accuracy significantly.

In the experiments, we compared the NS and NM fusion methods. In addition, the end-to-end CNN [8] and the ML-Net + EB (multi-level network + equator bias) [11] are used as the state-of-the-art benchmark algorithms to evaluate the performance of the proposed algorithm. Two metrics are used to evaluate the

Table 2. Experimental results.

Image index	CC			KL		
	NS	NM (ours)	CNN	NS	NM (ours)	CNN
1	0.739	0.787	0.507	0.461	0.421	0.589
8	0.747	0.776	0.655	0.475	0.472	0.484
16	0.669	0.703	0.610	0.344	0.333	0.249
19	0.525	0.547	0.512	0.635	0.619	0.384
20	0.574	0.532	0.497	1.072	1.219	0.618
26	0.747	0.735	0.596	0.369	0.397	0.276
48	0.706	0.662	0.266	0.440	0.516	0.622
50	0.531	0.530	0.59	1.116	1.188	0.479
60	0.523	0.553	0.671	0.500	0.489	0.273
65	0.781	0.808	0.531	0.518	0.502	0.475
69	0.664	0.696	0.686	1.181	1.313	0.433
71	0.595	0.592	0.388	1.198	1.209	0.404
72	0.604	0.512	0.271	0.489	0.606	0.459
73	0.782	0.811	0.536	0.492	0.514	0.364
74	0.663	0.678	0.656	0.907	0.927	0.385
78	0.620	0.651	0.581	0.768	0.771	0.514
79	0.798	0.828	0.818	0.628	0.563	0.360
85	0.705	0.748	0.714	0.669	0.645	0.351
91	0.604	0.695	0.628	0.777	0.769	0.498
93	0.671	0.713	0.488	0.519	0.475	0.496
94	0.576	0.595	0.468	1.046	1.182	0.583
95	0.639	0.680	0.624	0.801	0.848	0.455
96	0.629	0.699	0.722	1.628	1.442	0.582
97	0.581	0.580	0.369	1.328	1.492	0.471
98	0.327	0.345	0.309	0.718	0.765	0.652
Mean	0.640	0.658	0.548	0.763	0.787	0.458
Standard deviation	0.103	0.113	0.141	0.331	0.349	0.110

performance of the algorithms: KL (Kullback Leibler Dlivergence) and CC (Pearson Correlation Coefficient) [15]. The value of CC ranges from -1 to 1. The larger the absolute value of CC, the better the performance of the algorithm. The range of KL value is 0 to infinity. The smaller the KL value, the better the performance of the algorithm.

The experimental results of all test images for our algorithm compared to the CNN algorithm are listed in Table 2. The average CC and KL values for our algorithm with NM fusion are 0.65 and 0.78, respectively, compared to 0.63

of CC and 0.76 of KL for our algorithm with NS fusion. It indicates that the NM fusion can obtain slightly better performance than the NS fusion. Hence, we adopt the NM fusion in our algorithm.

In Table 2, the CC value of our algorithm is about 10% higher than the end-to-end CNN algorithm on average. Especially, for some images with center-surround contrast feature, our algorithm can obtain superior performance to the end-to-end CNN algorithm. For example, the image with index of 48, our algorithm can obtain the CC value of 0.66 compared to the CC value of 0.26 for end-to-end CNN algorithm. It is because that our algorithm extracts the visual frequency feature saliency in the viewport domain and it can capture the visual saliency more accurately. Comparably, the end-to-end CNN algorithm detects saliency directly on the panoramic image. Regarding the KL metric, the end-to-end CNN algorithm is superior to our algorithm. This is because the end-to-end CNN algorithm divides the panoramic image into slices. This particular operation refines the saliency detection that can capture the inward context information fully in the panoramic image. Correspondingly, it obtains more similar saliency distribution shape with that of ground-truth.

In the recent literature, the deep learning with head movement statistics (ML-Net + EB) is used to predict the saliency map of panoramic image. In ML-Net + EB algorithm, only the average mean and standard deviation of CC score is provided as 0.49 ± 0.11. The average mean and standard deviation of CC scores of our algorithm is 0.658 ± 0.11. Obviously, the performance of our algorithm is much better than that of the ML-Net + EB algorithm.

5   Conclusion

In this paper, we proposed a panoramic image saliency detection algorithm based on visual frequency feature and the context information of human viewing behavior. By extracting visual frequency feature in viewport domain, more accurate saliency cues are captured. And also, with considering the user's visual habit for browsing the panoramic image, a user's behavior function is built to model the probability of a user fixating on the viewport at specific latitude. Finally, these two kinds of saliency cues are combined to predict a final saliency map. Experimental results show that the proposed algorithm is superior to the state-of-the-art algorithms in CC metric. In the future work, we will optimize the proposed algorithm and extend it to the panoramic video saliency detection.

References

1. Serrano, A., Sitzmann, V., Ruiz-Borau, J., Wetzstein, G., Gutierrez, D., Masia, B.: Movie editing and cognitive event segmentation in virtual reality video. ACM Trans. Graph. **36**(4), 47 (2017)
2. Itti, L., Koch, C., Niebur, E.: A model of saliency-based visual attention for rapid scene analysis. IEEE Trans. Pattern Anal. Mach. Intell. **V20**(11), 1254–1259 (1998)

3. Cheng, M.M., Zhang, G.X., Mitra, N.J., Torr, P.H.S., Hu, S.M.: Global contrast based salient region detection. In: IEEE Conference on Computer Vision and Pattern Recognition, pp. 409–416 (2011)
4. Liu, T., Sun, J., Zhang, N., Tang, X., Shum, H.: Learning to detect a salient object. In: IEEE Conference on Computer Vision and Pattern Recognition, pp. 530–549 (2007)
5. Kienzle, W., Wichmann, F., Scholkopf, B., Franz, M.O.: A nonparametric approach to bottom-up visual saliency. In: NIPS, pp. 686–689 (2007)
6. Bogdanova, I., Bur, A., Hugli, H.: Visual attention on the sphere. IEEE Trans. Image Process. **17**(11), 2000–2014 (2008)
7. Maugey, T., Meur, O.L., Liu, Z.: Saliency-based navigation in omnidirectional image. In: International Workshop on Multimedia Signal Processing, pp. 1–6. IEEE (2017)
8. Monroy, R., Lutz, S., Chalasani, T., Smolic, A.: SalNet360: saliency Maps for omnidirectional images with CNN (2017)
9. Abreu, A.D., Ozcinar, C., Smolic, A.: Look around you: saliency maps for omnidirectional images in VR applications. In: Ninth International Conference on Quality of Multimedia Experience. IEEE (2017)
10. Upenik, E., Ebrahimi, T.: A simple method to obtain visual attention data in head mounted virtual reality. In: IEEE International Conference on Multimedia & Expo Workshops, pp. 73–78. IEEE (2017)
11. Sitzmann, V., et al.: Saliency in VR: how do people explore virtual environments? IEEE Trans. Vis. Comput. Graph., 1633–1642 (2018)
12. Rai, Y., Callet, P.L.: A dataset of head and eye movements for 360 degree images. In: ACM on Multimedia Systems Conference, pp. 205–210. ACM (2017)
13. Rai, Y., Callet, P.L., Guillotel, P.: Which saliency weighting for omni directional image quality assessment? In: Ninth International Conference on Quality of Multimedia Experience, pp. 1–6. IEEE (2017)
14. Rodieck, R.W.: Quantitative analysis of cat retinal ganglion cell response to visual stimuli. Vis. Res. **5**(12), 583–601 (1965)
15. Bylinskii, Z., Judd, T., Oliva, A., Torralba, A., Durand, F.: What do different evaluation metrics tell us about saliency models? IEEE Trans. Pattern Anal. Mach. Intell., **PP**(99), 1 (2016)

Coupled Learning for Image Generation and Latent Representation Inference Using MMD

Sheng Qian[1], Wen-ming Cao[1], Rui Li[1], Si Wu[2], and Hau-san Wong[1(✉)]

[1] Department of Computer Science, City University of Hong Kong,
Hong Kong, China
cshswong@cityu.edu.hk
[2] Department of Computer Science, South China University of Technology,
Guangzhou, China

Abstract. For modeling the data distribution or the latent representation distribution in the image domain, deep learning methods such as the variational autoencoder (VAE) and the generative adversarial network (GAN) have been proposed. However, despite its capability of modeling these two distributions, VAE tends to learn less meaningful latent representations; GAN can only model the data distribution using the challenging and unstable adversarial training. To address these issues, we propose an unsupervised learning framework to perform coupled learning of these two distributions based on kernel maximum mean discrepancy (MMD). Specifically, the proposed framework consists of (1) an inference network and a generation network for mapping between the data space and the latent space, and (2) a latent tester and a data tester for performing two-sample tests in these two spaces, respectively. On one hand, we perform a two-sample test between stochastic representations from the prior distribution and inferred representations from the inference network. On the other hand, we perform a two-sample test between the real data and generated data. In addition, we impose structural regularization that the two networks are inverses of each other, so that the learning of these two distributions can be coupled. Experimental results on benchmark image datasets demonstrate that the proposed framework is competitive on image generation and latent representation inference of images compared with representative approaches.

Keywords: Image generation · Latent representation inference
Coupled learning · Maximum mean discrepancy

1 Introduction

These years have witnessed the significant progress of deep generative models. Among these models, VAE [14] and GAN [8] are the two most popular frameworks. Both of them exploit deep neural networks to parameterize their models and can use stochastic backpropagation rules for model optimization. They

© Springer Nature Switzerland AG 2018
R. Hong et al. (Eds.): PCM 2018, LNCS 11165, pp. 430–440, 2018.
https://doi.org/10.1007/978-3-030-00767-6_40

have shown powerful capability in various computer vision applications, such as image generation [14,25], image translation [11,12], video prediction [22], shape prediction [29] and so on. Recently, researchers have paid considerable attention to simultaneously perform image generation and latent representation inference based on these two frameworks. Despite their successes, some shortcomings inherently exist in them as follows.

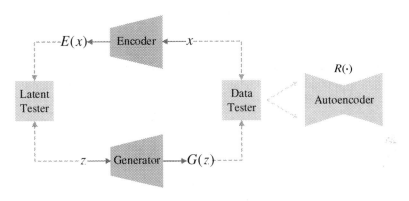

Fig. 1. The architecture of the proposed CoupledMMD.

Specifically, VAE possesses both capabilities of modeling the data distribution and latent representation distribution, however it tends to make inferred latent representations less meaningful for downstream tasks due to the evidence lower bound training criterion [3,32]. Because of the adversarial learning, the training stability and the generation performance of GAN are easily affected by different training criteria and model configurations [2]. Besides, GAN does not have the ability to infer latent representations.

For improving image generation and latent representation inference, various methods have been proposed [5,6,16]. Briefly speaking, [6] further extends the GAN framework by learning a generator consisting of an encoder and a decoder, and trains a discriminator to distinguish joint latent/data samples from the generator. Differently, [16] combines the VAE and GAN frameworks by collapsing VAE and GAN into one model. Since these two representative methods are based on the VAE and GAN frameworks, they still suffer from the above-mentioned inherent shortcomings to some extent.

In this work, we propose an unsupervised learning framework based on MMD to model both the distributions of the data and latent representation. In other words, the proposed framework can simultaneously perform image generation and latent representation inference. For simplicity, we refer to the proposed framework as **CoupledMMD**, as shown in Fig. 1. Specifically, CoupledMMD simultaneously learns an inference network (**Encoder**) and a generation network (**Generator**) through performing two two-sample tests within the MMD framework, and imposing structural regularization on these networks. Not only do we perform a two-sample test between stochastic latent representations from the prior distribution and inferred latent representations from the encoder; but we also perform a two-sample test between the real data and generated data. Moreover, in order to couple the learning of two distributions, we explicitly impose structural regularization to ensure that these networks are inverses of each other. The main contributions of this work are as follows:

- We propose a MMD-based framework for modeling both the distributions of the data and latent representation, which is formulated as a minimization problem. Compared with GAN, it avoids the difficult minimax problem and does not need elaborate model configurations. In addition, the training process is stable and the experimental performance is competitive.
- To the best of our knowledge, currently no existing research works have applied two-sample tests of MMD in both the data space and latent space simultaneously. We propose this approach and combine it with structural regularization, such that these two distributions can be jointly learnt, and the performance of both generation and inference can be effectively improved.

In the experiments on image datasets, we mainly focus on image generation, latent representation inference, sample interpolation and training stability. We have also performed comparisons to representative approaches, such as VAE, GAN-based models and MMD-based models. The experimental results demonstrate that the proposed CoupledMMD can generate images with good visual information fidelity, and simultaneously is capable of inferring meaningful and semantical latent representations. Moreover, it also shows good training stability.

2 Related Work

In this section we briefly introduce some related works and mainly focus on the following two types of frameworks.

GAN: As straightforward variants of GAN, [25] introduces well-applied architecture guidelines of CNN, and [23] restricts the GAN freedom through imposing auxiliary information. Besides using a single GAN, several works have tried to combine multiple GANs. [4,30] stack multiple GANs to synthesize images with the multi-scale strategy. To model the relationship of two domain distributions, [19] imposes weight-sharing over the high-level layers of GANs, while [12] makes GANs intersect each other. Besides researching on the architectural design, the training criteria of GANs also have a great influence on the stability of learning process. Similar to [21], [24] generalizes GAN in the variational divergence estimation framework. Differently, [1] applies the Earth-Mover distance as a better alternative of divergences, [10] trains a smooth discriminator to make the generator model the real data based on mode matching, and [31] reformulates adversarial learning as a process of shaping the energy surface.

MMD: Different from the GAN framework, recent approaches based on the MMD framework for modeling the data distribution also have attracted extensive attention. Among these approaches, [7] perform the MMD test in the data space in a straightforward way, but its model expressiveness is limited. Similarly, [18] leverages the auto-encoder to perform a MMD test between representations of the auto-encoder and those generated representations, and [32] identifies the basic defect of VAE and proposes a alternative MMD regularization. To maximize the testing power, [26] considers the variance of empirical MMD distances, and proposes generalized kernels in the GAN setting. Differently, [17] directly introduces adversarial learning into MMD, and proposes to learn adversarial kernels for the MMD test.

3   Coupled Learning

First we make a brief introduction about MMD, and then explain the framework of CoupledMMD in detail. Specifically, the architecture of CoupledMMD includes two types of components, as shown in Fig. 1. The first type of components consists of **Encoder** and **Generator** for mapping between the data space and the latent space. The second type of components consists of **Latent Tester** and **Data Tester** for performing two-sample tests in these two spaces, respectively. For coupled learning of the data distribution and latent representation distribution, we impose structural regularization that the encoder and the generator are inverses of each other.

3.1   Preliminary for MMD

Maximum-Mean Discrepancy (MMD) [9] is a framework to quantify the distance between two distributions $\mathbb{P}$ and $\mathbb{Q}$. It can be efficiently implemented using the kernel trick. However, only finite samples can be used to estimate the MMD distance in practice. Formally, given $X = \{x_1, x_2, ..., x_n\} \sim \mathbb{P}(x)$ and $Y = \{y_1, y_2, ..., y_m\} \sim \mathbb{Q}(y)$, the square of MMD distance based on the kernel $k(\cdot, \cdot)$ can be approximated by $L_{mmd}(X, Y)$:

$$L_{mmd}(X, Y) = \mathbb{E}[k(x_i, x_{i'})] + \mathbb{E}[k(y_j, y_{j'})] - 2\mathbb{E}[k(x_i, y_j)]. \tag{1}$$

3.2   MMD Testing in Data Space and Latent Space

Data Tester: Considering that the original images can be high-dimensional with complex patterns, if a two-sample test with fixed kernels is directly performed between real images and synthetic images, there is high possibility that this kind of tester does not have enough testing power to well distinguish between these two types of images. Hence, we adopt an indirect approach to design a powerful data tester. We expect to obtain low-dimensional and semantic representations of images with which we will actually perform a two-sample test. Through extracting the representations, on the one hand, the two-sample test can concentrate on the critical semantic patterns and ignore other trivial patterns; on the other hand, the two-sample test can be performed more efficiently due to dimension reduction.

To effectively obtain low-dimensional and semantic representations, we turn to an auto-encoder trained on the current dataset straightforwardly. Generally, an auto-encoder is used for data reconstruction. If trained properly with some regularization [27], an auto-encoder has the ability to learn a data manifold in an unsupervised learning manner, and its hidden layers can be used to extract primary statistical information of images. Hence, we can exploit the hidden layers of an auto-encoder to extract representations of images for performing the two-sample test, as illustrated in the right part of Fig. 1. Formally, given $X_{dt} = \{x_1, x_2, ..., x_n\} \sim p_{data}(x)$, $Z_{dt} = \{z_1, z_2, ..., z_m\} \sim p_z(z)$, and the hidden layers of an auto-encoder $R(\cdot)$, the square of MMD distance L_{dt} is given by

$$L_{dt} = \mathbb{E}[k(R(x_i), R(x_{i'}))] + \mathbb{E}[k(R(G(z_j)), R(G(z_{j'})))] - 2\mathbb{E}[k(R(x_i), R(G(z_j)))]. \tag{2}$$

Latent Tester: In order to perform image generation from the noise distribution, we should make latent representations of real images match this prior distribution of the latent space, we also perform a two-sample test in the latent space. Different from the data space, the latent space is presupposed to be low-dimensional with semantic information. Hence, a two-sample test with fixed kernels can be directly performed to well distinguish between stochastic latent samples from the latent space and inferred latent representations from the encoder. Formally, given $X_{lt} = \{x_1, x_2, ..., x_n\}$ and $Z_{lt} = \{z_1, z_2, ..., z_m\}$, the square of MMD distance L_{lt} is given by

$$L_{lt} = \mathbb{E}[k(E(x_i), E(x_{i'}))] + \mathbb{E}[k(z_j, z_{j'})] - 2\mathbb{E}[k(E(x_i), z_j)]. \tag{3}$$

If we only perform two two-sample tests in the data space and the latent space separately, we will then get a trivial solution to a large extent. Specifically, the encoder and the generator will be independently dominated by two two-sample tests. In other words, the encoder and generator do not bear any relationship. If latent representations inferred by the encoder are fed into the generator, the generator cannot generate realistic reconstructions. It means latent representations fail to capture primary statistical information of real images, which does not meet our expectations. In the next subsection, we will describe structural regularization to avert this situation.

3.3   Structural Regularization

For the purpose of avoiding trivial solutions, we explicitly impose structural regularization. It requires that the encoder and generator are inverses of each other, such that they are tightly coupled and the generated reconstructions could well approximate to real data with respect to the appearance and structure of images. Formally, given $X_{sr} = \{x_1, x_2, ..., x_n\}$, the structural regularization loss L_{sr} is given by

$$L_{sr} = \mathbb{E}[L_{rec}(D(E(x_i)), x_i)], \tag{4}$$

where L_{rec} measures the reconstruction error between a data sample x_i and its reconstruction sample $D(E(x_i))$.

In order to improve reconstruction, we propose to combine multiple choices of L_{rec}. Generally, a straightforward choice is the $\mathbf{L}^p$ difference loss function $\ell_{lp}(x, \hat{x}) = \|x - \hat{x}\|_p$. To further highlight significant edges and structural distortion for sharpening the image, we exploit the gradient difference loss (GDL) [22] and the structural similarity (SSIM) [28]. For simplicity, we denote them as $\ell_{gdl}(x, \hat{x})$ and $\ell_{ssim}(x, \hat{x})$ respectively, and more details can refer to the original papers. Overall, the reconstruction error of L_{rec} is given by

$$L_{rec} = \ell_{lp}(x, \hat{x}) + \alpha \ell_{gdl}(x, \hat{x}) + \beta(1 - \ell_{ssim}(x, \hat{x})), \tag{5}$$

where α and β are hyperparameters for adjusting the weights of these choices. α and β are set as 2.0 and 0.5 by experience, respectively.

3.4   Combining Loss

To learn the joint distribution, we combine the above-mentioned losses. Formally, the combined loss L is a weighted summation of L_{dt}, L_{lt} and L_{sr}:

$$L = \lambda_{dt} L_{dt} + \lambda_{lt} L_{lt} + L_{sr}, \tag{6}$$

where λ_{dt} and λ_{lt} are predefined parameters for adjusting the weights of losses. Both λ_{dt} and λ_{lt} are set as 8.0 by experience.

The whole training process of CoupledMMD includes two stages: (1) Train an auto-encoder for **Data Tester**; (2) Train the encoder and the generator. In stage (1), we also apply the above-mentioned L_{rec} to train the auto-encoder. For stage (2), we list key details in Algorithm 1.

Algorithm 1. Training the encoder and the generator

Use *Adam* optimizer with the learning rate η;

Initialize parameters $\theta_{eg} = \theta_{enc} \bigcup \theta_{gen}$, where θ_{enc} for Encoder, θ_{gen} for Generator.

repeat

    Step 1: MMD testing in the data space and latent space.

    (a) sample (X_{dt}, Z_{dt}) and (X_{lt}, Z_{lt}); and compute L_{dt} and L_{lt}

    (b) update $\theta_{gen} \leftarrow \theta_{gen} - \eta \lambda_{dt} \cdot Adam(\theta_{gen}, \nabla_{\theta_{gen}} L_{dt})$

    (c) update $\theta_{enc} \leftarrow \theta_{enc} - \eta \lambda_{lt} \cdot Adam(\theta_{enc}, \nabla_{\theta_{enc}} L_{lt})$

    Step 2: Structural regularization.

    (a) sample X_{sr}; and compute L_{sr}

    (b) update $\theta_{eg} \leftarrow \theta_{eg} - \eta \cdot Adam(\theta_{eg}, \nabla_{\theta_{eg}} L_{sr})$

until Convergence

---

4   Experiments

In the experiments, we evaluate the propose framework on the following benchmark datasets: CIFAR [15] and CelebA [20]. We focus on four aspects: image generation, latent representation inference, sample interpolation and training stability, and making comparisons to representative approaches. In addition, some key configurations of experimental implementation are listed as follows. **Network architecture:** We follow architecture design guidelines adopted by [25] and properly adjust the configuration of subnetwork layers. **Kernel design:** Since L_{dt} and L_{lt} are influenced by the choice of kernels, we follow the above-mentioned works and chose a mixture of K-RBF kernels [26]. **Hyperparameters:** We use Adam optimizer [13] with tuned hyperparameters recommended by [25]. For the dimensionality of the latent space, we set 64 for CIFAR and 32 for CelebA by experience according to the complexity of the dataset, respectively. The batch size is set as 256 for all datasets.

4.1   Image Generation

For image generation on CIFAR and CelebA, we mainly compare CoupledMMD with the following approaches: VAE, WGAN [1], ALI [6] and MMDGAN [17]. The generated data samples are illustrated in Fig. 2 (*for a better view by zooming in*). Compared with VAE, CoupledMMD can generate clearer and more realistic images on both datasets. Specifically, image regions seem more like objects in terms of boundary and shape on CIFAR, and face regions have more complete structures and more detailed textures on CelebA. Although images generated by WGAN, ALI and MMDGAN are sharper than those by CoupledMMD on both datasets, CoupledMMD can better capture the global structural feature of images. In particular, we can observe that faces generated by these GAN-based approaches suffer from more structural distortion on CelebA. Some faces are severely twisted and disfigured, and hair tends to be in a mess, especially in WGAN and MMDGAN. While faces from CoupledMMD show more reasonable global structures with tidily arranged hair. Although images generated by CoupledMMD on CIFAR are a little blurry, they have more obvious outline

(a) VAE   (b) WGAN   (c) ALI   (d) MMDGAN   (e) **Our**

(f) VAE   (g) WGAN   (h) ALI   (i) MMDGAN   (j) **Our**

Fig. 2. Data samples generated on CIFAR and CelebA as shown in the two rows.

of object-like regions with neat backgrounds. Overall, considering both image sharpness and structural distortion, CoupledMMD is competitive on data generation compared with these representative approaches. Nevertheless, this does not invalidate the promise of the GAN-based frameworks, and suggests our exploration is in line with our expectation.

4.2   Latent Representation Inference

To verify the effectiveness of latent representation inference, we inspect data reconstructions for qualitative evaluation and compare CoupledMMD with ALI. From Fig. 3, we observe that the reconstructions of ALI (*the left subfigure*) are realistic, but these reconstructions are different from the original faces in terms of the overall structure and textures. While the reconstructions of CoupledMMD (*the right subfigure*) are more similar to the original faces and basically keep the overall structure with losing some detailed textures. Although ALI tries to learn meaningful latent representations of data, the unsatisfactory reconstructions indicate that ALI may suffer from its proposed joint adversarial training scheme. Differently, we formulate the distribution learning as a minimization problem and propose structural regularization, which can provoke CoupledMMD to learn more meaningful latent representations and make it much easier to train.

Fig. 3. Reconstructions based on inferred latent representations. Each pair consists of a image and its reconstruction.

4.3   Sample Interpolation

We first encode a pair of image samples x_1 and x_2 to latent representations $E(x_1)$ and $E(x_2)$ via the encoder. We then generate synthetic images based on the interpolated representations between $E(x_1)$ and $E(x_2)$ via the generator. For comparison, we also directly perform the interpolation in the image space. As shown in Fig. 4, the interpolation from the image space (*the left two rows*) is almost the linear superposition. While the interpolation from the latent space (*the right two rows*) transits with gradual change of high-level characteristics, and can still show good visual information fidelity. Specifically, the faces are slowly changing in terms of face structure, hair style and face color. Moreover, the difference between the two types of interpolations is more obvious in the middle part. It indicates that CoupledMMD can achieve non-linearly mapping between the data space and latent space, and learn high-level semantic latent representations.

Fig. 4. Sample interpolation from the data space and the latent space respectively.

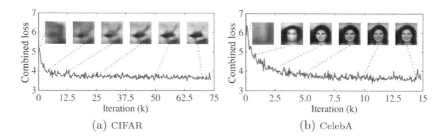

(a) CIFAR   (b) CelebA

Fig. 5. The training curves of the combined loss and generated samples.

4.4 Training Stability

Furthermore, we discuss the relationship between the combined loss and the generation quality, and analyze training stability qualitatively. We focus on the evolution of the combined loss during training on CIFAR and CelebA, as shown in Fig. 5. Besides, we exhibit samples generated from the fixed stochastic noise at six specific iterations. From the figure we can observe that CoupledMMD can converge quickly on both datasets, and the quality of generated samples can be gradually improved with further training. In Fig. 5(a), the basic structure of object regions does not change. The boundary becomes sharper and the textures become richer with clearer background. In Fig. 5(b), the global structure of faces keep stable and the appearance of faces becomes more obvious with increasing textures. In general, the combined loss and the image quality are negatively correlated and generated samples show a stable evolvement, which indicates that CoupledMMD shows good training stability. We attribute the stability to formulating the distribution learning as a minimization problem.

5 Conclusion

We propose an unsupervised learning framework (CoupledMMD) to perform coupled learning of the data distribution and the latent representation distribution based on kernel MMD. CoupledMMD consists of two networks for mapping between the data space and latent space, and two testers for performing two-sample tests in these two spaces. We then combine them with structural regularization, such that these two distributions can be jointly learnt. Compared with representative approaches on benchmark image datasets, CoupledMMD can achieve competitive performance in image generation. Through our experiments, we verify that CoupledMMD can infer high-level semantic latent representations of images effectively. Moreover, we also demonstrate that CoupledMMD shows good training stability on all the datasets used in our experiments.

References

1. Arjovsky, M., Chintala, S., Bottou, L.: Wasserstein GAN. CoRR arXiv:1701.07875 (2017)
2. Bojanowski, P., Joulin, A., Lopez-Paz, D., Szlam, A.: Optimizing the latent space of generative networks. CoRR arXiv:1707.05776
3. Chen, X., et al.: Variational lossy autoencoder. CoRR arXiv:1611.02731 (2016)
4. Denton, E.L., Chintala, S., Szlam, A., Fergus, R.: Deep generative image models using a laplacian pyramid of adversarial networks. In: NIPS, pp. 1486–1494 (2015)
5. Dosovitskiy, A., Brox, T.: Generating images with perceptual similarity metrics based on deep networks. In: NIPS, pp. 658–666 (2016)
6. Dumoulin, V., et al.: Adversarially learned inference. CoRR arXiv:1606.00704 (2016)
7. Dziugaite, G.K., Roy, D.M., Ghahramani, Z.: Training generative neural networks via maximum mean discrepancy optimization. In: UAI, pp. 258–267 (2015)
8. Goodfellow, I.J., et al.: Generative adversarial nets. In: NIPS, pp. 2672–2680 (2014)
9. Gretton, A., Borgwardt, K.M., Rasch, M.J., Schölkopf, B., Smola, A.J.: A kernel method for the two-sample-problem. In: NIPS, pp. 513–520 (2006)
10. Grewal, K., Hjelm, R.D., Bengio, Y.: Variance regularizing adversarial learning. CoRR arXiv:1707.00309
11. Isola, P., Zhu, J., Zhou, T., Efros, A.A.: Image-to-image translation with conditional adversarial networks. In: CVPR, pp. 5967–5976 (2017)
12. Kim, T., Cha, M., Kim, H., Lee, J.K., Kim, J.: Learning to discover cross-domain relations with generative adversarial networks. In: ICML, pp. 1857–1865 (2017)
13. Kingma, D.P., Ba, J.: Adam: a method for stochastic optimization. CoRR arXiv:1412.6980
14. Kingma, D.P., Welling, M.: Auto-encoding variational bayes. CoRR arXiv:1312.6114
15. Krizhevsky, A., Hinton, G.: Learning multiple layers of features from tiny images (2009)
16. Larsen, A.B.L., Sønderby, S.K., Larochelle, H., Winther, O.: Autoencoding beyond pixels using a learned similarity metric. In: ICML, pp. 1558–1566 (2016)
17. Li, C., Chang, W., Cheng, Y., Yang, Y., Póczos, B.: MMD GAN: towards deeper understanding of moment matching network. CoRR arXiv:1705.08584
18. Li, Y., Swersky, K., Zemel, R.S.: Generative moment matching networks. In: ICML, pp. 1718–1727 (2015)
19. Liu, M., Tuzel, O.: Coupled generative adversarial networks. In: NIPS, pp. 469–477 (2016)
20. Liu, Z., Luo, P., Wang, X., Tang, X.: Deep learning face attributes in the wild. In: ICCV, pp. 3730–3738 (2015)
21. Mao, X., Li, Q., Xie, H., Lau, R.Y.K., Wang, Z.: Multi-class generative adversarial networks with the L2 loss function. CoRR arXiv:1611.04076 (2016)
22. Mathieu, M., Couprie, C., LeCun, Y.: Deep multi-scale video prediction beyond mean square error. CoRR arXiv:1511.05440 (2015)

23. Mirza, M., Osindero, S.: Conditional generative adversarial nets. CoRR arXiv:1411.1784
24. Nowozin, S., Cseke, B., Tomioka, R.: f-GAN: training generative neural samplers using variational divergence minimization. In: NIPS, pp. 271–279 (2016)
25. Radford, A., Metz, L., Chintala, S.: Unsupervised representation learning with deep convolutional generative adversarial networks. CoRR arXiv:1511.06434
26. Sutherland, D.J., et al.: Generative models and model criticism via optimized maximum mean discrepancy. CoRR arXiv:1611.04488
27. Vincent, P., Larochelle, H., Bengio, Y., Manzagol, P.: Extracting and composing robust features with denoising autoencoders. In: ICML, pp. 1096–1103 (2008)
28. Wang, Z., Bovik, A.C., Sheikh, H.R., Simoncelli, E.P.: Image quality assessment: from error visibility to structural similarity. IEEE Trans. Image Process. **13**(4), 600–612 (2004)
29. Wu, J., Zhang, C., Xue, T., Freeman, B., Tenenbaum, J.: Learning a probabilistic latent space of object shapes via 3D generative-adversarial modeling. In: NIPS, pp. 82–90 (2016)
30. Zhang, H., et al.: StackGAN: text to photo-realistic image synthesis with stacked generative adversarial networks. CoRR **abs/1612.03242** (2016)
31. Zhao, J.J., Mathieu, M., LeCun, Y.: Energy-based generative adversarial network. CoRR arXiv:1609.03126
32. Zhao, S., Song, J., Ermon, S.: InfoVAE: information maximizing variational autoencoders. CoRR arXiv:1706.02262

Enhanced Discriminative Generative Adversarial Network for Face Super-Resolution

Xi Yang[1], Tao Lu[1(✉)], Jiaming Wang[1], Yanduo Zhang[1], Yuntao Wu[1], Zhongyuan Wang[2], and Zixiang Xiong[3,4]

[1] School of Computer Science and Engineering, Hubei Key Laboratory of Intelligent Robot, Wuhan Institute of Technology, Wuhan 430073, China
lutxyl@gmail.com
[2] School of Computer Science, Wuhan University, Wuhan 430072, China
[3] Department of ECSE, Monash University, Melbourne, Australia
[4] Department of ECE, Texas A&M University, College Station, TX 77024, USA

Abstract. Recently, some generative adversarial network (GAN)-based super-resolution (SR) methods have progressed to the point where they can produce photo-realistic natural images by using a generator (G) and discriminator (D) adversarial scheme. However, vanilla GAN-based SR methods cannot achieve good reconstruction and perceptual fidelity on real-world facial images at the same time. Because of D loss, them are hard to converge stably, which may cause the model collapse. In this paper, we present an Enhanced Discriminative Generative Adversarial Network (EDGAN) for SR facial recognition to achieve better reconstruction and perceptual fidelities. First, we discover that a versatile D boosts the adversarial framework to a preferable Nash equilibrium. Then, we design the D via dense connections, which brings more stable adversarial loss. Furthermore, a novel perceptual loss function, by reusing the intermediate features of D, is used to eliminate the gradient vanishing problem of Gs. To our knowledge, this is the first framework to focus on improving the performance of the D. Quantitatively, experimental results show the advantages of EDGAN on two widely used facial image databases against the state-of-the-art methods with different terms. EDGAN performs sharper and realistic results on real-world facial images with large pose and illumination variations than its competitors.

Keywords: Face super-resolution · Generative adversarial network
Densely connect · Feature reuse

T. Lu—This work is supported by the National Natural Science Foundation of China (61502354, 61501413, 61671332, 61771353, 41501505), the Natural Science Foundation of Hubei Province of China (2012FFA099, 2012FFA134, 2013CF125, 2014CFA130, 2015CFB451), Scientific Research Foundation of Wuhan Institute of Technology (K201713).

R. Hong et al. (Eds.): PCM 2018, LNCS 11165, pp. 441–452, 2018.
https://doi.org/10.1007/978-3-030-00767-6_41

1   Introduction

Face super-resolution refers to generating a high-resolution (HR) image from its low-resolution (LR) counterpart. This is a fundamental problem in computer vision and has a wide range of applications. In generally, facial image super-resolution is taken as a preprocessing step for facial recognition and video surveillance.

The aim of this paper is to improve on the quality and understanding of very-low resolution facial images. Existing Convolutional Neural Network (CNN)-based super-resolution methods usually focus on minimize the mean squared error (MSE) between the super-resolved (SR) image and the ground truth that also maximizes the Peak Signal-to-Noise Ratio (PSNR). The reconstruction results of these CNN-based methods always tend to be blurry and over-smoothing [1,2]. Recently, Ledig *et al.* [3] proposed an SRGAN using perceptual loss function and adversarial networks [4] to generate more realistic images. However, SRGAN doesn't perform well on real-world, low-resolution facial images [5]. As shown in Fig. 1, the facial features of the SRGAN reconstruction result are very confusing and have several artifacts.

Fig. 1. Visual results of Labeled Faces in the Wild (LFW) dataset [6] with a scaling factor of 4. (a) SRGAN trained with 20,000 iterations. (b) SRGAN trained with 200,000 iterations. (c) EDGAN. (d) ground truth. From (a) and (b), we can observe that the performance of SRGAN becomes worse when the iterations increase.

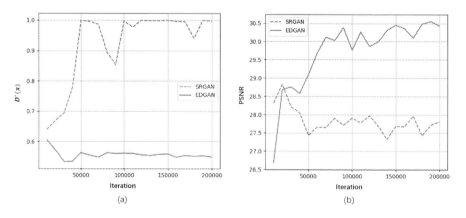

Fig. 2. Convergence Analysis of SRGAN and our method, all models are trained with an LFW dataset. According to Goodfellow *et al.* [4], if the generator G and discriminator D have enough capacity, then each can theoretically reach a point where both cannot improve because when HR = SR, D is unable to differentiate between the HR and SR, i.e., $D^*(x) = \frac{p_{HR}(x)}{p_{HR}(x)+p_{SR}(x)} = \frac{1}{2}$. (Color figure online)

The red dotted line in Fig. 2 visualizes the SRGAN convergence process after it is trained with about 50,000 iterations. The PSNR score dropped by about 1 dB during late-training and it never returned to its original level. Meanwhile, $D^*(x)$ quickly increased to around 1.0, which means that $p_{SR}(x)$ is much smaller than $p_{HR}(x)$, i.e., the reconstruction result of the generator is so poor that discriminator can identify it accurately. This is because SRGAN uses a loss function that is based on JS divergence, which converges too fast for real-world facial image datasets, this loss function also causes modal collapse and ultimately ensures that the adversarial process fails.

Designing a better loss function for GAN has been widely discussed since GAN was proposed. Variants of GAN, such as WGAN [7], LSGAN [8] and BEGAN [9] have proposed different loss functions. Recent work [10] uses WGAN for face super-resolution, but its image reconstruction step does not eliminate artifacts. In this paper, we discussed a new perspective of how to ensure generator achieves better performance in during the super-resolution facial reconstruction step by enhancing the capability of the discriminator. The blue solid line in Fig. 2 indicates that our method restores the normality of the adversarial process and the quality (of the generated facial images) increases with each iteration.

The contributions of EDGAN are highlighted as: (1) A dense discriminator block (DDB) that can read states from the preceding DDB via a contiguous memory (CM) mechanism. (2) An enhanced discriminator network (EDN), which is a deeper discriminator that improves the performance of adversarial network. (3) We design a novel perceptual loss function by reusing the intermediate features of D.

2   Related Work

Recently, deep convolution neural networks have been successfully applied to image super-resolution. Dong *et al.* proposed the SRCNN [11], this is the first paper that applies CNN to single-image super-resolution problems. Inspired by VGG-net [12], Kim *et al.* [13] uses a deeper convolutional network for accurate image super-resolution, which increases the network depth by stacking more convolutional layers and outperforms SRCNN with a large margin PSNR score. The face super-resolution problem is a special case of image super-resolution, which requires more informative structure priors and suffers from more challenging blurring. Early techniques [14] propose a deep cascaded bi-network that contains two functionality-specialized branches to recover different levels of texture details. Song *et al.* [15] present a two-stage method for facial hallucination. First, they generate facial components of the input image using CNNs, then they synthesize fine-grained facial structures from high resolution training images. After [16,17], researchers began to use residual learning and densely connected networks to design deeper network structures and develop state-of-the-art methods. [3,18–20]. These methods are all based on pixel loss, such as *l1* and *l2* loss.

Although the high-resolution images constructed by these models have high PSNR scores, they always tend to be blurry and lack high-frequency detail. More recently, Ledig *et al.* [3] presented a GAN-based approach that consists of a G based on deep residual networks and a D that differentiates between the super-resolved and the original HR images. The approaches also combine the perceptual loss function [21] (based on the VGG-net) and GAN model to generate more realistic image. However, the perceptual loss function based on VGG classification networks will introduce extra computations. To fix this issue, Wu *et al.* [22] designed a more robust perceptual loss function by reusing the features extracted by the discriminator, which can ensure higher perceptual fidelity. Bulat *et al.* [5] introduces a system that incorporates structural information in GAN-based super-resolution algorithms by integrating a sub-network for facial alignment (via heatmap regression) and optimizing a novel heatmap loss function.

In this paper, we describe an improved GAN-based architecture for face super-resolution, which requires a powerful discriminator, which can distinguish between a super-resolved image and a high-resolution original image, and contain a novel perceptual loss function.

3   Method

3.1   Network Architecture

In this section, we describe the proposed EDGAN, which is illustrated in Fig. 3. The generator is a super-resolution network used to reconstruct HR images from LR images. The improved discriminator needs to accomplish two tasks: the first is to distinguish between the super-resolved and the original high-resolution images; the second is to extract the features from the super-resolved images and the original high-resolution images.

Adversarial Loss. Following Goodfellow *et al.* [4], we optimize the alternating process between the G and D to solve the adversarial min-max problem,

$$\min_{G} \max_{D} E_{y \sim P_{data}(y)}[\log D(y)] + E_{z \sim P_{data}(z)}[\log(1 - D(G(z)))], \qquad (1)$$

In this formula, z is the LR image, which has a 50×50 (pixel) resolution, and y is the corresponding HR image, which has a 200×200 (pixel) resolution. $G(z)$ represents the image reconstructed by the G. In the training phase, D tries to minimize this objective function and G tries to maximize it. So, we optimize D with

$$l_D(G, D) = E_{y \sim P_{data}(y)}[\log D(y)] + E_{z \sim P_{data}(z)}[\log(1 - D(G(z)))], \qquad (2)$$

and we optimize G with

$$l_G(G, D) = -E_{z \sim P_{data}(z)}[\log(D(G(z)))], \qquad (3)$$

This approach encourages our generator to generate a solution that lies on the manifold of the HR images by trying to fool the discriminator.

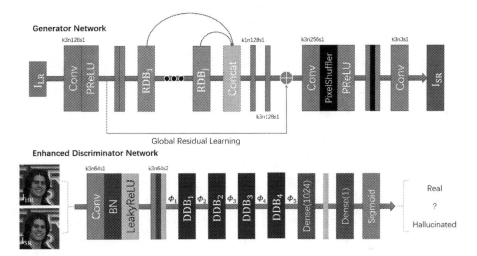

Fig. 3. The architecture of our proposed EDGAN, with corresponding kernel size (k), number of feature maps (n) and stride (s), is indicated for each convolutional layer. The term Φ_i represents the feature map computed by the convolution layer (after the activation function layer is introduced).

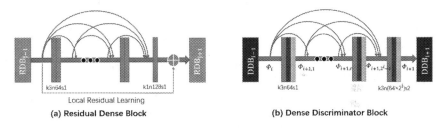

Fig. 4. The architecture of RDB (left) and DDB (right). In RDB, the number of convolution layers (followed by ParametricReLU) is 6, In DDB_i, the number of convolution layers (followed by batch-normalization layers and LeakyReLU) is $2^{i-1} + 1$.

3.2 Generator

The goal of the generator G is generate a high-resolution image that is as similar as possible to the high resolution image of the ground truth. We use two convolutional layers to extract shallow features and retain the output of the first convolutional layer for global residual learning. Then we continue learning the features of the image by using 16 residual dense blocks (RDB) [19]. RDB uses local feature fusion and local residual learning that make full use of the hierarchical features of the original LR images. The details of the RDB process can be found in Fig. 4a. After all RDBs, we use a global feature fusion layer to fuse the output of all RDBs. Inspired by [3], we increase the resolution of the input image via two trained sub-pixel convolution layers. The final convolution layer has 3 output channels, as we output the color HR images. We set the size of all convolutional layers to be 3×3, except for those in the local and global feature fusion (their kernel size is 1×1).

Pixel Loss Function. We utilize an $L1$ loss function to minimize the distance between the high-resolution and the super-resolved images, which has been demonstrated to be more powerful for performance and convergence [18]. The function is defined as follows:

$$l_{pixel} = E_{z,y \sim P_{data}(z,y)}(\|y - G(z)\|), \tag{4}$$

3.3   Enhanced Discriminator

To discriminate actual HR images from generated SR samples, we propose an enhanced discriminator network (EDN). Our EDN use local feature fusion to retain the hierarchical features from the image; its architecture is shown in Fig. 3. Our EDN contains a shallow feature extraction net, four dense discriminator blocks (DDB), two dense layers and a sigmoid activation function. We use a shallow feature extraction layer to get 64 feature maps when the image reaches the EDN, then we use the convolutions, with a stride of 2, to reduce the image resolution. We denote Φ_1 as the output of the Shallow feature extraction net.

The details about our proposed DDB in Fig. 4b. DDB$_i$ contains 2^{i-1} feature extraction layers and a local feature fusion layer. The local feature fusion layer and feature extraction layer are both composed of convolution layers, batch-normalizations and LeakyReLUs. The difference is that in the feature extraction layer, the stride of the convolution layer is 1, whereas the output of the convolution layer has 64 feature maps. In the local feature fusion layer, the stride of the convolution layer is 2 and the output of the convolution layer have 64×2^i feature maps. Let Φ_i and Φ_{i+1} be the input and output of the DDB$_i$, then the output of c-th convolution layer of DDB$_i$ can be formulated as:

$$\Phi_{i+1,c} = \sigma(W_{d,c}[\Phi_i, \Phi_{i+1,1}, \ldots, \Phi_{i+1,c-1}] + b_i), \tag{5}$$

where σ denotes the LeakyReLU activation function. $W_{d,c}$ and b_i is the weight and bias of the c-th convolution layer, respectively. The terms $[\Phi_i, \Phi_{i+1,1}, \ldots, \Phi_{i+1,c-1}]$ refer to the concatenation of the feature maps produced by the DDB$_i - 1$ convolutional layers $1, \ldots, (c-1)$ in the c-th DDB$_i$.

After all DDBs, we obtain 1024 feature maps. The resulting 1024 feature maps will be enter to two dense layers and a final sigmoid activation function to obtain a probability for sample classification.

Perceptual Loss Function. The pixel loss function can be helpful to recover the low frequency details and to get high PSNR score; however, we also need high-frequency details. Ledig *et al.* [3] uses a VGG loss function based on the ReLU activation layers of a pre-trained 19-layer VGG network to deal with this issue. Wu *et al.* [22] reuses the intermediate features in each convolution layer of the discriminator to build the perceptual loss function, which can reduce the computation budget. In our experiment, we use the outputs of shallow feature extraction networks and DDBs to build the perceptual loss function. We can define the perceptual loss function as:

$$l_{feature} = \sum_{i=1}^{n} E_{z,y \sim P_{data}(z,y)}(||\Phi_i(y) - \Phi_i(G(z))||), \tag{6}$$

Here n is 5 and Φ_i is the feature map extracted by D as shown in Fig. 3.

3.4   Final Loss Functions

Based on the individual loss functions shown above, we can define the final loss functions for discriminator D and generator G. The formulas are defined as follows:

$$l_g = l_{pixel} + \lambda_1 l_G(G, D) + \lambda_2 l_{feature}, \tag{7}$$
$$l_d = -l_D(G, D) + \lambda l_{feature}, \tag{8}$$

λ_1, λ_2 and λ are the corresponding weights, in our experiment, λ_1 and λ is 0.01, λ_2 is 0.02.

4   Experiments

4.1   Databases

Extensive experiments are evaluated on Labeled Faces in the Wild (LFW) [6] and CelebFaces Attributes (CelebA) [23] datasets.

LFW dataset contains 13,233 face images collected from the web, these images contain a variety pose variations and facial expressions. The LFW dataset has four different sets of images, including the original and three different types of aligned images. In our experiment, we use the original one. When generating the LR and HR pairs, we resize the original images (250×250 pixels) to be 200×200 pixels and label the resulting resized image an HR image. We obtained the LR images by down-sampling the HR images (RGB, C = 3) using a bicubic kernel with a down-sampling factor r = 4. We use 9526 images for training and the remaining 3707 images are used for testing.

CelebA is a large-scale facial attributes dataset with 10,177 unique identities and 202,599 face images in total. In our experiments, we only used a randomly selected subset of 1140 faces for testing.

The HR images are only available during training. We adopt the widely used Peak Signal-to-Noise Ratio (PSNR), structural-similarity (SSIM), feature-similarity (FSIM) [24] and Visual Information Fidelity (VIF) [25] on Y channel of transformed YCbCr space as our evaluation measures.

4.2   Training Details

Following the settings of [3], for each training batch, we randomly select 16 image patches as the high-resolution patches. Each patch has a 96×96 pixel size. To avoid overfitting, we perform random image flipping. For the optimization step, we use the ADAM algorithm [26] with $\beta_1 = 0.9$ for both G and D. Our EDGAN are trained with 10^5 update iterations at a learning rate of 10^{-4} and another 10^5 iterations at a lower rate of 10^{-5}.

Training via EDGAN, using 200,000 iterations, roughly takes 17 h with a NVIDIA GeForce GTX 1080Ti GPU. Unlike other GAN-based methods, such as SRGAN [3] and Super-FAN [5], our EDGAN doesn't require pre-training. We implemented the proposed networks with the tensorflow framework.

4.3   Qualitative Comparisons

We compare the proposed method with pre-existing state-of-the-art methods and show the super-resolution results in Fig. 5. The reconstruction results of CNN-based methods, such as CBN [14] and SRResnet [3], lack high-frequency detail; the edge detail of the resulting images produces a blurring effect. The SRGAN has high-frequency details, but it is limited by the ability of discriminator (only 8 Conv layers), the generated facial landmarks are more disordered. In comparison, our EDGAN can reconstruct more authentic facial detail that are almost indistinguishable from original.

4.4   Investigation of Models

We have shown a variety of networks and loss functions used for super-resolution that are all evaluated in this subsection. These methods are as follows:

Ours-pixel: this is a CNN-based super-resolution network that only uses a generator and trains with Eq. (4).

EDGAN-pixel: this GAN-based super-resolution network improves upon the Ours-pixel method by additionally training with the GAN loss function (Eqs. (2) and (3)) but doesn't use the perceptual loss function (Eq. (6)).

EDGAN: this is our final network, based on EDGAN-pixel and training using Eqs. (7) and (8).

We illustrated the reconstruction results produced by these models in Fig. 5. By including different types of loss functions, the reconstruction results have higher information entropy and the images are more realistic (Table 1).

Fig. 5. Visual reconstruction results [4× upscaling] and the corresponding reference of the HR image in the LFW dataset, All methods were trained with LFW dataset. We have highlighted the sub-regions, which are rich in facial detail. We magnify the sub-regions in the right boxes to show more detail. From the sub-region images, we can see that our method has a stronger ability to recover high frequency detail and sharp edges.

Table 1. The Entropy of our different models. A higher entropy signifies that an image has more information.

Databases	LFW			CelebA		
Methods	Ours-pixel	EDGAN-pixel	EDGAN	Ours-pixel	EDGAN-pixel	EDGAN
Entropy	7.1954	7.2150	7.2251	7.1481	7.1720	7.1754

4.5 Quantitative Comparisons

Our quantitative examination of LFW and CelebA datasets are shown in Table 2. On the LFW dataset, Ours-pixel achieved the highest score in terms of PSNR, SSIM, FSIM and VIF metrics. On the CelebA dataset, Ours-pixel and SRResnet achieved best PSNR score, in terms of SSIM and FSIM. The best-performing overall method is the EDGAN-pixel method. Ours-pixel also achieved the best performance on the VIF metric. From the resulting comparison of metrics, the best method is difficult to isolate. By visually inspecting the reconstruction results (see Fig. 5), the sharper and more realistic facial image reconstructions are by far produced by EDGAN-pixel and EDGAN.

4.6   Discussion of EDN

In this section, we further explored the role of EDN in SRGAN. We constructed a new adversarial model by combining EDN and SRResnet, the generator of SRGAN. Then, we compared the reconstruction results of SRResnet+EDN, SRGAN-MSE and SRGAN-VGG54. Visual results are shown in Fig. 6; these results are associated with data from Table 3. We can see SRResnet+EDN can produces more legible results than original SRGAN, and achieved higher score in four different metrics. This confirms that our proposed EDN is better than original discriminator, and not only allows our generator to produce realistic facial markers, but is also effective in other GAN-based super-resolution frameworks.

Table 2. PSNR-, SSIM-, FSIM- and VIF-based super-resolution performance on LFW and Celeb-A datasets (higher is better, red is highest in this table). The results are not indicative of visual quality. See Fig. 5.

Methods	LFW				CelebA			
	PSNR	SSIM	FSIM	VIF	PSNR	SSIM	FSIM	VIF
bicubic	29.18	0.8648	0.8754	0.4959	28.62	0.8329	0.8665	0.4700
VDSR	30.85	0.8882	0.9161	0.5264	29.12	0.8593	0.8976	0.5023
SRResnet	32.75	0.9302	0.9415	0.6414	30.56	0.8971	0.9226	0.5761
SRGAN	30.54	0.8891	0.9202	0.5279	28.96	0.8536	0.9044	0.4774
Ours-pixel	32.79	0.9314	0.9429	0.6472	30.56	0.8775	0.9194	0.5795
EDGAN-pixel	32.06	0.9130	0.9376	0.6089	29.98	0.8983	0.9239	0.5444
EDGAN	31.73	0.9093	0.9357	0.5996	29.72	0.8731	0.9181	0.5357

SRGAN-MSE   SRGAN-VGG54   SRResnet+EDN   HR

Fig. 6. SRGAN-MSE and SRGAN-VGG54 are models proposed by Ledig *et al.* [3]. **SRResnet+EDN** is the SRGAN that replaced the original discriminator with EDN.

Table 3. Reconstruction results of SRGAN-MSE, SRGAN-VGG54 and SRRes-net+EDN. Compared to SRGAN-MSE and SRGAN-VGG54, SRResnet+EDN achieved the best performance on all indicators.

Method	LFW				CelebA			
	PSNR	SSIM	FSIM	VIF	PSNR	SSIM	FSIM	VIF
SRGAN-MSE	30.32	0.8742	0.9146	0.5222	28.82	0.8437	0.9001	0.4771
SRGAN-VGG54	30.54	0.8891	0.9202	0.5279	28.96	0.8536	0.9044	0.4774
SRResnet+EDN	30.90	0.8946	0.9254	0.5645	29.09	0.8566	0.9074	0.5044

5   Conclusion

In this paper, we have highlighted some limitations of GAN-based super-resolution methods in the super-resolution facial recognition problem. The largest limitation we review is the instability of the discriminator. To solve this issue, we introduce EDGAN, which augments the discrimination process by using densely connected convolutional networks. We designed individual loss functions and combined them to form the objective functions of discriminators and generators, respectively. Our method surpasses other state-of-the-art methods on different metrics, and we also show good results with higher perceptual fidelity on real-world low resolution face images.

References

1. Dahl, R., Norouzi, M., Shlens, J.: Pixel recursive super resolution. arXiv preprint arXiv:1702.00783 (2017)
2. Sønderby, C.K., Caballero, J., Theis, L., Shi, W., Huszár, F.: Amortised map inference for image super-resolution. arXiv preprint arXiv:1610.04490 (2016)
3. Ledig, C., et al.: Photo-realistic single image super-resolution using a generative adversarial network. arXiv preprint (2016)
4. Goodfellow, I., et al.: Generative adversarial nets. In: Advances in Neural Information Processing Systems, pp. 2672–2680 (2014)
5. Bulat, A., Tzimiropoulos, G.: Super-fan: integrated facial landmark localization and super-resolution of real-world low resolution faces in arbitrary poses with gans. arXiv preprint arXiv:1712.02765 (2017)
6. Huang, G.B., Ramesh, M., Berg, T., Learned-Miller, E.: Labeled faces in the wild: a database for studying face recognition in unconstrained environments. Technical report 07-49, University of Massachusetts, Amherst, October 2007 (2007)
7. Arjovsky, M., Chintala, S., Bottou, L.: Wasserstein gan. arXiv preprint arXiv:1701.07875 (2017)
8. Mao, X., Li, Q., Xie, H., Lau, R.Y., Wang, Z., Smolley, S.P.: Least squares generative adversarial networks. In: 2017 IEEE International Conference on Computer Vision (ICCV), pp. 2813–2821. IEEE (2017)
9. Berthelot, D., Schumm, T., Metz, L.: Began: boundary equilibrium generative adversarial networks. arXiv preprint arXiv:1703.10717 (2017)

10. Chen, Z., Tong, Y.: Face super-resolution through wasserstein gans. arXiv preprint arXiv:1705.02438 (2017)

11. Dong, C., Loy, C.C., He, K., Tang, X.: Learning a deep convolutional network for image super-resolution. In: Fleet, D., Pajdla, T., Schiele, B., Tuytelaars, T. (eds.) ECCV 2014. LNCS, vol. 8692, pp. 184–199. Springer, Cham (2014). https://doi.org/10.1007/978-3-319-10593-2_13

12. Simonyan, K., Zisserman, A.: Very deep convolutional networks for large-scale image recognition. arXiv preprint arXiv:1409.1556 (2014)

13. Kim, J., Kwon Lee, J., Mu Lee, K.: Accurate image super-resolution using very deep convolutional networks. In: Proceedings of the IEEE Conference on Computer Vision and Pattern Recognition, pp. 1646–1654 (2016)

14. Zhu, S., Liu, S., Loy, C.C., Tang, X.: Deep cascaded bi-network for face hallucination. In: Leibe, B., Matas, J., Sebe, N., Welling, M. (eds.) ECCV 2016. LNCS, vol. 9909, pp. 614–630. Springer, Cham (2016). https://doi.org/10.1007/978-3-319-46454-1_37

15. Song, Y., Zhang, J., He, S., Bao, L., Yang, Q.: Learning to hallucinate face images via component generation and enhancement. arXiv preprint arXiv:1708.00223 (2017)

16. He, K., Zhang, X., Ren, S., Sun, J.: Deep residual learning for image recognition. In: Proceedings of the IEEE Conference on Computer Vision and Pattern Recognition, pp. 770–778 (2016)

17. Huang, G., Liu, Z., Weinberger, K.Q., van der Maaten, L.: Densely connected convolutional networks. In: Proceedings of the IEEE Conference on Computer Vision and Pattern Recognition, vol. 1, p. 3 (2017)

18. Lim, B., Son, S., Kim, H., Nah, S., Lee, K.M.: Enhanced deep residual networks for single image super-resolution. In: The IEEE Conference on Computer Vision and Pattern Recognition (CVPR) Workshops, vol. 1, p. 3 (2017)

19. Zhang, Y., Tian, Y., Kong, Y., Zhong, B., Fu, Y.: Residual dense network for image super-resolution. arXiv preprint arXiv:1802.08797 (2018)

20. Tong, T., Li, G., Liu, X., Gao, Q.: Image super-resolution using dense skip connections. In: 2017 IEEE International Conference on Computer Vision (ICCV), pp. 4809–4817. IEEE (2017)

21. Johnson, J., Alahi, A., Fei-Fei, L.: Perceptual losses for real-time style transfer and super-resolution. In: Leibe, B., Matas, J., Sebe, N., Welling, M. (eds.) ECCV 2016. LNCS, vol. 9906, pp. 694–711. Springer, Cham (2016). https://doi.org/10.1007/978-3-319-46475-6_43

22. Wu, B., Duan, H., Liu, Z., Sun, G.: Srpgan: perceptual generative adversarial network for single image super resolution. arXiv preprint arXiv:1712.05927 (2017)

23. Liu, Z., Luo, P., Wang, X., Tang, X.: Deep learning face attributes in the wild. In: Proceedings of International Conference on Computer Vision (ICCV) (2015)

24. Zhang, L., Zhang, L., Mou, X., Zhang, D.: FSIM: a feature similarity index for image quality assessment. IEEE Trans. Image Process. **20**(8), 2378–2386 (2011)

25. Sheikh, H.R., Bovik, A.C.: Image information and visual quality. IEEE Trans. Image process. **15**(2), 430–444 (2006)

26. Kingma, D.P., Ba, J.: Adam: a method for stochastic optimization. arXiv preprint arXiv:1412.6980 (2014)

Image Splicing Detection Based
on the Q-Markov Features

Hongda Sheng[1,2], Xuanjing Shen[1,2(✉)], and Zenan Shi[1,2]

[1] College of Computer Science and Technology, Jilin University,
Changchun, China
xjshen@jlu.edu.cn
[2] Key Laboratory of Symbolic Computation and Knowledge Engineering
of Ministry of Education, Jilin University, Changchun, China

Abstract. Recently, image splicing tamper detection has become an increasingly significant challenge, because of which all color information of color images and low detection rate of existing algorithms cannot be exploited. To overcome the shortcomings of this method, we propose a model, which employs difference matrix in quaternion domain (Q-DIFF) and Markov in quaternion domain (Q-Markov) in the quaternion discrete cosine transform domain (QDCT) for encoding tampering traces and quaternion back propagation neural network (QBPNN) for decision making. Furthermore, by introducing Q-DIFF and Q-Markov in the proposed model, the entire architecture of the algorithm is accumulated in the four-dimensional frequency domain (i.e., all color channels of color images are utilized). Moreover, the experimental results on public domain benchmark datasets demonstrate that the proposed model is superior to the other state-of-the-art splicing detection methods. Based on the experimental results, we suggest the direction that designs image tamper detection model, which invite all the processing in the model to operate in four-dimensional space (i.e. quaternion space).

Keywords: Image splicing detection · Markov model · Quaternion domain

1 Introduction

In recent years, computer vision has received more and more attention, such as image categorization [1–3], image fusion [4], image enhancement [5] and especially image detection [6, 7]. Image splicing, which transfers one or more objects or regions from one image and pastes them into another image, is often used as simple entertainment or the initial step of a photomontage. In addition, it is also a fundamental and extensive tampering method among the existing image tampering methods. However, spliced images used for malicious purposes can lead people to believe that digital image is not a reliable way to transfer daily information. This is because image splicing can be easily implemented through image editing tools, such as "Adobe Photoshop," and it is difficult for humans to visually detect such modifications. The approach of passive or blind image forgery detection approach, without any prior information has become the next research direction; it discriminates spliced images by exploiting the traces/artifacts

© Springer Nature Switzerland AG 2018
R. Hong et al. (Eds.): PCM 2018, LNCS 11165, pp. 453–464, 2018.
https://doi.org/10.1007/978-3-030-00767-6_42

(discontinuities and inconsistencies in the form of noise, lighting conditions, statistical pixels, etc. by modeling the artifacts of forgery) left by the tampering process.

Since Ng et al. [6, 7] proposed benchmark datasets and the basic framework of the image forgery detection model, researchers have proposed a variety of effective detection models in the past decade. In the following paragraphs, we present some representative detection models, and the main discrimination among these models is that the way encoded the artifacts due to the splicing operation are different.

He et al. [8] proposed the effective Markov feature detection model in DCT (discrete cosine transform) and DWT (Discrete Wavelet Transform) frequency domain, and the accuracy was up to 93.42% on the Columbia image dataset [7]. Inspired by the research in [8] for constructing the Markov detection model, Zhang et al. [9] extracted the intra-block Markov features by considering the different frequency ranges in DCT domain and the inter-block Markov features to handle color images in Contourlet domain. The model obtained detection accuracies of 94.10% and 96.69% on Columbia dataset and IFS-TC (Information Forensics and Security Technical Committee) dataset [10, 11] respectively.

Alahmadi et al. [12] proposed a detection model using the mean and standard deviation of DCT coefficient in LBP (local binary pattern) domain, and the SVM (Support vector machine) is used as a classifier. The method achieved 97%, 97.5%, and 97.77% accuracies on CASIA V1.0, CASIA V2.0 [13], and Columbia datasets respectively. Isaac et al. [11] proposed a detection method for extracting LPQ (Local Gabor Phase) features from Gabor images. The method obtained the accuracies of 96.3%, 99.33% and 85% on the CASIA v2.0, CASIA v1.0 and IFS-TC datasets respectively.

However, the color image need to be separated Cb or Cr channel before using the above algorithm to extract color image features, which can miss most of the color information. To overcome the shortcoming, Li et al. [14] extracted the intra-block and inter-block Markov features in QDCT domain. The entire image is subjected to 8×8 block QDCT to obtain a quaternion matrix of an image in the QDCT domain. And, the method obtained the accuracies of 95.217% and 92.38% on CASIA V1.0 and CASIA V2.0 respectively. In this paper, the quaternion domains and QDCT are introduced to preserve all color information of color images in the preprocessing step. But, the quaternion matrix is still converted to one real matrix due to the two-dimensional DIFF (difference matrix) and Markov features, which causes the color information to still be partially lost before extracting the final features.

Through the above techniques, we realize that the current stage of research leaves much space to improve. Furthermore, the main factor that contributes to all the deficiencies of the above algorithm is that the all color information of color image cannot be fully exploited, resulting in low detection performance of the previous research.

To overcome this shortcoming, this paper provides a novel passive image splicing detection method based on the Q-DIFF and Q-Markov features in the QDCT domain. The contributions of the work can be summarized as follows.

- In order to compensate for the loss of image information caused by using traditional DIFF, we introduce the Q-DIFF and achieve excellent performance.

- To construct a complete splicing image quaternion detection model, this paper extends the Markov feature to quaternion domains and we provide a new direction for the four-dimensional Markov.
- The proposed image forgery detection model is very robust and effective in recent state-of-the-art image forgery detection models.

The organization of the paper is as follows. Section 2 describes details about the proposed technique, including QDCT and Q-DIFF and Markov in quaternion domains. In Sect. 3 describes the experimental environment and analyzes the experimental results. Finally, Sect. 4 summarizes this work and emphasizes the creativity and contribution of this work.

2 Proposed Image Forgery Detection Model

In the detection model, all calculations of the algorithm are performed in quaternion domain space, QDCT and Q-DIFF which is to magnify structural changes that occurred in images due to forgery and Q-Markov features which is to capture this change. The architectural diagram of proposed approach is shown in Fig. 1.

(1) Firstly, the quaternion is used to represent the three channels of the original image and divided into non-overlapping 8×8 blocks.
(2) Secondly, the QDCT is applied to each image block, and all QDCT block matrices need to reassemble according to the site of blocking to construct an overall QDCT domain graphics [14].
(3) Thirdly, the difference matrix of QDCT coefficient matrix is calculated using the Q-DIFF.
(4) Fourthly, the Markov features of the Q-DIFF matrix is extracted in the four directions by using the Q-Markov.
(5) Finally, the training set and the test set of the Q-Markov features are fed to the QBPNN classifier.

2.1 QDCT

In the preprocessing step of most previous researches, the DCT is used to convert images from the digital domain to the frequency domain to focus the falsification information of the image. But before the DCT, only one-color channel of the three channels of the color image can be selected and processed due to the DCT can only process one color channel at a time. In order to overcome the dilemma, Li et al. [14] introduced the QDCT (designed by Feng Wei and Hu Bo [15] in 2008) to process color images, and all three color channels are used to generate a quaternion matrix. In the proposed model, the QDCT is also introduced, and the formula is shown as follows:

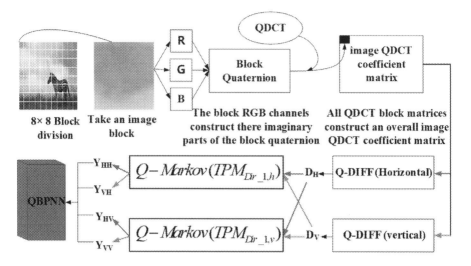

Fig. 1. The system design of the proposed Algorithm ($Dir\_1 = D_H$, D_V and dim = dimensionality)

$$q(u,v) = \alpha(u_1)\alpha(v_1) \sum_{u=0}^{M-1} \sum_{v=0}^{N-1} u_q \times f_q(u,v) \times H(u_1,v_1,u,v) \qquad (1)$$

Where,

$$f_q(u,v) = f_r(u,v)i + f_g(u,v)j + f_b(u,v)k \qquad (2)$$

$$\alpha(u_1) = \begin{cases} \frac{1}{\sqrt{M}}, & u_1 = 0 \\ \sqrt{\frac{2}{M}}, & otherwise \end{cases} \qquad \alpha(v_1) = \begin{cases} \frac{1}{\sqrt{N}}, & v_1 = 0 \\ \sqrt{\frac{2}{N}}, & otherwise \end{cases} \qquad (3)$$

$$H(u_1,v_1,u,v) = \cos\frac{\pi u_1(2u+1)}{2M}\cos\frac{\pi v_1(2v+1)}{2N}$$

$$u_q^2 = 1 \qquad (4)$$

Where, $f_q(u,v)$ is a two-dimensional M × N quaternion matrix. $f_r(u,v)$, $f_g(u,v)$ and $f_b(u,v)$ are three imaginary parts of a quaternion and represent the red, green and blue components of an RGB image respectively. And u and v indicate the location of pixels in rows and columns in the matrix, and the following u and v are also this meaning. u_1 and v_1 are the sampled values in the quaternion frequency domain, and M and N represent the dimensions of the rows and columns of the image matrix. u_q is a unit pure quaternion, and it conforms to the Eq. (4). According to previous examination [12, 14, 16], to utilize the spatial location of region of interests (ROI), before QDCT, the quaternion matrix of the image needs to be blocked, and the size of the block is 8 × 8.

2.2   Q-DIFF

In the existing research on image splicing detection models, the calculation of the difference matrix in the quaternion domain is a vacancy that needs urgent improvement. Due this shortcoming, Li et al. [14] reluctantly introduced the square root to convert a four-dimensional quaternion matrix of QDCT coefficient to a two-dimensional matrix. This method, which undoubtedly affects the consistency and frequency information of the four parts of the quaternion matrix and algorithm architecture, cannot be used on the quaternion domain. To solve the issue, this paper adopts the Q-DIFF [17], which can obtain the difference matrix of QDCT coefficients in the quaternion domain. The defined formula is follows ($q1$ and $q2$ are quaternions, $q1 = a + bi + cj + dk$, and $q2 = w + xi + yj + zk$):

$$DIFF = (a - w) + (b - x)i + (c - y)j + (d - z)k \tag{5}$$

In our model, to better decrease the correlation of the image content and highlight the impact of the splicing operation on the original image, the horizontal and vertical difference matrices of QDCT coefficients are all calculated. The derivation formulas are as follows (denoted by the obtained matrices $\mathbf{D_H}$ and $\mathbf{D_V}$):

$$D_H = q(u, v) - q(u + 1, v) \tag{6}$$

$$D_V = q(u, v) - q(u, v + 1) \tag{7}$$

By using the Q-DIFF, the data and logical relation between one real part and three imaginary parts of the QDCT quaternion coefficient matrix of the image can be completely utilized. In our model, the data of Q-DIFF must be preprocessed (round the QDCT coefficient matrix the nearest integer and take absolute value) because the input and output of the Q-DIFF are the QDCT coefficient matrices and the extracted features of Q-Markov, respectively. To reduce the computational complexity of extracting Q-Markov features, thresholding must be performed to the quaternion matrix obtained after Q-DIFF. And, threshold $T_H = 4$ was chosen in this study. If an element of $\mathbf{D_H}$ and $\mathbf{D_V}$ is either larger than T_H or smaller than $-T_H$, it will be represented by T_H or $-T_H$ correspondingly, applying Eq. (8). Where x is an element of $\mathbf{D_H}$ and $\mathbf{D_V}$.

$$x = \begin{cases} T_H, & x \geq T_H \\ -T_H, & x \leq -T_H \\ x, & otherwise \end{cases} \tag{8}$$

2.3   Q-Markov

After thresholding, the transition probability matrix of the Markov random process is applied to the threshold array as a tool to describe the correlation among QDCT coefficients [9]. However, the calculation of the Markov feature in the quaternion domain has always been a neglected issue in literature. When the quaternion matrix directly applies its four-arithmetic operation (primary division rule, i.e., matrix division) to extract Markov features, the matrix division will inevitably destroy the nature of the transition probability of the difference matrix that the Markov feature desires to obtain. Thus, the basic horizontal and vertical transition probability matrices (denoted by the obtained arrays TPM_{hh}, TPM_{hv}, TPM_{vh}, and TPM_{vv}) of $\mathbf{D_H}$ and $\mathbf{D_V}$ are still applied and are calculated by applying Eq. (9)–(12), respectively. In this paper, based on previous research such as the quaternion pseudo-Zernike [18] and QDCT [14], we attempted to design the Q-Markov algorithm flow that is different from the past. The details are given in Algorithm 1. Finally, $(2T+1) \times (2T+1) \times 4$ (i.e., 324) dimensionality Q-Markov features were obtained.

$$TPM_{hh}(i,j) = \frac{\sum_{u=1}^{S_u-2} \sum_{v=1}^{S_v} \delta(D_H(u,v) = i, D_H(u+1,v) = j)}{\sum_{u=1}^{S_u-2} \sum_{v=1}^{S_v} \delta(D_H(u,v) = i)} \tag{9}$$

$$TPM_{hv}(i,j) = \frac{\sum_{u=1}^{S_u-1} \sum_{v=1}^{S_v-1} \delta(D_H(u,v) = i, D_H(u,v+1) = j)}{\sum_{u=1}^{S_u-1} \sum_{v=1}^{S_v-1} \delta(D_H(u,v) = i)} \tag{10}$$

$$TPM_{vh}(i,j) = \frac{\sum_{u=1}^{S_u-1} \sum_{v=1}^{S_v-1} \delta(D_V(u,v) = i, D_V(u+1,v) = j)}{\sum_{u=1}^{S_u-1} \sum_{v=1}^{S_v-1} \delta(D_V(u,v) = i)} \tag{11}$$

$$TPM_{vv}(i,j) = \frac{\sum_{u=1}^{S_u} \sum_{v=1}^{S_v-2} \delta(D_V(u,v) = i, D_V(u,v+1) = j)}{\sum_{u=1}^{S_u} \sum_{v=1}^{S_v-2} \delta(D_V(u,v) = i)} \tag{12}$$

Where $i, j \in \{-T, -T+1, \ldots, 0, \ldots, T-1, T\}$, S_u and S_v denote the width and height of the input image. And, $\delta(\cdot) = 1$ only when all conditions are satisfying, otherwise $\delta(\cdot) = 0$.

Algorithm 1: The process of Q-Markov

Input: quaternion matrix $\mathbf{Y}$ after QDIFF and thresholding

Output: Q-Markov features

Procedure:

1. Extract four parts of the quaternion matrix $\mathbf{Y}$: Y_0, Y_i, Y_j, Y_k
2. For $z \in \{hh, hv, vh, vv\}$

      For each part Y_x, $x = 0, i, j, k$

         Calculate transition probability matrices (Eq. (9)–(12)): $\mathbf{Y}_{zx}$

      end

   end
3. Assemble all $\mathbf{Y}_{zx}$, $z = hh, hv, vh, vv$, $x = 0, i, j, k$ to construct two-direction Q-Markov features $Y_{hh}, Y_{hv}, Y_{vh}, Y_{vv}$

---

3   Experimental Results

3.1   Image Dataset

The proposed algorithm is evaluated and compared using 2 public benchmark datasets: CASIA V1.0 and CASIA V2.0 (CASIA version 1.0 and 2.0). Their detailed information is listed in Table 1 [19]. Figure 2 shows some sample images selected from the above 2 datasets.

Table 1. Detailed information of CASIA V1.0 and CASIA V2.0

Dataset	No. of images			Image type	Image size
	Authentic	Tampered	Total		
CASIA V1.0	800	921	1721	Jpg	$384 \times 256, 256 \times 384$
CASIA V2.0	7491	5123	12614	Jpg, tif, bmp	$240 \times 160 – 900 \times 600$

3.2   Evaluating Indicator

According to previous research work [16, 18, 20], the Accuracy, TPR (True Positive Rate), TNR (True Negative Rate), FPR (False Positive Rate) and AUC (Area Under Curve) value are introduced to verify the proposed model.

Accuracy is the proportion of all instances that are correctly classified.

$$Accuracy = \frac{(TP + TN)}{(TP + FP + TN + FN)} \times 100\% \tag{13}$$

TPR indicates probability that the tampered ones are identified as tampered ones.

$$TPR = \frac{TP}{(TP + FN)} \tag{14}$$

TNR indicates probability that the authentic ones are identified as authentic ones.

$$TNR = \frac{TN}{(TN + FP)} \tag{15}$$

FPR indicates probability that the authentic ones are identified as tampered ones.

$$FPR = 1 - TNR \tag{16}$$

The AUC is the area under the ROC (receiver operating characteristic curve), and the horizontal and vertical axes are FPR and TPR respectively. Where TN (True Negative) is the number of authentic images classified as authentic ones; FP (False Positive) is the number of authentic images classified as tampered ones; TP (True Positive) is the number of tampered images classified as tampered ones; and FN (False Negative) is

Fig. 2. Sample images selected from CASIA V1.0 and CASIA V2.0

the number of tampered images classified as authentic ones. And, the larger the value of Accuracy, AUC, TPR and TNR, the algorithm platforms better.

3.3   Parameter Initialization

In our model, size of QDCT block and T_H can directly affect complexity and performance. To select the appropriate size of block and T_H, 5 values of size and T_H are tested on CASIA V1.0, the corresponding experimental results are shown in Table 2.

Table 2. The results with different parameters (size of QDCT block and T_H)

The parameters	Accuracy (%)	AUC	TPR	TNR	Feature extraction time (s)
4×4, $T_H = 4$	98.5423	0.9857	0.9929	0.9767	5.52
16×16, $T_H = 4$	98.7755	0.9866	0.9908	0.9843	4.23
8×8, $T_H = 3$	94.6647	0.9517	0.9005	1	3.16
8×8, $T_H = 5$	99.4169	0.9945	0.9962	0.9918	5.7
8×8, $T_H = 4$	99.2128	0.9928	0.9929	0.9912	4.87

From Table 2 and the previous research [12, 14, 16], when size of blocks is 8×8 and $T_H= 4$, the performance and complexity of the model are acceptable.

In order to build a complete four-dimensional detection model, the QBPNN [18] is introduced as a classifier. According to work [18], the target output of the real image and spliced image are $(1+i+j+k, 0+0i+0j+0k)$ and $(0+0i+0j+0k, 1+i+j+k)$ respectively. The initial weights and thresholds are generated randomly between -0.3 and 0.3. The target error is 0.01, and the maximum number of epochs was 1000. In order to reduce the time of training the networks and overcome the local optimum parameter values, the additional momentum method and the learning rate adaptation method are introduced to the QBPNN. And, the momentum factor is 0.08.

3.4 Experimental Results Analysis

Contributions of Q-DIFF and Q-Markov. To prove the necessity of the quaternion framework, we conducted comparative experiments. The CASIA V1.0 was used for comparative investigations, and the experimental evaluation environment is described in Sects. 3.1, 3.2 and 3.3.

As the conventional DIFF and Markov can only be calculated in two-dimensional space and only process one part of QDCT coefficient matrix at a time, in the proposed model, the Q-DIFF and Q-Markov fully utilize the four parts of QDCT coefficient matrix and construct the entire framework of the model in the quaternion field. Further, to confirm its effect, we conducted the comparative experiment based on the conventional DIFF and Markov. In the experiment, after QDCT, a real part and three imaginary parts of QDCT coefficient matrix were separated and fed into the conventional DIFF and Markov, respectively, and the BP (back propagation) neural network was used to classify the features obtained. The corresponding experimental results are shown in Fig. 3.

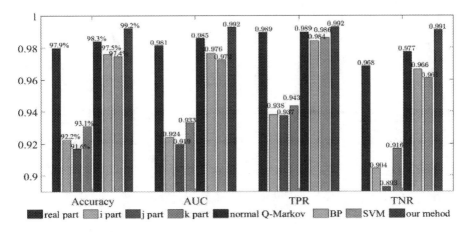

Fig. 3. Effects of Q-DIFF and Q-Markov on CASIA V1.0

Then, to verify the effectiveness of the designed Q-Markov, we compared all evaluating indicators of the method using the Q-Markov with the method using the normal Q-Markov (quaternion Markov calculated using the four-arithmetic operation of quaternion). Finally, to illustrate the effectiveness of introducing QBPNN, we adopt the BP and SVM classifiers to process the standard deviation of four parts of Q-Markov feature vector. Figure 3 shows the detailed experimental result.

From Fig. 3 it can be seen that, the Accuracies (98.3%, 99.2%), AUC values (0.985, 0.992), TPR values (0.989, 0.992) and TNR values (0.977, 0.991) of the normal Q-Markov and our method (by using our proposed Q-Markov) are the best two results. And we can conclude that the idea of constructing the image detection algorithm in quaternion domain (by using the proposed Q-DIFF) can fully exploit the data and logic of all color channels of color images and obtain better performance relative to real number domain models. Especially, because of that our proposed Q-Markov is superior to the normal Q-Markov method, the designed Q-Markov can contribute to improve the performance of the algorithm and have a certain theoretical significance. The results (97.5%, 97.4%, 0.976, 0.972 et al.) of using the BP and SVM are lower than those using the QBPNN classifier. This shows that the QBPNN can help us construct the quaternion image detection algorithm and improve the capability of our algorithm.

Comparison with Other Algorithms. To evaluate the algorithm comprehensively, the algorithm is compared with the following 3 state-of-the-art image splicing detection models. (1) The algorithm of Alahmadi et al. [12]: This model combines the mean and standard deviation of DCT coefficients in LBP domain, and obtain a prominent detection performance among the recently proposed models. (2) The algorithm of Li et al. [14]: This paper provides a fusion strategy (the square root) for the coefficient matrix obtained after QDCT, and the Markov are still used to extract discriminative features. (3) The algorithm of Shen et al. [11]: This paper proposed a novel textural feature based on the Gray Level Co-occurrence Matrices, which is a recent (2017) work. And, the results of the comparison experiments are shown in Table 3.

Table 3. The results of the comparison experiments

Dataset	Models	Accuracy (%)	AUC	TPR	TNR
CASIA V1.0	The proposed model	99.2128	0.9928	0.9929	0.9912
	Alahmadi et al. [12]	97.5	0.97	0.9675	0.9824
	Li et al. [14]	95.958	-	0.9544	0.9647
	Shen et al. [19]	98.54	-	0.9748	0.9951
CASIA V2.0	The proposed model	98.6686	0.9859	0.9870	0.9865
	Alahmadi et al. [12]	97.50	0.97	0.9845	0.9684
	Li et al. [14]	92.377	-	0.8918	0.9556
	Shen et al. [19]	97.73	-	0.9772	0.9780

From Table 3 we can summarize two conclusions: (1) Relative to the way of introducing the square root to process the QDCT coefficient matrix in [14], the paper proposes the Q-DIFF and perfects the Q-Markov. And, the experimental results proved that the Q-DIFF and Q-Markov all can improve the performance of the algorithm. (2) Relative to the two-dimensional real number domain model in [12] and [19], our algorithm is performed in quaternion domain space. And, the experimental data demonstrate that the four-dimensional detection algorithm has better performance than the two-dimensional detection algorithm.

Finally, the Table 3 shows that although the TPR on CASIA V1.0 of the Shen et al. [19] is 0.9951 which is higher than the proposed model, the accuracies (99.2128% and 98.6686%), AUCs, and TNRs of the proposed model significantly exceed the other three algorithms on CAISA V1.0 and CASIA V2.0 image datasets. This shows that this model has practical significance in the research of image splicing detection algorithms.

4 Conclusion

In this paper, to improve the detection rate of the image tamper detection model, the Q-DIFF and Q-Markov in QDCT domain were introduced. First, the 8×8 block QDCT is applied to the quaternion representation of the original color image to obtain QDCT coefficient matrix. Second, the horizontal and vertical difference matrix of QDCT coefficient matrix was calculated using the Q-DIFF. Finally, Q-Markov was used to calculate the Markov features of the difference matrix in the quaternion domain, and the QBPNN was used for classification. The experimental results demonstrate that by introducing Q-DIFF and Q-Markov, the proposed model is superior to most existing algorithms with accuracies of 99.2128% and 98.6686% on CASIA V1.0 and CASIA V2.0, respectively. Therefore, the proposed algorithm has a certain practical significance. In the future, we will focus on localizing the tampering region.

Acknowledgments. This research is supported by Key Projects of Jilin Province Science and Technology Development Plan (20180201064SF).

References

1. Zhang, L.: Detecting densely distributed graph patterns for fine-grained image categorization. IEEE Trans. Image Process. **25**(2), 553–565 (2016)
2. Zhang, L.: Image categorization by learning a propagated graphlet path. IEEE Trans. Neural Netw. Learn. Syst. **27**(3), 674–685 (2017)
3. Zhang, L.: Large-scale aerial image categorization using a multitask topological codebook. IEEE Trans. Cybern. **46**(2), 535–545 (2017)
4. Zhang, L.: Probabilistic skimlets fusion for summarizing multiple consumer landmark videos. IEEE Trans. Multimed. **17**(1), 40–49 (2014)
5. Hong, R.: Unified photo enhancement by discovering aesthetic communities from flickr. IEEE Trans. Image Process. **25**(3), 1124–1135 (2016)
6. Farid, H.: Detecting Digital Forgeries Using Bispectral. Analysis MIT AI Memo AIM-1657, pp. 1–9 (1999)
7. Ng, T.: A data set of authentic and spliced image blocks. Columbia University ADVENT Technical Report 203 (2004)
8. He, Z.: Digital image splicing detection based on Markov features in DCT and DWT domain. Pattern Recogn. **45**(12), 4292–4299 (2012)
9. Zhang, Q.: Joint image splicing detection in DCT and Contourlet transform domain. J. Vis. Commun. Image Represent. **40**, 449–458 (2016)
10. Wang, D.P.: A forensic algorithm against median filtering based on coefficients of image blocks in frequency domain. Multimed. Tools Appl. **4**, 1–17 (2018)
11. Isaac, M.: Multiscale local gabor phase quantization for image forgery detection. Multimed. Tools Appl. **1**, 1–22 (2017)
12. Alahmadi, A.: Passive detection of image forgery using DCT and local binary pattern. Signal Image Video Process. **11**(1), 1–8 (2016)
13. Dong, J.: CASIA image tampering detection evaluation database. In: 1st IEEE China Summit and International Conference on Signal and Information Processing, pp. 422–426. IEEE, Beijing (2013)

14. Li, C.: Image splicing detection based on Markov features in QDCT domain. Neurocomputing **228**, 29–36 (2017)
15. Feng, W.: Quaternion discrete cosine transform and its application in color template matching. In: 2008 Congress on Image and Signal Processing, Sanya, pp. 252–256. IEEE (2008)
16. Han, J.G.: Quantization-based Markov feature extraction method for image splicing detection. Mach. Vis. Appl. **29**(3), 543–552 (2018)
17. Qtfm Homepage. http://qtfm.sourceforge.net/. Accessed 14 Nov 2017
18. Chen, B.: Quaternion pseudo-Zernike moments combining both of RGB information and depth information for color image splicing detection. J. Vis. Commun. Image Represent. **49**, 283–290 (2017)
19. Shen, X.: Splicing image forgery detection using textural features based on the grey level co-occurrence matrices. IET Image Process. **11**(1), 44–53 (2017)
20. Hong, R.: Flickr circles: aesthetic tendency discovery by multi-view regularized topic modeling. IEEE Trans. Multimed. **18**(8), 1555–1567 (2016)

Snapshot Multiplexed Imaging Based on Compressive Sensing

Ying Fu, Chen Sun, Lizhi Wang, and Hua Huang[✉]

Beijing Institute of Technology, Beijing 100081, China
{fuying,sunchen,lzwang,huahuang}@bit.edu.cn

Abstract. Multiplexed imaging methods have been proposed to extend the field of view (FoV) of the imaging devices. However, the nature of multiple exposures hinders its application in time-crucial scenarios. In this paper, we design a snapshot multiplexed imaging system for wide FoV imaging. In the system, the scene is first spatially encoded by a mask, and then the coded scene is optically divided into multiple sub-regions which are finally superimposed and measured on a sensor array. We model the demultiplexing as a compressive sensing (CS) reconstruction problem and introduce two methods, one is based on Total Variation (TV) constraint and the other is based on sparsity constraint, to reconstruct the scene. Simulation results demonstrate the effectiveness of the proposed system.

Keywords: Multiplexed imaging · Snapshot · Wide FoV · CS

1 Introduction

Nowadays, wide FoV imaging is attracting more and more attention within the field of reconnaissance, surveillance, remote sensing, panoramic imaging and so on. In traditional optical design, image resolution is proportional to the number of pixels and is inversely proportional to FoV. So increasing FoV without suffering loss of resolution requires either more pixels or more exposures. Wide FoV imaging methods based on traditional optical design usually stitch together multiple images of different fields of view or scan across the wide field scene. However, the former requires multiple lens and multiple sensors which are costly and make the imaging system cumbersome, while the latter requires significant efforts in mechanical stability.

Multiplexed imaging systems [18,19], in which a single pixel can observe multiple different regions simultaneously, breaks the limit in traditional optical design that one pixel can only observe a small continuous region of the scene. Thus, multiplexed imaging systems could extend FoV but do not decrease resolution without the need to expand the size of sensor. Recently researchers proposed a shift-encoded optically multiplexed imaging system [16,17,21]. In this system, multiplexing assembly, which consists of an array of mirrors and is placed near the entrance pupil, directs different fields of view into a single objective lens to

© Springer Nature Switzerland AG 2018
R. Hong et al. (Eds.): PCM 2018, LNCS 11165, pp. 465–475, 2018.
https://doi.org/10.1007/978-3-030-00767-6_43

form a multiplexed image. Image coding is performed by shifting individual layer of the multiplexed image via tilting the mirrors. At each exposure, only a single layer is encoded. So, for a multiplexed imaging system with L channels, at least L exposures are needed to well reconstruct the scene. The scene is supposed to be static during the multiple exposures of the system. However, this assumption fails when capturing dynamic scenes, and these systems can not be used.

We propose an improved multiplexed imaging system in this paper. Our system can capture a wide field scene with only a single exposure. In our system, the scene is first spatially encoded by a mask, and then the encoded scene is divided into multiple sub-regions which are finally superimposed and measured on a sensor array. We model the demultiplexing as a CS reconstruction problem and introduce two methods, which are based on TV constraint and sparsity constraint respectively, to reconstruct the scene.

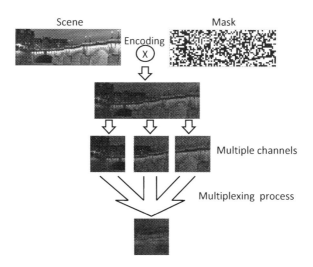

Fig. 1. Flow diagram of the multiplexed imaging process.

2   The Proposed System Model

Based on the imaging principle of the proposed system, as shown in Fig. 1, we depict the optical design of the system in Fig. 2. First, an objective lens is used to capture the scene, and on it's image plane, we place a mask [10,20] to encode the scene. Then, we use the multiplexed imaging system proposed in [16] to capture the encoded scene. As shown in Fig. 2, the encoded scene acts as the object of the second objective lens, and is optically divided into multiple channels to image. Multiplexing assembly [16,17,21] directs all these channels into the second objective lens to form a superimposed image on a common sensor array. The proposed system can be implemented by integrating the multiplexed imaging system proposed in [16] with an objective lens and a mask immediately.

And if there are L channels, and the size of the sensor is $S \times S$, the size of the mask should be $S \times LS$. We can capture scenes in different distance by just adjusting the focal length of the first objective lens without changing other parts of the system, even though the size of the sensor can not change.

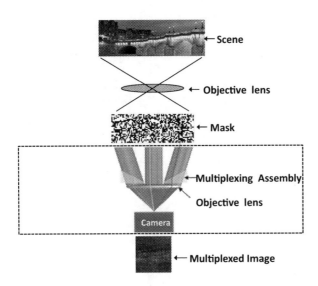

Fig. 2. The optical design of the multiplexed imaging system, and the part inside the dashed line box is the multiplexing system proposed by [16].

The imaging process can be formulated as a linear transform,

$$y = Ax + \epsilon, \tag{1}$$

where $y \in R^{n \times 1}$ represents the superimposed image measured on the image sensor, $A \in R^{n \times m}$ represents the transform matrix, $x \in R^{m \times 1}$ is a discretized m-pixel representation of the scene, and $\epsilon \in R^{n \times 1}$ represents the additive gaussian noise corrupting each pixel measurement. The transformation matrix A includes two operations: encoding and multiplexing. For the multiplexed imaging system with L channels, the transformation matrix A can be divided into L parts, each representing a channel encoding information, formulated as: $A = [a_1, a_2, ..., a_i, a_{i+1}, ..., a_L]$, where $a_i^{n \times n} \in \{0, 1\}$ is a diagonal matrix whose diagonal elements are partial values of the mask corresponding to i-th channel, and $[..., a_i, a_{i+1}, ...]$ arranged in the column direction indicates the multiplexing operation of multiple channels.

For $A \in R^{n \times m}$, $m = L * n$, Eq. 1 must be an under-determined linear problem in the case of a single exposure, which can't be directly solved by traditional linear solution methods. Since CS is a framework for solving the under-determined linear problem [3,4,9], we might regard the demultiplexing as a CS problem. Regularization term based on prior knowledge of the image is added to the reconstruction procedure, and the reconstruction problem can be further expressed as

$$\hat{x} = \arg\min_{x}\{\|y - Ax\|_2 + \lambda\Phi(x)\}, \tag{2}$$

where $\Phi(x)$ is a regularization term based on prior knowledge and used to constrain the solution so that the solution is as approximate as possible to the true solution, and λ denotes the regularization parameter balancing the fidelity term and the regularization term. Although there are many methods or algorithms to solve CS problem, most of them fall into three distinct categories: greedy algorithms, l_1 minimization algorithms [22], TV minimization algorithms [2,12,15]. Among them, greedy algorithms do not always guarantee an optimal solution, so we choose TV minimization algorithms and l_1 minimization algorithms to solve this problem, corresponding to TV regularization term and sparse regularization term respectively.

2.1   TV Constraint

Because image is piecewise smooth in the spatial representation, we take TV constraint as the regularization term, Eq. 2 can be further transformed into

$$\hat{x} = \arg\min_{x}\{\|y - Ax\|_2 + \lambda \cdot TV(x)\}, \tag{3}$$

The regularization term is formulated as

$$TV(X) = \sum_{i} \sqrt{(\Delta_i^h X)^2 + (\Delta_i^v X)^2}, \tag{4}$$

where $\Delta_i^h X$ and $\Delta_i^v X$ respectively represent horizontal and vertical first-order difference operators at pixel i. In general, for solving Eq. 4, Two-Step Shrinking/Iterative Threshold Algorithm (TwIST) [1] can usually derive an accurate and visually satisfactory result, therefore we choose TwIST to solve this optimization model.

The solving details of TwIST can be formulated as follows,

$$x_1 = \Gamma_\lambda(x_0), \tag{5}$$
$$x_{t+1} = (1 - \alpha)x_{t-1} + (\alpha - \beta)x_t + \beta\Gamma_\lambda(x_t), \tag{6}$$

where x_0 is the initial estimation, $\Gamma_\lambda(x) = \Psi_\lambda(x + A^T(y - Ax))$ and $\Psi_\lambda(\cdot)$ is a TV de-noising operation [5], x_{t+1} is not only depending on x_t but also depending on x_{t-1}, α and β are the parameters ensuring convergence.

2.2   Sparsity Constraint

Sparsity constraint can be described to be that an image x has a sparse representation on a set of basis D [11,14], which can be expressed as

$$\hat{\alpha} = \arg\min_{\alpha} \|\alpha\|_0, \ s.t. \ x = D\alpha, \tag{7}$$

Thus, during the process of reconstruction, the optimized solution should stay sparsity on D as also, which can be formulated as

$$\hat{\alpha} = \arg\min_{\alpha} \|\alpha\|_0, \ s.t. \ y = AD\alpha, \tag{8}$$

where $\|\alpha\|_0$ is a pseudo norm counting the number of non-zero entries of α. Since l_0 minimization is an NP-hard combinatorial optimization problem, it is usually relaxed to the convex l_1 minimization [7,8,13]. Take l_1 minimization as the regularization term, the reconstruction problem can be formulated in the following Lagrangian form:

$$\hat{\alpha} = \arg\min_{\alpha} \|y - AD\alpha\|_2 + \lambda\|\alpha\|_1, \tag{9}$$

where λ is a balance between the approximation error of x and the sparsity of α, based on which the solution of Eq. 9 can be split into two steps: updating x and sparse denoising [6–8].

– step1: sparse denoising

$$\hat{\alpha}^{(j+1/2)} = \arg\min_{\alpha} \|x^{(j)} - D\alpha\|_2 + \lambda\|\alpha\|_1, \tag{10}$$

– step2: updating x

$$x^{(j+1)} = x^{(j+1/2)} + \delta A^T(y - Ax^{(j+1/2)}), \tag{11}$$

where Eq. 10 can be optimally solved by TwIST. The connection between Eqs. 10 and 11 is established with $x^{(j+1/2)} = D\hat{\alpha}^{(j+1/2)}$.

3   Simulations

To verify the effectiveness of the proposed system and reconstruction methods, we conduct extensive simulations on different images in this section. We assume that the imaging system has $L = 3$, 4, 5, or 6 channels and do simulation respectively. Regardless of the number of channels, we just take one exposure. TV minimization algorithms and l_1 minimization algorithms are both used to reconstruct the scene.

The size of the imaging sensor in our simulation is 512×512. So for a imaging system with L channels, the size of the finally reconstructed image is $512 \times 512L$.

We first determine whether our proposed system can achieve accurate wide field scene reconstruction under ideal conditions. Sine the presence of noise can not be avoided during scene shooting, we also test whether the reconstruction methods is noise insensitive. Finally, we experimentally demonstrate that the shift-encoded multiplexed imaging system proposed in [16] can not be used to capture wide field scene when only a single exposure is allowed.

Table 1. The PSNR values of the proposed system reconstruction results respectively by TV minimization algorithms and l_1 minimization algorithms without noise.

Image	number of channels							
	3 channels		4 channels		5 channels		6 channels	
	TV	l_1	TV	l_1	TV	l_1	TV	l_1
	23.4454	23.8523	23.1368	23.4544	22.5320	22.8514	21.3284	21.5874
	28.8497	29.2802	28.1382	28.3380	27.4899	27.7676	27.0852	27.2909
	26.8259	27.7130	26.6491	27.2535	25.5900	25.9829	25.1507	25.4895
	24.7434	24.9674	24.4952	24.6679	23.8980	24.1946	23.9976	24.1549
	26.5671	26.9317	26.8843	27.0937	27.2770	27.5467	27.7629	27.9140
	35.7702	36.1012	33.8013	33.8654	33.1262	33.2489	32.8210	32.9570
	41.2256	41.3976	39.9155	40.0693	38.9248	38.9994	36.1631	36.2160
	34.3623	34.7880	32.6727	32.8658	31.0686	31.2131	29.5730	29.6774
	28.5038	28.6618	27.2511	27.3012	26.5902	26.6623	26.5591	26.5757
	31.6613	32.2164	30.4549	30.7873	29.1150	29.2886	28.4905	28.7282

3.1 Validation of the Proposed System

In this section, we select a set of pictures with the size of 512×3072. If the imaging system has L ($L <= 6$) channels, we will select parts of these images $(1 : 512, 1 : 512 * L)$ as the wide field scenes for simulations.

The reconstruction results of ten images are reported in Table 1. Form Table 1, we can see that both TV method and l_1 method achieve highly demultiplexing performance. In term of PSNR values, l_1 method performs better than TV method.

In Figs. 3 and 4 we visually show the reconstruction results and magnification of local details of the second image and the third image for 6 channels. Compared with the results of TV method, l_1 method can further restore the internal details of the image, which is visually more pleasant.

To verify whether our reconstruction methods is noise insensitive, we conduct simulation experiments with moderate noise corruption ($\sigma = 0.02$). Figures 5 and 6 visually demonstrate the reconstructed results for 6 channels in the case of noise. Compared with the noise-free case the reconstruction results decrease not only in PSNR values but also in visual effects. Although the results with noise are visually over-smooth, to some extent they are still acceptable and can reconstruct the rough structure of the scene.

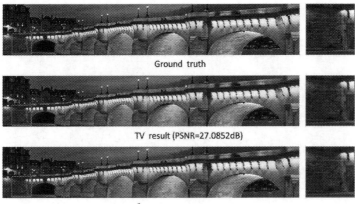

Fig. 3. Demultiplexing performance on the second image for the proposed system without noise.

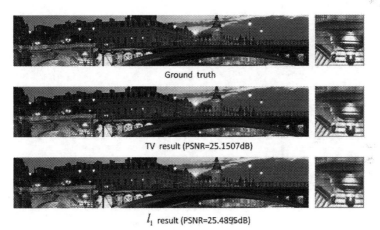

Fig. 4. Demultiplexing performance on the third image for the proposed system without noise.

3.2   Comparison with the Shift-Encoded Multiplexed Imaging System

Originally, the shift-encoded multiplexed imaging system proposed in [16] requires at least L exposures if the FoV scene is divided into L channels. However, it cannot apply for dynamic scenarios. A brute-force solution is to limit the shift-encoded multiplexed imaging system with only one exposure, then reconstruct the underlying FoV scene with the measurements from the one exposure. In such case, the only difference between our proposed system and the single-exposure shift-encoded multiplexed imaging system is the pattern of the masks, where the former follows a binary (i.e., 0 or 1) random distribution, and the latter is all 1.

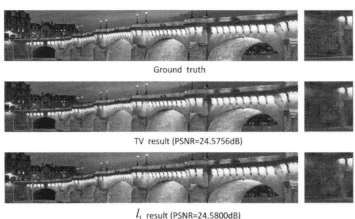

Fig. 5. Demultiplexing performance comparison on the second image for the proposed system with noise ($\sigma = 0.02$).

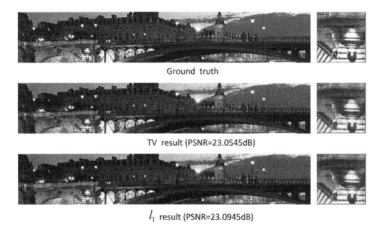

Fig. 6. Demultiplexing performance comparison on the third image for the proposed system with noise ($\sigma = 0.02$).

Table 2 shows the reconstruction results of the shift-encoded multiplexed imaging system in the case of one exposure. The PSNR values are always equal or less than 20. Figure 7 visually shows the TV reconstruction result of the second image, which is almost a tile of the measured image. Both quantitative and qualitative evaluations indicate the shift-encoded multiplexed imaging system is unable to demultiplex the superimposed image with one exposure.

Table 2. The PSNR values of the shift-encoded multiplexed imaging system reconstruction results respectively by TV minimization algorithms and l_1 minimization algorithms without noise.

Image	number of channels							
	3 channels		4 channels		5 channels		6 channels	
	TV	l_1	TV	l_1	TV	l_1	TV	l_1
	17.7497	17.7508	16.9604	16.9608	16.4889	16.4891	15.9056	15.9057
	14.8541	14.8546	14.0636	14.0638	13.6510	13.6511	13.2682	13.2683
	14.3994	14.3998	13.3569	13.3571	13.3421	13.3422	13.5285	13.5286
	12.7610	12.7614	11.9580	11.9581	11.6039	11.6040	11.7748	11.7749
	18.0290	18.0301	17.5745	17.5750	17.4512	17.4515	16.4867	16.4869
	15.4390	15.4393	14.2485	14.2487	14.0055	14.0056	14.0351	14.0352
	19.4046	19.4049	18.3601	18.3602	18.2972	18.2973	17.5195	17.5195
	15.4486	15.4491	15.4094	15.4097	15.2068	15.2070	13.7475	13.7475
	15.1411	15.1419	14.7725	14.7729	13.9344	13.9347	13.4551	13.4552
	17.2135	17.2144	16.4178	16.4182	16.1412	16.1414	15.2325	15.2326

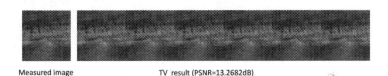

Measured image   TV result (PSNR=13.2682dB)

Fig. 7. The TV reconstruction result of the second image for the shift-encoded multiplexed imaging system without noise.

4   Conclusion

In this paper, we propose a snapshot multiplexed imaging system for wide field imaging by improving an existing multiplexed imaging system. Since demultiplexing is an under-determined problem, we model it as a CS reconstruction problem and solve it based on TV constraint and sparsity constraint respectively. Our simulation demonstrates the feasibility and effectiveness of this multiplexed imaging system. Although reconstruction methods used in this paper can achieve good reconstruction results, maybe, we could further optimize the demultiplexing model for better reconstruction performance.

Acknowledgements. This work was supported by the National Natural Science Foundation of China (Grant Nos. 61425013, 61701025, 61672096).

References

1. Bioucas-Dias, J.M., Figueiredo, M.A.: A new twist: two-step iterative shrinkage/thresholding algorithms for image restoration. IEEE Trans. Image Process. **16**(12), 2992–3004 (2007)
2. Brezis, H., Nguyen, H.M.: Non-local functionals related to the total variation and connections with image processing. Ann. PDE **4**(1), 9 (2018)
3. Candès, E.J., Romberg, J., Tao, T.: Robust uncertainty principles: exact signal reconstruction from highly incomplete frequency information. IEEE Trans. Inf. Theory **52**(2), 489–509 (2006)
4. Candes, E.J., Romberg, J.K., Tao, T.: Stable signal recovery from incomplete and inaccurate measurements. Commun. Pure Appl. Math. **59**(8), 1207–1223 (2006)
5. Chambolle, A.: An algorithm for total variation minimization and applications. J. Math. Imaging Vis. **20**(1), 89–97 (2004)
6. Dong, W., Li, X., Zhang, L., Shi, G.: Sparsity-based image denoising via dictionary learning and structural clustering. In: 2011 IEEE Conference on Computer Vision and Pattern Recognition (CVPR), pp. 457–464. IEEE (2011)
7. Dong, W., Shi, G., Li, X., Ma, Y., Huang, F.: Compressive sensing via nonlocal low-rank regularization. IEEE Trans. Image Process. **23**(8), 3618–3632 (2014)
8. Dong, W., Zhang, L., Shi, G., Li, X.: Nonlocally centralized sparse representation for image restoration. IEEE Trans. Image Process. **22**(4), 1620–1630 (2013)
9. Donoho, D.L.: Compressed sensing. IEEE Trans. Image Process. **52**(4), 1289–1306 (2006)
10. Gehm, M., John, R., Brady, D., Willett, R., Schulz, T.: Single-shot compressive spectral imaging with a dual-disperser architecture. Opt. Express **15**(21), 14013–14027 (2007)
11. Hua, O., Yu, W.: Compressed sensing to short duration voltage variation signal sparse in redundant dictionary. In: Chinese Automation Congress (CAC), pp. 656–660. IEEE (2017)
12. Li, C.: An efficient algorithm for total variation regularization with applications to the single pixel camera and compressive sensing. Ph.D. thesis, Rice University (2010)
13. Lin, X., Liu, Y., Wu, J., Dai, Q.: Spatial-spectral encoded compressive hyperspectral imaging. ACM Trans. Graph. (TOG) **33**(6), 233 (2014)
14. Rauhut, H., Schnass, K., Vandergheynst, P.: Compressed sensing and redundant dictionaries. IEEE Trans. Inf. Theory **54**(5), 2210–2219 (2008)
15. Ren, C., He, X., Nguyen, T.Q.: Single image super-resolution via adaptive high-dimensional non-local total variation and adaptive geometric feature. IEEE Trans. Image Process. **26**(1), 90–106 (2017)
16. Shah, V., Rachlin, Y., Shepard, R.H., Shih, T.: Shift-encoded optically multiplexed imaging. Opt. Eng. **56**(4), 041314 (2017)
17. Shepard, R.H., Rachlin, Y., Shah, V., Shih, T.: Design architectures for optically multiplexed imaging. Opt. Express **23**(24), 31419–31435 (2015)
18. Treeaporn, V., Ashok, A., Neifeld, M.A.: Increased field of view through optical multiplexing. Opt. Express **18**(21), 22432–22445 (2010)
19. Uttam, S., et al.: Optically multiplexed imaging with superposition space tracking. Opt. Express **17**(3), 1691–1713 (2009)

20. Wagadarikar, A., John, R., Willett, R., Brady, D.: Single disperser design for coded aperture snapshot spectral imaging. Appl. Opt. **47**(10), B44–B51 (2008)
21. Rachlin, Y., Shah, V., Hamilton Shepard, R., Shih, T.: Dynamic optically multiplexed imaging (2015). https://doi.org/10.1117/12.2188756
22. Zhang, Z., Xu, Y., Yang, J., Li, X., Zhang, D.: A survey of sparse representation: algorithms and applications. IEEE Access **3**, 490–530 (2015)

Partially Annotated Gastric Pathological Image Classification

Yanping Cui[1], Zhangcheng Wang[1], Guanzhen Yu[2(✉)], and Xinmei Tian[1(✉)]

[1] CAS Key Laboratory of Technology in Geo-Spatial Information Processing
and Application Systems, University of Science and Technology of China,
Hefei, China
{cuiyp,wzc1}@mail.ustc.edu.cn, xinmei@ustc.edu.cn
[2] Department of oncology, Longhua Hospital affiliated to Shanghai University
of Traditional Chinese Medicine, Shanghai 201203, China
qiaoshanqian@aliyun.com

Abstract. Previous works mainly address the medical datasets with image-wise labels or pixel-wise labels. However, it is difficult to train a model with only image-wise labels, and pixel-wise labels commonly refer to the high expense of annotations. A feasible solution is to make a compromise between data annotation and the performance. In this paper, we propose a cascaded convolutional neural network framework to classify partially annotated pathological images. A segmentation model is trained with the partially annotated samples to detect cancer regions, which are re-identified by a patch-wise classification network. Finally, the segmentation and classification results are combined to make the final image-wise classification. Several experiments are conducted on a landmark medical image dataset with partial annotations. We obtain a classification accuracy of 99.51%, which significantly outperforms other existing methods.

Keywords: Image classification · Pathological images · Convolutional neural network · Partial annotation

1 Introduction

Gastric cancer is one of the leading causes of cancer death worldwide [18]. Traditionally, pathologists must traverse through the entire pathological image to find lesions, but this process is time-consuming and fallible. Thus, computer-aided diagnostic systems are urgently required to reduce the burden of pathologists and improve the accuracy of diagnoses.

In recent years, deep learning has achieved remarkable success in the field of natural images [2,6,10,12], and it usually has a better performance than some traditional methods [7–9] in computer vision tasks. Considering the success of deep learning, many researchers, such as [1,4,5,16], attempted to apply it to the field of medical images. The performances of these applications largely rely

© Springer Nature Switzerland AG 2018
R. Hong et al. (Eds.): PCM 2018, LNCS 11165, pp. 476–486, 2018.
https://doi.org/10.1007/978-3-030-00767-6_44

on accurately annotated samples. However, it is often difficult to obtain a large amount of annotated samples in the field of medical images because of the high cost of accurate annotations.

To address these problems, a feasible solution is to make a compromise between data annotation and the performance. Thus, recent works proposed to use partially annotated samples in the medical image segmentation task [13]. The partially annotated samples only require the pathologists to annotate a part of the lesion areas in a medical image, which significantly reduces the cost of annotations and provides more detailed supervision information than the image-wise labels. However, in [13], they just iteratively generated new positive samples with high confidence to fine-tune the segmentation model. They ignored the rich information that spurious samples can provide which hardly affects the segmentation but greatly affects the classification. Therefore, we propose a new cascaded framework to classify partially annotated gastric pathological images in this paper. A segmentation model is trained with the partially annotated samples to detect cancer regions, which will be re-identified by a patch-wise classification network. Finally, the results of the segmentation network and classification network are combined to make the final image-wise classification. The major contributions of our work are as follows:

(1) We propose a new cascaded convolutional neural network framework to address the medical image classification problem with partial annotations; this framework is more efficient and practical than those of previous works;
(2) We propose a new training strategy to train the segmentation model by alternately using two complementary methods to sample the patches and considerably improve the classification performance;
(3) We further develop a medical image classification problem based on the segmentation problem by selecting patches that are difficult to distinguish by a segmentation model; then, we train the patch-wise classification model with these fallible patches. Thus, we combine the results of the segmentation model with the supervision information of the partial annotations to improve the performance in the image-wise classification task.

The remainder of our paper is organized as follows. Section 2 provides an introduction to closely related works. We provide the details of our proposed method in Sect. 3. Various experiments are conducted on a benchmark medical image dataset with partial annotations, and the results are reported in Sect. 4 which demonstrate the effectiveness of our method. In Sect. 5, we conclude our work and discuss the future works.

2   Related Work

Currently, the approaches of medical image classification can be divided into two main categories according to the datasets that they use: the approaches of image-wise labeled datasets and the approaches of pixel-wise labeled datasets. In the first category, the approaches often cannot obtain a good model because of limited supervision information [14,15]. In the second category, they first generated patches according to the pixel-wise labels, which are used to train a classifier [17]. The classifier is further modified to handle the input of entire images and provide an image-wise classification result. Nevertheless, this method requires sufficient supervision information of pixel-wise annotations, which is of high expense. To address these problems, partially annotated data have also been used in medical image segmentation, where new positive samples are iteratively generated with high confidence to fine-tune the segmentation network, which won the 2017 China Big Data & Artificial Intelligence Innovation and Entrepreneurship Competitions[1]. These new positive samples can provide rich information, which are very helpful for segmentation. However, some spurious samples will be inevitably introduced in the process of generating new positive samples, which hardly affects the segmentation but greatly affects the classification. Therefore, we propose to integrate the information of partial annotations into the generating process of patches. Moreover, we propose to make use of the spurious samples in training our classification models. In the following section, we present a detailed introduction to the proposed approach.

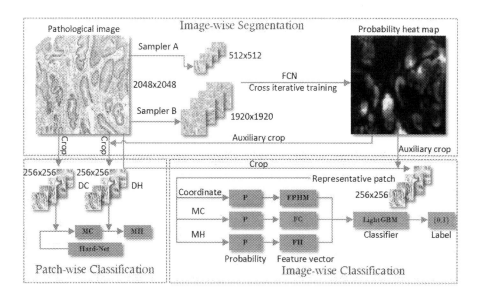

Fig. 1. Framework of the proposed approach. First, a segmentation model is trained with the partially annotated samples to detect cancer regions, which will be re-identified by a patch-wise classification network. Finally, we generate features from the segmentation and classification results and train a classifier to make the final decision.

---

[1] The challenge is held by Shanghai Big Data Alliance and Center for Applied Information Communication Technology (CAICT), the home page is http://www.datadreams.org/racerace3.html.

3   Proposed Approach

In this section, we describe the proposed method in details in three steps. The overall framework of our method is illustrated in Fig. 1. First, a segmentation network is trained with the partially annotated samples to obtain a heat map of the entire image, which provides the foundation for subsequent operations. We propose a new training strategy, which simultaneously considers two complementary sampling methods. Second, we develop a medical image classification problem based on the segmentation problem by generating the fallible patches from the segmentation heat map to train the patch-wise classification network. Finally, we present the image-wise classification process, which includes extracting features and training a classifier to obtain the final image-wise classification.

3.1   Image-Wise Segmentation

The main objective of the segmentation network is to make a pre-estimation of a given image. We use the existing segmentation network and propose a new training strategy to improve its performance. There are many classic segmentation networks such as [2,12,16]. In this paper, we use the FCN based on the pre-trained VGG16 model [19] as our segmentation network (**Seg-Net**) because pre-trained models can accelerate the convergence process and do not have stringent requirements for the segmentation details.

The cancer area of the original image is partially annotated, and the unannotated area is considered as a negative sample by default, which results in a large quantity of false-negative samples. We can address these obstacles by controlling the proportion of positive and negative samples. A smaller proportion of negative samples corresponds to a smaller risk of introducing false-negative samples. Meanwhile, adequate negative samples are required to train the Seg-Net. Considering these two factors, we propose two types of sampling methods.

For the first method (Sampler A), we sample the patches from the original images with the area ratio greater than 0.5. The area ratio P is defined as follows:

$$P = \frac{R}{R_{max}} \tag{1}$$

where R is the annotated cancer area in one patch, and R_{max} is the maximum among all R that we can obtain from the current image. Here, we use the area ratio to filter the patches instead of a fixed proportion of positive and negative samples as used in the previous work [13]. The reason is that the patches are unlikely sampled from those images with a small annotated cancer area with a fixed proportion of positive and negative samples, which are notably important to maintain the diversity of the samples. Even so, some positive samples of the annotated cancer area remain unused to train the segmentation network, which degrades the performance.

To address this problem, we propose the second sampling method (Sampler B) as a complement to the first method. Given an image from the partially

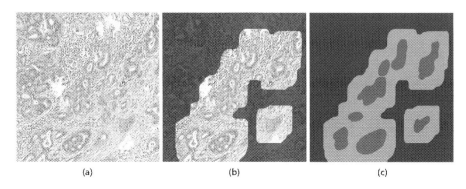

(a)   (b)   (c)

Fig. 2. An example generated by Sampler B. In the original image (a), the area circled by the green line is the cancer region, but there are some cancer regions in the remainder of the image. The highlighted area in image (b) is the valid area that contributes to the loss of the segmentation network. In image (c), the pixels in the red region are positive samples, and the pixels in the green region are negative samples. (Color figure online)

annotated dataset, the annotated cancer area is circled by the green line in the original image (Fig. 2(a)), which corresponds to the red area in Fig. 2(c); we denote this annotated area as R_{red}. However, some cancer areas remain unlabeled in the remainder of the image. The unannotated area near the annotated cancer area is less likely to contain false-negative samples. Considering these factors, we define the unannotated area near the annotated cancer areas as the security zone as follows:

$$|p_{sec} - p_{red}| < k \qquad (2)$$

where p_{red} are the pixels in R_{red}, p_{sec} are the pixels in the security zone, and k is a parameter to control the proportion of negative samples. The pixels in the security zone are considered negative samples, and k is determined to satisfy the following:

$$\frac{1}{3} < \frac{R_{red}}{R_{green} + R_{red}} < \frac{2}{3} \qquad (3)$$

where R_{green} is the area of the security zone (green region in Fig. 2(c)). The original image is further used as the input to train the segmentation network. We dismiss the pixels in R_{blue} when calculating the loss. We use a maximum filter with a kernel size $k \times k$ to generate an approximate security zone in implementation.

We propose a cross iterative training strategy that alternately uses two types of sampling methods in the training process. Thus, we make full use of the annotation information in the training process and can obtain a better performance.

3.2   Patch-Wise Classification

In this section, we introduce how to further develop a medical image classification problem based on the segmentation problem. First, we generate patches from the original images to train the patch-wise classification network (**Hard-Net**). Then, we use the pre-estimation from Seg-Net to generate fallible patches to fine-tune the Hard-Net. Hence, we make an amendment about the samples indistinguishable by the Seg-Net.

First, we introduce the method to train the Hard-Net. Since the image is partially labeled, we only sample 256×256 patches whose centroids are located in the annotated cancer areas of image as positive samples. For negative samples, we randomly sample patches from non-cancer images. We do not sample patches from the unannotated areas of the cancer images because some cancer regions may remain unlabeled in these areas, which results in a large number of false-negative samples. Meanwhile, pathologists only label typical cancer nests, which results in potential false-positive samples. To address this problem, those patches are excluded, where the number of cells is lower than a threshold N_{cell} because more cells in the patch corresponds to a higher confidence that the patch is a positive sample. However, we cannot accurately calculate the number of cells in a patch. We approximate the number of cells using the area of cells. Since the cell densities vary with different images, it is difficult to find a fixed N_{cell} for all images. To overcome this obstacle, we simultaneously sample multiple patches from one image and only select the patch with the largest area of cells as the positive sample. Finally, we denote those patches as (**DC**). We select resnet50 [6] as our base model and obtain the Hard-Net by modifying the dimension of the output layer from 1000 to 2. We pre-train the Hard-Net on the DC and denote this model as **MC**. Then, we generate the patches that are indistinguishable by the segmentation network to fine-tune MC. Generally, the maximum value P_{max} of the probability heat map from the Seg-Net can represent the probability of cancer. However, we find a strange phenomenon that P_{max} is relatively small for some cancer images but relatively large for some non-cancer images, which causes difficulty in the final classification and urgently requires re-identification.

We first select non-cancer images with P_{max} greater than 0.20 and cancer images with P_{max} less than 0.70. Then, we sample the patches from these selected images. We sample the patches from non-cancer images according to the probability heat map in descending order as negative samples. Specifically, we sample the patches from the annotated cancer area according to the probability heat map in an ascending order as positive samples. However, it does not work well. We analyze this phenomenon and attribute it to the impurity of the images; the positive samples from the annotated cancer area with a relatively low P_{max} are likely the impurity part of the cancer nest. Therefore, we also sample positive samples based on the probability heat map in descending order. We denote those patches as **DH**, fine-tune the Hard-Net on DH and denote this model as **MH**.

3.3 Image-Wise Classification

In this section, we introduce the method to extract features from the high-confidence patches and the training details of the classifier to implement the final classification.

First, we sample N patches from the original images based on the probability heat map in descending order as representative patches with the constraint that the overlap between the patches cannot exceed 50%. Then, we obtain N probability values through MH and construct the following features:

$$f_m = \sum_{n=1}^{N} Sign\left(n, \frac{m}{M+1}\right) \quad 1 \leqslant m \leqslant M \tag{4}$$

$$Sign(n,t) = \begin{cases} 1 & p_n \geq t \\ 0 & p_n < t \end{cases} \tag{5}$$

where f_m is the m-th element in the feature vector, p_n is the n-th probability value, and we can obtain a feature vector with length M. M is set to 16 in our experiments. For a specific patch, we can obtain three probability values P_1, P_2 and P_3 from **Seg-Net**, **MC** and **MH**, respectively. More specifically, P_1 is a probability of the heat map, where we sample the representative patch. Therefore, we can obtain three feature vectors, **FPHM**, **FC** and **FH**, which correspond to three probability values respectively. Finally, we merge the three feature vectors into one feature vector with a length of 48, which is further applied to train the image-wise classifier by LightGBM [11].

4   Experiments

In this section, we conduct various experiments to verify the effectiveness of our proposed method. All experiments are implemented on the Keras framework [3]. We first briefly introduce the partially annotated dataset in the experiments. Then, we analyze the effect of the proposed training strategy, different feature combinations and number of patches. Finally, we compare our proposed method with previous works.

4.1   Gastric Tumor Dataset

The Gastric Tumor Dataset was provided by 2017 China Big Data & Artificial Intelligence Innovation and Entrepreneurship, which contains 2100 images; 80% of the dataset is cancer images, whose cancer area is partially labeled. The images are available as 2048 × 2048 pixel TIFF images. We divide our dataset into training, validation and test sets with a split ratio of 3:1:3. To avoid randomness, we repeat the random splits for 4 times and report the average performance in our experiments. All parameters are selected on the validation set.

4.2   Effect of the Number of Representative Patches and Proposed Training Strategy

The number of representative patches that we use to extract features is a key parameter in our experiments. As shown in Fig. 3(a), a moderate accuracy is

Table 1. Comparison of different feature combinations in terms of mean accuracy ± standard deviation (%).

Methods	Accuracy	Dimension
FPHM	95.44 ± 0.56	16
FC	97.36 ± 0.39	16
FH	99.48 ± 0.22	16
FPHM + FC	97.41 ± 0.44	32
FPHM + FH	99.48 ± 0.21	32
FC + FH	99.50 ± 0.21	32
FPHM + FC + FH	**99.51 ± 0.24**	48

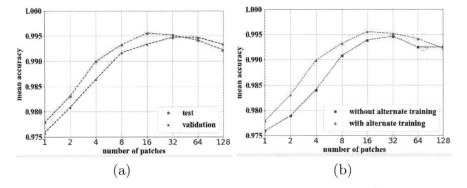

(a)   (b)

Fig. 3. The horizontal axis shows the number of representative patches sampled from one image, and the vertical axis shows the corresponding mean accuracy. In (a), the red (blue) line is the accuracy curve of the test (validation) data. In (b), the blue line is the accuracy curve without alternate training (only with Sampler A), and the red line stands for the accuracy curve with proposed alternate training (Samplers A and B). (Color figure online)

obtained with 32 representative patches; then, the accuracy decreases. The reason is that when there are too few patches, the collected information is insufficient, and when we select too many patches, the information is submerged by noise. A tradeoff must be made between the information and noise. From the results, we select 32 as the number of representative patches in our experiments according to the accuracy curve of the validation data.

We propose a cross iterative training strategy that alternately trains our segmentation model with two types of sampling methods. Fig. 3(b) shows us a comparison between different training strategies. From the result, we can easily conclude that the proposed training strategy improves the classification performance in most cases, which shows the effectiveness of the proposed training strategy.

4.3   Effect of Different Feature Combinations

This experiment analyzes the effect of different feature combinations. As mentioned, for a specific image, we can obtain three feature vectors: **FPHM**, **FC** and **FH**. Table 1 shows us the comparison between different combinations of these three feature vectors. We can conclude that FH outperforms the other two single features which is in consistent with our target. FH outperforms FPHM, which demonstrates that the re-identification of candidate cancer areas can help us improve the classification of the entire images. FH outperforms FC, which demonstrates the effectiveness of introducing fallible patches into training the patch-wise classification model. The fusion of features has a better performance, demonstrates that information from different perspectives can help us obtain a better performance of the final image-wise classification.

Table 2. Quantitative results of our proposed method and other recently published deep-learning-based methods in terms of mean accuracy ± standard deviation (%).

Methods	Accuracy	Precision	Recall	F1-score
WS [15]	95.22 ± 0.34	95.28 ± 0.46	98.70 ± 0.29	96.96 ± 0.25
End2End [17]	94.26 ± 1.02	95.69 ± 2.32	97.14 ± 1.34	94.38 ± 0.70
P-SEG [13]	97.89 ± 0.98	98.89 ± 0.83	98.49 ± 0.77	98.69 ± 0.61
Ours	$\mathbf{99.51} \pm 0.24$	$\mathbf{99.62} \pm 0.29$	$\mathbf{99.76} \pm 0.21$	$\mathbf{99.69} \pm 0.16$

4.4   Comparison with Existing Works

In this section, we compare our proposed method with three state-of-the-art methods (**WS** [15], **P-SEG** [13] and **End2End** [17]) to demonstrate the effectiveness of our method. The results are shown in Table 2. Our proposed method outperforms other methods in terms of mean accuracy, precision, recall, and F1-score. WS works with medical datasets with image-wise labels, and it does not make full use of the supervision information. End2End is applied to datasets with pixel-wise labels, and its performance largely relies on the quantity of pixel-wise labeled training samples. For the partially annotated datasets, they will suffer from the lack of labeled samples, which tremendously degrades the performance. P-SEG handles the segmentation problem of partially annotated datasets and iteratively generates new positive samples with high confidence to fine-tune the patch-wise classification model. However, some spurious samples may be introduced in the process of generating new positive samples. In our proposed method, we develop a medical image classification problem based on the segmentation problem by selecting patches that are indistinguishable by the segmentation model; then, we fine-tune the classification model with these fallible patches. Thus, we make full use of the supervision information of the partial annotations and obtain an outstanding performance with our method.

5   Conclusions

In this paper, we propose a cascaded convolutional neural network framework to classify partially annotated images. The proposed method achieves an outstanding performance on partially annotated medical image datasets. The proposed approach can prominently reduce the limitations of histopathology research, which will further promote the research of biomedical histopathology classification problems.

Acknowledgments. This work was supported by National Key Research and Development Program of China 2017YFB1002203, NSFC No. 61572451, and No. 61390514, Fok Ying Tung Education Foundation WF2100060004 and Youth Innovation Promotion Association CAS CX2100060016.

References

1. Chen, H., Qi, X., Yu, L., Heng, P.A.: DCAN: deep contour-aware networks for accurate gland segmentation. In: Proceedings of the IEEE Conference on Computer Vision and Pattern Recognition, pp. 2487–2496 (2016)
2. Chen, L.C., Papandreou, G., Kokkinos, I., Murphy, K., Yuille, A.L.: DeepLab: Semantic image segmentation with deep convolutional nets, atrous convolution, and fully connected CRFs. IEEE Trans. Pattern Anal. Mach. Intell. **40**(4), 834–848 (2018)
3. Chollet, F.: Keras. https://github.com/fchollet/keras (2015)
4. Cireşan, D.C., Giusti, A., Gambardella, L.M., Schmidhuber, J.: Mitosis detection in breast cancer histology images with deep neural networks. In: Mori, K., Sakuma, I., Sato, Y., Barillot, C., Navab, N. (eds.) MICCAI 2013. LNCS, vol. 8150, pp. 411–418. Springer, Heidelberg (2013). https://doi.org/10.1007/978-3-642-40763-5_51
5. Dhungel, N., Carneiro, G., Bradley, A.P.: The automated learning of deep features for breast mass classification from mammograms. In: Ourselin, S., Joskowicz, L., Sabuncu, M.R., Unal, G., Wells, W. (eds.) MICCAI 2016. LNCS, vol. 9901, pp. 106–114. Springer, Cham (2016). https://doi.org/10.1007/978-3-319-46723-8_13
6. He, K., Zhang, X., Ren, S., Sun, J.: Deep residual learning for image recognition. In: Proceedings of the IEEE conference on computer vision and pattern recognition, pp. 770–778 (2016)
7. Hong, R., Hu, Z., Wang, R., Wang, M., Tao, D.: Multi-view object retrieval via multi-scale topic models. IEEE Trans. Image Process. **25**(12), 5814–5827 (2016)
8. Hong, R., Zhang, L., Tao, D.: Unified photo enhancement by discovering aesthetic communities from flickr. IEEE Trans. Image Process. **25**(3), 1124–1135 (2016)
9. Hong, R., Zhang, L., Zhang, C., Zimmermann, R.: Flickr circles: aesthetic tendency discovery by multi-view regularized topic modeling. IEEE Trans. Multimed. **18**(8), 1555–1567 (2016)
10. Huang, G., Liu, Z., Weinberger, K.Q., van der Maaten, L.: Densely connected convolutional networks. In: Proceedings of the IEEE Conference on Computer Vision and Pattern Recognition, vol. 1, p. 3 (2017)
11. Ke, G., et al.: LightGBM: a highly efficient gradient boosting decision tree. In: Advances in Neural Information Processing Systems, pp. 3149–3157 (2017)

12. Long, J., Shelhamer, E., Darrell, T.: Fully convolutional networks for semantic segmentation. In: Proceedings of the IEEE Conference on Computer Vision and Pattern Recognition, pp. 3431–3440 (2015)
13. Nan, Y., et al.: Partial labeled gastric tumor segmentation via patch-based reiterative learning. arXiv preprint arXiv:1712.07488 (2017)
14. Paeng, K., Hwang, S., Park, S., Kim, M.: A unified framework for tumor proliferation score prediction in breast histopathology. In: Cardoso, M.J., et al. (eds.) DLMIA/ML-CDS -2017. LNCS, vol. 10553, pp. 231–239. Springer, Cham (2017). https://doi.org/10.1007/978-3-319-67558-9_27
15. Rakhlin, A., Shvets, A., Iglovikov, V., Kalinin, A.A.: Deep convolutional neural networks for breast cancer histology image analysis. arXiv preprint arXiv:1802.00752 (2018)
16. Ronneberger, O., Fischer, P., Brox, T.: U-Net: convolutional networks for biomedical image segmentation. In: Navab, N., Hornegger, J., Wells, W.M., Frangi, A.F. (eds.) MICCAI 2015. LNCS, vol. 9351, pp. 234–241. Springer, Cham (2015). https://doi.org/10.1007/978-3-319-24574-4_28
17. Shen, L.: End-to-end training for whole image breast cancer diagnosis using an all convolutional design. In: Neural Information Processing Systems 2017 Workshop on Machine Learning for Health (2017)
18. Siegel, R.L., Miller, K.D., Jemal, A.: Cancer statistics. CA: Cancer J. Clin. **66**(1), 7–30 (2016)
19. Simonyan, K., Zisserman, A.: Very deep convolutional networks for large-scale image recognition. arXiv preprint arXiv:1409.1556 (2014)

Facial Expression Recognition Based on Local Double Binary Mapped Pattern

Chunjian Yang[1,2], Min Hu[1,2(✉)], Yaqin Zheng[1,2], Xiaohua Wang[1,2], Yong Gao[1,2], and Hao Wu[1,2]

[1] School of Computer and Information of Hefei University of Technology,
Hefei, China
jsjxhumin@hfut.edu.cn
[2] Anhui Province Key Laboratory of Affective Computing and Advanced
Intelligent Machine, Hefei 230009, China

Abstract. Local feature descriptors play an important role in facial expression recognition. Local Binary Pattern (LBP) only considers the signal information of the difference between the gray value of the center pixel and the neighbor pixel. It does not take the magnitude information into consideration and has poor robustness. Local Mapped Pattern (LMP) is not ideal for discriminating differences between different textures and experimental result is not excellent. This paper proposes a novel feature descriptor based on gray-level difference mapping, called Local Double Binary Mapped Pattern (LDBMP).This new approach is an improvement over the previous LBP and LMP, not only retains the advantages of LBP and LMP but also preserves the information of magnitude and captures nuances that occur in the image. In our experiments, the new descriptor performs favorably.

Keywords: Local double binary mapped pattern · Feature fusion
Facial expression recognition

1 Introduction

With the rapid development of such areas as emotional computing, facial expression recognition as a branch of affective computing has also been widely used in many aspects. The wide application prospects of facial expression recognition make it a research hotspot for many scholars and have made huge progress in recent years [1].

The main steps of facial expression recognition include three parts: the acquisition and preprocessing of face images, feature extraction of images, and classification recognition. Among them, the feature extraction of face images is a key step to measure the quality of the final recognition [2]. Different features also have their advantages and disadvantages [3]. Currently, feature extraction methods with high frequency of use mainly include shape features based on face deformation, texture features based on pixel information, and multi-feature fusion methods. Among them, the extraction methods based on shape features mainly include active shape model (ASM) [4] and active appearance model (AAM) [5]. Extraction methods based on texture features include: local binary pattern (LBP) [6–8] and a series of improved methods, Gabor

© Springer Nature Switzerland AG 2018
R. Hong et al. (Eds.): PCM 2018, LNCS 11165, pp. 487–497, 2018.
https://doi.org/10.1007/978-3-030-00767-6_45

wavelet, HOG descriptor and so on. In order to make full use of the advantages of various methods, researchers began to fuse the extracted features and experiments in order to demonstrate they can also achieved good results [9–11].

The feature extraction algorithm used in this paper is based on texture features. The LBP operator proposed by Ojala et al. [6, 13] has gray invariance and rotation invariance, and it can be operated simply and easily. However, LBP only considers the signal information of the difference between the center pixel and the gray value of the neighboring pixels. It ignores the magnitude information of the original image and it is not robust. After that, a large number of feature extraction operators which have been improved based on the LBP also performed well. CLBP [12] not only extracts the signal feature of difference but also extracts the magnitude information and center pixel grayscale feature, but signal feature contains the most of the texture information and it has a large number of feature vectors and high computational complexity. Ferraz et al. [14] proposed a local mapped pattern (LMP) for image classification, which can classify different rotated and scaled images according to textures and obtain the difference information between pixels effectively. Through LMP, it can make the extracted texture features more robust but the discriminated differences between different textures is not apparently. Aiming at the problem of how to express the face texture effectively and identify it under the condition of low computational complexity, this paper proposes a face texture description named Local Double Binary Mapped Pattern (LDBMP) and its corresponding recognition method based on gray-level difference mapping.

The rest of this paper is organized as followed. In Sect. 2, we briefly describe the methods of LBP and LMP. Section 3 give details about the proposed approach LDBMP. The experimental evaluation and the results are presented in Sect. 4. In the end, we make a conclusion of this paper in Sect. 5.

2 Related Work

2.1 Local Binary Pattern (LBP)

LBP [6] is a simple but very effective texture descriptor, which is encoded by comparing the gray values of the center pixel and the neighboring pixels. The specific encoding method is shown in formula (1).

$$\text{LBP} = \text{round}\left(\frac{\sum_{k=1}^{w}\sum_{l=1}^{1}\left(f_{g(i,j)}p(k,l)\right)}{\sum_{k=1}^{w}\sum_{l=1}^{w}p(k,l)}\right)(B-1) \tag{1}$$

Through formula (1), each pattern defined by a w * w neighborhood. In this paper, we select 3 * 3 rectangular blocks and round () indicates that the result of calculation is rounded off. $p(k,l)$ is a weighting matrix of predefined values for each pixel position within the neighborhood, and B is the number of the histogram bins. $g(k,l)$ represent the gray value of the neighborhood pixel, $g(i,j)$ represent the gray value of the center pixel. $f_{g(i,j)}$ is a mapping function, which is shown in formula (2).

$$f_{g(i,j)} = H[g(k,l) - g(i,j)] \tag{2}$$

$$\text{Where} \quad H[g(k,l) - g(i,j)] = \begin{cases} 0, & g(k,l) - g(i,j) < 0; \\ 1, & g(k,l) - g(i,j) \geq 0. \end{cases} \tag{3}$$

$$\text{Weighting matrix: } p(k,l) = \begin{bmatrix} 1 & 2 & 4 \\ 128 & 0 & 8 \\ 64 & 32 & 16 \end{bmatrix}.$$

2.2   Local Mapped Pattern (LMP)

The mapping function of LBP is a discrete binary function. It only roughly compares the relation of magnitude between the gray value of the center pixel and the neighboring pixel, but ignores the specific difference information between the center pixel and the neighboring pixel. In LMP [14], the mapping function is a continuous function whose range is the (0, 1) interval. Selecting a continuous function can make the difference between the neighboring pixel and the central pixel be mapped to different values. The threshold is no longer either 0 or 1. That is, the proposed LMP operator can capture the nuances that occur in the image and it can preserve more texture information. The most commonly used mapping function is the sigmoid function. The specific encoding method is shown in formula (4).

$$LMP = \text{round}\left(\frac{\sum_{p=0}^{p-1} f_g(p)}{p}(B-1)\right) \tag{4}$$

In formula (4), P denotes the number of neighboring pixel points. The experimental value of this paper is 8. B is the number of the histogram bins. $f_g(p)$ is a mapping function, which is shown in formula (5).

$$f_g(p) = \frac{1}{1 + e^{-\frac{(g_c - g_p)}{\beta}}} \tag{5}$$

Here, g_c is the gray value of the center pixel, g_p is the gray value of the neighborhood point (p = 0, 1, 2,...7), β is the curve slope of the mapping function.

3   Local Double Binary Mapped Pattern (LDBMP)

Inspired by CLBP [12], this paper proposes a local double binary mapped Pattern LDBMP. CLBP decompose the local difference into sign(S) and magnitude (M). Moreover, to take advantage of global texture feature information, the difference between the central pixel and the mean value of the entire image's intensity is calculated by another operator named CLBP_C. So, CLBP contains three operators: CLBP_S, CLBP_M, CLBP_C. Similar to CLBP, the proposed LDBMP also can be decomposed into two operators: LDBMP_B, LDBMP_M, which are improvements over LBP and LMP respectively. Finally, these two codes are combined to form the final LDBMP.

The LDBMP_B consists of two operators: LBP_S and LBP_M. LBP_S compares difference the gray value of between the neighborhood pixel and the center pixel. The former is not greater than the latter then the value is 0, otherwise is 1. This step preserves the signal information of the LBP. LBP_M aim to preserves the LBP operator's magnitude information, it retains the absolute value of the difference between the gray value of neighborhood pixel and the center pixel, that is the magnitude's value, and then compares the obtained magnitude's value with the center pixel value. The process of operation is similar to LBP and then will get LBP_M code eventually. After generating coding map of LBP_S and LBP_M, we obtain the LDBMP_B by combining these two operators with weighting. Three LDBMP_B, LBP_S, LBP_M operators are defined in the formula (6), (7) and (8) respectively.

$$LDBMP\_B = \lambda LBP\_S + (1 - \lambda)LBP\_M \tag{6}$$

$$LBP\_S = \left(\frac{\sum_{k=1}^{w} \sum_{l=1}^{1} (f_{b1}(p)p(k,l))}{\sum_{k=1}^{w} \sum_{l=1}^{w} p(k,l)} \right) (B - 1) \tag{7}$$

$$LBP\_M = \left(\frac{\sum_{k=1}^{w} \sum_{l=1}^{1} (f_{b2}(p)p(k,l))}{\sum_{k=1}^{w} \sum_{l=1}^{w} p(k,l)} \right) (B - 1) \tag{8}$$

Where weighting matrix: $p(k, l) = \begin{bmatrix} 1 & 2 & 4 \\ 128 & 0 & 8 \\ 64 & 32 & 16 \end{bmatrix}$, B is the number of the histogram bins. λ is the term of trade-off between LBP_S and LBP_M. $f_{b1}(p)$ and $f_{b2}(p)$ are two different mapping function, g_c is the gray value of the center pixel, g_p is the gray value of the neighborhood point (p = 0, 1, 2,...7), β is the curve slope of the mapping function. $f_{b1}(p)$ and $f_{b2}(p)$ are shown in formula (9) and (10).

$$f_{b1}(p) = \begin{cases} 0, & [(g_p - g_c) < 0]; \\ 1, & [(g_p - g_c) \geq 0]. \end{cases} \tag{9}$$

$$f_{b2}(p) = \begin{cases} 0, & [(M - g_c) < 0]; \\ 1, & [(M - g_c) \geq 0]. \end{cases} \tag{10}$$

Where $M = |g_p - g_c|$. In this experiment, λ is 0.95.

The LDBMP_M operator is proposed based on the LMP. The formula is (11):

$$LDBMP\_M = \frac{\sum_{k=1}^{w} \sum_{l=1}^{1} (f_g(p)p(k,l))}{\sum_{k=1}^{w} \sum_{l=1}^{w} p(k,l)} (B - 1) \tag{11}$$

Where, $f_g(p)$ is mapping function, and $f_g(p) = \dfrac{1}{1 + e^{\frac{-(gc-gp)}{\beta}}}$. β is the curve slope of the mapping function. B is the number of the histogram bins. We set it is 0.1. $P(k,l)$ is weighting matrix, and $p(k,1) = \begin{bmatrix} 1 & 2 & 4 \\ 128 & 0 & 8 \\ 64 & 32 & 16 \end{bmatrix}$.

Compared with previous LMP, the LDBMP_M operator fixed the code on the [0, 255] interval by adding a weighting matrix, which can enlarge the difference of the gray value between central pixel and neighboring pixel. The comparison results are shown clearly in Fig. 1.

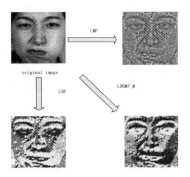

Fig. 1. Comparison results of three different descriptors under same image

As can be seen from Fig. 1, LDBMP_M operator extract the most of texture information and express the difference more clearly. So it is conducive to classify the category of facial expression.

After two LDBMP_B and LDBMP_M operators were used to extract local texture features from the facial expression grayscales, the above two features were combined with weighted to form the final LDBMP feature, which is presented in formula (12).

$$LDBMP = round(tLDBMP\_B + (1 - t)LDBMP\_M) \qquad (12)$$

Where t is term of trade-off between LDBMP_B and LDBMP_M.

4 Experiments

In this section, we evaluated our proposed method's performance by using two public facial expression databases and compared with other methods. All experiments were implemented by using Visual Studio 2013 and OpenCV 2.4.9.

4.1   Datasets Construction

In order to verify the effectiveness of our approach, we conducted experiments on the JAFFE and Cohn-Kanade (CK) databases respectively. The JAFFE database consists of 213 Japanese women's face images. It is recorded by 10 people for seven expressions: angry, disgust, fear, scare, happy, neutral, surprise. The ratio of training set and test set taken in this database is 2:1. Cohn-Kanade database consists of 2,105 images of 182 subjects in the age range of 18–30 years for six expressions: angry, disgust, fear, scare, happy, surprise. The ratio of training set and test set taken in this database is 1:1. In order to ensure the reliability of the experimental results, this paper use three cross-validations and the final results were averaged.

4.2   Experimental Procedure

(1) **Preprocess.** The image preprocessing process in this paper can be mainly divided into three parts: (a) Using Haar-like and Adaboost [15] to detect the position of the human eye in the image, using the coordinates of the two eyes to transform the face geometrically to decrease the impact of posture. (b) Detect and crop the face areas of interest, normalized to 96×96. (c) In order to diminish the influence of noise and improve the accuracy of classification, Gaussian filter processing is performed on the face image.

(2) **Feature extraction and combination.** In this paper, we use LDBMP_B and LDBMP_M to extract the local texture features of images and then two features were combined with weighted to form the final LDBMP feature.

(3) **Classification.** Using SVM [16] to classify features. Polynomial kernel function (POLY) is chosen as the SVM kernel function and training function automatically in order to determine optimal parameters.

4.3   Experimental Results and Analysis

In the process of feature extraction, the number of image blocks will affect the subsequent descriptors' performance. Too few blocks will lead to insufficient extraction of facial expression features in the local area. However, too many blocks will increase feature vectors' dimensions and time complexity. Therefore, the experiment of selecting the number of blocks is presented in this paper. The experimental results on the JAFFE and CK are shown in Fig. 2.

As can be seen from Fig. 2, for the CK, when block is 6, the corresponding average accuracy reaches a maximum of 99.33% then starts to decrease. For the JAFFE, the average accuracy is increasing along with the number of block, and when block is 8, the corresponding average accuracy reaches a maximum of 99.52%. Then it begin to decline, and is more obvious. Therefore, in this experiment, the number of image's block for JAFFE is 8 and CK is 6.

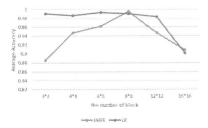

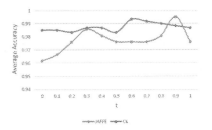

Fig. 2. Average accuracy for the number of block in JAFFE and CK

Fig. 3. Average accuracy for different t in JAFFE and CK

In order to study the term of trade-off to LDBMP operators, this paper conducted experiments on different values of t.

As can be seen from Fig. 3, t have an effect on the final performance of recognition. For JAFFE, the average accuracy shows an upward trend with the increase of t as a whole. When t is about 0.9, the corresponding average accuracy reaches a maximum of 0.9952. For CK, The overall average accuracy is above 98%. When the value of t is about 0.6, the maximum accuracy up to 99.33%.

There are several ways to combine features. In order to verify the effectiveness of the proposed approach, we conducted some comparative experiments in this paper to illustrate this issue. There are the following combination ways: (1) LBP_S only, (2) LBP_M only, (3) LBP_S cascades LDBMP_M; (4) LBP_M cascades LDBMP_M; (5) LBP_S, LBP_M cascade LDBMP_M. (6) LBP_S cascades LBP_M; (7) LBP_S weights LBP_M; (8) LBP_S weights LBP_M, then cascades LDBMP_M; (9) LBP_S weights LBP_M, then weights LDBMP_M. In this paper, the corresponding experiments were performed on the JAFFE and CK. The experimental results are shown in Table 1.

Table 1. Comparison of experimental results of different methods for combination (%)

Feature combination	JAFFE	CK
LBP_S only	97.60	98.51
LBP_M only	68.42	75.69
LBP_S cascades LDBMP_M	96.18	98.67
LBP_M cascades LDBMP_M	97.60	99.00
LBP_S cascade LBP_M cascade LDBMP_M	96.66	99.00
LBP_S cascades LBP_M	97.13	99.00
LBP_S weights LBP_M	98.09	98.51
LBP_S weights LBP_M, then cascades LDBMP_M	97.60	98.84
LBP_S weights LBP_M, then weights LDBMP_M	99.52	99.33

From Table 1, it can be seen that the average accuracy of using LBP_M is lower than others. Because LBP_M only reflects the magnitude information of difference and the extracted features are not complete for facial expression recognition. However, while combining it with other two features, the overall accuracy will increase. Compared to cascade, LBP_S weighted LBP_M performed well to obtain LDBMP_S. On this basis, it is verified that the LDBMP_M features weighted former can also improve performance. The effectiveness of the proposed algorithm is verified.

In order to study the effect of LDBMP for each expression, Table 2 shows the average accuracy of each expression in JAFFE and CK.

Table 2. Average accuracy for different expression in JAFFE and CK (%)

Expression	JAFFE	CK
Anger	100	95.56
Disgust	100	100
Fear	100	100
Happy	96.67	100
Neutral	100	-
Sad	100	100
Surprise	100	100
Average	99.52	99.33

In JAFFE, a few happy expressions were recognized as neutral mistakenly, a example image is shown in Fig. 4(a). In CK, a few angry expressions were recognized as surprise and sad mistakenly, as shown in Fig. 4(b). All of those mistaken recognition images have some similarities with mistaken expression.

(a)   (b)

Fig. 4. Two misrecognized samples

For Fig. 4(a), the woman's eyebrows are down and the mouth is opened slightly. This is similar to the neutral expression. For Fig. 4(b), the woman's eyebrows are tight and her mouth is open to a great degree. It somewhat similar to normal expression of surprise. The expression of these images is difficult to give a definite classification relatively.

In order to illustrate the reliability of LDBMP, this paper compared it with other papers. The results are shown in Table 3. As can be seen from following table, the proposed method is not only better than those traditional feature descriptors but also superior to deep learning [17, 18] because deep learning have some limitations.

For example: it requires a large number of samples to train neural network to ensure networks more precise. All in all, the proposed descriptor shows a certain degree of effectiveness and superiority in facial expression recognition.

Table 3. Comparison of different papers' average accuracy in JAFFE and CK (%)

Method	JAFFE	Method	CK
Feature co-clustering [19]	96.25	Curvelet transform + RBF [23]	95.17
Hierarchical classification [20]	97.00	Hierarchical classification [20]	98.00
Deep CNN [21]	97.71	LBP + 3DHLLBP + LNBPs [24]	98.80
LBP + LPQ + Gabor [22]	98.57	Gabor + PCA + SRC [25]	99.12
Proposed method	99.52	Proposed method	99.33

5 Conclusion

In this paper, we propose a novel local feature descriptor, which is local double binary mapped pattern (LDBMP). This descriptor is obtained by combining two operators: LDBMP_B and LDBMP_M. Compared with other local texture feature descriptors, the proposed method not only retains the advantages of LBP and LMP but also preserves the information of magnitude between neighboring pixel and central pixel and captures the difference of neighboring pixel and central pixel and enlarge nuances that occur in the image, which makes texture features extraction more easily and draw more detailed texture features. Experimental results showed that the proposed LDBMP outperformed some other methods. Since many facial expressions in real life are micro-expressions, in the future, we will continue study how to recognize them accurately.

Acknowledgements. This research has been partially supported by National Natural Science Foundation of China (Grant No. 61672202, 61502141), State Key Program of NSFC-Shenzhen Joint Foundation (Grant No. U1613217) and State Key Program of National Natural Science of China (61432004).

References

1. Durand, K., Gallay, M., Seigneuric, A., et al.: The development of facial emotion recognition: the role of configural information. J. Exp. Child Psychol. **97**(1), 14 (2007)
2. Abouyahya, A., Fkihi, S.E., Thami, R.O.H., et al.: Features extraction for facial expressions recognition. In: International Conference on Multimedia Computing and Systems, pp. 46–49. IEEE (2017)
3. Pratt, W.K.: Image feature extraction. In: Digital Image Processing: PIKS Inside, 3rd edn., pp. 509–550. Wiley (2002)
4. Cootes, T.F., Taylor, C.J.: Active shape models. In: Proceedings for the British Machine Vision Conference, pp. 266—275 (1992)

5. Cheon, Y., Kim, D.: Natural facial expression recognition using differential-AAM and manifold learning. Pattern Recogn. **42**(7), 1340–1350 (2009)
6. Ojala, T., Harwood, I.: A comparative study of texture measures with classification based on feature distributions. Pattern Recogn. **29**(1), 51–59 (1996)
7. Heikkilä, M., Pietikäinen, M., Schmid, C.: Description of interest regions with local binary patterns. Pattern Recogn. **42**(3), 425–436 (2009)
8. Ahonen, T., Matas, J., He, C., et al.: Rotation invariant image description with local binary pattern histogram fourier features. In: Proceedings of Image Analysis, Scandinavian Conference, Scia 2009, Oslo, Norway, June 15–18, pp. 61–70. DBLP (2009)
9. Fromont, E.: Discriminative feature fusion for image classification. In: IEEE Conference on Computer Vision and Pattern Recognition, pp. 3434–3441. IEEE Computer Society (2012)
10. Zhang, Z., Fang, C., Ding, X.: A hierarchical algorithm with multi-feature fusion for facial expression recognition. In: International Conference on Pattern Recognition, pp. 2363–2366 (2012)
11. Turan, C., Lam, K.M.: Region-based feature fusion for facial-expression recognition. In: IEEE International Conference on Image Processing, pp. 5966–5970. IEEE (2015)
12. Guo, Z., Zhang, L., Zhang, D.: A completed modeling of local binary pattern operator for texture classification. IEEE Trans. Image Process. **19**(6), 1657–1663 (2010)
13. Ojala, T., Pietikäinen, M., Mäenpää, T.: Multiresolution gray-scale and rotation invariant texture classification with local binary patterns. IEEE Trans. Pattern Anal. Mach. Intell. **24**(7), 971–987 (2000)
14. Ferraz, C.T., Pereira Jr., O., Gonzaga, A.: Feature description based on center-symmetric local mapped patterns. In: Proceedings of the 29th Annual ACM Symposium on Applied Computing, pp. 39–44. ACM, March 2014
15. Rätsch, G., Onoda, T., Müller, K.R.: Soft margins for AdaBoost. Mach. Learn. **42**(3), 287–320 (2001)
16. Ghimire, D., Jeong, S., Lee, J., et al.: Facial expression recognition based on local region specific features and support vector machines. Multimed. Tools Appl. **76**(6), 7803–7821 (2017)
17. Bazrafkan, S., Nedelcu, T., Filipczuk, P., et al.: Deep learning for facial expression recognition: a step closer to a smartphone that knows your moods. In: IEEE International Conference on Consumer Electronics, pp. 217–220 (2017)
18. Jan, A., Ding, H., Meng, H., Chen, L., Li, H.: Accurate Facial Parts Localization and Deep Learning for 3D Facial Expression Recognition. In: 2018 13th IEEE International Conference on Automatic Face and Gesture Recognition (FG 2018), pp. 466–472. IEEE, May 2018
19. Khan, S., Chen, L., Zhe, X., et al.: Feature selection based on co-clustering for effective facial expression recognition. In: International Conference on Machine Learning and Cybernetics, pp. 48–53. IEEE (2017)
20. Hu, M., Jiang, H., Wang, X., et al.: A hierarchical classification method of expressions based on geometric and texture features. Acta Electron. Sin. **45**(1), 164–172 (2017)
21. Nwosu, L., Wang, H., Lu, J., et al.: Deep convolutional neural network for facial expression recognition using facial parts. In: Dependable, Autonomic and Secure Computing, International Conference on Pervasive Intelligence and Computing, International Conference on Big Data Intelligence and Computing and Cyber Science and Technology Congress. IEEE (2018)
22. Zhang, B., Liu, G., Xie, G.: Facial expression recognition using LBP and LPQ based on Gabor wavelet transform. In: IEEE International Conference on Computer and Communications, pp. 365–369. IEEE (2017)

23. Uçar, A., Demir, Y., Güzeli, C.: A new facial expression recognition based on curvelet transform and online sequential extreme learning machine initialized with spherical clustering. Neural Comput. Appl. **27**(1), 131–142 (2016). https://doi.org/10.1007/s00521-014-1569-1

24. Moeini, A., Faez, K., Sadeghi, H., et al.: 2D facial expression recognition via 3D reconstruction and feature fusion. J. Vis. Commun. Image Represent. **35**, 1–14 (2016)

25. Lu, X., Kong, L., Liu, M., Zhang, X.: Facial expression recognition based on Gabor feature and SRC. Biometric Recognition. LNCS, vol. 9428, pp. 416–422. Springer, Cham (2015). https://doi.org/10.1007/978-3-319-25417-3_49

Pixel-Copy Prediction Based Lossless Reference Frame Compression

Weizhe Xu📵, Fangfa Fu📵, Binglei Lou📵, Yao Wang📵,
and Jinxiang Wang(✉)📵

Microelectronics Center, Harbin Institute of Technology, Harbin, China
xwz.xc@163.com, fff1984292@163.com, loubinglei@126.com, 375390859@qq.com ,
jxwang@hit.edu.cn

Abstract. With the increasing demands of high definition and high resolution for video applications, the bandwidth and power consumption of
accessing external memory storing reference frames during motion estimation bring serious pressure on practical video coding systems. Lossless
reference frame compression is a proper method to decrease memory size
and access bandwidth without any quality loss. This paper proposed
a pixel-copy prediction based lossless reference frame compression. The
method predicts current pixel by copying adjacent reconstructed pixels adaptively based on the estimations of the differences between the
original pixel and the left and upper reconstructed samples. Then residuals are encoded by Huffman encoding to generate bit stream. Realized
with HEVC reference software HM-16.5, experimental results show that
our method achieves averagely 67.45% data reduction ratio (DRR) for
luminance component that outperforms pervious works on computation
complexity and compression efficiency.

Keywords: Motion estimation · Lossless
Reference frame compression · Pixel-copy prediction

1 Introduction

As videos are applied widely in people's daily life, video compression is a necessary tool to eliminate spatial and temporal redundancies employing intra and
inter prediction. With the increasing requirements of higher definition and resolution, the demand of memory size storing reference frames in a practical video
codec grows rapidly. To decrease the on-chip cost of hardware resources, it's more
appropriate to introduce external memory to store reference frames. However,
the huge amount of access bandwidth and power consumption with off-chip
memory bring serious pressure for portable devices with limited computation
resources and power capacity. As previous researches show, the bandwidth cost

This work was supported by a grant from National Natural Science Foundation of
China (NSFC, No. 61504032).

of a real video encoder is about 878MB/s [1] when compressing a 1080P sequence and the power consumption is around 50% of the whole system [2].

To release the bandwidth and power pressure, many strategies are proposed including data reuse [3–5] and reference frame compression. Data reuse decreases the data fetching operations by keeping the overlapped pixels during the search window moves. However, this method requires extra on-chip caches and only reading access bandwidth can be reduced. Reference frame compression compresses the reference frames before writing to the external memory and decompresses them when they are required by motion estimation. This scheme not only reduces the writing and reading access bandwidth, but decreases the size of off-chip memory.

Reference frame compression can be classified into two types that are lossy and lossless. For lossy reference frame compression, transform-based [6], block-based [7,8], and hierarchical-based [9] compression methods are proposed which achieve outstanding compression performance and saves bandwidth significantly. However, the quality loss caused by this coding process leads to serious performance decrement for video coding system especially when the reference frame compression is different on encoder and decoder sides. Besides, the complex computation of transform-based and hierarchical-based methods increases the coding time negatively.

Lossless reference frame compression can maintain image quality and won't affect coding performance for practical video codec. Reference [10] presented a block-based prediction method with 5 prediction modes that supplied over 60% data reduction. A similar compression method was proposed in [11] obtaining 57.6% bandwidth savings averagely. A frequency hierarchical coding approach was introduced in [12] supplying not only lossless but lossy compression. This method could reduce 62.4% data amounts in lossless mode with only 2mW power consumption. Although the data reduction ratio (DRR) of the above algorithms is satisfactory, the computation complexity is too high. The authors of [13] proposed a compression method that encoded the original pixels directly with entropy encoding. The computation is simple, but the data reduction ratio is only 35%. To further improve the compression efficiency, prediction was employed before entropy coding [14] that saved averagely 60% data amount supplying a good balance between compression ratio and computation complexity. However, only one prediction mode was applied causing that the prediction was not accurate enough. To further improve the prediction, [15] introduced an extra difference calculation for predicted residuals which achieved 69% DRR significantly.

In general, lossless reference frame compression is more appropriate to be implemented in video coding systems. And to be realizable, the embedded coding algorithm must be in low complexity and high efficiency. In this paper, to overcome the problems of high complexity and low accuracy, we propose a pixel-copy based prediction for lossless reference frame compression. The method copies the left or upper adjacent reconstructed pixel to be the prediction based on the mode decision result adaptively for each pixel. Then residuals are calculated and encoded by Huffman encoding to generate bit stream. Implementing this method with HEVC reference software HM-16.5, experimental results show that the data reduction ratio can be reached by 67.45% on average for luminance component.

The rest of this paper is organized as follows. Section 2 analyzes the relationship between prediction mode and neighboring pixels. In Sect. 3, pixel-copy based prediction is presented along with the whole compression process. Experimental results and conclusion are shown in Sects. 4 and 5.

2   Correlation Analysis Between Prediction Mode and Adjacent Pixels

For all the compression methods, entropy encoding is employed to remove data statistic redundancies which is sensitive with the distribution of pixel values. And it is known to all that the more concentrated the samples distribute, the more effective the coding can reach. Thus, it is necessary to analyze the distribution of residuals for different prediction methods to design a more accurate prediction. To obtain the correlation, 50 frames from each test sequence in different scenes are encoded by HEVC reference software HM-16.5 in the configuration of low delay with TZ search on and the QPs are 22, 27, 32 and 37 separately.

Figures 1 and 2 present the distributions of original reference frame pixels and residuals of different predictions for sequences of *Basketball* and *Tennis* when QP is 22. As can be seen in Figs. 1(a) and 2(a), the distributions of the original reference frame pixels are not concentrated and it's difficult to design an effective entropy encoding to compress the pixels with different distribution characteristics. Although the horizontal and vertical predictions defined by Eqs. 1 and 2 that copy adjacent reconstructed samples directly are the most common used modes by reference frame compression methods, for the two test sequences, only one prediction mode cannot achieve the best prediction result as shown in Figs. 1 and 2. In the equations, *Diff(i, j)* is the predicted residual, *Pixel(i, j)* is the current pixel to be predicted, and $Recon(i, j-1)$ and $Recon(i-1, j)$ are the left and upper adjacent reconstructed samples separately. To obtain better prediction, we designed an optimal prediction combined horizontal and vertical predictions defined by Eq. 3. As shown in Figs. 1(d) and 2(d), the new introduced optimal prediction can supply the best prediction accuracy for both of the two sequences. Although the optimal prediction supplies the best residual concentration, it is impossible for a decoder to obtain the same prediction mode without adding extra mode decision bits into bit stream, otherwise the compression efficiency will deteriorate. Hence, adaptive prediction mode decision should be taken into consideration to balance compression ratio and prediction accuracy.

$$Diff_{hor}(i,j) = Pixel(i,j) - Recon(i, j-1) \qquad (1)$$
$$Diff_{ver}(i,j) = Pixel(i,j) - Recon(i-1, j) \qquad (2)$$

$$Diff_{opt}(i,j) = \begin{cases} Diff_{hor}(i,j), |Diff_{hor}(i,j)| \leq |Diff_{ver}(i,j)| \\ Diff_{ver}(i,j), others \end{cases} \tag{3}$$

$$if \quad i \geq 0, j \geq 0$$

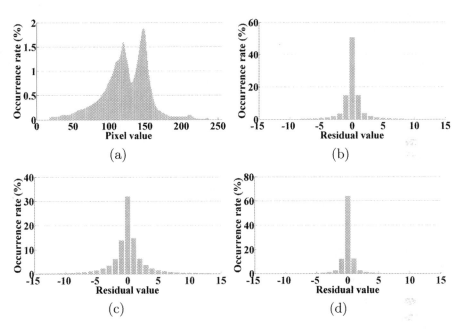

(a)   (b)

(c)   (d)

Fig. 1. The distributions of original reference frame pixels, and residuals of different prediction modes for sequence *Basketball*: (a)–(d) show the original pixels, horizontal, vertical and optimal prediction separately

In Eq. 3, the selection between horizontal and vertical modes is based on the premise that current pixel $Pixel(i, j)$ is already known. To satisfy the requirement of adaptive mode selection, we introduce the upper-left adjacent reconstructed pixel $Recon(i - 1, j - 1)$ to estimate the values of $\overline{Diff}_{hor}(i, j)$ and $\overline{Diff}_{ver}(i, j)$ as presented in Eq. 4 where $\overline{Diff_{hor}(i,j)}$ and $\overline{Diff_{ver}(i,j)}$ are the estimated values. And the mode selection can be modified into Eq. 5 in which $Diff_{adv}$ is the residual of the proposed adaptive prediction.

$$\begin{cases} \overline{Diff_{hor}(i,j)} = Recon(i - 1, j) - Recon(i - 1, j - 1) \\ \overline{Diff_{ver}(i,j)} = Recon(i, j - 1) - Recon(i - 1, j - 1) \end{cases} \tag{4}$$

$$Diff_{adv}(i,j) = \begin{cases} Diff_{hor}(i,j), |\overline{Diff_{hor}(i,j)}| \leq |\overline{Diff_{ver}(i,j)}| \\ Diff_{ver}(i,j), others \end{cases} \tag{5}$$

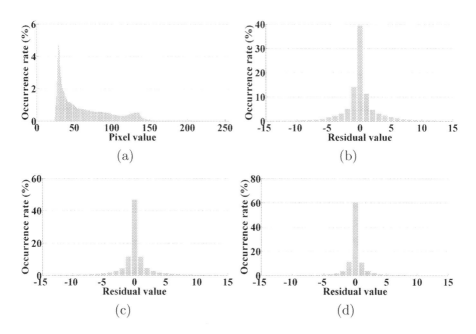

Fig. 2. The distributions of original reference frame pixels, and residuals of different prediction modes for sequence *Tennis*: (a)–(d) show the original pixels, horizontal, vertical and optimal prediction separately

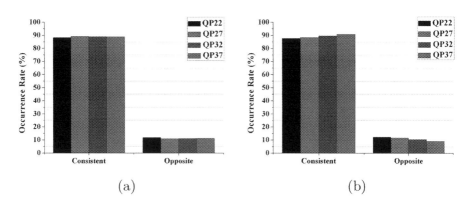

Fig. 3. The occurrence rate of the consistent mode estimation for different QPs: (a) *Basketball* and (b) *Tennis*

To evaluate the accuracy of the estimation, the consistence for different QPs of the optimal prediction mode and estimated prediction mode is analyzed which is shown in Fig. 3. The consistence is defined as that when the estimated mode equals to the optimal one, the result returns as consistency, otherwise the result is opposite. It can be seen that, for both of the two sequences with the 4 different QPs, the occurrence rate of consistency are above 88%. That is to say, for most of the pixels in one frame, the estimated prediction mode is identical to the optimal mode. It can be explained by that, the adjacent pixels has the similar variation trend in a local region.

3 Pixel-Copy Prediction Based Compression Algorithm

Based on the analysis above, the prediction can be adjusted adaptively for each pixel. Besides, in order to satisfy the random accessibility, the proposed prediction and embedded compression utilizes 32×32 block-based processing scheme. Unlike the traditional block-based compression algorithms, in our scheme, each pixel is predicted independently with the selected mode followed by Huffman encoding to generate coded bits.

3.1 Pixel-Copy Prediction Based Encoder

Figure 4 presents the process of pixel-copy prediction based encoder, the pixels in a block are processed in the raster scan order by different prediction modes based on their locations. For the first pixel of the block, the prediction is "0" as no reference sample is valid and the difference is equal to the original pixel value. While for the pixels in the first line and the leftmost column, horizontal and vertical predictions are assigned to be the unique prediction separately. When predicting the rest of the pixels, left or upper adjacent reconstructed samples are selected as prediction adaptively based on the mode decision process, during which the direction with smaller estimated value is specified to be the prediction mode. The processing above can be described in Eq. 6.

$$Diff(i,j) = \begin{cases} Pixel(i,j), \, if & i = 0, j = 0 \\ Pixel(i,j) - Recon(i, j-1), \, if & i = 0 \\ Pixel(i,j) - Recon(i-1, j), \, if & j = 0 \\ Pixel(i,j) - Recon(i, j-1), \, if & \overline{|Diff_{hor}(i,j)|} \leq \overline{|Diff_{ver}(i,j)|} \\ Pixel(i,j) - Recon(i-1, j), \, others \end{cases}$$

(6)

After prediction, entropy encoding will encode the residuals into coded bits. To achieve a good trade-off between encoding complexity and efficiency, Huffman encoding is more suitable. Usually, the Huffman table should be changed dynamically according to the statistics of the samples. However, the table calculation is so complex that it's difficult to reach real-time compressing. Besides, as the coding table is unknown for the decoder side, it needs to be transmitted to the decoder along with the bit stream. Thus, a constant Huffman table is more

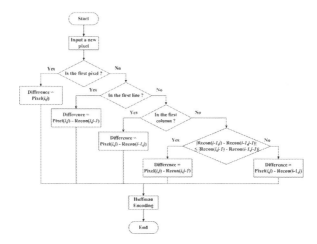

Fig. 4. The encoding process of the proposed compression method based on pixel-copy prediction

proper to be employed in both of the encoder and decoder sides to reduce coding complexity. To adapt to variety coding scenes, frames with different texture and motion characteristics are tested to generate an optimal coding table as shown in Table 1. The table has 32 items for residuals ranging from -15 to 15, and to encode residuals out of this range, a prefix named "others" is put in front of the residual value.

Table 1. The Huffman table for predicted residuals ranging from -15 to 15.

Residual	Coded bits	Residual	coded bits	Residual	Coded bits	Residual	Coded bits
-15	10111110001	-7	10110000	1	110	9	101111001
-14	10111110111	-6	10111111	2	1000	10	101010100
-13	1011000101	-5	1010111	3	101101	11	1011111001
-12	1011000111	-4	101001	4	101000	12	1011000110
-11	1011111010	-3	101110	5	1010110	13	1011000100
-10	101010101	-2	1001	6	10111101	14	10111110110
-9	101111001	-1	111	7	10101011	15	10111110000
-8	10101001	0	0	8	10101000	others	1011001

3.2 Pixel-Copy Prediction Based Decoder

The process of pixel-copy prediction based decoder is shown in Fig. 5. Firstly, the length of bits remained in decoding word registers is checked to decide whether to input a new coding word from bit stream. When the remained bits are less than coding word length equaling to 16 that is the maximum bit length in Huffman table, a new coding word will be fetched. Then the bits in the decoding word registers can be decoded directly by Huffman decoding to generate the residuals used for reconstruction. Meanwhile, the prediction of the current pixel is generated based on the mode decision process. By summing the prediction and decoded residual, the reconstructed pixel is calculated according to Eq. 7.

$$
Recon(i,j) = \begin{cases} Diff, if & i = 0, j = 0 \\ Recon(i, j-1) + Diff, if & i = 0 \\ Recon(i-1, j) + Diff, if & j = 0 \\ Recon(i, j-1) + Diff, if & \overline{|Diff_{hor}(i,j)|} \le \overline{|Diff_{ver}(i,j)|} \\ Recon(i-1, j) + Diff, others \end{cases}
$$

$$(7)$$

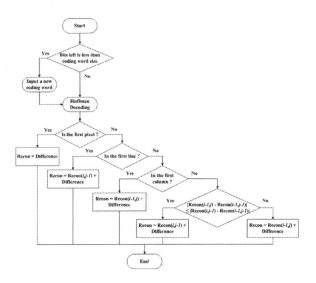

Fig. 5. The decoding process of the proposed compression method based on pixel-copy prediction

4   Experimental Results

To evaluate the efficiency of the proposed pixel-copy prediction based reference frame compression, HEVC reference software HM-16.5 was integrated with the proposed algorithm. 50 frames from 8 1080P sequences were encoded by the modified HM-16.5 under the configuration of low delay and TZ search on with the QPs of 22, 27, 32 and 37 separately. Table 2 presents the data reduction ratio calculated by Eq. 8 of the 3 components. In the table, the first 4 sequences on the left column are used to generate the Huffman table. It can be seen that the average DRR of different QPs for luminance component is 67.45% which is much lower than that of the chroma component that is 78.07%. The reason is that, the pixels from chroma component have a more concentrated distribution than luminance component and the residuals calculated by the proposed prediction distribute relatively more concentrated too.

Table 3 compares the DRR and computation complexity of the proposed compression method with previous works. It can be concluded that, the DRR of our proposed method is only lower than Reference [15] which has a higher complexity than ours. Though Reference [13] employs the least computation,

Table 2. The DRR of 8 test sequences.

Sequences		QP22	QP27	QP32	QP37	Sequences		QP22	QP27	QP32	QP37
Basketball	Y	69.69%	71.39%	71.95%	72.67%	Kimono1	Y	65.51%	65.28%	65.93%	67.35%
	U	81.01%	81.45%	81.91%	82.22%		U	75.46%	77.42%	79.16%	80.78%
	V	79.83%	80.21%	80.72%	80.99%		V	78.36%	80.41%	82.28%	83.38%
bluesky	Y	64.17%	64.20%	64.16%	64.64%	station2	Y	64.31%	65.75%	67.23%	68.86%
	U	68.94%	70.06%	71.35%	72.84%		U	79.44%	81.33%	82.86%	83.63%
	V	74.06%	75.17%	76.61%	77.83%		V	78.76%	80.97%	82.77%	83.81%
Tennis	Y	66.75%	68.07%	69.96%	72.30%	tractor	Y	58.96%	59.32%	60.23%	61.54%
	U	80.04%	81.22%	82.31%	82.88%		U	67.88%	69.44%	71.13%	72.85%
	V	80.60%	82.05%	83.30%	83.76%		V	69.52%	70.70%	71.94%	73.47%
pedestrian_area	Y	72.23%	72.70%	73.37%	73.83%	sunflower	Y	69.37%	68.97%	69.11%	69.51%
	U	81.38%	82.44%	83.23%	83.72%		U	69.66%	70.15%	70.93%	71.55%
	V	82.24%	82.80%	83.24%	83.61%		V	74.36%	75.13%	76.06%	76.70%
Average	Y	66.37%	66.96%	67.74%	68.71%						
	U	75.48%	76.69%	77.86%	78.81%						
	V	77.22%	78.43%	79.62%	80.44%						

Table 3. The DRR and complexity comparisons between the proposed compression and previous works.

Method	DRR	Complexity
Reference [10]	61.5%	5 DC + 1 TBP
Reference [11]	57.6%	4 MDA + 15 RGM + 1 SFL
Reference [12]	62.4%	2 DWT + 1 SPIHT
Reference [13]	35.46%	1 VLC
Reference [14]	60.8%	1 DC + 1 VLC
Reference [15]	69%	2 DC + 1 VLC
Proposed	67.45%	1 DC + 1 VLC

the compression efficiency is the worst. Thus, our method offers the best balance between complexity and compression ratio.

$$DRR = (1 - \frac{Length_{compressed}}{Length_{original}}) \times 100\% \qquad (8)$$

5 Conclusion

In this paper, we proposed a pixel-copy prediction for lossless reference frame compression. The method copies the left or upper adjacent reconstructed samples adaptively as the prediction based on the estimations of the differences between original pixel and the two adjacent samples. Then residuals are encoded by Huffman encoding to generate bit stream. Experimental results show that, the proposed lossless reference frame compression supplies an outstanding data reduction ratio that is 67.45% compared with previous works and offers a good balance between computation complexity and compression efficiency.

In future work, corresponding hardware architecture should be designed.

References

1. Lin, C.C., Chen, J.W., Chang, H.C.: A 160K gates/4.5 KB SRAM H.264 video decoder for HDTV applications. IEEE J. Solid State Circuits **42**(1), 170–182 (2006)
2. Budagavi, M., Zhou, M.: Video coding using compressed reference frames. IEEE International Conference on Acoustics, Speech and Signal Processing (ICASSP), pp. 1165–1168. IEEE, Las Vegas, USA (2008)
3. Tuan, J.C., Chang, T.S., Jen, C.W.: On the data resuse and memory bandwidth analysis for full-search block-matching VLSI architecture. IEEE Trans. Circuits Syst. Video Technol. **12**(1), 61–72 (2002)
4. Li, D.X., Zheng, W., Zhang, M.: Architecture design for H.264/AVC integer motion estimation with minimum memory bandwidth. IEEE Trans. Consum. Electron. **53**(3), 1053–1060 (2007)
5. Chen, C.Y., Huang, C.T., Chen, Y.H.: Level C+ data reuse scheme for motion estimation with corresponding coding orders. IEEE Trans. Circuits Syst. Video Technol. **16**(4), 553–558 (2006)
6. Said, A., Pearlman, W.A.: A new, fast, and efficient image codec based on set partitioning in hierarchical trees. IEEE Trans. Circuits Syst. Video Technol. **6**(3), 243–250 (1996)
7. Song, L., Zhou, D.J., Jin, X., et al.: An adaptive bandwidth reduction scheme for video coding. In: IEEE International Symposium on Circuits and Systems (ISCAS), pp. 401–404. IEEE, Paris, France (2010)
8. Lin, C.-L.: A low latency coding scheme for compressing reference frame in video codec. In: 2017 International Conference on Applied System Innovation (ICASI), pp. 1957–1960. IEEE, Sapporo, Japan (2017)
9. Ma, Z., Segall, A.: Frame buffer compression for low-power video coding. In: IEEE International Conference on Image Processing (ICIP), pp. 757–760. IEEE, Brussels, Belgium (2011)

10. Kim, J., Kyung, C.M.: A lossless embedded compression algorithm for high definition video coding. In: IEEE International Conference on Multimedia and Expo, pp. 193–196. IEEE, New York, USA (2009)
11. Guo, L., Zhou, D.J., Goto, S.: Lossless embedded compression using multi-mode DPCM & averaging prediction for HEVC-like video codec. In: Proceedings of the 21st European Signal Processing Conference (EUSIPCO), pp. 1–5. IEEE, Marrakech, Morocco (2013)
12. Cheng, C.C., Tseng, P.C., Chen, L.G.: Multimode embedded compression codec engine for power-aware video coding system. IEEE Trans. Circuits Syst. Video Technol. **19**(2), 141–150 (2009)
13. Silveira, D., Povala, G., Amaral, L., et al.: Memory bandwidth reduction for H.264 and HEVC encoders using lossless reference frame coding. In: IEEE International Symposium on Circuits and Systems (ISCAS), pp. 2624–2627. IEEE, Melbourne VIC, Australis (2014)
14. Silveira, D., Povala, G., Amaral, L., et al.: A new differential and lossless reference frame variable-length coder: an approach for high definition video coders. In: IEEE International Conference on Image Processing (ICIP), pp. 5641–5645, France, Paris (2014)
15. Silveira, D., Povala, G., Amaral, L. et al.: Efficient reference frame compression scheme for video coding systems: algorithm and VLSI design. J. R. Time Image Process., 1–21 (2015)

Robust Underwater Fish Classification Based on Data Augmentation by Adding Noises in Random Local Regions

Guanqun Wei[1], Zhiqiang Wei[1,2(✉)], Lei Huang[1,2], Jie Nie[1], and Huanhuan Chang[1]

[1] Ocean University of China, Qingdao 266000, China
wzqouc@foxmail.com
[2] Qingdao National Laboratory for Marine Science and Technology, Qingdao 266000, China

Abstract. Underwater fish classification is in great demand, but the unrestricted natural environment makes it a challenging task. The monitor placed underwater gets a lot of low-quality and hard-to-mark marine fish images. These images suffer from various illumination, complex background etc. At the same time, there are many high-quality and easy-to-mark marine fish pictures on the Internet. In this paper, we propose an effective data augmentation approach for improving the classification accuracy of low-quality marine fish images. In our method, unlike the existing global image method, random local regions are proposed for simulating local occlusion and fuzziness in various underwater environment. In addition, four types of noise are incorporated for augmenting training data set. Experimental results demonstrate that our approach can significantly enhance the classification performance of low-quality marine fish images under various challenging conditions when using high-quality marine fish images as training sets.

Keywords: Underwater vision · Fish classification
Convolution neural network · Data augmentation

1 Introduction

Underwater object recognition is in great demand and fish classification is one of the most important tasks for ocean observation. There are many practical applications using fish classification as the basic technology, e.g. fish farming and marine ecology monitoring.

Although there is a lot of previous work on fish classification, it is still a challenging research issue. The challenges mainly due to the following aspects: Firstly, the quality of most marine fish images is low because of the extreme conditions in the sea. Secondly, the distinctions between fish species is small, so the classification of marine fish images require expertise from marine biologists. According to above reasons, the classification of low-quality marine fish images is difficult. However, the Internet contains a large number of high-quality marine fish images captured by professional shooting. High-quality images contain more detailed information about fish, and more

© Springer Nature Switzerland AG 2018
R. Hong et al. (Eds.): PCM 2018, LNCS 11165, pp. 509–518, 2018.
https://doi.org/10.1007/978-3-030-00767-6_47

fish characteristics can be extracted. How to bridge the classification gap between high-quality marine fish images and low-quality marine fish images is a significant issue, as is shown in Fig. 1.

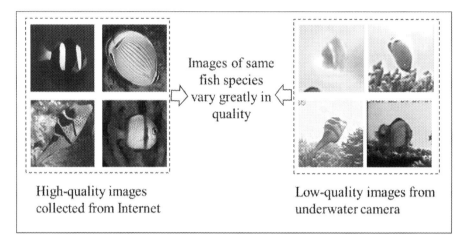

Fig. 1. The gap between images of different quality

In this paper, we aim to find a solution to improve the classification accuracy of low-quality fish images by using high-quality fish images downloaded from the Internet as training sets. In the existing research, there are ways to improve the applicability of classifiers by adding noise data. Reference [1] showed that the existing networks are susceptible to these quality distortions, particularly to blur and noise. Reference [2] used gray squares to block the different parts of the input image to investigate whether the model really recognizes the position of the object in the image. These methods had got satisfactory results in some tasks, but they are not suitable for underwater fish classification scenarios for that they treated the entire image. However, the underwater environment is complex, and the image quality changes are localized, so the existing methods cannot well simulate the real underwater situation. The contributions of our work are: (1) Random selected regions are proposed for simulating local occlusion and fuzziness in various underwater environment. (2) Four types of noise are incorporated for augmenting training data set.

2   Related Work

The previous fish classification researches are mainly based on constrained environment. Reference [3] used contour matching to recognize fishes in fish tanks. In [3] fish image features were extracted using morphometry and classified with three stage classifier models. Nearest Neighbor algorithm was used as classifier, and 25 fish species were classified with accuracy of about 99%. Reference [4] introduce an automatic classification approach for the Nile Tilapia fish using support vector machines (SVMs) algorithm in conjunction with feature Transform. Reference [5] derived shape and texture features from an appearance model, classified with LDA, and tested on a data set containing 108 images of three fish species under constrained condition (caught from the sea), achieving an accuracy of 76%. Reference [6] combined texture features and shape features and tested on a database containing 360 images of ten fish species, achieving an average accuracy of 92%. Previous studies primarily used handcrafted features and the recognized image is taken in a fixed scene. There are also studies that apply convolution neural networks to marine fish classification. Reference [7] combined PCANet with spatial pyramid pooling for underwater live fish classification and the foreground extraction is performed first in their works.

Data augmentation is an explicit form of regularization is also widely used in the training of deep CNN. Various translations such as, rotation, flipping, adding noise are used to enlarge training dataset. Adding noise is a common method in CNN training. Random Erasing [8] add noise to image and have better performance in image classification, object detection and person re-identification than without-noise method. "DisturbLabel" [9] is introduced by adding noise at the loss layer. But for the previous work, the training data and the test data usually have similar image quality, but the actual situation is not the case. For ocean fish classification, we can find a lot of well-marked marine fish pictures captured by professional shooting on the Internet, and their image quality is usually high. However, in reality, fish pictures taken from underwater often have low quality, low contrast, and even occlusions. Under this condition, the previous research encountered difficulties. In this paper, we try to bridge the classification gap between high-quality marine fish images and low-quality marine fish images.

3   Fish Classification

In order to make the trained classifier apply to the real underwater environment and improve the accuracy of fish image classification, we propose a new image data augmentation method. We introduce two characteristics in real underwater environment into the training set, and the two characteristics are: (1) Submarine environments involve interactions between seawater and light, and more complex etch, beam, and shadows occur. (2) Submarine scenes will be affected by the submarine topography, sea water depth, sea water scattering, etc. There is a variety of light and color areas in an image. And most interactions are in local regions in the images.

To solve this problem, we propose a data set expansion method based on random selected local regions. The method can simulate a variety for local noise/fuzziness. In the following sections, we will present our method in detail. Figure 2 give an overview of our paper.

We compare the high-quality marine fish images and the low-quality ocean images and try to add noise to high-quality images to make up the difference between high-quality and low-quality. Observing that the global noise is not appropriate for a complex regional environment from a clear image to a cross-data set of blurred images, we use the method based on local noise to cross-dataset learning.

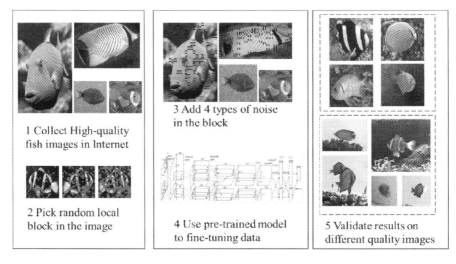

Fig. 2. The process of experiment

3.1 Random Local Region

The main difference between our method and the previous method is that we treat the image as random selected regions for adding noises. The method of random selection of areas is as follows.

Firstly, we normalize the high quality marine fish image to 500-by-500 pixels. Secondly, each image is divided into 20 * 20 blocks, i.e., (25 * 25 pixels for each block). Thirdly, as shown in Fig. 3, we select different proportions of blocks as local regions for adding noises.

(a)   (b)   (c)   (d)

Fig. 3. Random local region selection: (a) shows the original image, (b) shows the 10% of image is selected as the noise region. (c) 30% of the image is selected as the noise region. (d) 50% of the image is selected as the noise region.

3.2 Adding Noise in Local Regions

In this paper, we design four kinds of local noise: horizontal stripes noise, vertical stripes noise, Gaussian-noise and salt-pepper noise. The details of these noises are as follows.

Horizontal stripes noise: In the randomly selected regional area, a width of 5-pixel horizontal stripes is generated for each 10-pixels rectangle. The RGB value of the stripe is set to (255, 255, 255). The local region size in our work is 25 by 25 pixels. Horizontal stripes are presented in Fig. 4 (left).

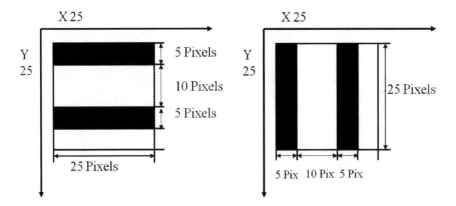

Fig. 4. Stripe noise generation: In every random selected local region, the black stripes were added to the image.

Vertical stripes noise: In the randomly selected regional area, a width of 5-pixel vertical stripes is generated for each 10-pixels rectangle. The RGB value of the stripe is set to (0, 0, 0). The local region size in our work is 25 by 25 pixels and vertical stripes are presented in Fig. 4 (right). Horizontal and Vertical stripes noise is considered as a same method to generate a stripe-noise data set.

Salt-pepper noise: Salt and pepper noise is a typical noise. The white and black dots are represented on image with a certain probability. In the random region, we set the signal-noise ratio to 0.8, and randomly select the pixel and the pixel's RGB value is set to [255, 255, 255] or [0, 0, 0].

Gaussian noise: Gaussian noise is a kind of noise that the probability density function of noise value obeys the Gaussian distribution. In this experiment, the Gaussian noise is generated by following formula:

$$P(i,j)_{noise} = P(i,j)_{original} + \delta G(seed)$$
$$where \quad G \sim N(seed)\, \delta = 0.8 \tag{1}$$

We use system time as a seed to generate a Gaussian distribution random number and add this Gaussian noise number to the original pixel value of the image. Salt-pepper and Gaussian noise is considered as a same method to generate a point-noise data set.

Using the noise generation strategy described above, we finally generated three training data sets: high-quality data set and high-quality with region noise data set. Figure 4 gives an example of noise images generated by noise generation strategy.

3.3 Fish Classification

In this paper, we use "AlexNet" [10] for fish classification. It has five convolutional layers, three fully connected layers, including a 1000-dimensional output layer, and has over 60 million parameters. We initialize the network with the representation learned on ImageNet, transferring all layers except the output layer, i.e. conv1-fc7. We change the output fc8 layer from 1000 dimensions to the number of fish categories in the data set and initialize the weights with a standard Gaussian distribution. The input image size is 227-by-227 pixels.

4 Experiments

4.1 Data Sets and Baseline Method

We use a high-quality marine fish image data set and a low-quality marine fish image data set to evaluate the proposed method. Table 1 give the distribution of two data sets.

Low-Quality Data Set
We use **Fish4Knowledge** [11] data set as our low-quality data set. This underwater live fish data set is acquired from a live video data set captured from the open sea. There are totally 27,370 verified fish images of 23 clusters and each cluster is presented by a representative species. The fish species are manually labeled by marine biologists. The distribution of the fish species in the data set is shown in Table 1. The second row of Fig. 5 shows the example of low-quality fish image. All low-quality images are used for testing.

High-Quality Data Set
We created a high-quality marine fish image data set by collecting fish images from the Internet. We use the same 23 kinds of the **Fish4Knowledge** data set. To construct this data set, we use "Google" as search engine and the fish kind name as keywords to download images first, then we remove the wrong images manually and cut the fish area using bounding boxes. The fish areas are normalized to 500 * 500 pixels. The number of image in our high-quality marine fish data set is 3,886, some samples are shown in the first row of Fig. 5. All high-quality images are divided into two subsets: 70% for training and 30% for test.

Baseline
We fine-tuned the high-quality data set based on the model pre-trained on a large image data set and use the classification accuracy of the low-quality test images and the high-quality test images on this fine-tuned model as our baseline. The classification accuracy of low-quality test data set on high-quality fine-turn models is 17.14%, and the high-quality test data set accuracy is 97.24%.

Table 1. The species distribution of data set

Species	Number of low-quality	Number of high-quality
Dascyllus reticulatus	12,112	181
Plectroglyphidodon dickii	2,683	52
Chromis chrysura	3,593	134
Amphiprion clarkii	4,049	287
Chaetodon lunulatus	2,534	260
Chaetodon trifascialis	190	98
Myripristis kuntee	450	189
Acanthurus nigrofuscus	218	226
Hemigymnus fasciatus	241	87
Neoniphon sammara	299	131
Abudefduf vaigiensis	98	241
Canthigaster valentini	147	377
Pomacentrus moluccensis	181	165
Zebrasoma scopas	90	146
Hemigymnus melapterus	42	77
Lutjanus fulvus	206	90
Scolopsis bilineata	49	135
Scaridae	56	309
Pempheris vanicolensis	29	91
Zanclus cornutus	21	188
Neoglyphidodon nigroris	16	68
Balistapus undulatus	41	273
Siganus fuscescens	25	81
Total number	27,370	3,886

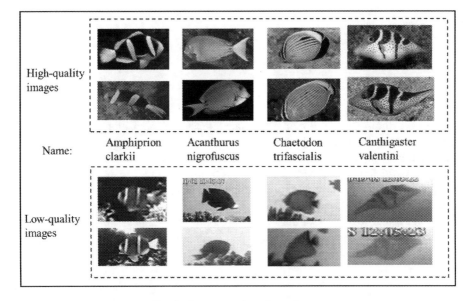

Fig. 5. Some samples of our data set

4.2   Experimental Results

Due to the limited number of images, we use the pre-trained model for fine-tuning. We implement the architecture with Caffe [12] framework. Stochastic gradient descent with a mini-batch size of 25 is used to update the parameters. The base learning rate is set to 0.001, in conjunction with a momentum term of 0.9 and weight decay of 0.0005. The weight and bias learning rate of last layer (a fully connected layer) is ten times of other layers. When the validation error turns to plateaus, the learning rate is decreased by a factor of 10. Based on the established regional noise generation strategy, nine high-quality data sets with regional noise levels ranging from 10% to 90% were generated and fine-tuned using a model pre-trained. 100% is adding global noise to image rather than random local region.

Figure 6 shows the results of low-quality images classification accuracy of our method, and it can be seen that our method can get promoted results, and the performance of vertical and horizontal stripe noise is obviously better than Gaussian & salt-pepper noise. With the increasing noise area, the data with dot noise is close to the baseline, and the performance of stripe noise is better than that of point noise. The best results (accuracy: 39.65%) were obtained at ratio of 30%, 22.50% higher than baseline. The data augmentation based on random local regions is better than adding noise to image globally.

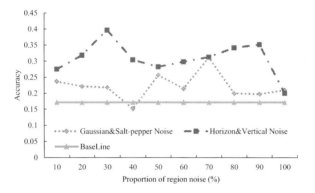

Fig. 6. Experimental results in low-quality image test dataset

Figure 7 shows the results of high-quality images classification accuracy of our method. We find that when the proportion of local noise area is small (<30%), the classification accuracy of stripe noise and Gaussian & salt-pepper noise is similar to baseline. As the proportion increases, the classification accuracy of point noise method is similar to baseline. The stripe noise method classification accuracy declined, so we reject all the stripe noise methods when area proportion is bigger than 40%. At last, we propose to add 20% or 30% of the vertical and horizontal stripes to images, and it can enhance the model without reducing the accuracy of the image and improving the accuracy of low-quality images.

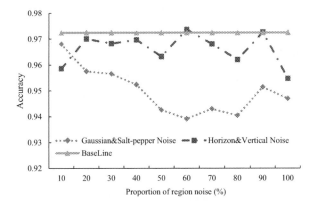

Fig. 7. Experimental results in high-quality image test dataset

5 Conclusion

In this paper, we propose a novel data augment method by adding noises in random local regions to bridge the gap between high-quality and low-quality images for underwater fish classification. Experiments show that the proposed method is robust for various challenging environments and can significantly improve the classification accuracy of low-quality fish images when using high-quality training data set.

Acknowledgements. This work is supported by the Aoshan Innovation Project in Science and Technology of Qingdao National Laboratory for Marine Science and Technology (No. 2016ASKJ07); National Natural Science Foundation of China (No. 61672475, 61402428, 61702471); Qingdao Science and Technology Development Plan (No. 16-5-1-13-jch).

References

1. Dodge, S., Karam, L.: Understanding How Image Quality Affects Deep Neural Networks (2016)
2. Zeiler, M.D., Fergus, R.: Visualizing and understanding convolutional networks. In: Fleet, D., Pajdla, T., Schiele, B., Tuytelaars, T. (eds.) ECCV 2014. LNCS, vol. 8689, pp. 818–833. Springer, Cham (2014). https://doi.org/10.1007/978-3-319-10590-1_53
3. Lee, D.J., Xu, X.: Contour matching for a fish recognition and migration-monitoring system. In: Proceedings of SPIE - The International Society for Optical Engineering, vol. 5606, pp. 37–48 (2004)
4. Fouad, M.M.M., Zawbaa, H.M., El-Bendary, N., Hassanien, A.E.: Automatic Nile Tilapia fish classification approach using machine learning techniques. In: International Conference on Hybrid Intelligent Systems, pp. 173–178 (2013)
5. Larsen, R., Olafsdottir, H., Ersbøll, B.K.: Shape and texture based classification of fish species. In: Salberg, A.-B., Hardeberg, J.Y., Jenssen, R. (eds.) SCIA 2009. LNCS, vol. 5575, pp. 745–749. Springer, Heidelberg (2009). https://doi.org/10.1007/978-3-642-02230-2_76

6. Spampinato, C., Giordano, D., Salvo, R.D., Chen-Burger, Y.H.J., Fisher, R.B., Nadarajan, G.: Automatic fish classification for underwater species behavior understanding. In: ACM International Workshop on Analysis and Retrieval of Tracked Events and Motion in Imagery Streams, pp. 45–50 (2010)
7. Qin, H., Li, X., Liang, J., Peng, Y., Zhang, C.: DeepFish: accurate underwater live fish recognition with a deep architecture. Neurocomputing **187**, 49–58 (2016)
8. Zhong, Z., Zheng, L., Kang, G., Li, S., Yang, Y.: Random Erasing Data Augmentation (2017)
9. Xie, L., Wang, J., Wei, Z., Wang, M., Tian, Q.: DisturbLabel: Regularizing CNN on the Loss Layer (2016)
10. Krizhevsky, A., Sutskever, I., Hinton, G.E.: ImageNet classification with deep convolutional neural networks. In: International Conference on Neural Information Processing Systems, pp. 1097–1105 (2012)
11. Boom, B.J., Huang, P.X., He, J., Fisher, R.B.: Supporting ground-truth annotation of image datasets using clustering. In: International Conference on Pattern Recognition, pp. 1542–1545 (2012)
12. Jia, Y., et al.: Caffe: Convolutional Architecture for Fast Feature Embedding, pp. 675–678 (2014)

Study on User Experience of Panoramic Images on Different Immersive Devices

Shilin Wu, Zhibo Chen$^{(\boxtimes)}$, Ning Liao, and Xiaoming Chen

CAS Key Laboratory of Technology in Geo-spatial Information Processing
and Application System, University of Science and Technology of China,
Hefei, China
chenzhibo@ustc.edu.cn

Abstract. Recently there has been a promising increment on the research and development of various virtual reality (VR) applications and devices. However, it is still an open issue to combine panoramic images in quality of experience research with different immersive devices. In this paper, we analyze five factors: the sense of immersion, sense of involvement, comfort factor, resolution factor and content factor presented by three typical types of panoramic display devices: CAVE, Oculus rift, and web panorama player. We compare the strength and the weakness of each device. In conclusion, the Oculus is best in immersion and involvement and CAVE is second best, while the web player is best in comfort and resolution. This will be useful for the development of emerging immersive technology.

Keywords: Virtual reality · Immersive device · Panorama
Quality of experience

1 Introduction

In the recent years, there has been a promising increment on the research and development of virtual reality applications and those applications are gradually coming into peoples daily live [1,2] . Nowadays, there are usually 3 categories of viewing environment: CAVE (Cave Automatic Virtual Environment) [3], HMD (Head Mounted displays, e.g. Oculus, HTC vivo [4]) and web panorama application, as shown in the Fig. 1.

Recently, the research on quantitative evaluation of panoramic image quality degradation caused by mosaicing techniques [5–7] and on quality assessment of panorama sequences caused by image compression [8,9] has made a lot progress. On the other hand, the research on virtual environment like evaluating the importance of multisensory input on memory and the sense of presence in virtual environments is also ongoing [10,11].

However, it is still an open issue to combine panoramic images in quality of experience research with different immersive devices. In this paper, we analyze how much the sense of immersion and the involvement are presented by

© Springer Nature Switzerland AG 2018
R. Hong et al. (Eds.): PCM 2018, LNCS 11165, pp. 519–527, 2018.
https://doi.org/10.1007/978-3-030-00767-6_48

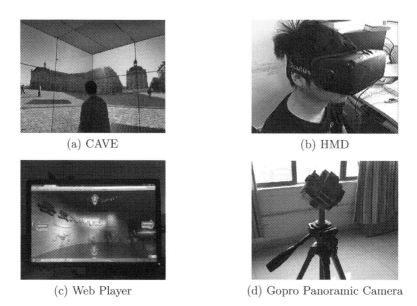

(a) CAVE

(b) HMD

(c) Web Player

(d) Gopro Panoramic Camera

Fig. 1. Experiment environment

panoramic display device. The research on the various factors to quality of experience of different immersive devices is important for the development of new immersive technology.

In the previous literatures, there are some research on the Sense of Presence (SoP) [12,13] in certain VR applications. They considered four factors that are thought to influence the sense of presence: control factor, sensory factor, distraction factor and realism factor. They designed several questions for each factor in questionnaire and calculated the Pearson correlation of each factor with the SoP. In [14], authors further considered the difference between subjects. However these works did not study the factors to the different user experience between virtual reality devices. Besides, they did not consider the influence of different image content on different devices. In this paper, we discuss the strengths and weaknesses of three typical devices systematically.

Specifically, we analyze the Sense of Presence in three different types of devices (CAVE, HMD, Web). From the experiment, we indicate the different user experience among three devices and compare the strength and weakness of each viewing devices, which will be useful for research field of immersive media processing.

Rest of the paper is organized as follows. In Sect. 2, we present the design of subjective experiment; we discuss the result of the experiment in Sect. 3, and present the conclusions and our future work in Sect. 4.

2   Design of Subjective Experiment

2.1   Display Devices for Panoramic Scenes

As shown in Fig. 1, for CAVE, a six side cubic screen is used to create a virtual environment. It is a fully immersive CAVE and a subject can watch the content displayed in each side. For the HMD device, Oculus rift DK2 is used as the experiment device. For the web player, the web panorama player available at [17] is used, and the screen size we use is 1920 * 1080. The display is View Sonic vs2370 s.

2.2   Source Content

Considering the VR applications such as visiting sceneries and museum, we classified the panoramic scenes into two categories: outdoor scenes and indoor scenes, as shown in Fig. 2. In the outdoor category, there are campus, park, street, buildings, and nature scenery such as snow, seaside, and night scenery. In the indoor category, there are house, shopping mall and museum. The total number of the panoramic images is 131.

Since subjects may feel boring to watch same images on different devices, we partitioned these images into three distinct parts, but each of which equally covers all the above content categories. Specifically, 51 images was used for CAVE, 40 images for Oculus and 40 images for web player. Moreover, on each display device, all the images are displayed in a random order for different viewers, so as to eliminate the recency effect.

All the images that we captured or collected have a resolution of more than 5k * 10k pixels. For CAVE, all the images shown on every side of the walls are of 1920 * 1200 pixels. For HMD, every panoramic image is of 2048 * 1096 pixels. For web displayer, the resolution is 2048 * 1096. As shown in Fig. 1(d), we captured the images through the panoramic camera which is composed of six Gopro hero4 cameras. The pictures in six directions were taken separately and we stitched six separate images to one panorama image.

Fig. 2. Content

Table 1. Questionnaire

Item stems	Factors
1. How natural did your interactions with the environment when you look around?	Involvement Factor
2. How aware were you of events occurring in the real world around you?	Immersion Factor
3. How much did your experiences in the virtual environment seem consistent with your real world experiences?	Involvement Factor/Immersion Factor
4. Were you able to anticipate what would happen next in response to the actions that you performed?	Immersion Factor
5. How completely were you able to actively survey or search the environment using vision?	Immersion Factor
6. How compelling was your sense of moving around inside the virtual environment?	Involvement Factor
7. How closely were you able to examine objects?	Involvement Factor
8. How much did you still feel immersive in the environment at the end of the experimental session?	Immersion Factor
9. How quickly did you adjust to the virtual environment experience?	Immersion Factor
10. How involved were you in the virtual environment experience?	Involvement Factor
11. How much did the resolution of the image influence you?	Resolution Factor
12. How much did the comfort of the device influence you?	Comfort Factor
13. How much did the the image content involve you? Indoor or Outdoor	Content Factor

2.3 Subjective Experiment Method

We adapt the questionnaire method [12] to investigate the potential factors to the quality of user experience. There are totally 13 questions as shown in Table 1, which reflects five types of factors, namely, the involvement factor, the immersion factor, the resolution factor, the comfort factor and the content factor. Because involvement and immersion are more abstractive than other factors, so we user more question to describe it. For each question, viewers will give their opinion on a five-point scale, where 1 point means strong disagreement and 5 points mean strong agreement.

The subjects were not asked to watch the panoramic images on the display devices in a fixed time. The participants can stop watching when they feel boring or uncomfortable, and we recorded the duration when they got immersive in the display environment. The watching order of the device is random for different subjects. The exposure duration is an important measure in immersive virtual environment testing [15,16]. We believe that the watch period may also be a good measure of the subjects' sense of immersion which will be experimentally analyzed in Sect. 3 later on.

16 subjects who have good color vision, participated in the subjective test. Before the subjective experiment, they were instructed about the categories of the image content, the meaning of the questions and the usage of the devices. The order of the display devices watched by the subjects is random and all images are displayed randomly for each subject. On each display device, the images are displayed one by one without any interruption and each image is displayed for

20 s. The images are displayed by slide show in CAVE, by Autopano software in Oculus, and by Web Panoramas player in the Web player case. After watching all images with one display device, the subject was required to rest for 10 min and to finish the questionnaire shown in Table 1. For the 13th question, the content factor, subjects were asked to pick the content category of the best experience from the list: campus, park, street, buildings, nature scenery, house, shopping mall and museum. The design of this content factor is supposed to tell how image content can affect the sense of immersion and the sense of involvement.

After watching all images with all the devices, the subjects were asked to give an overall score of the three devices separately. A five point score, where 1 point means lowest overall experience and 5 points mean best overall experience, was given.

After the subjective test, we screened out 2 outlier subjects, the correlation of whose scores with the average scores was lower than 0.75.

3   Results and Analysis

3.1   Mean Score and Standard Deviation

In Table 2, for each device, we first calculated the mean score of all the questions. Then we calculated the standard deviation of valid subjects' scores for each question, and got average standard deviation over 13 questions. The mean score of Oculus (HMD) is obviously higher than the CAVE and the web player (p-value $\ll 0.01$), which means Oculus provides better comprehensive experience for subjects.

The average of the standard deviation of the CAVE is almost higher than other devices, which means the user experience presented by the CAVE is controversial to the subjects.

Table 2. Mean score and Standard Deviation

Mean opinion score (MOS)		Standard deviation	
CAVE	3.35	CAVE	1.06
Oculus	4.28	Oculus	0.66
Web player	2.57	Web player	0.88

3.2   Factor Analysis

In the questionnaire, the question is divided into five kinds of factors: immersion factor, involvement factor, comfort factor, resolution factor and content factor (see Table 1). Immersion presents subjective immersive feeling that people immerse in the virtual environment and do not realize the real world. Involvement means that people get involved in the virtual environment and can interact

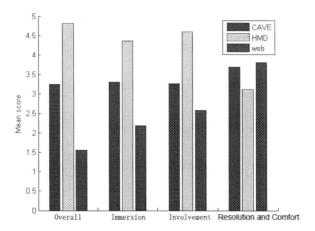

Fig. 3. Factor analysis

with the environment. Comfort factor and resolution factor present the subjective feelings affected by device resolution and comfort.

In Fig. 3, the bars list the average of all viewers' overall score, the average score of the immersion factor, the involvement factor, the resolution and comfort factor. Oculus rift is the best with respect to the overall experience, immersion and involvement(p-value<<0.01). And people can hardly get immersive and involved with the web player as indicated by very low immersion and involvement score. However, it notes that the web player gets highest score in terms of resolution and comfort factors, which means that though the web player is not an immersive devices, it is a comfortable panoramic media player.

The CAVE and Oculus still need some improvement in high resolution and comfort. From the detail information we gathered from the subjects, the resolution of every side screen of the CAVE is not high enough (1920*1200) and this cubic environment requires viewers stand at a fixed position. If they deviate from this position slightly, viewers can easily find some distortion in the edges of the screens. For Oculus, its low resolution and unfriendliness wearing manner for shortsighted people, who have to adjust Oculuss sight distance before looking, have great negative effect. Therefore, the devices still need some improvement in resolution and comfort.

3.3 Immersion Duration Analysis

In Fig. 4, we separately added up the viewers' immersion duration distribution of three devices and calculated cumulative frequency. As discussed in part 2, this chart shows how long in time the subjects get immersed in the device. The duration of immersion time in the web is obviously lower than the CAVE and Oculus. Most participants stay less than 10 min in the experiment of web player. The Pearson Correlation Coefficient between time and immersion score is 0.7546, demonstrating the strong relationship between immersion experience and watching time.

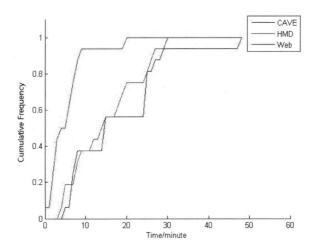

Fig. 4. Immersion duration analysis

3.4 Content Analysis

To investigate the influence of panorama contents on the QoE delivered by different devices, the participants were asked to choose the image content of best experience after all the experiments. In Fig. 5, we gathered the votes for the preferred content category and classified it into outdoor content or indoor content. The term "all good" means participants choose all the categories and "all bad"

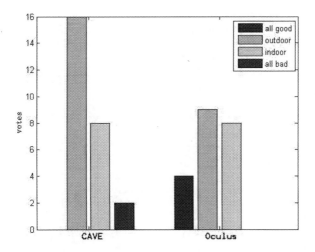

Fig. 5. Content analysis

means participants choose nothing. Because most viewers chose "all bad" for the web player, this part is not shown on the chart.

As shown in Fig. 5, for CAVE, outdoor contents were preferred more than indoor content, which means that the CAVE delivers better user experience in the open sight than close sight. Referring to the information we gather from the subjects, the reason is that the close sight may cause people confused of sight distance in the edge of the cubic box, and the outdoor scenery is brighter than indoor scenery. However, for Oculus, there is not much difference between outdoor and indoor content because of the sphere panorama structure in Oculus.

4   Conclusion

This paper mainly discuss about the user experience in three difference panoramic display devices: CAVE, HMD and web player. We design the subjective assessment method, and study the factors that affect quality of experience in different devices. The Oculus, used as a HMD device, get the best user experience in the overall experience. The CAVE gets the second best, and the outdoor scenery has better result than the indoor content. Both Oculus and CAVE still need some improvement in high resolution and comfort in the future. They all can be called immersive device. The web player does not provide any sense of immersion and involvement, but it is good as a panorama displayer in terms of resolution and comfort.

Acknowledgments. This work was supported in part by the National Key Research and Development Program of China under Grant No. 2016YFC0801001, the National Program on Key Basic Research Projects (973 Program) under Grant 2015CB351803, NSFC under Grant 61571413, 61632001,61390514, and Intel ICRI MNC.

References

1. Jung, K., Morris, K.C., Lyons, K.W.: Using formal methods to scope performance challenges for Smart Manufacturing Systems: focus on agility. Presence Concurr. Eng. Res. Appl. **23**(4), 343–354 (2015)
2. Amiri Atashgah, M.A., Malaek, S.M.B.: An integrated virtual environment for feasibility studies and implementation of aerial MonoSLAM. Presence Virtual Real. **16**, 215–232 (2012)
3. Manjrekar, S., Gondhalekar, M., Sandilya, S.: CAVE: an emerging immersive technology - a review, pp. 1–6 (2014)
4. Eberhardt, B.: Head mounted displays. Seminar Ambient Computing, Universitaet zu Luebeck, SS (2012)
5. Weibo, Z., Jianxun, L., Zhi, Z.: Performance evaluation approach for image mosaicing algorithm. In: Control and Decision Conference (CCDC) (2013)
6. Ghosh, D., Park, S., Kaabouch, N., Semke, W., Fevig, R.A.: Quantitative evaluation of image mosaicing in multiple scene categories. In: Conference on Image Processing - Algorithms and Systems X and Parallel Processing for Imaging Applications II (2012)

 7. Qureshi, H.S., Khan, M.M., Hafiz, R., Cho, Y.: Quantitative quality assessment of stitched panoramic images. IET Image Process. **6**, 1348–1358 (2012)
 8. Leorin, S., Lucchese, L., Cutler, R.G.: Quality assessment of panorama video for videoconferencing applications. In: Multimedia Signal Processing (2005)
 9. Ries, M., Nemethova, O., Badic, B., Rupp, M.: Assessment of H.264 coded panorama sequences. In: 1st International Conference on Multimedia Services Access Networks (MSAN05) (2005)
10. Dinh, H.Q., Walker, N., Hodges, L.F., Song, C., Kobayashi, A.: Evaluating the importance of multisensory input on memory and the sense of presence in virtual environments. In: IEEE Virtual Reality Conference (1999)
11. Lugrin, J.L., Cavazza, M., Charles, F.: Immersive FPS games: user experience and performance. In: ACM International Workshop on Immersive Media Experiences, ImmersiveMe 2013 (2013)
12. Witmer, B.G., Singer, M.J.: Measuring presence in virtual environments: a presence questionnaire. Presence Teleoperators Virtual Environ. **7**, 225–240 (1998)
13. Lessiter, J., Freeman, J., Keogh, E., Davidoff, J.: A crossmedia presence questionnaire: the ITC-sense of presence inventory. Presence **10**, 282–297 (2001)
14. Slater, M.: Measuring presence: a response to the Witmer and Singer presence questionnaire. Presence Teleoperators Virtual Environ. **8**, 560–565 (1999)
15. Stanney, K., Kingdon, K., Graeber, D., Kennedy, R.: Human performance in immersive virtual environments: effects of exposure duration, user control, and scene complexity. Hum. Perform. **15**, 339–366 (2002)
16. So, R.H.Y., Lo, W.T., Ho, A.T.K.: Effects of navigation speed on motion sickness caused by an immersive virtual environment. Hum. Factors **43**, 452–461 (2001)
17. Kolor Eyes VR Player. http://www.kolor.com/kolor-eyes/

Environmental Sound Classification Based on Multi-temporal Resolution Convolutional Neural Network Combining with Multi-level Features

Boqing Zhu[1], Kele Xu[1,2], Dezhi Wang[3(✉)], Lilun Zhang[3], Bo Li[4],
and Yuxing Peng[1]

[1] Science and Technology on Parallel and Distributed Laboratory,
National University of Defense Technology, Changsha, China
zhuboqing09@nudt.edu.cn, pengyuxing@aliyun.com

[2] College of Information Communication, National University of Defense Technology,
Wuhan, China
kelele.xu@gmail.com

[3] College of Meteorology and Oceanography, National University of Defense
Technology, Changsha, China
wang_dezhi@hotmail.com, zll0434@163.com

[4] Beijing University of Posts and Telecommunications, Beijing, China
deepblue.lb@gmail.com

Abstract. Motivated by the fact that characteristics of different sound classes are highly diverse in different temporal scales and hierarchical levels, a novel deep convolutional neural network (CNN) architecture is proposed for the environmental sound classification task. This network architecture takes raw waveforms as input, and a set of separated parallel CNNs are utilized with different convolutional filter sizes and strides, in order to learn feature representations with multi-temporal resolutions. On the other hand, the proposed architecture also aggregates hierarchical features from multi-level CNN layers for classification using direct connections between convolutional layers, which is beyond the typical single-level CNN features employed by the majority of previous studies. This network architecture also improves the flow of information and avoids vanishing gradient problem. The combination of multi-level features boosts the classification performance significantly. Comparative experiments are conducted on two datasets: the environmental sound classification dataset (ESC-50), and DCASE 2017 audio scene classification dataset. Results demonstrate that the proposed method is highly effective in the classification tasks by employing multi-temporal resolution and multi-level features, and it outperforms the previous methods which only account for single-level features.

Keywords: Audio scene classification · Multi-temporal resolution
Multi-level · Convolutional neural network

© Springer Nature Switzerland AG 2018
R. Hong et al. (Eds.): PCM 2018, LNCS 11165, pp. 528–537, 2018.
https://doi.org/10.1007/978-3-030-00767-6_49

1   Introduction

Audio classification aims to predict the most descriptive audio tags from a set of given tags determined before the analysis. Generally, it can be divided into three main sub-domains: environmental sound classification, music classification and speech classification. Environmental sound signals are quite informative in characterizing environmental context in order to achieve a detailed understanding of the acoustic scene itself [1–3]. And a wide range of applications can be found in [4,5]. Environmental sound classification (ESC) is also very important for machines to understand the surroundings, but it is still a challenging problem, which has attracted extensive interest recently. In particular, the deep-learning based methods using more complex neural networks [6–8] have shown great potential and significant improvement in this field. Due to the capability of learning hierarchical features from high-dimensional raw data, convolutional neural networks (CNNs) based approaches have become a choice in audio classification problem.

Time-frequency representation and its variants, such as spectrograms, mel-frequency cepstral coefficients (MFCCs) [9,10], mel-filterbank features [11,12], are the most popular input for CNN-based architectures. However, the hyper-parameters (such as hop size or window size) of short time Fourier transform (STFT) in the generation of these spectrogram-based representations is normally not particularly optimized for the task, while environmental sounds actually have different discriminative patterns in terms of time-scales and feature hierarchy [13–15]. To avoid exhausting parameter search, this issue may be addressed by applying feature extraction networks that directly take raw audio waveforms as input. There are a decent number of CNN architectures that learns from raw waveforms [16,17]. The majority of them employed large-sized filters in the input convolutional layers with various sizes of stride to capture frequency-selective responses, which are carefully designed to handle their target problems. There are also a few works that used small filter and stride sizes in the input convolution layers [18,19] inspired by the VGG networks in image classification that use very small filters for convolutional layers.

Inspired by the fact that the different environmental sound tags have different performance sensitivity to different time-scales, a multi-scale convolutional neural network named WaveMsNet [20] was proposed to extract features by filter banks at multiple scales. It uses the waveform as input and facilitates learning more precise representations on a suitable temporal scale to discriminate difference of environmental sounds. After combining the representations of the different temporal resolutions, the proposed method claimed that superior performance can be achieved with waveform as input on the environmental sound classification datasets ESC-10 and ESC-50 [21]. Unlike previous attempts focusing on the adjustment the CNN architectures to enable the feature extraction at multi time-scales; in this paper, we explore to extend the approach to handle the multi-level features from hierarchical CNN layers together with multi-scale audio features in order to even further improve the current performance on the ESC problem. Similar to the network setup of DenseNet [22], the concatenation of multi-level features is implemented by direct connections between convolutional layers in a feed-forward fashion, which accounts for more hierarchical features and also allows convolutional networks to be more efficient to train with the help of similar mechanism of skip connections [23]. Moreover, our method is also evaluated on another benchmark dataset from the DCASE 2017 audio scene classification task to demonstrate the generalization capability.

In this study, we have following contributions: (1) a novel CNN-based architecture is designed that is capable of comprehensively combining the audio features with multi-temporal resolutions from raw waveforms and the multi-level features from different CNN hierarchical layers. (2) Comparatively studies are conducted to demonstrate the effect of multi time-scale and multi-level features on the classification performance of environmental sounds. (3) Explore to visualize the learned multi-temporal resolution and multi-level audio features to explain the physical meaning of what the model has really learned.

The rest of the paper is organized as follows. Section 2 discusses related work. In Sect. 3, we describe our proposed architecture with implementation details. The experimental setup and results are given in Sect. 4, while Sect. 5 concludes this paper.

2   Related Work

Due to the rapid development in signal processing and machine learning domains, there is an extensive surge of interest in applying deep learning approaches for the audio classification (or audio tagging) task. Most of the approaches with good performance [11,24] for the environmental sound classification related tasks of the DCASE 2017 challenge [2] utilize deep learning models such as CNNs, which have already become the most popular method. The frequency based features are commonly used as input of CNN models in the environmental sound classification. The frequency based features are also replaced with raw audio waves as the input for the classifiers in some studies. This kind of end-to-end learning approach has been successfully used in speech recognition [16], music genre recognition [25] and so on. Recently, a raw waveform-based approach so-called Sample-CNN model [26] shows comparable performance to the spectrogram-based CNN model in music tagging by using sample-level filters to learn hierarchical audio characteristics.

Most of the previous studies in environmental sound classification utilize only one level or one scale of features for the classification, which is typically adopted in image classification. However, this kind of method ignores that, for the audio, discriminative features are generally positioned in different levels or time-scales in a hierarchy. This issue is addressed in some work by comparing or combining multi-layer or multi-scale audio features [27,28]. The combination of different resolutions of spectrograms in terms of time-scale [27] is extensively studied for the prediction of audio tags. This idea is also further improved by using Gaussian and Laplacian pyramids [28]. Instead of concatenating the multi-scaling features only on the input layer, attempts are also made to combine audio features from different levels [15], which is believed to provide a superior performance. To increase temporal resolutions of Mel-spectrogram segments for acoustic scene classification, an architecture [29] consisting of parallel convolutional neural networks is presented where it shows a significant improvement compared with the best single resolution model. In the paper [30], mixup method is explored to provide higher prediction accuracy and robustness.

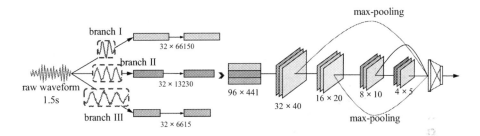

Fig. 1. Network architecture for environmental sound classification.

3   Proposed Method

In this section, we investigate the combination of multi-temporal resolution and multi-level features, for the problem of environmental sound classification.

3.1   Overview

The proposed network architecture is presented in the Fig. 1. The network is designed as an end-to-end system which takes the wave signal as input and class label as output. When training the network, we randomly select 1.5 s from the original training raw waveform data and input it into the network. The selected section is different in each epoch, and we use the same training label regardless of the selected section. When testing, we classify testing data based on probability-voting. That is, we divide the testing audio into multiple 1.5 s sections and input each of them into the network. We take the sum of all the output probabilities after softmax and use it to classify the testing data.

3.2   Multi-temporal Resolution CNN

The architecture is composed of a set of separated parallel 1-D time domain convolutional layers with different filter sizes and strides, in order to learn feature representations with multi-temporal resolutions. Specifically, to learn high-frequency features, filters with a short window are applied at a small stride. Low-frequency features, on the contrary, employ a long window that can be applied at a larger stride. Then feature maps with different temporal resolution are concatenated along frequency axis and pooled to the same dimension on the time axis.

In our experiments, we apply three branches of separated parallel 1-D convolutional layers (branch I: (size 11, stride 1), branch II: (size 51, stride 5), branch III: (size 101, stride 10)). Each branch has 32 filters. Another time-domain convolutional layer is followed to create invariance to phase shifts with filter size 3 and stride 1. We aggressively reduce the temporal resolution to 441 with a max pooling layer to each branch s feature map. Then we concatenate three feature map together to get the multi-temporal resolution features to represent the audios.

3.3   Multi-level Feature Concatenation

Next, we apply four convolutional layers for the multi-temporal resolution feature map. The two dimensions of the feature map correspond to $frequency \times time$. There are 64, 128, 256, 256 filters in each convolutional layer respectively with a size of 3×3, and we stride the filter by 1×1. We leverage non-overlapping max pooling to down-sample the features to the corresponding size as shown in Fig. 1. The outputs of the four convolutional layers are concatenated and then delivered to the full connection layers. Before the concatenation, the dimensions of the outputs are reduced to 4×5 by max pooling. In the experimental section, we investigate the effect of the concatenated layers in multi-level features. The input size of full connection layer adjusts to the dimensionality of the concatenated feature maps. For instance, when we pick features from last 3 layers, the model will have $(128 + 256 + 256) \times 4 \times 5$ dimensional feature maps.

4   Experiments

In this section, details of the DCASE 2017 ASC dataset and ESC-50 dataset used in the experiment are first introduced. Then the model parameters and experimental setup are presented for the comparison of performance between the proposed model and the previous models. We performed 5-fold cross-validation five times on the dataset. Finally, the conclusion is drawn based on the experimental results.

4.1   Dataset

We use two datasets: 2017 DCASE challenge dataset for audio scene classification task and ESC-50 dataset to validate the performance of proposed method. DCASE challenge dataset [2,3] is established to determine the context of a given recording through selecting one appropriate label from a pre-determined set of 15 acoustic scenes such as cafe/restaurant, car, city center and so on. Each scene contains 312 recordings with a length of 10 s, a sampling rate of 44.1 kHz and 24-bit resolution in stereo in the development dataset. Totally there are 4680 audio recordings in the development dataset which is provided at the beginning of the challenge, together with ground truth. Besides, an evaluation dataset is also released with 1620 audio recordings in total after the challenge submission is closed. A four-fold cross-validation setup is provided so as to make results reported strictly comparable. The evaluation dataset is used to evaluate the performance of classification models.

ESC-50 [21] dataset which is public labeled sets of environmental recordings are also used in our experiments. ESC-50 dataset comprises 50 equally balanced classes, each clip is about 5 s and sampled at 44.1 kHz. The 50 classes can be divided into 5 major groups: animals, natural soundscapes and water sounds, human non-speech sound, interior/domestic sounds, and exterior/urban noises. Datasets have been prearranged into 5 folds for comparable cross-validation and other experiments [18] used these folds. The same fold division is employed in our evaluation. The metric used is classification accuracy, and the average accuracy across the five folds is reported for comparison.

4.2   Experimental Details

For the network training, cross-entropy loss is used. To optimize the loss, the momentum stochastic gradient descent algorithm is applied with momentum 0.9. We use Rectified Linear Units (ReLUs) to implement nonlinear activation functions. A batch size of 64 is applied. All weight parameters are subjected to ℓ_2 regularization with coefficient 5×10^{-4}. We train models for 160 epochs until convergence. Learning rate is set as 10^{-2} for first 60 epochs, 10^{-3} for next 60 epochs, 10^{-4} for next 20 epochs and 10^{-5} for last 20 epochs. The weights in the time-domain convolutional layers are randomly initialized. The models in experiment are implemented by PyTorch [31] and trained on GTX Titan X GPU cards. We randomly select a 1.5 s waveform as input when training the model. In testing phase, we use the probability-voting strategy.

Table 1. Comparison of Multi-temporal resolution and single-temporal resolution.

Temporal resolution	Filter number			Mean accuracy (%)	
	Branch I	Branch II	Branch III	ESC-50	DCASE 2017
Low	96	0	0	69.1 ± 2.63	70.3 ± 3.63
Middle	0	96	0	68.2 ± 2.29	71.6 ± 3.78
High	0	0	96	68.4 ± 3.13	71.3 ± 4.02
Multi	32	32	32	**71.6** ± 2.58	**73.1** ± 3.34
Baseline [2,32]	-			64.5	61.0

4.3   Results

Effect of Multi-temporal Resolution. We compare the performances with constant filter size at three different temporal resolutions, low temporal-resolution (Low), middle temporal-resolution (Middle) and high temporal-resolution (High). These three models remain only one corresponding branch (Low remains branch I, Middle remains branch II and High remains branch III). As we reduce the number of convolution filters, it may cause performance degradation. So we use triple filters in time-domain convolution in single temporal resolution models for fair comparison. These three variant models are trained separately. Table 1 demonstrates the mean accuracy and standard error using multi-temporal resolution features and single-temporal resolution features. Both single-temporal resolution CNNs and multi-temporal resolution CNN showed better performance against the baseline. Our multi-temporal resolution model achieves average improvement of 3.0% and 2.0% compared with the single-temporal resolution models on ESC-50 and DCASE2017 dataset respectively.

Table 2. Performance with multi-level feature.

Last N layers feature	Mean accuracy (%)	
	ESC-50	DCASE 2017
$N = 1$	71.6 ± 2.58	73.1 ± 3.34
$N = 2$	71.8 ± 2.79	73.2 ± 3.27
$N = 3$	73.0 ± 2.19	73.9 ± 2.95
$N = 4$	**73.2** ± 2.90	**74.7** ± 2.46

Effect of Multi-level Features. Next, we demonstrate the effectiveness of multiple-level features. We down-sample and stack the feature map of the last N ($N = 1, 2, 3, 4$) layers of the network. They are delivered to the full connection layers. When $N = 1$, single-level features are used. As demonstrated in Table 2, the accuracies consistently increase on both datasets. Further, the performances are always benefited from the increase of N. When $N = 4$, that is, concatenating features of each layer of 2D convolution layer, we got the best result on ESC-50 and DCASE2017.

Analysis of the Results. Here, we present the analysis of the multi-temporal resolution features. The technique of visualizing the filters at different branches [26] can provide deeper understanding of what the networks have learned from the raw waveforms. Figure 2 shows the responses of the multi-temporal resolution feature maps. Most of the filters learn to be band-pass filters while the filters are sorted by their central frequencies from low to high as shown in the figure. Branch I has learned more dispersed bands across the frequency that can extract the features from all frequencies. But the frequency resolution

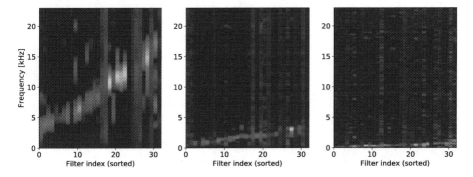

Fig. 2. Frequency response of the multi-temporal resolution feature maps. Left shows the frequency response of feature map product by branch I. Middle corresponds to branch II. Right corresponds to branch III.

is lower. On the contrary, branch III has learned high-frequency resolution bands and most of them locate at the low-frequency area. Branch II behaves between branch I and III. This indicts that different branch could learn discrepant features, and the filter banks split responsibilities based on what they efficiently can represent. This explains why multi-temporal resolution models get a better performance than the single-temporal resolution model shown in Table 1.

5   Conclusion

In this article, we proposed an effective CNN architecture integrating the networks for multi-temporal resolution analysis and multi-level feature extraction in order to achieve more comprehensive feature representations of audios and tackle the multi-scale problem in the environmental sound classification. Through the experiments, it is shown that combining the multi-level and multi-scale features improves the overall performance. The raw waveforms are directly taken as the model input, which enables the proposed approach to be applied in an end-to-end manner. The frequency response of learned filters at the model branches with different temporal resolutions is visualized to better interpret the multi time-scale effect on filter characteristics. In future, we would like to evaluate the performance of our method on a large-scale dataset of Google AudioSet for the general-purpose audio tagging task.

Acknowledgments. This study is funded by the National Basic Research Program of China (973) under Grant No. 2014CB340303 and the Scientific Research Project of NUDT (No. ZK17-03-31).

References

1. Marchi, E., et al.: Pairwise decomposition with deep neural networks and multi-scale kernel subspace learning for acoustic scene classification. In: Proceedings of the Detection and Classification of Acoustic Scenes and Events 2016 Workshop (DCASE 2016), pp. 65–69 (2016)
2. Mesaros, A., et al.: DCASE 2017 challenge setup: tasks, datasets and baseline system. In: DCASE 2017-Workshop on Detection and Classification of Acoustic Scenes and Events (2017)
3. Stowell, D., Giannoulis, D., Benetos, E., Lagrange, M., Plumbley, M.D.: Detection and classification of acoustic scenes and events. IEEE Trans. Multimed. **17**(10), 1733–1746 (2015)
4. Bugalho, M., Portelo, J., Trancoso, I., Pellegrini, T., Abad, A.: Detecting audio events for semantic video search. In: Tenth Annual Conference of the International Speech Communication Association (2009)
5. Eyben, F., Weninger, F., Gross, F., Schuller, B.: Recent developments in opensmile, the munich open-source multimedia feature extractor. In: Proceedings of the 21st ACM International Conference on Multimedia, pp. 835–838. ACM (2013)
6. Han, Y., Lee, K.: Acoustic scene classification using convolutional neural network and multiple-width frequency-delta data augmentation. arXiv preprint arXiv:1607.02383 (2016)
7. Fonseca, E., Gong, R., Bogdanov, D., Slizovskaia, O., Gómez Gutiérrez, E., Serra, X.: Acoustic scene classification by ensembling gradient boosting machine and convolutional neural networks. In: Virtanen, T., et al. (eds.) Detection and Classification of Acoustic Scenes and Events 2017 Workshop (DCASE2017), 16 Nov 2017, Munich, Germany, Tampere (Finland): Tampere University of Technology 2017, pp. 37–41 (2017)
8. Abeßer, J., Mimilakis, S.I., Gräfe, R., Lukashevich, H., Fraunhofer, I.: Acoustic scene classification by combining autoencoder-based dimensionality reduction and convolutional neural networks (2017)
9. Valero, X., Alias, F.: Gammatone cepstral coefficients: biologically inspired features for non-speech audio classification. IEEE Trans. Multimed. **14**(6), 1684–1689 (2012)
10. Boddapati, V., Petef, A., Rasmusson, J., Lundberg, L.: Classifying environmental sounds using image recognition networks. Procedia Comput. Sci. **112**, 2048–2056 (2017)
11. Jung, J.W., Heo, H.S., Yang, I.H., Yoon, S.H., Shim, H.J., Yu, H.J.: DNN-based audio scene classification for DCASE 2017: dual input features, balancing cost, and stochastic data duplication. System **4**(5) (2017)
12. Deng, J., et al.: The university of passau open emotion recognition system for the multimodal emotion challenge. In: Tan, T., Li, X., Chen, X., Zhou, J., Yang, J., Cheng, H. (eds.) CCPR 2016. CCIS, vol. 663, pp. 652–666. Springer, Singapore (2016). https://doi.org/10.1007/978-981-10-3005-5_54
13. Lee, J., Park, J., Kim, K.L., Nam, J.: Sample-level deep convolutional neural networks for music auto-tagging using raw waveforms. arXiv preprint arXiv:1703.01789 (2017)
14. Xu, Y., Huang, Q., Wang, W., Plumbley, M.D.: Hierarchical learning for dnn-based acoustic scene classification. arXiv preprint arXiv:1607.03682 (2016)
15. Lee, J., Nam, J.: Multi-level and multi-scale feature aggregation using pretrained convolutional neural networks for music auto-tagging. IEEE Signal Process. Lett. **24**(8), 1208–1212 (2017)

16. Sainath, T.N., Weiss, R.J., Senior, A., Wilson, K.W., Vinyals, O.: Learning the speech front-end with raw waveform CLDNNs. In: Sixteenth Annual Conference of the International Speech Communication Association (2015)
17. Dai, W., Dai, C., Qu, S., Li, J., Das, S.: Very deep convolutional neural networks for raw waveforms. In: 2017 IEEE International Conference on Acoustics, Speech and Signal Processing (ICASSP), pp. 421–425. IEEE (2017)
18. Tokozume, Y., Harada, T.: Learning environmental sounds with end-to-end convolutional neural network. In: IEEE International Conference on Acoustics, Speech and Signal Processing, pp. 2721–2725 (2017)
19. Palaz, D., Magimai.-Doss, M., Collobert, R.: Analysis of CNN-based speech recognition system using raw speech as input. Technical report, Idiap (2015)
20. Zhu, B., Wang, C., Liu, F., Lei, J., Lu, Z., Peng, Y.: Learning environmental sounds with multi-scale convolutional neural network. arXiv preprint arXiv:1803.10219 (2018)
21. Piczak, K.J.: ESC: dataset for Environmental Sound Classification. In: Proceedings of the 23rd Annual ACM Conference on Multimedia, pp. 1015–1018. ACM Press (2015). https://doi.org/10.1145/2733373.2806390. http://dl.acm.org/citation.cfm?doid=2733373.2806390
22. Huang, G., Liu, Z., Weinberger, K.Q., van der Maaten, L.: Densely connected convolutional networks. In: Proceedings of the IEEE Conference on Computer Vision and Pattern Recognition, vol. 1, p. 3 (2017)
23. He, K., Zhang, X., Ren, S., Sun, J.: Deep residual learning for image recognition. In: Computer Vision and Pattern Recognition, pp. 770–778 (2016)
24. Han, Y., Park, J., Lee, K.: Convolutional neural networks with binaural representations and background subtraction for acoustic scene classification (2017)
25. Dieleman, S., Schrauwen, B.: End-to-end learning for music audio. In: 2014 IEEE International Conference on Acoustics, Speech and Signal Processing (ICASSP), pp. 6964–6968. IEEE (2014)
26. Lee, J., Park, J., Kim, K.L., Nam, J.: SampleCNN: end-to-end deep convolutional neural networks using very small filters for music classification. Appl. Sci. 8(1), 150 (2018)
27. Hamel, P., Bengio, Y., Eck, D.: Building musically-relevant audio features through multiple timescale representations. In: ISMIR, pp. 553–558 (2012)
28. Dieleman, S., Schrauwen, B.: Multiscale approaches to music audio feature learning. In: 14th International Society for Music Information Retrieval Conference (ISMIR-2013), pp. 116–121. Pontifícia Universidade Católica do Paraná (2013)
29. Schindler, A., Lidy, T., Rauber, A.: Multi-temporal resolution convolutional neural networks for the dcase acoustic scene classification task (2017)
30. Xu, K., et al.: Mixup-based acoustic scene classification using multi-channel convolutional neural network. arXiv preprint arXiv:1805.07319 (2018)
31. Paszke, A., et al.: Automatic differentiation in pytorch (2017)
32. Piczak, K.J.: Environmental sound classification with convolutional neural networks. In: 2015 IEEE 25th International Workshop on Machine Learning for Signal Processing (MLSP), pp. 1–6. IEEE (2015)

Discriminative Dictionary Learning Based on Sample Diversity for Face Recognition

Yuhong Wang[1,2,3], Shigang Liu[1,2,3], Yali Peng[1,2(✉)], and Han Cao[2,3]

[1] Key Laboratory of Modern Teaching Technology, Ministry of Education,
Xi'an 710062, China
pengyl_sx@163.com
[2] Engineering Laboratory of Teaching Information Technology
of Shaanxi Province, Xi'an 710119, China
[3] School of Computer Science, Shaanxi Normal University,
Xi'an 710119, China

Abstract. Dictionary learning algorithms applied to face recognition always suffer from the following problems. It is difficult for face databases to provide sufficient training samples, which hinders algorithms from extracting reliable atoms. Then, facial images are susceptible to external conditions. And even the facial images from the same individual vary with facial poses and expressions. That is the reason why most of the dictionary algorithms are not robust. Moreover, the discrimination of algorithm is limited due to ignoring the locality characteristics. We proposed a novel discriminative dictionary learning algorithm framework based on sample diversity to solve the above problems. First of all, the skillfully generated virtual face images can enrich sample diversity and alleviate the small sample size problem. Secondly, new error constraints are added to the elaborate objective function to restrain outliers from the training samples and make algorithm robust. Thirdly, the local information of atoms is preserved via the graph Laplacian matrix of dictionary instead of directly using the training samples, which aims to enhance the discrimination of the dictionary and reduce the influence of noise. Experimental results show that the proposed dictionary learning algorithm framework can achieve higher performance level than some previous state-of-the-art algorithms.

Keywords: Dictionary learning · Face recognition · Locality constrained
Sample diversity

1 Introduction

Recently, dictionary learning has attracted more and more attention and also obtained some remarkable achievements for face recognition [1–3]. However, the applications of dictionary learning for face recognition are very limited in practice. On the one hand, face recognition is a typical small sample size problem [4–6]. Insufficient samples are unfavorable for algorithm to adequately extract facial information from high-dimensional images and to recognize faces. On the other hand, face images are not stable [7–9]. The face images are not easily disturbed by external factors only, but the face images have many changes in expressions and postures. Therefore, it is very

© Springer Nature Switzerland AG 2018
R. Hong et al. (Eds.): PCM 2018, LNCS 11165, pp. 538–546, 2018.
https://doi.org/10.1007/978-3-030-00767-6_50

necessary for face recognition to find a dictionary learning algorithm which is robust and insensitive to expression, illumination, and noise [10–14].

In order to solve these problems, researchers have proposed some learning algorithms [15–19]. However, these algorithms could not solve the problems fundamentally, such as small sample size and various face images. Moreover, ignoring the local structures [20–22] of the training samples weakens discrimination of the learned dictionary.

In this paper, we propose a new discriminative dictionary learning algorithm based on sample diversity. Firstly, we allow both the original samples and virtual samples generated elaborately as training samples. Secondly, we construct locality constraint via the similarities between atoms to inherit the local geometric characters of data and to enhance the discrimination. In addition, we add new noise constraint term to constraints of the virtual samples by modeling; that is, we not only preserve the sample diversity but also further limit the noise caused by the virtual training samples.

As for the general structure of this paper, the proposed algorithm framework is explained in Sect. 2. Section 3 demonstrates the experimental results and analyses. The overall conclusion is given in the last section.

2   The Proposed Algorithm Framework

Given N training samples $Y = [y_1, y_2 \ldots y_N] \in \Re^{n \times N}$ and C classes, let the labels of all training samples be recorded in label matrix $H = [h_1, h_2 \ldots h_N] \in \Re^{C \times N}$. Let $D = [d_1, d_2 \ldots d_K] \in \Re^{n \times K}$ denote dictionary matrix where d_i is the i th atom, and K is the number of the atoms in the dictionary. Coding coefficients matrix is $X = [x_1, x_2 \ldots x_N] \in \Re^{K \times N}$.

2.1   Locality Constraint of Atoms

According to [23, 24], we use the local structure information of the atoms instead of the original training samples. k-nearest neighbor graph M is constructed to describe the structure of the atoms in dictionary,

$$M_{i,j} = \begin{cases} \exp\left(-\frac{\|d_i - d_j\|_2}{\sigma}\right), & \text{if } d_j \in k - NN(d_i) \\ 0, & \text{else} \end{cases}, \tag{1}$$

where $k\text{-}NN(d_i)$ denotes the set that contains k nearest atoms of d_i, σ is parameter, and $M_{i,j}$ is the similarity between d_i and d_j.

Zheng et al. [25] discovers profiles [26] can also represent the intrinsic geometry of the data distribution in manifold learning. Thereby we hope to employ both atoms and profiles to construct local constraints. According to [25], we need to minimize the following formulas.

$$\frac{1}{2} \sum_{i=1}^{K} \sum_{j=1}^{K} \|\hat{x}_i - \hat{x}_j\|_2 M_{i,j} = Tr(X^T L X), \quad L = T - M, \tag{2}$$

where L is the Laplacian matrix, M can be calculated by Eq. (1), and $T = diag(t_1, t_2 \ldots t_K)$, $\quad t_i = \sum_{j=1}^{K} M_{i,j}$.

2.2 The Objective Function

Many dictionary learning algorithms for face recognition are often not robust due to sensitive facial images and noises included in training samples. So this paper provides the following objective function based on restrained sample diversity and local information of atoms:

$$\min_{D,X,L,E} ||Y - DX||_2^2 + \lambda_1 ||Y_{alter} - DX - E||_2^2 + \lambda_2 Tr(X^T L X) + \lambda_3 ||X||_2^2 + \lambda_4 ||E||_2^2$$

$$s.t. ||d_i||^2 = 1, i = 1, 2 \ldots, K, \tag{3}$$

where Y_{alter} denotes the alternative training samples generated by flexible and elaborate approaches, E represents noises and outliers matrix of the alternative training samples, L is the Laplacian matrix, and λ_1, λ_2, λ_3, λ_4 are all controllable parameters.

As for Eq. (3), the first term of the objective function is the reconstruction error term of the original training samples; the second term is the reconstruction error term of the alternative training samples; the third term is the local constraint which is designed to promote the discrimination of the algorithm; the fourth one guarantees the numerical stability of X; the fifth one ensures the differences between the original training samples and the alternative training samples in a reasonable range.

2.3 Optimization of Objective Function

In this subsection, it optimizes our framework and solves each variable by fixing the remaining variables.

Firstly, dictionary D and coefficient matrix X can be initialized by using the K-SVD algorithm [27], and Laplacian matrix L is initialized according to Eq. (2).

Then, we fix D, X, and L in order to obtain coefficient matrix X, and the objective function has a closed form solution,

$$X = (D^T D + \lambda_1 D^T D + \lambda_2 L + \lambda_3 I)^{-1}(D^T Y + \lambda_1 D^T Y_{alter} - \lambda_1 D^T E). \tag{4}$$

Similarly, we can calculate E and D as follows.

$$E = (\lambda_1 Y_{alter} - \lambda_1 DX)/(\lambda_1 + \lambda_4), \tag{5}$$

$$D = (YX^T + \lambda_1 Y_{alter} X^T - \lambda_1 EX^T)(XX^T + \lambda_1 XX^T)^{-1}. \tag{6}$$

Moreover, we can update Laplacian matrix L according to Eq. (2).

2.4   The Classification Procedure

Besides, linear classifier is applied to the framework. The classifier parameter matrix $W \in \Re^{C \times K}$ can be derived as follows according to [21].

$$W = HX^T(XX^T + I)^{-1}. \tag{7}$$

In addition, we solve the sparse coding model via the OMP algorithm [28] and get sparse coding vector x^{test} of test sample y^{test}. Finally, y^{test} is classified by using label vector $g^{test} = Wx^{test}$. Hence, our dictionary learning framework and entire classification process are reflected in Table 1.

Table 1. The proposed dictionary learning framework.

Input: Training sample matrix and alternative training sample matrix; λ_1, λ_2, λ_3, λ_4 ; test sample; label matrix of training samples.

Output: Dictionary; coding coefficients matrix; sparse coding vector; label of the test sample.

Step 1. Initialization: obtaining initial dictionary and coding coefficients matrix via the K-SVD algorithm;

Step 2. While not converged **do**:

Step 3. Update X using Eq. (4) .

Step 4. Update E using Eq. (5) .

Step 5. Update D using Eq. (6) .

Step 6. Update L using Eq. (2) .

Step 7. end while

Step 8. Calculate W using Eq. (7).

Step 9. Calculate sparse coding vector x^{test} via the OMP algorithm.

Step 10. Let $g^{test} = Wx^{test}$, $c^{test} = \arg\max_i(g_i)$, then test sample is classified in the c^{test} th class.

In general, differing from previous dictionary learning algorithms, our dictionary learning framework has the following characteristics:

(1) Our algorithm invokes double training samples to improve the diversity of the training samples, which makes it possible to recognize multi-pose test samples.

(2) More training samples may introduce some unavoidable noise and outliers. In order to resolve the problem, the noise model is added to the proposed algorithm. So that it can clearly constrain the differences between the dual samples and make the algorithm more robust.

(3) Our algorithm brings in local constraints of atoms, which can effectively improve the discrimination of the dictionary.

(4) Our algorithm is just a framework. There are various ways to flexibly generate alternative samples, such as the corrupted images of the original training samples. This paper invokes samples corrupted by Gaussian noise and Salt & Pepper noise. We call these algorithms GN-DDL and SPN-DDL respectively.

3   Experimental Results and Analysis

In this section, we arrange a large number of experiments on the Labeled Faces in the Wild (LFW) [29] and AR [30] face databases. In order to demonstrate the performance of our algorithm, the paper compares the algorithm with other four algorithms at the same time, which are LLC [31], LC- KSVD [32], GN-DL [20], SPN-DL [20].

3.1   Experimental Settings

The "imnoise" function of MATLAB produces the corrupted samples. The mean and variance of the Gaussian noise are set to 0 and 0.1, respectively. The noise density of Salt and Pepper noise is set to 0.1. According to [32], the sparsity is set to 30. As for the GN-DL and SPN-DL algorithms, we use the parameters in [20]. And the algorithms perform classification with the linear classifier method [21].

3.2   Experimental Results on the AR Face Database

The AR face database is selected from 3120 images as data. And each image is resized to 40×50. We use 7 face images in the first session and one corrupted face image of each class as training samples (the first sunglasses image in sessions are not used as training samples). All algorithms are repeated 10 times, and then the average results are recorded in Table 2. The number of atoms is equal to 720. Meanwhile, the optimal parameters of our algorithm are $\lambda_1 = 0.001$, $\lambda_2 = 0.00001$, $\lambda_3 = 0.001$, $\lambda_4 = 0.001$.

As for Table 2, the average recognition rate of our algorithm reaches the highest value among all algorithms. And we use different numbers of atoms ($K = 120$, $240, \ldots, 600, 720$) to evaluate the algorithms. The experimental results are recorded in Fig. 1. We can find that average recognition rates of our algorithm are still better than other algorithms, and GN-DDL gets the highest average recognition rate when the number of atoms is equal to 720.

Table 2. The average recognition rates, training time, and testing time on the AR face database.

Algorithm	Average recognition rates (%) ± standard deviation	Training time (s)	Testing time (s)
LLC	72.3 ± 0.060	-	8.51e−2
LC-KSVD	75.1 ± 0.058	81.7	1.63e−3
GN-DL	80.2 ± 0.067	46.5	1.69e−3
SPN-DL	79.1 ± 0.066	46.5	1.69e−3
GN-DDL	82.1 ± 0.056	46.7	1.69e−3
SPN-DDL	80.8 ± 0.063	46.7	1.69e−3

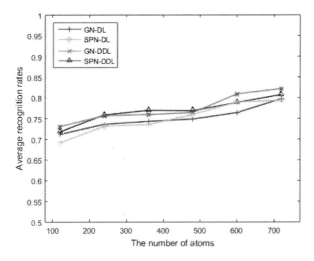

Fig. 1. The average recognition rates with different numbers of atoms on the AR face database.

3.3 Experimental Results on the LFW Face Database

According to [23], a subset which includes 86 people totaling 1215 images is generated as dataset, and each image is normalized to the size of 32×32. We randomly select 8 face images of each class (including the first 5 images) as the training samples. We repeat all algorithms 10 times, and results are saved in the Table 3. The number of atoms is equal to 516. In addition, $\lambda_1 = 0.1$, $\lambda_2 = 0.0001$, $\lambda_3 = 0.001$, $\lambda_4 = 0.001$.

Table 3. The average recognition rates, training time, and testing time on the LFW face database.

Algorithm	Average recognition rates (%) ± standard deviation	Training time (s)	Testing time (s)
LLC	26.3 ± 0.013	-	3.9e−2
LC-KSVD	28.7 ± 0.012	9.91	1.81e−3
GN-DL	33.9 ± 0.015	8.03	1.99e−3
SPN-DL	33.6 ± 0.020	8.03	1.99e−3
GN-DDL	35.5 ± 0.012	8.15	1.95e−3
SPN-DDL	35.6 ± 0.015	8.15	1.95e−3

According to Table 3 we can find that our algorithm framework can obtain higher average recognition rate and smaller standard deviation than other algorithms where $K = 516$. At the same time, we run the algorithms with different numbers of atoms

($K = 86, 172, 258, 344$) and the experimental results are shown in Fig. 2. It can be seen that GN-DDL achieves the best experimental result.

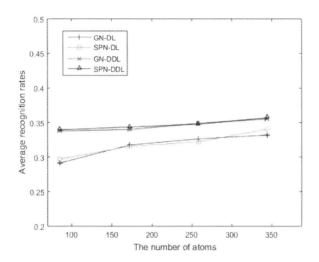

Fig. 2. The average recognition rates with different numbers of atoms on the LFW face database.

3.4   Experimental Analysis

We have the following discoveries regarding to the experimental results.

In Tables 2 and 3, our algorithm can obtain better experimental results than LLC algorithm. LLC algorithm uses training samples to classify tests directly, which potentially introduces noise from the training samples. Compared with the LC-KSVD algorithm, our algorithm has more outstanding performance. It may due to that LC-KSVD algorithm lacks the usage of local information and available sample diversity. In other words, the local constraint and sample diversity of our algorithm overcome these disadvantages to some extent. Compared with the GN-DL and SPN-DL algorithms, our algorithms either obtain higher average recognition rates or are comparable to them in all cases, which indicates that the added constraints can make the algorithm more discriminative and robust.

Figures 1 and 2 indicate that the average recognition rates are improved with the increasing number of atoms. At the same time, the curves also show that GN-DDL and SPN-DDL algorithms are always superior to the GN-DL and SPN-DL algorithms, as the number of atoms grows.

Tables 2 and 3 show that GN-DL and SPN-DL algorithms and our algorithm framework require approximately the same training time and LC-KSVD algorithm requires more. Since LC-KSVD, GN-DL, SPN-DL, and our algorithms all use the same linear classifier to perform the classification, the testing time is nearly equal and shorter than LLC algorithm.

4 Conclusion

We propose an innovative discriminative dictionary learning algorithm framework based on sample diversity. The main advantage of this algorithm is that it can maintain the sample diversity while enhancing the robustness and discrimination of the dictionary. In addition, this algorithm is also a flexible framework that allows various techniques to produce alternative samples according to different requirements. The experimental results also show that our algorithm framework is superior to some previous state-of-the-art dictionary learning and sparse coding algorithms.

Acknowledgements. This work is supported by the National Natural Science Foundation of China (Nos. 61672333, 61703096, 41471280), the Natural Science Foundation of Shaanxi Province of China (No. 2018JM6050), China Postdoctoral Science Foundation (No. 2017M611655), the Program of Key Science and Technology Innovation Team in Shaanxi Province (No. 2014KTC-18), the Key Science and Technology Program of Shaanxi Province (No. 2016GY-081), the National Natural Science Foundation of Jiangsu Province (No. BK20170691), the Fundamental Research Funds for the Central Universities (Nos. GK201803059, GK201803088), Interdisciplinary Incubation Project of Learning Science of Shaanxi Normal University.

References

1. Peng, Y., Li, L., Liu, S., Li, J., Wang, X.: Extended sparse representation based classification method for face recognition. Mach. Vis. Appl. (2018). https://doi.org/10.1007/s00138-018-0941-z
2. Li, L., Liu, S., Peng, Y., Sun, Z.: Overview of principal component analysis algorithm. Optik Int. J. Light Electron Opt. **127**(9), 3935–3944 (2016)
3. Hong, R., Zhang, L., Zhang, C., Zimmermann, R.: Flickr circles: aesthetic tendency discovery by multi-view regularized topic modeling. IEEE Trans. Multimed. **18**(8), 1555–1567 (2016)
4. Zhang, X., Hu, W., Xie, N., Bao, H., Maybank, S.: A robust tracking system for low frame rate video. Int. J. Comput. Vis. **115**(3), 279–304 (2015)
5. Liu, Z., Qiu, Y., Peng, Y., Pu, J., Zhang, X.: Quaternion based maximum margin criterion method for color face recognition. Neural Process. Lett. **45**(3), 913–923 (2017)
6. Xu, Y., Zhong, A., Yang, J., Zhang, D.: LPP solution schemes for use with face recognition. Pattern Recogn. **43**(12), 4165–4176 (2010)
7. Hong, R., Zhang, L., Tao, D.: Unified photo enhancement by discovering aesthetic communities from Flickr. IEEE Trans. Image Process. **25**(3), 1124–1135 (2016)
8. Zhang, X., Hu, W., Chen, S., Maybank, S.: Graph-embedding-based learning for robust object tracking. IEEE Trans. Ind. Electron. **61**(2), 1072–1084 (2014)
9. Peng, Y., Li, L., Liu, S., Lei, T., Wu, J.: A new virtual samples-based CRC method for face recognition. Neural Process. Lett. (2017). https://doi.org/10.1007/s11063-017-9721-4
10. Ke, J., Peng, Y., Liu, S., Li, J., Pei, Z.: Face recognition based on symmetrical virtual image and original training image. J. Mod. Opt. **65**(4), 367–380 (2018)
11. Xu, Y., Roy-Chowdhury, A., Patel, K.: Pose and illumination invariant face recognition in video. In: IEEE Conference on Computer Vision & Pattern Recognition, pp. 1–7 (2007)

12. Liu, S., Zhang, X., Peng, Y., Cao, H.: Virtual images inspired consolidate collaborative representation based classification method for face recognition. J. Mod. Opt. **63**(12), 1181–1188 (2016)

13. Zhang, X., Li, C., Tong, X., Hu, W., Maybank, S., Zhang, Y.: Human pose estimation and tracking via parsing a tree structure based human model. IEEE Trans. Syst. Man Cybern. Syst. **44**(5), 580–592 (2014)

14. Li, L., Peng, Y., Qiu, G., Sun, Z., Liu, S.: A survey of virtual sample generation technology for face recognition. Artif. Intell. Rev. **50**(1), 1–20 (2018)

15. Zhang, K., Peng, Y., Liu, S.: Discriminative face recognition via kernel sparse representation, multimedia tools and applications (2018). https://doi.org/10.1007/s11042-018-6110-6

16. Zhang, X., Peng, Y., Liu, S., Wu, J., Ren, P.: A supervised dimensionality reduction method based sparse representation for face recognition. J. Mod. Opt. **64**(8), 799–806 (2017)

17. Ke, J., Peng, Y., Liu, S., Wu, J., Qiu, G.: Sample partition and grouped sparse representation. J. Mod. Opt. **64**(21), 2289–2297 (2017)

18. Peng, Y., Li, L., Liu, S., Lei, T.: Space-frequency domain based joint dictionary learning and collaborative representation for face recognition. Sig. Process. **147**, 101–109 (2018)

19. Hong, R., Hu, Z., Wang, R., Wang, M., Tao, D.: Multi-view object retrieval via multi-scale topic models. IEEE Trans. Image Process. **25**(12), 5814–5827 (2016)

20. Xu, Y., Li, Z., Zhang, B., Yang, J., You, J.: Sample diversity, representation effectiveness and robust dictionary learning for face recognition. Inf. Sci. **37**, 171–182 (2017)

21. Xu, Y., Li, Z., Yang, J., Zhang, D.: A survey of dictionary learning algorithms for face recognition. IEEE Access **5**(99), 8502–8514 (2017)

22. Xu, Y., Zhu, Q., Fan, Z., Zhang, D., Mi, J.: Using the idea of the sparse representation to perform coarse-to-fine face recognition. Inf. Sci. **238**(7), 138–148 (2013)

23. Peng, Y., Zhang, Y., Liu, S., Wang, S., Guo, M.: Kernel negative ε dragging linear regression for pattern classification. Complexity (2017). 14 pages, Article ID 2691474

24. Liu, S., Peng, Y., Ben, X., Yang, W., Qiu, G.: A novel label learning algorithm for face recognition. Sig. Process. **124**, 141–146 (2016)

25. Zheng, M., et al.: Graph regularized sparse coding for image representation. IEEE Trans. Image Process. **20**(5), 1327–1336 (2011)

26. Peng, Y., Liu, S., Lei, T., Li, J., Guo, M.: Negative ε dragging technique for pattern classification. IEEE Access **6**(1), 488–494 (2018)

27. Aharon, M., Elad, M., Bruckstein, A.: K-SVD: an algorithm for designing overcomplete dictionaries for sparse representation. IEEE Trans. Signal Process. **54**(11), 4311–4322 (2016)

28. Zhang, J., Yan, K., He, Z.: Improved OMP selecting sparse representation used with face recognition. In: IEEE International Conference on Software Engineering and Service Since, pp. 589–592 (2014)

29. Huang, G., Ramesh, M., Berg, T., Learned-Miller, E.: Labeled faces in the wild: a database for studying face recognition in unconstrained environments, Univ. Massachusetts, Amherst, MA, USA, Tech. Rep. 07-49 (2007)

30. Martinez, A., Benavente, R.: The AR face database, CVC, New Delhi, India, Tech. Rep. #24 (1998)

31. Wang, J., Yang, J., Yu, K., Lv, F., Huang, T.: Locality-constrained linear coding for image classification. Comput. Vis. Pattern Recognit. **119**(5), 3360–3367 (2010)

32. Jiang, Z., Lin, Z., Davis, L.S.: Learning a discriminative dictionary for sparse coding via label consistent K-SVD. In: Proceedings of IEEE Conference Computer Vision Pattern Recognition, pp. 1697–1704 (2011)

Spatial Attention Network for Head Detection

Rongchun Li$^{(\boxtimes)}$, Biao Zhang, Zhen Huang, Xiang Zhao, Peng Qiao,
and Yong Dou

National Laboratory for Parallel and Distributed Processing,
National University of Defense Technology, Changsha 410073, China
{rongchunli,huangzhen,pengqiao,yongdou}@nudt.edu.cn
zhtbiao@163.com, yoobright@163.com

Abstract. Human head detection is widely used in computer vision. However, in practical applications, human head detection is likely to cause false alarms because of the angle, light condition, and cameras. This paper proposes a novel spatial attention network (SAN) which adopts the saliency module to exploit the environmental information beyond the proposal which is ignored in the Faster-RCNN. At the meantime, the class score and saliency score are fused together through a suitable strategy to effectively suppress false positive samples. In order to train and test our model, this paper has established a dataset including 55,802 images. We have evaluated our method and the final experimental results show that our model is significantly superior to the Faster-RCNN model.

Keywords: Head detection · CNN · Spatial Attention Network

1 Introduction

In computer vision research, face detection and pedestrian detection are two important research directions in recent years and have made great progress. Among them, pedestrian detection is a key component in computer vision tasks such as pedestrian identification, motion recognition, pedestrian attribute analysis, and automatic driving. However, in the actual scene application, face detection and pedestrian detection still remain to be very challenging. Face detection requires frontage face which means that a person who turns his back to the camera can not be detected. In pedestrian detection, many of the pedestrians are obstructed by objects and most part of body are not visible. In order to detect people more accurately, head detection is a more effective pedestrian detection method and can be applied to in-door scenarios.

Because of the different scales of the human head, too many similar objects, and obstructions, there are so many false alarms that the accuracy of the current human head detection can not satisfy the requirement for the practical applications. How to extract human head features and effectively distinguish between

© Springer Nature Switzerland AG 2018
R. Hong et al. (Eds.): PCM 2018, LNCS 11165, pp. 547–557, 2018.
https://doi.org/10.1007/978-3-030-00767-6_51

heads and backgrounds, especially those similar to human heads, is a key issue worthy of study.

Recently, the Convolutional Neural Networks (CNN) have brought great progress in the computer vision, especially the image classification and object detection. In particular, many CNN-based object detection frameworks, such as Faster-RCNN [1], R-FCN [2], YOLO [3], and SSD [4], can achieve significant gains. However, these models treat all the objects as independent ones with the bounding boxes. The global information in the scene beyond objects also can supply contextual cues for the object detection and recognition.

This paper uses the saliency network SAN to consider the contextual information around the object proposals which is ignored in the Faster-RCNN.

2   Related Work

Due to the different size and appearance of the head, how to effectively use the extracted features to locate the head and distinguish it from the background is still a big problem. Many previous methods used different levels of deep convolutional networks generated by multi-scale features. SSD [4] seeks to use multi-scale features to estimate class probabilities and bounding box coordinates. Hariharan et al. [5] encode different levels of connected rescaled feature maps into vectors for each location, called super-columns. In the paper [6], a top-down framework is proposed for constructing high-level semantic feature maps and predicting feature maps at different scales. Other methods, such as HyperNet [7], concatenate multiple layers in the CNN to predict the final scores. All the works above demonstrate that the multi-scale features can be use in the object detection.

There are few works related to the head detection. Vu et al. [8] proposed a joint CNN framework which combines the local, global and pairwise models. In the global model, they leverage person-scene relations to predict positions and scales of heads directly from the full image. In the pairwise model, they explicitly model pairwise relations among objects. However, this paper is presented based on the R-CNN, which is obsolete model for the object detection. At the meantime, their work do not solve the problem of the false alarm of head proposals. As a result, this model is not suitable for the practical applications.

In this paper, we propose a multi-scale feature network named Spatial Attention Network (SAN) to precisely locate the human head in the image by concatenating the multi-scale feature maps together. As a consequence, the SAN can learn the target salient features and information about the location of the environment nearby, which raises the probability of the head.

At the meantime, the current head detection datasets, such as the Brainwash [9] and Hollywood [8], only focus on the particular scenes where each human head is independent of each other and there is no obstructions for the heads. So the existing head datasets can not satisfy the requirement of practical application. we also collected and marked a head detection dataset named NUDT-HEAD, which promotes the diversity of data and can enhance the generalization ability of the proposed network.

In summary, the main contributions of this article are as follows:

(1) We propose a novel attention network SAN by adding environmental information around the human heads, which promote the accuracy for head detection.
(2) We present an automatic filtering and fusion method between SAN and
Faster-RCNN to integrate our environmental information into the Faster-
RCNN prediction results.
(3) We propose a new dataset to effectively expand the application and test
scenarios of head detection.

3 Method

3.1 Overall Architecture

Our framework is shown in Fig. 1. It mainly consists of three parts: Faster-RCNN,
SAN, and Fusion Processing. Firstly, feature maps are obtained by using the deep
convolution network. Then the feature maps are both sent to the Faster-RCNN
and SAN. On the one hand, Faster-RCNN performs the human head detection
to obtain the scores and boxes of the head proposals. On the other hand, SAN
performs the saliency detection of the human head, and obtains feature map
scores. Finally, Fusion Processing mainly integrates the detection results of the
two branches of Faster-RCNN and SAN to obtain the final prediction result.
Among them, SAN carries out saliency detection in the level of feature map.
Because Faster-RCNN can extract information such as semantic information
and geographical location in the picture, we directly extract Feature Extraction
as the input of SAN. As a result, the SAN shares feature extraction module with
the Faster-RCNN. Then, SAN makes the saliency predictions of the human head
proposal in the image.

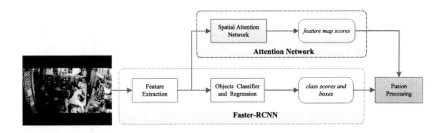

Fig. 1. The proposed architecture for head detection

The authors in the previous work [10] figure out that the head size distribution of more than 80% of the target size is between 16 and 400. We designed SAN
to consider the importance of different sizes and the surrounding environment of
the target. Compared to the fact that the Faster-RCNN network performs target
detection independently and does not consider the environment information, the

SAN can learn the salient features and the environmental information around the object. The environmental information adds weight to the head proposal and raises the probability of the proposal.

We now describe the network structure of SAN. As shown in Fig. 2: First, we use the three 3×3 convolution layers each with 256 filters to further extract Feature features, each layer has 256 convolution kernels. Then, the information of different feature layers is concatenated, in which we use 1×1 convolution layer with 256 filters to extract Feature. Finally, each pixel of the $1 \times 1 \times 1024$ concatenated output information is to judge whether the receptive field of the original image corresponding to this pixel contains the key information of the object. In this step, we use 1×1 convolution layer and sigmoid activation to obtain the saliency detection result. According to the statistical analysis of the paper [10], the 80% of the size of the face and the head are distributed in the range of 16 to 400 pixels. The pixel of feature maps is 16*16 for the receptive field. As a result, the shallow and deep 4-layer features are combined to detect the significant position of the target.

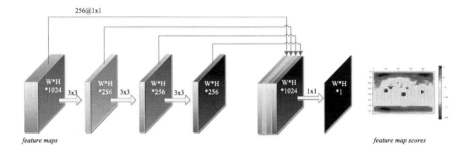

Fig. 2. The structure of spatial attention network

Because the human head object is generally a relatively regular shape, we can use the ground truth box to label the pixels of the output layer feature map. We obtain the corresponding receptive field in the original image according to the coordinates of the feature map pixel in the output layer, and then compare the receptive field with ground truth. When more than 30% of the receptive field is in the ground truth box, mark the pixel as the positive sample. Through this aggressive way, we can get more positive samples. The loss function of the saliency module is defined as follows:

$$Loss = \sum_{k=1}^{n} -label_k \cdot log(x_k) - (1 - label_k) \cdot log(1 - x_k) \qquad (1)$$

where x_k represents the kth proposal and $label_k$ denotes label of x_k When $label_k = 1$, it means the proposal x_k contains the human head. The results obtained by the SAN need to be integrated with the Faster-RCNN network to effectively improve the detection accuracy of the Faster-RCNN. In the fusion processing part, we perform the post-processing of the output of Faster-RCNN detection results to obtain candidate proposal and prediction scores. Then, the saliency score of the candidate proposal is obtained according to the score of the SAN by exploiting several methods, such as median, average, center zone average, etc. In this step, we exploit the average value of the 70% center area of the feature map within the proposal. Finally, we fuse the Fast-RCNN prediction scores and saliency scores of the proposal to obtain the final prediction results.

3.2   Fusion Strategy

The fusion strategy is mainly to integrate the SAN scores with the Faster-RCNN scores through suitable strategies to achieve an effective trade-off between the false positive (FP) ratio and the false negative (FN) ratio, so as to obtain the appropriate proposal filtering threshold and improve the overall detection accuracy. We can achieve the following three types of fusion strategies.

Strategy 1: From the perspective of reducing FP, we set strategy 1 as $f_s > s_1$ and $a_s > s_2$, where f_s and a_s are the thresholds of Faster-RCNN scores and SAN scores, respectively. s_1 and s_2 are obtained by statistically counting the lower 5% distribution value of the positive samples for Faster-RCNN proposals and 2% for the SAN proposals, respectively.

Strategy 2: From the perspective of reducing FN, we set strategy 2 as $f_s > s_3$ and $a_s > s_4$, where s_3 is the same as s_1 in Strategy 1. s_4 is obtained by statistically counting the lower 5% distribution value of the positive samples.

Strategy 3: Strategies 1 and 2 require a priori statistical analysis. However, there are no data labels in the actual application scene and thereby the statistical analysis of the data cannot be performed for the scenario. For this reason, it is necessary to predict the statistics distribution to automatically give out the appropriate thresholds. In order to solve these problems, we adopt an automatic fusion strategy, using two types of score fusion for classification prediction.

Common auto-fusion strategies include linear classifiers, Bayesian classifiers and SVM classifiers. In this paper, we use an automatic classifier based on SVM which includes two steps. First, a 4-dimensional input vector is constructed as $f = (f_s, a_s, f_s + a_s, f_s * a_s, \|f_s - a_s\|)$. In this equation, we fully considered f_s, a_s and their relative values when building the SVM model. The last three relationship values, such as addition, product and absolute value of difference, are added to enhance the distinguish effect of the SVM classifier. After this, the SVM classifier model is trained. Finally, in the detection, we only need to use the trained SVM classifier to automatically filter out the proposals to generate the final prediction result.

4   Experiments

In this section, we describe our datasets and experimental setups to indicate the advantages and detection performance of SAN and fusion strategies. First, we briefly introduce the related human head datasets, the evaluation index and training techniques. Then, we analyze the impact of different fusion strategies. Finally, we test our method on our own dataset and public datasets.

4.1 Datasets

Currently, the public dataset for human head detection includes Brainwash, Hollywood, and other datasets. The Brainwash dataset comes from a surveillance video in a caffe shop with a fixed camera angle and little change in the target distribution. The Hollywood dataset is mainly taken from 21 movie pictures which generally pay attention to do some atmosphere modification and character focus, making the head take up most part of the picture. In the practical applications, the scenes of head detection include not only the above scenarios, but also the complicated scenes with more severe obstructions, more similar objects and more densely people. In order to reflect the robustness of head detection in different scenarios, we constructed a human head detection dataset called NUDT-HEAD, which includes 21,405 images and 369,846 marked head targets. It is mainly collected from networks and surveillance cameras, including classrooms, laboratories, halls, corridors, and supermarkets, which is shown in Fig. 3. At the same time, a large number of obstructed objects with similar heads, such as hats, school bags, and horns, are similar to the human head. NUDT-HEAD dataset can better reflect the diversity of human head detection in multiple scenarios.

Fig. 3. The images in the NUDT-HEAD dataset

To prove that our model has generalization ability on multiple datasets, We build a head detection dataset based on different scenarios which combines the above three datasets as the training set, validation set, and test set in our experiments. It has 55,802 images, of which 11,769 were taken from brainwash, 22,628 from Hollywood, and 21,405 from NUDT-HEAD. The dataset used in this paper is distributed as follows (Table 1):

Table 1. The distribution of proposed head detection datasets

	Training set	Validation set	Test set	Overall
Brainwash	10769	500	500	11769
Hollywood	20394	932	1302	22628
NUDT-HEAD	18337	1068	2000	21405
Merge	49500	2500	3802	55802

4.2 Measure Index

In practical applications, there are high requirements for recall and precision, average precision (AP) can not indicate the above two important indexes. In addition, AP cannot point out the prediction accuracy of the images with no false positive samples or false negative samples. Due to more precisely indicate the accuracy of the number of detecting head in the image, we use the FP and FN which can better evaluate the accuracy of the head detection. We define a weighted harmonic mean index based on precision and recall as below:

$$F_\beta = ((1 + \beta) * P * R)/((\beta^2 * P) + R) \tag{2}$$

This index is an evaluation comprehensively considering both recall and precision, in which P and R represents the precision and recall, and β measures the relative importance of the recall rate on the precision rate. When $\beta > 1$, the recall rate has a greater impact. When $\beta < 1$, the precision rate has a greater impact. When $\beta = 1$, the recall and the accuracy are equally important. When F is larger, the detection performance is better and vice versa.

4.3 Training Details

We used a stochastic gradient descent training (SGD) as the training algorithm, in which the Momentum and weight decay were 0.9 and 0.001, respectively. The specific training process is based on the pretrained model of the COCO dataset. The Faster-RCNN module is trained first, after which the SAN module is trained. Finally the fusion training is performed.

4.4 Experiments for Fusion Strategy

In order to verify and compare proposed three fusion strategies, we set up the following experiments to perform the comparison of Faster-RCNN and our fusion strategies based on the verification dataset.

First of all, we train the SAN based on Resnet-50, and get $mAP = 0.933$ while Faster-RCNN get $map = 0.935$, which indicates SAN is approximately equivalent to Faster-RCNN. According to Sect. 3.2, we can obtain $s_1 = 0.8$, $s_2 = 0.2$, and $s_3 = 0.5$. As a result, the index value of the Faster-RCNN and our methods can be obtained, as shown in Table 2, where P stands for presicion and R for recall.

Comparing post-processing strategy 1 with Faster-RCNN, our method is better than the Faster-RCNN with a single threshold. We can see that the recall decline declined by 0.4%, precision and F increases by 1.8% and 0.8%, respectively. Strategy 2 has a smaller increase in recall while precision and F decreases. Strategy 3 decreases by 2.1% in recall, while precision and F increase by 4.2% and 0.8%, respectively. As a result, Strategy 3 gets the best results with F as the measure index.

Table 2. The comparison of different fusion strategy

	P	R	F
Faster RCNN	0.919	0.895	0.907
Our Strategy 1	0.937	0.891	0.913
Our Strategy 2	0.888	0.917	0.902
Our Strategy 3	0.961	0.874	0.915

4.5   Experiment for Different Backbone Networks

This part is mainly to compare and analyze head detection performance between different backbone networks based on the NUDT-HEAD test set using the third fusion strategy. As shown in Table 3, our method has a significant improvement over the Faster-RCNN in VGG16, Resnet-50, and Resnet-101 networks. In the three types of the backbone networks, the Resnet-50 network can obtain the best performance for head detection, which is chosen as the final backbone network in the next experiment.

Table 3. The comparison of different backbone networks

	VGG16			Resnet-50			Resnet-101		
	P	R	F	P	R	F	P	R	F
Faster RCNN	0.726	0.912	0.808	0.804	0.931	0.863	0.841	0.935	0.885
Proposed (Faster RCNN + SAN)	0.891	0.858	0.874	0.892	0.912	0.902	0.923	0.911	0.917

Table 4. The comparison of different datasets

	Hollywood			Brainwash			NUDT-HEAD		
	P	R	F	P	R	F	P	R	F
Faster RCNN	0.833	0.856	0.844	0.941	0.846	0.891	0.804	0.931	0.863
Proposed (Faster RCNN + SAN)	0.891	0.84	0.874	0.964	0.832	0.893	0.892	0.912	0.902

4.6 Experiment for Different Datasets

Based on the above two experiments, we select Resnet-50 as the backbone network and Strategy 3 as the fusion strategy. In the following experiment, we perform an evaluation experiment of proposed head detection network and Faster-RCNN based on different datasets. The result is as follows:

From the results in Table 4, the proposed network achieves higher detection performance than the Faster-RCNN network based on three datasets. The proposed network not only can get better precision and F value in the scenes with single observe angle (Brainwash) or no obstruction (Hollywood), but also can get better precision and F value in the complicated scenario that has the obstructions and objects similar to human heads, only with a slightly decline of recall, which proves that our approach has a strong generalization ability across different actual application scenes.

We also evaluate the proportion of images with no false alarms in all the images on different datasets. The result is shown in Table 5. This experiment shows that our method can achieve 3.1%–21.0% lower number of images with no false positive samples than the Faster-RCNN on the three datasets, which means the proposed network can effectively solve the problem of false alarms in head detection.

Table 5. The proportion of images with no FP samples in all the images under different datasets

	Hollywood	Brainwash	NUDT-head
Faster RCNN	0.732	0.600	0.583
Proposed (Faster RCNN + SAN)	0.763	0.750	0.793

Figure 4 shows the effect of our detection method on suppressing false positive head proposals under different datasets. By exploiting the SAN, we can carries out saliency detection in the level of feature map. As a result, SAN makes the saliency predictions of the human head proposal in the image. Finally, it can eliminate the wrong head proposals which do not exist in the saliency heat map through the fusion processing.

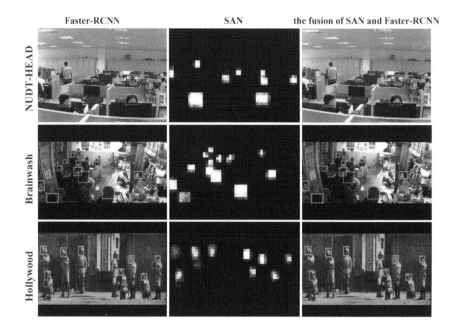

Fig. 4. The visual result of the proposed network

We also compare our method and other detection networks. The mAP for head detection in [8] is 0.727, while ours is 0.898. Other detection methods such as SSD, yolo and so on are not ideal for small target detection.

5   Conclusions

In this paper, we proposed a saliency attention network SAN, which can perform saliency detection in the level of feature map and a fusion strategy to automatically merge the SAN and Faster-RCNN to promote the detection accuracy of human heads. Compared to a separate Faster-RCNN network, SAN can learn about the context information around the proposal. Experiments show that the proposed head detection network has better robustness and performance than Faster-RCNN, especially for the scenes with similar human head and obstructions.

Acknowledgments. This work was supported by National Key Research and Development Program of China under No. 2018YFB1003405.

References

1. Ren, S., He, K., Girshick, R., Sun, J.: Faster R-CNN: towards real-time object detection with region proposal networks. In: Advances in Neural Information Processing Systems, pp. 91–99 (2015)

2. Dai, J., Li, Y., He, K., Sun, J.: R-FCN: object detection via region-based fully convolutional networks. In: Advances in Neural Information Processing Systems, pp. 379–387 (2016)
3. Redmon, J., Farhadi, A.: YOLO9000: better, faster, stronger, arXiv preprint arXiv:1612.08242 (2016)
4. Liu, W., et al.: SSD: single shot MultiBox detector. In: Leibe, B., Matas, J., Sebe, N., Welling, M. (eds.) ECCV 2016. LNCS, vol. 9905, pp. 21–37. Springer, Cham (2016). https://doi.org/10.1007/978-3-319-46448-0_2
5. Hariharan, B., Arbelez, P., Girshick, R., Malik, J.: Hypercolumns for object segmentation and fine-grained localization. In: Proceedings of the IEEE Conference on Computer Vision and Pattern Recognition, pp. 447–456 (2015)
6. Lin, T.-Y., Dollr, P., Girshick, R., He, K., Hariharan, B., Belongie, S.: Feature pyramid networks for object detection. In: Proceedings of the IEEE Conference on Computer Vision and Pattern Recognition, vol. 1, no. 2, p. 4 (2017)
7. Kong, T., Yao, A., Chen, Y., Sun, F.: Hypernet: towards accurate region proposal generation and joint object detection. In: Proceedings of the IEEE Conference on Computer Vision and Pattern Recognition, pp. 845–853 (2016)
8. Vu, T.H., Osokin, A., Laptev, I.: Context-aware CNNs for person head detection. In: Proceedings of the IEEE International Conference on Computer Vision, pp. 2893–2901 (2015)
9. Stewart, R.: Brainwash dataset. Stanford Digital Repository (2015). http://purl.stanford.edu/sx925dc9385
10. Stewart, R., Andriluka, M., Ng, A.Y.: End-to-end people detection in crowded scenes. In: Proceedings of the IEEE Conference on Computer Vision and Pattern Recognition, pp. 2325–2333 (2016)

Arbitrary Perspective Crowd Counting
via Multi Convolutional Kernels

Minghui Yu[1], Teng Li[1(✉)], Jun Zhang[1], Jiaxing Li[1], Feng Yuan[2],
and Ran Li[2]

[1] Anhui University, Hefei 230601, China
liteng@ahu.edu.cn
[2] Cloud Computing Center, Chinese Academy of Sciences, Dongguan, China

Abstract. Cross-scene crowd counting plays a more and more important role in intelligent scene monitoring, and it is very important in the safety of personnel and the scene scheduling. The traditional estimation of crowd counting is mainly dependent on the simple background of scenes, which is not conducive to the complex background. To address this problem, in this paper, we propose a multi convolutional kernels net for crowd counting, which discards the subjectivity and the occasionality of the traditional manual feature extraction. Firstly, we label dataset for convolution output features. Then we use the fully convolutional network to create the density map at the end of the network with multi convolutional kernels. Finally, we perform integral regression on density maps to estimate the crowd counting. The dataset that we used is a set of publicly available datasets, which are the Shanghaitech dataset, the UCF_CC_50 dataset and the UCSD dataset. The experiments based on video images show that the proposed method is more effective than traditional methods in terms of robustness and accuracy.

Keywords: Crowd counting · Deep learning · Multi Convolutional Kernels

1 Introduction

With the increase of social activities and urban population, In some public places, the increase in population density is a typical scene of the crowd, such as train stations, subway, and large shopping malls. There are two social problems when the density of people reaches a certain quantity. First, human life safety can not be guaranteed because it is easy to cause a stampede. Second, it is not conducive to the supervision and management of personnel management. Fortunately, with the development of modern digital and image processing technology, the research of automatic intelligent crowd density analysis has become practical.

There are two problems to be solved for the crowd counting with an arbitrary camera perspective, the first one is how to get the crowd density map, and the other is how to count the crowd. Density map mainly reflects the intensity of the crowd, we can get the degree of crowd density by thermography. However, we can only reflect the true density map by the location of pedestrians if the density map is difficult to generate by the method of detection. At the same time, the camera's perspective is an important

© Springer Nature Switzerland AG 2018
R. Hong et al. (Eds.): PCM 2018, LNCS 11165, pp. 558–569, 2018.
https://doi.org/10.1007/978-3-030-00767-6_52

factor affecting the density map. For the dense crowd, it is difficult to detect the number of people because of occlusion, and it is not wise to calculate the number by foreground segmentation due to the randomness of foreground segmentation.

In recent years, there have been more and more researches on crowd counting by domestic and foreign scholars, but most of them are based on single perspective. For example, the front crowd in A (see Fig. 1), the profile in B (see Fig. 1), the back crowd in C (see Fig. 1). This single Angle has great limitations and is not suitable for real-life situations.

A   B   C

Fig. 1. People in different perspective.

In order to solve the above problems, in this paper, we propose a new framework based on convolutional neural network (CNN) for crowd counting in an arbitrary perspective crowd image. We propose the network with Multi Convolutional Kernels (hereinafter referred to as MECK-Net), which includes 4 different efficient input kernels, Because different sizes of convolution kernels the depth of image feature extraction is different, we use the $3 \times 3, 5 \times 5, 7 \times 7$ and 9×9 as the convolution kernels for the feature extraction of the input image. We tested that the combination of those four convolution kernels is effective for arbitrary perspective crowd counting. At the same time, these four kinds of convolution kernels can extract the features of the input image comprehensively, which can correctly reflect the information of the image. The output of the network is a crowd density map whose integral regression gives the overall crowd count.

2 Related Work

Some works have been done for crowd counting. The [10] use a method of detection for crowd counts. Based on the appearance and motion features, the scan detector reads two consecutive frames of video sequences to estimate the number of people. [8, 10] have used a similar detection framework to estimate the number of pedestrians.

In [4], the authors use motion patterns to estimate the count of moving objects. Rabaud et al. use the foreground segmentation [2] to obtain the moving object and the number of people by regression according to the characteristics of the moving object, such as the edge feature, area feature, gradient histogram feature, etc. These methods rely on the moving objects, which can be obtained only in the case of a continuous video stream. However, they cannot be used for still images.

The algorithm proposed by Idrees et al. [5] realize that it is not able to obtain the accurate number of people relying solely on single feature. To address this problem, they combine a variety of image features, such as HOG features, Fourier transform, region of interest calibration, etc. Post-processing is performed by using multi-scale Markov random fields. However, there is a certain impact on the accuracy of the results based on the characteristics of manual extraction with the requirements of the illumination, video perspective, as well as the severe occlusion.

There are also some methods that focus on crowd counts from still images. The reference [5] proposes the use of multiple sources of information to count crowd in a very dense scene from a single image. Idrees extract features from neural networks, and use support vector machines (SVM) to generate the crowd counts from still images. Wang et al. [6] utilize a deep learning network to get the number of people, but the result is mainly to generate the population count. Zhang et al. [13] also use a deep network to analyze the crowd count, and the method can generate the model by analyzing the density map from the perspective of the scene.

3   Proposed Method

3.1   Multi Convolutional Kernels

There will be a nonlinear perspective in the actual scene taken by the camera because of the different viewing angles of the camera equipment. The larger the target projection near the camera, the more the number of pixels in the area of the target in the image. On the contrary, the smaller the target projection near the camera, indicates that the less the number of pixels in the area of the target in the image. The unit pixels in different positions of the image have different contribution to the target foreground because of the existence of the perspective effect, so we analyze the number of people in the image through the semantic level image [14]. We use different kernels to convolute the input image, which can cover the whole pixel of the image, and analyze the different characteristics of the image, so as to improve the effect of perspective. An architecture of MECK-Net for generating crowd density map is shown (see Fig. 2). In the following, we describe the network in detail.

The images of scene contain person heads of different sizes because of perspective distortion, so the filters have different sizes of kernels. These kernels can be used to perceive the overall information of the image through the receptive fields of different sizes, including the head of different scales. For example, the larger receptive field of the filter corresponds to the larger header information.

As shown (see Fig. 2), we present four different sizes of convolution kernels, 3×3, 5×5, 7×7 and 9×9 (If we just use 9 * 9 convolutional kernel, the recipe filed is too much to fit heads of different sizes.), to separate regions of different sizes. There are six volumes of each line, and the number of output channels are the same to meet the similar learning of network. Each of the two convolutional layers is embedded with a layer of pooling, and the characteristics of the pool is the way of MAX for image

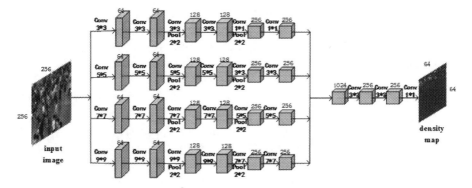

Fig. 2. The architecture of MECK-Net for generating crowd density map.

dimensionality reduction, but the convolutional layer does not participate in the image dimension reduction. For each line of the last two volumes, we have reduced the size of the filter to ensure that the characteristics of the image without distortion. The activation function used in this method is ReLU [15] to prevent the over fitting of the iteration. In order to obtain the more accurate crowd density map, we expanded the dimension of the convolution feature map from 64 to 256. For the output layer of the network, we use the convolution kernel of 1×1, and the dimension of the output is 1. Then we use Euclidean distance as the loss function, which is defined as follows:

$$L(\theta) = \frac{1}{4N} \sum_{k}^{N} ||F(X_k; \theta) - F_k||_2^2 \tag{1}$$

where L is the loss between estimated density map $F(X_k; \theta)$ and the ground truth density map F_k. θ is the set of parameters of the MECK-Net model and N is the number of training samples. X_k is the input image and F_k is the ground truth density map of image X_k. $F(X_k; \theta)$ represents the density map of the MECK-Net generated by the parameter θ for sample X_k.

There are two important points should be noted. First, the corresponding output density map of MECK-Net should be the size of the input image 1/4 times due to the use of two layers of the pooling. Second, because our network is designed to be a fully convolutional network, the size of the input image is arbitrary, which will not cause image distortion.

3.2 Density Map Generating

In order to estimate the number of people in the scene image, we can understand the image directly by establishing density map. The output phase of our network corresponds to the density map. So the output layer can determine the density map of the scene image. For density maps, we can visually determine where the scene is dense, and it is clear to determine where the accident prone. At the same time, we can accurately get the distribution of the scene density according to the phenomenon of scene perspective, which has certain advantages for the MECK-Net learning.

There is density map for output label corresponding to the input image according to the corresponding principle of the network label. We generate the density map of head label on the label of the head for getting the label of the output of the network. The density function is defined as follows:

$$M(x) = \sum_{i=1}^{N} \delta(x - x_i) * G_p(x) \tag{2}$$

Where x_i is a head point, and $\delta(x - x_i)$ is represented a delta function. $G_p(x)$ is a Gaussian kernel [11]. In order to adapt to the perspective, our Gauss kernel is adaptive kernel. The size of the Gauss kernel is different according to the proportion of pedestrian window, which means the Gauss kernel increases with the increase of the Y axis of the image, so p is the adaptive kernel. $M(x)$ is the density map.

We generate the density map of the two kind of Shanghaitech dataset [1] according to the density function, as shown (see Fig. 3), and the other two datasets do the same. Part_A and Part_B are some ground truth and its corresponding density map. We can see that the higher the density, the more obvious the red alert, which really reflects the distribution of the density of the scene.

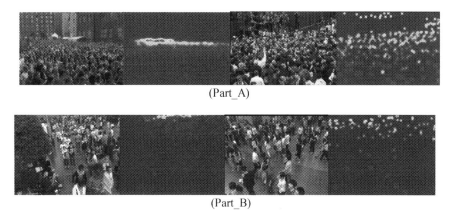

(Part_A)

(Part_B)

Fig. 3. (Part_A) Representative density map of Part_A ground truth. (Part_B) Representative density map of Part_B ground truth.

3.3 Data Augmentation

The data distribution of the original Shanghaitech dataset [1] is shown in Table 1, and the other two datasets do the same.

Table 1. The data distribution of the original Shanghaitech dataset.

Dataset		NUM	TOTAL	MAX	MIN	AVG	SIZE
Shanghaitech	Part_A	482	241,677	3139	33	501.4	various
	Part_B	716	88,488	578	9	123.6	1024 × 768

Where NUM is the number of images. TOTAL is the number of labels for all images. MAX is the maximum number of labels. MIN is the minimum number of labels. AVG is the average number of all labels. SIZE is the size of images. As we can see from Table 1, the number of images and labels can meet the convolutional neural network learning, but in order to make the it to obtain more accurate results, we need a large amount of data as the training set. Moreover, we should consider these two problems, one is that the data cover the range of each density, and the two is not to make the image distortion. To achieve these order, we used three data augmentation methods with rotation, mirroring and cropping.

We rotate the image to four different angles with 0°, 90°, 180°, 270°, respectively, At the same time we can mirror the image horizontally and vertically. We can see that the training data can be extended 6 times through the rotation and mirror. Then we cut the image of 2 × 2 and 4 × 4, each block corresponds to the label of block. We can also divide the number of labels in different degrees by cropping, which can make the network more robust.

3.4   Optimization of MECK-Net

As can be seen from Formula (1), the network we have learned is achieved by stochastic gradient descent and back propagation. We need to try to fine tune the network approach for the not very large dataset, so that the network will not enter the gradient of the state of the disappearance.

Our approach is divided into two steps. First, multi-column of the training network are trained according to the training set, and the output is used as the feature map to be classified in front of the merge layer, and ground truth is used as the label for feedback training. Second, the four network model generated by the first step is fine tuned through the network architecture (see Fig. 2). Figure 4 show that the proposed method effectively reduces the probability of over fitting.

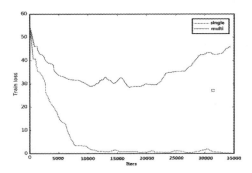

Fig. 4. Demonstrate the different results obtained by different training methods. *single* represents a single training for the network, *multi* represents multi-training for the network.

4 Experiments

The dataset that we used is a set of publicly available datasets, which are the Shanghaitech dataset [1], the UCF_CC_50 dataset [5] and the UCSD dataset [3]. The Shanghaitech dataset [1] that we mainly used consists two parts: Part A and Part B. Images in Part_A come from the Internet, most of them contain a large number of people. Image in Part_B are from the commercial street in Shanghai, with the number of people ranging in different number. Both parts of these images have been annotated. Figure 5 shows some samples of the dataset. The UCF_CC_50 dataset [5] contains 50 images from the Internet, the average number of person from per image is about 1280, the total number of people from the dataset is 63974. It is a very challenging dataset because there are a lot of people in each image. At the same time, the number of people from each image range from 94 to 4543, which led to the uncertainty of the number of people. The UCSD dataset [3] contains 2000 images from UCSD campus, the average number of people from images is 25, the image size is 158×238, while the dataset provides a ROI area for each image. At the same time we do the data augmentation.

(Part_A)

(Part_B)

Fig. 5. (Part_A) Representative images of Part_A contain a large number of people from Shanghaitech dataset [1]. (Part_B) Representative images of Part_B range in different number.

We used two evaluation policies [16] for different methods to test with both the mean absolute error (MAE) the mean squared error (MSE), which are defined as follows:

$$MAE = \frac{1}{N} \sum\nolimits_1^N |z_i - y_i| \tag{3}$$

$$MSE = \sqrt{\frac{1}{N} \sum\nolimits_1^N (z_i - y_i)^2} \tag{4}$$

Where N is the number of test images, z_i is the actual number of people in the ith image, and y_i is the estimated number of people in the i-th image. Roughly speaking, MAE indicates the accuracy of the estimates, and MSE indicates the robustness of the estimates. The crowd counts are derived from the integral regression of density maps. We obtain the density map accurately from the last output layer of the network, and we can get the exact number of people by the integral regression of the pixel value. The integral regression function is defined as follows:

$$N_p = REG\left(\sum_{i=1}^{N} X_i\right) \tag{5}$$

Where N_p is the number of people, REG represents linear regression using least squares. X_i is the value of pixel i. N represents the number of all pixel values for an image.

4.1 Experimental Results

The experimental results of the MECK network are shown (see Fig. 6), which shows examples of ground truth density maps and estimated density maps of images in Part_A and Part_B from the Shanghaitech dataset. The experimental results are very similar to the ground truth image, which indicates that the experimental results of this data set are good.

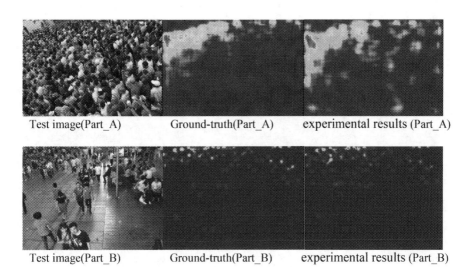

| Test image(Part_A) | Ground-truth(Part_A) | experimental results (Part_A) |
| Test image(Part_B) | Ground-truth(Part_B) | experimental results (Part_B) |

Fig. 6. (Part_A) Representative a corresponding estimated crowd density map of testing image. (Part_B) Representative a corresponding estimated crowd density map of testing image.

Finally, to show the effective of our method, we also compared with many different algorithms. Table 2 shows the comparison results. Among them, the first (Rodriguez et al. [7]) to the eight (LBP + RR) are typical traditional algorithms. The last two

(Zhang et al. [6] and MCNN [1]) are advanced deep learning algorithms. As we can see, for our proposed, both MAE and MSE have a lower error rate. Even for some algorithms there is a significant reduction. Therefore the accuracy and the robustness of the estimates are better than the previous algorithm. It is further explained that great progress has been made in our experimental results.

Table 2. Comparing performances of different methods on Shanghaitech dataset [1], the UCF_CC_50 dataset and the UCSD dataset.

| Method | The Shanghaitech dataset | | | | The UCF_CC_50 dataset | | The UCSD dataset | |
| | Part_A | | Part_B | | | | | |
	MAE	MSE	MAE	MSE	MAE	MSE	MAE	MSE
Rodriguez et al. [7]	—	—	—	—	655.7	697.8	—	—
Lempitsky et al. [11]	—	—	—	—	493.4	**487.1**	—	—
Idrees et al. [5]	—	—	—	—	419.5	541.6	—	—
Kernel Ridge Regression [8]	—	—	—	—	—	—	2.16	7.45
Ridge Regression [2]	—	—	—	—	—	—	2.25	7.82
Gaussian Process Regression [3]	—	—	—	—	—	—	2.24	7.97
Cumulative Attribute Regression [9]	—	—	—	—	—	—	2.07	6.86
LBP + RR [12]	303.2	371.0	59.0	81.7	—	—	—	—
Zhang et al. [6]	181.8	277.7	32.0	49.8	467.0	498.5	1.60	3.31
MCNN [1]	110.2	173.2	26.4	41.3	337.6	509.1	1.07	1.35
MECK-Net	**105.3**	**169.5**	**24.0**	**40.5**	**336.5**	510.2	**1.03**	**1.30**

4.2 Analysis

In this section, we analyze the experimental results from four aspects.

MECK-Net: We use the MECK-Net to carry on the deep level expression through the effective four initial convolution to estimate the input image. And the network covers targets of different sizes, which causes the network to be more robust to the scene information. We also tried to analyze the scene through two or three columns of the network layer, and even tried to use more than five columns to evaluate the network. The experimental results showed that only four columns network can meet the requirements of our experiments, two or three columns network can not get more accurate results, and more than five columns network will be over-fitting, which is not conducive to the conduct of the experiment while the amount of training parameters is too big. Details are shown in Table 3.

Table 3. Different experimental results

	MECK-Net							
1 columns?	☑							
2 columns?		☑						
3 columns?			☑					
4 columns?				☑	☑	☑	☑	☑
1 time?				☑				
1/2 times?					☑			
1/3 times?						☑		
1/4 times?	☑	☑	☑				☑	☑
Data Augmentatio?	☑	☑	☑	☑	☑	☑		☑
MSE of testing The Shanghaitech dataset	184.5	178.9	174.8	171.4	180.3	176.3	178.6	**169.5**

Loss Function: We use the loss function as shown in Formula (1). Euclidean distance is a commonly used method to calculate the loss function, but there is a difficult problem that how to get the loss function which meets the requirements of the experiment. We have tried 1 times [6], 1/2 times [1] and 1/3 times loss function calculation method, and the experimental results show increasing state. So we try to use 1/4 times loss function, the experimental results also prove that the 1/4 times loss function is the best. Details are shown in Table 3. Thus we analyze that the network is entered by 4 columns, each column has different effects on the input image and the experimental results, so we get the best experimental results according to the law.

Data Augmentation: The dataset that we used are the UCF_CC_50 dataset [5] and the UCSD [3] dataset. The data distribution of the original Shanghaitech dataset [1] is shown in Table 1. As can be seen from Table 1, the amount of data used for density analysis is not large enough. And it is well known that the requirement of deep learning is that the number of training images must be large. Therefore, we extend the training dataset in different degree. The experimental results show that our method can be used for reference. Details are shown in Table 3.

Optimization of MECK-Net: During the experiment, we found that if we trained directly on MECK-Net, the experimental results were random. The final model may have a good effect on a certain scene, but if there are different perspective, the experimental results can not achieve good results. So we consider whether the network can be layered training, the results of each layer are obtained by fine-tuning to share the weight, which is the source of our training ideas. In the end, we train the network in different degrees, and get the results as shown in Table 2 and Fig. 6.

5   Conclusion

In this paper, we present fully convolutional network to create the density map at the end of the network with MECK-Net. The network mainly takes into account the size distribution of the head, uses MECK-Net to extract the features of the image, which is more able to express the image information. At the same time, the density map is generated in the output layer of network, and the number of people is obtained by integral regression. We test our proposed method in the Shanghaitech dataset [1], the UCF_CC_50 dataset [5] and the UCSD [3] dataset, Moreover, the experimental results show that the accuracy and the robustness of our method are better than the state-of-art crowd counting methods. Moreover, our MECK-Net does not produce over fitting state during training because of the fine tuning of the network, which indicates that the generality of our model is powerful. The advantage of our method should be because of the change of the loss function, the expansion of the dataset and the optimization of the network training. They solve the problem of non-robustness caused by the small amount of data and the problem of over-fitting.

In addition, the experiment also shows that the proposed method is time-consuming on test data. In future, we will try to study how to compress the network or reduce the layer without affecting the accuracy.

Acknowledgments. This work is supported by the National Natural Science Foundation (NSF) of China (No. 61572029, No. 61702001).

References

1. Zhang, Y., Zhou, D., Chen, S., et al.: Single-Image crowd counting via multi-column convolutional neural network. In: CVPR, pp. 589–597 (2016)
2. Chen, K., Loy, C.C., Gong, S., Xiang, T.: Feature mining for localised crowd counting. In: BMVC, vol. 1, p. 3 (2012)
3. Chan, A.B., Liang, Z.S.J., Vasconcelos, N.: Privacy preserving crowd monitoring: counting people without people models or tracking. In: CVPR, pp. 1–7 (2008)
4. Brostow, G.J., Cipolla, R.: Unsupervised Bayesian detection of independent motion in crowds. In: CVPR, pp. 594–601 (2006)
5. Idrees, H., Saleemi, I., Seibert, C., et al.: Multi-source multi-scale counting in extremely dense crowd images. In: CVPR, pp. 2547–2554 (2013)
6. Zhang, C., Li, H., Wang, X., et al.: Cross-scene crowd counting via deep convolutional neural networks. In: CVPR, pp. 833–841 (2015)
7. Rodriguez, M., Laptev, I., Sivic, J., et al.: Density-aware person detection and tracking in crowds. In: ICCV, pp. 2423–2430 (2011)
8. An, S., Liu, W., Venkatesh, S.: Face recognition using kernel ridge regression. In: CVPR, pp. 1–7 (2007)
9. Chen, K., Gong, S., Xiang, T., Loy, C.C.: Cumulative attribute space for age and crowd density estimation. In: CVPR, pp. 2467–2474 (2013)
10. Viola, P., Jones, M.J., Snow, D.: Detecting pedestrians using patterns of motion and appearance. Int. J. Comput. Vis. **63**(2), 153–161 (2005)

11. Lempitsky, V., Zisserman, A.: Learning to count objects in images. In: Advances in Neural Information Processing Systems, ICONIP, pp. 1324–1332 (2010)
12. Ojala, T., Pietikäinen, M., Harwood, D.: Performance evaluation of texture measures with classification based on Kullback discrimination of distributions. In: ICPR, pp. 582–585. IEEE (1994)
13. Wang, C., Zhang, H., Yang, L., et al. Deep people counting in extremely dense crowds. In: Proceedings of the 23rd Annual ACM Conference on Multimedia Conference, pp. 1299–1302 (2015)
14. Long, J., Shelhamer, E., Darrell, T.: Fully convolutional networks for semantic segmentation. In: PAMI, pp. 640–651. IEEE (2017)
15. Krizhevsky, A., Sutskever, I., Hinton, G.E.: ImageNet classification with deep convolutional neural networks. In: International Conference on Neural Information Processing Systems, ICONIP. CAI, pp. 1097–1105 (2012)
16. Zhang, C., Li, H., Wang, X., Yang, X.: Cross-scene crowd counting via deep convolutional neural networks. In: CVPR, pp. 833–841. IEEE (2015)

3D Shape Co-segmentation by Combining Sparse Representation with Extreme Learning Machine

Hongyan Li[1,2], Zhengxing Sun[1(✉)], Qian Li[1], and Jinlong Shi[1]

[1] State Key Laboratory for Novel Software Technology,
Nanjing University, Nanjing 210046, China
szx@nju.edu.cn
[2] Institute of Computing and Software,
Nanjing College of Information Technology, Nanjing 210023, China

Abstract. Unsupervised shape co-segmentation is proposed to segment a set of 3D shapes into meaningful parts without any labelled data. At the same time, a correspondence is created between the segmented parts. Usually, there are two main steps: correlation analysis and representation learning. In this paper, we propose an affinity matrix construction method based on parameter-free and high-efficiency simplex sparse representation to analysis correlation. This construction avoids the blindness of parameter setting. Based on the affinity matrix, we propose a co-segmentation approach via an unsupervised extreme learning machine to train a transform network for feature representation. This representation learning could attain good performance in lower embedding dimension. Therefore, co-segmentation can be implemented by clustering on lower dimensions embedding space. So the execution is more efficient. Moreover, once the transform network is trained, it can be applied to the data representation acquisition process without re-computing the affinity matrix. Experiments validate the method proposed in this paper. The method is unsupervised and can perform efficient and effective co-segmentation. Moreover, it also can deal with incremental co-segmentation when the data set is expanded.

Keywords: Unsupervised co-segmentation · Simplex sparse representation
Extreme learning machine

1 Introduction

Shape segmentation that partitions a 3D shape into meaningful parts has become more and more important to the digital geometric processing and analysis field. Previous methods pay attention to segmenting an individual-shape based on a limited number of geometric features, such as convexity [1], curvature [2], shape diameter function [3]. Unfortunately, there is no perfect all-encompassing segmentation algorithm and an individual shape cannot provide sufficient geometric diversity to identify its meaningful parts [6].

R. Hong et al. (Eds.): PCM 2018, LNCS 11165, pp. 570–581, 2018.
https://doi.org/10.1007/978-3-030-00767-6_53

To utilize the geometric cues, various methods have been proposed to segment a set of shapes consistently and to take advantage of the correlation between shapes. Based on the availability of labelled data, this type of shape segmentation can be classified as either supervised segmentation [4, 7–12] or unsupervised segmentation [13–18]. Supervised methods tend to obtain better segmentation results. In particular, deep learning methods have proven very valuable [7–12]. However, these methods require substantially-labelled shapes. In shape segmentation task, annotation should be done on each triangular face, which is very tedious and time-consuming.

To address the problem of lack of labels, we focus on shape co-segmentation in this paper, e.g., unsupervised consistent segmentation. Such an unsupervised segmentation directly extracts appropriate knowledge inherent to the shape set, where shapes belong to a common family, share the same functionality and exhibit a general form. Early attempts at co-segmentation are mainly based on shape matching. Matching-based methods [13] usually reformulate the co-segmentation of a shape to a graph clustering problem, based on shape correspondence. Due to its high-dependence of spatial alignment, these methods can only handle limited shape types. Some recent co-segmentation approaches perform clustering based on correlation analysis in a descriptor-space [14–18]. These clustering-based methods can handle more variable shape types independent of their orientation, location and cardinality. Most of them adopt spectral-based clustering and follow two processes: correlation analysis and clustering. In the correlation analysis process, an affinity matrix is usually created, via pairwise similarities between segments or patches, to infer correspondence between the shapes. One example of this process is to apply a Gaussian kernel to the segment/patch similarities to construct pairwise affinities, which need to be fine-tuned via a scaling parameter [14, 18]. Usually, the neighbouring similarities are more reliable than the far-away ones, then another way of constructing pairwise affinities is in terms of k-nearest neighbour graph, which is very sensitive to the number of neighbours [17]. In other words, existing methods require carefully choosing parameters that may be restrictive. During the clustering process, clustering is performed according to the embedded feature distribution. Existing methods mainly obtain data representations via a direct data embedding [14–18]. That means the eigenvectors of the affinity matrix are directly used for data representation. Thus, when a new shape appears and adds to the shape set, affinity matrix need be re-computed and the whole co-segmentation process should be repeated. Therefore, for shape co-segmentation, adapting to the variable scale of the data set remains a challenging complication.

To better address these challenges, we propose a novel shape co-segmentation by Combining Sparse Representation (SSR) with Extreme Learning Machine (US-ELM). US-ELM provides a method of representation learning for unlabeled shapes, which can achieve the data reduction and embedding by constructing a transform network. Based on the consistency hypothesis of co-segmentation, when a new shape of the same category appears, the transform network can be used to obtain its feature representation without redoing the entire process again. The solution of network parameters is based on its affinity matrix; therefore, a calculation method in term of SSR is proposed in our approach. Our method focuses on representing every data element using other data directly; therefore we don't need to worry about how to measure similarity using appropriate metrics.

In summary, our contributions are threefold.

1. Inspired by sparse coding, we propose a correlation analysis method based on SSR to avoid the blindness of parameter setting in the calculation of affinity matrices. It can effectively reduce the sensitivity of the method to parameters, thus avoiding the influence of parameter selection on segmentation results.
2. Based on manifold regularization, we recast co-segmentation as a generalized eigendecomposition problem and constructs a transform network for representation learning. Based on the transform network, co-segmentation can be implemented by clustering on lower dimensions, so the execution is more efficient.
3. Because the obtained similarities are used to compute the parameters of the network instead of the data representations directly, the ELM network can be applied to any shape in the original input space once it is trained. Therefore, our shape co-segmentation method can deal with the problem of incremental segmentation when the data set is expanded.

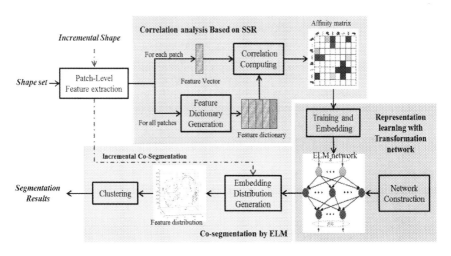

Fig. 1. Overview of our approach for shape co-segmentation.

2 Proposed Method

An overview of our approach is illustrated in Fig. 1. Our approach includes three aspects: pre-processing, correlation analysis, representation learning and co-segmentation. In the pre-processing stage, each shape is over-segmented to patches and the features of each patch are extracted. Based on the feature vectors, each patch can be represented by the feature dictionary to complete the correlation analysis. In other words, the sparse affinity matrix is calculated to measure similarities between patches by introducing parameter-free and high-efficiency SSR. In representation learning process, a transform network is constructed and trained with Laplacian of affinity matrices to obtain a transform approach for shape representation. Based on the transform network, we can perform dimension reduction and embed original data into the embedded space simultaneously. Finally, k-means clustering is performed in the embedded space to achieve co-segmentation of the shape set. Moreover, once the transform network is trained, it can be applied to map new shape patches to the embedded space. In where these new patches are assigned to the closest cluster. Thus, a consistent and incremental co-segmentation is achieved. The proposed method will be described in detail in the following subsections.

2.1   Pre-processing

Feature Descriptors. In accordance to the study of feature selection in the learning approach [8], we select four shape descriptors including shape context (SC), Gaussian curvature (GC), average geodesic distance (AGD), and shape diameter function (SDF). All these feature descriptors are defined and computed on mesh triangles and then concatenated together.

Over-segmentation. Given shapes $M_1, M_2, \cdots M_n$ from the same category, each shape M_i is decomposed into p_i patches via normalized cuts [20]. The number of patches-per-shape is set to $p_i = 50$ in our implementation. Thus, there are $N = \sum_{i=1}^{n} p_i$ patches in the same set. Then, we compute the histograms that capture the distribution of the feature measurements found on the triangles of a single patch. In this paper, the number of bins for the histogram is set to 100. After that, we have a feature matrix $X = \{x_i\}_{i=1}^{N}$ for all the patches in the original feature space.

2.2   Correlation Analysis

To describe the correlation, affinity matrices are usually calculated by pairwise similarities between patches. Recently, sparse representation assumes that each data point can be reconstructed as a linear combination of additional data points. This routine has demonstrated its efficacy in various applications like face recognition and image classification [22, 27]. Therefore, an efficient SSR [23] is adopted to optimize the combination coefficient of the dictionary base.

Given the feature matrix $X = (x_1, x_2, \cdots, x_N) \in \mathbb{R}^{d \times N}$, each patch can be written as a sparse combination of all other patches. Then, the objective function is defined as:

$$\min \|X_{-i}\alpha_i - x_i\|_2^2$$
$$s.t. \ \alpha_i \geq 0, \alpha_i^T \mathbf{1} = 1 \tag{1}$$

where, $X_{-i} = [x_1, \cdots, x_{i-1}, x_{i+1}, \cdots, x_N] \in \mathbb{R}^{d \times (N-1)}$ is the feature matrix without column i. The similarity is denoted as $\alpha_i = [\alpha_{i1}, \cdots, \alpha_{i,i-1}, \alpha_{i,i+1}, \cdots, \alpha_{iN}]^T \in \mathbb{R}^{N-1}$, which is between the i-th feature and other features to satisfy $x_i \approx X_{-i}\alpha_i$. We use an accelerated project gradient method to optimize combined coefficients. The similarity matrix is then defined as $S = [\hat{\alpha}_1, \hat{\alpha}_2, \cdots, \hat{\alpha}_N]$. Here, $\hat{\alpha}_i$ is the result of inserting coefficient 0 for its i-th of α_i, i.e., $\hat{\alpha}_i = [\alpha_{i1}, \cdots, \alpha_{i,i-1}, 0, \alpha_{i,i+1}, \cdots, \alpha_{iN}]^T$. Obviously, the similarity matrix S is not necessarily symmetric. Therefore, we get an average of S and its transpose to calculate the final affinity matrix $W = \frac{S+S^T}{2}$. Normalized Laplacian matrix is calculated as $L = I - D^{-1/2}WD^{-1/2}$, where D is the degrees diagonal matrix.

2.3   Representation Learning

We formalized co-segmentation as a generalized eigendecomposition problem. The obtained eigenvectors are then used as the parameters of the network instead of the data representation directly. Thus, a transform network could be constructed for shape representation learning and co-segmenting. Generally, the construction consists of two stages: hidden layer construction and parameters weight solution. Given the patch set $X = \{x_i\}_{i=1}^{N}$, we construct the hidden layer. Rectified Linear Unit (ReLU) is chosen as the activation function for each input feature, i.e., $G(a, b, x) = \max(0, ax + b)$. Where the parameter a_i is the weight connecting the i-th hidden-node and the input-nodes and b_i is the bias of i-th hidden node. Both parameters are initialled to a uniform distribution. N_h randomly generates hidden neurons map the data from the original input space into an N_h-dimensional embedded space. And the output vector of the hidden layer is $h(x) \in \mathbb{R}^{1 \times N_h}$. We use $\beta \in \mathbb{R}^{N_h \times N_o}$ for the output weights that connect the hidden layers with the output layers. Thus, the output of the ELM network is $f(x) = h(x)\beta$.

In the second stage, unsupervised learning is built to solve the output weight. Since it is difficult to compute the conditional probability, the cost function of unsupervised learning is approximately formulated as

$$L_m = \frac{1}{2} \sum_{i,j} w_{ij} \left\| y_i - y_j \right\|^2 \tag{2}$$

where, $w_{i,j}$ is the pair-wise similarity between the patch x_i and x_j, y_i and y_j are the category predictions of x_i and x_j, respectively. Equation (2) can be expressed as a matrix form $L_m = T_r(Y^T LY)$. L is known as the normalized Laplacian graph of the affinity matrix. $Y = H\beta$ is the output of the network and denotes the feature distribution in the embedded space, $T_r(\cdot)$ denotes the trace of the matrix. Therefore, the unsupervised objective function of ELM is defined as

$$\min_{\beta} \|\beta\|^2 + \lambda T_r(\beta^T H^T LH\beta) \tag{3}$$

$$\text{s.t.}\quad (H\beta)^T H\beta = I_{N_o},$$

where, $H = \left[h(x_1)^T, \cdots, h(x_N)^T \right]^T \in \mathbb{R}^{N \times N_h}$ and λ is a tradeoff parameter. The solution of Eq. (3) is found by choosing matrix β, whose columns are the eigenvectors corresponding to the smallest eigenvalues.

$$\left(I_{N_h} + \lambda H^T LH \right) v = \gamma H^T Hv, \tag{4}$$

where, $\gamma_1, \gamma_2, \cdots, \gamma_{N_o+1}$ are the (N_o+1) smallest eigenvalues of Eq. (4) and $v_1, v_2, \cdots, v_{N_o+1}$ are their corresponding eigenvectors. Because the first eigenvector of Eq. (4) leads to small variations in embedding; we discard the first one and solve the output weights β via $\beta = [\tilde{v}_2, \tilde{v}_3, \cdots, \tilde{v}_{N_o+1}]$, where, $\tilde{v}_i = v_i / \|Hv_i\|$ are the normalized eigenvectors that satisfy the constraint in Eq. (3). When the number of patches is fewer than the number of hidden neurons, i.e., $N \leq N_h$, Eq. (4) is underdetermined. The following alternative formulation can be found using known information $(I_u + \lambda L H H^T)u = \gamma H H^T u$. Similarly, if $u_1, u_2, \cdots, u_{N_o+1}$ are the eigenvectors corresponding to the (N_o+1) smallest eigenvalues, then output weights β is given by $\beta = H^T[\tilde{u}_2, \tilde{u}_3, \cdots, \tilde{u}_{N_o+1}]$, where, $\tilde{u}_i = u_i / \|HH^T u_i\|$ are the normalized eigenvectors. After β is found, the transform network is constructed and could be used for learning data representations.

2.4 Co-segmentation

While training the ELM network, the patches can be mapped into the embedded space and obtain a more effective representation. We adopt litekmeans [25] to perform multi-cluster clustering in the embedded space. Therefore, the resulting co-segmentation of the shape set from the same class is achieved.

Once the ELM network is trained, it can be applied to any out-of-sample data in the original feature space. For out-of-sample shapes, we first over-segment them and then extract patch features. Given the patch set of out-of-sample x, we construct the hidden layer with the same input weight and bias parameters of the trained networks. The feature distribution $f(x) = h(x)\beta$ is generated via known output-weight parameters of the trained network. Next, we assign patches from the closest cluster in the embedded space to the existing co-segmentation shape set. Finally, we label each patch with the same category as the nearest cluster to achieve consistent incremental co-segmentation.

3 Experiments

3.1 Dataset and Evaluation

The proposed method is evaluated on dataset Princeton Segmentation Benchmark (PSB) [6] and COSEG [14]. Because our approach is entirely unsupervised, the ground truth is only used for statistical evaluation. The co-segmentation is implemented for each category. All the results presented in this paper were produced via parameters $N_h = 2000$, $N_o = 3$, $p_i = 50$, and $\lambda = 0.1$. To evaluate our results, we use the accuracy measure as in [8], which measures the correctly-labeled shape area as: $Accuracy(l, t) = (\sum_i a_i \delta(map(l_i) - t_i))/(\sum_i a_i)$ where, a_i is the area of face i, l_i is the category label computed by our co-segmentation, t_i is the ground-truth label, $\delta(x)$ is a function that equals to 1 if and only if x equals 0, and $map(\cdot)$ is an optimal permutation function that maps each cluster label to a category label via Hungarian algorithm [26].

3.2   Sensitivity to Parameter Selection in Affinity Matrix Construction

To verify the availability of affinity matrix construction, we compare the segmentation results using two different construction methods: SSR and k-nearest neighbors. For comparison, after the affinity matrix is obtained, we exploit the same unsupervised shape representation and clustering algorithm. The comparison between the k-nearest neighbors and our method is illustrated by segmentation accuracy in Fig. 2. Obviously, the segmentation accuracies that use different k change widely. For example, the highest average segmentation accuracy of the Teddy category is 93.72% with $k = 10$, while the lowest one is only 70.7% with $k = 2$. It means that affinity matrix construction is sensitivity to the selection of k. In contrast, our method does not need selecting any parameters in the process of affinity matrix construction and has stable and better segmentation performance.

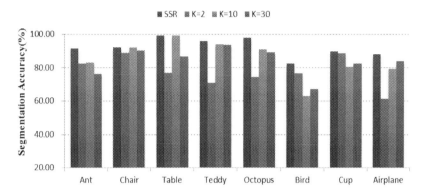

Fig. 2. Segmentation accuracy in different affinity matrix construction.

3.3   The Effectiveness of Data Representation in Embedded Space

To compare the effectiveness of data representations computed by US-ELM and other nonlinear embedding methods, we construct an affinity matrix via SSR that above description and perform clustering in different embedded spaces. Figure 3 shows the average segmentation accuracy of US-ELM, diffusion maps, LLE and Kernel-PCA for different dimensions of the embedded space on several categories, such as Table in Fig. 3(a), Hand in Fig. 3(b), Teddy in Fig. 3(c) and Chair in Fig. 3(d). From these results, we can see the highest segmentation accuracy is usually achieved by US-ELM. Moreover, the US-ELM attains its best performance in very low dimensional embedding, which means it is an efficient method since fewer eigenvectors are required to be computed while finding more accurate clustering method.

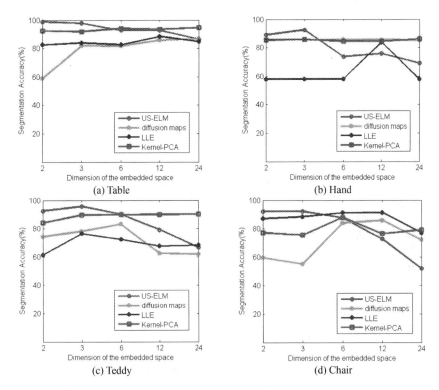

Fig. 3. Segmentation accuracy in different dimensions of the embedded space.

3.4 Co-segmentation Result

Figure 4 shows the co-segmentation results of our approach on several categories. These examples demonstrate that our approach can segment shapes into consistent corresponding parts and is insensitive to orientation, location and shape variation.

Comparisons to State-of-the-Art. We make comparisons of our approach with three state-of-the-art methods [14, 16, 17] on PSB dataset with the results shown in Fig. 5. Notice that Hu's method [16] and Wu's method [17] get higher accuracies than our method for several shape categories. That is possible because, in complex shape sets, the feature-fusion strategy can avoid some feature conflicts than the concatenated feature vector used in our method. However, they both lead to an extremely complex optimization problem. The average accuracies of [14, 16, 17] and ours performed on the concatenated feature vector (CFV) are 78.68%, 80.96%, 87.55% and 89.43%, respectively. Obviously, our approach has better segmentation accuracy on average; and, compared with the other methods, our approach does not use any post-processing operation to refine the results. As mentioned in the method by Sidi et al. [14], if the fine-tuning process is added, there will also be about 4% improvement from the initial to the refined co-segmentation. In recent years, segmentation methods based on deep learning have achieved a lot of success and usually has higher segmentation precision

[7–12]. However, these methods are all supervised, which means they rely on large amount of labelled data. The preparation of the training set is tedious and time-consuming. By contrast, our approach does not require any labelled data, and can segment shape directly according to the correlation of the shape set.

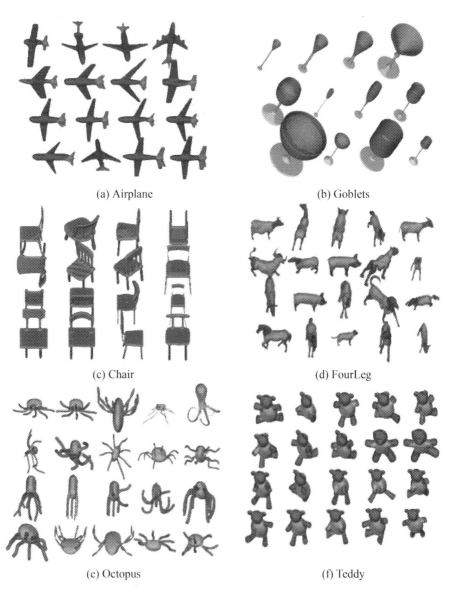

(a) Airplane

(b) Goblets

(c) Chair

(d) FourLeg

(e) Octopus

(f) Teddy

Fig. 4. Co-segmentation results on various categories produced by our approach. (Corresponding parts in each category are shown with the same color.). (Color figure online)

Performance. All the experiments are performed on a machine with Intel(R) Core™ 3.6 GHz CPU and 16 GB RAM. The total running time of our co-segmentation algorithm is very fast and is about half a minute on a set of 20 shapes (excluding the time of feature extraction). In an incremental co-segmentation setting, instead of handling the whole co-segmentation process as in spectral clustering method, we only implement feature embedding and clustering to the out-of-sample shapes. Thus, the time consumed for segmenting one extra shape is about a second and a half.

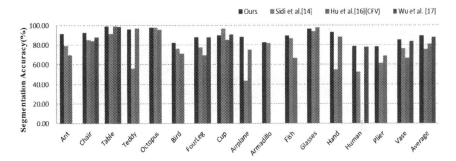

Fig. 5. Comparison of average segmentation accuracy between our approach with three state-of-the-art co-segmentation methods.

4 Conclusion

In this paper, we present an unsupervised method for co-segmentation a set of 3D shapes. The method has three advantages. First, we exploit SSR to construct the affinity matrix, which is parameter-free and leads to a significant reduction in computational cost. Second, we propose a new unsupervised method for shape co-segmentation via US-ELM, which implements network training and using for co-segmentation simultaneously. Third, based on the transform network, our approach can deal with the problem of incremental co-segmentation when the data set is expanded. Experimental results show our method's favorable performance when compared to state-of-the-art co-segmentation algorithms.

However, there are still problems that need to be addressed in unsupervised shape co-segmentation. If there are different semantic parts with high geometric similarity, they may be incorrectly clustered to the same semantic part. In the future, we plan to look for semantic-level feature descriptors to further improve the discrimination between the geometric components. In addition, we will further explore incremental, flexible and extensible co-segmentation methods.

Acknowledgements. This work was supported by Key Natural Science Fund of Nanjing College of Information Technology (No. YK20170401), National High Technology Re-search and Development Program of China (No. 2007AA01Z334), National Natural Science Foundation of China (Nos. 61321491 and 61272219), Program for New Century Excellent Talents in University of China (NCET-04-04605), the China Postdoctoral Science Foundation (Grant No. 2017M621700) and Innovation Fund of State Key Lab for Novel Software Technology (Nos. ZZKT2013A12, ZZKT2016A11 and ZZKT2018A09).

References

1. Agathos, A., Pratikakis, I., Perantonis, S., Sapidis, N., Azariadis, P.: 3D mesh segmentation methodologies for CAD applications. Comput. Aided Des. **4**(6), 827–841 (2007)
2. Shamir, A.: A survey on mesh segmentation techniques. Comput. Graph. Forum **27**(6), 1539–1556 (2008)
3. Shapira, L., Shamir, A., Cohen-Or, D.: Consistent mesh partitioning and skeletonisation using the shape diameter function. Vis. Comput. **24**(4), 249–259 (2008)
4. Guo, K., Zou, D., Chen, X.: 3D mesh labeling via deep convolutional neural net-works. ACM Trans. Graph. **35**(1), 3:1–3:12 (2015)
5. Shu, Z.-Y., et al.: Unsupervised 3D shape segmentation and co-segmentation via deep learning. Comput. Aided Geom. Des. **43**, 39–52 (2016)
6. Chen, X., Golovinskiy, A., Funkhouser, T.: A benchmark for 3D mesh segmentation. ACM Trans. Graph (SIGGRAPH) **28**(3), 73 (2009)
7. Le, T., Bui, G., Duan, Y.: A multi-view recurrent neural network for 3D mesh segmentation. Comput. Graph. **66**, 103–112 (2017)
8. Kalogerakis, E., Averkiou, M., Maji, S., Chaudhuri, S.: 3D shape segmentation with projective convolutional networks. In: Proceedings of CVPR, vol. 1, no. 2, pp. 3779–3788 (2017)
9. Yi, L., Su, H., Guo, X., Guibas, L.: SyncSpecCNN: synchronized spectral CNN for 3D shape segmentation. In: Proceedings of CVPR, pp. 2282–2290 (2017)
10. Wang, P., Gan, Y., et al.: 3D shape segmentation via shape fully convolutional networks. Comput. Graph. **70**, 128–139 (2017)
11. Qi, C.R., Su, H., Mo, K., Guibas, L.: J. PointNet: deep learning on point sets for 3D classification and segmentation. In: Proceeding of CVPR, pp. 77–85 (2017)
12. Xu, H., Dong, M., Zhong, Z.: Directionally convolutional networks for 3D shape segmentation. In: Proceedings of the IEEE International Conference on Computer Vision, pp. 2717–2726 (2017)
13. Golovinskiy, A., Funkhouser, T.: Consistent segmentation of 3D models. Comput. Graph. **33**(3), 262–269 (2009)
14. Sidi, O., Van Kaick, O., Kleiman, Y., Zhang, H., Cohen-or, D.: Unsupervised co-segmentation of a set of shapes via descriptor-space spectral clustering. ACM Trans. Graph. **30**(6), 126:1–126:10 (2011)
15. Luo, P., Wu, Z., Xia, C., Feng, L.: Co-segmentation of 3D shapes via multi-view spectral clustering. Vis. Comput. **29**(6–8), 587–597 (2013)
16. Hu, R., Fan, L., Liu, L.: Co-segmentation of 3D shapes via subspace clustering. Comput. Graph. Forum **31**(5), 1703–1713 (2012)
17. Wu, Z., Wang, Y., Shou, R., Chen, B., Liu, X.: Unsupervised co-segmentation of 3D shapes via affinity aggregation spectral clustering. Comput. Graph. **37**, 628–637 (2013)
18. Meng, M., Xia, J., Luo, J., He, Y.: Unsupervised co-segmentation for 3D shapes using iterative multi-label optimization. Comput. Aided Des. **45**(2), 312–320 (2013)
19. Kalogerakis, E., Hertzmann, A., Singh, K.: Learning 3D mesh segmentation and labeling. ACM Trans. Graph **29**, 102:1–102:12 (2010)
20. Shi, J., Malik, J.: Normalized cuts and image segmentation. IEEE Trans. Pattern Anal. Mach. Intell. **22**(8), 888–905 (2000)
21. Peng, C., Kang, Z., Yang, M., Cheng, Q.: Feature selection embedded subspace clustering. IEEE Signal Process. Lett. **23**(7), 1018–1022 (2016)
22. Reuter, M., Wolter, F.E., Peinecke, N.: Laplace-Beltrami spectra as 'shape-DNA' of surfaces and solids. Comput. Aided Des. **38**(4), 342–366 (2006)

23. Huang, J., Nie, F.-P., Huang, H.: A new simplex sparse learning model to measure data similarity for clustering. In: Proceedings of the Twenty-Fourth International Joint Conference on Artificial Intelligence, pp. 3569–3575 (2015)
24. Huang, G., Song, S., Gupta, J.N.D., Wu, C.: Semi-supervised and unsupervised extreme learning machines. IEEE Trans. Cybernet. **44**(12), 2405–2417 (2014)
25. Cai, D.: Litekmeans: the fastest matlab implementation of kmeans. http://www.zjucadcg.cn/dengcai/Data/Clustering.html
26. Papadimitriou, C.H., Steiglitz, K.: Combinatorial Optimization: Algorithms and Complexity. Courier Dover Publications, Mineola (1998)
27. Zhang, Z., Member, S., Xu, Y., Member, S.: A Survey of Sparse Representation: Algorithms and Applications, vol. 3 (2015)
28. Belkin, M., Niyogi, P., Sindhwani, V.: Manifold regularization: a geometric framework for learning from labeled and unlabeled examples. J. Mach. Learn. Res. **7**, 2399–2434 (2006)

RS-MSSF Frame: Remote Sensing Image Classification Based on Extraction and Fusion of Multiple Spectral-Spatial Features

Hanane Teffahi[1,2] and Hongxun Yao[1(✉)]

[1] School of Computer Science and Technology, Harbin Institute of Technology, Harbin, China
{hteffahi89, h.yao}@hit.edu.cn
[2] Algerian Space Agency, Algiers, Algeria
hteffahi@asal.dz

Abstract. Classifying remote sensing images with high spectral and spatial resolution became an important topic and challenging task in computer vision and remote sensing (RS) fields because of their huge dimensionality and computational complexity. Recently, many studies have already demonstrated the efficiency of employing spatial information where a combination of spectral and spatial information in a single classification framework have attracted special attention because of their capability to improve the classification accuracy. Shape and texture features are considered as two important types of spatial features in various applications of image processing. In this study, we extracted multiple features from spectral and spatial domains where we utilized texture and shape features, as well as spectral features, in order to obtain high classification accuracy. The spatial features considered in this study are produced by Gray Level Co-occurrence Matrix (GLCM) and Extended Multi-Attribute Profiles (EMAP), while, the extraction of deep spectral features is done by Stacked Sparse Autoencoders. The obtained spectral-spatial features are concatenated directly as a simple feature fusion and are fed into the Support Vector Machine (SVM) classifier. We tested the proposed method on hyperspectral (HS) and multispectral (MS) images where the experiments demonstrated significantly the efficiency of the proposed framework in comparison with some recent spectral-spatial classification methods and with different classification frameworks based on the used extractors.

Keywords: Remote sensing image classification · Multiple features
Stacked sparse autoencoder · Gray level co-occurrence matrix
Extended multi-attribute profiles · Support vector machine

1 Introduction

Nowadays, with the development of remote sensing sensors, high-resolution satellite imagery in both the spectral and spatial domains can be acquired and used in various domains (e.g. urban mapping, environmental management and military purpose). New generation of HS and MS sensors can collect images with very high spectral and spatial

© Springer Nature Switzerland AG 2018
R. Hong et al. (Eds.): PCM 2018, LNCS 11165, pp. 582–595, 2018.
https://doi.org/10.1007/978-3-030-00767-6_54

resolution where the spatial resolution of the panchromatic image can be as high as 0.3 m. While, the spatial resolution for multispectral and hyperspectral images can be as high as 1.6 m to 3 m, respectively (e.g. World View satellites, Geoeye, ROSIS, CASI…) [1] and the spectral resolution is between eight bands until two hundred bands.

The high spectral and spatial dimensionality of RS images produces Hughes phenomenon making the processing of these images difficult. In particular, classification of the hyperspectral and multispectral image with high resolution is a challenging task. However, the highly improved remote sensing image classification suffers from the amount of detailed ground information which increases the intra-class variation and decreases the interclass variation and leads to classification difficulty in the spectral domain [2]. Based on these issues, many spectral-spatial classification frameworks were developed. To well characterize a pixel, it is important to find effective and robust features that can well represent spectral and spatial characteristics [1]. In this way, some spatial extractors were investigated for the classification of remote sensing images as Grey-Level Co-occurrence Matrix [3], Gabor texture features [4], morphological features [5], attribute profiles, Multi-attribute profiles [6, 7] and Extinction profiles [8]. These spatial feature extraction techniques are proposed to describe the texture, structural and shape features such as they were successfully applied for remote sensing images. In the last decay, deep learning (DL) methods proved their capability to compute high-level features such as it is commonly recognized that high-level features are more efficient and robust when dealing with image classification of different remote sensing data (e.g. VHR multispectral and hyperspectral images, RADAR "SAR" images, LIDAR images). However, the most popular deep learning methods for remote sensing image classification are Stacked Autoencoders (SA) [9–12] and Convolutional Neural Network (CNN) [13–15].

This paper is the presentation of a new contribution for spectral-spatial classification of remote sensing images. In this work, in addition to the deep spectral features extracted using Stacked Sparse Autoencoders (SSAE), two different types of spatial features are extracted based on "crisp neighborhood" and "Adaptive neighborhood" systems exposed in [1] for spatial extraction. In our approach, Adaptive neighborhood system used for extracting geometric-shape features based on Extended Multi-Attribute Profiles (EMAP) and crisp neighborhood used for extracting texture features based on Fast-Grey Level Co-occurrence Matrix (FGLCM) which is introduced in [16]. Therefore, the fused features are the input of the SVM classifier. The fusion of these high-level features is a good and robust representation of remote sensing image classification achieving high classification accuracy. The proposed feature extraction and fusion process is applied to different types of remote sensing images.

2 Feature Extraction Techniques

2.1 Spectral Feature Extraction Based on Stacked Sparse Autoencoders

A shallow sparse autoencoder introduces a specific kind of neural network containing input, hidden, and reconstruction layers that can be employed to train the high-level feature representations in an unsupervised manner [17]. Sparse autoencoder estimates a reconstruction function that maximizes the similarity score of decoding and input layer functions (i.e. minimizing the difference between input and its reconstruction). The Sparse Autoencoder (SAE) is defined based on the basic autoencoder and KL-divergence regularization [11, 18, 19]. A basic SAE contains three layers, that is, one input layer, one hidden layer, and one reconstruction layer (output layer). However, according to [9, 11, 20, 21], we expose sparse autoencoder as follows: the input X describes by $\{x^{(1)}, x^{(2)}, \ldots, x^{(n)}, \ldots, x^{(N)}\}$, where $x^i \in \mathbb{R}^N$ presents the spatial information from origin the al panchromatic image for a single SAE. However, the input vector X for the first SAE of a deep autoencoder applied for multispectral image, presents the spectral information from the original MS image. However, the input vector X for the first SAE of a deep autoencoder applied for hyperspectral image presents the spectral information from the fourth first principal components (PC) of the HS image. In the following, we utilize x for the input and h for the hidden when explaining the SAE. In the training of SAE, we have two steps "encoding" and "decoding. During the encoding step, an input vector $x \in \mathbb{R}^N$ is processed by applying a linear mapping and a nonlinear activation function to the network: $h = f(W_h x + b_h)$ where $W_h \in \mathbb{R}^{N \times K}$ is a weight matrix with K features, $b_h \in \mathbb{R}^K$ is the encoding bias, and f is the logistic sigmoid function $f(x) = 1/1 + e^{-x}$. We decode a vector using a separate linear decoding matrix: $z = f(W_z h + b_z)$ where $W_z \in \mathbb{R}^{K \times N}$ is a weight matrix and $b_z \in \mathbb{R}^N$ is the decoding bias. W_h and W_z denote the input-to-hidden and the hidden-to-output weights, respectively. For rendering the parameterizations identical, we restrain $W = W_h = W_z^T$. b_h and b_z denote the bias of hidden and output units respectively. By employing the back-propagation algorithm, the spectral features are extracted by minimizing the difference between input and its reconstruction, and then the features are encapsulated in weight matrix W and bias vector b and the objective function of autoencoder is: $Arg\ min_{W, b_h, b_z}[L(x, z)]$. The reconstruction error can be measured as the traditional squared error:

$$L(x, z) = \|x - z\|^2 + \lambda \|W\|^2 \tag{1}$$

The reconstruction error $L(x,z)$ is defined as the Euclidean distance between x and z where the reconstruction z is forced to approximate the input data $x \left(\|x - z\|^2 \to 0 \right)$. The parameters of the autoencoder (W, b_h and b_z) are generally optimized by using mini-batch stochastic gradient descent (SGD). After training the network, the reconstruction layer is removed and the learned feature lies in the hidden layer, which can be used for classification. The objective function of SAE architecture with a weight decay term and a sparsity constraint term defined as follows [11, 18]:

$$L(x, z) = \|x - z\|^2 + \lambda \|W\|^2 + \beta\ KL(x\|z) \tag{2}$$

The first term (on the right side) of the Eq. (2) is an average sum-of-square errors term represented the gap between the input x and the output z. The second term is the weight decay term, such as it is utilized for reducing the autoencoder from overfitting by monitoring the amplitude of the weights, where λ is a weight decay parameter. The third term represents a sparsity penalty term, such as β controls the weight of the term, and KL is a Kullback-Leibler divergence, formulated as $\text{KL}\,(x\|z) = \sum_{j=1}^{K}[x_j \log z_j + (1 - x_j) \log (1 - z_j)]$ where K is the number of neurons in the hidden layer, and the index j is summing over the hidden units in our network. Stacked Sparse Autoencoder (SSA) is a layer-wise encoding neural network in which multiple layers of Sparse Autoencoders are stacked and pre-trained layer by layer [22]. Stacked Sparse Autoencoder finds optimal features for a training set by multiple hidden layers [19] such as SSAE makes a deep representation of input data at the output of the last layer. After finishing training one specific layer of parameters, the next layer is trained according to the output of its previous layer. Stacking these input-to-hidden layers sequentially constructs a stacked autoencoder [9]. The learning process of SSAE is the same as the SAE architecture, where the objective function is minimizing the reconstruction error. In this paper, we use stacked sparse autoencoder (SSAE) to compute deep representations of spectral data by extracting high-level features from the original MS and HS datasets.

2.2 Extraction of Shape Features Using Extend Multi-attribute Profiles

As defined in [6, 7], Attribute Profiles are based on Attribute filters which operate on the connected components that compose an image, according to a specific criterion evaluated on each connected component CC of the image f and associated to thickening and thinning transformations. However, the AP is defined as computing an attribute A for every connected component CC of an image f for a given reference value λ. For each connected component C_i of the image, whether the attribute satisfies a predefined criterion, afterward the region remains unaffected; otherwise, it is set to the radiometric value of the adjacent region with the nearest value, so C_i can be merged to the adjacent connected component. When the region is merged to the adjacent region of a lower (or higher) grey level, the process achieved is a thinning (or a thickening) [21]. When, we consider a sequence of thresholds $\{\lambda_1, \lambda_2 \ldots, \lambda_n\}$, an AP is obtained by making a successive attribute thinning and attribute thickening as follows [23]: AP $(f) = \{\phi_n(f), \ldots, \phi_1(f), f, \gamma_1(f), \ldots, \gamma_n(f)\}$ such as φ_i and γ_i denote respectively the thickening and thinning transformations, with n morphological attributes thickening (φ^T) and attributes thinning (γ^T). Therefore, each pixel p of an image f can be configured and typified using the values resulting from the successive filtering operations. The concept of AP was extended to multispectral and hyperspectral images with the definition of Extended Attribute Profiles. An EAP [6] is computed by concatenating the APs considering the same attribute of the r principal components PCs, extracted from the original image:

$$\text{EAP} = \{\text{AP}(\text{PC}_1), \text{AP}(\text{PC}_2), \ldots, \text{AP}(\text{PC}_r)\} \tag{3}$$

Where $\text{PC}_i(1 \leq i \leq r)$ are the rth first principal components obtained after applying PCA for the original multispectral image. PCA uses performing attribute filtering on the rth first PC in order to reduce computational complexity. Extended Multi-Attribute Profiles are obtained by the concatenation of EAPs. Since the dimensionality of the features is increased, the EMAP has a much greater capability to extract the spatial information from the remote sensing images than a single EAP [7]. In various approaches using EMAP for multispectral and hyperspectral images, four EAPs were used. These attributes are [7]: "a" area of the regions, "d" the diagonal length's of the box bounding the region, "i" the moment of inertia, "s" the standard deviation of pixels radiometric values in the regions.

2.3 Texture Feature Extraction Based on Fast-Gray Level Co-occurrence Matrix

GLCM matrices and their statistics were introduced as a tool for extracting texture features from images. In this work, we use a specific GLCM technique named Fast-GLCM developed by Mirzapour and Ghassemian for extracting efficient texture features. The Fast Gray Level Co-occurrence Matrix (FGLCM) is defined as follows [16]: For each pixel with (x_p, y_p) coordination in the original single-band image, a neighbourhood system (usually a square-shaped window) is considered. Then, a GLCM matrix is extracted from this window and is assigned to the center pixel [24]. The (i,j)-element of the $\text{GLCM}^{(x_p,y_p)}$ matrix is the number of pixels in the neighborhood window $\text{W}^{(x_p,y_p)}$ having the grey level of j, occurring at an angular offset of (d, θ) from another pixel with the grey level of i, where all these pixels are located in the $\text{W}^{(x_p,y_p)}$ neighborhood window. We can formulate this process as follows [16, 24]:

$$\text{GLCM}^{(x_p,y_p)}(i,j) = \sum_{(x_1,y_1) and (x_2,y_2) \in \text{W}^{(x_p,y_p)}} \begin{cases} 1 \; if f(x_1,y_1) = i \, and f(x_2,y_2) = j \\ 0; else \end{cases}$$
$$(x_2,y_2) = (x_1 + dcos\theta, y_1 + dsin\theta)$$

$$(4)$$

Where f referred to the Kronecker delta function δ applied to the grey level image I [25], (d, θ), is the angular distance. Traditional GLCM matrices are extracted from the input image re-quantized by 5 bits where its offset parameters (d, θ) are set to $(1, 0)$ and a square neighborhood window of size (11×11) is used. After calculating a GLCM matrix for each pixel using Eq. (11), 16 statistical features as described in [27] are extracted from the matrix and assigned to the pixel [24]. We have implemented the Fast-GLCM algorithm described in [25] retained only some useful features as: *mean, variance, homogeneity, contrast, entropy, angular second moment, correlation.*

3   Feature Extraction and Fusion Scheme

In this section, the whole plan of our work is illustrated in Fig. 1. As shown in Fig. 1, to extract spectral features from the two datasets we use Stacked Sparse Autoencoders such as for the MS data, the input of SSAE is the whole image but for the HYP data, the input of SSAE is the fourth first principal components of the hyperspectral image.

The spatial extracted features are two kinds: shape and texture features. The shape features are extracted by EMAP. On the other hand, texture features are given from segmentation based Gray Level Co-occurrence Matrix (GLCM). For each dataset, the extracted spectral and multi-spatial features are concatenated directly as a simple fusion features.

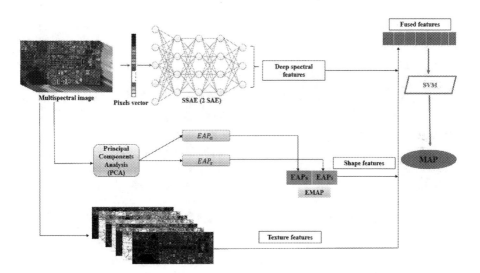

Fig. 1. The overall architecture of the proposed approach.

Feature Stacking is used for feature fusion and defined as is a simple approach to integrate extracted features from each dataset. In this manner, let X_{Spect}, X_{Shape}, and $X_{Texture}$ denote the spectral features, shape features and texture features respectively extracted from MS and HYP data. In this case, the Feature Stacking approach simply concatenates the spectral and spatial features (shape and texture), i.e., $X_{Fusion} = [X_{Spect} \quad X_{Shape} \quad X_{Texture}]$ where x, y, and z are three feature vectors extracted from input images $(x = X_{Spect}, y = X_{Shape}, z = X_{Texture})$ and D, K and Q are the dimensions of the spectral and spatial features vectors respectively, then the fused feature vector is $k = X_{Fusion}$ with size equal to $(D + K + Q)$. The fused features are integrated into Support Vector Machine (SVM) for classification.

4   Datasets in Investigation

4.1   Houston Data

The data is a hyperspectral image. This dataset was acquired by the compact airborne spectrographic imager over the University of Houston campus and the neighboring urban area on June 23, 2012, and was distributed for the 2013 GRSS data fusion contest. The size of the data is 349 × 1905 pixels with the spatial resolution of 2.5 m and consists of 144 spectral bands ranging from 0.38 to 1.05 μm.

The ground truth map contains 15 classes of interests. Figure 2 shows a color composite representation of the HS data and the corresponding ground truth data. Table 1 gives information about the number of training and test samples for different classes of interests [14].

Fig. 2. (Up) a color composite representation of the Huston data using bands 70, 50, and 20, as R, G, and B, respectively; (Down) Ground truth data.

Table 1. Ground reference classes of Huston data

Class number	Class name	Training samples	Testing samples
1	Grass healthy	198	1053
2	Grass stressed	190	1064
3	Grass synthetic	192	505
4	Tree	188	1056
5	Soil	186	1056
6	Water	182	143
7	Residential	196	1072
8	Commercial	191	1053
9	Road	193	1059
10	Highway	191	1036
11	Railway	181	1054
12	Parking lot 1	192	1041
13	Parking lot 2	184	285
14	Tennis court	181	247
15	Running track	187	473

4.2   Washington D.C Data

Washington D.C (USA) data: was acquired with the WorldView-2 satellite on 9 February 2016 (downloaded from www.digitalglobe.com). As shown in Fig. 3, the analyzed dataset is available with 8 spectral bands and spatial resolution 1.6 m for multispectral. The dimension of the image is 2438 × 896 pixels. The ground data consisted of eight classes of interest as described in Table 2. This data was used for the first time in [21].

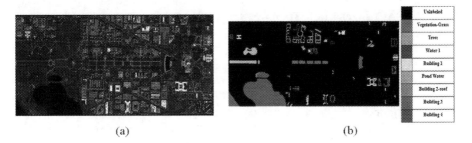

(a)   (b)

Fig. 3.   (a) Washington D.C image with true colors/(b) Ground Truth image with eight classes.

Table 2.   Eight ground reference classes of Washington D.C data.

Class number	1	2	3	4	5	6	7	8
Total samples	14550	12846	148872	18153	22859	19229	3545	12468

5   Experiments and Discussion

In the multispectral dataset, we divide the labeled parts of the images into two sets: training samples and testing samples. We randomly choose 0.01 as the training-rate (1%) from the labeled samples from each class and the rest is defined as the testing and validating set. The multispectral image "Washington DC" contains eight (8) bands; a single pixel can be represented as an 8-dimensional vector. Window structure can increase the discriminant information. Therefore, a pixel can be represented by a box with $w \times w \times 8$ pixels ($w = 28$). The hyperspectral image "Huston" contains 144 bands; a single pixel can be represented as a 144-dimensional vector. Window structure can increase the discriminant information. Therefore, a pixel can be represented by a box with $w \times w \times 144$ pixels ($w = 64$). In the proposed framework, SSAE is employed to extract deep spectral features from of the original multispectral and hyperspectral images. The SSAE for MS image is composed of two autoencoders; however, SSAE for the hyperspectral image is composed of four autoencoders. We experimented the SSAE for MS and HS data according to the variation of the hidden size as in Tables 3 and 4 where the tables list the overall classification accuracy (OA) and kappa coefficient of SSAE-SVM method for "Washington DC" and "Huston" data respectively, versus the number of neurons in a hidden layer. For these two datasets, the proposed method was

Table 3. Classification accuracies for MS image of Washington DC data using SSAE.

Measure	$h = 50$	$h = 100$	$h = 200$	$h = 300$	$\boldsymbol{h = 400}$	$h = 500$
OA (%)	98.3949	98.8054	99.0122	99.0822	**99.1099**	99.0690
Kappa	0.9743	0.9808	0.9842	0.9853	**0.9857**	0.9851
AA (%)	96.85	97.53	97.74	97.96	**98.11**	98.05

trained by stochastic gradient descent, with a batch size of 28 for MS image and 64 for HYP image, a maximum number of iteration is 1000 for SSAE and the sparsity penalty value of SSAE is equal to 0.001 for MS and HYP images. In addition, there is no important change in the classification accuracy using a big number of neurons for the same training rate.

For extracting efficient shape-spatial features we use EMAP. For multispectral data "Washington DC", EMAP is composed of two EAPs {EAP_a, EAP_s} using two reference values λ for building each of the two EAPs where each EAP leads to 15_dimensional profiles and the λ values considered are the following: λ_a= [100 500], λ_s= [25 125]. For the hyperspectral data "Huston", EMAP is composed of four EAPs {EAP_a, EAP_d, EAP_i, EAP_s} using four reference values λ for building each of the four EAPs where each EAP leads to 45_dimentional profiles (composed by four Aps of 15 levels computed on the 4 PCs). The λ values considered are the following: λ_a = [100 500 1000 5000], λ_d= [10 25 50 100], λ_i = [0.2 0.3 0.4 0.5], λ_s = [20 30 40 50].

We have implemented the Fast GLCM algorithm described in [25]. The main drawback of GLCM is its slow nature, which is addressed and managed by the fast GLCM where we retained only some useful features as mean, variance, homogeneity, contrast, entropy, angular second moment, correlation. The parameters of GLCM are initialized as the angular distance $(d, \theta) = (1, 0)$ and the size of neighborhood window (w) is equal to (11×11) for MS image and (33×33) for HS image. G is the number of gray levels of the image after re-quantization where $G = 32$ for HS data and $G = 16$ for MS data.

Table 4. Classification accuracies for HYP-Huston data using SSAE.

Measure	$h = 50$	$h = 100$	$h = 200$	$h = 300$	$h = 400$	$\boldsymbol{h = 500}$
OA (%)	95,5001	95.8051	96.0221	96.0452	97.1080	**97,8902**
Kappa	0.9513	0.9538	0.9582	0.9585	0.9666	**0.9742**
AA (%)	94.8231	95.0801	95.7280	95.7102	96.3980	**97.2003**

The best classification performance for Washington DC data can be obtained with $h = 400$ for MS-SSAE fused with the shape features extracted by EMAP and the texture features extracted by GLCM. On the other hand, the best classification performance for Huston data can be obtained with $h = 500$, for HYP-SSAE fused with the shape features extracted by EMAP (constructed from the fourth attributes) and the texture features extracted by GLCM. The classification is performed by using SVM based on One

Against One strategy (OAO) and SVM is done by using LIBSVM through its MATLAB interface. In Table 5, the accuracy assessment of the proposed framework is demonstrated by the overall accuracy (OA), the kappa coefficient and the average accuracy (AA). In another hand, the two measures producer's accuracy (PA) and user's accuracy (UA) evaluated the individual class accuracies (Tables 6 and 7).

Table 5. Classification accuracies of the proposed method.

Data	OA	Kappa	AA
Washington	99.85	0.9942	99.28
Huston	99.02	0.9861	98.43

Table 6. Individual classification accuracies of Washington DC data.

	C1	C2	C3	C4	C5	C6	C7	C8
PA	0.9881	0.9888	**0.9969**	**0.9988**	0.9961	**0.9986**	**0.9966**	**0.9987**
UA	0.9975	0.9886	**0.9978**	**0.9994**	0.9930	**0.9980**	**0.9988**	**0.9979**

Table 7. Individual classification accuracies of Huston data.

Class	PA	UA
C1	0.9745	0.9758
C2	0.9881	0.9885
C3	0.9780	0.9801
C4	0.9810	0.9825
C5	0.9888	0.9881
C6	0.9903	0.9909
C7	**0.9935**	**0.9971**
C8	**0.9942**	**0.9951**
C9	**0.9964**	**0.9970**
C10	**0.9901**	**0.9900**
C11	**0.9889**	**0.9883**
C12	0.9880	0.9887
C13	0.9886	0.9889
C14	0.9788	0.9801
C15	0.9797	0.9805

The classification accuracy after feature fusion [SSAE ⊕ EMAP ⊕ GLCM feature fusion fed into SVM] is exposed in Table 5 and compared with different frameworks applied to the same datasets. These comparative methods are exposed as follow: **F1**: Stacked Sparse Autoencoder for the two Remote Sensing images (SSAE) using two SAE for Washington DC where hidden size $h_1 = h_2 = 400$ and three SAE for Huston

Table 8. Comparison of the overall classification accuracies (OA%)

Frameworks	Washington DC	Huston
F1	99,10	97,89
F2	98.20	96.80
F3	97.90	96.50
F4	99.56	98.10
F5	99.64	98.70
F6	98.79	95.20
F7	96,01	95,40
F8	98,02	96,80
Proposed F	**99,85**	**99,02**

data where hidden size $h_1 = h_2 = h_3 = 500$ (Tables 3 and 4); **F2:** PCA-SAE: spectral feature extraction for MS and HYP images; **F3:** GLCM texture feature extraction technique with SVM for the two datasets (PCA-GLCM + SVM) such as we used 4 PCs for the HYP image and 2 PC for MS image. The parameters of GLCM are $(d, \theta) = (1, 0)$, $w_{MS} = (11 \times 11)$, $w_{HS} = (33 \times 33)$, $G_{HS} = 32$, $G_{MS} = 16$ and the parameters of SVM are: the semi-radius of the kernel function (RBF) $g = 1$ and penalized parameters $c = 100$ (the parameters of SVM are the same for all frameworks); **F4:** EMAP + SAE + SVM [21]; **F5:** Spectral-EMAP-SAE such as it is started by building EMAP with 4-EAPs for HS image and 2-EAPs for MS image. Next, we combined the spectral information from the original image (original spectral features) with EMAP. After that, we used SAE for feature extraction and dimensionality reduction where. Finally, classification is performed on the reconstruction layer of SAE by using SVM, the parameters of SAE are the same as in (F4) [28]; **F6:** SVM for hyperspectral and multispectral images where the kernel function is radial basis function (RBF), the semi-radius of the kernel function $g = 1$ and penalized parameters $c = 100$; **F7:** Fusion of (SSAE) with (GLCM) for the two images; **F8:** GLCM + SAE + SVM (the parameters of the methods are the same as in F3 and F4).

According to the results presented in Table 8, we notice that the proposed method gave the best accuracy for both multispectral and hyperspectral images. In addition, the results of the proposed classification framework show that extracting powerful and complementary features can produce satisfactory results even in a simple combinational framework. It is worth noting that EMAP and GLCM capture diverse spatial features of an image (the size and shape and texture of the elements of the image, respectively), and are able to complement each other very well. Moreover, the simple fusion of these spatial features with deep spectral features (high-level features) obtains great results in solving the classification problem of Multispectral and Hyperspectral images according to our experiments.

6 Conclusion

In this paper, we presented a new approach for multispectral and hyperspectral image classification. The proposed method is based on feature extraction and fusion techniques. The fusion of different types of spatial features with deep spectral features proved its efficiency for remote sensing image classification.

Exploring the classification results, we saw that including spatial information in the classification process could improve the results, as was expected. Nonetheless, we came to some useful conclusions, among them: (1) we saw that the fusion of EMAP and FGLCM for extracting spatial features may represent well the spatial information, specifically for urban data. (2) Extracting complementary and informative features is an important step in classification tasks where this process can improve the classification performance significantly even when we used a simple feature fusion technique such as stacking. (3) Finally, we compared our method with some other spectral-spatial classification methods and classification methods using only spectral feature extractors. The comparative results show the good capabilities of the proposed method to classify the VHR-MS and hyperspectral images.

Acknowledgments. This work is supported by the National Natural Science Foundation of China (61472103) (61772158).

References

1. Zhao, W., Guo, Z., Yue, J., Zhang, X., Luo, L.: On combining multiscale deep learning features for the classification of hyperspectral remote sensing imagery. Int. J. Remote Sens. **36**(13), 3368–3379 (2015)
2. Huang, X., Zhang, L.: An SVM ensemble approach combining spectral, structural, and semantic features for the classification of high-resolution remotely sensed imagery. IEEE Trans. Geosci. Remote Sens. **51**(1), 257–272 (2013)
3. Agüera, F., Aguilar, F.J., Aguilar, M.A.: Using texture analysis to improve per-pixel classification of very high-resolution images for mapping plastic Greenhouses. ISPRS J. Photogrammetry Remote Sens. **63**(6), 635–646 (2008)
4. Reis, S., Taş Demir, K.: Identification of hazelnut fields using spectral and gabor textural features. ISPRS J. Photogrammetry Remote Sens. **66**(5), 652–661 (2011)
5. Benediktsson, J.A., Palmason, J.A., Sveinsson, J.R.: Classification of hyperspectral data from urban areas based on extended morphological profiles. IEEE Trans. Geosci. Remote Sens. **43**(3), 480–491 (2005)
6. Dalla Mura, M., Benediktsson, J.A., Bruzzone, L.: Classification of hyperspectral images with extend attribute profiles and feature extraction techniques. In: IEEE Geosciences and Remote Sensing Symposium, pp. 76–79 (2010)
7. Ghamisi, P., Dalla Mura, M., Benediktsson, J.A.: A survey on spectral-spatial classification techniques based on attribute profiles. IEEE Trans. Geosci. Remote Sens. **53**(5), 2335–2353 (2015)
8. Ghamisi, P., Souza, R., Benediktsson, J.A., Zhu, X.X., Rittner, L., Lotufo, R.A.: Extinction profiles for the classification of remote sensing data. IEEE Trans. Geosci. Remote Sens. **54**(10), 5631–5645 (2016)

9. Chen, Y., Lin, Z., Zhao, X., Wang, G., Gu, Y.: Deep learning-based classification of hyperspectral data. IEEE J. Sel. Topics Appl. Earth Obs. Remote Sens. 7(6), 2094–2107 (2014)

10. Tao, C., Pan, H., Li, Y., Zou, Z.: Unsupervised spectral-spatial feature learning with stacked sparse autoencoder for hyperspectral imagery classification. IEEE Geosci. Remote Sens. Lett. 12(12), 2438–2442 (2015)

11. Wang, L., Zhang, J., Liu, P., Choo, K.K.R., Huang, F.: Spectral-spatial multi feature-based deep learning for hyperspectral remote sensing image classification. Soft Comput. 21(1), 213–221 (2016)

12. Shao, Z., Zhang, L., Wang, L.: Stacked sparse autoencoder modelling using the synergy of airborne LiDAR and satellite optical and SAR data to map forest above-ground biomass. IEEE J. Sel. Topics Appl. Earth Observations Remote Sens. 10(12) (2017)

13. Zhao, W., Du, S.: Spectral-spatial feature extraction for hyperspectral image classification: a dimension reduction and deep learning approach. IEEE Trans. Geosci. Remote Sens. 54(8), 4544–4554 (2016)

14. Chen, Y., Li, C., Ghamisi, P., Jia, X., Gu, Y.: Deep fusion of remote sensing data for accurate classification. IEEE Geosci. Remote Sens. Lett. 14(8), 1253–1257 (2017)

15. Wang, A., He, X., Ghamisi, P., Chen, Y.: LiDAR data classification using morphological profiles and convolutional neural networks. IEEE Geosci. Remote Sens. Lett. 15(5), 774–778 (2018)

16. Mirzapour, F., Ghassemian, H.: Fast GLCM and gabor filters for texture classification of very high-resolution remote sensing images. Int. J. Inf. Commun. Technol. Res. 7(3), 21–30 (2015)

17. Abdi, G., Samadzadegan, F., Reinartz, P.: Spectral-spatial feature learning for hyperspectral imagery classification using deep stacked sparse autoencoder. J. Appl. Remote Sens. 11(4), 1–15 (2017)

18. LeCun, Y., Bengio, Y., Hinton, G.E.: Deep learning. Nat. Int. Weekly J. Sci. 521, 436–444 (2015)

19. Vincent, P., Larochelle, H., Bengio, Y., Manzagol, P.A.: Extracting and composing robust features with denoising autoencoder. In: The 25th International Conference on Machine Learning, pp. 1096–1103 (2008)

20. Ng, A.: "Sparse autoencoder", CS294A Lecture notes, Stanford University (2010)

21. Teffahi, H., Yao, H., Belabid, N., Chaib, S.: Feature extraction based on extend multi-attribute profiles and sparse autoencoder for remote sensing image classification. In: MIPPR 2017: Multispectral Image Acquisition, Processing, and Analysis, Proceedings of SPIE, vol. 10607 (2018)

22. Wan, X., Zhao, C., Wang, Y., Liu, W.: Stacked sparse autoencoder in hyperspectral data classification using spectral-spatial, higher order statistics and multifractal spectrum features. Infrared Phys. Technol. 86, 77–89 (2017)

23. Song, B., et al.: Remotely sensed image classification using sparse representations of morphological attribute profiles. IEEE Trans. Geosci. Remote Sens. 52(8), 5122–5136 (2014)

24. Mirzapour, F., Ghassemian, H.: Improving hyperspectral image classification by combining spectral, texture, and shape features. Int. J. Remote Sens. 36(4), 1070–1096 (2015)

25. Mirzapour, F., Ghassemian, H.: Using GLCM and Gabor filters for classification of PAN images. In: IEEE 21st Iranian Conference on Electrical Engineering (ICEE), pp. 1–6 (2013)

26. Aferi, F.D., Purboyo, T.W., Saputra, R.E.: Cotton texture segmentation based on image texture analysis using Gray Level Co-occurrence Matrix (GLCM) and euclidean distance. Int. J. Appl. Eng. Res. 13(1), 449–455 (2018)

27. Haralick, R.M., Shanmugam, K., Dinstein, I.H.: Textural features for image classification. IEEE Trans. Syst. Man Cybern. **3**(6), 610–621 (1973)
28. Teffahi, H., Yao, H., Belabid, N., Chaib, S.: A Novel Spectral-Spatial Classification Technique of Multispectral Images Using Extend Multi-Attribute Profiles and Sparse Autoencoder, December 2017. (Accepted, Submitted to Remote Sensing Letters)

Hierarchical Image Segmentation Based on Multi-feature Fusion and Graph Cut Optimization

Anqi Hu, Zhengxing Sun$^{(\boxtimes)}$, Yuqi Guo, and Qian Li

State Key Laboratory for Novel Software Technology, Nanjing University,
Nanjing, China
szx@nju.edu.cn

Abstract. The task of hierarchical image segmentation attempts to parse images from coarse to fine and provides a structural configuration by the output of a tree-like structure. To deal with the challenges of keeping semantic consistency in each level caused by the variable scale of different objects in image, this paper proposes a hierarchical image segmentation approach guided by multi-feature fusion and energy optimization. We transform the image into a region adjacency graph (RAG) by superpixels and design a bottom-up progressive merging framework based on graph cut for a hierarchical region tree. A multiscale structural edge is designed as a feature map for mapping to the hierarchical levels, while we conduct salient map and object window as a weakly-supervised prior during the optimization process. Experimental results demonstrate that our approach gets a better performance in semantic consistency while has an encouraging performance compared with some state-of-the-arts.

Keywords: Hierarchical segmentation · Semantic consistency
Graph cut · Multiscale

1 Introduction

Image segmentation plays an important role in many computer vision researches, e.g., object detection, image retrieval and image style transfer. General image segmentation problem is defined as: to divide an image into different regions with different labels, and the pixels with same labels in the image show similar feature. Actually, the image segmentation in single-scale can't satisfy different requirements at the same time because of the diverse applications. For example, object detection generally obtains the contour of targets while geological image analysis searches texture details besides boundary information for better image parsing. Therefore, many current researches attempt to deal with the problem of image segmentations in different scales. This task is also called hierarchical image segmentation which can be considered as the combination of several image segmentations from coarse to fine to some extent. The result of hierarchical image segmentation is generally represented as an Ultrametric Contour Map (UCM) [1], which is a convenient representation of the structural tree since the segmentation at all scales can be retrieved by thresholding the UCM.

© Springer Nature Switzerland AG 2018
R. Hong et al. (Eds.): PCM 2018, LNCS 11165, pp. 596–606, 2018.
https://doi.org/10.1007/978-3-030-00767-6_55

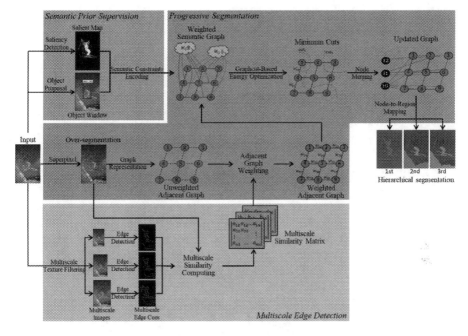

Fig. 1. The overview of proposed approach.

Most of the existing works [1–3] achieve the goal of hierarchical image segmentation by progressive merging or progressive division. These progressive processes normally generate different flat partitions based on threshold division approaches with low-level features. Specially, the approach [4] performs hierarchical segmentation at different scale and combines them into a single result for multiscale information. However, changing the size of input makes each segmentation process independent, and we could furtherly exploit advantages of multiscale strategy if it is used to con-duct the computing process instead of combine after segmentation.

On the other hand, how to keep the segmentations hierarchical both in region scale and semantic scale is still a challenge in this field. The hierarchies guided by low-level features hardly keep the semantic consistency in each threshold level, and may detect objects which belong to the same classes in different flat partitions because they only force on the scale of the regions. The improved approach [5] attempts to deal with this problem by aligning the hierarchical structure of existing segmentations. Actually, this method can be seen as a post-processing of hierarchies and we try to design an end-to-end framework in our work.

This paper introduces a hierarchical image segmentation method to overcome these limitations above. The pipeline of our method is illustrated in Fig. 1. Starting with superpixels, we transform the image into a region adjacency graph (RAG) so that the problem of hierarchical segmentation can be formulated as energy optimization based on graph theory. Then multiscale structural edge is extracted as a feature map to guide the progressive optimization. Finally, we conduct salient map and object window as constraint condition to make a weak supervision during hierarchical segmentation and overcome the influence of different object scales. This strategy of multi-feature prior can keep the semantic consistency in each level.

The remainder of this paper is organized as follows: Sect. 2 introduces the progressive framework of our hierarchical segmentation. Then the details of multiscale structural edge and the semantic prior constraint of salient map and object window are presented in Sect. 3. Section 4 describes the extended segmentation algorithm for region merging based on graph cut. Section 5 includes the experimental results and Sect. 6 makes a conclusion of our work.

2   Graph-Based Hierarchical Segmentation Model

The proposed approach starts with an over-segmentation of image I. According to [6], all pixels P of image I can be segmented into a set of regions, which is also called superpixels. We donate these superpixels as $S = \{s_i\}$, where s_i is the i-th segmented region of I, and all segments in S form the whole image. Superpixels are popular as a pre-processing for many image segmentation works and they can significantly increase running efficiency compared with pixel-wise algorithm while maintain the performance.

The over-segmentation of image I can be seen as a region adjacency graph (RAG): $G = (V, E, W)$, in which each node $v_i \in V$ represents the special region $s_i \in S$, an edge $e_i \in E$ corresponds to the adjacency relationship of node v_i and v_j in image I, and $w_{ij} \in W$ represents the weight of e_{ij} measured via similarity matrixes. The similarity matrixes measured by multiscale feature directly determines the performance of segmentation and will be introduced in Sect. 3.1 in details.

The task of image segmentation is transformed into finding a coarse-to-fine combination of node s_i based on the weighted graph G. In this paper, we use a bottom-up region merging algorithm following for a hierarchical tree to show the structure of image segmentation. The RAG G_L constituted by superpixels could be seen as the finest partitions at scale L and we iteratively perform the merging operation below until the newly generated map shows a coarsest segmentation: each time, all nodes v_i^l in graph G_l are clustered into graph G_{l-1} via energy optimization algorithm and $l - 1$ represents the label of newly generated graph and the level of corresponding segmentation. Therefore, the final hierarchical tree-like structure $T = \{G_1, \ldots, G_l, \ldots, G_{L-1}\}$ is a family of all RAGs.

3   Multi-feature Fusion

In this section we describe the image vision cues used in segmentation process, where the multiscale structural edge measures the similarity between a pair of adjacent regions and the semantic prior including saliency detection and object window guides the hierarchical structure as a weak supervision.

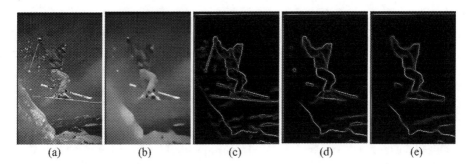

    (a)   (b)   (c)   (d)   (e)

Fig. 2. Example of multiscale edge cue. (a) An input image, (b) the structure element of image after structure-texture decomposition, (c) the structural edge map in 1.0 scale, (d) the structural edge map in 0.6 scale, and (e) the structural edge map in 0.2 scale.

3.1 Multiscale Edge Cue

The multiscale edge clue is used to provide the feature information for graph-based hierarchical segmentation approach, which is the basis to ensure the rationality of the segmentation structure. In order to extract coarse-fine structure at different scales, we firstly use structure-texture decomposition to decompose an image to structure elements and texture details. Then structural edge detector is used to find the main structure based on the structure elements part of structure-texture decomposition.

An image can be decomposed as $I = St + Te$ by structure-texture decomposition, where I, St and Te represent the input image, structure elements and texture details, separately. In this paper, we use a novel filtering-based method [7] for structure-texture decomposition to find the multiscale image structures. Obviously, the structure elements at lower scale which represent the coarser segmentations relevantly ignore more texture information than the structure element at higher scale.

Therefore, edge detector based on structure element can find edge information from coarse to fine effectively. The edge possibility map [8] is used to detect structural edges in different scales. Compared with typical edge detection where each pixel is labeled with a binary variable, edge possibility map contains more edge information of image structure and can be translated into binary edge mask by threshold division.

The edge possibility maps extracted in multi scales are shown in Fig. 2 and we measure the similar of two adjacency nodes according to these structural edge features in our work. Following [9], we define the similar matrix $f = \{f_1, f_2, \ldots, f_K\}$ as:

$$f_k(s_i, s_j) = 1 - avg(\{EPM_k(x) | x \in B(s_i, s_j)\}) \tag{1}$$

$$B(s_i, s_j) = (s_i \cap N(s_j)) \cup (s_j \cap N(s_i)) \tag{2}$$

where $EPM_k(x)$ is the value of pixel x in the edge possibility map at scale k, and $N(s_i)$ represents the set of neighbor pixels of s_i.

3.2   Semantic Prior Constraint

In this subsection, we introduce the constraint and formulation of semantic priors which is used to supervise the progressive merging process. Actually, the hierarchical segmentation of image should not only focus on the differences in the scale of regions, but also keep the segmentations of each level as consistent as possible. According to human's cognitive process of the real world, people normally tend to achieve segmentation of the foreground and background in the high-level partitions, distinguish individual instances from object regions in the middle-level partitions and explore more details of the targets in low-level partitions. The low-level local features or middle-level statistical features can hardly accomplish this task, therefore we use the salient map and the object windows as semantic information to constrain the hierarchy structure and make it more reasonable.

Salient Map. We use the saliency detection algorithm in work [10] to search the salient region in image. The constraint of salient map is based on a phenomenon that the salient objects tend to obtain a finer segmentation with the granularity of the partition becomes smaller. Thus, we define a saliency weight to amend the similarity matrix $f(s_i, s_j)$:

$$fsm(s_i, s_j) = Wsm(s_i, s_j) \cdot f(s_i, s_j) \tag{3}$$

where the weight $Wsm(s_i, s_j)$ is determined by the mean saliency of a pair of adjacent regions s_i and s_j; the similarity should be enlarged if at least one of regions is salient and narrowed if two regions are both non-salient:

$$Wsm(s_i, s_j) = \begin{cases} 1, sm(s_i) < t_l \text{ and } sm(s_j) < t_l \\ l, otherwise \end{cases} \tag{4}$$

ℓ and t_l represent positive correlation parameter and saliency threshold separately, $sm(s_i)$ represents the mean saliency of region s_i; the value of t_l change constantly according to the iteration counter l because of the dynamic amendment strategy which will be described in Sect. 4 in details.

Object Window. We use object window to overcome the limitation that object region at small scale may be ignored in coarse segmentations. This semantic prior is detected via Faster-RCNN [11] which is efficient and wide-used in computer vision and we further modify the similarity matrix based on Eq. (3):

$$fsm(s_i, s_j) = Wsm(s_i, s_j) \cdot f(s_i, s_j) + \alpha_{ij} \cdot Wow(s_i, s_j) \tag{5}$$

where $Wow(s_i, s_j)$ indicates weight of region s_i and s_j in object prior and it is defined as:

$$Wow(s_i, s_j) = \begin{cases} 1, s_i \in s_{ow} \text{ and } s_j \in s_{ow} \\ 0, otherwise \end{cases} \tag{6}$$

here, s_{ow} denotes the pixels within object windows; value a_{ij} is determined by the distance between $f(s_i, s_j)$ and the mean similarity of object windows.

4 Graphcut-Based Segmentation

According to the proposed framework described in Sect. 2, each node merging operation generated an updated graph and we define it as:

$$G_{l-1} = C(l, G_l) \tag{7}$$

in which l represents the threshold level of graph in hierarchical tree.

The task of merging operation is to find an optimal cut in the graph and the divided parts could minimize the cost of energy function. The existing method called Graph Cuts [12] which is well-known for single image segmentation solved the graph cut problem via "max-flow" algorithm and achieved an outperformance. Therefore, we improve the design of energy function according to the hierarchical framework and the extracted multi-feature and use "max-flow" algorithm to find the optimal solution following Graph Cuts.

Given the G_l, the basic energy function is formulated as:

$$E(A) = \lambda \sum R_i(A_i) + \sum B_{ij} \tag{8}$$

where R_i and B_{ij} represent the region term and boundary term separately. In order to encode the feature constraints in $E(A)$, we convert the formulation as:

$$E(A) = \lambda \sum R_i(A_i) + \sum \exp(-\frac{fsm(s_i, s_j)^2}{2\sigma^2}) \tag{9}$$

where $f(s_i, s_j)$ used in $fsm(s_i, s_j)$ by Eq. (5) is determined by the multiscale edge cue $\{f_1, f_2, \ldots, f_K\}$ and it is defined as:

$$f(s_i, s_j) = \sum_{k \in K} Wf_k \cdot f_k(s_i, s_j) \tag{10}$$

here, the similarity weight Wf_k obeys a normal distribution in a merging operation and the peak has a deviation from f_1 to f_K with the change of threshold level according to the *dynamic amendment strategy*. This strategy is based on the phenomenon that the multiscale edge cue should have mapping relation with hierarchical segmentation, and edge cues at small scale have a greater influence on similarity matrix than large scale as the bottom-up merging process goes on. The value t_l in Eq. (4) also follows the dynamic amendment strategy while it obeys an arithmetic progression.

5   Experiment

We compare our approach with other state-of-the-art hierarchical segmentation methods for the validity of the novel approach, and analyze performance of the results generated by different methods for the effective influence of multi-feature fusion. Experiment results that our approach gets a better performance in semantic consistency while achieves the state-of-the-art accuracy.

5.1   Experimental Settings

We conduct experiments on the BSDS500 dataset [1] and compare our results with several popular and state-of-the art hierarchical segmentation approaches. The BSDS500 consists of 500 natural images in different scenes. It includes 200 images for training, 200 images for testing, 100 images for validation, and each image is annotated with multiple groundtruth.

Following [1], we use the Segmentation Covering (SC), the probabilistic Rand Index (PRI) and the Variation of Information as evaluation measures of hierarchical segmentation because of their wide use in related work. The Segmentation Covering computes the sum of overlapping ratios between each proposed segment and the true segment in groundtruth. Each true segment is matched with an optimal proposed segment which has the largest overlapping ratio weighted by segment sizes. The probabilistic Rand Index measures the accuracy of clustering algorithm originally and could also be used in evaluation of segmentations. It computes the ratio of the number of pixel pairs that have identical label in segmentations and groundtruth over the number of all pixel pairs. The Variation of Information measures the relative entropy between segmentation and the groundtruth by the conditional image entropies.

Each image is evaluated at both Optimal Dataset Scale (ODS) and Optimal Image Scale (OIS) according to the previous work [1]. ODS searches for the optimal threshold level of all images in dataset which has the best evaluation, while OIS selects different threshold levels for each test image.

5.2   Comparisons with Other Methods

Table 1 shows the results of our approach compared with other existing methods [1, 4, 13], which are called UCM, MCG and PMI for short. In order to verify the effectiveness of our multi-feature fusion strategy, we also evaluate the performances of our work without salient map or object window which are called Ours-NS and Ours-NO separately in Table 1. The complete approach in last line has a great improvement compared with Ours-NS and Ours-NO, especially on ODS than OIS. Actually, the evaluations on OIS need to search for the optimal segmentation in hierarchical structure for each image and limit the effect of semantic prior, while ODS reduces the cost of labeling threshold level and could be more effective in real applications. The comparison to other related hierarchical segmentation methods also indicates that our approach attains state-of-the-art while parts of evaluations have an outperformance.

Table 1. The results of our approach with a comparison to the related methods. The bold data represent optimal results in our work.

	SC		PRI		VI	
	ODS	OIS	ODS	OIS	ODS	OIS
MCG [4]	0.61	0.67	0.81	0.86	1.55	1.37
UCM [1]	0.59	0.65	0.83	0.86	1.69	1.48
PMI [13]	0.53	0.59	0.76	0.81	2.03	1.80
Ours-NS	0.56	0.59	0.75	0.79	1.65	1.56
Ours-NO	0.53	0.56	0.72	0.74	1.72	1.60
Ours	**0.63**	0.64	**0.84**	0.85	**1.52**	1.43

5.3 Qualitative Analysis

The examples in Fig. 3, from left to right, show the segmentation at different levels respectively, which correspond to high-level, middle-level and low-level. Hierarchical results prove that our approach can better reflect the semantic information in structural segmentations. For instance, the input image of riding bikes displayed in the fourth line tends to merge people and trees into one region from middle-level to high-level if we only use low-level feature in segmentation process. Therefore, the result of our approach which divides image into people and background in coarse segmentation is more reasonable, and this kind of advantages sometimes can't be reflected in quanti-tative assessment of ODS and OIS because the performance of high-level and low-level segmentations may be ignored while only the optimal result is evaluated.

5.4 Running Time

Table 2 shows the running time in total for each image, where the MCG and SCG represent the multiscale segmentation algorithm and the single-scale segmentation algorithm in work [4]. The average time of our approach is approximate 2.3 s per image which doesn't include the parts of semantic prior (salient map and object window) detection. All times of our approach are measured on the same standard Windows machine.

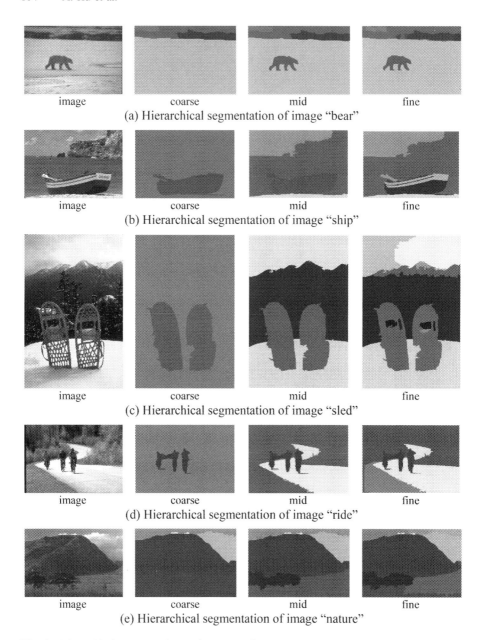

image   coarse   mid   fine
(a) Hierarchical segmentation of image "bear"

image   coarse   mid   fine
(b) Hierarchical segmentation of image "ship"

image   coarse   mid   fine
(c) Hierarchical segmentation of image "sled"

image   coarse   mid   fine
(d) Hierarchical segmentation of image "ride"

image   coarse   mid   fine
(e) Hierarchical segmentation of image "nature"

Fig. 3. Hierarchical segmentation performance of our approach on BSDS500 dataset. From left to right, we display the original image, the coarse segmentation result, the mid segmentation result and the fine segmentation result respectively for each image.

Table 2. Time comparison in seconds per image with the related methods.

	MCG [4]	SCG [4]	SAA [5]	Ours
Average time (s)	25	2.7	3	2.3

6 Conclusion

In this work, we presented a hierarchical image segmentation approach based on multi-feature fusion and graphcut-based segmentation. We formulated the scale prediction for the segments in a hierarchy as a graph label problem, which is solved by progressive energy optimization. Multiscale edge cue extracts more information in different scales in input image and is mapped to each threshold level of hierarchical via our progressive process and dynamic amendment strategy. With the semantic-prior as constraints, we encode them into the energy function of graph cut to keep the semantic consistency in structural segmentations. The experiments on BSDS500 compared with three existing methods also show the improvement of our work and the advantages of progressive merging framework and multi-feature fusion.

Acknowledgments. This work was supported by National High Technology Re-search and Development Program of China (No. 2007AA01Z334), National Natural Science Foundation of China (Nos. 61321491 and 61272219), Program for New Century Excellent Talents in University of China (NCET-04-04605), the China Postdoctoral Science Foundation (Grant No. 2017M621700) and Innovation Fund of State Key Lab for Novel Software Technology (Nos. ZZKT2013A12, ZZKT2016A11 and ZZKT2018A09).

References

1. Arbelaez, P., Maire, M., Fowlkes, C., Malik, J.: Contour detection and hierarchical image segmentation. IEEE Trans. Pattern Anal. Mach. Intell. **33**(5), 898–916 (2011)
2. Kim, T.H., Lee, K.M., Lee, S.U.: Learning full pairwise affinities for spectral segmentation. IEEE Trans. Pattern Anal. Mach. Intell. **35**(7), 1690–1703 (2013)
3. Salembier, P., Garrido, L.: Binary partition tree as an efficient representation for image processing, segmentation, and information retrieval. IEEE Trans. Image Process. **9**(4), 561–576 (2000)
4. Pont-Tuset, J., Arbelaez, P., Barron, J., Marques, F., Malik, J.: Multiscale combinatorial grouping for image segmentation and object proposal generation. IEEE Trans. Pattern Anal. Mach. Intell. **39**(1), 128–140 (2016)
5. Chen, Y., Dai, D., Pont-Tuset, J., Van Gool, L.: Scale-aware alignment of hierarchical image segmentation. In: Computer Vision and Pattern Recognition, pp. 364–372. IEEE (2016)
6. Achanta, R., Shaji, A., Smith, K.: SLIC superpixels compared to state-of-the-art superpixel methods. IEEE Trans. Pattern Anal. Mach. Intell. **34**(11), 2274–2282 (2012)
7. Lee, H., Jeon, J., Kim, J., Lee, S.: Structure-texture decomposition of images with interval gradient. Comput. Graph. Forum **36**(6), 262–274 (2017)
8. Dollar, P., Zitnick, C.L.: Fast edge detection using structured forests. IEEE Trans. Pattern Anal. Mach. Intell. **37**(8), 1558–1570 (2015)

9. Liu, T., Seyedhosseini, M., Tasdizen, T.: Image segmentation using hierarchical merge tree. IEEE Trans. Image Process. **25**(10), 4596–4607 (2016)
10. Kim, J., Han, D., Tai, Y.-W., Kim, J.: Salient region detection via high-dimensional color transform. In: Computer Vision and Pattern Recognition, pp. 883–890. IEEE (2014)
11. Ren, S., He, K., Girshick, R.: Faster R-CNN: towards real-time object detection with region proposal networks. IEEE Trans. Pattern Anal. Mach. Intell. **39**(6), 1137–1149 (2017)
12. Boykov, Y.Y., Jolly, M.-P.: Interactive graph cuts for optimal boundary & region segmentation of objects in ND images. In: International Conference on Computer Vision, p. 105. IEEE (2001)
13. Isola, P., Zoran, D., Krishnan, D., Adelson, E.H.: Crisp boundary detection using pointwise mutual information. In: Fleet, D., Pajdla, T., Schiele, B., Tuytelaars, T. (eds.) ECCV 2014. LNCS, vol. 8691, pp. 799–814. Springer, Cham (2014). https://doi.org/10.1007/978-3-319-10578-9_52

A Sound Image Reproduction Model Based on Personalized Weight Vectors

Jiaxi Zheng[1,2], Weiping Tu[1,2(✉)], Xiong Zhang[1,2], Wanzhao Yang[1,2], Shuangxing Zhai[1,2], and Chen Shen[1,2]

[1] National Engineering Research Center for Multimedia Software,
Wuhan University, Wuhan 430072, China
{jiaxi0720,15002708424,zhao_26,
zhaishuangxing,shenchenZHJQ}@163.com,
tuweiping@whu.edu.cn
[2] School of Computer Science, Wuhan University, Wuhan 430072, China

Abstract. Many perceptual models for audio reconstruction have been proposed to create the virtual sound, but the direction of the virtual sound maybe deviate from the desired direction due to the distortion of binaural cues. In this paper, a binaural cues' equation for real sound and virtual one reproduced by dual loudspeakers is established to derive weight vectors based on the head-related transfer function (HRTF). After being filtered by the weight vectors, sound signals emitted from the loudspeakers can deliver an accurate spatial impression to the listener. However, the HRTFs change with listeners, by which the weight vectors calculated also vary from person to person. Therefore, a radial basis function neural network (RBFNN) is designed to personalize weight vectors for each specific listener. Compared with the three methods including vector base amplitude panning (VBAP), the HRTF-based panning (HP) and the band-based panning (BP), the method in this paper can reproduce binaural cues more accurately, and subjective test also indicates that there is no significant difference in perception between real sound and virtual sound based on the proposed methods.

Keywords: Weight vector · Loudspeaker · Personalize
Sound image reproduction

1 Introduction

The virtual auditory display (VAD) is a significant technology in virtual reality, and usually implemented by headset or loudspeakers. The key to reproduce sound image by the headset is to filter the sound signal with the head-related transfer function (HRTF) which varies with directions and subjects [1, 2], but it's inconvenient to communicate for the users wearing headphones. However, loudspeaker array can avoid the drawbacks

This work is supported by National Nature Science Foundation of China (No. 61671335); Technological Innovation Major Project of Hubei Province (No. 2017AAA123).

R. Hong et al. (Eds.): PCM 2018, LNCS 11165, pp. 607–617, 2018.
https://doi.org/10.1007/978-3-030-00767-6_56

of headset when reproducing the sound-images. The mainstream VAD technologies based on loudspeakers mainly include wave field synthesis (WFS), Ambisonics technology and amplitude panning technology (AP). WFS and Ambisonics technology can restore the physical properties of the sound field but demand for many loudspeakers with fixed arrangement [3, 4]. Meanwhile these two methods with the high computational complexity are difficult to be applied in practice. On the contrary, AP methods are simpler and more flexible to achieve sound spatialization because of the free loudspeaker placement, such as vector-base amplitude panning (VBAP) and multiple-direction amplitude panning (MDAP) [5–7]. First, the AP methods establish a simple geometric model which consists of several loudspeakers and one listening point and calculate the gain of each loudspeaker by vector decomposition theory. Then, the difference in gains of loudspeakers generates an auditory perception that the sound seems to be emitted from the virtual position among the actual loudspeakers.

However, there are obvious defects in AP. Firstly, spatial psychoacoustics indicates that interaural time difference (ITD) and interaural level difference (ILD) are two key binaural cues of sound localization in the horizontal plane, and ITD works at low frequency (below about 1500 Hz) while ILD works at high frequency [8, 9]. Although the differences between/among gains of loudspeakers are translated into ITD and ILD at the listener's ear drums, ITD is not considered by AP methods and it is not controllable and may be distorted. Secondly, the AP technique creates a simple model which regards the listener as a listening point and ignores the filtering effect of the head and torso on the sound so that the reproduced ILD is inconsistent with the ideal ILD. Therefore, the perceived direction of the virtual sound reproduced by the AP methods deviates from the desired direction due to the distortion of binaural cues.

Many studies have been conducted on the problems of AP technology. In 2013, Jeroen [10] compares ILDs of real source and virtual sound created by pair-wise amplitude panning in an anechoic environment. The study indicated that the translation of the amplitude of loudspeakers' signals to ILDs depends on the source frequency, the individual's HRTFs, the loudspeaker angle, and the source direction angle. Although some methods considering frequency dependency and HRTF dependency have been proposed, there are still some problems. First, HRTF-based panning (HP) method, such as individualized crosstalk cancellation (CTC) system, requires precise crosstalk cancellation modules which are designed by individualized HRTFs [11]. Many methods are proposed for HRTF individualization, but there are always some errors between individualized HRTF and real one. After calculating the crosstalk cancellation modules by the HRTF, the errors are amplified, and the binaural cues reproduced by the loudspeakers deviate from the ideal binaural cues. Second, the band-based panning method (BP) has been proposed in [12] where each frequency band corresponds to a different gain instead of a common gain for all frequency bands in AP. However, non-personalized gain in each frequency band in FP may lead to more serious distortion of perception.

In this paper, to deliver the accurate spatial impression to different listeners by two loudspeakers, a non-sophisticated sound reproduction model that can solve the problem of both frequency dependency and individual dependency is created. For frequency dependency, weight vectors are obtained by a binaural signals' equation for real sound and virtual sound created by loudspeaker pair, and each element of weight vectors represents the amplitude of each frequency band of the audio signal in a channel. For individual dependency, a rather simple radial base function neural network (RBFNN) which can predict time series precisely and rapidly is designed to personalize weight vectors for each specific listener.

The remainder of the paper is organized as follows. In Sect. 2, the approach for reproducing virtual sound image by weight vectors is introduced. In Sect. 3, the process of personalizing weight vectors by RBFNN is described. In Sect. 4, the objective and subjective comparison and analysis are made on the proposed method, VBAP method, BP method and HP method. Finally, we conclude in Sect. 5.

2   Proposed Approach for Reproduction of Virtual Sound Image

2.1   Fundamental Principle of the Approach

The goal of this paper is to give the listeners the same direction perception of the virtual sound as that of the real sound in an anechoic environment on the horizontal plane. Spatial psychoacoustics also indicates that the auditory system uses binaural cues (ITD and ILD) for sound source localization in the horizontal plane [8]. Therefore, when the binaural cues formed by a single loudspeaker are identical to that created by dual loudspeakers, the listener will have the same orientation perception between them.

Figure 1 describes the process of generating binaural signals with single and dual loudspeakers respectively. In order to precisely reproduce a virtual sound image with the same direction as the real sound, there should be the same binaural cues in Fig. 1(a) and (b). The binaural cues are derived from the binaural signals, so a stricter constraint that the binaural signals are identical in Fig. 1(a) and (b) is proposed in this paper. In such a constraint, the reconstructed virtual sound image will be exactly the same as the real sound image.

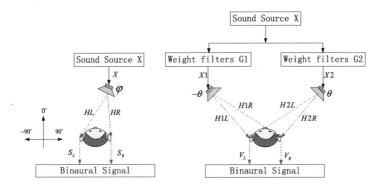

Fig. 1. A schematic diagram of the process of generating binaural signals in the ears: (a) binaural signals formed by one loudspeaker; (b) binaural signals formed by two loudspeakers.

2.2   Derivation of the Ideal Weight Vectors

According to the previous section, the weight vectors are derived from the binaural signals' equation of the real sound and the virtual one generated by the loudspeaker pair. The derivation process is described below.

As shown in Fig. 1, the human head center is regarded as the coordinate center, the direction in front of the head is set to $0°$, and the azimuths of the left and right ears are respectively defined as $-90°$ and $90°$. All the data in this article are expressed in the form of frequency domain. HL, HR, $H1L$, $H1R$, $H2L$ and $H2R$ separately express the HRTF of the left and right ears corresponding to the 3 loudspeakers in Fig. 1.

In Fig. 1(a), a target sound signal X is multiplied by HL and HR respectively to get binaural signals $S_L(\omega)$ and $S_R(\omega)$.

As shown in Fig. 1(b), the weight filters of the loudspeakers are set as unknown vectors $G1(\omega)$ and $G2(\omega)$. $X1(\omega)$ and $X2(\omega)$ are the results of the signal $X(\omega)$ filtered by $G1(\omega)$ and $G2(\omega)$, and then are loaded to the dual loudspeakers. According to the summing localization criterion [13], the binaural signals generated by the dual loud-speakers are equivalent to the superposition of the binaural signals generated by two separate loudspeakers signals ($X1(\omega)$ and $X2(\omega)$) in the frequency domain. Therefore, the binaural signals of the virtual sound generated by the loudspeaker 1 and 2 are $V_L(\omega)$ and $V_R(\omega)$.

$$X(\omega) \cdot G1(\omega) \cdot H1L(\omega) + X(\omega) \cdot G2(\omega) \cdot H2L(\omega) = V_L(\omega)$$
$$X(\omega) \cdot G1(\omega) \cdot H1R(\omega) + X(\omega) \cdot G2(\omega) \cdot H2R(\omega) = V_R(\omega) \tag{1}$$

If the binaural signals of the virtual sound image are equal to that of the target sound image, $G1(\omega)$ and $G2(\omega)$ can be solved as:

$$G1(\omega) = \frac{g1(\omega)}{Com(\omega)}, \quad G2(\omega) = \frac{g2(\omega)}{Com(\omega)} \tag{2}$$

In formula (2), there are common part Com and different molecules $g1$, $g2$ between $G1(\omega)$ and $G2(\omega)$, and the corresponding expression is as follows.

$$Com(\omega) = H1L(\omega) \cdot H2R(\omega) - H1R(\omega) \cdot H2L(\omega)$$
$$g1(\omega) = HL(\omega) \cdot H2R(\omega) - HR(\omega) \cdot H2L(\omega) \tag{3}$$
$$g2(\omega) = HR(\omega) \cdot H1L(\omega) - HL(\omega) \cdot H1R(\omega)$$

The vector that consists of Com, $g1$ and $g2$ is called the weight vector. The binaural signals of the virtual sound image reconstructed based on the calculated weight vectors are equal to that of the target sound image, which is a ideal situation in the sound image reproduction system. Therefore, we call the calculated weight vectors as ideal weight vectors.

2.3   Calculation of Weight Vectors by Using CIPIC Database

The HRTFs describe the diffraction and reflection effect of listener's head, pinna, and torso on sound waves before the sound reaches outer ear drums. Therefore, to obtain the weight vectors conveniently, a public high spatial-resolution database of HRTFs, measured and provided by CIPIC of California University [14], is used to simulate the process of generating virtual sound images based on single loudspeaker and dual loudspeakers respectively.

If the listener is a subject in the CIPIC database, the ideal weight vectors can be calculated according to Eq. (3). After filtered by the ideal weight vectors, sound signals are loaded into the corresponding loudspeaker pair and then are emitted from the loudspeakers, so the perceived direction of virtual sound image will be same as the desired direction for the listener.

3   The Personalized Reproduction of Virtual Sound Image

Therefore, it can be seen from the above process that the precondition of obtaining ideal weight vectors is to use the HRTFs of the listener. For a listener not in the CIPIC database, the weight vectors used to reproduce the virtual sound image cannot be calculated directly. Although the HRTFs can be measured, it is time consuming and requires specific anechoic chamber [15]. To personalize the HRTFs first and then calculate the corresponding weight vectors can avoid the above problems, but the complexity is greatly improved. Therefore, a method of personalizing the weight vectors is proposed below.

Firstly, the correlation between the weight vectors and the corresponding anthropometry parameters in the CIPIC library is analyzed based on the Pearson coefficient, and the anthropometry parameters with the larger correlation coefficient is screened out. Secondly, a rather simple RBFNN is chosen as a prediction model and the selected parameters and the corresponding weight vectors are combined into a set of training samples to train the RBFNN model. Finally, the RBFNN with a definite structure is used to predict weight vectors of a new listener.

3.1   Anthropometry Parameters Selection

A set of 27 anthropometry measurements are recorded in the CIPIC database, including parameters of head and torso ($X_1 \sim X_{17}$) as well as the parameters of the pinna ($D_1 \sim D_8, \theta_1, \theta_2$). A model to determine the mapping between the weight vectors and anthropometric parameters is established, but it is obvious unreasonable to introduce all of parameters into the model because there are different correlations between these parameters and the weight vectors. In order to simplify the model and reduce the number of anthropometry parameters to be measured when predicting weight vectors, the several key parameters will be screened out.

Pearson correlation coefficient, a measure of the linear correlation between two variables, is used to analyze the correlation between the weight vectors and anthropometry parameters in this paper. Then several parameters with larger correlation coefficient are selected. The distribution of correlation coefficients that greater than 0.45 is shown in Fig. 2 where the colorbar corresponds to the scale of the correlation coefficient. The final parameters $X_3, X_6, X_9, X_{12}, X_{14}, X_{15}, X_{16}, X_{17}$ are selected as independent variables of the neural network models.

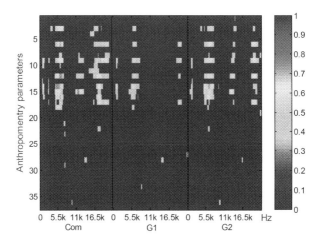

Fig. 2. The distribution of Pearson correlation coefficient between the weight vectors and anthropometry parameters. The abscissas represents the frequency of the three components (*Com*, *g*1, and *gr*) that belong to the weight vectors, respectively. The ordinate includes 37 anthropometry parameters. The first 17 parameters represent $X_1 \sim X_{17}$, and the last 20 parameters respectively represent $D_1 \sim D_8$, θ_1 and θ_2 of left and right pinna. (Color figure online)

3.2   Generation of Training Samples for RBFNN

Through the above correlation analysis, we find the weight vectors have a strong correlation with the 8 selected anthropometry parameters. In addition, the structure of the RBFNN is simple, and the learning rate and convergence rate are also superior to the back propagation neural network. Thus, a three-layer RBF neural network is used to establish a mapping model of selected anthropometry parameters and weight vectors. RBFNN are typically trained from pairs of input and target values, so an experimental sample data set that contains input and output should be established.

The CIPIC database includes HRTFs for 45 subjects, but records only 35 subjects' anthropometry parameters fully, so the maximum number of experimental samples is 35. As shown in Fig. 3, under the fixed loudspeaker angles and sound source direction angle, ideal weight vectors of a subject in a CIPIC library are calculated according to an equation of binaural signals generated by real sound and the virtual one created by loudspeaker pair. Then an experimental sample consisting of weight vectors and anthropometry parameters is obtained. And experimental samples of the rest 34 subjects are obtained in the same way. In order to predict the weight vectors by RBFNN, ideal weight vectors are used as the target values and the anthropometry parameters are used as the inputs of the RBFNN.

Although *Com*, *g*1 and *g*2 are complex values, only the magnitudes of weight vectors are trained and predicted by RBFNN, while the phase is uniformly set to be the average of phases of the weight vectors of 35 subjects. HRTF in this paper is 64-point symmetrical and the valid length of HRTF is 32. Therefore, the length of the weight vectors is 96.

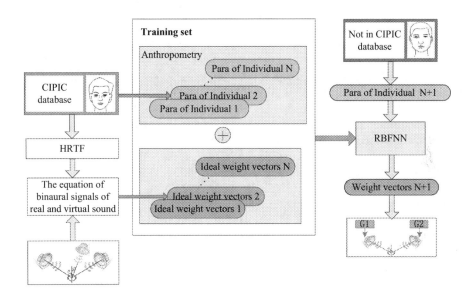

Fig. 3. Block diagram of personalize weight vectors

3.3   Personalization of Weight Vectors Based on RBFNN

In this paper, a RBFNN model is chosen to establish the mapping between anthropometry parameters and weight vectors. As shown in Fig. 3, when the configuration of the loudspeaker pair and the target sound source are fixed, 35 experimental samples including ideal weight vectors and anthropometry parameters can be obtained, where 31 samples are selected as training sets. By learning the training samples, the neural network will have a definite structure. Therefore, in the case of the same loudspeaker configuration and the same target orientation, if the anthropometry parameters of the listener are measured, the weight vectors can be predicted by the established RBFNN. After filtered by the right and left weight filters derived from weight vectors, sound signals are loaded into the corresponding loudspeaker pair, and then the virtual sound are reproduced to the listeners through loudspeakers.

4   Experiment and Evaluation

4.1   Comparison of Binaural Cues

In a loudspeaker configuration of $[-45°\ 45°]$, 9 RBFNNs corresponding to 9 orientations of the sound source are established respectively, where the 31 randomly selected training sets are identical and the 9 orientations are from $0°$ to $40°$ in $5°$ increments. The IDs of the 4 test subjects in the CIPIC library in this experiment are 40, 61, 134 and 154. The test signal is Gaussian white noise with a sampling frequency of 44.1 kHz and a length of 20 ms.

The generations and comparisons of the target signals, experimental signals, VBAP signals, HP signals, and BP signals are as follows. The above 5 signals are all binaural signals. Target signals are obtained by convolving Gaussian white noise with head-related impulse response (HRIR) of the target azimuth, where HRIR is the time domain representation of HRTF. Based on a summing localization criterion, experimental signals are synthesized by filtering Gaussian white noise with the predicted weight vectors, and VBAP signals are also synthesized from the Gaussian white noise multiplied by the weights of VBAP [16]. HP signals are obtained by filtering Gaussian white noise with 4 CTC filters described in [17], where the HRTFs are personalized by manifold learning approach [18]. The BP signals are similar to the HP signals, but the CTC filters are calculated by the HRTFs of the artificial head model in CIPIC instead of the personalized HRTFs. Since the HRTFs are fixed, the CTC module and the binaural signal generation module are combined to obtain two weight filters where the coefficients of filters corresponding to the frequency bands of the signals [19], so the binaural signals obtained by this method are BP signals.

In this paper, the reconstruction of the virtual sound image in horizontal plane is studied, so the ILD and ITD of the binaural signals are only analyzed. The calculation of ILD and ITD is described in [20]. The ILDs and ITDs of the experimental signals, ideal signals, VBAP signals, HP signals and BP signals for 4 test samples under 9 different target positions are calculated and compared. The result is shown in Fig. 4.

The first column of Fig. 4 shows that the ITDs of experimental signals and HP signals are close to the ILDs of the target signals for every subject, but the ITDs of VBAP signals and BP signals are greatly different from the target signals. The second column indicates little difference of the ILD values among the target signals, VBAP signals and experimental signals, while there are relatively large differences between target signals and HP signals as well as BP signals. The above results manifest that, compared with the VBAP, HP and BP, the binaural cues of the virtual sound image reproduced by the personalized weight vectors are closer to that of the target sound image on the horizontal plane. Therefore, the method proposed in this paper has a best positioning effectiveness than VBAP, HP and BP methods.

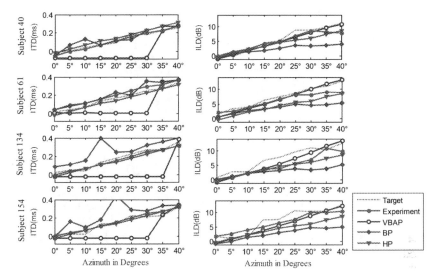

Fig. 4. The proposed method is compared with the VBAP, BP, HP. The angle of the loudspeaker pair is set to [−45° 45°], so the abscissas of all subfigures indicates the target sound image angles of 0°, 5°, 10°, …, 35°, and 40°.

4.2   Subjective Test

In addition to calculating binaural cues, a subjective test is performed to evaluate the proposed method. The subjective test consists of 4 listening sequences, each consisting of two 0.5 s non-silent sound segments, one of which is the target signal and the other of which is one of the 4 signals including experimental signal, VBAP signal, the HP signal, and the BP signal. In each listening sequence, 1-s silent signal is inserted into two non-silent segments. Then the listener perceives the sound through the earphone and evaluates the differences of azimuths of the two signals. The scoring criteria is as follows. When the listener cannot distinguish the orientation of the two signals, the score is 10. When the listener can distinguish them, the score is given according to the azimuth difference between them, and the bigger the perceived difference is, the lower the score is.

As shown in Fig. 5, all the scores are high, and the scores of experimental signals are greater than the scores of VBAP signals, BP signals and HP signals. It can be seen from the subjective evaluation that, compared with the three methods, the direction of the reproduced virtual sound image based on proposed method is closer to that of the target sound image.

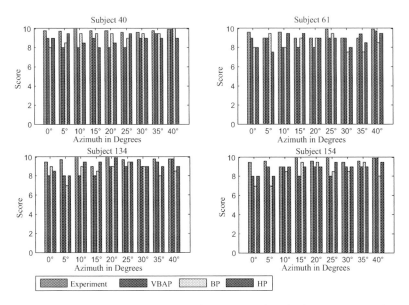

Fig. 5. The score of the subjective test of the proposed method, VBAP, BP and HP.

5   Conclusion

By analyzing the weight vectors and the anthropometric parameters, a novel method based on RBFNN is proposed to personalize a new listener's weight vectors which is used to reproduce the target sound in this study, and the performance of the proposed method was evaluated by both objective simulation experiment and subjective listening experiment. Compared with the three methods including VBAP, HP and BP, the difference of the binaural cues between the target sound and the virtual sound reconstructed by the proposed method is smaller, and there is no significant difference in perception between real sound and virtual sound based on the method in this paper.

In future, the anthropometry parameters of the actual listeners will be measured to estimate the corresponding weight vectors and the signals filtered by the weight vectors will be emitted from loudspeakers in an anechoic room instead of using HRTF in CIPIC database to simulate the process. Aiming to reproduce the orientation of the original sound image more accurately, we will also improve the proposed method as well as the prediction performance of neural networks and reduce the complexity of calculating the weight vectors.

References

1. Wenzel, E.M., Arruda, M., Kistler, D.J., et al.: Localization using nonindividualized head-related transfer functions. J. Acoust. Soc. Am. **94**(1), 111 (1993)
2. Zotkin, D.N., Duraiswami, R., Davis, L.S.: Rendering localized spatial audio in a virtual auditory space. IEEE Trans. Multimed. **6**(4), 553–564 (2004)

3. Poletti, M.A.: Three-dimensional surround sound systems based on spherical harmonics. J. Audio Eng. Soc. **53**(11), 1004–1025 (2005)
4. Ward, D.B., Abhayapala, T.D., Member, S.: Reproduction of a plane-wave sound field using an array of loudspeakers. IEEE Trans. Speech Audio Process. **9**, 697–707 (2001)
5. Pulkki, V.: Localization of amplitude-panned virtual sources I: stereophonic panning. Audio Eng. Soc. **49**(9), 739–752 (2001)
6. Pulkki, V., Karjalainen, M.: Multichannel audio rendering using amplitude panning [DSP applications]. Signal Process. Mag. IEEE **25**(3), 118–122 (2008)
7. Pulkki, V.: Uniform spreading of amplitude panned virtual sources. In: 1999 IEEE Workshop on Applications of Signal Processing to Audio and Acoustics, pp. 187–190. IEEE (2002)
8. Blauert, J., Butler, R.A.: Spatial hearing: the psychophysics of human sound localization by Jens Blauert. J. Acoust. Soc. Am. **77**(1), 334–335 (1996)
9. Macpherson, E.A.: A computer model of binaural localization for stereo image measurement. J. Audio Eng. Soc. **39**(9), 604–622 (1989)
10. Breebaart, J.: Comparison of interaural intensity differences evoked by real and phantom sources. J. Audio Eng. Soc. **61**(11), 850–859 (2013)
11. Choi, T., Park, Y., Youn, D., et al.: Virtual sound rendering in a stereophonic loudspeaker setup. IEEE Trans. Audio Speech Lang. Process. **19**(7), 1962–1974 (2011)
12. Laitinen, M.V., Vilkamo, J., Kai, J., et al.: Gain normalization in amplitude panning as a function of frequency and room reverberance. In: International Conference: Spatial Sound, pp. 85–94. AES (2014)
13. Pulkki, V., Hirvonen, T.: Localization of virtual sources in multichannel audio reproduction. IEEE Trans. Speech Audio Process. **13**(1), 105–119 (2004)
14. Algazi, V.R., Duda, R.O., Thompson, D.M., et al.: The CIPIC HRTF database. In: 2001 IEEE Workshop on the Applications of Signal Processing to Audio and Acoustics, pp. 99–102. IEEE (2001)
15. Gardner, W.: HRTF measurement of a KEMAR. J. Acoust. Soc. Am. **97**, 3907–3908 (1995)
16. Wang, J., Wang, X., Tu, W., Chen, J., Wu, T., Ke, S.: The analysis for binaural signal's characteristics of a real source and corresponding virtual sound image. In: Zeng, B., Huang, Q., El Saddik, A., Li, H., Jiang, S., Fan, X. (eds.) PCM 2017. LNCS, vol. 10736, pp. 626–633. Springer, Cham (2018). https://doi.org/10.1007/978-3-319-77383-4_61
17. Majdak, P., Masiero, B., Fels, J.: Sound localization in individualized and non-individualized crosstalk cancellation systems. J. Acoust. Soc. Am. **133**(4), 2055–2068 (2013)
18. Grijalva, F., Martini, L., Florencio, D., et al.: A manifold learning approach for personalizing HRTFs from anthropometric features. IEEE/ACM Trans. Audio Speech Lang. Process. **24**(3), 559–570 (2016)
19. MingFang: Application of minimum-phase approximation on the signal processing of virtual auditory display. A dissertation of the degree of Master, South China University of Technology, Guangzhou, China (2012)
20. Breebaart, J., Steven, V.D.P., Kohlrausch, A., et al.: Parametric coding of stereo audio. EURASIP J. Adv. Signal Process. **2005**(9), 1–18 (2005)

Reconstruction of Multi-view Video Based on GAN

Song Li, Chengdong Lan$^{(\boxtimes)}$, and Tiesong Zhao

School of Physics and Information Engineering,
Fuzhou University, Fuzhou, China
lancd@fzu.edu.cn

Abstract. There is a huge amount of data in multi-view video which brings enormous challenges to the compression, storage, and transmission of video data. Transmitting part of the viewpoint information is a prior solution to reconstruct the original multi-viewpoint information. They are all based on pixel matching to obtain the correlation between adjacent viewpoint images. However, pixels cannot express the invariability of image features and are susceptible to noise. Therefore, in order to overcome the above problems, the VGG network is used to extract the high-dimensional features between the images, indicating the relevance of the adjacent images. The GAN is further used to more accurately generate virtual viewpoint images. We extract the lines at the same positions of the viewpoints as local areas for image merging and input the local images into the network. In the reconstruction viewpoint, we generate a local image of a dense viewpoint through the GAN network. Experiments on multiple test sequences show that the proposed method has a 0.2–0.8-dB PSNR and 0.15–0.61 MOS improvement over the traditional method.

Keywords: Hybrid resolution · SRGAN · Virtual view reconstruction
EPI · Multi-view video

1 Introduction

In recent years, with the rapid development of computing and multimedia technology, immersive video has also made great progress in order to satisfy users' increasing demand for high-quality visual experience [1]. In 2009, the largest and most technologically advanced 3D movie, Avatar, was popular among people [2]. Later, with the success and popularity of 3D digital movies, 3D TVs have begun to reach the public. Moreover, VR has been applied in education [3], entertainment, health [4] and other fields due to its good development prospects. Currently, Facebook, Sony, and HTC have launched new products for consumers. To meet peoples' thirsty for high-quality visual experience, an effective reconstruction technology is required in multi-view videos.

© Springer Nature Switzerland AG 2018
R. Hong et al. (Eds.): PCM 2018, LNCS 11165, pp. 618–629, 2018.
https://doi.org/10.1007/978-3-030-00767-6_57

1.1   Related Work

Considering the transmission of data, the currently promising reconstruction technologies in Multi-view Video are mainly hybrid resolution and Depth Image Based Rendering (DIBR).

Hybrid Resolution. Due to the limitations of data transmission and capacity storage, how to transmit high-quality super resolution viewpoints to users is a huge challenge in the related field [5]. Therefore, a super resolution technique under a multi-view hybrid resolution framework [6] has been proposed, which mainly uses the high-frequency part of the high-resolution graph to increase the quality of adjacent low-resolution viewpoints. In [7], a hybrid resolution scheme that intercepts low-resolution and full-resolution viewpoints is proposed. In [8], an algorithm based on the displacement compensation high-frequency synthesis method to correct the projection error was proposed. In [9], the adjacent images are interleaved and complementarily downsampled at the coding end, and the missing pixels are interpolated and restored by the virtual viewpoint at the decoding end.

DIBR. DIBR mainly includes: 3D-Warping, pixel interpolation and hole filling. On the one hand, due to the inaccuracy of the depth map used in DIBR technology, on the other hand, there are still some void areas in the virtual viewpoint images after pixel interpolation, and the void needs to be filled. For inaccurate depth maps, the entire depth map is smoothed using a Gaussian filter and an asymmetric Gaussian low-pass filter [10]. This technique causes geometric distortion in other without void regions; therefore, an adaptive edge smoothing filters [11] were successively proposed. Filling holes: In the spatial domain, in [12] uses the neighborhood interpolation method; in [13], the method of image fusion is proposed; in [14], the pixels are filled according to the priority of the void boundary. In the time domain, background updating techniques are used in [15]. In [16], we use it as a reference algorithm, the technique needs the adjacent view-points' texture images and corresponding depth maps. Then, a large number of Gaussian mixture models are used to separate the foreground pixels from the background pixels, and the pixel intensity is modified accordingly; and adaptive weighted mean generation of pixels is used to restore the missing pixels of the background image, correcting errors in Warping.

However, using the pixel value as a similarity criterion to match the content of adjacent viewpoints, the pixel change caused by the different viewpoints of the adjacent viewpoints will cause the inaccuracy of the depth information calculation, so the calculation based on the pixel level will directly affect the final reconstruction result.

1.2   Contribution

The main contributions are as follows:

- We improve input of the network based on multi-view video and take the hybrid resolution local images as input.
- We propose a multi-view reconstruction framework based on SRGAN. VGG is used to extract the high-dimensional features of the content and location of local images and use them to generate local images with dense viewpoints.

We talk about the proposed network architecture in Sect. 2. The experimental process and experimental results are described in Sects. 3 and 4 respectively. In the final Sect. 5, the work of this article is summarized and prospected.

2   Proposed Multi-view Reconstruction

This section describes the proposed multi-view reconstruction method based on the SRGAN network [17]. We have taken the hybrid resolution local images as input which contain information from different viewpoints. To solve the inaccuracy based on pixel-matching similarity, the VGG19 network [18] is used to extract high-dimensional features and used for the representation of correlation. In the network training, the mapping relation between the low-resolution and the high-resolution local image is learned.

2.1   Adversarial Network Architecture

As shown in Fig. 1, the adversarial network contains two networks, a generator and a discriminator. G generates a high-resolution Epipolar Plane Image (HREPI) from a low-resolution Epipolar Plane Image (LREPI). D is used to determine if the distribution is similar. D can help G achieve better performance by loss. Specifically expressed as:

$$\min_{\theta_G} \max_{\theta_D} E_{I^{HREPI} \sim p_{train}(I^{HREPI})}[\log D_{\theta_D}(I^{HREPI})]$$
$$+ E_{I^{LREPI} \sim p_G(I^{LREPI})}[\log(1 - D_{\theta_D}(G_{\theta_D}(I^{LREPI})))] \tag{1}$$

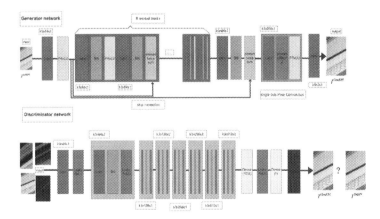

Fig. 1. GAN in multiview reconstruction, the figure above is the generate network, the figure below is the discriminator network.

In the generation network, there are B residual blocks which have the same distribution. Each residual block has two convolutional layers. After the convolutional layer, batch-normalization is added, and ReLU is used as the activation function. The convolution layers are all with 3×3 convolutional kernels and have 64 feature maps. In this network, the extracted viewpoint information is synthesized by training subpixel convolutional layer and the resolution is improved. We have used single sub-pixel convolution as shown in generator. The function loss is the same as the SRGAN.

2.2 Relevance Representation

Inaccurate Representations of Correlation Redundancy Between Adjacent Views. In the process of transmission, the conventional method includes not only the original image and the low-resolution image but also the data of the part of the depth map. The depth map information represents the redundancy correlation between the viewpoints. The similarity of information is measured based on the error between pixel values. Therefore, it only uses the low-dimensional features to represent the correlation between viewpoints, resulting in inaccuracy and large redundancy.

The SRGAN Network Uses High-Dimensional Features to Represent Correlations. In the SRGAN network, VGG19 can extract the high-dimensional features of the generated image and high digital image, as shown in Fig. 2. If $filter_i$ is used to extract a certain feature i, and p_i is a receptive filed with the high-dimensional feature of the filter, then there is

$$G_{i_n} = P_{i_n} * fliter_{i_n} \tag{2}$$

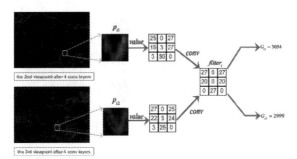

Fig. 2. The 2nd point-view of lovebird1 is convolved with 4 layers, the information of the "V" collar p_{i_1} is extracted and assumed to be quantified into a $3 * 3$ matrix with a $3 * 3$ $fliter_i$, which is specifically used to extract "V" collars, then when p_{i_1} is convoluted with $fliter_i$, a large value G_{i_1} will be obtained. Similarly, G_{i_2} can be obtained by quantifying and convolving with four layers.

Where G_{i_n} represents the convolution result of the feature i of the n-th viewpoint. If the value is large, it means that the curve in the input content may activate the filter. If $\varepsilon > 0$ exists:

$$|G_{i_j} - G_{i_k}| < \varepsilon \tag{3}$$

Where ε is a small positive number relative to the G value. Then the high-dimensional feature i of the j-th viewpoint and the k-th viewpoint have high similarity. Therefore, we can use a kind of high-dimensional feature to express the same content information in multi-viewpoints. Then, the content information of the occlusion can be generated from the high dimensional features extracted from other corresponding locations.

In order to use high-dimensional features to represent the relevance of multi-view video, we did the following:

Synthesis EPI. In order to input multi-viewpoint information by single image, we introduce Epipolar Plane Image (EPI) to multi-viewpoint signals. Due to the principle of polar plane, the same scene object captured by different viewpoints will appear on a diagonal line of the EPI image. The slope is related with the disparity and directly depends on the depth of field between the object and the viewpoint of the photographing. Therefore, information on corresponding objects in different viewpoints can be gathered to the same image by using EPI [19], so the reconstruction can more easily utilize the correlation between the viewpoints. Based on the above analysis, we chose EPI as a representation of the input signal in the multi-view reconstruction framework for deep learning. Compared with natural images, EPI has a specific diagonal texture, as shown in Fig. 3. The EPI construction method is as follows.

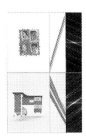

Fig. 3. From left to right, original image, EPI and its spectrum diagram

Let K view images be $I_1, I_2 \ldots I_K$ respectively. The definition matrix is a matrix whose m-th row is 1 and all other rows are 0. The size of the matrix A_m is equal to the image size and is expressed as follows:

$$A_m = \begin{bmatrix} 0 & 0 & \cdots & 0 & 0 \\ \vdots & \vdots & \vdots & \vdots & \vdots \\ 1 & 1 & \ldots & 1 & 1 \\ \vdots & \vdots & \vdots & \vdots & \vdots \\ 0 & \ldots & \ldots & \ldots & \ldots \end{bmatrix} \begin{matrix} 1 \\ \vdots \\ m \\ \vdots \\ M \end{matrix} \tag{4}$$

Then, the EPI can be expressed as:

$$E_m = \sum_{i=1}^{K} (I_i. \times A_m)^T \qquad (5)$$

Where T represents the matrix transpose. m represents the m-th row of the multi-view image, and K represents the total number of multi-view images.

However, the above-mentioned single-row pixel-level EPI is not suitable for the network; because a single-row pixel-level EPI is not conducive to extract the high-dimensional features. Thus, we extend the matrix whose n rows are 1, and obtain the local image (EPI consists of multiple lines of pixels), as shown in step 1 of Fig. 4, which can get hybrid resolution local images more easily and efficiently.

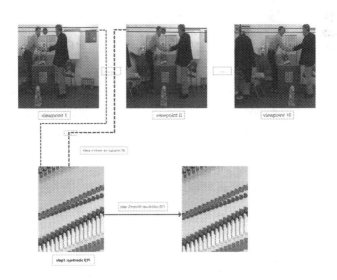

Fig. 4. The book arrival sequence with 16 viewpoints takes n rows of pixel synthesis local images

Capture Hybrid Resolution Local Images. As shown step 2 in Fig. 4, we reduce the resolution to 1/2 for all even-numbered columns of viewpoint information while leaving the information for the odd-numbered columns completely preserved. Then, we achieve hybrid resolution local images whose even-numbered columns of viewpoint are low resolution. Finally, we put hybrid resolution local images into SRGAN to train.

2.3 Multiview Reconstruction

Content Information. From Sect. 2.2, it can be seen that the content information in the purple rhombus frame in the viewpoint 2 need to be synthesized in Fig. 5 and can be directly expressed by the high-dimensional features extracted by the viewpoints 1

and 3 because the content information of the multi-viewpoints has high similarity. Moreover, Low-resolution information in even-numbered view-points can assist adjacent view-points to finish reconstruction.

Location Information of Content. Because the EPI has a specific slash texture compared to the natural image, as shown by the two red slashes in the right figure in Fig. 5, we can reconstruct location information of content more accurately with obvious slash features. In Sect. 3, we also designed related experiments to verify this slash feature.

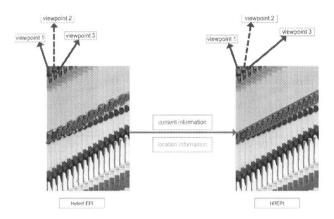

Fig. 5. Reconstruction of multiple views, where purple represents content information, and green represents location information of content. (Color figure online)

Block Effect. Blocking is mainly due to the fact that our input is obtained by splicing EPI blocks. During training, the network is likely to learn the segmentation features of the boundary, leading to significant block effects; we perform post-processing by overlapping and simply filtering operations.

3   Experiment

3.1   Dataset

In the multi-view transmission task, all of our information is known, but we artificially determine which part of the data is transmitted in order to reduce the amount of data, and to reconstruct high-quality images on the decoding side. Therefore, we just use different content in the same scenario to test. Our datasets are from Nagoya University, etc. as following: *Newspaper*, *Lovebird1*, *Lovebird2*, *Book_arrival* and *Balloon*. Firstly, we obtain the images at intervals of 10 frames from the first half of each multi-view video, and then compose the images into local images as a training set. The test set consists of the frames with a large difference in the content information extracted from the latter half of the video.

3.2   Training Details and Parameters

All of our networks are trained on Nvidia GPU:1080Ti using 1080 local images made from Nagoya et al.'s multi-view dataset. We crop the 384 * 384 sub-local images of different HREPI images from the leftmost side. In order to optimize the network, an Adam gradient algorithm was used, where $\beta_1 = 0.9$. Only the pixel mean square error is used when initializing the generator, and the training is performed 100 times with a learning rate of 1e−4 to avoid unnecessary local optimum. During training, 250 epochs were trained using 1e−4, then the learning rate was reduced to the initial 10%, and 250 epochs were trained with 1e−5. Our experiment is based on tensorflow and tensorlayer.

3.3   Slash Features of EPI

In Fig. 6, Since the slash features saved by the network training are: relative to the previous given viewpoint, the latter reconstruction viewpoint shifts upward with a fixed slope, so the dense view EPI map with the overall upward slash texture feature can be well reconstructed in the right figure. What is interesting is that in the left figure, the reconstruction view downward with a fixed slope occurs when we deliberately reversed the order of the view-points. Therefore, after the network is trained, the slash-like features of the EPI maps are effectively learned. Moreover, the two slopes are opposite numbers to each other.

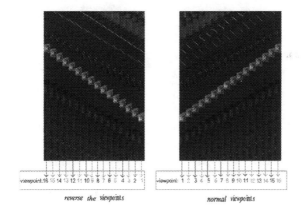

reverse the viewpoints   normal viewpoints

Fig. 6. The green part represents the reconstructed viewpoint. During test, we deliberately reversed the local image's view order in the left image, while the right image is the local image reconstructed from the normal viewpoint order. (Color figure online)

3.4   Experimental Result

Figure 7 illustrates the subjective quality for *Newspaper* video sequence. Figure 7(a) shows the original images, i.e. 10th original frame of the virtual view and green rectangular boxes are utilized to mark the cropped and zoomed portion which is shown in

Fig. 7(b). Similarly, Fig. 7(c), (e), (g), (i) shows the view synthesis by bicubic, Gaussian mixture model and the proposed technique and Fig. 7(b), (d), (f), (h), (j) shows corresponding cropped and zoomed images. As can be seen from the Fig. 7, the bicubic which generate from the network has a remarkable block effect, and the proposed hybrid resolution is more realistic than the Gaussian in terms of hair texture.

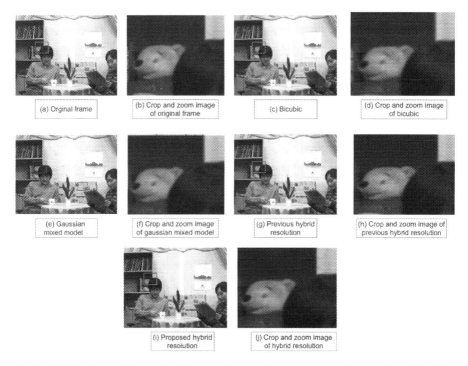

Fig. 7. Original image (a), synthesis images (c, e, g, i), crop and zoom images (b, d, f, h, j) for *Newspaper* video sequence by the proposed method and three traditional methods.

Figure 8 has demonstrated that the proposed method brings a large improvement in the subjective quality when the network learned the well mapping between the LREPI and the HREPI. What's more, high-dimensional features can express more accurately in the content information without ghosting.

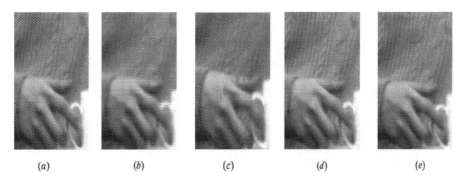

(a)   (b)   (c)   (d)   (e)

Fig. 8. (a) Original image (b) Image after bicubic (c) Image after learning inverse mapping in Gaussian mixed model (d) Previous hybrid resolution (e) Proposed hybrid resolution

Tables 1 and 2 show the average PSNR and MOS comparison on *Newspaper*, *Lovebird1*, *Book_arrival* and *Balloon* four sequences. Proposed technique improves 0.8 dB, 0.2 dB 0.2 dB on average for *Newspaper*, *Lovebird1* and *Balloon*, but there is a 0.9 dB drop on *Book_arrival*. Similarly, in MOS score, proposed technique improves 0.61, 0.15 0.32 on average for *Newspaper*, *Lovebird1* and *Balloon*, but there is a 0.04 drop on *Book_arrival*. We extracted 10 synthesis images in each sequence, then we invited 10 students and teachers who major in multiview video scoring from 1 to 5 points.

Table 1. Average PSNR comparison for 10 frames with 4 kinds of methods on 4 sequences

	Newspaper	Lovebird1	Book_arrival	Balloon
Bicubic	31.9	26.5	22.6	25.2
Gaussian mixture model	32.3	28.1	**25.2**	25.7
Previous hybrid resolution	32.5	28.1	24.2	25.7
Proposed hybrid resolution	**33.3**	**28.3**	24.3	**25.9**

Table 2. Average MOS comparison for 10 frames with 4 kinds of methods on 4 sequences

	Newspaper	Lovebird1	Book_arrival	Balloon
Bicubic	2.36	2.28	2.23	2.72
Gaussian mixture model	3.46	4.08	**4.05**	3.97
Previous hybrid resolution	3.71	4.11	3.98	3.93
Proposed hybrid resolution	**4.32**	**4.26**	4.01	**4.25**

4   Discussion and Feature Work

The hybrid resolution framework proposed in this paper can reconstruct high quality images with relatively little moving background. When the moving amplitude is large, the network performance is not robust enough. If the movement is not very strong, the proposed method can still provide better results. In the later work, we intend to completely remove the content information of the even-numbered viewpoints in the local images, and try to use the information of the neighboring viewpoints and the slash feature of the local images to achieve the reconstruction.

5   Conclusion

In this paper, we show a method for multi-view reconstruction based on SRGAN network. In our proposed method, we use a single two-dimensional image to input multi-viewpoint information; in the reconstruction of multi-viewpoints, we effectively use the slash feature of the EPI to recover the location information of content, and the content information is reconstructed through the high-dimensional features expression. Our method can effectively reduce the mismatch of pixel-level correlation representations. The original 3D information can be reproduced to the users more non-destructively.

Acknowledgement. This research was funded by the National Natural Science Foundation of China (No. 61671152, No. 61471124); the Natural Science Foundation of Fujian Province of China (No. 2014J01234, No. 2017J01757).

References

1. Schreer, O., Thomas, G., Niamut, O.A.: Format-agnostic approach for production, delivery and rendering of immersive media. In: Proceedings of the 4th ACM Multimedia Systems Conference, pp. 249–260 (2013)
2. Seguin, D.: 3D at the B.O: avatar has changed everything. Canada's Broadcast Prod. J. 2(5), 43–44 (2010)
3. Żmigrodzka, M., Wiśniowski, W.: Development of virtual reality technology in the aspect of educational applications. Mark. Sci. Res. Organ. (2017)
4. Chirico, A., Lucidi, F., Milanese, C.: Virtual reality in health system: beyond entertainment. A mini-review on the efficacy of VR during cancer treatment. J. Cell. Physiol. 231(2), 275–287 (2016)
5. Xiao, J., Hannuksela, M.M., Tillo, T., Gabbouj, M., Zhu, C., Zhao, Y.: Scalable bit allocation between texture and depth views for 3-D video streaming over heterogeneous networks. IEEE Trans. Circuits Syst. 25(1), 139–152 (2015)
6. Diogo, C., Garcia, G., Camilo Dorea, C., Ricardo, L.: Super resolution for multiview images using depth information. IEEE Trans. Circuits Syst. Video Technol. 20(3), 132–135 (2012)
7. Aflaki, P., Hannuksela, M.M., Hakkinen, J., Lindroos, P., Gabbouj, M.: Subjective study on compressed asymmetric stereoscopic video. In: Proceedings of the 17th IEEE International Conference on Image Processing (ICIP), vol. 2, pp. 4021–4024. IEEE (2010)

8. Richter, T., Seiler, J., Schnurrer, W., Kaup, A.: Robust super-resolution for mixed-resolution multiview image plus depth data. IEEE Trans. Circuits Syst. Video Technol. **26**(5), 814–828 (2016)

9. Jin, Z., Tillo, T., Xiao, J., Zhao, Y.: Multiview video plus depth transmission via virtual-view-assisted complementary down/upsampling. EURASIP J. Image Video Process. **2016**, 19 (2016)

10. Horng, Y.-R., Tseng, Y.C., Chang, T.-S.: Stereoscopic images generation with directional Gaussian filter. In: Proceedings of the IEEE International Symposium on Circuits Systems, pp. 2650–2653. IEEE (2010)

11. Lee, P.J., Effendi, : Nongeometric distortion smoothing approach for depth map preprocessing. IEEE Trans. Multimed. **13**(2), 246–254 (2011)

12. Do, L., Zinger, S.: Quality improving techniques for free-viewpoint DIBR. In: IS&T/SPIE Electronic Imaging, pp. 75240I–75240I-10. SPIE Press, Boston (2010)

13. Rahaman, D.M.M., Paul, M.: Free view-point video synthesis using Gaussian mixture modelling. In: Proceedings of the IEEE Conference on Image and Vision Computing New Zealand, pp. 1–6. IEEE (2015)

14. Oliveira, A., Fickel, G., Walter, M., Jung, C.: Selective hole-filling for depth-image based rendering. In: Proceedings of the IEEE International Conference on Acoustics, Speech and Signal Processing, pp. 1186–1190 (2015)

15. Yao, C., Zhao, Y., Xiao, J., Bai, H., Lin, C.: Depth map driven hole filling algorithm exploiting temporal correlation information. IEEE Trans. Broadcast. **60**(2), 394–404 (2014)

16. Dmm, R., Paul, M.: Virtual view synthesis for free viewpoint video and multiview video compression using gaussian mixture modelling. IEEE Trans. Image Process. **27**(3), 1190–1201 (2018)

17. Simonyan, K., Zisserman, A.: Very deep convolutional networks for large-scale image recognition. arXiv preprint arXiv:1409-1556 (2014)

18. Ledig, C., Theis, L., Huszar, F., et al.: Photo-realistic single image super-resolution using a generative adversarial network. In: Proceedings of the IEEE Conference on Computer Vision and Pattern Recognition, Honolulu, HI, USA, pp. 105–114 (2016)

19. Wu, G., Zhao, M., Wang, L., Dai, Q., Chai, T., Liu, Y.: Light field reconstruction using deep convolutional network on EPI. In: Proceedings of the IEEE Computer Vision and Pattern Recognition, pp. 6317–6327 (2017)

Contextual Attention Model for Social Recommendation

Hongfeng Bao, Le Wu[✉], and Peijie Sun

Hefei University of Technology, Hefei, China
lewu@hfut.edu.cn

Abstract. Recently, with the emergence of a large number of social platforms, more and more works have been explored for social recommendation. On a social platform, social scientists converged that there exists social influence among users. Thus, accurately modeling the social influence could alleviate the data sparsity issue in Collaborative Filtering (CF). Most of the methods simply define the social influence with the normalized constant weights. However, this is not accuracy enough, which requires more reliable modeling. Besides, many studies have adopted neural network with CF in various recommendation tasks due to the effective ability of neural network for representation. In this paper, we attempt to apply attention mechanism based neural network structure for social recommendation. Specifically, social attention can weigh the contribution of social influence in the form of scores from each neighbor, and then generates each user's social context. Finally, extensive experimental results confirm the feasibility and effectiveness of our proposed model.

Keywords: Deep learning for recommender system
Attention model · Social network

1 Introduction

With the rapid development of the Internet and the explosive growth of information, recommender systems have been increasingly applied to help users extract the information they need efficiently. Collaborative Filtering is one of the most commonly used method due to its making recommendation based on the collaborative behaviors of users without requiring the explicit user and item profiles [1]. However, CF also has several weaknesses, especially it achieves poor performance when the user-item interaction data is very sparse. Thus, many researchers have explored additional auxiliary information to alleviate this issue of CF such as social information.

It is well known that users in a social network are correlated, and users' interests are similar to or influenced by his social connections apart from their own interests [2]. Therefore, many works are proposed to analyse how to learn social influence from social network effectively, and then incorporate it in the classical latent factor based models [3–5]. Although several methods successfully used

© Springer Nature Switzerland AG 2018
R. Hong et al. (Eds.): PCM 2018, LNCS 11165, pp. 630–641, 2018.
https://doi.org/10.1007/978-3-030-00767-6_58

the social information and solved the data sparsity problem of CF, they modeled the social influence between one user and his neighbor with a normalized constant weight. This is obviously unreasonable, thus requiring a more appropriate approach to represent social influence.

The wide population of deep learning techniques gives us more chances to further to enhance the recommendation performance for social recommendation. The neural network has a strong ability of feature representation to better represent users' preferences and items' contents from available data. Hence, several researchers actively explored the approach to combining deep learning techniques with CF models for various recommendation tasks [6–8]. These efforts applied deep neural networks to learn the complex user-item relationship from training data and make tremendous progress toward improving the effectiveness of recommender system.

In this paper, we make an attempt to model social influence through an attention based neural network. Attention mechanism can be regarded as a resource allocation model, which has a wide range of applications in many fields of machine learning [9–11]. In our model, we use the attention mechanism to quantize the social influence between a user and his neighbors in the form of probability, and then create a context. As a consequence, our objective prediction function is composed of a classical latent factor based model and a social context generated by an attention model, which models the users' interests and social influence respectively. Finally, we conduct extensive experiments on two real-world datasets to demonstrate the superiority of our proposed model.

2   Related Work

2.1   Collaborative Filtering and Social Recommendation

CF aims at making recommendations or predictions of the unknown preferences for a group of users by learning the known other users' historical behavior [1]. So far many CF techniques have been developed and the latent factor based models are undoubtedly the most popular one among them due to their relatively high performance [12,13]. In some real world applications, only binary data can be available, which reflect a user's action or inaction. Such recommendation with implicit feedback is called the one-class problem, where only positive feedback is available [14]. In this paper, we use common approach of sample based learning samples negative feedback from the missing data [7,13,14]. Since there are many challenges to CF systems such as data sparsity [1,15], many efforts were done for social recommendation. Several researchers completed a series of attempts of fusing users' preference with social links and achieve desirable results [4,5,16,17]. In addition, more and more deep learning techniques are applied to social recommendation now [18]. Some of them processed analysed user-provided content in social media to understand social information [19]. These models make better use of social information thus showing improved performance over classical recommendation models.

2.2   Deep Learning and Attention Mechanism

Recently, many data scientists have put their focus on deep learning in the field of recommender systems [20]. Several researches used autoencoder to learn hidden user representations from the rating matrix and improved the result of recommendation based on collaborative filtering techniques [6, 21–23]. NeuMF combines a generalization of matrix factorization and a multilayer perceptron [7]. Except for these, attention model is also increasingly applied to the framework of neural network for recommendation and successfully captures various contextual information [24]. The main idea of attention mechanism is that human recognition only focuses on selective parts of a target scenario instead of the whole perception space. IARN learns different scores from all time steps in user-item interaction history to capture user and item dynamics jointly [25]. Besides, Chen et al. [26] and Seo et al. [27] proposed a hierarchical attention network and a dual attention network respectively for recommendation. Our work extends attention mechanism in social network. Specifically, we score social influence of each user's all connections through attention model for more reliable modeling.

3   The Proposed Model

In a social recommender system, we have a userset U with N users and an itemset I with M item. Let $\mathbf{S} \in \mathbb{R}^{N \times N}$ denotes social network matrix and S_u represents all the followees of user u, i.e., if $S_{uv} > 0$, then $v \in S_u$. Since the social contextual information is made up of social influence, the social context C_{S_u} can be computed by social influence from each user u's social neighbors S_u. Besides, The ratings expressed by users on items are given in a user-item interaction matrix $\mathbf{R} \in \mathbb{R}^{N \times N}$ with $R_{ui} = 1$ if user u shows preference for item i, otherwise it equals 0. Given the sparse social matrix $\mathbf{S}$ and user-item interaction matrix $\mathbf{R}$, our goal is to recommend the potential top-N items each user may like based on various social contextual information.

3.1   General Framework

In this subsection, we describe our proposed Contextual Attention Model for Social Recommendation. Figure 1 shows the overall architecture of our model, which is composed of two parts. Attention model part models user's preference score with respect to the social contextual information from social influence. Given a user u, we use a_{uv} to denote the degree of a user is influenced by his neighbors S_u. Besides, in the part of social embedding, we apply an autoencoder to learn the social embedding of each user with the social network structure. Then we take it as another input of attention model, which makes the training of attention model consider richer social information.

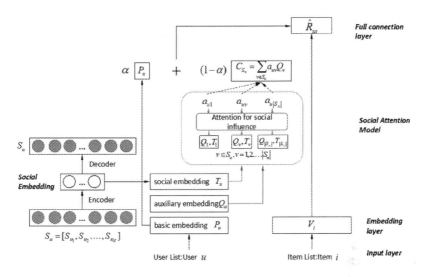

Fig. 1. The overall architecture of the proposed model.

Objective Prediction Function. In a social recommender system, a user u's preference score on item i can be predicted by the user-item interaction and social contextual information. The former is interpreted by the user u's favor on item i, and the latter represents the user u's potential favor on item i affected by his social neighbors S_u's favors. For the input, we parameterize each user u with a base embedding P_u and each item i with a base embedding V_i as general latent factor based models. The base embedding denotes a user's base latent interest or the basic item latent vector of an item. In the other part, we use attention mechanism to estimate the social influence strength of one user to another. Then we get:

$$\hat{R}_{ui} = (\alpha P_u + (1 - \alpha) C_{S_u}) \times V_i \tag{1}$$

In the above function, the basic latent factor based model and the social context are smoothed by the parameter α, which controls the effects of social neighbors on the predicted rating. Specifically, the part of the social context C_{S_u} is an attention subnetwork and use learned attentive weights to model social influence. The realization of the attention part is stated in detail in the following parts. In addition, if we replace all attentive weights with normalized weights $\frac{1}{|S_u|}$, our model turns to a RSTE model [4].

Social Network Embedding. To train the attention model by making more use of social information, we add the social embedding for each user in the alignment model. The social embedding part aims at gaining another representation for each user from another different low-dimensional space. In this paper, we choose the autoencoder to construct such a social embedding space directly among many useful methods [28,29].

Autoencoder is one of the feed-forward neural network. Its basic idea is compressing the high-dimensional input data into hidden representation and then reconstructing the input. The whole structure of autoencoder consists of an encoder part and a decoder part, which is shown in the left part in Fig. 1. In our model, we let a sparse social matrix $\mathbf{S}$ as the input and expect to gain a low-dimensional social embedding space $T \in \mathbb{R}^{N \times D}$ from two parts:

$$T_u = f\left(\mathbf{W_1} \times S_u + b\right),$$
$$\hat{S}_u = f'\left(T_u\right) = f'\left(\mathbf{W_1}' \times T_u + b'\right),$$
(2)

where $f(x)$ and $f'(x)$ are both nonlinear functions. $\hat{S}_u$ a reconstructed vector and $S_u \in \mathbb{R}^N, T_u \in \mathbb{R}^D$. Its parameter set is $\theta_1 = \{\mathbf{W_1} \in \mathbb{R}^{D \times N}, b \in \mathbb{R}^D, \mathbf{W_1}' \in \mathbb{R}^{N \times D}, b' \in \mathbb{R}^N\}$.

Social Attention Model. In this part, we describe how to use attention mechanism to learn influence weights between users from social network matrix $\mathbf{S}$. Specifically, we use e_{uv} to denote the social influence strength of v to u if user u follows v. Then, the social attention model generates a social contextual vector in accordance with the contribution of social influence from each user's social connections. The attention model can be simply realized by a single multilayer perceptron such that:

$$e_{uv} = w_3 \times \sigma\left(\mathbf{W_2} \times [P_u, P_v, T_u, T_v]\right),$$
(3)

where we choose the sigmoid function $\sigma(x)$ as activation function. $\theta_2 = [\mathbf{W_2}, w_3]$ are the parameters in the social attention model, with $\mathbf{W_2}$ denotes the matrix parameter and w_3 is the parameters of the activation function. P_u and P_v denote the base embeddings of the user u and one of his neighbor v respectively. Similarly, T_u and T_v are the social embeddings of them which represent their own social network structures.

Then, we obtain the final attentive social influence scores a_{uv} by normalizing the above attentive scores using Softmax function:

$$a_{uv} = \frac{exp\left(e_{uv}\right)}{\sum_{v \in S_u} exp\left(e_{uv}\right)}.$$
(4)

The normalized attentive social influence scores a_{uv} above can be interpreted as the contribution of the preference of user u's all neighbors to user u, so we calculate the social context C_{S_u} of user u as:

$$C_{S_u} = \sum_{v \in S_u} a_{uv} P_v.$$
(5)

Here we make an improvement of adding an auxiliary embedding vector Q_u to distinguish the user embedding in two parts as showed in Fig. 1:

$$R_{ui} = \left(\alpha P_u + (1 - \alpha) \sum_{v \in S_u} a_{uv} Q_v \right) \times V_i,$$

$$\text{where } a_{uv} = \frac{exp\left(e_{uv}\right)}{\sum_{v \in S_u} exp\left(e_{uv}\right)}, e_{uv} = w_3 \times \sigma\left(\mathbf{W_2} \times [Q_u, Q_v, T_u, T_v]\right). \quad (6)$$

In the new network, the base embedding of user P_u only participates in the first half of equation, and the social influence score in the part of attention model is calculated by the auxiliary embeddings and the social embeddings. With this change, the model is better trained, thus our results ending up with a lot of improvement.

3.2 Model Learning

As similar as most of the latent factor based models [12], we apply the regularized squared loss to the optimization function:

$$\arg \min_{P,V,\theta_2} \sum_{(u,i) \in D} \left(R_{ui} - \hat{R}_{ui} \right)^2 + \lambda \left(\|P\|^2 + \|V\|^2 \right), \quad (7)$$

where P and V denote the base embedding matrices and θ_2 is the parameters of social attention model. D represents the set that the users have the observed ratings on items in training data. λ controls the strength of regularization.

During actual model training, the social embedding part was pretrained. Besides, our datasets are processed for implicit feedback of users, which has no negative values. Therefore, for each observed positive value, we randomly sample 5 negative values from missing data at each iteration of the training process [14]. At last, we choose TensorFlow to train our model using mini-batch Adam.

4 Experiments

4.1 Experimental Settings

Datasets. We choose two publicly accessible datasets to complete our experiments: Flixster [3] and Douban. At first, due to high sparsity of datasets, we filter the datasets by retaining users that have at least 5 rating records and 5 social links. We also filter out items that have been rated less than 5 times. Then, since we are concerned about recommendation with implicit feedback, we relabel the rating matrix with 1 or 0 indicating whether the user has rated the item [7,26]. The characteristics of the two datasets after adjustment are showed in Table 1. In data splitting process, we adopt the widely used leave-one-out procedure in [7,13,26]. Specifically, for each user, we hold-out his latest rating record as the test data, and the remaining data are used for training, 5% of which form the validation dataset.

Table 1. Statistics of the two evaluation datasets.

Dataset	Users	Items	Ratings	Links	Rating density
Douban	6,739	17,902	840,828	201,014	0.697%
Flixster	9,874	10,978	283,503	120,306	0.262%

Evaluation Metrics. We select two widely adopted metrics for top-K ranking performance in several recent works for recommendation: the Hit Ratio (HR) and Normalized Discounted Cumulative Gain (NDCG) [7,26]. HR measures the percentage of top-K ranked items that are rated by users in the test data. NDCG gives a higher score when the hit items are ranked higher in a ranking list. As it is too time-consuming to complete top-K ranking evaluation for all users, we follow the common strategy: for each test user, we randomly select 1000 unrated items in the training data. Then we mix them with the records in both validation and test data to generate top-K results, and we repeat this process for 10 times and average the results [7,26].

Baselines. We choose the following baselines compared with our model:

- *PMF*. It is a basic competitive method for recommendation [12].
- *SocialMF*. This model incorporates the social information into the matrix factorization framework. It models the social influence with the equal normalized weights, which is the improvement of RSTE model [3,4].
- *SR*. It is another state-of-the-art model for social recommendation [5].
- *AutoRec*. It provides an autoencoder based neural network structure for recommendation [6].
- *NeuMF*. It is an excellent method as a competitive neural network based model for recommendation [7].

Parameter Settings. In view of the effectiveness and efficiency, we finally set the batch size to 2048 and the learning rate to 0.001. Besides, the latent feature dimension is set as 16 and the parameter α is set as optimal value in the following parts if not specified. For the initialization of these methods, we initialized these models with the parameters of PMF model.

4.2   Performance Comparison

Figure 2 only shows the performance of NDCG@10 with respect to the number of latent factors for page limit. We can observe that our proposed model always achieves the best performance on both datasets as the number of predictive factors grows. This means that attention mechansim can capture more accurate social influence to improve social recommendation significantly. What's more, we find that most models have the largest increases from the factor of 8 to 16.

Thus, we set the number of predictive factors as 16 in the following experiments, although our proposed model has better results with a larger factor.

Next, the performance of Top-K recommendation lists on the two datasets are presented in Fig. 3. We can see that our proposed method always performs the best on both datasets across all positions. In particular, our model is greatly superior to socialMF, which indicates that social influence can be modeled more reasonably with attentive weights than the equal normalized weights. Besides, other four baselines outperform PMF. This shows that encoding social network information into the latent factor based model or applying deep learning techniques really help to enhance the performance of recommendation.

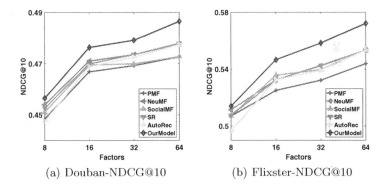

(a) Douban-NDCG@10   (b) Flixster-NDCG@10

Fig. 2. Performance of HR@10 and NDCG@10 w.r.t. the number of predictive factors on the two datasets.

4.3   Impact of Parameter α

In our proposed model, parameter α balances how much information from the users' own characteristics and their neighbors' preference can be utilize in social recommendation, similar as that in RSTE [4]. If $\alpha = 1$, it turns to a PMF model, which does not encode any social information. If $\alpha = 0$, we train the recommendation model only by social information learned from attention mechanism.

Figure 4 shows the metric of NDCG@10 on two datasets with different values of α. To show the difference clearly, we remove the minimum value when $\alpha = 0$ here. We observe that two curves have a similar trend, and the optimal value of α is around 0.9 and 0.7 respectively. This means that our method learns effective social information to improve the performance of recommendation.

4.4   Effect on Attention Input

Since our model introduces many parameters to learn social influence from social matrix, it inevitably leads to overfitting. In this paper, dropout is applied to

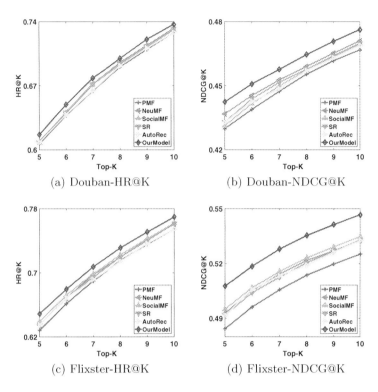

(a) Douban-HR@K   (b) Douban-NDCG@K

(c) Flixster-HR@K   (d) Flixster-NDCG@K

Fig. 3. Performance of Top-K item recommendation where K ranges from 5 to 10 on the two datasets.

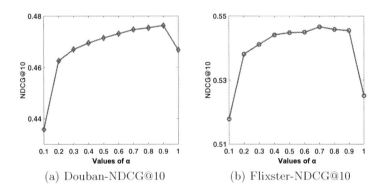

(a) Douban-NDCG@10   (b) Flixster-NDCG@10

Fig. 4. Impact of parameter α

reduce overfitting to enhance the performance of model [30]. Table 2 shows the results with different dropout ratio ρ. We choose the dropout ratio ρ as 0.2 and 0.3 respectively for their best performances. Compared with the performance when $\rho = 0$, there are great improvements on both datasets.

In Table 3, we show whether the addition of auxiliary embeddings for users can make the model better trained. We observe that the model with two different embeddings for users performs better on both datasets. Especially, on Flixster, the second way relies more on social information with a smaller value of α, which means it learns more reliable social influence.

Table 2. HR@10 and NDCG@10 with different dropout ratio.

Dropout ratio	0	0.1	0.2	0.3	0.4	0.5
Douban						
HR	0.7325	0.7369	**0.7372**	0.7333	0.7310	0.7289
NDCG	0.4722	0.4756	**0.4763**	0.4736	0.4714	0.4691
Flixster						
HR	0.7612	0.7634	0.7673	**0.7694**	0.7682	0.7659
NDCG	0.5348	0.5377	0.5428	**0.5466**	0.5450	0.5411

Table 3. The best results of our model with two ways for users.

Model	Optimal α	Dropout ratio	HR	NDCG
Douban				
$P = Q$	0.9	0.2	0.7341	0.4735
$P \neq Q$	0.9	0.2	0.7369	0.4763
Flixster				
$P = Q$	0.9	0.3	0.7663	0.5449
$P \neq Q$	0.7	0.3	0.7694	0.5466

5   Conclusion

In this paper, we proposed an attention mechanism based neural network architecture for social recommendation. Specifically, we attempted to capture social influence from social network through attention mechanism and learn more real attentive weights to generate social context. Obviously, the attentive weights learned are not all the same between one user and different neighbors. Then, we encoded the learned social information in a general matrix factorization model. The extensive experiments were conducted on two real-world datasets, which demonstrated the effectiveness of our proposed model. After analysing the results of comparative experiments, it is proved that attention model can learn social influence more effectively than those models with the simple modeling likes the normalized constant weights for social recommendation.

Acknowledgments. This work was supported in part by the National Key Research and Development Program of China (Grant No. 2017YFC0820604), the National Natural Science Foundation of China (Grant No. 61602147, No. 61632007), the Open Project Program of the National Laboratory of Pattern Recognition (NLPR Grant No. 201700017), and the Fundamental Research Funds for the Central Universities (Grant No. JZ2018HGTB0230).

References

1. Su, X., Khoshgoftaar, T.M.: A survey of collaborative filtering techniques. Adv. Artif. Intell. **2009**(4), 2 (2009). https://doi.org/10.1155/2009/421425
2. Anagnostopoulos, A., Kumar, R., Mahdian, M.: Influence and correlation in social networks. In: SIGKDD, pp. 7–15. ACM, New York (2008). https://doi.org/10.1145/1401890.1401897
3. Jamali, M., Ester, M.: A matrix factorization technique with trust propagation for recommendation in social networks. In: RecSys, pp. 135–142. ACM, New York (2010). https://doi.org/10.1145/1864708.1864736
4. Ma, H., King, I., Lyu, M.R.: Learning to Recommend with Social Trust Ensemble. In: SIGIR, pp. 203–210. ACM, New York (2009). https://doi.org/10.1145/1571941.1571978
5. Ma, H., Zhou, D., Liu, C., Lyu, M.R., King, I.: Recommender systems with social regularization. In: WSDM, pp. 287–296. ACM, New York (2011). https://doi.org/10.1145/1935826.1935877
6. Sedhain, S., Menon, A.K., Sanner, S., Xie, L.: Autorec: autoencoders meet collaborative filtering. In: WWW, pp. 111–112. ACM, New York (2015). https://doi.org/10.1145/2740908.2742726
7. He, X., Liao, L., Zhang, H., Nie, L., Hu, X., Chua, T.-S.: Neural collaborative filtering. In: WWW, pp. 173–182, Perth, Australia (2017). https://doi.org/10.1145/3038912.3052569
8. Cheng, H.-T., Koc, L., Harmsen, J., et al.: Wide & deep learning for recommender systems. In: Proceedings of the 1st Workshop on Deep Learning for Recommender Systems, pp. 7–10. ACM, New York (2016). https://doi.org/10.1145/2988450.2988454
9. Bahdanau, D., Cho, K., Bengio, Y.: Neural machine translation by jointly learning to align and translate. arXiv:1409.0473 (2014)
10. Xu, K., et al.: Show, attend and tell: Neural image caption generation with visual attention. In: ICML, pp. 2048–2057. JMLR.org (2015)
11. Yang, Z., Yang, D., Dyer, C., He, X., Smola, A., Hovy, E.: Hierarchical attention networks for document classification. In: Proceedings of the 2016 Conference of the North American Chapter of the Association for Computational Linguistics: Human Language Technologies, pp. 1480–1489 (2016)
12. Mnih, A., Salakhutdinov, R.: Probabilistic matrix factorization. In: NIPS, pp. 1257–1264. Curran Associates Inc., New York (2007)

13. Rendle, S., Freudenthaler, C., Gantner, Z., Schmidt-Thieme, L.: BPR: Bayesian personalized ranking from implicit feedback. In: UAI, pp. 452–461. AUAI Press Arlington, Virginia (2009)
14. Pan, R., et al.: One-class collaborative filtering. In: ICDM, pp. 502–511. IEEE, New York (2008). https://doi.org/10.1109/ICDM.2008.16
15. Adomavicius, G., Tuzhilin, A.: Toward the next gen- eration of recommender systems: a survey of the state-of-the-art and possible extensions. TKDE **17**(6), 734–749 (2005). https://doi.org/10.1109/TKDE.2005.99
16. Ma, H., Yang, H., Lyu, M.R., King, I.: SoRec: social recommendation using probabilistic matrix factorization. In: CIKM, pp. 931–940. ACM, New York (2008). https://doi.org/10.1145/1458082.1458205
17. Wu, L., et al.: Modeling the evolution of users preferences and social links in social networking services. TKDE **29**(6), 1240–1253 (2017). https://doi.org/10.1109/TKDE.2017.2663422
18. Deng, S., Huang, L., Xu, G., Wu, X., Wu, Z.: On deep learning for trust-aware recommendations in social networks. TNNLSS **28**(5), 1164–1177 (2017). https://doi.org/10.1109/TNNLS.2016.2514368
19. Li, Z., Tang, J.: Weakly Supervised deep matrix factorization for social image understanding. IEEE Trans. Image Process. **26**(1), 276–288 (2017). https://doi.org/10.1109/TIP.2016.2624140
20. Zhang, S., Yao, L., Sun, A.: Deep learning based recommender system: a survey and new perspectives. arXiv:1707.07435 (2017)
21. Vincent, P., Larochelle, H., Lajoie, I., Bengio, Y., Manzagol, P.-A.: Stacked denoising autoencoders: learning useful representations in a deep network with a local denoising criterion. JMLR **11**(3), 3371–3408 (2010)
22. Wu, Y., DuBois, C., Zheng, A.X., Ester, M.: Collaborative denoising autoencoders for top-n recommender systems. In: WSDM, pp. 153–162. ACM, New York (2016). https://doi.org/10.1145/2835776.2835837
23. Wang, H., Wang, N., Yeung, D.-Y.: Collaborative deep learning for recommender systems. In: SIGKDD, pp. 1235–1244. ACM, New York (2015). https://doi.org/10.1145/2783258.2783273
24. Sun, P., Wu, L., Wang, M.: Attentive recurrent social recommendation. In: SIGIR, pp. 185–194. ACM, New York (2018). https://doi.org/10.1145/3209978.3210023
25. Pei, W., Yang, J., Sun, Z., Zhang, J., Bozzon, A., Tax, D.M.J.: Interacting attention-gated recurrent networks for recommendation. In: CIKM, pp. 1459–1468. ACM, New York (2017). https://doi.org/10.1145/3132847.3133005
26. Chen, J., Zhang, H., He, X., Nie, L., Liu, W., Chua, T.-S.: Attentive collaborative filtering: multimedia recommendation with item-and component-level attention. In: SIGIR, pp. 335–344. ACM, New York (2017). https://doi.org/10.1145/3077136.3080797
27. Seo, S., Huang, J., Yang, H., Liu, Y.: Interpretable convolutional neural networks with dual local and global attention for review rating prediction. In: RecSys, pp. 297–305. ACM, New York (2017). https://doi.org/10.1145/3109859.3109890
28. Chang, S., Han, W., Tang, J., Qi, G.-J., Aggarwal, C.C., Huang, T.S.: Heterogeneous network embedding via deep architectures. In: SIGKDD, pp. 119–128. ACM, New York (2015). https://doi.org/10.1145/2783258.2783296
29. Wang, D., Cui, P., Zhu, W.: Structural deep network embedding. In: SIGKDD, pp. 1225–1234. ACM, New York (2016). https://doi.org/10.1145/2939672.2939753
30. Srivastava, N., Hinton, G.E., Krizhevsky, A., Sutskever, I., Salakhutdinov, R.: Dropout: a simple way to prevent neural networks from overfitting. JMLR **15**(1), 1929–1958 (2014)

Cracked Tongue Recognition Based on Deep Features and Multiple-Instance SVM

Yushan Xue[1], Xiaoqiang Li[2($\boxtimes$)] , Qing Cui[1], Lu Wang[1], and Pin Wu[1]

[1] School of Computer Engineering and Science,
Shanghai University, Shanghai, China
wupin@shu.edu.cn
[2] Shanghai Institute for Advanced Communication and Data Science,
Shanghai University, Shanghai, China
xqli@shu.edu.cn

Abstract. Cracked tongue can provide valuable diagnostic information for traditional Chinese Medicine doctors. However, due to similar model of real and fake tongue crack, cracked tongue recognition is still challenging. The existing methods make use of handcraft features to classify the cracked tongue which leads to inconstant performance when the length or width of crack is various. In this paper, we pay attention to localized cracked regions of the tongue instead of the whole tongue. We train the Alexnet by using cracked regions and non-cracked regions to extract deep feature of cracked region. At last, cracked tongue recognition is considered as a multiple instance learning problem, and we train a multiple-instance Support Vector Machine (SVM) to make the final decision. Experimental results demonstrate that the proposed method performs better than the method extracting handcraft features.

Keywords: Cracked tongue recognition · Suspected region · Feature extraction
Multiple instance learning · Tongue diagnosis

1 Introduction

Traditional Chinese Medicine (TCM) is a system that consists of the clinical and theoretical investigation on the physiology and pathology of organs and functions [1]. According to the TCM concept of holism, human body is interconnected "whole" comprised of mind, body and spirit. On the principles of TCM, zang-fu theory explains the physiological functions and pathological changes of the internal organs, which are determined by observing the outward manifestations of the body [2]. Tongue diagnosis is a simple, non-invasive and valuable tool that belongs to the method of inspection. As the only organ that could be observed by baked eyes, tongue has abundant symptoms to reflect the pathological changes of the internal organs. Based on these above, TCM practitioner could find potential diseases by analyzing the patient's symptoms.

Over the past decades, tongue diagnosis has widely developed with digital imaging technology. Computer system aimed to aid tongue diagnosis has been established [3] and specific tongue image acquisition device has been designed, which could get high

© Springer Nature Switzerland AG 2018
R. Hong et al. (Eds.): PCM 2018, LNCS 11165, pp. 642–652, 2018.
https://doi.org/10.1007/978-3-030-00767-6_59

quality tongue images [4]. Some researches have been done on tongue characteristic detection. [5] used geometric and tongue-mouth relation features extracted from tongue image to analyze the shape of tongue body. [6] used a method named wide line detection for tongue lines extraction from tongue images.

Tongue cracks, also called tongue lines, refer to the surface of the tongue covered with lines in deep or shallow shape, which can provide valuable diagnostic information for TCM doctors. But not every line or fissure on the tongue is the tongue crack. Many algorithms have been developed and applied in different areas. [7] extracted multiple features from images to detect timber defects like knots and cracks on wood surface. [8] first proposed the method named wide line detector (WLD) which used isotropic nonlinear filtering. This method performs better on detecting lines of different widths and extracts them entirely. [9, 10] used the WLD algorithm and improved WLD algorithm to detect and extract tongue cracks. These two methods are valuable on tongue lines detection but they didn't pay attention to cracked tongue classification. Hu [10] proposed a wide line detection method to extract lines by using the water flow method, which didn't concern if the tongue is cracked tongue.

The cracked tongue has at least one cracked region on the tongue and normal tongue includes non-cracked region. As the organ of human body, the tongue has abundant information on the surface. Not every line on the surface of tongue is tongue crack. Therefore, it will be very significant that classifying cracked tongue automatically in aiding tongue diagnostic. Li [11] proposed a method by using improved WLD method to detect tongue lines and extracted features of these lines to train a binary support vector machine for cracked tongue classification. This method is impactful to distinguish cracked tongues which have long or salient cracks from normal tongue. However, it is not good enough to find out cracked tongues that have short cracks. Instead of using complex handcrafted features as proposed in [11], the proposed method firstly detect tongue lines by the aid of wide line detector and then extract a feature vector for the region that contains suspected tongue cracks by using the convolutional neural network (CNN) to. Then we group feature vectors from each tongue into bag and train a multiple-instance SVM to make a final decision.

The paper is organized as follows. Training AlexNet on Imagenet with Caffe and the training data are described in Sect. 2. In Sect. 3, the multiple-instance SVM and the classification process will be produced. Experiment and result discussion will be presented in Sect. 4. Finally, we will make a conclusion and discuss the future work in Sect. 5.

2   Network Training and Dataset

2.1   Training AlexNet on Imagenet with Caffe

As a deep learning framework, Caffe has already powers academic research projects, startup prototypes and even large scale industrial application in vision, speech and multimedia [12]. Because the training dataset we built is not large enough to train a useful model, we fine-tuned Alexnet that have pre-trained on Imagenet with Caffe. The network performs not good enough without pre-trained: lower accuracy and higher loss. Before training the net, every image in the training dataset was resized to 256 * 256, and a fixed-size 227 * 227 sub-image is randomly cropped and flipped it horizontally by using the same strategy described in [13]. In our research, stochastic gradient descent was used to fine-tune our network with a batch size of 128, learning rate of 0.0001, momentum of 0.9 and weight decay of 0.0005. In this paper, we use the trained network to extract deep features of suspected cracked region.

2.2   Dataset

In our research, the original image includes tongue and part of face (Fig. 1(a)). At the first, we segment tongue from the original image, Fig. 1(b) shows a tongue image. The tongue image dataset used in this paper is provided and acquired by Shanghai Daosh Medical Technology Company, Ltd. It is a small dataset, which contains more than 700 tongue images (both normal tongues and cracked tongues) in total acquired in three different time. The image size is 4896 × 3672 pixels. The tongue images are labeled by TCM practitioners. However, whether a tongue is a cracked tongue is not absolute, the boundary can be very blur when the symptoms is not serious, which is the reason why this is a challenging task. So the label of a tongue image is voted by multiple TCM practitioners.

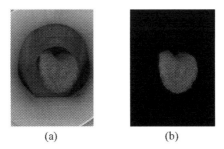

(a)   (b)

Fig. 1. (a) original tongue image with part of face; (b) tongue image segment from (a).

What we import into the network for training and testing is part of the tongue image, named as cracked region. There are various cracks on the tongue: a clear rip on smooth tongue, deep shape line with thick tongue fur and shallow line on tongue surface. In contrast, the characteristics of normal tongue are: smooth tongue surface, lines caused by broken coating, exfoliation of tongue coating, and some shallow shape lines. Figure 2 shows different types of regions from cracked tongue and normal tongue. All of regions generated from cracked tongue and normal tongue are not using a fixed size. To build a useful dataset, we applied data augmentation method such as rotation and flip on every region generated. The dataset we built is 11682 regions including 2064 cracked regions and 9038 normal regions.

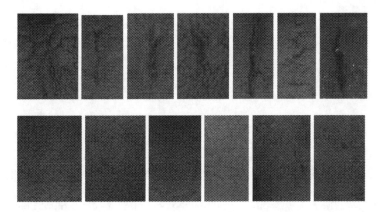

Fig. 2. All kinds of region selected from cracked tongue and normal tongue. First row is cracked regions and second row is normal regions.

3 The Proposed Method

In this section, we will describe proposed algorithm in detail. The proposed method regards a tongue as a bag in multiple-instance learning and cracked regions as instances.

The method follows three steps during classification processing. First, finding and separating regions that have suspected crack on the tongue. Then, using the net to extract feature vector of the region. At last, grouping feature vectors from a tongue to a bag and using a multiple-instance SVM classify the tongue based on the bag.

3.1 Generating and Separating Suspected Cracked Regions

The suspected cracked regions generated from a cracked tongue can include both real cracked regions and healthy non-cracked regions, but at least one cracked region should be included, just like the assumption of multiple-instance binary classification that a positive bag should contains at least one positive instance. The suspected cracked regions generated from a healthy non-cracked tongue should include only healthy regions. We generate a bounding box for every suspected cracked region.

In this study, we tried to use two strategies for generating regions. One is sliding a window to select ROIs on the tongue image. Another one is using improved WLD method described in [9, 11] to detect suspected tongue cracks and separate the regions that contain them. In this method, we make sure every suspected line on the center of the region and also get regions with suspected lines on non-cracked tongues. Due to the tongue crack always appear at the middle area of tongues, we cropped the tongue image before using these two methods. The formula used in cropped tongue images is defined as,

$$
\begin{cases}
x_c = 0.15 \times W_t \\
y_c = 0.1 \times H_t \\
w_c = 0.7 \times W_t \\
h_c = 0.8 \times H_t
\end{cases}
\tag{1}
$$

where, W_t and H_t are width and height of the tongue image, (x_c, y_c) is the left-top coordinate of the cropped image in tongue image and w_c, h_c are width and height of the cropped image. Figure 3 shows the result of getting ROIs from tongue images by using two strategies. The process in use WLD method could find the location of lines directly and get the region but the sliding window method should get many regions to make sure some of them contain suspected lines. For non-cracked tongues, the WLD also detect suspected lines that are not tongue cracks in naked eye observation.

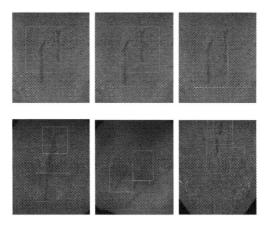

Fig. 3. Select regions on tongue with two different methods. First row is part regions on a cracked tongue by using sliding window; the second row is regions select from different cracked tongue by using WLD method.

3.2   Classification Based on Multiple-Instance SVM

In this part, we train a multiple-instance SVM to classify the tongue images based on suspected cracked region. The multiple-instance SVM used in this paper was proposed by [14]. They introduce two forms of multiple-instance SVM in their work, mi-SVM and MI-SVM. In this paper, we describe the basic idea of MI-SVM. The input of a multiple-instance SVM is a bag B_I that represents the Ith tongue in our case. And the instances in the bag are the feature vectors $B_I = \{x_i : i \in I\}$ we extracted from the cracked regions of the tongue. Instead of explicitly associating a label y_i with each instance, a label Y_I is associated with a bag B_I. If $Y_I = -1$, then $y_i = -1$ for all $i \in I$; if $Y_I = 1$, then at least one instance $x_i \in B_I$ is a positive example.

MI-SVM: The functional margin of a bag is defined by,

$$\gamma_I = Y_I \max_{i \in I}(\langle \omega, x_i \rangle + b) \tag{2}$$

The task in MI-SVM for finding the optimal hyperplane is to minimize the objective function as follow,

$$\min_{(\omega, x, \xi)} \frac{1}{2}\|\omega\|^2 + C \sum_I \xi_I$$
$$s.t. \forall I : Y_I \max_{i \in I}(\langle \omega, x_i \rangle) \geq 1 - \xi_I, \ \xi_I \geq 0 \tag{3}$$

where, ξ_i is a slack parameter that allows classification errors [15]. As can be seen from the definition, instead of taking every instance into account, MI-SVM only looks at parts of the instance. For a positive bag, the margin is defined by the most positive instance, while the margin of a negative bag is defined by the least negative instance of the bag.

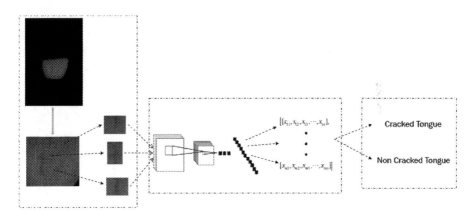

Fig. 4. The framework of the method. There are three stages: Preprocessing for tongue images to get suspected regions, feature extraction and classification by using MI-SVM.

The framework of the method is shown in Fig. 4. Suspected regions are generated by using the algorithm described in Sect. 3.1 first. Suspected regions from one tongue image are grouped into one bag and a feature vector is extracted from each suspected region using the fixed fine-tuned Alexnet. Then, a multiple-instance SVM is used to make the final decision.

4   Experiment

In this section, we will present the experiment result by using different cracked tongue classification methods including the method named WLDF proposed in [11] and our method with two strategies generating suspected cracked regions.

Before presenting results, there are three metrics in our research for evaluating them that are accuracy (ACC), true positive rate (TPR) and true negative rate (TNR), formulations of them are defined as follow:

$$ACC = \frac{TP + TN}{TP + TN + FP + FN} \tag{4}$$

$$TPR = \frac{TP}{TP + FN} \tag{5}$$

$$TNR = \frac{TN}{TN + FP} \tag{6}$$

4.1   Result of WLDF Method

The method proposed in [11] extracts nine features from every tongue image. In the method, nine features are extracted mainly around the three largest MAX-distance of detected areas. We select 79 images (32 cracked tongue images) to train a standard SVM classifier and select other 56 images (29 cracked tongue images) for evaluating the classifier. The result of predicting shown in Table 1.

Table 1. The Result of WLDF method.

Method	ACC	TPR	TNR
WLDF	85.7%	79.31%	92.59%

The WLDF method performs better on cracked tongue that has conspicuous cracks but it is difficult to distinguish non-cracked tongue with ample texture and cracked tongue with short and less obvious cracks. The performance of this method is very dependent on the crack length. Figure 5 shows three tongue images that difficult to distinguish.

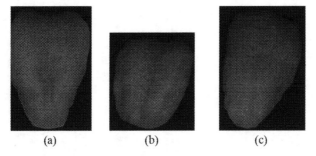

(a)   (b)   (c)

Fig. 5. Images that difficult to distinguish by WLDF. (a) is a non-cracked tongue with ample texture, (b) and (c) are cracked tongues.

4.2   Result of the Proposed Method

In this part, the result of the proposed method is presented. We use sliding window and the improved WLD method to get regions respectively. Then, the proposed method tested on the data of 79 samples as training set and the data of 56 samples as testing set.

As mentioned in Sect. 3.1, the tongue images were cropped by using the formula (1) before selecting regions. Sliding the window with fixed size on the cropped images will get too larger regions. In this paper, the size of sliding window and step size are determined by the experiment. We define that $Roi.h$ and $Roi.w$ are height and width of the sliding window. $Roi.h$ and $Roi.w$ were defined as,

$$\begin{cases} Roi.h = a * w_c \\ Roi.w = b * h_c \end{cases} \tag{7}$$

where, a and b are parameters to determine height and width of sliding window. In this paper, the range of a and b is [0.2, 0.6] with the change step of 0.05. We fix the step size as [200, 200] and test the effect of different values of a and b on the tongue classification result. Figure 6 shows the change trend of the tongue classification result with different combinations of a and b. In the figure, we set the combination of a and b as $a * b$.

Figure 6 shows when the value of a is 0.55 and b is 0.5 with the step size [200, 200], the result of classification performs best. So, they are the parameters we used in our research. The result by using this method to select regions shows in Table 2.

Table 2. Result of sliding window for pre-processing.

Method	ACC	TPR	TNR
Slide window	87.50%	79.31%	96.29%

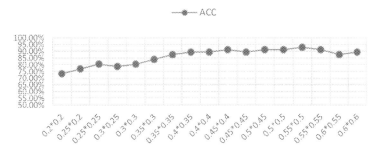

Fig. 6. The tongue classification result with different combinations of *a* and *b*.

Using this method, the cracked tongue with clearly cracks such as long and sharp cracks could be classified easily but it also could not classify the cracked tongue with shallow cracks (like Fig. 7(a)). From the result, it proves that the classification with this method couldn't recognize cracked tongue with chapped thick tongue coating (Fig. 7(b)) and non-cracked tongue with lines caused by thick tongue coating (Fig. 7(c)). Compared to WLDF method, classification in this way seemingly does not rely on the length of cracks.

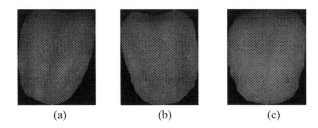

| (a) | (b) | (c) |

Fig. 7. Images are difficult to distinguish based on sliding window. (a) is a cracked tongue with shallow crack; (b) is a cracked tongue with thick tongue coating; (c) is a non cracked tongue with lines caused by thick tongue coating.

Before using WLD method to select regions, the tongue images also were cropped by using the formula (1). The WLD method detect lines on cropped images at first, then generating the suspected cracked region based on the size of the detected line. It is not necessary to use the best parameters in WLD method that would cost long time for processing. What we want is to make sure the line located in the middle of the region. The result of evaluating the classifier trained on regions selected by WLD method shows in Table 3.

Table 3. Result of WLD method for pre-processing.

Method	ACC	TPR	TNR
WLD	89.28%	79.31%	100%

The WLD method could detect suspected cracks accurately but classification performs not good enough on cracked tongue classifying, especially the cracked tongue with shallow cracks as shown in Fig. 8.

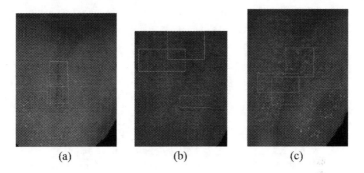

(a)   (b)   (c)

Fig. 8. Cracked tongue images with shallow cracks.

We changed the number of suspected lines selected from cropped images. Choosing 1, 2, 3, 4, 5, 6, 7, 8, 9 suspected regions based on suspected lines in our research and conduct some experiments based on these regions respectively. The result of these nine conditions shows in Table 4.

Table 4. Result of different number of suspected cracked regions.

Number	ACC	TPR	TNR
1	87.50%	79.31%	96.29%
2	91.07%	86.20%	96.29%
3	89.28%	79.31%	100%
4	83.92%	75.86%	92.59%
5	82.14%	79.31%	85.18%
6	85.71%	79.31%	92.59%
7	83.92%	79.31%	88.88%
8	82.14%	75.86%	88.88%
9	82.14%	72.41%	92.59%

From a series of experiments, the classification based on two suspected lines performs better than others but it still difficult to classify cracked tongue with shallow lines.

5   Discussion

In this paper, we proposed a method for cracked tongue recognition that aim to avoid using handcrafted features. We use pre-trained AlexNet on ImageNet and fine-tune it on cracked regions to extract deep features of cracked region. Then, the classification of tongue images is regard as a multiple-instance learning problem, and MI-SVM is used for final decision based on bag of feature vectors. Experiments show that our method performs better than the method based on sliding window or based on handcrafted features. However, the proposed method does not achieve a high accuracy. The future work is to look for new representation learning method for cracked tongue region and take more tongue samples.

References

1. Nestler, G., Dovey, M.: Traditional Chinese medicine. Clin. Obstet. Gynecol. **44**(4), 801–813 (2001)
2. Lao, L., Xu, L., Xu, S.: Traditional Chinese Medicine. pp. 125–135 (2012)
3. Zhang, H.Z., Wang, K.Q., Zhang, D., et al.: Computer aided tongue diagnosis system. In: 27th Annual International Conference of the Engineering in Medicine and Biology Society, IEEE-EMBS 2005, pp. 6754–6757. IEEE (2005)
4. Wang, X., Zhang, D.: A high quality color imaging system for computerized tongue image analysis. Expert Syst. Appl. **40**(15), 5854–5866 (2013)
5. Cui, Q., Li, X., Li, J., Zhang, Y.: Geometric and tongue-mouth relation features for morphology analysis of tongue body. In: Chen, E., Gong, Y., Tie, Y. (eds.) PCM 2016. LNCS, vol. 9917, pp. 490–497. Springer, Cham (2016). https://doi.org/10.1007/978-3-319-48896-7_48
6. Liu, L.L., Zhang, D., Kumar, A., et al.: Tongue line extraction. In: 19th International Conference on Pattern Recognition, ICPR 2008, pp. 1–4. IEEE (2008)
7. Hittawe, M.M., Muddamsetty, S.M., Sidibé, D., et al.: Multiple features extraction for timber defects detection and classification using SVM. In: IEEE International Conference on Image Processing (ICIP), pp. 427–431. IEEE (2015)
8. Liu, L., Zhang, D., You, J.: Detecting wide lines using isotropic nonlinear filtering. IEEE Trans. Image Process. **16**(6), 1584–1595 (2007)
9. Liu, L.L., Zhang, D.: Extracting tongue cracks using the wide line detector. In: Zhang, D. (ed.) ICMB 2008. LNCS, vol. 4901, pp. 49–56. Springer, Heidelberg (2007). https://doi.org/10.1007/978-3-540-77413-6_7
10. Hu, Y., Zhang, W., Lu, H., et al.: Wide line detection with water flow. In: 2016 IEEE International Conference on Bioinformatics and Biomedicine (BIBM), pp. 1353–1355. IEEE (2016)
11. Li, X., Wang, D., Cui, Q.: WLDF: effective statistical shape feature for cracked tongue recognition. J. Electr. Eng. Technol. **12**(1), 420–427 (2017)
12. http://caffe.berkeleyvision.org/
13. Krizhevsky, A., Sutskever, I., Hinton, G.E.: Imagenet classification with deep convolutional neural networks. In: Advances in Neural Information Processing Systems, pp. 1097–1105. (2012)
14. Andrews, S., Tsochantaridis, I., Hofmann, T.: Support vector machines for multiple-instance learning. In: Advances in Neural Information Processing Systems, pp. 577–584 (2003)
15. Doran, G., Ray, S.: A theoretical and empirical analysis of support vector machine methods for multiple-instance classification. Mach. Learn. **97**(1–2), 79–102 (2014)

Multitask Learning for Chinese Named Entity Recognition

Qun Zhang[✉], Zhenzhen Li, Dawei Feng, Dongsheng Li, Zhen Huang,
and Yuxing Peng

National University of Defence Technology, Changsha, China
{zhangqun16,lizhenzhen14,dsli,huangzhen,pengyuxing}@nudt.edu.cn,
davyfeng.c@gmail.com

Abstract. Named Entity Recognition (NER) for Chinese corpus such as social media text and medical records is a grand chanllenge as the entity boundary is not easy to be accurately clarified. In this work, we describe and evaluate a character-level tagger for Chinese NER, which incorporates multitask learning, self-attention and multi-step training methods to exploit richer features and further improve the model performance. The proposed model has achieved 90.52% strict F1 on the Electronic medical records dataset (CCKS-NER 2017), which is the best single model at present. In addition, we also conducted experiments on a Chinese Social Media dataset and the CCKS-NER 2018 dataset, whose results illustrate the effectiveness of the proposed method for Chinese Named Entity Recognition task.

Keywords: Named entity recognition · Multitask learning
Electronic medical records · Social media

1 Introduction

Named entity recognition (NER) is a challenging task in NLP, which aims to identify entities from raw text, and classifies the detected entities into one of predefined categories such as person, organization, location, etc. It is a core component of a wide range of downstream applications, such as information extraction and knowledge base construction. In Chinese named entity recognition, Social Media and electronic medical records datasets have appeared one after another. Recently, much progress has been made in NER using Deep Neural Network (DNN) based approaches, the main method converts the NER problem into a sequence labeling problem and uses the encoder-decoder model for labeling. In general, sequence labeling models consist of three steps: Firstly, current neural models generally make use of word embeddings, which allow them to learn similar representations for semantically or functionally similar words. Secondly, use LSTM, CNN, or GRU to encode the pre-trained word embeddings (character embeddings) and extract features. Finally, sentence level tagging is performed according to the hidden layer features obtained by the encoder layer.

© Springer Nature Switzerland AG 2018
R. Hong et al. (Eds.): PCM 2018, LNCS 11165, pp. 653–662, 2018.
https://doi.org/10.1007/978-3-030-00767-6_60

To improve the accuracy of the model, many improvements aim at the step 2: Designing modified coding methods or rich vector features for specific datasets. The generalization of these methods is weak, because there are also differences between languages. For example, when we train a Chinese NER model, it is more difficult to determine the boundaries of entities, such as: "师范大学附中" and "无症状性心肌缺血病史" (We label "asymptomatic myocardial ischemic disease" as "disease"). The former is difficult to determine the entity boundary due to abbreviation and entity nesting problems; and the latter is difficult because of the overlap between the conventional vocabulary. These problems can be well solved if a rule-based approach is employed, however it's difficult for a deep learning model to fit well alone.

In this work, we propose three improvements to overcome the problem of inaccurate Chinese entity boundary: (1) A multitask training framework is introduced to improve the LSTM encoding without increasing labeling data. We use the current character to predict its previous and next character, and further excavate the relevant information between characters. (2) A self-attention mechanism is conducted to align the encoded sequence after the encoder layer, such that the local features of the sequence is obtained. (3) A multi-step training is adopted to balance the positive and negative sample proportions, which makes the model more focused on the entity and its surroundings at the same time.

Our contributions are two-fold: (1) We propose a framework that incorperates multitask training, self-attention and multi-step training to effectively improve the performance of Chinese named entity recognition task; (2) Experiment results on several Chinese NER datasets achieve the best performing accuracy of single models. The remainder of this paper is organized as follows. Sect. 2 discusses related work. Sect. 3 gives an overview of baseline model, and improvements to the model. Sect. 4 presents the experiments and evaluation results. Finally, Sect. 5 concludes this paper and discusses directions for future work.

2   Related Work

After Collobert et al. [4,5] applied the deep neural network to the named entity recognition task, neural networks replace traditional method gradually. In this section, we introduce the research progress on named entity recognition technique and some works that inspire us.

Deep Learning for Named Entity Recognition. In [9], Lample et al. introduced hierarchy in the architecture by replacing hand-engineered features in prior works with additional bidirectional LSTM encoders and became the main method of NER for deep learning. Subsequently, Chiu and Nichols [3] improved model by using LSTM-CNN and achieved the best results on the CoNLL-2003 and OntoNotes 5.0 datasets. Previous research [20] also proved that CNN could effectively extract feature information. In other related work, [13,14] respectively tried to improve decoding with RNN and LSTM. However, experimental results [3,9] show that CRF decoding is better. Ma et al. (2016) [12] pointed out that it is beneficial to consider the corelations between labels in neighborhoods and jointly decode the best chain of labels for a given input sentence. Shen et al. [21] proposed a more efficient model and joined with active learning.

Self-Attention. At present, most reading comprehension models have joined the self-attention mechanism [2,8]. Vaswani et al. [23] replaced CNN and LSTM with attention in machine translation, and improved the accuracy and efficiency in experiments. In addition, self-attention is used in semantic role labelling (Tan [20]), text classification (Yang [24]), natural language interaction tasks (Parikh [15]) and sentence-level embedding (Lin [10]). Compared with LSTM, attention can record the dependencies between words within a long context. Our experiments show that self-attention is also effective in other sequence labeling tasks.

Multitask Learning. Although multitask learning is an important direction for deep learning, only a small part of the work has studied how to use multitask learning more effectively to train neural network (Ruder [19]). At present, there are few works use multitask learning methods in named entity recognition. The representative work includes: Luong et al. [11] based on the sequence labeling model; Peng and Dredze [16,17] combined Chinese word segmentation and Chinese named entity recognition for joint training, and used the feature information of word segmentation to enhance the effect of NER. They added auxiliary task to the sequence labeling problem which is very relevant to the main objective function, so the model can learn a better feature representation during training.

3   Named Entity Recognition Model

In this paper, we choose the BiLSTM-CRF [9] as the baseline mode, since it is the most widely studied model and has shown its competitive performance in various NER dataset evaluations. This section consists of three parts: data representation, the BiLSTM-CRF model and our contributions to the baseline model. Components of the BiLSTM-CRF method as well as the proposed modules are illustrated in Fig. 1.

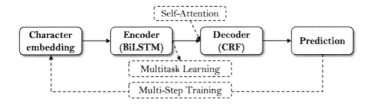

Fig. 1. Illustration of BiLSTM-CRF model, and the improvements, i.e. multitask learning, self-attention module and multi-step training, we add to it.

3.1   Data Representation

The data representation method refers to Shen et al. [21]. We use special [BOS] and [EOS] tokens at the beginning and the end of the sentence, respectively, and add [PAD] tokens at the end of sentences to make their lengths uniform inside a bucket. Different from existing English model, in the Chinese model, we only use the Chinese character-level embedding, as it is equivalent to the English word-level embedding. In this way, the error caused by word segmentation will not propagate into the following model training and testing phases.

3.2   Bidirectional LSTM Model

Character-Level Encoder. We use BiLSTM (Dyer et al. [6]) to encode character vectors. Formally, let S represents a Chinese sequence with n tokens, then its character-level embeddings can be given as following:

$$S = (c_1, c_2, ...c_n). \tag{1}$$

Here c_t is a vector standing for a d dimensional character embedding for the t-th Chinese character in the sentences. Character embedding will then be used as input to two LSTM encoders to generate a specific contextual representation. And we concatenate each $\overrightarrow{h_t}$ with $\overleftarrow{h_t}$ to obtain a hidden state h_t.

$$\overrightarrow{h_t} = LSTM(c_t, \overrightarrow{h_{t-1}}) \qquad \overleftarrow{h_t} = LSTM(c_t, \overleftarrow{h_{t+1}}) \qquad h_t = [\overrightarrow{h_t}; \overleftarrow{h_t}]. \tag{2}$$

For simplicity, we note all the n h_ts as H, each h_t is a $2d$ dimension vector.

$$H = (h_1, h_2, ..., h_n). \tag{3}$$

Tag Decoder. Although result in [21] indicates that the LSTM decoder is better than the CRF, CRF is still the main method for tag decoding. Given that y is a sequence of labels $y = (y_1, y_2, ..., y_n)$, then the CRF score for this sequence can be calculated as:

$$S(y) = \sum_{t=0}^{T} A_{t,y_t} + \sum_{t=1}^{T} B_{y_t,y_{t+1}}, \tag{4}$$

$$A_{t,y_t} = W_{t,y_t} h_t + b. \tag{5}$$

The CRF consists of two parts: the confidence matrix A_{t,y_t} obtained by the hidden layer and the state transition matrix $B_{y_t,y_{t+1}}$ between the labels y_t and y_{t+1}. The parameters W and b can be learned during training.

3.3 Model Improvements

Multitask Learning. Inspired by a simple strategy from Rei [18], we add an auxiliary task to the existing language model by predicting the next word of each character in the sentences. The implementation is simple and does not require additional sample annotation. Feature representation improves during the training of the model. Firstly, the hidden representations from forward and backward-LSTMs are mapped to a new space using a non-linear layer, and then we use a softmax layer to predict the preceding and the following word:

$$\overrightarrow{m_t} = tanh(\overrightarrow{W}_m \overrightarrow{h_t}) \qquad P(w_{t+1}|\overrightarrow{m_t}) = softmax(\overrightarrow{W}_q \overrightarrow{m_t}), \tag{6}$$

$$\overleftarrow{m_t} = tanh(\overleftarrow{W}_m \overleftarrow{h_t}) \qquad P(w_{t-1}|\overleftarrow{m_t}) = softmax(\overleftarrow{W}_q \overleftarrow{m_t}). \tag{7}$$

The objective function for both components is then constructed as a regular language mode objective, by calculating the negative loglikelihood of the next word in the sequence. Finally, we combine the named entity recognition training target with the objective functions based on forward and backward prediction to a new multi-task learning goal.

$$\tilde{E} = E + \gamma(\overrightarrow{E} + \overleftarrow{E}). \tag{8}$$

Here E represents the objective function of the sequence labelling problem, and γ is a hyperparameter.

Self-Attention. Self-attention will generate the feature representation of each Chinese character based on the sequence, which further improves the recognition accuracy of the entity boundary. Our inspiration comes from [1,23]. In contrast, we do not use the multi-headed attention module given in [23], because it brings more parameters that are not suitable for the entity recognition model. Here we use $H = (h_1, h_2, ..., H_n)$ to represent the hidden layer feature matrix output by LSTM, then the self-attention matrix A of the sequence H is calculated as following:

$$A = softmax(tanh(HWH^T)). \tag{9}$$

We compute the entire sequence representation M by multiplying the self-attention matrix A and LSTM hidden states H.

$$M = AH. \tag{10}$$

Finally, the hidden states H and M are concatenated and fed into CRF. The entire model architecture is shown in Fig. 2.

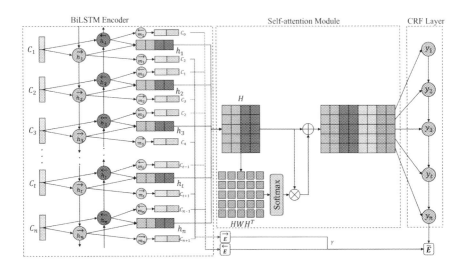

Fig. 2. Main architecture of the network. c_t is character-level embedding. We use BiL-STM to obtain the context features of the sequence, and then feed into self-attention module. Finally, we use CRF for decoding.

Multi-step Training. In named entity recognition task, the proportion of positive and negative samples is uneven, and the entity labels are sparse. For example: On the Conll 2003 entity identification data set [22] only 17% of words represent entities. Unlike machine translation or reading comprehension tasks where the long-term information plays an important role, in the named entity recognition task it is more concerned with local information around a character. Therefore, we consider a two-step model to complete this task.

- Step I uses a simple BiLSTM-CRF model to identify the boundaries of entities;
- Step II uses the boundary results of Step I as an input, with a slightly expanded entity boundary(e.g. adding several characters before and after the detected entity boundary). In this way, model in Step II can learn more accurate position and entity-type information.

4   Experiments and Results

4.1   Dataset

[7,16,17] describe the details of the Chinese Social Media data set. The corpus is composed of Sina Weibo messages annotated for NER. In CCKS-CNER 2017 dataset, there are 400 labeled electronic medical records (EMR), and 2205 unlabeled records. Table 1 shows the statistics of the EMR. We divided training set and the test set with a ratio of 9 : 1. In CCKS-CNER 2018, there are 600 labeled records with 5 types of entities as the same in CCKS-CNER 2017. Among them, 480 records are chosen for training, 60 for validation and the other 60s for testing.

Table 1. Statistics of entity on different categories in CCKS-NER 2017.

Training set	Body	Symptom	Treatment	Disease	Exam
General items (300)	181	558	0	74	1
Medical history (300)	6373	4608	138	570	5902
Diagnosis & Treatment (299)	875	547	902	74	794
Discharge summary (299)	3290	2118	8	4	2849
Summary	10719	7831	1048	722	9546

4.2 Implementation Details

In the experiment, we utilize the BIOES tags for entity boundaries. The evaluations are conducted on the 2017 CCKS-CNER dataset, which output micro-average precisions (Prec.), recalls (Rec.) and F1-scores (F1) under two criteria: Strict Metrics - checks whether the boundary and category of an entity is exactly matched with a gold one; Relaxed Metrics - only considers the boundary of an entity is overlapped with a gold one of same category "strict" is the primary one. Experiments mainly refer to strictly F1. We use 300-dimensional word2vec to represent each Chinese character. Characters not encoded by word2vec are replaced with [UNK].

In order to batch the computation, sentences with similar length are grouped together into buckets. Specific experimental parameters are as follows: The LSTM layer size was set to 300 in each direction for character-level components, and sentences were grouped into batches of size 20. In addition, we add a few characters before and after the results of Step I in multi-step training to ensure that the entity is completely covered, and the best character number is 2 as shown in experiments.

4.3 Results

Table 2 shows the result of different modules on CCKS-NER 2017. The first row is a baseline model presented in [9].

- In second row, multitask learning improves the model by 0.66% obviously.
- From the third row, a slight improvement can be observed on the self-attention mechanism. The effectiveness of a direct use of the self-attention on the entire sentence is not obvious, as the NER relies more on the local feature representation.
- The fourth row shows the performance of multi-step training, where a notable improvement of 0.73% is demonstrated. Although, the Prec. is slightly decreased, the Rec. improves significantly.
- Overall, our model achieves 90.52% on the test set (the best single model result). Table 3 shows the result of other methods.

Table 2. Performance comparison of different modules on CCKS-NER 2017 dataset. Evaluation index include precisions, recalls and F1-scores (F1) under two criteria: Strict Metrics and Relaxed Metrics

Method	Strict(%)			Relaxed(%)			Increase
	Prec.	Rec.	F1	Prec.	Rec.	F1	
BiLSTM-CRF(Baseline)	88.57	89.33	88.95	94.21	95.02	94.61	
+mult-task	88.91	90.32	89.61	94.38	95.51	94.94	0.66
+self-attention	88.68	89.36	89.02	94.15	95.05	94.59	0.07
+mult-step	88.42	90.98	89.68	93.95	95.34	94.63	0.73
+mult-task+mult-step	89.02	91.52	90.25	94.47	95.62	95.04	1.30
+mult-task+mult-step+self-att	89.24	91.83	90.52	94.51	95.72	95.11	1.57

Table 3. The main method test on the CCKS-NER dataset, mt, att and ms denote multitask learning, self-attention and multi-step training, respectively. Consistency Check: Manually verify the result.

	Method	Strict F1(%)
Jianglu Hu et al.	BI-LSTM-FEA (Ensemble)	91.03
Jinhang Wu et al.	ReSeg+LSTM-CRF (Consistency Check)	90.82
En Ouyang et al.	Bi-RNN-CRF (Ensemble)	90.39
Yuhang Xia et al.	Bi-LSTM-CRF (Ensemble)	89.88
Yanxu Chen	CRF (Feature-rich)	89.74
Our	Bi-LSTM-CRF (mt+att+ms)	90.52

Table 4 shows the results on the Chinese Social Media test data. Our method outperforms Peng et al. [17], which increases by 2.72% compared to baseline. It is worth mentioning that the Precision and Recall of our model are close to each other.

Table 5 is the result of CCKS-NER 2018. We can see that multitask learning works well, and self-attention and multi-step training shows a slight increase. To sum up, multitask learning exhibits consistent improvement to NER on all three datasets, and the other two methods are beneficial as well.

5   Conclusion

In this paper, we propose a character-level tagger for Chinese NER, that incorporates multi-task learning, self-attention and multi-step training methods to exploit richer features. The proposed model has achieved 90.52% strict F1 (the best single model at present) on the CCKS-NER 2017. Further, experiment results on a Chinese Social Media dataset and the CCKS-NER 2018 dataset also demonstrate the effectiveness of the proposed method for Chinese Named Entity Recognition task. As for future work, it is possible to combine the Chinese word

Table 4. NER results for named and nominal mentions on Chinese Social Media test data. Since the entities in named and nominal mentions are similar, (*) are averaged by named and nominal mentions results.

Method	Precision	Recall	F1
BiLSTM-MMNN [He et al. 2017]	68.08*	42.08*	51.44
Word Seg (LSTM)+NER (CRF) [Peng et al. 2016]	63.33	39.18	48.41
CRF [Peng et al. 2015]	63.84	29.45	40.38
Bi-LSTM-CRF (Baseline)	52.85	44.00	48.02
Bi-LSTM-CRF (mt+att+ms)	58.76	44.65	50.74

Table 5. NER results on CCKS-NER 2018 training data.

Method	Precision	Recall	F1
Bi-LSTM-CRF (baseline)	87.37	88.56	87.96
+mult-task	88.88	88.45	88.66
+mult-task+mult-step	89.13	88.52	88.82
+mult-task+mult-step+self-att	89.38	88.49	88.93

segmentation and the NER to train a multi-task learning model, as the word segmentation feature as well as the character-level feature will benefit each other model through training.

References

1. Bahdanau, D., Cho, K., Bengio, Y.: Neural machine translation by jointly learning to align and translate. arXiv preprint arXiv:1409.0473 (2014)
2. Cheng, J., Dong, L., Lapata, M.: Long short-term memory-networks for machine reading. In: Proceedings of the 2016 Conference on Empirical Methods in Natural Language Processing, pp. 551–561 (2016)
3. Chiu, J.P.C., Nichols, E.: Named entity recognition with bidirectional LSTM-CNNs. Trans. Assoc. Comput. Linguist. **4**, 357–370 (2016)
4. Collobert, R., Weston, J.: A unified architecture for natural language processing: deep neural networks with multitask learning. In: International Conference on Machine Learning, pp. 160–167 (2008)
5. Collobert, R., Weston, J., Bottou, L., Karlen, M., Kavukcuoglu, K., Kuksa, P.: Natural language processing (almost) from scratch. J. Mach. Learn. Res. **12**(Aug), 2493–2537 (2011)
6. Dyer, C., Ballesteros, M., Ling, W., Matthews, A., Smith, N.A.: Transition-based dependency parsing with stack long short-term memory. arXiv preprint arXiv:1505.08075 (2015)
7. He, H., Sun, X.: A unified model for cross-domain and semi-supervised named entity recognition in Chinese social media. In: AAAI, pp. 3216–3222 (2017)
8. Hu, M., Peng, Y., Qiu, X.: Reinforced mnemonic reader for machine comprehension. CoRR, abs/1705.02798 (2017)

9. Lample, G., Ballesteros, M., Subramanian, S., Kawakami, K., Dyer, C.: Neural architectures for named entity recognition. In: Proceedings of NAACL-HLT, pp. 260–270 (2016)
10. Lin, Z., et al.: A structured self-attentive sentence embedding. arXiv preprint arXiv:1703.03130 (2017)
11. Luong, M.-T., Le, Q.V., Sutskever, I., Vinyals, O., Kaiser, L.: Multi-task sequence to sequence learning. arXiv preprint arXiv:1511.06114, 2015
12. Ma, X., Hovy, E.: End-to-end sequence labeling via bi-directional LSTM-CNNs-CRF. In: Proceedings of the 54th Annual Meeting of the Association for Computational Linguistics, vol. 1, pp. 1064–1074 (2016). Long Papers
13. Mesnil, G., He, X., Deng, L., Bengio, Y.: Investigation of recurrent-neural-network architectures and learning methods for spoken language understanding. In: Interspeech, pp. 3771–3775 (2013)
14. Nguyen, T.H., Sil, A., Dinu, G., Florian, R.: Toward mention detection robustness with recurrent neural networks. arXiv preprint arXiv:1602.07749 (2016)
15. Parikh, A.P., Täckström, O., Das, D., Uszkoreit, J.: A decomposable attention model for natural language inference. arXiv preprint arXiv:1606.01933 (2016)
16. Peng, N., Dredze, M.: Named entity recognition for Chinese social media with jointly trained embeddings. In: Proceedings of the 2015 Conference on Empirical Methods in Natural Language Processing, pp. 548–554 (2015)
17. Peng, N., Dredze, M.: Improving named entity recognition for chinese social media with word segmentation representation learning. arXiv preprint arXiv:1603.00786 (2016)
18. Rei, M.: Semi-supervised multitask learning for sequence labeling. arXiv preprint arXiv:1704.07156 (2017)
19. Ruder, S.: An overview of multi-task learning in deep neural networks. arXiv preprint arXiv:1706.05098 (2017)
20. Santos, C.D., Zadrozny, B.: Learning character-level representations for part-of-speech tagging. In: Proceedings of the 31st International Conference on Machine Learning (ICML-14), pp. 1818–1826 (2014)
21. Shen, Y., Yun, H., Lipton, Z.C., Kronrod, Y., Anandkumar, A.: Deep active learning for named entity recognition. In: ACL 2017, p. 252 (2017)
22. Tjong, E.F., Sang, K., De Meulder, F.: Introduction to the CoNLL-2003 shared task: language-independent named entity recognition. In: Proceedings of the Seventh Conference on Natural Language Learning at HLT-NAACL 2003-Volume 4, pp. 142–147. Association for Computational Linguistics (2003)
23. Vaswani, A., et al.: Attention is all you need. In: Advances in Neural Information Processing Systems, pp. 6000–6010 (2017)
24. Yang, Z., Yang, D., Dyer, C., He, X., Smola, A., Hovy, E.: Hierarchical attention networks for document classification. In: Proceedings of the 2016 Conference of the North American Chapter of the Association for Computational Linguistics: Human Language Technologies, pp. 1480–1489 (2016)

Sparse-Region Net: Local-Enhanced Facial Depthmap Reconstruction from a Single Face Image

Haoqian Wang[1,2]([⊠]), Shuhao Zhang[1], Xingzheng Wang[1], and Yongbing Zhang[1]

[1] Graduate School at Shenzhen, Tsinghua University, Shenzhen 518055, China
wanghaoqian@tsinghua.edu.cn
[2] Shenzhen Institute of Future Technology, Shenzhen 518071, China

Abstract. In this paper, we propose a novel end-to-end deep neural network for region-enhanced depthmap reconstruction from a single face image. Unlike most of popular depthmap reconstruction methods, the proposed network fully takes the region information of RGB images into consideration, and thus results in more accurate correspondences. Specially, we treat the depthmap reconstruction task as an independent problem, investigating to improve facial depth regression through multi-task learning (facial landmark detection and mask detection), and then a attention region training technique to utilize the rich texture information in some semantic regions to infer local detailed depth information. Next, we describe a zigzag dilated convolution framework to sparse the kernel size of our network, which will enlarge the network's receptive field and simultaneously avoid gridding artifacts. In training, we also propose a new method for setting the learning rates, called discrete cyclical learning rates, to improve the results. Experimental results illustrate that the proposed Sparse-Region Net (SRN) outperforms the state-of-the-art baseline methods by a large margin.

Keywords: Facial depth prediction · Dilated convolution
Multi-task learning · Sub-region attention enhancing
Facial reconstruction

1 Introduction

Depthmap reconstruction is a classic and fundamental problem in computer vision. For some important and emerging domains such as industrial robots, VR, AR and medical 3D reconstruction, accurate depth information would be a key for understanding high-level hierarchy structures. In the industry, depth sensors (e.g. LiDARs, Kniect, and stereo camera) are expensive or less accurate. In the academic field, traditional methods extract depth cues directly from multi-view, multi-lighted or multi-focusing images [1,8], using hand defined features which lack of quality of results and speed. Recently, with the fast development of deep neural networks, the large amount of existing data make this problem

© Springer Nature Switzerland AG 2018
R. Hong et al. (Eds.): PCM 2018, LNCS 11165, pp. 663–673, 2018.
https://doi.org/10.1007/978-3-030-00767-6_61

better solved on the condition that the correlation between 2D and 3D is learned from large training data [4–7]. However, these methods pay little attention to the study of detail features of a image, which would restrict them to get better results.

In this paper, we propose a novel deep neural network called "Sparse-Region Net (SRN)" for facial depthmap reconstruction. In real-life, facial image is an important category in the daily recorded images and its depth information contributes to many tasks like face reconstruction, expression tracking and recognition. In view of this problem, our network fully considers the detail features and learns to directly predict depthmap, landmark location and plane segmentation masks from a single facial RGB image, which simultaneously perceives the global and edge information of a facial image. Next, inspired by the popular attention model [23], we propose to take the local organs like eyes, nose and mouth as our attention regions, as these regions are more discriminative to the human vision system to judge an expression or identity a person. Based on the predicted landmark location, we train on these special attention regions for depthmap prediction to improve the final result. On the network structure, unlike [4–7], we sparse the kernel size of our network to enlarge the receptive field and simultaneous enhance the connection between non-zero kernels, which makes our network sense more tiny feature. Furthermore, following a recent work on setting learning rates [16], we define a discrete cyclic learning rate that is fully symmetric and easy to deploy.

Our main contributions can be summarized as follows. First, we propose a novel Sparse-Region Net (SRN) for depthmap reconstruction. "sparse" means that we deploy zigzag dilated convolution framework to sparse the convolution kernels. "region" means that we fully consider detail features by proposing to fuse multi-task learning with attention region learning. Second, we propose discrete cyclic learning rates for training our SRN to get the better results. Third, we propose a novel training loss that enforces the fitting between depth and local region information.

2   Related Work

Depth Prediction. For depth prediction from single image, Saxena et al. [3] propose the first supervised learning-based solution, where hand-crafted features in multiple scales are required to predict depth. Based on deep neural networks, Eigen et al. [4] propose a multi-scale network on the basis of their previous work to refine the depth results, but their results are low resolution and their network is shallow. Cao et al. [5] bring forward a new theory that depth prediction is also capable of being regarded as a classification problem, and they modify their loss function for training. Xu et al. [6] make use of Conditional Random Field(CRF) in the network architecture to refine results. And Laina et al. [7] present a ResNet-based architecture that contains several kinds of highway-architecture. In an unsupervised manner, Godard et al. [8] propose a left-right consistency framework to predict depth, but their local information lacks accuracy.

Single View Face Reconstruction. Reconstructing 3D face from single image is an ill-posed problem. Some methods relay on an initial template model, followed by the Shape from Shading (SFS) technique which deforms the template to generate more detailed 3D faces [9]. Recently, deep learning techniques are used in face reconstruction. Laine et al. [10] train a network that can directly map a face image to a 3D face represented in a low dimensional (160 dimensions) space. Liu et al. [11] reconstruct a dense face mesh, where high frequency facial features are preserved. Richardson et al. [12] use the SFS technique to construct the loss functions in the training, and thus facial details could also be reconstructed. Jackson et al. [13] utilize hourglass network to directly map 2D facial image to 3D facial model, but their results lack texture features.

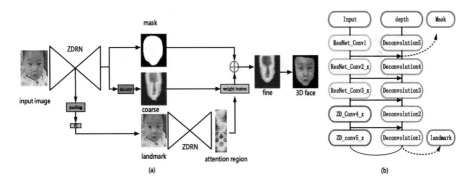

Fig. 1. a: The overview of SRN that predicts depth map, landmark location and segmentation mask. ZDRN means zigzag dilated residual network. b: the description of the proposed ZDRN.

3 Sparse-Region Net

3.1 ZDRN (Zigzag Dilated Residual Network)

In Fig. 1, the proposed SRN contains two ZDRN, the first one aims at multi-task training, while the other aims at attention region training. All of them are fed with 384×384 RGB images. The encoding part, which is based on ResNet-101 and initialized with the pretrained weights, captures global information and decreases the input image resolution to obtain $2048 \times 12 \times 12$ feature maps. The decoding part consists of five 4×4 transposed convolutional layers in which each layer doubles the feature map, thus guiding the network to predict the final depth map.

Zigzag Dilated Convolution Framework: Dilated convolutions recently go down well with semantic segmentation [14]. Actually, dilated convolution is a powerful tool to enlarge the convolution kernel's receptive field and maintain the resolution by thinning convolution kernels with zeros. Specially, if the input feature map x is sampled by a dilated convolution with a w filter, for each location i on the output y can be expressed as:

$$y\{i\} = \sum_k x\{i + r \cdot k\} w\{k\} \tag{1}$$

where r is the dilated rate, therefore, one can change the receptive field of the kernel by setting different dilated rates without loss of resolution.

However, for the encoding part, simply using dilated convolution will cause "gridding problem" (Fig. 2 middle) in deeper network, because zero-padded convolution kernel will only sample those pixels related to non-zero locations of the kernel, losing precious information among adjacent pixels. Generally, the gridding problem becomes increasingly fierce along with the dilation rate grows in high layers as the convolutional kernel in there is too sparse to blanket any local information, leading to severe information loss. Here, in Table 1, inspired by [15], we process the "conv4_x" and "conv5_x" of ResNet-101 by zigzag dilated rates (r=1,2,5,9 for the 3×3 convolutional kernels of "conv4_x" repeatedly and $r = 5, 9, 17$ for the 3×3 convolutional kernels of "conv5_x"). The dilation rates are selected by "maximum distance [15]" between two nonzero values, which is defined as:

$$M_i = max\{M_{i+1} - 2r_i, r_i, -(M_{i+1} - 2r_i)\} \tag{2}$$

For N convolutional layers with kernel size K x K dilated by $[r_1,...,r_i,...,r_n]$, let M_n equal to r_n. The zigzag dilated rates are designed to let $M_2 < K$.

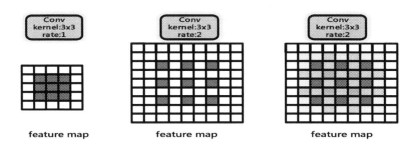

Fig. 2. Left: the standard convolution. Middle: dilated convolution with rate = 2. Right: zigzag dilated convolution with rate = 2.

Table 1. The zigzag dilated convolution framework for ResNet-101.

ResNet-101	conv4_x			conv5_x
Kernel size	3 × 3	3 × 3	3 × 3	3 × 3
Location	(res3b3) to (res4b2)	(res4b3) to (res4b18)	(res4b19) to (res4b21)	(res5a) to (res5c)
Rate	(2, 2, 5, 9)	(1, 2, 5, 9)*4	(1, 2, 5)	(5, 9, 17)

3.2 Multi-task Training

The proposed multi-task training consists three independent branches for the following specific missions:

The first one is facial depth branch. This branch stars with the ZDRN mentioned above, followed by a deconvolution layer to produce a one-channel depthmap. we define the loss for the branch as the sum of squared depth differences between the predicted depthmap and ground-truth:

$$L^d = \frac{1}{I_d} \sum_{p \epsilon I_d} M^{(p)}(D^{(p)} - D^{*(p)})^2 \tag{3}$$

where M denotes the facial mask indicating the face area in the RGB image, I_d is the total number of pixels of M. For each pixel p, $D^{(p)}$ and $D^{*(p)}$ denote the depth value and the corresponding ground truth value.

The second branch is facial landmark. Different from [4–7], we further explore the high-level prior knowledge by involving additional tasks. This branch starts with the end of ZDRN's encoding part, followed by a pooling layer and a fully connected layer to produce coordinates of the landmark location. The encoding part for landmark feature extraction is shared with the aforementioned basic network. For this branch, we define the loss as the sum of squared coordinates differences:

$$L^l = \frac{1}{I_l} \sum_{p \epsilon I_l} [(X^{(p)} - X^{*(p)})^2 + (Y^{(p)} - Y^{*(p)})^2] \tag{4}$$

$X^{(p)}$ and $Y^{(p)}$ denote the predicted coordinates, while $X^{*(p)}$ and $Y^{*(p)}$ denote the ground truth coordinates (obtained by the landmark detection model[1]), I_l is the total number of landmarks.

The third one is facial mask branch. We incorporate a facial mask detection task in the proposed multi-task model, this is very useful for handling real recorded images as it can be used to eliminate the wrong depth values assigned to the background. The network structure of the facial mask detection is shared with the facial depth regression network expect the last transposed convolutional layer. The loss is defined as follow:

$$L^m = \frac{1}{I_m} \sum_{p \epsilon I_m} (M^{(p)} - M^{*(p)})^2 \tag{5}$$

---

[1] https://github.com/ageitgey/face_recognition.

where $M^{(p)}$ denotes the predicted probability of pixel p belongs to facial region. $M^{*(p)}$ is the ground truth. I_m is the total number of pixels of mask area. Therefore, our overall training loss can be summarized as follow:

$$L = L^d + \alpha L^l + \beta L^m \tag{6}$$

where α and β control the relative importance between L^d, L^l and L^m. In training, the parameters α and β are 1.

3.3 Attention Region Training

Different from previous depthmap reconstruction networks [4–7], which rely on global region training. We combine this problem with attention region. Attention region training actually simulates the human brain's attention model. For example, when we look at a picture, the whole picture can be seen, but when we look deeply and carefully, the eyes are focused only on a small piece. That is to say, at this time, the attention of the human brain to the whole picture is not balanced, and there is a certain weight distinction.

Here, we approximate the attention region as a square using the landmark location predicted by the multi-task training discussed above. This process can be represented by:

$$[c_x, c_y, h_l] = f(W_c * X) \tag{7}$$

where W_c denotes the mapping parameters (map the input image X to landmark location) of the first ZDRN. $f(\cdot)$ represents the fully-connected layer (Fig. 1) which outputs the coordinates (x, y). Based on (x, y), we can get the center coordinates c_x, c_y and the square's half side length given by h_l. Therefore, the attention region can be represented as:

$$X^{(i)} = X \odot R(c_x, c_y, h_l) \qquad i \in att \tag{8}$$

where $R(\cdot)$ denotes the attention mask to generate the attention region $X^{(i)}$ which belongs to att (eyes, mouth, nose). Therefore, we define the loss for attention training as follow:

$$L^a = \frac{1}{I} \sum_{i \epsilon att} \sum_{p \epsilon I} (D_i^{(p)} - D_i^{*(p)})^2 \tag{9}$$

where I denotes the total number of pixels in corresponding attention region.

4   Experiments

4.1   Discrete Cyclical Learning Rates

For training the neural network, the setting of learning rates is crucial. Some methods [4–7] usually give an initial value, and then decrease it constantly with the increment of iterations. However, this way is difficult do decide when to drop the learning rates. To solve this problem, we propose the discrete cyclical learning rates (DCLR) based on [16] to train our network aforementioned. As shown in Fig. 3, unlike [16], our DCLR policy is fully symmetric easier to deploy. We set the learning rates for training according to:

$$lr(t) = L_b cos(\frac{\alpha}{2b\pi}t) + 2L_b \tag{10}$$

where t denotes the epoch number which belongs to positive integers, and b denotes the batch size for each epoch. α reflects the oscillation of the whole learning rates. In our training, the L_b for multi-task training is 10^{-5} and the L_b for attention region training is 10^{-7}. The batch size is set to 50.

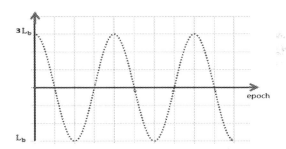

Fig. 3. The discrete cyclical learning rates policy. The longitudinal axes denotes the learning rates. Transverse axes denotes epoch.

4.2   Dataset

In order to train our proposed model, a large dataset of facial images and the corresponding depth maps are required. Therefore, we use 3DMM [17] to generate synthetic data with different perspectives and illumination conditions. See Fig. 4, we use BU-3DFE database and synthetic data for both training, validation and testing. For quantitative evaluation, we compare the ground-truth depth with our predicted depth and compute the mean-square error (MSE) as well as mean-absolute error (MAE) in centimeters (cm).

Fig. 4. The example of our dataset, which contains facial RGB images and depth maps.

4.3   Ablation Study

In Fig. 5 and Table 2, we compare the performance of different key components, containing basic network, multi-task training, attention region training and discrete cyclical learning rates. In Table 2, we see that the ZDRN described in Sect. 3.1 is beneficial and outperforms other architecture by a large margin. Specifically, although our model has shorter depth than Resnet152, the resultant decrease in error far outstrips (mae decreased by 19.7% and mse decreased by 32.3%) the Resnet152 model. Then, we further evaluate the multi-task training and the DCLR policy. It can be observed that the idea of combining 3D landmark detection, facial mask detection in our multi-task learning and training with DCLR policy works well. We then further optimize the final results by fusing the sub-reg depth maps shown in Fig. 5(b). A qualitative comparison between attention and no attention is shown in Fig. 5(c). The results show that training 'SRN' based on 'DCLR' works best.

Table 2. Ablation on the structure. we compare the depth error of different models according to the same evaluation protocols on BU-3DFE dataset.

Evaluation on various key									
Epoch = 200	Key components								
Error	FCN32s	FCN8s	VGG-16	ResNet-50	ResNet-101	ResNet-152	ZDRN	ZDRN + multi-task	SRN + DCLR
MSE	0.534	0.529	0.390	0.221	0.197	0.193	0.165	0.155	**0.139**
MAE	0.471	0.461	0.257	0.085	0.070	0.065	0.047	0.044	**0.036**

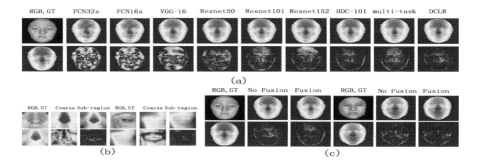

(a)

(b)   (c)

Fig. 5. Qualitative results. a: comparison of the various key components. The first column illustrates the facial RGB images and the corresponding ground truth depth maps, the other columns evince the prediction results and errors of various components. b: results of attention region training. c: results fused with attention region.

4.4   Comparison with State-of-the-Art Methods

We report the obtained experiments and results compared with the state-of-the-art methods, utilizing the open source codes provided by the authors or the results in their papers.

We first compare with HPEN [18], CMSCA [4], FCRN [7] and DRN [20] both qualitatively and quantitatively. It is clear to see that our model outstrips these methods by showing lower errors and more accurate facial depth information visually, such as the second, fourth and sixth lines in Fig. 6(a). As shown in Table 3, our mean absolute error (mae) is lower than DRN-A-50 [20] by 19.2%, lower than FCRN [7] by 37.1%, meanwhile, as for the mean square error(mse), the decrease is 30.8% and 57.6% respectively.

Using the same evaluation dataset, we also compare our method against several competitive methods [2,19,21,22]. As shown in Fig. 6(b), the proposed method has fewer error on the error heat map. This benefits from the multi-task learning and the attention training which utilize the global and local face priors respectively. Figure 6(c) illustrates the results on CelebA dataset.

Table 3. Comparison with state-of-the-art methods

Epoch = 200	Methods				
Error	CMSCA [4]	FCRN [7]	DRN-C-26 [20]	DRN-A-50 [20]	Ours
MAE	0.364	0.221	0.238	0.172	**0.139**
MSE	0.130	0.085	0.098	0.052	**0.036**

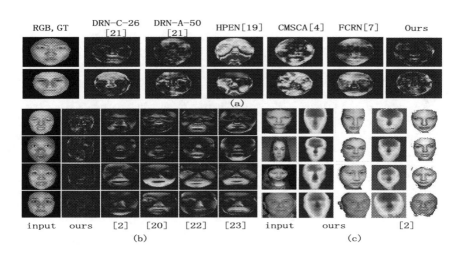

Fig. 6. Visual Comparisons of our methods and existing supervised methods.

672 H. Wang et al.

5   Conclusion

In this paper, we introduce a novel method for depthmap reconstruction from a single facial image, which combines multi-task model with attention region training to optimize results. We further apply zigzag dilated convolutions to residual block and propose discrete cyclical learning rates for training. Experimental results demonstrate the effectiveness of our method.

Acknowledgments. This work is supported by the Guangdong Provincial Science and Technology Project (2017B010110005), the Shenzhen Science and Technology Project under Grant (JCYJ20170817161916238, GGFW2017040714161462).

References

1. Karsch, K., Liu, C., Kang, S.B.: Depth extraction from video using non-parametric sampling. In: Fitzgibbon, A., Lazebnik, S., Perona, P., Sato, Y., Schmid, C. (eds.) ECCV 2012. LNCS, vol. 7576, pp. 775–788. Springer, Heidelberg (2012). https://doi.org/10.1007/978-3-642-33715-4_56
2. Sela, M., Richardson, E., Kimme, R.: Unrestricted facial geometry reconstruction using image-to-image translation. In: ICCV (2017)
3. Saxena, A., Chung, S.H., Ng, A.Y.: 3-D depth reconstruction from a single still image. Int. J. Comput. Vis. **76**, 53–69 (2008)
4. Eigen, D., Fergus, R.: Predicting depth, surface normals and semantci labels with a common multi-scale convolutional architecture. In: ICCV, pp. 2650–2658 (2015)
5. Cao, Y., Wu, Z., Shen, C.: Estimating depth from monocular images as classification using deep fully convolutional residual networks. IEEE Trans. Circuits Syst. Video Technol. **PP**(99), 1 (2016)
6. Xu, D., Ricci, E., Wuyang, W.: Multi-scale continuous CRFs as sequential deep networks for monocular depth estimation. In: CVPR (2017)
7. Laina, I., Rupprecht, C., Belagiannis, V.: Deeper depth prediction with fully convolutional residual networks. In: Fourth International Conference on 3D Vision, pp. 239–248 (2016)
8. Godard, C., Aodha, O.M., Brostow, G.J.: Unsupervised monocular depth estimation with left-right consistency. In: CVPR, pp. 6602–6611 (2017)
9. Kemelmacher-Shlizerman, I., Basri, R.: 3D face reconstruction from a single image using a single reference face shape. IEEE Trans. Pattern Anal. Mach. Intell. **33**, 394–405 (2011S)
10. Laine, S., Karras, T., Aila, T.: Facial performance capture with deep neural networks. arXiv preprint arXiv:1609.06536 (2016)
11. Liu, F., Zeng, D., Zhao, Q., Liu, X.: Joint face alignment and 3D face reconstruction. In: Leibe, B., Matas, J., Sebe, N., Welling, M. (eds.) ECCV 2016. LNCS, vol. 9909, pp. 545–560. Springer, Cham (2016). https://doi.org/10.1007/978-3-319-46454-1_33
12. Richardson, E., Sela, M., Or-El, R.: Learning detailed face reconstruction from a single image. In: CVPR, pp. 5553–5562 (2017)
13. Jackson, A.S., Bulat, A.: Large pose 3D face reconstruction from a single image via direct volumetric CNN regression. In: ICCV, pp. 1031–1039 (2017)
14. Yu, F., Koltun, V.: Multi-scale context aggregation by dilated convolutions. In: ICLR, pp. 1–10 (2016)

15. Wang, P., Chen, P., Yuan, Y.: Understanding convolution for semantic segmentation (2017)
16. Smith, L.N.: Cyclical learning rates for training neural networks. In: Computer Science, pp. 464–472 (2015)
17. Blanz, V., Vetter, T.: A morphable model for the synthesis of 3d faces. In: Proceedings of the 26th Annual Conference on Computer Graphics and Interactive Techniques, pp. 187–194 (1999)
18. Zhu, X., Lei, Z., Yan, D.: High-fidelity pose and expression normalization for face recognition in the wild. In: CVPR, pp. 787–796 (2015)
19. Ren, M., Cao, X., Sun, J.: Face alignment at 3000 fps via regressing local binary features. In: CVPR, pp. 1685–1692 (2014)
20. Yu, F., Koltun, V.: Dilated residual networks. In: CVPR (2017)
21. Jourabloo, A., Liu, X.: Large-pose face alignment via CNN-based dense 3D model fitting. In: CVPR (2016)
22. Zhang, Z., Luo, P., Loy, C.C., Tang, X.: Facial landmark detection by deep multi-task learning. In: Fleet, D., Pajdla, T., Schiele, B., Tuytelaars, T. (eds.) ECCV 2014. LNCS, vol. 8694, pp. 94–108. Springer, Cham (2014). https://doi.org/10.1007/978-3-319-10599-4_7
23. Fu, J., Zheng, H., Mei, T.: Look closer to see better: recurrent attention convolutional neural network for fine-grained image recognition. In: CVPR, pp. 4476–4484 (2017)

Entropy Based Boundary-Eliminated Pseudo-Inverse Linear Discriminant for Speech Emotion Recognition

Dongdong Li[1,2], Linyu Sun[1], Zhe Wang[1,2(✉)], and Jing Zhang[1]

[1] Department of Computer Science and Engineering, East China University of Science and Technology, Shanghai 200237, People's Republic of China
wangzhe@ecust.edu.cn
[2] Provincial Key Laboratory for Computer Information Processing Technology, Soochow University, Jiangsu 215006, People's Republic of China

Abstract. Remarkable advances have achieved in speech emotion recognition (SER) with efficient and feasible models. These studies focus on the ability of the model itself. However, they ignore the potential distributed information of speech data. Actually, emotion speech is imbalanced due to the expression of human being. To overcome the imbalanced problems of speech data, the ongoing work furthers our previous study of the Boundary-Eliminated Pseudo-Inverse Linear Discriminant (BEPILD) model through introducing the information entropy that contributes to describing the distribution of the speech data. As a result, an Entropy-based Boundary-Eliminated Pseudo-Inverse Linear Discriminant model (EBEPILD) is proposed to generate more robust hyperplanes to tackle the speech data with high class uncertainty. The experiments conducted on the Interactive Emotional Dyadic Motion Capture (IEMO-CAP) database with four emotion states show that the EBEPILD has outstanding performance compared with other algorithms.

Keywords: Speech emotion recognition · Imbalanced problems
Information entropy · Pseudo-Inverse Linear Discriminant

1 Introduction

In recent years, with the increasing demand for human-computer interaction, speech emotion recognition (SER) has attracted much attention. It is well-know that emotional information can be analyzed from a variety of sources, among speech is the indispensable way to carry emotional information. In addition, speech not only conveys semantic content, but also contains full emotions. Therefore, for future friendly human-computer interaction and discovering customer sentiment in service industry, it is essential to automatically analyze and distinguish emotions from human voice.

In the past decade, researchers have used various models on SER [3]. For instance, Hu et al. proposed Gaussian Mixture Model (GMM) supervector based

ⓒ Springer Nature Switzerland AG 2018
R. Hong et al. (Eds.): PCM 2018, LNCS 11165, pp. 674–685, 2018.
https://doi.org/10.1007/978-3-030-00767-6_62

Support Vector Machine (SVM) [6] outperforming standard GMM on SER. The Deep Neural Network (DNN)-Hidden Markov Model (HMM) [7] proposed by Li et al. got superior performance compared with GMM-HMM. Han et al. [9] used DNN to extract utterance-level features which can be put into Extreme Learning Machine (ELM) attaining the state-of-the-art result. But most researchers only focus on the ability of the model itself, without paying attention to potential distributed information of samples. In general, the sample distribution of speech emotional data is imbalanced. For example, in the INTERSPEECH 2009 Emotion Challenge [13], the AIBO corpus is designated as a game-specific speech database including various emotions such as joyful, irritated, angry, and neutral. Shockingly, there are 39169 sentences in neutral emotion, but there have only 3 sentences in helpless emotion. The most mainstream multi-modal emotional database, Interactive Emotional Dyadic Motion Capture (IEMOCAP) [2], recorded by the University of South California is also imbalanced. It can be seen that in the real situation, the emotions that human conveys are imbalanced. Corresponding to the task of emotional recognition, the number of certain kinds of emotional utterances is more than that of other kinds of emotional utterances.

Although there are some tricks for dealing with imbalanced problems on SER. For example, some researchers chose the cost sensitive method [10] to solve the non-equilibrium problem or mixed the same attributes of emotional utterances to balance the data [5]. However, more attentions should be paid to the pertinent algorithms that could copy with the imbalanced problems. In our previous study, a novel algorithm named Boundary-Eliminated Pseudo-Inverse Linear Discriminant (BEPILD) [15] was proposed to handle the imbalanced data [8], which was rapid and highly-efficient. In this paper, we further this study and introduce it to SER. An improved version of BEPILD is proposed by employing the information entropy, which is named Entropy based Boundary-Eliminated Pseudo-Inverse Linear Discriminant (EBEPILD). In original BEPILD, the two hyperplanes are determined by the centroids of the two classes samples. To get more robust decision hyperplanes, the information entropy is introduced to assign the weight of each sample, which represents its class certainty. According to the weights of samples, the samples are selected to construct a tighter decision interval. In detail, the low weighted samples will fall into stable areas as much as possible. Conversely, the samples with high weight will fall into confused area as much as possible.

The outline of this paper is organized as follows: Sect. 2 introduces the architecture of SER with the proposed EBEPILD. Section 3 gives the details of Pseudo-Inverse Linear Discriminant(PILD), the calculation method of information entropy and the proposed EBEPILD. The experimental protocol and comparison is presented in Sect. 4. Finally, we give a conclusion in Sect. 5.

2   Framework Overview

Figure 1 shows the architecture of SER in our research. Four emotions (anger, happy, neutral, sadness) are chosen from IEMOCAP to verify our model. Speech features are extracted from both training sets and testing sets. "One vs One"

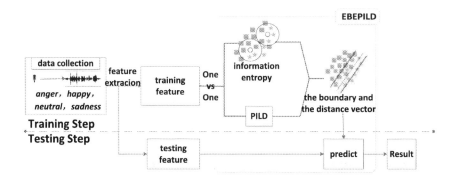

Fig. 1. Entropy based Boundary-Eliminated Pseudo-Inverse Linear Discriminant for SER framework

strategy is employed since the EBEPILD is a binary-class classifier. The proposed EBEPILD combines Pseudo-Inverse Linear Discriminant (PILD) [14] and information entropy theory to obtain more robust hyperplanes. Finally, the testing samples are predicted by the proposed model.

3   Method

3.1   Pseudo-Inverse Linear-Discriminant (PILD)

The PILD [14] is the linear classifier which is used to calculate the optimal w and b in order to form the robust hyperplane, and its decision function can be built as:

$$g(x) = w^T x + b \begin{cases} > 0, x \in \text{class one} \\ < 0, x \in \text{class two} \end{cases}. \tag{1}$$

For the sake of the convenient calculation, weight w and bias b are combined into augmented vectors $\widetilde{w} = [b, w^T]^T$ and the training sample matrix X_{full} is extended to $\widetilde{X_{full}} = [1_{N \times 1}, X_{full}]$. Finally, the linear classification model can be modeled as:

$$\widetilde{y} = \widetilde{X_{full}}\widetilde{w}. \tag{2}$$

If the training matrix is not singular, it is easy to know that the optimal weight vector $\widetilde{w}$ can be directly obtained from the follow equation:

$$\widetilde{w} = \widetilde{X_{full}}^{-1}\widetilde{y}. \tag{3}$$

However, the matrix is often singularity in practical problems. Therefore, the MSE criterion loss function of linear classification model can be defined as follows:

$$L(\widetilde{w}) = \left\| \varphi - \widetilde{y} \right\|_2^2, \tag{4}$$

where φ is the theoretical class labels. In order to make the sum of squared errors approach to zero, gradient descent method is employed to obtain optimal $\widetilde{w}$:

$$\frac{\partial L(\widetilde{w})}{\partial \widetilde{w}} = 2\widetilde{X_{full}}^T (\varphi - \widetilde{X_{full}\widetilde{w}}) = 0. \tag{5}$$

Finally, the optimal weight vector $\widetilde{w}$ can be computed as:

$$\widetilde{w} = (\widetilde{X_{full}}^T \widetilde{X_{full}})^{-1} \widetilde{X_{full}}\varphi. \tag{6}$$

3.2   Information Entropy Based on K Nearest Neighbor

Information entropy [11] is the common metric for measuring purity of data set. Furthermore, the higher entropy represents that the information of samples are more uncertain. So we can take advantage of the information entropy to determine class certainty of data set. Suppose that in the data set D, the proportion of sample belonging to class r is p_r. The information entropy of D can be computed as:

$$E(D) = - \sum_{r=1}^{r} p_r \log_2 p_r, \tag{7}$$

where r is the total number of categories. If the value of $E(D)$ is high, the class certainty of D is low. It is obviously that magnitude of the dataset D determines the value of information entropy. In this paper, we utilize k nearest neighbor to divide the data set for calculating entropy. Because our algorithm is used for handling two-binary imbalanced data, information entropy is also calculated from two classes. In our research, the majority class is defined as

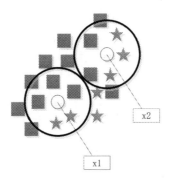

Fig. 2. Method of calculating entropy according to k nearest neighbor. Squares represent negative classes, and Pentagrams represent positive classes. Selecting k nearest neighbor samples of x1 is first step. Secondly, the number of positive class samples and negative class samples are counted in nearest neighbor. Then, we calculate the probabilities of x1 belonging to positive class and negative class. Finally, the entropy is computed from the probabilities. The method of calculating the entropy of x2 is the same as x1.

negative class and the minority class is defined as positive class. As shown in the Fig. 2, for the given training samples $\{x_j, y_j\}_{j=1}^{N}$, $y_j \in \{-1, +1\}$, we firstly choose the k nearest neighbors. Then, the number of the positive samples and negative samples, which are denoted as num_{+j} and num_{-j}, are counted in selected k samples. The probabilities of x_j, which belong to positive class and negative class, can be computed as:

$$p_{+j} = \frac{num_{+j}}{k},$$ (8)

$$p_{-j} = \frac{num_{-j}}{k}.$$ (9)

Finally, according to the probabilities of positive class and negative class, the entropy of x_j can be calculated as:

$$E(x_j) = -p_{+j} \log_2(p_{+j}) - p_{-j} \log_2(p_{-j}).$$ (10)

After obtaining the entropy of each sample, the class certainty is determined by entropy. Considering that the samples with low class certainty contain more information which is conducive to classification, these samples are kept as much as possible. When it comes to the mathematical formula, we set the weights for the each sample. The weight W_j of sample x_j can be built as follow:

$$W_j = \begin{cases} 1, \text{ if } E(x_j) > \text{threshold} \\ 0, \text{ if } E(x_j) \leq \text{threshold} \end{cases}.$$ (11)

As mentioned above, if the entropy $E(x_j)$ of sample x_j is larger than the set threshold, these samples will be retained, else will be discarded.

3.3 Entropy Based Boundary-Eliminated Pseudo-Inverse Linear Discriminant (EBEPILD)

As shown in the Fig. 3, this method is divided into two main steps: computing hyperplanes and generating distance discrimination parameters. In the first step, a traditional linear classification model is trained to generate the hyperplane l_o. It is evident that the computational complexity of this step depends on the selected model. Therefore, in order to improve the parsing speed, we choose the PILD method to obtain the suitable hyperplane l_o. The hyperplane l_o can reflect the preliminary distribution of training samples. Then, the samples with high class uncertainty are selected in the positive and negative samples using entropy. Assume that the number of positive class samples is N_{pos} and the number of negative class samples is N_{neg} and the total number of samples is N, where $N = N_{pos} + N_{neg}$. The centroids of the selected positive samples and the selected negative samples are computed as:

$$\begin{cases} C_{pos} = \frac{\sum_{j=1}^{N_{pos}} W_j x_j}{\sum_{j=1}^{N_{pos}} W_j} \\ C_{neg} = \frac{\sum_{j=N_{pos}+1}^{N} W_j x_j}{\sum_{j=N_{pos}+1}^{N} W_j} \end{cases},$$ (12)

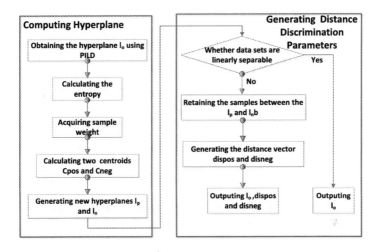

Fig. 3. Training Flow Chart of EBEPILD

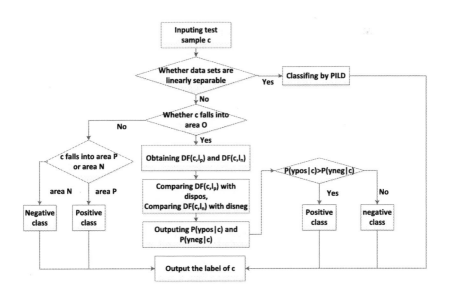

Fig. 4. Testing Flow Chart of EBEPILD

where x_j is the jth sample of X_{full} and the calculation method of W_j is illuminated in Sect. 2. Then, the two hyperplanes paralleled to l_o are formed. Furthermore, the locations of two hyperplane are determined by C_{pos} and C_{neg}. In detail, one hyperplane which through C_{pos} is defined as l_p, and the other hyperplane which through C_{neg} is defined as l_n. As shown in Fig. 5, we can get the three spaces through the l_p and l_n. In the second part, the training samples locating at area O are chosen to constitute the new data set ND. Suppose that the number of positive samples is $\widetilde{N}_{pos}$ and the number of negative samples is $\widetilde{N}_{neg}$ in ND. The samples belonging to ND can be defined as p_k. Afterwards, the distance function $DF(p_k, l)$ is described as the distance from the sample p_k to hyperplane l. Furthermore, the distances from all the positive sample $\sum_{k=1}^{\widetilde{N}_{pos}} p_k$ to l_p, which can be put into vector defined as $dispos = [dp_1, dp_2, ..., dp_{\widetilde{N}_{pos}}] \in \mathbb{R}^{\widetilde{N}_{pos} \times 1}$. Similarly, the distances from all negative samples $\sum_{k=\widetilde{N}_{pos}+1}^{\widetilde{N}_{pos}+\widetilde{N}_{neg}} p_k$ to l_n are calculated and the result is stored in the vector $disneg = [dn_1, dn_2, ..., dn_{\widetilde{N}_{neg}}] \in \mathbb{R}^{\widetilde{N}_{neg} \times 1}$. We define the maximum distance from $dispos$ as $disposmax$ and define the maximum distance from $disneg$ as $disnegmax$. If $disposmax \geq disnegmax$, it is indicated that the data set are linearly separable. Thus, the PILD can be used to solve classification problems directly. If $disposmax < disnegmax$, it means that the data have overlap phenomenon.

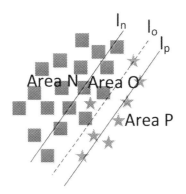

Fig. 5. Three spaces decided by l_p and l_n

As shown in Fig. 4, in testing phase, if the data are linearly inseparable, we will calculate the location of the testing sample c firstly. If c does not fall into the area between l_p and l_n, its category is the same as the class of area which c fall into. Otherwise, the $DF(c, l_p)$ and $DF(c, l_n)$ are calculated. Then, the $\widetilde{M}_{pos}$

and $\widetilde{M}_{neg}$ can be computed as:

$$\widetilde{M}_{pos} = \sum_{j=1}^{\widetilde{N}_{pos}} dp_j = \begin{cases} 1, \text{if } dp_j > DF(c, l_p) \\ 0, \text{if } dp_j \leq DF(c, l_p) \end{cases}, \tag{13}$$

$$\widetilde{M}_{neg} = \sum_{j=1}^{\widetilde{N}_{neg}} dn_j = \begin{cases} 1, \text{if } dn_j > DF(c, l_n) \\ 0, \text{if } dn_j \leq DF(c, l_n) \end{cases}. \tag{14}$$

The probability that the testing sample c is predicted to be a positive class can be calculated as:

$$P(y_{pos}|c) = \frac{\widetilde{M}_{pos}}{\widetilde{N}_{pos}}, \tag{15}$$

and the probability that the testing sample c is predicted to be a negative class can be defined as:

$$P(y_{neg}|c) = \frac{\widetilde{M}_{neg}}{\widetilde{N}_{neg}}. \tag{16}$$

Finally, compared $P(y_{neg}|c)$ with $P(y_{pos}|c)$, the class of sample c is determined by the larger probability. It's remarkable that if $P(y_{neg}|c) == P(y_{pos}|c)$, the sample can be classified by hyperplane l_o. In addition, if the training samples are linearly separable, the testing sample c can be directly predicted by the original hyperplane l_o. The objective equation of PILD is shown as follow:

$$g_{pild}(x_j) = w^T x_j + b. \tag{17}$$

For simple calculation, the distance calculation formula $DF(\cdot)$ can be defined as:

$$DF(x_j, l) = g_{pild}(x_j). \tag{18}$$

4   Experiment

4.1   Speech Emotion Database

In order to verify the effectiveness of the algorithm, we use Interactive Emotional Dyadic Motion Capture (IEMOCAP) database [2]. This emotion corpus contains five sessions. Each session contains two performances, scripted scenarios and improvised scenarios, which was performed by two actors that one is female and the other is male. We choose categorical labels in our research and we only take into account four emotional categories: anger, happiness, neutral state and sadness. In our research, four sessions are used to train and the rest session is used to test. In order to adjust the hyper-parameters of the classifier, we randomly select 25% training set to validate.

4.2   Feature Extraction

The experiment applies the hand-crafted low-level descriptions (LLDs) for speech emotion recognition [1]. In detail, The specific features, which are employed in the INTERSPEECH 2009 Emotion Challenge [13], consist of Harmonics-to-Noiscratio (HNR), Zero-Crossing-Rate (ZCR), Root Mean Square (RMS) frame energy, 1–12 Mel-Frequency Cepstral Coefficients (MFCC), Fundamental Frequency (F0). And then these delta coefficients can be calculated to form the 32-dimensional features. Furthermore, the functionals (mean, minimum and maximum value, std, kurtosis, skewness, relative position, range, as well as offset, slope and their MSE) are used to flat the frame-level features for obtaining the 384-dimensional utterance-level features. For the interoperability of data, the open source toolkit called openSMILE [4] is applied to extract features in this paper.

4.3   Experiment Result

The Weighted Accuracy (WA) and Unweighted Accuracy (UA) are selected as evaluation metrics. WA is the number of the correctly classified samples on the whole test data. UA is the average classification accuracy in each emotion, which explains models performance on imbalanced classes. All experiments are conducted on 384-dimensional utterance-level features from IEMOCAP. In order to prove the advantage of the proposed algorithm, we choose the Cost-Sensitive SVM (CSSVM) with RBF kernel as a baseline system. In view of the success of deep learning in speech emotion recognition, deep learning [12] is considered as a comparison algorithm. In this paper, DNN consists of one input layer with 384 nodes and three fully-connected layers which have 256 nodes, 128 nodes and 64 nodes, respectively, and each fully-connected layer follows by 0.5 dropout, batchnorm and relu. Then, the last fully-connected layer with 64 nodes is connected with softmax layer for classification. BEPILD is a binary classification algorithm without parameter, so the "One vs One" strategy is applied to copy with multi-class classification case. The proposed EBEPILD has two hyper-parameters, number of nearest neighbors k and entropy value *threshold*. In our research, k is selected from $[5, 7, 10, 15, 20, 40, 80]$ and *threshold* is chosen from $[0.5, 0.6, 0.7, 0.8, 0.9]$.

Table 1. WA and UA of CSSVM, DNN, BEPILD, EBEPILD on IEMOCAP four-class task

Method	WA(%)	UA(%)
CSSVM	55.52	53.56
DNN	55.52	57.54
BEPILD	55.41	59.26
EBEPILD	**56.26**	**60.01**

Table 1 lists the WA and UA of CSSVM, DNN, BEPILD and EBEPILD. Compared with CSSVM and DNN, BEPILD has a remarkable advantage in solving imbalanced problem according to UA and it also performs fine capability in classification accuracy. It can be concluded that BEPILD is an effective algorithm to handle unbalanced speech data according to experiment result. In order to get better results, we refine the algorithm by adopting entropy. According to the results from Table 1, the proposed EBEPILD shows better performance compared with BEPILD when k is set as 10 and *threshold* is set as 0.9. As shown in the Fig. 6, *anger*(874) denotes the number of anger utterance is 874 in training set. It is visible that the happy class can be considered as the positive class. From the Fig. 6, CSSVM has advantages in identifying neutral class, but it does not work in identifying happy class. DNN achieves the best result in identifying anger and sadness. Considering that angry emotion and sad emotion have obvious mood swings, DNN can distinguish these samples primely through high-dimensional mapping. But DNN must have enough quantity of samples to learn the general characteristics of the sample, it is not sensitive to identify the positive class. Compared with CSSVM and DNN, BEPILD obtain the best recognition rate of happy emotion, because BEPILD classify the positive class and negative class samples with high class uncertain through the sample distribution instead of learning the intrinsic characteristics of samples. Furthermore, it can be seen that the BEPILD has more stable recognition accuracy for each class. Compared with BEPILD, the proposed EBEPILD has a few improvements in anger recognition accuracy, neutral recognition accuracy and sadness recognition accuracy, because of adding entropy to tighten the two hyperplanes can precisely classify the samples with high class certainty. In summary, from the average emotion accuracy, the proposed EBEPILD manifests excellent performance in dealing with imbalanced problems of speech emotion data.

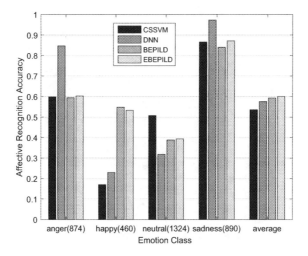

Fig. 6. The recognition rate of each classifier for each kind of emotion

5   Conclusion

In this paper, we introduce the BEPILD algorithm to overcome the imbalanced problems on SER. Due to its prominent performance on IEMOCAP, we make the improvement to BEPILD. The refined algorithm named EBEPILD can obtain more robust boundaries by utilizing entropy. The experiment results confirm that the proposed EBEPILD can achieve superior results compared with other algorithms. In order to handle the high-dimensional speech features, the kernel method will be introduced to the EBEPILD as our future work.

Acknowledgements. This work is supported by Natural Science Foundations of China under Grant No. 61672227, "Shuguang Program" supported by Shanghai Education Development Foundation and Shanghai Municipal Education Commission, and Action Plan for Innovation on Science and Technology Projects of Shanghai under Grant No. 16511101000.

References

1. Ayadi, M.E., Kamel, M.S., Karray, F.: Survey on speech emotion recognition: features, classification schemes, and databases. Pattern. Recogn. **44**, 542–587 (2011)
2. Busso, C., Bulut, M., Lee, C.C., Kazemzadeh, A., Mower, E., Kim, S.: Iemocap: interactive emotional dyadic motion capture database. Lang. Res. Eval. **42**(4), 335–359 (2008)
3. Calvo, R.A., D'Mello, S.: Affect detection: an interdisciplinary review of models, methods, and their applications. IEEE Trans. Affect. Comput. **1**(1), 18–37 (2010)
4. Eyben, F., Wöllmer, M., Schuller, B.: OpenSmile: the Munich versatile and fast open-source audio feature extractor. In: ACM International Conference on Multimedia, pp. 1459–1462 (2010)
5. Fayek, H.M., Lech, M., Cavedon, L.: Evaluating deep learning architectures for speech emotion recognition. Neural Netw. **92**, 60–68 (2017)
6. Hu, H., Xu, M.X., Wu, W.: GMM supervector based SVM with spectral features for speech emotion recognition. In: IEEE International Conference on Acoustics, pp. IV-413 - IV-416 (2007)
7. Li, L., Zhao, Y., Jiang, D., Zhang, Y., Wang, F., Gonzalez, I., et al.: Hybrid deep neural network-hidden markov model (DNN-HMM) based speech emotion recognition. In: Pun, T., Pelachaud, C., Sebe, N. (eds.) ACII 2013, vol. 7971, pp. 312–317. IEEE, Geneva (2013). https://doi.org/10.1109/ACII.2013.58
8. He, H., Garcia, E.A.: Learning from imbalanced data. IEEE Trans. Knowl. Data Eng. **21**(9), 1263–1284 (2009)
9. Han, K., Yu, D., Tashev, I.: Speech emotion recognition using deep neural network and extreme learning machine. In: INTERSPEECH (2014)
10. Mirsamadi, S., Barsoum, E., Zhang, C.: Automatic speech emotion recognition using recurrent neural networks with local attention. In: IEEE International Conference on Acoustics, Speech and Signal Processing(ICASSP), pp. 2227–2231. IEEE(2017)
11. Shannon, C.E.: A mathematical theory of communication. Bell Syst. Tech. J. **27**(4), 379–423 (1948)
12. Schmidhuber, J.: Deep learning in neural networks: an overview. Neural Netw. **61**, 85–117 (2014)

13. Schuller, B., Steidl, S., Batliner, A.: The Interspeech 2009 emotion challenge. In: INTERSPEECH 2009, Conference of the International Speech Communication Association, pp. 312–315 (2009)
14. Tian, T., Ji, W., Gao, D.Q.: Threshold optimization of pseudo-inverse linear discriminants based on overall accuracies. In: International Joint Conference on Neural Networks, pp. 1–6 (2015)
15. Zhu, Y.J., Wang, Z., Zha, H.Y., Gao, D.Q.: Boundary-eliminated pseudoinverse linear discriminant for imbalanced problems. IEEE Trans. Neural Netw. Learn. Syst. **99**, 1–14 (2017)

An Improved C-COT Based Visual Tracking Scheme to Weighted Fusion of Diverse Features

Lifang Wu, Qi Wang, Dezhong Xu, and Meng Jian[⊠]

Faculty of Information Technology,
Beijing University of Technology, Beijing 100124, China
mjian@bjut.edu.cn

Abstract. Visual tracking is one of hot researches in computer vision in recent years. C-COT [8] has obtained excellent results on many visual tracking benchmarks. However, it cannot exploit CNN features effectively because it gave the same weight for different CNN features. Furthermore, it updated model frame by frame, it possibly results in model drift. To address these problems, we propose an improved C-COT based visual tracking scheme to weighted fusion of diverse features. We present a weighted sum model that convolutional responses from different convolutional layers are weighted and summed to obtain the final response score. Secondly, we introduce a context based updating strategy for high confidence model update to avoid samples corruption and model drift. The experimental results on the challenging OTB dataset demonstrate that the proposed method is more competitive than state-of-the-art trackers.

Keywords: Diverse features · C-COT · Weighted fusion
Context based updating strategy

1 Introduction

In recent years, visual object tracking is one of hot topics in computer vision, which aims to estimate the trajectory of a target in a video given the initial position. Although some researches have made great success, it is still a challenging task because of occlusion, deformation, and scale variations.

Most state-of-the-art methods belong to tracking-by-detection schemes which learnt a discriminative appearance model of target object to handle visual tracking problem. Among discriminative tracking method, discriminative correlative filter based approaches achieve excellent performance in terms of accuracy and robustness on visual tracking benchmarks [1, 2]. These methods train a discriminative correlative filter (DCF) efficiently in the frequency domain by fast Fourier transform (FFT).

Since Bolme et al. [3] introduced the CF into the visual object tracking, several advancements have been made to improve DCF tracking framework. Danelljan et al. [4] introduced Color Names (CN) to represent color information of target object. Danelljan et al. [5] tackled scale changes estimation by learning a scale pyramid representation.

© Springer Nature Switzerland AG 2018
R. Hong et al. (Eds.): PCM 2018, LNCS 11165, pp. 686–695, 2018.
https://doi.org/10.1007/978-3-030-00767-6_63

Because of the periodic assumption of all circular shifts training samples, unexpected bounding effects is introduced into CF which degrade the discrimination of models. To handle this problem, Danelljan et al. [6] presented a Spatially Regularized Discriminative Correlation Filter (SRDCF), adding a spatial regularization to penalize coefficients of filter away from the spatial center of target. Due to rapidly development of deep convolutional neural networks, many DCF trackers attempted to combine with CNN features which bring rich feature representations for discriminate target from complicated background. According to [7], powerful feature representation is most important for visual tracking. Rich features can bring different target information leading to a robust estimation and better results. However, for discriminative correlative filters based method, the efficient use of CNN features from different convolutional layers has not been adequately studied.

C-COT [8] has obtained excellent results on many visual tracking benchmarks and ranked first in the VOT 2016 challenge. C-COT method gave the same importance to different CNN features; however, shallow layers and deep layers of CNN features have different contributions for the tracking performance. Therefore, C-COT method cannot exploit CNN features efficiently. Furthermore, C-COT updated model at each frame, which is at risk of model drift whenever the target is missing.

Motivated by above we propose an improved C-COT based visual tracking scheme to weighted fusion of diverse features. We first present a weighted sum model that convolutional responses from different convolutional layers are weighted and summed to obtain the final response score. Secondly, we introduce a context based updating strategy for high confidence model updates to avoid samples corruption and model drift. Experimental results show that the improved scheme outperforms the baseline.

The contributions of this paper are as follows:

1. We present a weighted sum model to combine the convolutional responses from different convolutional layers with different weight.
2. We propose a context based updating strategy for high confidence model update. It avoids samples corruption and model drift.

2   Baseline Approach

The C-COT learnt discriminative correlation filter with continuous convolution operator based on a set of M training samples $\{x_j\}_1^M \in \chi$ where χ denotes the sample space. Each sample is consist of D feature channels $x_j^1, x_j^2 \ldots, x_j^D$, extracted from the same image patch. On the contrary with conventional DCF trackers that require feature maps to have the same spatial resolution, C-COT breaks through these restriction. In the formulation, N^d denotes the number of spatial samples in x_j^d and each feature layer x_j^d has an independent resolution N^d. The feature channel $x_j^d \in \mathbb{R}^{Nd}$ is viewed as a function $x_j^d[n]$ indexed by the discrete spatial variable $n \in \{0, \cdots, N_d - 1\}$. In order to transfer the learning problem into continuous spatial domain, they introduce an interpolation model for training samples. the continuous interval $[0, T] \in \mathbb{R}$ is considered as the spatial support of feature map where scalar T denotes the size of support region. The interpolation operator J_d is followed as Eq. (1)

$$J_d\{x^d\}(t) = \sum_{n=0}^{N_d-1} x^d[n]b_d\left(t - \frac{T}{N_d}n\right) \tag{1}$$

Here, $b_d \in L^2(t)$, the Hilbert space, is an interpolation function with period T > 0. The interpolated sample $J_d\{x^d\}(t)$ is viewed as an interpolated feature layer of feature channel d.

Overall, this method is aim to train a linear convolutional operator which maps a sample $x \in \chi$ to a detection score function $S_f\{x\}(t)$ as Eq. (2),

$$S_f\{x\}(t) = f * J\{x\} = \sum_{d=1}^{D} f^d * J_d\{x^d\} \tag{2}$$

In this formulation, $J\{x\}$ denotes to the entire interpolated feature map; $f = (f^1, f^2, \cdots, f^D)$ is a set of convolution filters, specifically f^d is the filter of feature channel d; * is the circular convolution operator. The multi-channel convolution responses are summed to obtain the detection score $S_f\{x\}(t)$, defined in the continuous interval $[0, T] \in \mathbb{R}$, indicating the confidence score of target in the corresponding image region. Similar to other DCF trackers, the confidence score function with highest response is estimated position of target in the tracking stage.

For the standard correlation filters based methods, each sample is labeled by a Gaussian function representing the desire response output. In the continuous formulation, each sample $x_i \in \chi$ is labeled by confidence functions y_i which is the desired output of $S_f\{x\}(t)$ in continuous spatial domain. Filters are trained by minimizing the cost function and the cost function as Eq. (3):

$$E(f) = \sum_{j=1}^{M} \alpha_j \left\| S_f\{x_j\} - y_j \right\|_{L^2}^2 + \sum_{d=1}^{D} \left\| wf^d \right\|_{L^2}^2 \tag{3}$$

where the $\|g\|_{L^2}^2$ denotes L^2 norm of g; $\alpha_j \geq 0$ is the weight of each training sample x_j. w is an regularization weight to determine the significance of filter coefficient f^d. Coefficients f^d near the target region are emphasized by assigning smaller weight w and vice versa. In this way, correlation filters can be learned arbitrary size of image region different from standard DCF trackers that introduce unexpected boundary effect and reduce the discriminating ability of filters [6].

3   Our Scheme

In this section, we first introduce the weighted sum formulation. Next, a context based update strategy is proposed to prevent model shift.

3.1   Weighted Feature Fusion

C-COT method integrate feature maps of input RGB image, conv1 layer and conv5 layer. Table 1 shows the spatial size and dimension of feature map extracted from different convolutional layers. The input image is 224×224 while the spatial resolution of layer 1 and layer 5 are 109×109 and 13×13 respectively. If feature map concatenates with the same weight, larger feature map will dominate smaller feature map, resulting in some information loss for feature maps of layer 1 and layer 5. What's more, shallow layers have more outline information of target while deep layers have more semantic information and high level features. CNN features from different layers have different contributions for representing information of object target. In order to use CNN features efficiently, we propose a weighted features integration, assigning larger weights to the response score of layer 1 and layer 5.

Table 1. The spatial size and dimension of feature map extracted from different convolutional layers. Layer 0 is the input RGB image.

	Layer 0	Layer 1	Layer 2	Layer 3	Layer 4	Layer 5
Spatial size	224×224	109×109	26×26	13×13	13×13	13×13
Dimension	3	96	256	512	512	512

Similar to C-COT [8], feature maps from all layers are transferred to the spatial continuous domain by interpolation operation introduced in Sect. 2, and then convolve with corresponding convolutional filters to gain response map. Response maps are summed to obtain the confidence score as Eq. (4)

$$S_f(x) = W_0 \sum_{a=1}^{D_{layer0}} f^a * J_a\{x^a\} + W_1 \sum_{b=1}^{D_{layer1}} f^b * J_b\{x^b\} + W_2 \sum_{c=1}^{D_{layer5}} f^c * J_c\{x^c\}$$

$$(4)$$

Here, D_{layer0}, D_{layer1} and D_{layer5} donates the dimension of input image, conv1 feature map and conv5 feature map respectively. $W0$, $W1$ and $W2$ represent weight coefficient of each convolutional layer.

The loss function of convolutional filters are

$$E(f) = \sum_{j=1}^{m} \alpha_j \left\| W_0 \sum_{a=1}^{D_{layer0}} f^a * J_a\{x^a\} + W_1 \sum_{b=1}^{D_{layer1}} f^b * J_b\{x^b\} \right.$$
$$\left. + \sum_{c=1}^{D_{layer5}} f^c * J_c\{x^c\} - y_j \right\|^2 + \sum_{d=1}^{D} \left\| w f^d \right\|^2$$

$$(5)$$

where, α_j donates weight of each training sample; w donates regularization penalty term; $D = D_{layer0} + D_{layer1} + D_{layer5}$ represents the whole dimension. Convolutional filters $f = (f^1, f^2, \cdots, f^D)$ are trained by minimizing Eq. (5).

3.2   Context Based Update Strategy

Assuming estimated location is accurate, most existed trackers update model at each frame. In C-COT method, given estimated location, convolutional filters are optimized by Eq. (5) iteratively. However, this strategy is at risk of model drift once the target is estimated inaccurately, occluded or missing the target. Additionally, some recent works gain good performance without model update or updating under some conditions [9, 10]. These indicate updating model at each frame is redundant and increase some computation.

The peak value and fluctuation of response map reflect the confidence of tracking result to some extent. When the estimated location approaches ground truth, the desire response map has a sharp and high peak and smooth in the other area. The sharper and higher the peak are, the better estimated location is. On the contrary, if response map fluctuates or have many peaks with small value, the possibility of tracking failure is very high. Figure 1 shows the tracking results on a sequence from OTB dataset. (a) is an accurate tracking result whose response map (b) is low fluctuation and peak value is 0.70932 while (c) shows missing target result whose response map (d) is fluctuate and peak value reduces to 0.11661. If updating model in this situation, the tracking model will be corrupted.

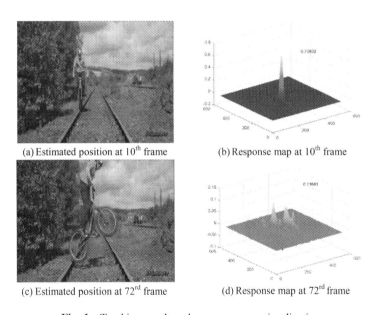

(a) Estimated position at 10th frame   (b) Response map at 10th frame

(c) Estimated position at 72rd frame   (d) Response map at 72rd frame

Fig. 1. Tracking result and response map visualization

[11] proposed a high confidence update, which defines the average peak-to-correlation energy (APCE) to measure fluctuation and represented as Eq. (6),

$$APCE = \frac{|F_{max} - F_{min}|^2}{mean\left(\sum_{w,h}\left(F_{w,h} - F_{min}\right)^2\right)} \qquad (6)$$

Here $F_{w,h}$ represents the wth row h^{th} column elements of response map; F_{max} and F_{min} donates the maximum and minimum of response map respectively. When the target is detected accurately, the response map only have one sharp peak and APCE will become larger. Otherwise, when estimated target is not accurate, APCE will decrease.

Our method proposes a context based updating strategy for high confidence model updates by two criteria: APCE as Eq. (6) and peak value of response map defined as Eq. (7)

$$F_{max} = F \tag{7}$$

Different from [11], our method updates model when two criteria are greater than the average of previous m frames with certain proportion γ_1 and γ_2, which is defined as Eq. (8). This the lower subscript t of $APCE_t$ and F_{max-t} donates current frame t.

$$\begin{cases} APCE_t > \gamma_1 \dfrac{\sum_{i=t-m}^{t-1} APCE_i}{m} \\ F_{max-t} > \gamma_2 \dfrac{\sum_{i=t-m}^{t-1} F_{max-i}}{m} \end{cases} \tag{8}$$

4   Experiments

In this section, the features are firstly compared, so that the best feature combination could be obtained. Then the coefficient of weighted feature fusion and parameters in model update are analyzed. Finally, the proposed scheme is compared with state-of-the-art methods.

4.1   Diverse Feature Comparison

CN and HOG are very popular handcraft features in visual tracking. However, with development of deep learning, CNN features present better performance. In order to explore powerful representation of CNN features, I evaluate the performance when using different convolutional layers and their combinations in VGG-M network.

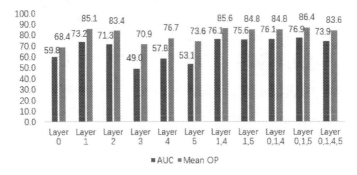

Fig. 2. Tracking results when using different convolutional layers in tracking framework.

Figure 2 shows mean overlap precision (OP) and area-under-the-curve (AUC) on OTB-2013 dataset. OP is defined as the percentage of frames in a video where the intersection-over-union overlap exceeds a threshold of 0.5. AUC is computed from the success plot. Results demonstrate that shallow layers have more outline information of target and feature maps have high spatial resolution which is helpful for locating target while deep layers have more semantic information and high level features which is helpful for discriminating target from complex background. Most convolutional layers are better than input image. The first convolutional layer achieves great improvement, however with convolutional layers deeper from conv1 to conv3, the performance is poorer for the reason that the spatial resolution of feature map gradually decreases, which is unfavorable for locating position. Whereas the fourth and fifth convolutional layers provide a performance improvement comparing with third convolutional layer. This is may be because the high level features have more semantic information and have stronger discrimination. Combining the shallow layer and deep layer gain great performance promotion while fusing too many layers can degrade performance. In summary, the best results are obtained when fusing the input RGB image, the first and fifth convolutional layers.

4.2   Analysis of the Model with Different Weights

Due to the small size of deep convolutional feature map, weighted sum feature integration should be into consideration. In this experiment, We select 55 videos from OTB-2015 dataset and get the results with different weight combination as Fig. 3. The W_0, W_1 and W_2 donate the weight coefficient of response score for layer 0, layer 1 and layer 5 respectively. The ratios in the figure represents W_0:W_1:W_2. Figure 3 indicates that assigning the response score of the first and fifth convolutional layers more weight can promote performance. The main reason is high level features have powerful discrimination with small feature maps. The baseline method C-COT obtains a Mean Overlap Precision 96.9%, an AUC 90.4%. When we assign weight W_0:W_1:W_2 = 1:2:4, our method obtained best results that the Mean Overlap Precision is 98.0% and the AUC is 92.0%.

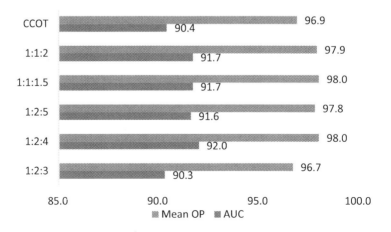

Fig. 3. Results of different weight coefficients

4.3   Analysis of Model Update Parameters

We use tracker with weighted sum feature integration as a baseline. The dataset used is the same as introduced dataset in Sect. 4.2. γ_1 and γ_2 are coefficients introduced in Sect. 3.2 and another significant parameter is historic number of frames m. I evaluate these parameters effects on tracking performance when m varies as 5, 6, 7, 8 corresponding different coefficients. Results are shown in Table 2 that when $m = 6$, $\gamma_1 = 0.3$, $\gamma_2 = 0.3$, our proposed method gain best results.

Table 2. Results of different model update parameters.

$m = 5$					$m = 6$			
γ_1	γ_2	AUC	Mean OP		γ_1	γ_2	AUC	Mean OP
0.2	0.3	92.0	97.9		0.2	0.3	92.0	98.0
0.3	0.3	92.0	98.0		0.3	0.3	92.0	98.4
0.3	0.4	92.0	98.0		0.3	0.4	91.8	98.0

(a)					(b)			
$m = 7$					$m = 8$			
γ_1	γ_2	AUC	Mean OP		γ_1	γ_2	AUC	Mean OP
0.2	0.3	91.8	98.0		0.2	0.3	91.5	98.0
0.3	0.3	91.9	98.4		0.3	0.3	91.6	98.0
0.3	0.4	91.6	98.1		0.3	0.4	91.2	98.2

(c)   (d)

4.4   Comparison with State-of-the-Art Schemes

Our tracker is evaluated on the same dataset as Sect. 4.2 with 10 state-of-the-art methods: ECO [12], C-COT [8], MD-Net [10], SA-Net [13], DeepSRDCF [14], MCPF [15], ACFN [16], Staple [17] and HDT [18].

Figure 4(a) shows the precision plot on OTB 55 video sequences. The SA-Net tracker, based on convolutional features representation, obtains a Mean OP of 97.9%. The baseline method C-COT achieves a Mean OP of 96.7%. Our method achieves the best result with a mean OP of 98.4%, outperforming C-COT by 1.5% and the second best method by 1%. Figure 4(b) shows results of the success plot. The DCF based methods Staple and MCPF obtain AUC score of 81.1% and 85.9%. Based on deep features trackers ECO, SA-Net, MD-Net, C-COT and DeepSRDCF provide excellent performance. Among them, the ECO, improved method based on C-COT, achieves best result of AUC score of 91.9%. Overall, our method obtains the best results, outperforming ECO method.

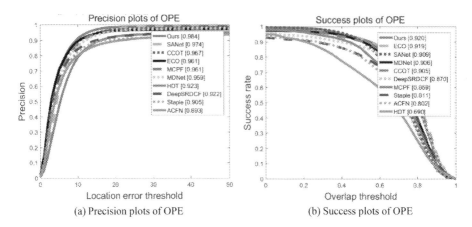

(a) Precision plots of OPE (b) Success plots of OPE

Fig. 4. Comparison results on OTB-55 dataset

5 Conclusion

In this paper, we propose an improved C-COT based visual tracking scheme to weighted fusion of diverse features. We explore the performance effect of CNN features from different layers so that our method fuses diverse CNN features efficiently. The proposed method presents a weighted sum model, assigning different weights to convolutional response from different layers. Weighted fusion of CNN features makes the proposed method have stronger discrimination. For model update, we propose a contextual high confidence update strategy that model updates when peak value and APCE are greater than the average of previous frames, efficiently reducing the possibility of model drift. Eventually, our experimental results and performance comparison with state-of the-art tracking methods on challenging benchmark tracking datasets shows that our method archives state-of-the-art performance.

Acknowledgement. We thank Beijing National Natural Science Foundation of China (61702022), Postdoctoral Research Foundation(2017-KZ-029), China Postdoctoral Science Foundation funded project(2018T110019), and Beijing university of technology "Ri Xin" cultivation project for partially supporting this work

References

1. Wu, Y., Lim, J., Yang, M.H.: Object tracking benchmark. TPAMI **37**(9), 1834–1848 (2015)
2. Kristan, M., Pflugfelder, R., Leonardis, A., Matas, J., et al.: The visual object tracking VOT2014 challenge results. In: Agapito, L., Bronstein, M.M., Rother, C. (eds.) ECCV 2014. LNCS, vol. 8926, pp. 191–217. Springer, Cham (2015). https://doi.org/10.1007/978-3-319-16181-5_14
3. Bolme, D.S., Beveridge, J.R., Draper, B.A., Lui, Y.M.: Visual object tracking using adaptive correlation filters. In: IEEE Conference on CVPR, pp. 2544–2550 (2010)

4. Danelljan, M., Shahbaz Khan, F., Felsberg, M., Van de Weijer, J.: Adaptive color attributes for real-time visual tracking. In: CVPR, pp. 1090–1097 (2014)
5. Danelljan, M., Hager, G., Khan, F., Felsberg, M.: Accurate scale estimation for robust visual tracking. In: BMVC Press (2014)
6. Danelljan, M., Hager, G., Shahbaz Khan, F., Felsberg, M.: Learning spatially regularized correlation filters for visual tracking. In: ICCV, pp. 4310–4318 (2015)
7. Wang, N., Shi, J., Yeung, D.-Y., Jia, J.: Understanding and diagnosing visual tracking systems. In: ICCV, pp. 3101–3109 (2015)
8. Danelljan, M., Robinson, A., Shahbaz Khan, F., Felsberg, M.: Beyond correlation filters: learning continuous convolution operators for visual tracking. In: Leibe, Bastian, Matas, Jiri, Sebe, Nicu, Welling, Max (eds.) ECCV 2016. LNCS, vol. 9909, pp. 472–488. Springer, Cham (2016). https://doi.org/10.1007/978-3-319-46454-1_29
9. Bertinetto, L., Valmadre, J., Henriques, J.F., Vedaldi, A., Torr, P.H.S.: Fully-convolutional siamese networks for object tracking. In: Hua, G., Jégou, H. (eds.) ECCV 2016. LNCS, vol. 9914, pp. 850–865. Springer, Cham (2016). https://doi.org/10.1007/978-3-319-48881-3_56
10. Nam, H., Han, B.: Learning multi-domain convolutional neural networks for visual tracking. In: CVPR, pp. 4293–4302 (2015)
11. Wang, M., Liu, Y., Huang, Z.: Large margin object tracking with circulant feature maps. In: CVPR, pp. 4800–4808 (2017)
12. Danelljan, M., Bhat, G., Khan, F.S., Felsberg, M.: ECO: efficient convolution operators for tracking. arXiv preprint arXiv:1611.09224 (2016)
13. Fan, H., Ling, H.: SANet: structure-aware network for visual tracking. In: Computer Vision and Pattern Recognition Workshops, pp. 2217–2224. IEEE (2017)
14. Danelljan, M., Hager, G., Shahbaz Khan, F., Felsberg, M.: Convolutional features for correlation filter based visual tracking. In: ICCV Workshops, pp. 58–66 (2015)
15. Zhang, T., Xu, C., Yang, M.H.: Multi-task correlation particle filter for robust object tracking. In: CVPR, pp. 4819–4827 (2017)
16. Choi, J., Chang, H.J., Yun, S., et al.: Attentional correlation filter network for adaptive visual tracking. In: CVPR, pp. 4828–4837 (2017)
17. Bertinetto, L., Valmadre, J., Golodetz, S., et al.: Staple: complementary learners for real-time tracking. In: Computer Vision and Pattern Recognition, pp. 1401–1409. IEEE (2016)
18. Qi, Y., Zhang, S., Qin, L., et al.: Hedged deep tracking. In: Computer Vision and Pattern Recognition, pp. 4303–4311. IEEE (2016)

Focal Liver Lesion Classification Based on Tensor Sparse Representations of Multi-phase CT Images

Jian Wang[1,2], Xian-Hua Han[3], Jiande Sun[1], Lanfen Lin[4], Hongjie Hu[5],
Yingying Xu[4], Qingqing Chen[5], and Yen-Wei Chen[2,4(✉)]

[1] School of Information Science and Engineering,
Shandong Normal University, Jinan, China
[2] College of Information Science and Engineering,
Ritsumeikan University, Kyoto, Japan
chen@is.ritsumei.ac.jp
[3] Faculty of Science, Yamaguchi University, Yamaguchi, Japan
[4] College of Computer Science and Technology,
Zhejiang University, Hangzhou, China
[5] Department of Radiology, Sir Run Run Shaw Hospital, Hangzhou, China

Abstract. The bag-of-visual-words (BoVW) method has been proved to be an effective method for classification tasks in both natural imaging and medical imaging. In this paper, we propose a multilinear extension of the traditional BoVW method for classification of focal liver lesions using multi-phase CT images. In our approach, we form new volumes from the corresponding slices of multi-phase CT images and extract cubes from the volumes as local structures. Regard the high dimensional local structures as tensors, we propose a K-CP (CANDECOMP/PARAFAC) algorithm to learn a tensor dictionary in an iterative way. With the learned tensor dictionary, we can calculate sparse representations of each group of multi-phase CT images. The proposed tensor was evaluated in classification of focal liver lesions and achieved better results than conventional BoVW method.

Keywords: Multi-phase CT · Tensor analysis · Sparse coding
Image classification · Focal liver lesion

1 Introduction

Liver cancer is one of the leading causes of death worldwide. Early detection of liver cancers by analysis of medical images is a helpful way to reduce death due to liver cancer. High-definition medical images produced by modern medical imaging devices provide more detailed descriptions of tissue structures and thus facilitate more accurate diagnoses. High-definition medical images and large unorganized medical datasets, however, post challenges to doctors from the viewpoint

ⓒ Springer Nature Switzerland AG 2018
R. Hong et al. (Eds.): PCM 2018, LNCS 11165, pp. 696–704, 2018.
https://doi.org/10.1007/978-3-030-00767-6_64

of analysis and review. Computer-aided diagnosis (CAD) systems will assist doctors by characterizing the focal liver lesion (FLL) images.

Based on clinical observations, different types of liver lesions exhibit different visual characteristics at various time points after intravenous contrast injection. To capture the visual feature transitions of liver tumors over time, multi-phase contrast-enhanced computer-tomography (CT) scanning is generally employed on patients who are thought to have liver problems. In the multi-phase contrast-enhanced CT scan procedure, four phases of images are obtained: noncontrast-enhanced (NC) phase images are obtained from scans before contrast injection, arterial (ART) phase scanned 25–40 s after contrast injection, portal venous (PV) phase 60–75 s after contrast injection, and delayed (DL) phase scanned 3–5 min after contrast injection.

Characterization of FLLs, including classification and retrieval, has attracted considerable research interest recently. Mir *et al.* [1] first presented texture analysis in liver characterization, which illustrated the importance of gray-level distribution for distinguishing normal and malignant tissue. Yu et al. in [2] developed a content-based image retrieval system to differentiate among three types of hepatic lesions by using global features derived from a nontensor product wavelet filter and local features based on image density and texture. Roy et al. [4] used four types of features, that is, density, temporal density, texture, and temporal texture, which are derived from four-phase medical images, to retrieve the most similar images of five types of liver lesions. Shape feature was adopted in [5] in combination with density and texture features for retrieving five types of FLLs. Comparing to low-level features introduced above, the middle-level feature bag-of-visual-words (BoVW) has been proved to be considerably more effective for classifying and retrieving natural images. Diamant *et al.* [8] learned BoVW representation of the interior and boundary regions of FLLs for classifying three types of FLLs from single-phase CT images. A variant of BoVW called bag of temporal co-occurrence words (BoTCoW) was proposed by Xu *et al.* [9]. In BoTCoW, BoVW was applied to temporal co-occurrence images, which were constructed by connecting the intensities of multi-phase images, to extract temporal features for retrieving five types of FLLs from triple-phase CT images. After a common codebook learning procedure, Diamant *et al.* [11] proposed a visual word selection method based on mutual information to select more meaningful visual words for each specific classification task. In addition to these variants and enhanced versions of BoVW based on the hard-assignment mechanism, Wang *et al.* [12] learned sparse representations of local structures, which is a soft-assignment BoVW method, of multi-phase CT scans for FLL retrieval. Research on learning high-level features by deep learning methods, in particular using the convolutional neural networks (CNNs), is growing rapidly. [13] surveyed the use of deep learning methods in medical image analysis tasks, such as image classification, object detection, segmentation, registration. Due to the difficulties in collecting professional marked medical images, current medical image databases are always too small for deep learning methods. Most of current approaches use pretrained CNNs to extract feature descriptors from medical images. We have not yet seen many applications of deep learning methods in medical image feature extraction, especially in classification of focal liver lesions. To our knowledge, Bag-of-Visual-Words is still the state-of-the-art method in this field.

However, the conventional vector-based BoVW methods, as mentioned above, analyze the multi-phase images separately, in which the temporal co-occurrence information is neglected. In this study, we explore a multilinear generalization of the soft-assignment BoVW, that is, the tensor sparse representation approach, for joint analysis of multi-phase CT images and apply the proposed method for classification of four classes of focal liver lesions.

2   Tensor Sparse Representation of Multi-phase Medical Images

2.1   Tensor Codebook Learning by the Proposed K-CP Algorithm

First, we introduce the notations used throughout this paper. A vector is denoted by a lowercase boldface letter, for example, $\boldsymbol{x}$. A matrix is denoted by an uppercase boldface letter, for example, $\boldsymbol{X}$. A tensor is denoted by a Lucida Calligraphy letter, for example, X. We define tensor multiplication in a way similar to that in [14].

Given a set of tensor training samples Y, we proposed a K-CP method to learn tensor codebook D. Implementation of the proposed K-CP method comprises two iterated stages: calculation of sparse coefficients, assuming that the codebook is fixed, and codeword update based on the calculated sparse coefficients.

The first stage can be solved easily by using the tensor generalization of Orthogonal Matching Pursuit (OMP) algorithm. The OMP algorithm is a greedy algorithm that finds sparse coefficients of vector-based signals using a given codebook, whose codewords (atoms) are also vectors. In tensor OMP, given a collection of samples $Y = [Y_1, Y_2, ..., Y_N]$, where $Y_i \in \mathbb{R}^{I_1 \times I_2 \times ... \times I_M}, i = 1, 2, ..., N$, is an Mth-order tensor and $Y \in \mathbb{R}^{I_1 \times I_2 \times ... \times I_M \times N}$ is an $(M+1)^{th}$-order tensor. Suppose a codebook D comprises of K tensor codewords $D_k \in \mathbb{R}^{I_1 \times I_2 \times ... \times I_M}$. Then, D is a $(M+1)^{th}$-order tensor. The tensor OMP can be formulated as follows:

$$i = 1, 2, ..., N \quad \min_{\boldsymbol{x}_i} || Y_i - D \bar{\times}_{(M+1)} \boldsymbol{x}_i ||_2^2, \tag{1}$$
$$s.t. \ || \boldsymbol{x}_i ||_0 \leqslant T, \forall i$$

where a column vector $\boldsymbol{x}_i$ in $\boldsymbol{X}$ represents a combination of the codewords that approximates a sample Y_i, and T is a sparsity measure.

In the codeword update stage, each tensor codeword is updated individually. To update codeword D_k, we first find the row vector $\boldsymbol{x}_k^T$ in $\boldsymbol{X}$, in which each entry corresponds to the coefficient of a sample in Y to D_k. Then, we define the approximation error without using codeword D_k as follows:

$$E_k = Y - \sum_{j \neq k}^{K} D_j \circ \boldsymbol{x}_j^T \tag{2}$$

The total reconstruction error can be written as follows:

$$\| Y - D \times_{(M+1)} \boldsymbol{X} \|^2 = \| E_k - D_k \circ \boldsymbol{x}_k^T \|^2 \tag{3}$$

Algorithm 1. Tensor codebook learning via TOMP & K-CP

---

1. Task: Find the optimal dictionary that represent the data samples $\{y_i\}_{i=1}^N$ as sparse compositions. Objective function is:

$$\min_{\mathcal{D},\mathbf{X}}\{\|\mathcal{Y}-\mathcal{D}\times_{(M+1)}\mathbf{X}\|_F^2\}, \quad \text{s.t.}\,\|\mathbf{x}_i\|_0 \leq T,\ \forall i$$

2. Initialize: Set dictionary matrix $\mathcal{D}^{(0)}$ with l^2 normed columns
3. **while** (stopping rule not met) **do**
4.   **Sparse coding stage**: Compute $\mathbf{x}_i$ to approximate example $\mathcal{Y}_i$ using any pursuit algorithm, by solving:

$$i=1,2,...,N \quad \min_{\mathbf{x}_i}\{\|\mathcal{Y}_i-\mathcal{D}\,\overline{\times}_{(M+1)}\,\mathbf{x}_i\|_2^2\}, \quad \text{s.t.}\,\|\mathbf{x}_i\|_0 \leq T$$

5.   **Codebook update stage**: Update each dictionary atom $\mathcal{D}_k$, $k=1,2,...,K$
6.   --Define the group of indices corresponding to atom $\mathcal{D}_k$,

$$\omega_k = \{i\,|\,1\leq i\leq N,\ \mathbf{x}_k^T(i)\neq 0\}$$

7.   --Compute the overall representation error:

$$\mathcal{E}_k = \mathcal{Y}-\sum_{j\neq k}\mathcal{D}_j\mathbf{x}_k^T$$

8.   --Obtain $\mathcal{E}_k^R$ by restricting $\mathcal{E}_k$ using ω_k
9.   --Apply CP on $\mathcal{E}_k^R$. Codeword $\mathcal{D}_k$ is updated by the rank one tensor in the decomposition result.
10. **end while**

Our aim is to find the optimal D_k that well approximates the reconstruction error E_k in Eq. (3), which can be solved easily by applying CP decomposition on E_k.

CP (CANDECOMP/PARAFAC decomposition) decomposes a Pth-order tensor D into a sum of rank-one tensors [14].

$$D \approx \sum_{r=1}^{R} \lambda_r(\boldsymbol{d}_r^1 \circ \boldsymbol{d}_r^2 \circ ... \circ \boldsymbol{d}_r^P) \tag{4}$$

where $\circ$ denotes the outer product. We suppose the vector $\boldsymbol{d}_r^p$ is normalized to unit length, and the weight of each rank-one tensor is λ_r.

However, applying CP on E_k directly would fill the coefficient vector $\boldsymbol{x}_k^T$, which means that the sparsity would be destroyed. Therefore, we construct a constraint vector $\boldsymbol{\omega}_k = (i|1 \leq i \leq N, \boldsymbol{x}_k^T \neq 0)$ that captures the nonzero entries of $\boldsymbol{x}_k^T$. According to $\boldsymbol{\omega}_k$, we must restrict E_k and $\boldsymbol{x}_k^T$ to E_k^R and $\boldsymbol{x}_k^R$, respectively. By applying CP to E_k^R with a rank-one tensor component, D_k can be updated by using the decomposition result and the coefficient vector $\boldsymbol{x}_k^T$ can be updated by zero-padding the weight λ, as in Eq. (4)

The process of applying the CANDECOMP/PARAFAC (CP) decomposition to the reconstruction residual tensor is executed K times to update each of the K tensor codewords in each iteration. Thus this method is called K-CP method.

The above two stages are iterated until a pre-specified reconstruction error is achieved or the maximum iteration number is reached. The details of the K-CP method for overcomplete tensor codebook learning are given in Algorithm 1.

2.2 FLL Classification Using Tensor Sparse Representations of Spatiotemporal Structures

For each patient in the dataset, there are triple-phase (NC/ART/PV) CT images, which is explained in detail in Sect. 3.1. Based on the structure of the dataset, spatiotemporal features are extracted by using the BoVW models, in which codebooks are learned by the proposed tensor sparse coding method.

To capture the temporal feature of multi-phase CT images, corresponding slices from triple-phase CT images were center-aligned according to the tumor masks and stacked to form three-layer volumes. By this operation, the temporal co-occurrence information is transformed into spatial information in the third dimension of the constructed volumes. A spatiotemporal codebook can be learned by applying our proposed method on the tensor training samples, which are local descriptors extracted from three-layer volumes. Spatiotemporal feature of each medical case can be then calculated by summarizing the representations of local descriptors using mean pooling method. Spatiotemporal feature of a query case can also be calculated based on the learned spatiotemporal codebook

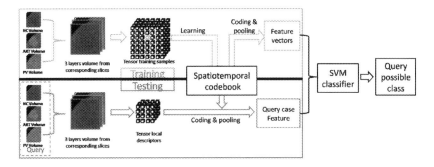

Fig. 1. Learning spatiotemporal features via the proposed tensor sparse coding method from multi-phase images

under the same mechanism. Features of the query and cases in the dataset were fed into a support-vector machine (SVM) classifier with a Radial basis function (RBF) kernel to predict the possible class that the query case may belong to. The workflow is shown in Fig. 1.

3   Experiments and Results

3.1   Multi-phase Medical Dataset

A multi-phase medical dataset was constructed with the help of radiologists to evaluate the performance of the proposed method. The dataset comprises four types of FLLs collected from 111 medical cases. For each medical case, triple-phase (NC/ART/PV) CT images were collected, with spacing of $(0.5 - 0.8) \times (0.5 - 0.8) \times (5/7)$ mm^3. The size of a CT slice was fixed to 512×512 pixels, while the number of CT slices was set depending on the region scanned (full body or only the abdomen). All tumors in each CT image were manually marked by an experienced medical doctor. In our experiments, however, only the major tumor, that is, the tumor with the largest volume, was considered. As a result, 111 FLLs were selected for use in our experiments, including 38 lesions of the cyst class, 19 cases of focal nodular hyperplasia (FNH), 26 cases of hepatocellular carcinoma (HCC), and 28 cases of hemangioma (HEM). Examples of the four types of FLLs are shown in Fig. 2.

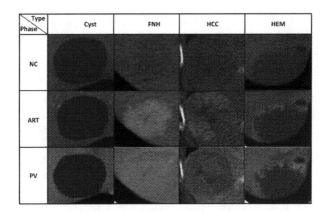

Fig. 2. Examples of each lesion type on 3 phases. Rows are images belong to same contrast phase, while columns are images from same lesion: cyst, FNH, HCC, HEM

3.2   Evaluation Method

Considering the constructed small dataset, the leave-one-out cross-validation method is used in performance evaluation. The classification accuracy are calculated for quantitative measurement, shown as follows:

$$Accuracy = TP/(TP + FP) \tag{5}$$

where, TP is number of correct classified cases, FP represents the number of miss classified cases. $(TP+FP)$ is the total number of cases in the corresponding FLL type.

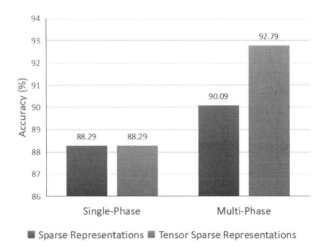

Fig. 3. Comparing classification performance by the proposed tensor sparse representation and the conventional sparse representation method using sing-/multi-phase CT images

3.3   Experimental Results

We compared the classification performance of the proposed tensor sparse representation method with the conventional sparse representation method over both single-/multi-phase medical images, as shown in Fig. 3. We used PV phase images in the single-phase experiments as most of related works do, since most liver lesion types can be visualized clearly on PV phase images. It's interesting that both the two methods got exactly the same results using single-phase images. The accuracy is more significantly improved, however, by the proposed tensor sparse representation method than the conventional one when using multi-phase images, which emphases that the proposed method is more effective in capturing the temporal information from multi-phase images. The detailed classification result of the proposed method is shown in Table 1. Due to the clear texture features and temporal enhancement features of Cyst and FNH, they are much easier to be classified from the others when using the temporal co-occurrence information captured by the proposed tensor sparse representation method.

A comparison of the performance of the proposed method with those of the state-of-the-art methods is given in Table 2. As mentioned in previous sections, considerable research effort has been invested to exploring variants and enhanced versions of the BoVW model for FLL characterization. Most of he state-of-the-art methods are based on the BoVW framework. Table 2 shows a comparison of the proposed method with a few other BoVW models. The proposed tensor sparse coding method outperforms the other methods by preserving spatiotemporal features captured from multi-phase CT images, especially for FNH that shows significant different contrast enhancement features in different phases.

Table 1. The performance of the proposed method

FLL type	Cyst	FNH	HCC	HEM	Accuracy (%)
Cyst	36	0	0	2	94.74
FNH	0	18	1	0	94.74
HCC	0	2	24	0	92.31
HEM	1	1	1	25	89.29

Table 2. Compare the classification accuracy (%) of the proposed method with those of state-of-the-art methods

Methods	Cyst	FNH	HCC	HEM	Averaged
Dual dictionary BoVW [8]	72.09	67.75	**96.08**	40.00	74.00
BoVW-MI [11]	93.23	90.53	66.67	68.00	79.80
TextureSpecific BoVW [10]	79.16	94.12	80.39	69.24	80.13
BoTCoW [9]	**95.83**	82.35	84.32	84.61	87.42
The proposed method	94.74	**94.74**	92.31	**89.29**	**92.79**

4   Conclusion

In this paper, we proposed the K-CP method to learn tensor sparse representations of multi-phase medical images. We learned tensor codebooks by using the proposed method and built BoVW models for extracting spatial features and temporal co-occurrence of multi-phase medical images. Experiments of applying the proposed method on focal liver lesion classification showed that the proposed method achieved more significant improvement from single-phase to multi-phase images than conventional sparse representation method and the proposed method outperforms the state-of-the-art methods in this task.

Acknowledgments. This research was supported in part by the Grant-in Aid for Scientific Research from the Japanese Ministry for Education, Science, Culture and Sports (MEXT) under the Grant No. 18H03267, and No. 18H04747, in part by the Key Science and Technology Innovation Support Program of Hangzhou under the Grant No.20172011A038, and in part by the National Key Basic Research Program of China under the Grant No. 2015CB352400.

References

1. Mir, A.H., Hanmandlu, M., Tandon, S.N.: Texture analysis of CT-images. IEEE Eng. Med. Biol. **5**, 781–786 (1995)
2. Yu, M., Lu, Z., Feng, Q., Chen, W.: Liver CT image retrieval based on non-tensor product wavelet. In: International Conference on Medical Image Analysis and Clinical Applications (MIACA), pp. 67–70 (2010)
3. Duda, D., Kretowski, M., Bezy-Wendling, J.: Texture characterization for hepatic tumor recognition in multiphase CT. Biocybern. Biomed. Eng. **26**(4), 15–24 (2006)
4. Roy, S., Chi, Y., Liu, J., Venkatesh, S.K., Brown, M.S.: Three-dimensional spatiotemporal features for fast content-based retrieval of focal liver lesions. IEEE Trans. Biomed. Eng. **61**(11), 2768–2778 (2014)
5. Xu, Y., et al.: Combined density, texture and shape features of multi-phase contrast-enhanced CT images for CBIR of focal liver lesions: a preliminary study. In: Chen, Y.-W., Toro, C., Tanaka, S., Howlett, R.J., Jain, L.C. (eds.) Innovation in Medicine and Healthcare 2015. SIST, vol. 45, pp. 215–224. Springer, Cham (2016). https://doi.org/10.1007/978-3-319-23024-5_20
6. Yu, M., Feng, Q., Yang, W., Gao, Y., Chen, W.: Extraction of lesion- partitioned features and retrieval of contrast-enhanced liver images. In: Computational and Mathematical Methods in Medicine (2012)
7. Yang, W., Lu, Z., Yu, M., Huang, M., Feng, Q., Chen, W.: Content-based retrieval of focal liver lesions using bag-of-visual- words representations of single- and multiphase contrast-enhanced CT images. J. Digit Imaging **25**, 708–719 (2012)
8. Diamant, I., et al.: Improved patch based automated liver lesion classification by separate analysis of the interior and boundary regions. IEEE J. Biomed. Health Inform. **20**(6), 1585–1594 (2016)
9. Xu, Y., et al.: Bag of temporal co-occurrence words for retrieval of focal liver lesions using 3D multiphase contrast-enhanced CT images. In: 2016 23rd International Conference on Pattern Recognition (ICPR 2016) (2016)
10. Xu, Y., et al.: Texture-specific bag of visual words model and spatial cone matching-based method for the retrieval of focal liver lesions using multiphase contrast-enhanced CT images. Int. J. Comput. Assist. Radiol. Surg. **13**(1), 151–164 (2018)
11. Diamant, I., Klang, E., Amitai, M., Konen, E., Goldberger, J., Greenspan, H.: Task-driven dictionary learning based on mutual information for medical image classification. IEEE Trans. Biomed. Eng. **64**(6), 1380–1392 (2017)
12. Wang, J., et al.: Sparse codebook model of local structures for retrieval of focal liver lesions using multi-phase medical images. Int. J. Biomed. Imaging. vol. 2017, Article ID 1413297, 13 pages (2017)
13. Litjens, G.: A survey on deep learning in medical image analysis. Med. Image Anal. **42**(9), 60–88 (2017)
14. Kolda, T.G., Bader, B.W.: Tensor decompositions and applications. SIAM Rev. **51**(3), 455–500 (2009)
15. Foruzan, A.H., Chen, Y.-W.: Improved segmentation of low-contrast lesions using sigmoid edge model. Int. J. CARS **11**(7), 1267–1283 (2016)

Joint Learning of LSTMs-CNN and Prototype for Micro-video Venue Classification

Wei Liu[1,2(✉)], Xianglin Huang[1], Gang Cao[1], Gege Song[1],
and Lifang Yang[1]

[1] Communication University of China, Beijing, China
cuclwl2@163.com
[2] Nanyang Institute of Technology, Nanyang, China

Abstract. Generally, venue category information of the micro-video is an important cue in social network applications, such as location-oriented applications and personalized services. In the existing micro-video venue classification methods, the discrimination becomes worse due to unsuitable convolutional filter and convolutional padding, and the robustness is not enough that is caused by the softmax layer. In order to alleviate such problems, we propose a novel learning framework which jointly learns LSTMs-CNN and Prototype for micro-video venue classification. Specifically, LSTMs-CNN with convolutional padding of the SAME type and small convolutional filter is used to extract spatio-temporal information. The Prototype is simultaneously learned to improve the robustness against softmax classification function. We adopt Euclidean distance loss function to train the whole network. Extensive experimental results on a real-world dataset show that our model significantly outperforms the state-of-the-art baselines in terms of both Micro-F and Macro-F scores.

Keywords: Micro-video venue classification · LSTMs-CNN
Prototype learning · Euclidean distance loss function

1 Introduction

Recently, with the popularity of smart mobile phones and social media platforms, such as Vine[1], Snapchat[2], and Instagram[3], users can conveniently shoot, capture and share micro-videos to record their own life stories vividly whenever and wherever possible. As a new form of user generated contents (UGCs), micro-videos are not only different from traditional long videos which contain lots of multi-scenes, but also distinct from some single images of missing sequence scene information. In spite of merely several seconds time spans, micro-videos have attracted extensive interests of smartphone users to record live events at some specific places, which benefit many potential location-oriented applications and personalized services, such as location recognition [1], landmark search [2] and commercial recommendations [3].

---

[1] https://vine.co/.
[2] https://www.snapchat.com/.
[3] https://instagram.com/.

© Springer Nature Switzerland AG 2018
R. Hong et al. (Eds.): PCM 2018, LNCS 11165, pp. 705–715, 2018.
https://doi.org/10.1007/978-3-030-00767-6_65

In the last few years, venue category estimation from micro-video has attracted the attention of some researchers. Zhang et al. [4] conducted a preliminary study on venue category estimation. Firstly, a large-scale public benchmark micro-video dataset collected from Vine was released. And then, they utilized the consensus information on visual, acoustic, and textual modalities of micro-videos and built a tree-guided multi-task multi-modal model (TRUMANN) to recognize venue information. Unfortunately, it is known from their experiments [4] that the acoustic modality demonstrates weaker capability than the visual modality in indicating venue information. This is because the quality of the acoustic signals is usually relatively lower. To alleviate the sparsity problem of the acoustic signals, Nie et al. [5] developed a deep transfer model (DARE) which jointly leveraged external sounds to strengthen the acoustic concept learning and the category similarity. In the above works, it is regrettable that characteristics of time-series from micro-video are not considered to estimate venue categories. In order to extract time-series information, Liu et al. [6] built an EASTERN model using three independent LSTMs to characterize the sequential structures of three modalities in parallel and using a convolutional neural network to learn their sparse and conceptual representations. However, because the sizes of convolutional filters are too large (namely n × 1) and the type of convolutional padding is VALID, these cause the model to become worse discriminative. Besides, because of the softmax function at the end of the architecture, the LSTMs and dictionary lack of robustness resulting in the reduction of final classification performance.

To address aforementioned problems, we propose an end-to-end joint learning of LSTMs-CNN and prototype model, as illustrated in Fig. 1. In particular, firstly, we also leverage three independent LSTMs to extract the sequential features of three modalities in parallel and project them into a common space by three distinct mapping functions. Secondly, for the three projected vectors with the same length, we then devise a CNN layer with filters of smaller size and padding of the SAME type to extract their sparse and conceptual spatio-temporal information. Thirdly, we assign multiple prototypes to represent different classes. It's easy to get the results of the classification by using Euclidean distance to find the nearest prototype. Finally, we design Euclidean distance loss functions for this whole model, making the feature representations extracted by LSTMs-CNN and prototypes being learned jointly. Therefore, the whole framework can be trained efficiently and effectively. We validate our model on a publicly accessible benchmark dataset, whereby each micro-video is labeled with one venue category. Extensive experiments demonstrate that our model can achieve comparable or even better classification accuracies compared with several state-of-the-art baselines.

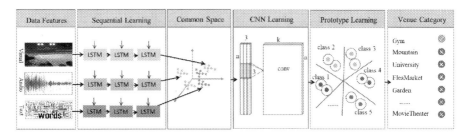

Fig. 1. Proposed framework using LSTMs-CNN and Prototype.

In this paper, our main contribution lies in the end-to-end jointly learning of LSTMs-CNN and prototype model to venue category estimation in the micro-video. Different from the prior work [6], our model applies the CNN with smaller filters and the SAME type of padding, followed by the directly learning of prototypes for each class. Prototype matching is used for decision making.

The rest of this paper is structured as follows. In Sect. 2, we detail our proposed model for micro-video venue classification. We conduct experiments and analyze the results in Sect. 3, followed by the conclusion drawn in Sect. 4.

2 Our Proposed Model

In this section, we present an end-to-end jointly learning of LSTMs-CNN and proto-type model for micro-video venue classification.

2.1 Notations

In this subsection, several notations are declared firstly. Bold capital letters (e.g., X) and bold lowercase letters (e.g., x) are employed to denote matrices and vectors, respectively. Non-bold letters (e.g., x) are used to represent scalars, and Greek letters (e.g., φ) as parameters. If not clarified, all vectors are in column form.

Suppose there are N micro-videos $X = \{x_i\}_{i=1}^{N}$. Each micro-video x is associated with one of c venue categories, namely a one-hot label vector y. We pre-segment each micro-video into three modalities $x = \{x^v, x^a, x^e\}$, whereinto the superscripts v, a and e respectively represents the visual, acoustic, and textural modality. We further denote $m \in \{v, a, e\}$ as modality indicator, and $x^m \in R^d$ as the d-dimensional feature vector over the m-th modality.

2.2 Sequential Feature Learning

The visual, acoustic, and textual modalities of micro-videos can be treated as sequential data. To capture the embedded sequential information, we can adopt LSTM suited for modeling sequential data. From prior work [6], for the three modalities of micro-videos, a parallel LSTMs can be devised as follow:

$$
\begin{cases}
i_t^m = \sigma\left(W_i^m x_t^m + U_i^m h_{t-1}^m + b_i^m\right) \\
f_t^m = \sigma\left(W_f^m x_f^m + U_f^m h_{t-1}^m + b_f^m\right) \\
o_t^m = \sigma\left(W_o^m x_t^m + U_o^m h_{t-1}^m + b_o^m\right) \\
g_t^m = tanh\left(W_g^m x_t^m + U_g^m h_{t-1}^m + b_g^m\right) \\
c_t^m = f_t^m \odot c_{t-1}^m + i_t^m \odot g_t^m \\
h_t^m = o_t^m \odot tanh\left(c_t^m\right) \\
m \in \{v, a, e\}
\end{cases}
\tag{1}
$$

where $\sigma(.)$ and $tanh(.)$ are the element-wise sigmoid and hyperbolic tangent functions, respectively; $\odot$ is the element-wise multiplication operator; for the visual, acoustic, and textual modalities, i_t^m, f_t^m and o_t^m are respectively treated as input, forget, and output gates, g_t^m represents a candidate cell state, c_t^m represents a memory cell vector at each time step; $x_t^m \in R^{d_v}$, $x_t^a \in R^{d_a}$ and $x_t^e \in R^{d_e}$ respectively represent the visual, acoustic, and textual input sequences at time t; W_j^m, U_j^m, and b_j^m for $j \in \{i, f, o, g\}$ are the parameters of the mth modality LSTM. The inputs of the three modalities are neither required to have the same dimensional feature space nor required to have the same time steps, namely the three LSTM networks are independent, they have different hidden units and different time steps. For micro-video venue classification, we extract respectively the final hidden representation from the last LSTMs step as the output of the parallel LSTMs.

2.3   Common Space Learning

As analyzed above, three independent feature vectors with different dimensions are outputted by the parallel LSTMs for each micro-video. These feature vectors are extracted from the same micro-video, so they are highly related. We can project them into common space to capture the intrinsic and latent characteristics. It is formally defined as:

$$\begin{cases} \tilde{x}_n^m = W_{nd}^m h_d^m + b_d^m \\ m \in \{v, a, e\} \\ d \in \{d_v, d_a, d_e\} \end{cases} \tag{2}$$

where $\tilde{x}_n^m$ is the mth modality embedding; W_{nd}^m is the mth modality embedding matrices; b_d^m is bias vector for the mth modalities; h_d^m is the final hidden representation from the last step of the mth LSTM; n is the dimension of the feature vectors in this learned common space; d represents the dimension of the hidden representation for the visual, acoustic, and textual embedding.

2.4   CNN Learning

From the preceding common space above, we obtain three feature vectors with equal length whereby each vector denotes one sequential feature of a modality from the micro-video. Such three feature vectors are come from the visual, acoustic and textual modalities of the same micro-video, so the vectors are not independent but highly correlated. Formally, for each micro-video, we can stack the three feature vectors into one feature map with size of $n \times 1 \times 3$ (i.e. n height, 1 width and 3 Channels) as:

$$\tilde{X} = [\tilde{x}^v, \tilde{x}^a, \tilde{x}^e] \tag{3}$$

where $\tilde{X} \in R^{n \times 1 \times 3}$ is a $n \times 1 \times 3$ matrix and represents the feature map, whereinto $\tilde{x}^v, \tilde{x}^a, \tilde{x}^e$ respectively denotes the embedding over the visual, acoustic, and textual modalities.

In our model, we aim to learn a sparse and semantic representation using Convolutional Neural Network. In particular, we devise a 2-D convolutional layer with k filters of size $[p, 1]$, a stride of $[1, 1]$ and SAME padding, where $p \leq n$. The feature map $\tilde{X}$ as input data. And ReLU is selected as the activation function. The formula is as follows:

$$f\left(\tilde{X}\right) = \max\left(0, W_{conv} * \tilde{X} + b_{conv}\right) \tag{4}$$

where W_{conv} and b_{conv} respectively represent the filters and bias, and $*$ denotes the convolution operation. Here, we apply k convolutions to extract the multi-modal feature, and each convolution has a kernel size p $\times$ 1. This is illustrated at the fourth part of Fig. 1. The output is composed of k feature maps, and the size of each feature map is $n \times 1$. The shape of the output is $n \times 1 \times k$, we finally reshape the output to a vector as a sparse representation of the three modalities.

2.5   Prototype Learning

After obtaining the multi-modal representation of micro-videos, we can denote it as $F(x; W)$, where x and W denote the raw input and parameters of the LSTMs-CNN, respectively. Motivated by [7], we also maintain and learn several prototypes on the sparse representation for each class and use prototype matching for classification. The prototypes are denoted as m_{ij} where $i \in \{1, 2, ..., c\}$ represents the index of the classes and $j \in \{1, 2, ..., p\}$ represents the index of the prototype in each class. Here we assume each class having equal number of p prototypes. The c is equal to the number of class label and the p is equal to the feature dimension. According to Sect. 2.4, the p value is $n \times 1 \times k$. Prototype of $c \times p$ can be denoted as M. This is illustrated at the center part of Fig. 2.

Figure 2 illustrates the distance representation learning procedure. In particular, assuming that we obtain a p-dimension feature vector of a micro-video via LSTMs-CNN. We aim to learn a distance representation based upon prototype M. Representation learning over prototype M with $c \times p$ is equivalent to get Euclidean distance applying p-dimension feature vector and c rows of the prototype M. The outputs are c-classes coefficients. The formula of Euclidean distance is as follows:

$$dis_i(x; W, M) = \left|\left|F(x; W) - M_{i,}\right|\right|_2^2 \tag{5}$$

For a general vector for classification, the probability that the sample x belongs to category y can be define:

$$x \in class\, y\ arg\ \max_{i=1}^{c} g_i(x) \tag{6}$$

However, we classify micro-video sample x to the category y according to the nearest prototype in distance representation comprised of the prototypes. That is to say,

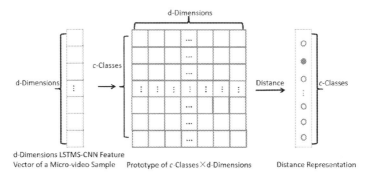

Fig. 2. Illustration of distance representation learning.

the smaller the distance, the greater the probability of the sample x belongs to a category y. Thus, the discriminant function $g_i(x)$ for category y can be formulated as:

$$g_i(x) = -\min_{i=1}^{c}||F(x; W) - M_{i,}||_2^2 \tag{7}$$

2.6   Euclidean Distance Loss Function

As aforementioned, the distance can be used to measure the similarity between the samples and the prototypes. The smaller the distance between them, the higher the similarity. For example, from the right part of Fig. 2, we can see that if the c-th distance is the smallest in the distance vector, the sample belongs to the c-th class. Therefore, the probability of a sample (x, y) can be measured by the distance between them:

$$p(y|x) \propto -||F(x; W) - M_{i,}||_2^2 \tag{8}$$

The probability of a sample (x, y) can be further given by softmax function as follow:

$$p(y|x) = softmax(F(x; W)) = \frac{e^{-\varphi dis\left(F(x;W), M_{i,}\right)}}{\sum_i^c e^{-\varphi dis\left(F(x;W), M_{i,}\right)}} \tag{9}$$

where φ is hyper-parameter to adjust ratio between p(y|x) and $dis\left(F(x; W), M_{i,}\right)$.

Therefore, based on the probability of p(y|x) cross entropy loss can be defined as follows:

$$loss((x, y); W, M) = -\frac{1}{N}y\log(p(y|x)) \tag{10}$$

This loss function is defined based on the Euclidean distance, so we call it Euclidean distance cross entropy loss. The loss function is applied in training stage of our model.

3   Experimental Results and Discussion

In this section, we conducted extensive experiments to verify our model.

3.1   Baselines, Dataset, Evaluation Metrics

To demonstrate the effectiveness of our proposed model, we chose the following several state-of-the-art methods as baselines:

TRUMANN: This is a tree-guided multi-task multi-modal learning method, which learns a common feature space from multi-modal heterogeneous spaces and utilizes the learned common space to represent each micro-video.

LSTMs + Late_Fusion: This is a model with three parallel LSTMs, which extract respectively modality-specific sequential features from visual, acoustic and textual modalities of each micro-video, and then the one feature vector is cascaded via the three extracted feature vectors and fed into a softmax classifier.

EASTERN: This is an end-to-end joint sequential-sparse model, which is capable of jointly capturing the sequential structures of visual, acoustic and textual modalities and sparsity of micro-videos.

LSTMs + CNN + FC and LSTMs + CNN + No_FC: The two models utilize three parallel LSTMs to extract modality-specific sequential features. And then the three sequential features are projected into a common space. After that the three projected vectors are fed into a CNN with SAME padding and small filters. Behind the architecture of the combined network, if fully connected layer is added, the one model is named as LSTMs + CNN + FC; if not, the other model is named as LSTMs + CNN + No_FC.

In order to validate our work, we leveraged a public benchmark micro-video dataset[4], which was crawled from Vine trough public API[5]. This dataset was released by [6] to estimate venue category. It consists of 20,093 micro-videos distributed in 22 Foursquare venue categories. This dataset is for sequence modeling task, so 11 frames are extracted, 6 audio clips are segmented and 30 text words are separated from each micro-video and its textual description. It is described by a rich set of features, namely, 4,096-D convolutional neural networks (CNN) visual features for each frame by AlexNet [8], 512-D acoustic features from each of these audio clips using Librosa[6] and 100-D features for each word in the textual description by Word2Vector [9].

Meanwhile, Macro-F [10] and Micro-F [11] were used to evaluate performance of our approach and the baselines from different angles. Macro-F gives equal weight to each class-label in the averaging process, while Micro-F gives equal weight to all

---

[4] https://acmmm17.wixsite.com/eastern.

[5] https://github.com/davoclavo/vinepy.

[6] https://github.com/librosa/librosa.

instances in the averaging process. Both Macro-F and Micro-F metrics reach their best value at 1 and worst at 0.

3.2   Experimental Settings

We randomly shuffled the dataset and split it into two parts: 18,000 micro-videos for training and 2,093 ones for testing. We trained our model and the baselines over the same training set and verified them over the same testing one. And we repeat the experiment ten times for each model and reported the average experimental results. Except TRUMANN model that was conducted based on Matlab2015[7], we implemented our model and baselines with the help of Tensorflow[8]. Thus, the GPU can be used for accelerating. All the experiments above were conducted on a 64-bit Ubuntu 16.04 Server with Intel Xeon(R) CPU E5-1603 v3 at 2.80 GHz ×4 on 20 GB RAM and NVIDIA GeForce GTX 1080 GPU support.

At the sequential feature learning and common space stages, according to the previous work [6], the time steps and hidden layer units of visual, acoustic and textual LSTMs were selected in [11, 6, 30] and [500, 300, 80], respectively. The dimension of common space was 150. At the CNN learning stage, we used 1 convolutional layer with convolutional padding of the SAME type and stride of 1. The filters size was set to be 3×1 and the number of convolutional filters was set to be 100. And at the prototype learning stage, we set the prototype size (i.e., M_{ij}), namely rows of the prototype as 22, columns of the prototype as $150 \times 1 \times 100$. During the training, all networks were trained using Adam optimizer with learning rate of 0.01 and learning rate decay of 0.25. Parameter φ in Eq. (9) was set to 1. Training stop if there is no improvement after 10 epochs with batch size of 100.

3.3   Performance Comparison Among Models

We summarized the performance comparison among different models in Table 1. From this table, we have the following observations: (1) The TRUMANN model achieves the worst performance, as compared to other learning approaches. This is because the method without using LSTMs cannot capture sequential information. (2) The EASTERN and LSTMs + CNN + No_FC models are better than LSTMs + CNN + FC model. This may be due to that the fully connected layer generates some redundancies. (3) the EASTERN and LSTMs + CNN + No_FC models outperform LSTMs + Later fusion model. This demonstrate the CNN can capture discriminate CNN features and such the features are beneficial to venue category classification. (4) The EASTERN performs weaker than the LSTMs + CNN + No_FC. This due to the fact that the EASTERN model used the VALID padding and filter of N $\times$ 1, while the LSTMs + CNN + No_FC model applied SAME padding and filter of 3×1. (5) Our proposed model is better than EASTERN and LSTMs + CNN + No_FC models. It demonstrates that prototype learning can achieve higher performance.

---

[7] https://ww2.mathworks.cn/.

[8] https://www.tensorflow.org.

Table 1. Performance comparison between our proposed method and several state-of-the-art baselines in terms of Micro-F and Macro-F.

Method	Micro-F (%)	Macro-F (%)
TRUMANN	58.58	18.87
LSTMs + Later fusion	61.47	31.03
LSTMs + CNN + FC	61.24	28.13
LSTMs + CNN + No_FC	62.24	32.72
EASTERN	62.00	30.69
Ours	62.73	32.93

Table 2. Performance comparison between our proposed method and EASTERN baseline with different modality combinations in terms of Micro-F and Macro-F.

Method	Modality	Micro-F (%)	Macro-F (%)
EASTERN	Visual	57.29	26.82
	Audio	44.55	11.09
	Text	41.38	8.60
	Visual + Audio	60.08	27.82
	Visual + Text	60.09	29.24
	Audio + Text	49.79	15.30
	ALL	62.00	30.69
Ours	Visual	58.03	28.57
	Audio	44.63	12.01
	Text	42.82	10.87
	Visual + Audio	60.50	29.95
	Visual + Text	60.72	31.04
	Audio + Text	50.66	17.00
	ALL	62.73	32.93

3.4 Comparison on Modality Combination Between Our Model and EASTERN

To further verify the effectiveness of our model, we also conducted experiments over different modality combinations of the visual, acoustic and textual modalities. Table 2 summarizes the multi-modal analyses and comparable results between our model and EASTERN. From this table, it is obvious that our model achieves better performance over different combinations of the three modalities, as compare to the EASTERN model. This further shows the effectiveness of prototype learning for LSTMs + CNN feature. Moreover, compared to the prior work [6], we can obtain the same conclusion that the more modalities we incorporate, the better performance we can achieve between our model and EASTERN model. This further demonstrates that one modality is insufficient and multimodalities are complementary to each other.

4 Conclusion

In this paper, we present a novel deep learning framework for micro-video venue classification by jointly learning between LSTMs-CNN and prototype. This approach is capable to model LSTMs-CNN to extract the spatio-temporal information and learn multiple prototypes for each class and then use prototype matching for decision making. To train our whole model, we applied the Euclidean distance loss function based on cross-entropy. This is an end-to-end scheme including four components: triple parallel LSTMs, common space projection, CNN learning and prototype learning. Experimental results demonstrate that our approach achieves much better classification accuracies and robustness, and thereby can be superior to several state-of-the-art baselines. In the future, we will try to joint prototype learning of different dimensions with deep model and apply more distance loss functions for the higher classification accuracies and more robustness.

Acknowledgments. We would like to thank the anonymous reviewers for their valuable comments. This work was supported by the National Natural Science Foundation of China (61772539), and the Fundamental Research Funds for the Central Universities (Nos. 3132017XNG1715, 3132018XNG1806).

References

1. Hays, J., Efros, A.A.: IM2GPS: estimating geographic information from a single image. In: Proceedings of IEEE Conference on Computer Vision and Pattern Recognition, pp. 1–8 (2008)
2. Zhu, L., Huang, Z., Liu, X., He, X., Song, J., Zhou, X.: discrete multi-modal hashing with canonical views for robust mobile landmark search. IEEE Trans. Multimed. **19**(9), 2066–2079 (2017)
3. Ye, M., Yin, P., Lee, W. C.: Location recommendation for location-based social networks. In: Proceedings of ACM SIGSPATIAL International Symposium on Advances in Geographic Information Systems, pp. 458–461 (2010)
4. Zhang, J., Nie, L., Wang, X., He, X., Huang, X., Chua, T.S.: Shorter-is-better: venue category estimation from micro-video. In: Proceedings of ACM International Conference on Multimedia, pp. 1415–1424 (2016)
5. Nie, L., Wang, X., Zhang, J., He, X., Zhang, H., Hong, R., et al.: Enhancing micro-video understanding by harnessing external sounds. In: Proceedings of ACM International Conference on Multimedia, pp. 1192–1200 (2017)
6. Liu, M., Nie, L., Wang, M., Chen, B.: Towards micro-video understanding by joint sequential-sparse modeling. In: Proceedings of ACM International Conference on Multimedia, pp. 970–978 (2017)
7. Yang, H., Zhang X., Yin F, Liu C.: Robust classification with convolutional prototype learning. In: Proceedings of IEEE Conference on Computer Vision and Pattern Recognition (2018)
8. Krizhevsky, A., Sutskever, I., Hinton, G.E.: Imagenet classification with deep convolutional neural networks. In: Proceedings of International Conference on Neural Information Processing Systems, pp. 1097–1105 (2012)

9. Mikolov, T., Sutskever, I., Chen, K., Corrado, G., Dean, J.: Distributed representations of words and phrases and their compositionality. In: Proceedings of International Conference on Neural Information Processing Systems, pp. 3111–3119 (2013)
10. Lepri, B., Mana, N., Cappelletti, A., Pianesi, F.: Automatic prediction of individual performance from thin slices of social behavior. In: Proceedings of ACM International Conference on Multimedia, pp. 733–736 (2009)
11. Sanden, C., Zhang, J.Z.: Enhancing multi-label music genre classification through ensemble techniques. In: Proceedings of International ACM SIGIR Conference on Research and Development in Information Retrieval, pp. 705–714 (2011)

Spatial Pixels Selection and Inter-frame Combined Likelihood Based Observation for 60 fps 3D Tracking of Twelve Volleyball Players on GPU

Yiming Zhao$^{(\boxtimes)}$, Xina Cheng, and Takeshi Ikenaga

Graduate School of Information, Production and Systems, Waseda University,
Kitakyushu, Japan
zhaoyiming@moegi.waseda.jp

Abstract. 3D players tracking plays an important role in sports analysis. Tracking of players contributes to high level game analysis such as tactic analysis and commercial applications such as TV contents. Many services like sports live and broadcasting have strict limitation on processing time, thus real-time implementation for 3D players tracking is necessary. This paper proposes a particle filter based 60 fps multi-view volleyball players tracking system on GPU platform. There are three proposals: body region constraint prediction, spatial pixels selection and inter-frame combined likelihood. The body region constraint prediction uses player's body region as limitation in prediction to increase tracking accuracy. The spatial pixels selection method selects pixels for likelihood calculating to reduce calculation amount in spatial space. The inter-frame observation method does particle filter algorithm with two frames each time to reduce calculation amount in temporal space. Our experiments are based on videos of the Final and Semi-Final Game of 2014 Japan Inter High School Games of Men's Volleyball in Tokyo Metropolitan Gymnasium. On the GPU device GeForce GTX 1080Ti, our tracking system achieves real-time on 60 fps videos and keeps the tracking accuracy higher than 97%.

Keywords: GPU acceleration · Particle filter
3D volleyball players tracking · Sports analysis

1 Introduction

Among sports analysis, 3D volleyball players tracking is a significant technology which extracts players' 3D position, trajectory and velocity. And these kinds of player information are useful for TV contents, player action recognition and tactic analysis. To meet requirements of applications such as volleyball game live and broadcasting, not only the accuracy but also the processing speed should be considered. As high quality live service needs fast processing system to analysis

© Springer Nature Switzerland AG 2018
R. Hong et al. (Eds.): PCM 2018, LNCS 11165, pp. 716–726, 2018.
https://doi.org/10.1007/978-3-030-00767-6_66

players performance, real-time tracking system is necessary. Also real-time player tracking is needed in efficient high level sports analysis such as player action recognition and player performance evaluation.

However, tracking algorithm with high accuracy usually needs large amount of calculation, which is the main difficulty to overcome when implementing real-time tracking system. Our research target is to implement a real-time 60 fps multi-view tracking system for twelve volleyball players in both teams with high accuracy.

Due to the rapid development of computer vision, a lot of efficient tracking systems appeared in recent years. Some tracking systems use KLT algorithm to track multiple targets by feature detection [1], but when it comes to volleyball players tracking, similar features of multiple players and frequent occlusion between players cause large amount of noises. There are also works on real-time tracking systems which use particle filter algorithm [2], but in their work only single person with simple movements is tracked. When it comes to volleyball players tracking the action and movements of players are rapid, random and complex. The number of players also influences tracking speed. There is a work on soccer players tracking using particle filter [6], but in volleyball games, players density is larger than soccer game, the occlusion problem is more complex and frequent. Some works use graph optimization to track multiple sports players [5], but they use some algorithm such as K-shortest paths (KSP) which has high data dependency between players, thus it is not friendly to acceleration on heterogeneous computation devices. Thus we choose to implement the real-time multi-view 3D players tracking system with particle filter algorithm that is robust enough for multiple volleyball players [3].

The chosen conventional work [3] tracks volleyball players based on particle filter and achieves high accuracy, but one team's players are tracked because of view limitation. The number of views is changed from 3 to 4 in our implementation to catch both two teams' information. To make it more robust for twelve volleyball players' situation, we propose one proposal, the description of this proposal in given in Sect. 2.

In the past years, heterogeneous computation device develops very fast, more and more people focus on powerful platforms such as GPU to implement high speed system. And particle filter is an algorithm suitable for parallel computation. There are also real-time tracking systems for volleyball on GPU using particle filter [4,7] which show high potential of GPU acceleration. Thus we select GPU as our platform.

The conventional volleyball players tracking method [3] runs on CPU and process each particles sequentially. In the chosen tracking system, the most complex and time-consuming part is the observation. Since in observation part, HSV, Sobel and number feature of players are calculated. What's more, large amount of particles and multiple camera views also cause numerous calculation amount. In the conventional real-time volleyball tracking system [4], the tracking target size is very small, less than 200 pixels are calculated in each camera view. The number of target to track is 1 and the feature used is HSV. But in volleyball players tracking [3], totally 12 players are needed to track and each player has over 5000 pixels in each camera view, and multiple features such as HSV and Sobel are calculated. Thus volleyball players tracking has much larger calculation amount than volleyball tracking.

To solve the large calculation amount problem, we proposed two proposals. The spatial pixels selection method reduces calculation amount in spatial space based on decreasing number of pixels used in feature calculation. The inter-frame combined likelihood method reduces calculation amount in temporal space based on separating two kinds of feature calculation into two adjacent frames. The detail descriptions for this two acceleration proposals are in Sect. 3.

The remaining part of this paper is arranged as follows: Sect. 2 introduces the framework of our 3D volleyball players tracking system and one proposal which makes algorithm more robust. Section 3 covers description of our two acceleration proposals. Section 4 is experiment results and analysis. The final section is conclusion.

2   3D Volleyball Players Tracking Framework

2.1   Tracking Algorithm Framework

We show framework of our twelve volleyball players tracking algorithm in Fig. 1, the input is 4 cameras' synchronous videos that catch all 12 players, and output is volleyball player's 3D position. The flow of conventional algorithm is divided into four parts: Prediction, Observation, Reweight and Resampling. We use only HSV and Sobel features since number feature has very small influence on tracking result but costs very large time. When we implement the tracking system on GPU, we propose the first proposal in prediction to make tracking algorithm more robust.

The body region constraint prediction model helps to predict players' next positions more reasonable and avoid the position exchange error between players in tracking system, we show the difference of our proposed prediction method in Fig. 2. In the conventional work [3], players are considered separately when distributing new particles of them, which means relationship between players are not considered. In this case when two players stay very close with each other, they are possible to exchange their position in tracking results. While in our proposed body region constraint method, we add the limitation caused by player's body region to avoid impossible distribution. We apply a cylinder region to each player as the body region, the cylinder size is constant after initialization. Then for each player we check whether the new particles for this player go into other players' body region or not. With player body region's limitation on particle distribution, we make our tracking algorithm performs more robust when multiple players stay very close with each other.

2.2   GPU Implementation Framework

Our GPU implementation framework is shown in Fig. 3. It is based on conventional GPU implementation work of ball tracking [4]. In our implementation, the

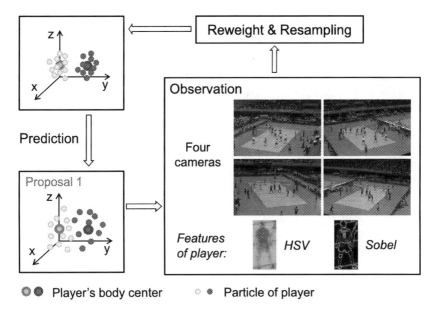

●◉ Player's body center   ○ ● Particle of player

Proposal 1: Body region constraint prediction model

Fig. 1. Overall framework of multiple players tracking algorithm.

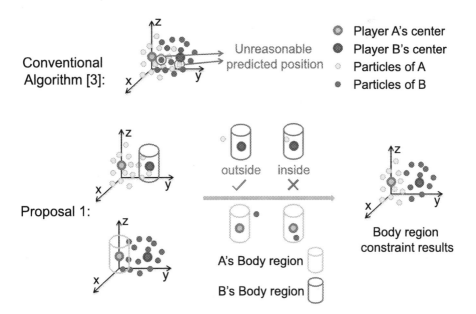

Fig. 2. Comparison of proposal 1: Body region constraint prediction.

initial position of players and video information are transferred from host computer to GPU device, after that the tracking algorithm are executed for every input frame on GPU. The tracked 3D positions of players are returned to host device to generate result videos. There are six kernels implemented on GPU, K1: Convert color, K2: Prediction, K3: HSV feature observation, K4: Sobel feature observation, K5: Resampling and K6: Reweight.

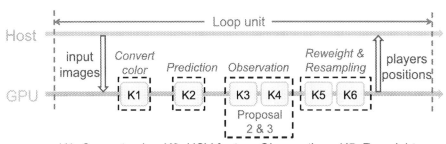

Fig. 3. Framework of GPU implementation.

In the first kernel, we convert the input RGB images into HSV and gray images. In observation part we use HSV values to calculate HSV likelihood and gray values to calculate Sobel likelihood.

In K2, we do prediction to distribute particles. This kernel mainly assigns each particles a 3D position as the predicted players center point. We distribute particles according to Gaussian distribution and Least Square Fitting method in conventional work [3], and add player region constraint model to make prediction more reasonable.

For observation task, we have 2 kernels K3 and K4 to calculate the 2D likelihood for each particles. HSV feature is used in K3 to calculate HSV likelihood and Sobel feature is used in K4 to obtain Sobel likelihood. In these two kernels, the thread count is 4 times of particles number, while 4 is cameras number. We fully utilize the shared memory in GPU device to help accelerate by loading as much as possible pixels around last player's center position to shared memory. The observation task has largest calculation amount and we propose two proposals to accelerate, they are explained in Sect. 3.

Finally is the reweight and resampling, K5 calculates 3D likelihood according to the results of observation. K6 estimates the 3D positions of players and then updates players information. Behind these 2 kernels, players positions are returned to host device as tracking results.

3   Acceleration on GPU

In our particle filter based volleyball players tracking algorithm, observation costs the most time. In observation we need to calculate likelihood for each particle that distributed in prediction. While in our tracking systems, for each player we use 1024 particles. When calculating likelihood for these particles, we need to calculate 2D likelihood on each view. Just as Fig. 4 shows, for each particle we need to use all pixels in players ROI (Region of interest) to calculate HSV and Sobel likelihood, the features we calculate for each ROI are summed into a histogram, then we calculate the similarity of histogram as 2D likelihood. In this progress, the number of pixels in players ROI (Usually more than 5000) is very large which brings large calculation. However, as likelihood focuses on similarity of features and our video has high resolution, we don not need to calculate all pixels in player's ROI. Thus we propose spatial pixels selection method to decrease the number of pixels used in observation.

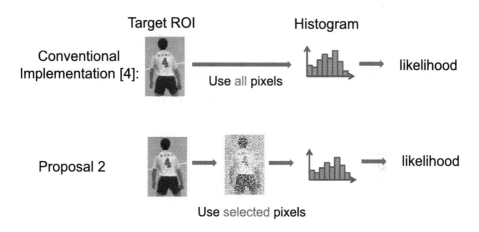

Fig. 4. Proposal 2: Spatial pixels selection in observation.

The selection strategy is described as follow. We divide players ROI into multiple 3×3 regions. Mark 9 pixels in a 3×3 regions as $p_i(i = 1, 2, \ldots, 9)$. We select 2 pixels (p_a and p_b) in each 3×3 region to use in feature calculation. Firstly we give each pixel a weight w_i which is determined by color distance between current frame and last frame. We mark the 9 pixels value in last frame as $q_i(i = 1, 2, \ldots, 9)$. We set the weight as

$$w_i = \frac{Distance(p_i, q_i)}{Max\_distance} + 1. \tag{1}$$

The distance function we use is Euclidean distance. After the determination of weights, we sum some pixels' weights together if they have same RGB values. And then we select the most two highest weighted pixels to use in feature calculation. If there are candidates with same weight, we select randomly. With the pixel selection, the number of pixels used in feature calculation decreases but similarity calculation between histogram still works well.

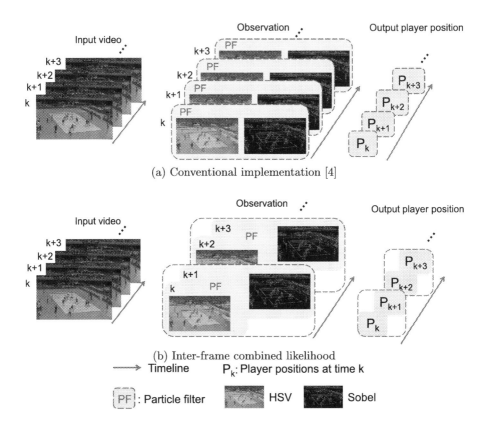

(a) Conventional implementation [4]

(b) Inter-frame combined likelihood

Fig. 5. Concept difference between conventional and proposal 3

In conventional tracking algorithm we do particle filter frame by frame to track players, but in 60 fps videos, the 3D positions change between adjacent frames is very small. Thus we consider it is possible to decrease the calculation amount on temporal space. We propose the inter-frame combined likelihood to do particle filter with two frames each time. As Fig. 5 shows, the conventional work does particle filter algorithm on each frame. But in our propose method, after we know players' positions on time t, we predict players' next position on time $t + 2$ and use the mid point between t and $t + 2$ as position on time $t + 1$. In observation Sobel feature is based on frame $t + 1$ and HSV feature is based on frame $t + 2$. In reweight and resampling part, we determine players position on time $t + 2$ first and use mid point between t and $t + 2$ as result of frame $t + 1$. Since the position change between adjacent frames is very small for volleyball players, our proposed method also catches players' positions with high accuracy.

4 Experiment

4.1 Experiment Environment

Our experiment uses one GPU device, the NVIDIA GeForce GTX 1080Ti. And for host computer, the CPU is Intel i7-6700K, 4.01 GHz, the memory size is 32 GB. For the third party tools we use OpenCV 2.4.11 and CUDA 8.0. Our program is written in C++ and CUDA C language.

4.2 Dataset and Evaluation Method

Our experiment is based on videos of an official volleyball game (Final and Semi-Final Game of 2014 Japan Inter High School Games of Mens Volleyball in Tokyo Metropolitan Gymnasium). There are totally 4 different teams in our sequences. And each sequence includes 4 cameras' synchronous videos. We set the 4 cameras separately at 4 corners of volleyball game playground to catch all twelve volleyball players. For the cameras we use, the shutter speed is 1000 per second, the resolution is 1920×1080 and the frame rate is 60 fps. In our sequences, there is no motion blur occurs on volleyball players.

For tracking accuracy evaluation, we follow the method in Huang's work [3]. In Huang's evaluation method, RU (Respond Unit) is defined. We divide each round of volleyball game into several RU and each RU is a period contains action series as: One team's player receives ball from opposite team, then passes ball, finally returns ball to opposite team by spiking. Under definition of RU, the success rate is defined as

$$Success\_rate = \frac{\Sigma_1^n Tracked\_players}{12 \times n}. \tag{2}$$

In this equation, n is the number of RUs and 12 is the number of volleyball players.

For evaluation on acceleration, we consider the time consuming T on each frame, which includes these processes: data transmission between CPU and GPU, tracking algorithm execution on GPU. The time for video reading and output video creating on CPU is not included. Speed up of our implementation is calculated as

$$Speedup = \frac{T_{CPU\ implementation}}{T_{GPU\ implementation}}. \tag{3}$$

4.3 Experiment Result and Consideration

When we show our experiment result, P1 is body region constraint prediction model, P2 is spatial pixels selection method, P3 is inter-frame combined likelihood method.

In experiment result part, we show the comparison on tracking accuracy in Table 1, then show comparison of system's tracking speed in Table 2.

Table 1. Tracking accuracy comparison.

		Game 1		Game 2		Total
		Team1	Team2	Team3	Team4	
$12 \times$ RU		372	372	348	348	1440
Conventional [3]	Tracked	370	353	330	340	1393
	Success rate	99.46%	94.89%	94.82%	97.70%	96.73%
P1	Tracked	370	361	342	344	1417
	Success rate	99.46%	97.04%	98.27%	98.85%	98.40%
P1 + P2 + P3	Tracked	363	356	338	340	1397
	Success rate	97.58%	95.69%	97.12%	97.70%	97.01%

In Table 1, there are 2 games including 4 teams in our dataset. The conventional work [3] achieves 96.7% tracking accuracy on our dataset. After adding P1 the tracking accuracy increases to 98.4%. With acceleration proposals P2 and P3, the implementation on GPU has tracking accuracy higher than 97%.

Table 2. Speedup comparison.

Time-consuming (ms/frame)	CPU	GPU			
		Conventional [4]	P1	P1 + P2	P1 + P2 + P3
K1: Convert color	130	0.5	0.5	0.5	0.5
K2: Prediction	20	0.65	0.66	0.66	0.34
K3: HSV observation	15530	117.56	117.56	17.14	8.65
K4: Sobel observation	8090	64.45	64.45	9.48	4.79
K5: Reweight	5	0.06	0.06	0.06	0.03
K6: Resampling	25	3.39	3.39	3.39	1.70
SUM	23800	186.62	186.62	31.23	16.01
Speedup	1	127.5	127.5	762.1	1486.5

In Table 2 we give a comparison between CPU and GPU implementation on execution time. The time cost on CPU is very large and far away from real-time, but when we implement it on GPU the time cost decreases a lot. As showed in the table, conventional implementation has low time cost but still not enough for real-time on 60 fps videos. After adding the first proposal to improve tracking accuracy, the time cost in K2 increased. And after adding spatial pixels selection method, the time cost decreases to lower than 31.23 ms/frame and accuracy is kept. And with the help of inter-frame combined likelihood method, the time cost per frame decrease to 16.01 ms/frame that is low enough for 60 fps tracking (less than 16.66 ms/frame).

And as the Table 2 shows, the most time costing part is observation part which includes HSV and Sobel feature observation. K3 and K4 calculate HSV and Sobel likelihood and have largest calculation amount. By adding spatial pixels selection, we decrease time cost in K3 and K4 a lot. And after applying inter-frame combined likelihood method to do particle filter with two frames each time, as the total execution times of particle filter algorithm decreases, time cost on most kernels reduces a lot.

5   Conclusion

In this paper we present a GPU based implementation of real-time 60 fps multi-view 3D tracking system for twelve volleyball players. We present one proposal in prediction to increase tracking accuracy and two acceleration proposals in observation to achieve real-time tracking with 60 fps videos. The body region constraint prediction avoids impossible particles distribution in order to improve accuracy of tracking algorithm. The spatial pixels selection method decreases the number of pixels used in likelihood calculation in each frame. The inter-frame combined likelihood method decreases calculation amount on temporal space by separating HSV feature and Sobel feature calculation into two adjacent frames. With experiment on single NVIDIA GeForce GTX1080Ti device, our tracking system's speed is faster than 16.01 ms/frame and accuracy is higher than 97%. Compared to CPU implementation runs on Intel i7-6700K 4.01 GHz, our system is 1486 times faster. Compared to conventional GPU implementation, our system is 11.65 times faster.

For future work, various efforts such as tracking recovery method are needed to apply real-time volleyball players tracking to applications. High level analysis such as tactic analysis and players action analysis are possible to be more efficient with real-time volleyball players tracking system.

Acknowledgements. This work was supported by KAKENHI (16K 13006) and Waseda University Grant for Special Research Projects (2018K-302).

References

1. Buddubariki, V., Tulluri, G.S., Mukherjee, S.: Multiple object tracking by improved KLT tracker over SURF features. In: 5th National Conference on Computer Vision, Pattern Recognition, Image Processing and Graphics (NCVPRIPG), pp. 1–4 (2015)
2. Tseng, T., Liu, A., Hsiao, P., Huang, C., Fu, L.: Real-time people detection and tracking for indoor surveillance using multiple top-view depth cameras. In: IEEE/RSJ International Conference on Intelligent Robots and Systems, pp. 4077–4082 (2014)
3. Huang, S., Zhuang, X., Cheng, X., Ikoma, N., Honda, M., Ikenaga, T.: Player feature based multi-likelihood and spatial relationship based multi-view elimination with least square fitting prediction for volleyball players tracking in 3D space. IIEEJ Trans. Image Electr. Vis. Comput. 4(2), 145–155 (2016)

4. Deng, Z., Hou, Y., Cheng, X., Ikenaga, T.: Vectorized data combination and binary search oriented reweight for CPU-GPU based real-time 3D ball tracking. In: Zeng, B., Huang, Q., El Saddik, A., Li, H., Jiang, S., Fan, X. (eds.) PCM 2017. LNCS, vol. 10736, pp. 508–516. Springer, Cham (2018). https://doi.org/10.1007/978-3-319-77383-4_50
5. Nishikawa, Y., Sato, H., Ozawa, J.: Multiple sports player tracking system based on graph optimization using low-cost cameras. In: IEEE International Conference on Consumer Electronics (ICCE), pp. 1–4 (2018)
6. Petsas, P., Kaimakis, P.: Soccer player tracking using particle filters. In: IEEE International Symposium on Signal Processing and Information Technology (ISSPIT), pp. 57–62 (2016)
7. Hou, Y., Cheng, X., Ikenaga, T.: Real-time 3D ball tracking with CPU-GPU acceleration using particle filter with multi-command queues and stepped parallelism iteration. In: International Conference on Multimedia and Image Processing (ICMIP), pp. 235–239 (2017)

An Interactive Light Field Video System with User-Dependent View Selection and Coding Scheme

Bing Wang[1]([✉]), Qiang Peng[1], Xiao Wu[1], Eric Wang[2], and Wei Xiang[2]

[1] Southwest Jiaotong University, Chengdu, China
iceice_wang@outlook.com
[2] James Cook University, Cairns, Australia

Abstract. Due to the sheer size of Light Field (LF) video, the traffic of LF video is many times larger than traditional multimedia, which brings new challenges on how to efficiently store and transmit this huge amount of complex data. In this paper, we propose an interactive LF video streaming system with user-dependent view selection and coding scheme. The user trajectory prediction model proposed in this paper can be used to calculate the viewing area of the user in a limited consecutive number of time slots. With the projection model, the system can determine the interested views of the user for transmission. Furthermore, we construct a special LF video sequence with only the selected view at a consecutive number of time slots. By reducing the number of views for transmission, the system can significantly reduce the traffic load in transmission and realize a low-latency real-time LF video application.

Keywords: Light Field video · Video transmission · View selection
User-dependent · Trajectory prediction

1 Introduction

Recent advances in computational photography have enabled us to explore Light Field from a digital perspective [1]. LF photography is able to capture both spatial and angular information, and enables new possibilities for digital imaging [2]. With the advent of commercial LF cameras, LF photography has been popularly adopted by professional photographers to create animated and multi-focus digital photos. Additionally, the depth perception available from LF has enabled LF technology to be implemented in a wide range of conventional industry applications [3,4].

One problem when dealing with LF video is the sheer size of data volume. Compared to conventional 3D data in a format of coarsely distributed multi-view images, LF data exhibits a more complex structure, consisting of densely distributed 2D image arrays. All current coding methods for LF video consider an LF video as a whole and compress all the views to achieve a high compression

© Springer Nature Switzerland AG 2018
R. Hong et al. (Eds.): PCM 2018, LNCS 11165, pp. 727–736, 2018.
https://doi.org/10.1007/978-3-030-00767-6_67

rate and save the bandwidth [5–8]. In some interactive or selective applications where the user is able to choose the interested area to be displayed, compressing and transmitting the whole LF video not only generates a very significant computation load, but also creates an enormous coding and transmission redundancy. It is hence difficult to employ them in low-latency streaming applications, such as telehealth consultation [4].

In [9], the authors proposed an user dependent scheme for multi-view video transmission. The proposed scheme makes use of the periodically feedback information to encode and transmit only the frames that are possibly to be displayed instead of all the frames, and is able to decrease the bit-rate for multi-view video transmission. To our knowledge, none work has been conducted to an interactive user-dependent scheme for LF video.

In this paper, we propose an interactive LF video streaming system using user-dependent coding and transmission scheme. The user trajectory prediction model proposed in this paper is able to adapt to multiple input devices, such as mouse/one finger, two fingers/hands, and user movement tracking. The predicted trajectory can be used to calculate the viewing area of the user in a limited consecutive number of time slots. The system can determine the interested views of the user for transmission by using the proposed prediction model. The remainder of the paper is organized as follows. The effective framework of the interactive LF video streaming system is presented in Sect. 2. We illustrate the structure for user-dependent view selection in Sect. 3. The conclusion and direction in the future research are provided in Sect. 4.

2   An Interactive Light Field System Model

In the LF video streaming system without user dependent LF video coding, all the views of an LF are transmitted as a whole, which creates a high complexity and generates a large burden for transmission. As a result, it is difficult to realize real-time coding and transmission for LF video streaming services. In this paper, we propose an interactive LF video streaming system with user-dependent view selection and coding scheme. As shown in Fig. 1, the captured LF video in raw LF format are first decoded, calibrated and rectified as an LF multi-view video sequence. In contrast to the algorithm proposed in [7], where the whole LF multi-view video sequence is compressed and transmitted directly via networks, an user interaction feedback scheme is adopted in this system to provide the user interaction information.

The position of the user or navigation destination are then translated via the corresponding projection model to obtain the selected views for different positions for each frame in a consecutive number of time frames. The selected view sequence is then compressed using an adaptive coding algorithm for LF video coding to a single video stream. The video stream is then transmitted through networks, which is then decoded to recover the selected view sequence. The rendering algorithm then render the selected views for display.

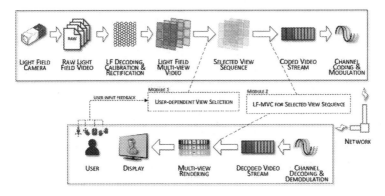

Fig. 1. System framework of interactive LF video streaming.

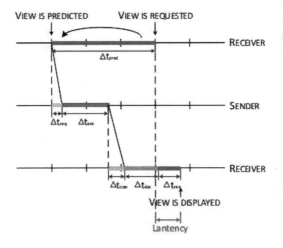

Fig. 2. User-dependent transmission protocol.

Figure 2 shows a diagram of the set of actions taken by the server and receiver under user-dependent LF video coding, scheduling and transmitting scenarios. In the user-dependent streaming system, the first step of the streaming process is predicting the view trajectory. Details of how view prediction is actually performed is given in Sect. 3. Essentially, it can be thought of as a way of knowing a desired view, approximately, ahead of time. To make this idea more precise, suppose that the user wants to view the trajectory $\mathbf{v} = \{v_1, v_2, ..., v_M\}$, where M is the predicted time blocks. It is assumed that the sender and receiver clocks are synchronized, and therefore, all times can be given in terms of single global reference clock. If view trajectory can be predicted ahead by Δt_{pred}, then view $\tilde{v}_i$ an approximate of the set of views v_i can be known at time $t_{v_i} - \Delta t_{pred}$ instead of time t_{v_i}.

The prediction of view trajectory is implemented at the receiver. In the user-dependent scenario, having predicted the view trajectory, the desired set of views can be requested. The view requests are transmitted from receiver to the sender, taking an interval of time Δt_{req}, and the server responds immediately by encoding the corresponding selected view sequence, which takes time interval Δt_{enc}. The encoded video stream takes time interval Δt_{tran} to arrive at the receiver. Once the client receives the encoded video stream, it decodes them and renders the view, taking times Δt_{dec} and Δt_{ren}, respectively. The selected view sequence in the view trajectory must arrive within a fixed period of time after the user selects it. Suppose that the maximum tolerable latency between having selected a view and it being rendered is L_{max}. Therefore, view set v_i must be completely rendered on the display screen by $t_{v_i} + L_{max}$.

Therefore, at a random streaming time t_s, in order to satisfy the latency requirement for streaming, the processing time for encoding, decoding and rendering $\Delta t_{proc} = \Delta t_{req} + \Delta t_{ec} + \Delta t_{tran} + \Delta t_{dec} + \Delta t_{ren}$ has to satisfy the following constraint,

$$t_s + \Delta t_{proc} - L_{max} < t_{v_i} \leq \Delta t_{pred}. \tag{1}$$

3   User-Dependent View Selection

In the user-dependent streaming system, the first step of the streaming process is predicting the view trajectory and determine the set of views for rendering and display. The framework of the user-dependent view selection method proposed in this paper is shown in Fig. 3, where we illustrate framework of using the input signals from users to determine the set of views for rendering and display.

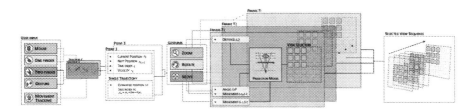

Fig. 3. User-dependent view selection framework.

The signals from user movement or gestures are first collected, which can be categorized into two groups, (1) single point input devices, such as mouse and one finger touch input; and (2) multiple point input devices, with the popular two point touch or gesture input. With the additional one extra input, these input devices is able to achieve a more complicated and richer user gestures, e.g., moving, rotating and zooming. The two interactive methods follow two completely different projection model, therefore the trajectory calculation and prediction for the two methods also follow two different scheme. In the following two sections, we will discuss the two scheme separately.

The following calculations are carried out in the horizontal direction, while the vertical direction navigation can be easily explored with the same procedures presented given below. First of all, we will make some notations for the coordinate system using in the two projection models.

Denoted by Z the axis along the depth in horizontal direction, and the camera array lies on the $z = 0$ plane. The depth range of the scene is from $-z_{max}$ to $-z_{min}$, with z_o as the center plane of the interested object in the scene. On the other hand, S is the axis along the field of view in horizontal direction. Assuming that the center of the object is aligned with the centers of the scene and camera array, which all lie on $s = 0$ plane. The distance between each camera in the camera array is d, and the angle of the field of view of each camera is β, while that of the user is α. The size of the camera array is denoted by $H \times V$, where H is the number of cameras in the horizontal direction, and V is the number of cameras in the vertical direction. Therefore, the positions of cameras in S axis $\{s_1^c, s_2^c, s_3^c, ..., s_V^c\}$ can be calculated as

$$s_i^c = \left(i - \frac{H}{2}\right)d - \frac{d}{2}, \forall H = 2k : k \in \mathbb{Z}, \tag{2}$$

$$s_i^c = \left(i - \frac{H-1}{2}\right)d, \forall H = 2k+1 : k \in \mathbb{Z}. \tag{3}$$

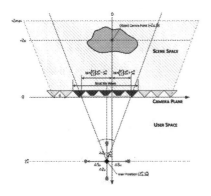

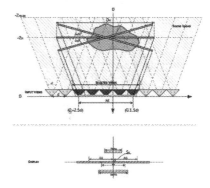

Fig. 4. Projection model for LF display with user movement tracking.

Fig. 5. Projection model for 2D display with gesture recognition.

3.1   User Tracking

In this interactive system, the user position and movement is tracked via a 2D/3D camera. Hence, with the current user location and the predicted trajectory of movement in a limited length of lookahead time period, the user position can be predicted for any time index ahead in a limited time window. Via the projection model, which maps the user position to the viewing area of the display, the sets of views for any time index ahead in this limited time window can be calculated, thus encoded and transmitted by the sender, and decoded and rendered by the receiver.

At the first transmission of the streaming service, the whole captured scene is displayed at the user side in disregard of user's position. The initial position of the user is denoted by $p^0 = \{z_U^0, s_U^0\}$, where z_U^0 is the user's distance to the display, as shown in Fig. 4, and s_U^0 is the distance of the user to the centre of camera array in S axis.

The position of user at time index k is denoted by $p_U^k = (z_U^k, s_U^k)$. For $k = 0, 1, 2, ...$, the estimated velocity of the user can be given by

$$v_U^k = \sigma(p_U^k - p_U^{k-1}) + (1 - \sigma)v_U^{k-1}, \tag{4}$$

where σ is a smoothness factor for velocity estimation.

If the user position is known up to time index k, then the velocity estimate v_U^k can be computed recursively, and the user position for time index $m > k$ is computed using

$$p_U^m = p_U^k + (m - k)v_U^k. \tag{5}$$

Denoted by $p_U^m = \{z_U^m, s_U^m\}$ the position of the user at time index m, the viewing region of the scene can be calculated as

$$\left\{-z_{max}, \tan\frac{\alpha}{2}(z_U^m + z_{max}) - s_U^m\right\}, \left\{0, \tan\frac{\alpha}{2}z_U^m - s_U^m\right\},$$

$$\left\{-z_{max}, \tan\frac{\alpha}{2}(z_U^m + z_{max}) + s_U^m\right\}, \left\{0, \tan\frac{\alpha}{2}z_U^m + s_U^m\right\}.$$

Similarly, the camera region for the rendering can be calculated as

$$\left\{0, \tan\frac{\alpha}{2}z_U^m - s_U^m\right\}, \left\{0, \tan\frac{\alpha}{2}z_U^m + s_U^m\right\}. \tag{6}$$

Therefore, the selected set of camera views $i^m \in \{i_{min}^m, i_{max}^m\}$ for rendering can be derived as

$$i_{min}^m = \frac{H}{2} - \left\lceil \frac{\tan\frac{\alpha}{2}z_U^m - s_U^m - \frac{d}{2}}{d} \right\rceil,$$

$$i_{max}^m = \frac{H}{2} + 1 + \left\lceil \frac{\tan\frac{\alpha}{2}z_U^m + s_U^m - \frac{d}{2}}{d} \right\rceil,$$

$$\forall H = 2k : k \in \mathbb{Z}, \tag{7}$$

and

$$i_{min}^m = \frac{H+1}{2} - \left\lceil \frac{\tan\frac{\alpha}{2}z_U^m - s_U^m}{d} \right\rceil,$$

$$i_{max}^m = \frac{H+1}{2} + \left\lceil \frac{\tan\frac{\alpha}{2}z_U^m + s_U^m}{d} \right\rceil,$$

$$\forall H = 2k + 1 : k \in \mathbb{Z}. \tag{8}$$

Moreover, from the projection model, it can be easily deduced that the set of camera views will move for one camera per d movement on either side, i.e.,

$$\{s_U^m - s_U^k = d\} \to \{i_{min}^m = i_{min}^k + 1, i_{max}^m = i_{max}^k + 1\}, \tag{9}$$

$$\{s_U^k - s_U^m = d\} \to \{i_{min}^m = i_{min}^k - 1, i_{max}^m = i_{max}^k - 1\}. \tag{10}$$

and the camera views will move for one camera per d movement in both sides for the zoom-in and zoom-out actions, i.e.,

$$\{z_U^m - z_U^k = d\} \to \{i_{min}^m = i_{min}^k - 1, i_{max}^m = i_{max}^k + 1\}, \tag{11}$$

$$\{z_U^k - z_U^m = d\} \to \{i_{min}^m = i_{min}^k + 1, i_{max}^m = i_{max}^k - 1\}. \tag{12}$$

3.2   Gesture Navigation

For this interactive method, users use different input device to navigate the scene displayed on the screen. For the single point navigation system, the calculation of the movement is similar to the calculation of the user position in the last subsection. Therefore, we will focus on analyzing the gestures using multiple points, such as two finger or two hands gesture control.

The initial state of the navigation system is defined by $\{D_{s0}, s_0 = 0\}$, where D_{s0} represent the width of the field of view at horizontal direction, and s_0 is the distance of the centre of display to the centre of cameras in S axis, as shown in Fig. 5. The projection factor is defined as the ratio between the displacement of the gesture against the displacement in the display scene in the object centre, as a, where $D_{v0} = 2Z_{max} \tan \frac{\beta}{2} + Hd$. Therefore, to determine the selected view, we need to calculate the position S_s according to the displacement of the fingers/hands, the viewing area D_s according to the displacements of two fingers, and the rotation angle θ_s according to the displacements of two fingers.

The positions of the left and right fingers (or hands) at time index k are denoted by $p_l^k = (z_l^k, s_l^k, t_l^k)$ and $p_r^k = (z_r^k, s_r^k, t_r^k)$, respectively. Denoted by

$$\begin{aligned} \boldsymbol{r} &= \overrightarrow{r}_z \hat{\mathbf{i}} + \overrightarrow{r}_s \hat{\mathbf{j}} + \overrightarrow{r}_t \hat{\mathbf{k}} \\ &= (z^k - z^{k-1})\hat{\mathbf{i}} + (s^k - s^{k-1})\hat{\mathbf{j}} + (t^k - t^{k-1})\hat{\mathbf{k}}, \end{aligned} \tag{13}$$

the vector for one single finger, we are able to determine the eight different gestures as shown in Table 1.

The positions of left and right fingers/hand at time index k are denoted by $p_l^k = (z_l^k, s_l^k, t_l^k)$ and $p_l^k = (z_l^k, s_l^k, t_l^k)$. Following the similar analyse for user tracking, the positions of left and right fingers/hand at time index $m > k$ can be computed using

$$p_l^m = p_l^k + (m - k)v_l^k, \tag{14}$$

$$p_r^m = p_r^k + (m - k)v_r^k. \tag{15}$$

For zoom-in and zoom-out actions,

$$D_v^m = D_v^k \mp \Delta D,$$

$$\Delta D = \max \left(|s_l^m - s_l^k|, |s_r^m - s_r^k| \right) a. \tag{16}$$

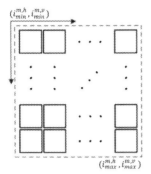

$(i_{min}^{m,h}, i_{min}^{m,v})$

$(i_{max}^{m,h}, i_{max}^{m,v})$

Fig. 6. The selected set of views for time index m.

Table 1. Gesture conditions

Gestures	Condition
Zoom-in	$\overrightarrow{r}_s^{\,l} < 0\,\&\,\overrightarrow{r}_s^{\,r} > 0$
Zoom-out	$\overrightarrow{r}_z^{\,l} > 0\,\&\,\overrightarrow{r}_z^{\,r} < 0$
Clockwise rotation	$\overrightarrow{r}_z^{\,l} > 0\,\&\,\overrightarrow{r}_z^{\,r} < 0$
Anti-clockwise rotation	$\overrightarrow{r}_z^{\,l} < 0\,\&\,\overrightarrow{r}_z^{\,r} > 0$
Moving left	$\overrightarrow{r}_s^{\,l} < 0\,|\,\overrightarrow{r}_s^{\,r} < 0$
Moving right	$\overrightarrow{r}_s^{\,l} > 0\,|\,\overrightarrow{r}_s^{\,r} > 0$
Moving up	$\overrightarrow{r}_t^{\,l} > 0\,|\,\overrightarrow{r}_t^{\,r} > 0$
Moving down	$\overrightarrow{r}_t^{\,l} < 0\,|\,\overrightarrow{r}_t^{\,r} < 0$

On the other hand, for clockwise and anti-clockwise rotation gestures,

$$\theta_v^m = \theta_v^k \mp \Delta\theta_v,$$
$$\Delta\theta_v = \frac{\max\left(|z_l^m - z_l^k|, |z_r^m - z_r^k|\right)}{b}, \tag{17}$$

where b is the step size for rotation actions, that is preset in a display system. The viewing region of the scene can be calculated as

$$\left\{-z_{max}, s_{l,r}^m - \left(z_{max} - z_o + \frac{D_v^m}{2}\sin\Delta\theta_v\right)\tan\frac{\alpha}{2} + \frac{D_v^m}{2}\cos\Delta\theta_v\right\},$$
$$\left\{0, s_{l,r}^m - \frac{D_v^m}{2}\cos\Delta\theta_v - \left(z_o - \frac{D_v^m}{2}\sin\Delta\theta_v\right)\tan\frac{\alpha}{2}\right\},$$
$$\left\{-z_{max}, \frac{D_v^m}{2}\cos\Delta\theta_v + \frac{D_v^m}{2}\sin\Delta\theta_v\tan\frac{\alpha}{2} + s_{l,r}^m\right\},$$
$$\left\{0, \frac{D_v^m}{2}\cos\Delta\theta_v + \left(z_o + \frac{D_v^m}{2}\sin\Delta\theta_v\right)\tan\frac{\alpha}{2} + s_{l,r}^m\right\}.$$

Similarly, the camera region for the rendering can be calculated as

$$\left\{0, s_{l,r}^m - \frac{D_v^m}{2}\cos\Delta\theta_v - \left(z_o - \frac{D_v^m}{2}\sin\Delta\theta_v\right)\tan\frac{\alpha}{2}\right\},$$
$$\left\{0, \frac{D_v^m}{2}\cos\Delta\theta_v + \left(z_o + \frac{D_v^m}{2}\sin\Delta\theta_v\right)\tan\frac{\alpha}{2} + s_{l,r}^m\right\}.$$

Therefore, the selected set of camera views $i^m \in \{i^m_{min}, i^m_{max}\}$ for rendering can be derived as

$$i^m_{min} = \frac{H}{2} - \left\lceil \frac{\frac{D^m_v \cos \Delta\theta_v H d}{2D_{v0}} - s^m_{l,r} - \frac{d}{2}}{d} \right\rceil,$$

$$i^m_{max} = \frac{H}{2} + 1 + \left\lceil \frac{\frac{D^m_v \cos \Delta\theta_v H d}{2D_{v0}} + s^m_{l,r} - \frac{d}{2}}{d} \right\rceil,$$

$$\forall H = 2k : k \in \mathbb{Z}, \tag{18}$$

and

$$i^m_{min} = \frac{H+1}{2} - \left\lceil \frac{\frac{D^m_v \cos \Delta\theta_v H d}{2D_{v0}} - s^m_{l,r}}{d} \right\rceil,$$

$$i^m_{max} = \frac{H+1}{2} + \left\lceil \frac{\frac{D^m_v \cos \Delta\theta_v H d}{2D_{v0}} + s^m_{l,r}}{d} \right\rceil,$$

$$\forall H = 2k+1 : k \in \mathbb{Z}, \tag{19}$$

Similarly, from the projection model, it can be easily deduced that the set of camera views will move for one camera per $\frac{d \cos \Delta\theta_v}{a}$ movement on either side, i.e.,

$$\left\{ s^m_{l,r} - s^k_{l,r} = \frac{d \cos \Delta\theta_v}{a} \right\}$$
$$\rightarrow \left\{ i^m_{min} = i^k_{min} + 1, i^m_{max} = i^k_{max} + 1 \right\}, \tag{20}$$

$$\left\{ s^k_{l,r} - s^m_{l,r} = \frac{d \cos \Delta\theta_v}{a} \right\}$$
$$\rightarrow \left\{ i^m_{min} = i^k_{min} - 1, i^m_{max} = i^k_{max} - 1 \right\}. \tag{21}$$

and the camera views will move for one camera per $\frac{d}{a}$ movement in both sides for the zoom-in and zoom-out actions, i.e.,

$$\left\{ \max \left(s^m_l - s^k_l, s^m_r - s^k_r \right) = \frac{d}{a} \right\}$$
$$\rightarrow \left\{ i^m_{min} = i^k_{min} - 1, i^m_{max} = i^k_{max} + 1 \right\}, \tag{22}$$

$$\left\{ \max \left(s^k_l - s^m_l, s^k_r - s^m_r \right) = \frac{d}{a} \right\}$$
$$\rightarrow \left\{ i^m_{min} = i^k_{min} + 1, i^m_{max} = i^k_{max} - 1 \right\}. \tag{23}$$

Equations (18) and (19) are derived only for the horizontal projection, while the vertical projection can be derived with the similar procedure here. Hence, with both horizontal and vertical view selections, the selected set of the views for time index m can be given by

$$i^{m,h} \in \{i^{m,h}_{min}, i^{m,h}_{max}\}, i^{m,v} \in \{i^{m,v}_{min}, i^{m,v}_{max}\}, \tag{24}$$

which is illustrated in Fig. 6.

4   Conclusion

In this paper, we have presented a full framework of interactive LF video sequence in order to realize real time coding and transmission for LF video streaming services. In the system, one of the most important tasks is to predict the view trajectory and determine the set of views for transmission and display. Hence, we discussed two different trajectory calculation and prediction schemes according to the movement of the user or navigation gesture. User-dependent view selection is the foundation of effective LF video coding and transmission, and the scheme plays an indispensable role in boosting the development of LF video applications.

In the future study, we will research and develop an adaptive LF video coding method for different LF video sequence based on the proposed user-dependent view selection, and then an complete interactive LF video scheme combined the user-dependent view selection and the adaptive coding method will be presented.

Acknowledgement. This work was supported in part by the National Natural Science Foundation of China (Grant No: 61772436), Sichuan Science and Technology Innovation Seedling Fund (2017RZ0015), and the Fundamental Research Funds for the Central Universities.

References

1. Levoy, M.: Light fields and computational imaging. Computer **39**(8), 46–55 (2006)
2. Yang, R., Huang, X., Li, S., Jaynes, C.: Toward the light field display: autostereoscopic rendering via a cluster of projectors. IEEE Trans. Vis. Comput. Graph. **14**(1), 84–96 (2008)
3. Xiang, W., Wang, G., Pickering, M., Zhang, Y.: Big video data for light-field-based 3D telemedicine. IEEE Netw. **30**(3), 30–38 (2015)
4. Wang, G., Xiang, W., Pickering, M.: A cross-platform solution for light field based 3D telemedicine. Comput. Methods Programs Biomed. **125**, 103–116 (2015)
5. Kovcs, P.T., Nagy, Z., Barsi, A., Adhikarla, V.K.: Overview of the applicability of H. 264/MVC for real-time light-field applications. In: 3DTV-Conference: The True Vision-Capture, Transmission and Display of 3D Video (3DTV-CON), pp. 1–4. IEEE, Budapest (2014)
6. Vetro, A., Wiegand, T., Sullivan, G.J.: Overview of the stereo and multiview video coding extensions of the H. 264/MPEG-4 AVC standard. Proc. IEEE **99**(4), 626–642 (2011)
7. Wang, G., Xiang, W., Pickering, M.: Light field multi-view video coding with two-directional parallel inter-view prediction. IEEE Trans. Image Process. **25**(11), 5104–5117 (2016)
8. Hong, R., Hu, Z., Wang, R., Wang, M.: Multi-view object retrieval via multi-scale topic models. IEEE Trans. Image Process. **25**(12), 5814–5827 (2016)
9. Pan, Z., Ikuta, Y., Bandai, M., Watanabe, T.: User dependent scheme for multi-view video transmission. In: 2011 IEEE International Conference on Communications (ICC), pp. 1–5. IEEE, Kyoto (2011)

Deep Residual Net Based Compact Feature Representation for Image Retrieval

Cong Bai[1(✉)], Jian Chen[1], Qing Ma[1,2], Zhi Liu[1], and Shengyong Chen[1]

[1] College of Computer Science, Zhejiang University of Technology, Hangzhou, China
congbai@zjut.edu.cn
[2] College of Science, Zhejiang University of Technology, Hangzhou, China

Abstract. Deep learning technology has been introduced into many multimedia processing tasks, including multimedia retrieval. In this paper, we propose a deep residual net (ResNet) based compact feature representation improve the content-based image retrieval (CBIR) performance. The proposed method integrates ResNet and hashing networks to convert the raw images into binary codes. The binary codes of images in query set and that of the database are compared using Hamming distance for retrieval. Comprehensive experiments are executed on three public databases. The results show that the proposed method outperforms state-of-the-art methods. Furthermore, the impact of the deep convolutional network (DCNN)'s depth on the performance is investigated.

Keywords: Content-based image retrieval · Residual Nets · Hashing
Depth of deep convolutional neural network

1 Introduction

With the huge accumulation of digital images and videos in the society, the demand of searching such kind of data is also increasing [8]. Traditional multimedia search engines usually use the surrounding meta data, such as titles and tags or manually annotated keywords as the index to retrieval the multimedia data, named keyword-based retrieval [18]. The main drawback of such kind of retrieval technology is the inconsistence between the textural information and visual content of multimedia data. So the content-based multimedia retrieval is proposed and makes great progress in the past decades [13]. In content-based multimedia retrieval, semantic gap is a challenging problem, which refers to the difference between the low level representations of images and the higher level concepts used by human beings to describe the images. To narrow this gap, extensive efforts have been made both from the academic and industry communities [9,10,17].

Over the past few years, deep learning has been witnessed as one of the most promising technology in computer vision as their outstanding performance in

© Springer Nature Switzerland AG 2018
R. Hong et al. (Eds.): PCM 2018, LNCS 11165, pp. 737–747, 2018.
https://doi.org/10.1007/978-3-030-00767-6_68

a series of vision related tasks, such as image classification [19], face recognition [29], image segmentation [27] and so on. Since the successful application of AlexNet [12] in computer vision, the instinct idea to get better feature representation is to use deep convolution neural networks (DCNN). For example, AlexNet contains 8 learned layers, VGGNet [21] has 19 learned layers and GoogLeNet [22] consists of 22 learned layers. However, going deeper means that training such network will become more difficult. Thus deeper but easy to be trained DCNN is proposed, namely, ResNet [6]. Inspired by these successful DCNN, deep learning has already been introduced into content-based image retrieval (CBIR) [2]. However, it is instinct to ask: whether using very deep DCNN will improve the performance of image retrieval, especially for large scale image database?

In order to answer the above question, we use a very deep DCNN, ResNet for image retrieval. However, features extracted by ResNet are high-dimensional, thus they are not compatible with large scale image database if they are used directly for retrieval. So the proposed method constitutes a framework for converting the raw images into binary codes for effectively large scale image retrieval. To do so, the raw images are firstly input into the ResNet to get the deep features. And then the deep features are converted into binary codes by a DCNN based hashing network. In summary, the contributions of this work are twofold:

(1) We investigate a new framework that could convert the raw images into binary codes for large scale image retrieval. This framework integrates ResNet and DCNN based hashing network.

(2) Extensive experiments are conducted for comprehensive evaluations of the proposed framework, especially with different depth of the DCNN.

The reminder of the manuscript is structured as follows: Sect. 2 describes the proposed framework, followed by the experiment results in Sect. 3. Finally, conclusion and perspectives are given in Sect. 4.

2   Proposal

The framework of our method is shown in Fig. 1. The inputs are the pixels of the raw image and the corresponding label information of the image (for training only) and the output are the binary codes of the images. Such codes could be used for image retrieval by comparing the Hamming distances between the query's codes and the codes of the gallery images. The proposed framework includes two kinds of DCNN. Deep Residual Network (ResNet) [6] is used to convert the raw images into deep features and hashing neural network (HNN) is used to convert the deep feature into binary codes. We name our proposal as ResHNN. Details will be explained in this section.

2.1   ResNet

ResNet won the 1st place on the ILSVRC 2015 classification task and is proved to be easily optimized. And it gains improvement from increased depth [6]. ResNet is composed by many stacked "Residual Units" and each unit could be expressed in the following form:

$$y_i = h(x_i) + F(x_i, W_l)$$
$$x_{i+1} = f(y_i) \tag{1}$$

where x_i and $x_i + 1$ are input and output of the i-th unit, and F is residual function. $h(x_i)$ is an identity mapping and f is a ReLU function [7]. The core idea of ResNet is to learn the additive residual function F with respect to $h(x_i)$, with a key choice of using an identity mapping $h(x_i) = x_i$. This is realized by attaching an shortcut connection that performs identity mapping and their outputs are added to the outputs of the stacked layers. The residual unit we used is as shown in Fig. 1. More details could be found in [6,7].

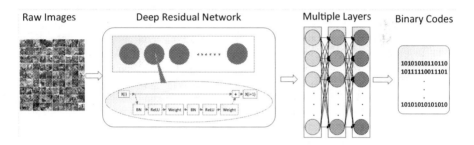

Fig. 1. The framework of transferring images into compact feature representation

2.2   HNN

The hashing layer is based on Nonlinear Discrete Hashing [3], which uses a multi-layer neural network to obtain the compact binary codes through nonlinear transformations. Let $X = [x_1, x_2, \ldots, x_n] \in R^{n \times d}$ denote the training set with n samples, where each sample $x_i \in R^d (1 \leq i \leq n)$ is a data point of d dimension. Assuming the m-th layer consists of $u^{(m)}$ units, the output of each layer is computed as:

$$h^{(1)}(x_i) = s(x_i W^{(1)} + c^{(1)}), i = 1, \cdots, n \tag{2}$$
$$h^{(m)}(x_i) = s(h^{(m-1)}(x_i)W^{(m)} + c^{(m)}), i = 1, \cdots, n \tag{3}$$

where $s(\cdot)$ is a nonlinear activation function such as the $tanh$ function, and the projection matrix $W^{(m)}$ and the bias vector $c^{(m)}$ are the parameters to be learned for the m-th layer of the network.

And for a I-layer network, we could have the output in the form of:

$$F(x) = h^{(I)}(x) \in R^{u^{(I)}} \tag{4}$$

where the mapping $F : R^d \rightarrow R^{u^{(I)}}$ is a parametric nonlinear function determined by $\{W^{(m)}, c^{(m)}\}_{m=1}^{I}$. We treat the sign of the output of the network as the binary code of these n samples and put the binary code of all the samples together as:

$$B = sgn(F(X)) \in \{-1, +1\}^{(n \times r)} \tag{5}$$

Specifically, we treat the 0 as +1. The formula is as follows:

$$sgn(b) = \begin{cases} 1, & b \geq 0 \\ -1, & b < 0 \end{cases} \tag{6}$$

The goal is to find a binary matrix that minimizes the value of loss function. The formula is as follows:

$$arg \min_B Q = Q(L, B) + Q(B, X) \tag{7}$$

where $Q(L, B)$ means the difference between the predicted labels through the hash code matrix B and the ground truth labels of all samples, and the $Q(B, X)$ means the information loss caused by transforming to binary code. Denoting the classifier weight matrix as C, the first term $Q(L, B)$ can be considered as:

$$Q_C(L, B) = ||L - CB^T||_F^2 \tag{8}$$

where the L is the ground truth label of the samples and $|| \cdot ||_F$ means the Frobenius norm.

$Q(B, X)$ measures the discrepancy between the binary codes and the data samples including the quantization loss term and the similarity preserving term

$$Q_F^{(I)}(B, X) = ||B - F^{(I)}||_F^2$$
$$+ \alpha \sum_{i=1}^n \sum_{j=1}^n S_{ij} ||F^{(I)}(i, :) - F^{(I)}(j, :)||_F^2$$
$$s.t. B \in \{-1, 1\}^{n \times r}, B^T B = nI_r \tag{9}$$

where S is the similarity matrix. To reduce the redundancy of information, $B^T B = nI_r$ is added. But the problem of constraint makes optimization difficult, so a real-valued matrix Y is introduced in $\Omega = \{Y \in \mathbb{R}^{n \times r} || Y^T Y = nI_r\}$ approaching to B. So the Eq. 10 is introduced to substitute the independent constraint.

$$Q_I(B) = ||B - Y||_F^2 \tag{10}$$

We notice that the loss function Eq. 7 only considers the outputs of the top layer of the network, but the hidden layers are not included. So the companion loss function is introduced as follows:

$$Q_F(B, X) = Q_F^{(I)}(B, X) + \sum_{m=1}^{I-1} \alpha^{(n)} h(Q_F^{(m)} - \tau^m) \tag{11}$$

where $h(x) = \max(x, 0)$ and $Q_F^{(m)} = \sum_{i=1}^n \sum_{j=1}^n S_{ij} ||F^{(m)}(i, :) - F^{(m)}(j, :)||_F^2$, $m = 1, 2, \ldots, I - 1$.

In consideration of all the mentioned above, the overall function is defined as follow:

$$\arg \min_{\mathbf{B},\mathbf{P},\{\mathbf{F}^{(m)}\}_{m=1}^{I},\mathbf{Y}} = Q_C + \lambda_1 Q_I + \lambda_2 Q_F + \lambda_3 Q_R$$

$$s.t. B \in \{-1,1\}^{n \times r} \tag{12}$$

where $Q_R = \|C\|_F^2 + \sum_{m=1}^{I} \|W^{(m)}\|_F^2 + \sum_{m=1}^{I} \|c^m\|_F^2$ contains the regularizer to control the scales of the parameters.

Since the above joint optimization problem is non-convex and difficult to solve. Sub-optimal problems with respect to one variable while keeping other variable fixed is used. So we could iterate each variable of optimal solution in sub-optimal problem one by one. And this problem could be solved by Singular Value decomposition (SVD) and Gram-Schmidt process. More details could be found in [3].

3 Experiments

In this section, we conduct experiments on three datasets: MNIST [4], CIFAR10 [11], and SUN379 [26], to evaluate the performance of the proposal and try to answer the question we posed in the introduction.

Table 1. Retrieval performance on MNIST with 16, 32 and 64 bit length of binary codes

Method	mAP (%)			Precision@500 (%)		
	16	32	64	16	32	64
LSH [1]	15.81	25.41	32.78	28.08	38.56	48.39
SMLSH [25]	31.68	38.28	43.42	41.93	49.16	55.14
ITQ [5]	38.11	42.13	43.63	54.35	60.15	62.03
SPLH [24]	48.67	49.38	48.71	59.69	60.57	63.06
CCA-ITQ [5]	58.61	60.34	62.51	67.95	69.37	71.42
FastH [15]	95.04	96.19	96.71	93.60	94.67	95.27
SDH [20]	92.28	93.74	94.81	91.45	92.07	92.88
DeepH [16]	70.91	74.10	76.34	76.75	79.13	81.55
NDH [3]	94.64	95.88	96.29	93.82	94.81	94.99
SSDH [28]	-	**98.20**	-	-	98.50	-
ResHNN-50	**98.01**	98.03	**98.07**	**98.60**	**98.62**	**98.63**

3.1 Databases

MNIST: It is a handwritten digit dataset consisting of 70000 images with the size of 28×28. Each image is associated with a digit from 0 to 9 and represented as a 784-dimensional gray-scale feature vector by concatenating all pixels [3]. It's a simple dataset, so we extract a 256 dimensional feature vector by ResNet for each image. Following the same setting in [24], 1000 images with 100 images per class are randomly selected from original test set to form the query set, and use the remaining 69000 images as gallery database.

CIFAR10: It is a set of 60000 manually labeled color images. They are from 10 classes, and each class has 6000 images. Each image is with the size of 32×32. ResNet is used to extract a 1024 dimensional feature vector for each image. Similar to the MNIST, we use 1000 images consist of 100 images per class from original test set as query set and construct the gallery database with the remaining images.

SUN397: This dataset contains 108754 images which are classified into 397 categories. It is bigger and more complex than the two mentioned above databases, it could be a challenge to retrieve semantic neighbors. Each image is represented by a 2048 dimensional feature vector extracted by ResNet. Following the same protocol of the referred methods, 8000 images are randomly sampled as query images and the remaining images are left to form the gallery database.

3.2 Evaluation Metric

All experiments are repeated 10 times and the averaged values are took as the final result. Two metrics are used to measure the performance of different methods: precision at N samples and mean Average Precision (mAP). Given top N returned samples, precision at N samples is calculated as the percentage of relevant retrieved images:

$$Precision@N = \frac{\sum_{k=1}^{N} rel(k)}{N} \tag{13}$$

where $rel(k) = 1$ if k-th image is a relevant retrieved image, otherwise, $rel(k) = 0$. The mean Average Precision (mAP) presents an overall measurement of the retrieval performance by computing the area under the precision-recall curve, which delivers good discrimination and stability. It is calculated as follows:

$$AveP = \frac{\sum_{k=1}^{N}(P(k) \times rel(k))}{number\ of\ relevant\ images},$$
$$MAP = \frac{\sum_{q=1}^{Q} AveP(q)}{Q} \tag{14}$$

where k is the rank in the sequence of retrieved documents, N is the number of retrieved images, $P(k)$ is the precision at cut-off k in the list, and $rel(k)$ is equal to 1 if the item at rank k is a relevant image, otherwise, it is equal to 0 [23]. Q is the number of the queries.

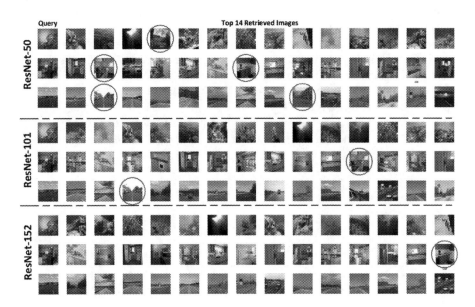

Fig. 2. Top 14 retrieved images from SUN397 dataset by different number of layers of ResNet with 128 bits binary codes. The results of ResHNN-50 are shown in the first three rows, the results of ResHNN-101 are shown in the middle three rows, and the results of ResHNN-152 are shown in the last three rows. The irrelevant images are marked by red circle. (Color figure online)

3.3   Results and Analysis

Result on MNIST: The training set used for hashing net is with the size of 5000 images by selecting 500 images from each class. The ResNet and HNN are trained separately and the depth of the ResNet we used in this database is 50. For the HNN, we take the tanh function as the nonlinear activation function and initialize the biases $c^{(m)}$ to be 0. Each element of $W^{(m)}$ is uniformly sampled from the range $\left[-\sqrt{\frac{6}{row(m)+col(m)}}, \sqrt{\frac{6}{row(m)+col(m)}} \right]$, where $row(m)$ is the number of rows of $W^{(m)}$ and $col(m)$ is the number of columns of $W^{(m)}$. The numbers R and L are set as 5 and 3. And we set $\alpha^{(1)}$ and $\alpha^{(2)}$ as 20, $\alpha^{(3)}$ as 100, $\tau^{(1)}$ and $\tau^{(2)}$ as 1000, λ_1 as 1e−3, λ_2 and λ_3 as 1e−5, learning rate η as 1e−3. And the same setting is adopted for all the other datasets. The experimental results are shown in Table 1. The results are compared on hash code with lengths of 16, 32, and 64 bits. As MNIST is a simple handwrite characters dataset, a lot of methods could achieve good performance, so does ResHNN. As the performance achieved in this database is quiet high, the impact of depth of the DCNN is difficult to estimate, so we did not conduct further experiments on it.

Result on CIFAR10: Similar as the setting of experiments on the MNIST dataset, the training set is constructed with 5000 images with 500 images per category. We compare the results in different depth of the ResNet with 50, 101 and 152 to evaluated their impacts on feature extracting and hashing. Results are shown in Table 2. It is obviously that our method ResHNN outperforms referred methods obviously in different kinds of bit length, both in the aspect of mAP and Precision@500 and ResHNN-152 is the best. And with the ResNet going deeper, the retrieval performance improves slightly. We believe that the reason is that deeper networks could extract features from images more efficiently, which is preserved in our hash layers.

Table 2. Retrieval performance on CIFAR-10 with 16, 32 and 64 bit length of binary codes

Method	mAP (%)			Precision@500 (%)		
	16	32	64	16	32	64
LSH [1]	12.63	13.70	14.62	15.32	17.23	19.36
SMLSH [25]	14.96	16.41	16.98	17.82	19.75	20.36
ITQ [5]	15.57	15.80	16.57	19.91	21.04	22.53
SPLH [24]	17.08	19.38	21.21	21.22	26.39	29.34
CCA-ITQ [5]	16.21	16.02	16.49	24.63	24.44	26.77
FastH [15]	27.94	33.09	36.55	37.74	43.13	46.84
SDH [20]	29.21	29.22	32.67	39.08	39.62	42.15
DeepH [16]	24.04	25.96	27.53	32.45	34.09	36.85
NDH [3]	33.75	35.93	37.90	43.58	46.67	48.24
LPMH [14]	67.54	72.17	73.59	-	-	-
SSDH [28]	-	81.20	-	-	82.80	-
ResHNN-50	93.04	93.31	93.68	92.81	92.98	93.40
ResHNN-101	93.69	94.32	94.46	93.45	94.10	94.28
ResHNN-152	**94.16**	**94.65**	**94.92**	**93.87**	**94.35**	**94.74**

Result on SUN397: In order to verify whether the proposed ResHNN works well under large and complex conditions, more experiments were conducted in SUN397 database. As this database is a larger collection, we evaluate the impacts of the different number of the layers of ResNet also, with respects of 50, 101 and 152. Results are shown in Table 3. We notice that the proposed ResHNN could

Table 3. Retrieval performance on SUN397 with 48, 64 and 128 bit length of binary codes respectively

Method	mAP (%)			Precision@2000 (%)		
	48	64	128	48	64	128
ITQ [5]	5.16	5.58	6.73	6.14	6.43	6.98
SPLH [24]	1.27	1.89	0.99	2.90	3.33	2.65
CCA-ITQ [5]	7.22	6.38	6.08	6.21	5.90	5.56
FastH [15]	2.71	4.98	8.28	2.90	3.90	5.22
SDH [20]	9.87	9.65	11.85	7.57	7.81	8.52
DeepH [16]	9.31	9.73	8.32	7.54	7.52	6.76
NDH [3]	**11.39**	**12.96**	13.86	**7.81**	**8.32**	9.05
ResHNN-50	9.96	10.41	16.61	6.67	7.01	9.36
ResHNN-101	10.12	11.32	18.95	6.74	7.44	10.09
ResHNN-152	10.23	11.67	**19.58**	6.96	7.70	**10.26**

achieve the comparable performance with referred methods, and outperforms in long length of bits. Furthermore, the increase of the depth of the ResNet could trigger the obvious improvements on the retrieval with the longer length of binary codes. This could be explained by the fact that the deeper of the DCNN layers, the more information of the visual content of the image could be extracted, and with the longer of the length of the binary codes, such information could be preserved better. The examples of retrieval results with different layers of ResNet are shown in Fig. 2. The wrong returned images are marked by red circles.

4   Conclusion and Perspective

In this paper, we propose a ResNet based compact feature representation for image retrieval, namely, ResHNN, which integrates Residual net and hashing neural networks to generate the binary code for CBIR. Extensive experimental results on three widely used public databases demonstrate the superiority of the proposed ResHNN. Furthermore, we explore the impact of ResNet's depth on the performance. The impacts of the different deep features and different hashing method will be discovered further.

Acknowledgement. This work is supported by the National Natural Science Foundation of China under Grants No. 61502424 and U1509207, Zhejiang Provincial Natural Science Foundation of China under Grant No. LY18F020032 and LY16F020033.

References

1. Andoni, A., Indyk, P.: Near-optimal hashing algorithms for approximate nearest neighbor in high dimensions. In: Proceedings - Annual IEEE Symposium on Foundations of Computer Science, FOCS, pp. 459–468 (2006)
2. Bai, C., Huang, L., Pan, X., Zheng, J., Chen, S.: Optimization of deep convolutional neural network for large scale image retrieval. Neurocomputing **303**, 60–67 (2018)
3. Chen, Z., Lu, J., Feng, J., Zhou, J.: Nonlinear discrete hashing. IEEE Trans. Multimedia **19**(1), 123–135 (2017)
4. Deng, L.: The MNIST database of handwritten digit images for machine learning research [best of the web]. IEEE Sig. Process. Mag. **29**(6), 141–142 (2012)
5. Gong, Y., Lazebnik, S., Gordo, A., Perronnin, F.: Iterative quantization: a procrustean approach to learning binary codes for large-scale image retrieval. IEEE Trans. Pattern Anal. Mach. Intell. **35**(12), 2916–2929 (2013)
6. He, K., Zhang, X., Ren, S., Sun, J.: Deep residual learning for image recognition. In: 2016 IEEE Conference on Computer Vision and Pattern Recognition (CVPR), pp. 770–778 (2016)
7. He, K., Zhang, X., Ren, S., Sun, J.: Identity mappings in deep residual networks. In: Leibe, B., Matas, J., Sebe, N., Welling, M. (eds.) ECCV 2016. LNCS, vol. 9908, pp. 630–645. Springer, Cham (2016). https://doi.org/10.1007/978-3-319-46493-0_38
8. Hong, R., Hu, Z., Wang, R., Wang, M., Tao, D.: Multi-view object retrieval via multi-scale topic models. IEEE Trans. Image Process. **25**(12), 5814–5827 (2016). https://doi.org/10.1109/TIP.2016.2614132
9. Hong, R., Zhang, L., Tao, D.: Unified photo enhancement by discovering aesthetic communities from flickr. IEEE Trans. Image Process. **25**(3), 1124–1135 (2016). https://doi.org/10.1109/TIP.2016.2514499
10. Hong, R., Zhang, L., Zhang, C., Zimmermann, R.: Flickr circles: aesthetic tendency discovery by multi-view regularized topic modeling. IEEE Trans. Multimedia **18**(8), 1555–1567 (2016). https://doi.org/10.1109/TMM.2016.2567071
11. Krizhevsky, A.: Learning multiple layers of features from tiny images. Technical report (2009)
12. Krizhevsky, A., Sutskever, I., Hinton, G.E.: Imagenet classification with deep convolutional neural networks. In: International Conference on Neural Information Processing Systems, pp. 1097–1105 (2012)
13. Lew, M., Sebe, N., Djeraba, C., Jain, R.: Content-based multimedia information retrieval: state of the art and challenges. ACM Trans. Multimedia Comput. Commun. Appl. **2**(1), 1–19 (2006)
14. Li, K., Qi, G.J., Hua, K.A.: Learning label preserving binary codes for multimedia retrieval: a general approach. ACM Trans. Multimedia Comput. Commun. Appl. **14**(1), 2:1–2:23 (2017)
15. Lin, G., Shen, C., Shi, Q., Van Den Hengel, A., Suter, D.: Fast supervised hashing with decision trees for high-dimensional data. In: Proceedings of the IEEE Computer Society Conference on Computer Vision and Pattern Recognition, pp. 1971–1978 (2014)
16. Liong, V.E., Lu, J., Wang, G., Moulin, P., Zhou, J.: Deep hashing for compact binary codes learning. In: Proceedings of the IEEE Computer Society Conference on Computer Vision and Pattern Recognition, 07–12 June 2015, pp. 2475–2483 (2015)
17. Liu, Y., Zhang, D., Lu, G., Ma, W.: A survey of content-based image retrieval with high-level semantics. Pattern Recogn. **40**(1), 262–282 (2007)

18. Mei, T., Rui, Y., Li, S., Tian, Q.: Multimedia search reranking. ACM Comput. Surv. **46**(3), 1–38 (2014)
19. Russakovsky, O., et al.: ImageNet large scale visual recognition challenge. Int. J. Comput. Vis. **115**(3), 211–252 (2015)
20. Shen, F., Shen, C., Liu, W., Shen, H.T.: Supervised discrete hashing. In: Proceedings of the IEEE Computer Society Conference on Computer Vision and Pattern Recognition, 07–12 June 2015, pp. 37–45 (2015)
21. Simonyan, K., Zisserman, A.: Very deep convolutional networks for large-scale image recognition. In: International Conference on Learning Representations (ICRL), pp. 1–14 (2015)
22. Szegedy, C., et al.: Going deeper with convolutions. In: Proceedings of the IEEE Computer Society Conference on Computer Vision and Pattern Recognition, 07–12 June 2015, pp. 1–9 (2015)
23. Turpin, A., Scholer, F.: User performance versus precision measures for simple search tasks. In: Proceedings of the 29th Annual International ACM SIGIR Conference on Research and Development in Information Retrieval, SIGIR 2006, pp. 11–18. ACM, New York (2006)
24. Wang, J., Kumar, S., Chang, S.F.: Semi-supervised hashing for large-scale search. IEEE Trans. Pattern Anal. Mach. Intell. **34**(12), 2393–2406 (2012)
25. Weng, L., Jhuo, I.H., Shi, M., Sun, M., Cheng, W.H., Amsaleg, L.: Supervised multi-scale locality sensitive hashing. In: Proceedings of the 5th ACM on International Conference on Multimedia Retrieval, ICMR 2015, pp. 259–266. ACM, New York (2015)
26. Xiao, J., Hays, J., Ehinger, K.A., Oliva, A., Torralba, A.: Sun database: large-scale scene recognition from abbey to zoo. In: 2010 IEEE conference on Computer vision and pattern recognition (CVPR), pp. 3485–3492. IEEE (2010)
27. Ye, L., Liu, Z., Li, L., Shen, L., Bai, C., Wang, Y.: Salient object segmentation via effective integration of saliency and objectness. IEEE Trans. Multimedia **19**(8), 1742–1756 (2017)
28. Zhang, J., Peng, Y.: SSDH: semi-supervised deep hashing for large scale image retrieval. IEEE Trans. Circ. Syst. Video Technol. **PP**(99), 1 (2017)
29. Zheng, J., Yang, P., Chen, S., Shen, G., Wang, W.: Iterative re-constrained group sparse face recognition with adaptive weights learning. IEEE Trans. Image Process. **26**(5), 2408–2423 (2017)

Sea Ice Change Detection from SAR Images Based on Canonical Correlation Analysis and Contractive Autoencoders

Xiao Wang[1,2], Feng Gao[1,2(✉)], Junyu Dong[1,2], and Shengke Wang[1,2]

[1] College of Information Science and Engineering, Ocean University of China, Qingdao, China
{gaofeng,dongjunyu,neverme}@ouc.edu.cn
[2] Qingdao Key Laboratory of Mixed Reality and Virtual Ocean, Qingdao, China
1195747610@qq.com

Abstract. In this paper, we proposed a novel sea ice change detection method for Synthetic Aperture Radar (SAR) images based on Canonical Correlation Analysis (CCA) and Contractive Autoencoders (SCAEs). To alleviate the effect of multiplicative speckle noise, structured matrix decomposition is utilized for difference image enhancement, and therefore, better difference image with less noisy spots can be obtained. In order to get good data representations in changed and unchanged pixels classification, CCA and SCAEs are combined to exploit more effective changed features. Experiments on two real sea ice datasets demonstrate the robustness and efficiency of the proposed method in comparison with three other state-of-the-art methods.

Keywords: Change detection · Synthetic aperture radar
Canonical correlation analysis · Contractive autoencoders
Remote sensing image

1 Introduction

Sea ice is an important part of the cryosphere that interacts continuously with the underlying oceans and the overlaying atmosphere. Global warming accelerates the loss of sea ice, which threatens animals living in the Polar Regions [1]. In addition, polar sea ice information is very important for safe navigation, since the amount of ice adversely impacts the friction against the hull of a vessel. Therefore, polar sea ice research has attracting increasing attentions and raised global security concerns these years.

More and more Synthetic Aperture Radar (SAR) sensors have been developed, and a large number of multitemporal image pairs have been acquired in

This work was supported by the National Natural Science Foundation of China (Grant Nos. 41606198 and 41576011) and in part by the Shandong Province Natural Science Foundation of China under Grant No. ZR2016FB02.

© Springer Nature Switzerland AG 2018
R. Hong et al. (Eds.): PCM 2018, LNCS 11165, pp. 748–757, 2018.
https://doi.org/10.1007/978-3-030-00767-6_69

the past decades. SAR images provides the capabilities for all-weather, day-and-night surveillance. The applications of SAR images has become a hot research topic in remote sensing communities. Among these applications, sea ice monitoring is an essential issue. However, sea ice change detection from SAR images exhibits difficulties due to the presence of multiplicative speckle noise [8]. How to alleviate the interference of speckle noise has become a critical issue in sea ice change detection from SAR images.

In general, change detection [2] consists of three main steps: (1) image preprocessing; (2) difference image (DI) generation from a pair of multitemporal images; (3) DI analysis to achieve the segmentation of the changed regions. The first step mainly includes geometric correction and denoising. Some methods have been employed to alleviate the effect to speckle noise, e.g., Gamma-MAP [3] and SRAD [4]. In the second step, the log-ratio operator is usually employed since it is considered to be robust to the speckle noise [5]. In this paper, the log-ratio operator is employed for DI generation. In the final step, clustering methods are very popular, since they do not require DI image distribution. Celik [6] used the principal component analysis (PCA) and k-means clustering for classification. Gao et al. [7] use Gabor wavelets and fuzzy c-means (FCM) algorithm to select samples for classification. Gong et al. [8] proposed a reformulated fuzzy local-information c-means clustering algorithm for classifying changed and unchanged regions in the generated DI. These clustering methods gain satisfying performance and are popular in the past decades. However, clustering methods are sensitive to initial values, and improper initial values may result in premature convergence on local optima.

In recent years, deep learning shows its dominant performance in many fields, such as image classification [9], voice recognition [10] and nature language processing, etc. In [11], Wang et al. demonstrated that convolution neural network (CNN) has been used to estimate ice concentration using SAR scenes captured during the melt season. In [12], a deep neural network is established for SAR image change detection. It is empirically verified that deep learning can further improve the change detection performance.

The above mentioned methods have gain good performance in SAR image change detection. However, there are still two problems to be considered: (1) The speckle noise should be suppressed in the process of DI generation; (2) Good data representations should be achieved from multitemporal images to identify changed regions. With respect to the first problem, in this paper, Structured Matrix Decomposition (SMD) is utilized for DI enhancement, and therefore better DI with less noisy spots can be obtained. To solve the second problem, we develop a classification model based on Canonical Correlation Analysis and Contractive Autoencoders (CCA-SCAEs) to exploit good representations from multitemporal SAR images.

The main contributions of the proposed method are listed as follows: First, SMD is introduced to suppress the speckle noise to obtain DI with less noisy spots. Saliency detection method can capture the distinctive patterns and suppress the background noise, but it is rarely considered in change detection.

Second, a novel classification model based on CCA and SCAEs is established to exploit change information from multitemporal SAR images. Experimental results on two real SAR datasets demonstrate the effectiveness of the proposed method.

The remainder of this paper is organized as follows: Sect. 2 presents the change detection problem statements and describes the proposed method in details. Section 3 shows the experimental results and the paper closes with a conclusion in Sect. 4.

2   Proposed Method

Given two multitemporal sea ice images I_1 and I_2, which are acquired at the same location but at different times t_1 and t_2. The aim of sea ice change detection is to generate a change map, which gives the interpretations about the changes occurred. The framework of the proposed method is illustrated in Fig. 1.

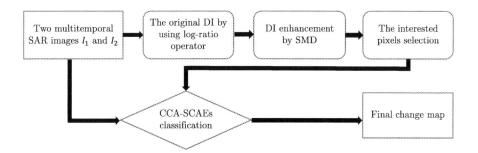

Fig. 1. Framework of the proposed change detection method

Our proposed method is illustrated in Fig. 1, which is mainly comprised of two steps:

Step 1 – DI generation and enhancement. Given two multitemporal images I_1 and I_2, the log-ratio operator is employed to obtain an initial DI. In order to remove speckle noise spots, we use the SMD method to enhanced the DI and suppress the speckle noise.

Step 2 – Training samples generation and classification. We need reliable samples for the classification model. Hierarchical FCM clustering algorithm [7] is used for reliable samples generation. Then, a classifier based on CCA and SCAEs will be built. The changed regions can be identified by feeding pixels from multitemporal images into the classifier. Then, the final change map can be obtained.

2.1   DI Generation and Enhancement

The log-ratio operator is chosen to obtain an initial DI. We choose log-ratio operator since it is empirically verified to be capable to reduce the speckle noise to some extent. The log-ratio operator is defined as DI $= \log(I_1/I_2)$. However, it is a challenging task to find the interested and distinctive areas from SAR images, which includes many speckle noise spots. Based on this, we should find a better way to improve DI robustness. Saliency map shows interested areas which has a strong contrast with the entire image. Inspired by this, we use saliency detection methods to enhance the DI. Figure 2 shows the similarity between the DI and the saliency map. Figure 2(a) and (b) are the original multitemporal images. Figure 2(c) is the original DI obtained by using the log-ratio operator, which contains many noisy spots. Figure 2(d) is the saliency map, which has less speckle noise spots compared with the original DI.

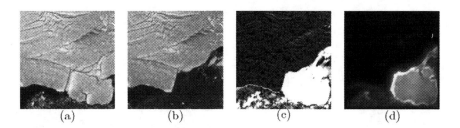

(a)   (b)   (c)   (d)

Fig. 2. The similarity between the original DI and the saliency map. (a) and (b) are two original multitemporal images. (c) The original DI obtained by using log-ratio operator. (d) The saliency map acquired by SMD.

In this paper, saliency detection methods based on Structured Matrix Decomposition [13] is utilized. The SMD is defined as follows:

$$\min_{L,S} \varphi(L) + \alpha\Omega(S) + \beta\theta(L,S) \quad s.t. \quad F = L + S, \tag{1}$$

where $\varphi(\cdot)$ is a low-rank constraint to allow identification of the intrinsic feature subspace of the redundant background patches, $\Omega(\cdot)$ is a structured sparsity regularization to capture the spatial and feature relations of patches in S, $\theta(\cdot)$ is an interactive regularization term to enlarge the distance between the subspaces drawn from L and S, and α, β are positive tradeoff parameters. Detailed information about SMD can be found in Peng's work [13].

Let I_D represents the initial DI, and I_{SMD} represent the saliency map obtained by the SMD method, the enhanced difference image I_E is computed by:

$$I_E = \exp(k \cdot I_{SMD}) \cdot I_D, \tag{2}$$

In our implementations, we found that the I_{SMD} is of great significance. Therefore, we use an exponential function to emphasize I_{SMD}. In our implementations, we use $k = 0.2$ as the scaling factor.

After obtaining I_E, we partition the pixels in DI into three groups by using the hierarchical FCM algorihtm [7]: changed class Ω_c, unchanged class Ω_u, and intermediate class Ω_i. Pixels belonging to Ω_c and Ω_u have high probabilities to be changed or unchanged. These pixels are selected as reliable samples for CCA-SCAEs. The classification model will be introduced in the next subsection in detail.

2.2   Classification via CCA-SCAEs Classifier

Deep learning is a hot topic in recent years and achieves extraordinary results in many fields. The motivation of deep learning is to establish and simulate the neural network of human brain for analysis and learning. It imitates the mechanism of the human brain to explain the data. In this paper, in order to acquire discriminative representation of changed features, neighborhood pixels features of multitemporal SAR images are fed into CCA-SCAEs to learn more effective changed features.

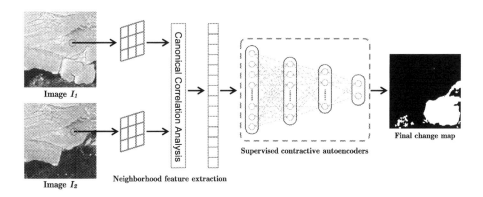

Fig. 3. The proposed classification framework based on Canonical Correlation Analysis and Contractive Autoencoders (CCA-SCAEs).

The framework of CCA-SCAEs is illustrated in Fig. 3. Neighborhood pixel features of multi-temporal SAR images are first processed by canonical correlation analysis (CCA). Then, the obtained features are fed into supervised contractive autoencoders (SCAEs) to learn more effective changed features. Suppose an L layer network, the front $L - 1$ layers consist of SCAEs and the top layer is sigmoid function. As for the lth layer with the parameters $\theta^l = \{W_1^l, W_2^l, b_1^l, b_2^l\}$, the training process can be divided into two stages: pretraining and updating [14]. The pretraining objective function is formulated as:

$$J_{\text{pre}}^l(\theta^l) = \frac{1}{2N} \sum_{i=1}^{N} \left\{ \|\hat{x}_i^l - h_i^{l-1}\|_2^2 + \lambda \sum_{j=1}^{n_l} (h_{ij}^l(1 - h_{ij}^l))^2 \|W_{1j}^l\|_2^2 \right\}$$
$$+ \frac{\delta}{2} \|W_1^l\|_F^2, \tag{3}$$

where the first term represent the reconstruction error, the second term denotes the contractive penalty to optimize the encoding function, and the third term stands for the weight decay of the parameters. After pretraining, multinomial logistic regression module is connected with the encoder to update the parameters of SCAE. The updating function is written as:

$$J_{\text{update}}^l(\theta^l) = -\frac{1}{N}\sum_{i=1}^{N} y_i \log p_i^l + \frac{\tau}{2}\|W_2^l\|_F^2 \tag{4}$$

where the first term is used to capture the relevant information between features and labels, and the second term denotes the weight decay. After training of each SCAE, the whole network is fine-tuned from the top layer to the final layer. The sigmoid function on the top layer is applied to classify each pixel into changed and unchanged class.

3   Experimental Results and Analysis

3.1   Datasets Description

In this section, we evaluate the performance of the proposed method on two real datasets, which are acquired from two large SAR images of the region of the Sulzberger Ice Shelf. Both of two SAR images, which have the size of 2263 × 2264 pixels, are provided by the European Space Agency's Envisat satellite on March 11 and 16, 2011. It is huge to us, so we selected a part of these images, which area is 256 × 256 pixels. Both datasets are shown in Figs. 4 and 5. (a) and (b) are the original images and (c) is the ground truth image, which are obtained by integrating prior information with photo interpretation.

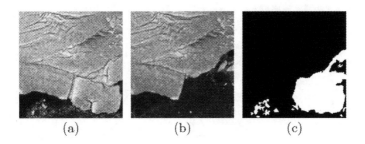

$$\text{(a)}\qquad\qquad\text{(b)}\qquad\qquad\text{(c)}$$

Fig. 4. Dataset I from Sulzerger Ice Shelf. (a) Image acquired in March 11 in 2011. (b) Image acquired in March 16 in 2011. (c) Ground truth image obtained by integrating prior information with photo interpretation.

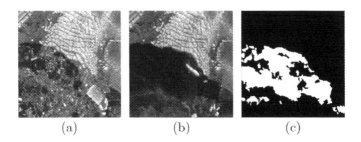

(a)   (b)   (c)

Fig. 5. Dataset II from Sulzerger Ice Shelf. (a) Image acquired in March 11 in 2011. (b) Image acquired in March 16 in 2011. (c) Ground truth image obtained by integrating prior information with photo interpretation.

3.2   Evaluation Criteria

In order to prove the effectiveness of the proposed method, we introduce five criteria to evaluate the performance of experimental results, such as false positives (FP), false negatives (FN), overall error (OE), percentage correct classification (PCC) and Kappa coefficient (KC). FP is the number of pixels belong to the unchanged class in the ground truth image but wrongly classified as the changed class. On the contrary, FN is the number of pixels belong to the changed class in the ground truth image but wrongly classified as the unchanged class. OE represents the sum of FP and FN. PCC is the percentage correct classification of the algorithm and its representation can be defined as follows:

$$PCC = \frac{N - FP - FN}{N} \times 100\% \tag{5}$$

where N is the total number of pixels in the images. The KC is a critical measurement of change detection accuracy, and it is a more persuasive coefficient because more detailed information is involved.

3.3   Experimental Results on Dataset I

In this paper, we compare our algorithm with other three state-of-the-art methods: PCAKM [6], GaborTLC [15] and GaborPCANet [7]. And the accuracy of the experiment results is showed in two ways: image change map and tabular form. We first evaluate our proposed method on the Dataset I. Figure 6 shows the final experimental results on Dataset I and Table 1 illustrate the evaluate criteria of different methods. From Fig. 6, we can find that the result of GaborTLC is polluted by many white noisy spots. Many unchanged pixels are wrongly classified into the changed class. Similar conditions can be found in the results of PCAKM and GaborPCANet. The PCC value of the proposed method achieves the best performance on this dataset. It can be concluded that the proposed method can efficiently suppress the multiplicative speckle noise by introducing saliency detection.

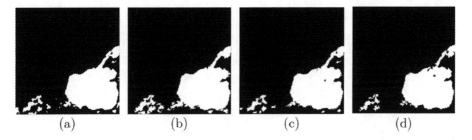

(a)   (b)   (c)   (d)

Fig. 6. Experimental results on Dataset I. (a) Result by PCAKM. (b) Result by GaborTLC. (c) Result by GaborPCANet. (d) Result by the proposed method.

Table 1. Change detection results of different methods on Dataset I.

Methods	FP	FN	OE	PCC (%)	KC (%)
PCAKM [6]	711	479	1190	98.18	87.13
GaborTLC [15]	1171	423	1594	97.57	92.39
GaborPCANet [7]	435	833	1268	98.07	93.73
Proposed method	232	930	1162	98.23	94.21

3.4 Experimental Results on Dataset II

We also test the proposed model on the Dataset II. Figure 7 shows the final sea ice change detection results of different methods. It can be observed that the results of PCAKM, GaborTLC and GaborPCANet are relatively inefficient in speckle noise suppression, which lead to the change maps containing many speckle noise spots. Table 2 shows the comparisons of different methods according to the five evaluation criterion. From Table 2, we find that the FP values of different methods are higher than that in the Dataset I. One important reason is that Dataset II contain very complicated background compared with Dataset I, which poses a challenge to change detection and needs the higher robust method

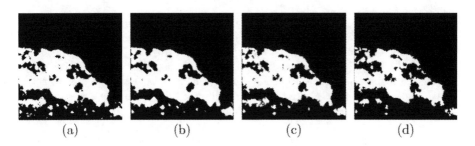

                (a)   (b)   (c)   (d)

Fig. 7. Experimental results on Dataset I. (a) Result by PCAKM. (b) Result by GaborTLC. (c) Result by GaborPCANet. (d) Result by the proposed method.

Table 2. Change detection results of different methods on Dataset II.

Methods	FP	FN	OE	PCC (%)	KC (%)
PCAKM [6]	3215	141	3356	94.88	87.13
GaborTLC [15]	2805	174	2979	95.44	88.48
GaborPCANet [7]	2237	599	2836	95.67	88.82
Proposed method	889	863	1861	97.16	92.44

to improve the accuracy. The proposed methods produces fewer false alarms and it suppresses the bad effects of multiplicative speckle noise. Both visual and data quantitative comparison on this dataset has demonstrated the effectiveness of the proposed method.

4   Conclusion

In this paper, we proposed a sea ice change detection method from SAR images based on Canonical Correlation Analysis and Contractive Autoencoders (CCA-SCAEs). On the one hand, the structured matrix decomposition can improve the difference image, which has less speckle noise spots for the classification. On the other hand, by using the hierarchical FCM algorithm, some reliable samples can be acquired to train the CCA-SCAEs classifier. Finally, we use the CCA-SCAEs classifier to classify pixels from the original SAR images into changed and unchanged class, and the final change map can be obtained. The proposed method is implemented on two real sea ice datasets. The experimental results have demonstrated the proposed method can efficiently reduce the speckle noise, while also preserving the fine details of change features.

References

1. Wang, L., Scott, K.A., Clausi, D.A.: Sea ice concentration estimation during freeze-up from SAR imagery using a convolutional neural network. Remote Sens. **9**, 409 (2017)
2. Zhang, H., Gong, M., Zhang, P., Su, L., Shi, J.: Feature-level change detection using deep representation and feature change analysis for multispectral imagery. IEEE Geosci. Remote Sens. Lett. **13**, 1666–1670 (2016)
3. Lopes, A., Nezry, E., Touzi, R., Laur, H.: Structure detection and statistical adaptive speckle filtering in SAR images. Int. J. Remote Sens. **14**, 1735–1758 (1993)
4. Gomez, L., Munteanu, C., Berlles, J., Mejail, M.: Evolutionary expert-supervised despeckled SRAD filter design for enhancing SAR images. IEEE Geosci. Remote Sens. Lett. **8**, 814–818 (2011)
5. Bazi, Y., Bruzzone, L., Melgani, F.: An unsupervised approach based on the generalized Gaussian model to automatic change detection in multitemporal SAR images. IEEE Trans. Geosci. Remote Sens. **43**, 874–887 (2005)

6. Celik, T.: Unsupervised change detection in satellite images using principal component analysis and k-means clustering. IEEE Geosci. Remote Sens. Lett. **6**, 772–776 (2009)

7. Gao, F., Dong, J., Li, B., Xu, Q.: Automatic change detection in synthetic aperture radar images based on PCANet. IEEE Geosci. Remote Sens. Lett. **13**, 1792–1796 (2016)

8. Gong, M., Zhou, Z., Ma, J.: Change detection in synthetic aperture radar images based on image fusion and fuzzy clustering. IEEE Trans. Image Process. **21**, 2141–2151 (2012)

9. Lin, K., Lu, J., Chen, C.-S., Zhou, J.: Learning compact binary descriptors with unsupervised deep neural networks. In: IEEE Conference on Computer Vision and Pattern Recognition (CVPR), pp. 1183–1192 (2016)

10. Sell, G., Garcia-Romero, D., McCree, A.: Speaker diarization with I-Vectors from DNN senone posteriors. In: Proceedings of Interspeech, pp. 3096–3099 (2015)

11. Wang, L., Scott, K.A., Xu, L., et al.: Sea ice concentration estimation during melt from dual-pol SAR scenes using deep convolutional neural networks: a case study. IEEE Trans. Geosci. Remote Sens. **54**, 4524–4533 (2016)

12. Gong, M., Zhao, J., Liu, J., Miao, Q., Jiao, L.: Change detection in synthetic aperture radar images based on deep neural networks. IEEE Trans. Neural Netw. Lear. Syst. **27**, 125–138 (2016)

13. Peng, H., Li, B., Ling, H., Hu, W., Xiong, W., Maybank, S.: Salient object detection via structured matrix decomposition. IEEE Trans. Patter. Anal. Mach. Intell. **39**, 818–832 (2017)

14. Geng, J., Wang, H., Fan, J., Ma, X.: Deep supervised and contractive neural network for SAR image classificaiton. IEEE Trans. Geosci. Remote Sens. **55**, 2442–2459 (2017)

15. Li, H., Celik, T., Longbotham, N.: Gabor feature based unsupervised change detection of multitemporal SAR images based on two-level clustering. IEEE Geosci. Remote Sens. Lett. **12**, 2458–2462 (2015)

Pedestrian Attributes Recognition in Surveillance Scenarios with Hierarchical Multi-task CNN Models

Wenhua Fang[1(✉)], Jun Chen[1], Tao Lu[2], and Ruimin Hu[1]

[1] National Engineering Research Center for Multimedia Software,
Computer School of Wuhan University, Wuhan 430072, Hubei Province, China
{fangwh,chenj,hrm}@whu.edu.cn
[2] Computer School of Wuhan Institute of Technology,
Wuhan 430205, Hubei Province, China
lutxyl@gmail.com

Abstract. Pedestrian attributes recognition is a very important problem in video surveillance and video forensics. Traditional methods assume the pedestrian attributes are independent and design handcraft features for each one. In this paper, we propose a joint hierarchical multi-task learning algorithm to learn the relationships among attributes for better recognizing the pedestrian attributes in still images using convolutional neural networks (CNN). We divide the attributes into local and global ones according to spatial and semantic relations, and then consider learning semantic attributes through a hierarchical multi-task CNN model where each CNN in the first layer will predict each group of such local attributes and CNN in the second layer will predict the global attributes. Our multi-task learning framework allows each CNN model to simultaneously share visual knowledge among different groups of attribute categories. Extensive experiments are conducted on two popular and challenging benchmarks in surveillance scenarios, namely, the PETA and RAP pedestrian attributes datasets. On both benchmarks, our framework achieves superior results over the state-of-the-art methods by 88.2% on PETA and 83.25% on RAP, respectively.

Keywords: Attributes recognition · CNN · Multi-task learning

1 Introduction

Visual recognition of pedestrian attributes, such as the estimation of gender, age and clothes colors or styles, is an emerging and important research topic in computer vision community, due to its great potential in the real surveillance system. For example, pedestrian attributes recognition has been used to assists pedestrian re-identification [1] and pedestrian detection [2], and has been proved to greatly improve other vision related tasks [3]. As a middle-level representation, pedestrian attributes may shorten the semantic gap between low-level features

© Springer Nature Switzerland AG 2018
R. Hong et al. (Eds.): PCM 2018, LNCS 11165, pp. 758–767, 2018.
https://doi.org/10.1007/978-3-030-00767-6_70

and human description [4]. In addition, they may play a critical role for the specific person search in practical applications for public security, such as the retrieval of the suspects in London underground bombing event.

Recognizing pedestrian attributes remains a challenging problem in video surveillance environments, because there are large intra-class variations [5]. In addition, pedestrian attributes are difficult to predict because of different camera viewpoints, pose variations, large time range, low image quality, different orientations and partial occlusion, etc. In surveillance scenarios, even the appearances of the same persons attributes will vary widely.

Pedestrian attributes recognition approaches can be divided into two categories according to feature representation: hand-crafted features based and deep learning features based. The former are artificially designed based on the statistical properties of the frame pixels. Early work was done on relatively small pedestrian attributes datasets. Daniel et al. [3] proposed a part based feature representation for human attributes including facial hair, eyewear, and clothing color in video surveillance. Lubomir et al. [6] tried to recognize 9 binary human attributes in personal photo album images by a poselet based approach [7]. To recognize more pedestrian attributes in surveillance environment, Zhu et al. [8] constructed the Attributed Pedestrians in Surveillance (APiS) database including 11 binary and 2 multi-class attribute annotations. Moreover, they proposed two baseline methods with color and texture features to predict the attributes. Deng et al. [5] extended the above work and released the first large scale pedestrian dataset including 19, 000 images with 61 annotated attributes. And they adopted multiple low-level color and texture features to present the attributes. However, the main drawback of such hand-crafted descriptors is that they lack enough discriminative capacity for pedestrian attributes recognition in surveillance scenarios [9].

On the other hand, deep learning features, such as deep Convolutional Neural Networks (CNN), has demonstrated superior performance in many computer vision tasks. And it has also been shown that CNN can generate robust and generic features [10]. Encouraged by the ability of feature learning of CNN, recently researchers have exploited the CNN features for pedestrian attributes recognition. Li et al. [9] proposed two CNN based models to recognize pedestrian attributes. On the one hand, they treated each attribute as an independent component and trained single attribute recognition model to recognize each attribute one by one. On the other hand, they treated the pedestrian attributes recognition as classic multiple classification task to exploit the relationship among attributes. Zhu et al. [11] proposed a multi-label convolutional neural network to predict multiple attributes together in a unified framework, in which a pedestrian image was roughly divided into multiple overlapping body parts, which were simultaneously integrated in the multi-label CNN model. Kai Yu et al. [12] formulated pedestrian attribute recognition in an attribute localization framework and proposed a weakly-supervised pedestrian attribute localization network to predict the attribute labels.

Given the aforementioned issues, in this paper we propose a CNN based hierarchical multi-task framework for the large scale pedestrian attributes recognition in the surveillance scenes, as shown in Fig. 2. The motivation is that there are inherent relationships among the pedestrian attributes which can benefit the recognition of each attribute. We explicitly model such relationships by splitting the attributes into local and global ones according to their location and semantics. In the first layer, each multi-task CNN is constructed for local attributes in each local groups and in the second layer, the auxiliary network on the top of last convolutional layers from all local multi-task CNNs is proposed for learning global attributes. Unlike the work in [13], we introduce the spatial and semantic based grouping information to obtain flexible global and local sharing and competition between groups. The contributions of this paper are summarized as follows:

- We find that there are relationships among pedestrian attributes that can boost the recognition of each attribute and propose the strategy to split the attributes into relative groups.
- We propose a CNN based multi-task learning framework for the large scale pedestrian attributes recognition in the surveillance scenes.
- We conduct sufficient experiments on two challenging large scale pedestrian attributes databases and achieve the state-of-the-art performance.

2 Proposed Approach

In this section, we will explain the details of the proposed approach of the multi-task CNN model. Figure 1 shows the overall structure of the proposed method, starting from raw images and ending with attribute classification. We start with the presentation of pedestrian attributes grouping strategy, then introduce the CNN based multi-task learning model for local attributes recognition in each group. Based on the learned local CNN features in each group, we discuss the global attributes recognition in detail.

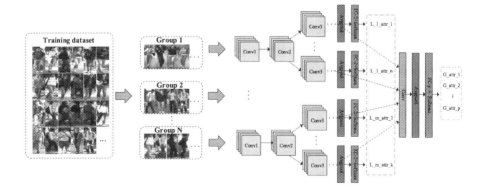

Fig. 1. CNN based multi-task learning framework for pedestrian attributes recognition

2.1 Pedestrian Attributes Grouping

Pedestrian attributes can be naturally split into different groups according to spatial location and semantic relation. For spatial location, pedestrian attributes can be divided into global and local. As the name implies, local attributes describe the head (glasses,long hair), upper body (T-shirt) and lower body (long pants) and global attributes describe the whole body (age, gender). For semantic relation, pedestrian attributes can be divided into correlated and non-correlated. For example, Long hair and red dresses may indicate that the gender of the pedestrian is female. So female is correlated with long hair and red dressed.

We follow the work of [13], and split the pedestrian attributes into the global and local ones. Because the number of the global attributes is relatively small, usually less than 10, we aggregate them into only one group. For the local attributes, according to the different locations, we split them into three categories: the head-shoulder, the upper body and the lower body. Moreover, according to the correlation among the local attributes, we divide them into different groups, such as color, texture, shape and action. Detailed attributes grouping tables are listed in Sect. 3.

2.2 Multi-task Learning for Local Attributes

In this section, we will explain the details of the proposed approach of the multi-task CNN model for local attributes. The local attributes recognition is described as illustrated in Fig. 2. Firstly, the images of training dataset are divided into different groups according to the attributes grouping mentioned above. And then Images from each group are fed into a multi-task CNN model for learning semantic attributes. Due to multiple attributes labels of each image, an image will belong to different groups. Emily et al. [14] proposed a multi-task model based on Alexnet for facial attributes recognition with two lower convolutional layers shared by all attributes, ignoring the semantics and spatial relationship. Unlike that, we use and modify the popular network structure VGG-16 to learn more discriminative features. In our multi-task framework, the lower four convolutional layers with two max pooling layers are shared for all attributes in the same group. This allows for learning of implicit relationships among attributes at a lower level. After that, feature maps are fed into different convolutional layers, max pooling layers, global average pooling layer and subsequent softmax layer for each attribute recognition. Different from standard VGG model, we not only remove the last two fully connected layers and add a 1×1 convolutional layer with global average pooling layer [13], but also replace some 3×3 convolutional layers with the 1×1 convolutional layers, as shown in Fig. 3. This can not only reduce the parameters by a large margin, and accelerate convergence in the training, but also reduce the risk of over-fitting without loss of performance. This improvement can boost the performance by about 1.2% in mean Accuracy (mA). Moreover, we use a weighted sigmoid cross-entropy loss for all attribute scores to facilitate training to solve the imbalance distribution of attributes. This

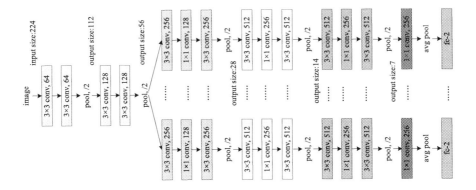

Fig. 2. Multi-task learning for local attributes recognition

can improve the performance by about 0.9% in mean Accuracy (mA). As pre-processing steps, the training mean is subtracted from the images. This helps the network to be robust to shifts in the input. Unlike other attributes classification methods, we do not perform any manual alignment or part extraction in the preprocessing stage. Both manual alignment and part extraction are expensive and error-prone processes, and so we save time and avoid problems associated with poor and misalignment by skipping these steps. Our method is also more applicable to real-world imagery for which alignment may be challenging.

2.3 Multi-task Learning for Global Attributes

Since the global attributes are related to the multiple local attributes of the pedestrian, the local CNN features are shared for all global attributes. Inspired by AUX architecture [14], we also design a novel AUX architecture. In our AUX architecture, we add a 1×1 convolutional layer and global average pooling on the top of the last convolutional layer of the multi-task CNN for local attributes. We also use the sigmoid cross-entropy loss for all global attribute scores to facilitate training like local attributes learning. This can boost the performance about 1%.

In Fig. 3, the purple dashed line shows the connection between multi-task CNN for local attributes learning and global attributes prediction. And the orange dashed lines show the multi-task CNN for local attributes learning. In AUX architecture [14], two fully connected layers are added on the output of the multi-task CNN to predict the global facial attributes. However, this may result in the risk of over-fitting (five fully connected layers for global attribute recognition). From the perspective of CNN based object classification, the convolutional layers and the fully connected layers are regarded as feature extraction and classifier respectively. Our model is quite different from AUX architecture. We use the all convolutional features as the input of our model to share the features directly, not the predicted vectors.

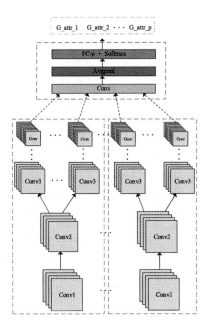

Fig. 3. Multi-task learning for global attributes recognition

2.4 Implementation

We preprocess all examples by warping the input image to 256×256 pixels. In order to learn robust features from CNNs, during training, we make use of several data augmentation strategies [14]. We resize the images of training database to 227×227, and randomly crop a 224×224 patch from them. Secondly, the cropped patches are horizontally flipped by random. Finally, we use a scale jittering strategy to help CNN to learn robust features. Furthermore, we crop a patch on three scales (1, 0.875, and 0.75), which yield the scaled patches of size 224×224, 196×196, and 168×168. The scaled patches are then resized to 224×224. In testing phase, we disuse the data augmentation strategy and just crop one 224×224 patch from the center of testing image. Multi-task CNN is pre-trained on ImageNet dataset and then the model parameters are finely tuned on pedestrian attributes recognition datasets. In the process of CNN training, the learning rate is set to 10^{-3}, decreases to its $1/10$ every $10K$ iterations, and stops at $50K$ iterations [8].

3 Experiments

In this section, we show the pedestrian attributes recognition performance of the proposed method. We firstly introduce the datasets, evaluation criterion and baselines for evaluating our proposed approach. Then we give the grouping lists on the datasets based on the grouping strategy. Finally, the proposed method is compared with the state-of-the-art methods.

Table 1. Pedestrian attributes grouping strategy on PETA

Category		Attributes		
Global		Gender, age, body shape, role, viewpoint, etc.		
Local	Head	Hair style, hat, etc.	Hair/hat color, etc.	Telephoning, talking, etc.
	Upper	Clothes style, etc.	Clothes color, etc.	Carrying backbag, pushing, etc.
	Lower	Clothes style, etc.	Clothes color, etc.	Walking, etc

3.1 Datasets

We evaluate the proposed approach on two popular large scale pedestrian attributes recognition datasets in surveillance scenes: PETA [14] and RAP [17]. The PETA dataset is collected from ten popular pedestrian datasets in surveillance scenes and consists of $19,000$ images with resolution ranging from 17×17 to 169×365 pixels. Each image in PETA is labeled with 61 binary and 4 multi-class attributes (color attributes). The binary attributes cover an exhaustive set of characteristics of interest, including demographics (e.g. gender and age range), appearance (e.g. hair style), upper and lower body clothing style (e.g. casual or formal), and accessories.

The RAP dataset is the largest collection of realistic multi-camera surveillance scenarios with long term collection, where data samples are annotated with not only fine-grained human attributes but also environmental and contextual factors. The dataset is composed of $41,585$ images, each of which is annotated with 72 attributes as well as viewpoints, occlusions, body parts information. We use the default training/testing splits in our experiment. We also train our CNN model in image classification task from ImageNet to initiate our model and fine-tune it in our target databases.

3.2 Attributes Grouping

Firstly, the pedestrian attributes are divided into two parts: the global and local. The global attributes usually describe the features of the whole body. Because of the relatively small volume, we aggregate them into one group. Secondly, the local attributes are divided into three parts (head-shoulder, upper body and lower body). Finally, each local part is split into two groups according to semantic appearance and action. Detailed attributes groups are listed in Table 1.

3.3 Performance Evaluation on PETA

In this section, we compare our method with four baselines on PETA dataset. The performance of the benchmarks are listed in [10,13]. As shown in Table 2, our multi-task CNN framework obtains the highest performance on two datasets in all four evaluation criterions. According to Table 2, we can learn that the mean accuracy, accuracy, precision, recall and F1 of our method are 88.20%, 78.31%, 86.23%, 89.21% and 87.69%, which are 2.7%, 1.33%, 2.16%, 3.43% and

Table 2. Pedestrian attributes evaluation on PETA

Methods	mA	ACC	Prec	Rec	F1
ELF [15]	75.21	43.68	49.45	74.24	59.36
ACN [16]	81.15	73.66	84.06	81.26	82.64
DeepMAR [10]	82.89	75.07	83.68	83.14	83.41
WPAL [13]	85.50	76.98	84.07	85.78	84.90
Ours	**88.20**	**78.31**	**86.23**	**89.21**	**87.69**

2.79% higher than that of the state-of-the-art approaches respectively. And the average improvement of the proposed method over the second best method is 3.5%. In general, the proposed multi-task CNN framework obviously improves the recognition accuracy of pedestrian attributes.

3.4   Performance Evaluation on RAP

Finally, we compare our method with the state-of-the-art on RAP dataset. As shown in Table 3, For RAP dataset, according to Table 3, we can learn that the mean accuracy, accuracy, precision, recall and F1 of our method are 83.25%, 63.13%, 82.52%, 81.65% and 82.08%, which are 2.0%, 0.53%, 2.30%, 3.26% and 6.1% higher than that of the state-of-the-art approaches. Our multi-task deep learning framework produces the highest performance on the two datasets. Both on the PETA and RAP datasets, some work with competitive results is based on multi-task learning CNN feature. Compared with the result of the original multi-task method in [16] (mA: 69.66%), our framework (mA: 83.25%) is much better with the additional multi-task learning and attributes grouping to explore the discriminative information. The recent multi-task CNN works [13] adopted different multi-task framework, so the results are not directly comparable. Our deep learning framework is similar to [13], but our results improved 2% compared to [13].

Table 3. Pedestrian attributes evaluation on RAP

Methods	mA	ACC	Prec	Rec	F1
ELF [15]	69.94	29.29	32.84	71.18	44.95
ACN [16]	69.66	62.61	80.12	72.26	75.98
DeepMAR [10]	73.79	62.02	74.92	76.21	75.56
WPAL [13]	81.25	50.30	57.17	78.39	66.12
Ours	**83.25**	**63.13**	**82.52**	**81.65**	**82.08**

4   Conclusions

In this paper, we have proposed a novel group based hierarchical multi-task CNN framework for pedestrian attributes recognition. The main idea is splitting the pedestrian attributes into different groups according to the spatial and semantic relationship and building the novel hierarchical multi-task CNN framework for local attributes and global attributes recognition by global sharing and competition between groups through learning deep discriminative features. The proposed method can significantly improve the performance of pedestrian attributes recognition. Experimental results on two public benchmark databases have demonstrated its superiority over the state-of-the-art methods. In our future work, the relationships among local and global attributes are considered to model by graphical models explicitly.

Acknowledgment. This research is based upon work supported by National Nature Science Founda- tion of China (No. U1736206),National Nature Science Foundation of China(61671336), National Nature Science Foundation of China(61671332),Technology Research Program of Ministry of Public Security (No. 2016JSYJA12),Hubei Province Technological Innovation Major Project(No. 2016AAA015),Hubei Province Technological Innovation Major Project2017AAA123),The National Key Research and Development Program of China(No.2016YFB0100901),Nature Science Foun- dation of Jiangsu Province (No. BK20160386) and National Nature Science Foundation of China(61502354).

References

1. Lin, Y.T., Liang, Z., Zeng, Z.D.: Improving person re-identification by attribute and identity learning, arXiv preprint arXiv:1703.07220 (2017)
2. Tian, Y., Luo, P., Wang, X.: Pedestrian detection aided by deep learning semantic task. In: Proceedings of IEEE Conference on Computer Vision and Pattern Recognition, pp. 5079–5087 (2015)
3. Vaquero, D.A., Feris, R.S., Duan, T.: Attribute-based people search in surveillance environments. In: Proceedings of IEEE Workshop on Applications of Computer Vision, pp. 1–8 (2009)
4. Deng, Y., Luo, P., Chen, C.: Pedestrian attribute recognition at far distance. In: Proceedings of ACM International Conference on Multimedia, pp. 789–792 (2014)
5. Zhu, J., Liao, S., Lei, Z.: Pedestrian attribute classification in surveillance: database and evaluation. In: Proceedings of the IEEE International Conference on Computer Vision Workshops, pp. 331–338 (2013)
6. Bourdev, L., Maji, S., Malik, J.: Describing people: a poselet-based approach to attribute classification. In: Proceedings of IEEE International Conference on Computer Vision, pp. 1543–1550 (2012)
7. Dangei, L., Zhang, Z., Tang, C.X.: A richly annotated dataset for pedestrian attribute recognition, arXiv preprint arXiv:1603.07054 (2016)
8. Layne, R., Hospedaes, T.M., Gong, S.G.: Person Re-identification by attributes. In: British Machine Vision Conference (2012)
9. He, K., Zhang, X., Ren, S.Q.: Deep residual learning for image recognition. In: Proceedings of IEEE Computer Vision and Pattern Recognition, pp. 770–778 (2016)

10. Bengio, Y., Courville, A., Vincent, P.: Representation learning: a review and new perspectives. IEEE Trans. Patt. Anal. Mach. Intell. **35**(8), 1798–828 (2013)
11. Jayaraman, D., Fei, S., Grauman, K.: Decorrelating semantic visual attributes by resisting the urge to share. In: Proceedings of IEEE Conference on Computer Vision and Pattern Recognition, pp. 1629–1636. IEEE Computer Society (2014)
12. Zhu, J., Liao, S., Yi, D.: Multi-label CNN based pedestrian attribute learning for soft biometrics. In: Proceedings of the IEEE International Conference on Biometrics, pp. 535–540 (2015)
13. Yu, K., Leng, B., Zhang, Z.: Weakly-supervised learning of mid-level features for pedestrian attribute recognition and localization, arXiv preprint arXiv:1611.05603 (2016)
14. Emily, M.H., Rama, C.B.: Attributes for improved attributes: a multi-task network utilizing implicit and explicit relationships for facial attribute classification. In: Proceedings of AAAI Conference on Artificial Intelligence, pp. 4068–4074 (2017)
15. Sudowe, P., Spitzer, H., Leibe, B.: Person attribute recognition with a jointly-Trained holistic CNN model. In: Proceedings of IEEE International Conference on Computer Vision Workshop, pp. 329–337 (2015)
16. Zhu, J., Liao, S., Lei, Z.: Multi-label convolutional neural network based pedestrian attribute classification. Image Vis. Comput. **58**, 224–229 (2016)

Re-Ranking Person Re-Identification with Forward and Reverse Sorting Constraints

Meibin Qi⬥, Yonglai Wei$^{(\boxtimes)}$⬥, Kunpeng Gao⬥, Jianguo Jiang⬥, and Jingjing Wu⬥

Hefei University of Technology, Hefei 230009, Anhui, China
qimeibin@163.com, wyl_hfut@163.com, gkp_hfut@163.com, jgjiang@hfut.edu.cn,
hfutwujingjing@mail.hfut.edu.cn

Abstract. Person re-identification task aims at matching pedestrian images across multiple camera views. Extracting more robust feature of the pedestrian images and finding more discriminative metric learning are the main research directions in person re-identification. The achieved results are provided in the form of a list of ranked matching persons. It often happens that the true match which should be in the first position is not ranked first. In order to correct some false matches and improve the accuracy of person re-identification, this paper proposes a re-ranking method with forward and reverse sorting constraints. The forward sorting constraint makes the image, which is in the front position of one forward sorting list, be backward in the position of other forward sorting lists; The reverse sorting constraint makes two images of the same pedestrian be in the front position of each other's sorting list. Experiments on four public person re-identification datasets, VIPeR, PRID450S, CUHK01 and CUHK03 confirm the simplicity and effectiveness of our method.

Keywords: Person re-identification · Re-ranking
Forward and reverse sorting constraints

1 Introduction

With the popularity of video surveillance, more and more video surveillance systems are employed in the public areas, such as shopping malls, airports and hospitals. The control center of this system is usually connected with multiple cameras which are distributed in different areas. The control center operator can find and track a specific pedestrian (such as a criminal suspect) by observing the cameras. However, with the increasing number of cameras in the surveillance

Supported by organization by the National Natural Science Foundation of China Grant 61632007 and Key Research and Development Project of Anhui Province, China 1704d0802183.

R. Hong et al. (Eds.): PCM 2018, LNCS 11165, pp. 768–779, 2018.
https://doi.org/10.1007/978-3-030-00767-6_71

system, it becomes increasingly difficult to manually find and track pedestrians. Therefore, monitoring system needs to automatically find and track pedestrians.

The key of this monitoring system is to match the same pedestrian under different cameras, which is known as person re-identification. The problem of person re-identification can be expressed as follows: supposed that the existing n pedestrians pass through camera A and camera B in turn, the pedestrian images captured by camera A and camera B are called the probe images and the gallery images, respectively. Each probe image could determine a sort of the gallery images. The top-ranked gallery images are considered to be more similar to the probe images, in other words, they are more likely to be the same pedestrian.

However, many factors affect the performance of the person re-identification, such as various camera viewpoints, illumination, occlusion and the limitation of metric function. In recent years, there are two key points of person re-identification methods, i.e. extracting robust features and learning discriminative metrics. For feature representation, Liao et al. [10] propose an efficient feature representation called Local Maximal Occurrence (LOMO), using color and Scale Invariant Local Ternary Pattern (SILTP) histograms to represent picture appearance in a high dimension. In [15], a method called Gaussian of Gaussian (GOG) descriptor is proposed based on a hierarchical Gaussian distribution of pixel features. Lisanti et al. [12] propose a kernel descriptor to encode person appearance and project the data into common subspace using Kernel Canonical Correlation Analysis (KCCA). Chen et al. [1] propose an Spatially Constrained Similarity function on Polynomial feature map as SCSP to divide an image into four non-overlapping horizontal stripe regions, and each stripe region can be described by four visual cues which are organized as HSV1/HOG, HSV2/SILPT, LAB1/SILPT and LAB2/HOG. For distance metric, methods are designed to maximize the inter-class similarity and minimize the intra-class similarity. KISS Metric Learning (KISSME) [5], Crossview Quadratic Discriminant Analysis (XQDA) [10], Metric Learning with Accelerated Proximal Gradient (MLAPG) [11], Top-push Distance Learning model (TDL) [21] are representative methods.

While the first image of the list is usually not the true matching image in existing person re-identification methods. In order to solve this problem and improve the accuracy of person re-identification, a large number of researches called re-ranking have emerged for person re-identification [2,3,6,7,13,20,22]. Some of these researches [13] require the supervision of tag information, while this paper prefers unsupervised automatic re-ranking research. The method proposed in [3] learns an unsupervised re-ranking model by jointly considering content and context information in a sorting list for effectively eliminating fuzzy samples and improving the performance of person re-identification. Lend et al. [20] propose a bidirectional ranking method which combines the similarity of the content with context, and uses the new similarity to correct the initial sorting list. Recently, it is increasingly popular to correct the initial sorting by using the k-reciprocal neighbors [23]. However, the method in [23] has no effect if there is only one positive for each identity in the gallery (single-shot). In order to solve this problem, this paper proposes a forward and reverse sorting constraints algorithm under single-shot to significantly improve pedestrian re-identification performance.

2   Proposed Approach

2.1   Problem Definition

Given a probe set with M pedestrian images $P = \{p_i|i = 1, 2, ..., M\}$ and a gallery set with N pedestrian images $G = \{g_j|j = 1, 2, ..., N\}$, the original distance between the pedestrian images p_i and g_j is represented as $d(p_i, g_j)$. Therefore, the initial ranking list $L(p_i, G) = \{g_{k,j}^i|i \in (1, 2, ...M); j, k \in (1, 2, ...N)\}$ can be obtained by calculating the distances between the probe image p_i and all gallery images G, where $g_{k,j}^i$ denotes the gallery g_j which is ranked k-th in the list of p_i. The goal of re-ranking is to modify the initial ranking list $L(p_i, G)$ so that more positive samples are ranked at the top of the list to improve the performance of person re-identification.

2.2   Forward Sorting Constraint

The matching of person re-identification is to calculate the distances between a probe image and all gallery images, then we sort the image of the gallery based on the distances. The first image of the list is considered to be the true match of the probe image. This sorting method is called forward sorting in this paper. However, the true matching results are not ranked first in the sorting list after the forward sorting. Inspired by [16], we propose the algorithm of forward sorting constraint.

As shown in Fig. 1(a), the left images are the probe images, and the right images are the sorting of gallery images for them. The values in parentheses of the figure indicate the distance values between the corresponding gallery images

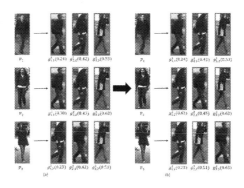

Fig. 1. Forward sorting constraint: (a) Initial ranking results. (b) Re-ranking results.

and the probe images. It can be seen from the figure, the gallery image g_1 is respectively rank-1 in the sorting list of p_1, rank-1 in the sorting list of p_2, and rank-2 in the sorting list of p_3. Therefore, according to the ranking, the possibilities that the gallery image g_1 and the probe images p_1, p_2 are the same pedestrian are higher than that of p_3. But it still cannot be judged that whether p_1 or p_2 is more likely to be the same pedestrian as g_1. Therefore, this paper combines the distance information with the constraint of the sorting information to judge. From the Fig. 1(a), because of $d(p_1, g_1) < d(p_2, g_1)$, the possibility that g_1 and p_1 are the same pedestrian is higher than the possibility that g_1 is the same pedestrian as p_2.

In summary, the possibility that the gallery image g_1 and the probe image p_1 are the same pedestrian is the highest. We assume that g_1 and p_1 are the same pedestrian, so g_1 and p_2, p_3 should be different pedestrian. Under the situation, the matching accuracy can be improved by "punishing" the distance between g_1 and p_2, p_3 (increasing the distance between them). The result of re-ranking after "punishment" is shown in Fig. 1(b). The ranking of correct matching is improved in the forward sorting list of p_2, p_3.

According to the above analysis, how to "punish" the distances of false matches becomes the primary problem. The algorithm of this paper gives the appropriate "punishment" based on the original distance value of mismatches. In short, the smaller (larger) original distance, the smaller (heavier) "punishment". In this way, some excessive punishment can be effectively avoided. The detailed algorithm is summarized in Algorithm 1.

Algorithm 1: Forward sorting constraint

Input: Initial distance matrix $dist(N, M)$
Output: Re-ranking distance matrix $outdist(N, M)$
for $j = 1, 2..., N$
    for $i = 1, 2, ..., M$
        $K(i) = g_j$ position in $L(p_i, G)$
    end
    $array\_k = argmin(K(i))$
    $k = array\_k$ lets $min(dist(p_{array\_k}, g_j))$
    for $i = 1, 2, ..., M$
        if $(i! = k)$ $outdist(j, i) = dist(j, i) + \lambda * dist(j, i)$
    end
end

2.3 Reverse Sorting Constraint

The forward sorting uses the distance values between the probe images and the gallery images to sort the gallery images. In turn, the distance values between them can also sort the probe images, and the sorting method is called the reverse sorting in this paper.

In Fig. 2, we show an example of the reverse sorting constraint. The true match (green box) of the probe image p_1 is incorrectly ranked, shown in Fig. 2(a). Figure 2(b) shows an reverse sorting list of the first two gallery images from the forward sorting list in Fig. 2(a). It is not difficult to find that the probe image p_1 is located in the third and first positions of reverse sorting list of the gallery images g_1 and g_2, respectively. There are some distinctions between Fig. 2(a) and 2(b). Figure 2(a) shows that the position of g_2 is relatively low-ranking in the forward sorting list of p_1. In Fig. 2(b), in the reverse sorting list of g_2, the position of p_1 is relatively high-ranking. It inspires us to use the reverse sorting information to constrain the initial sorting.

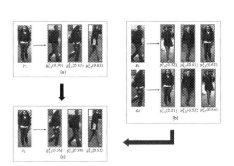

Fig. 2. Reverse sorting constraint: (a) Probe image and its forward sorting list. (b) Gallery images g_1, g_2 and their reverse sorting lists. (c) Re-ranking results.

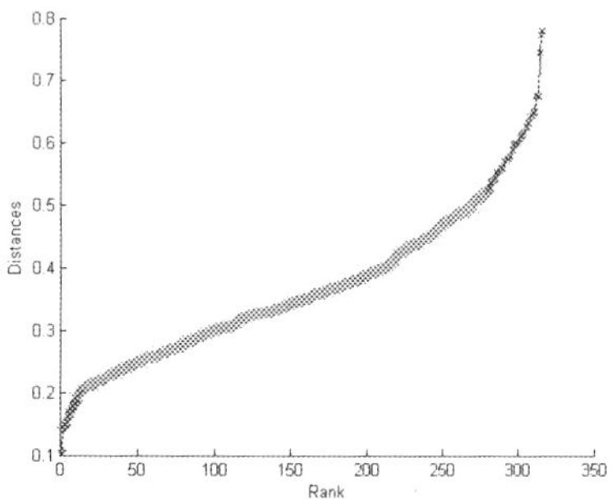

Fig. 3. An example of the results obtained by applying clustering algorithm to the distance computed between a probe and all the gallery images.

According to the above analysis, in order to efficiently use the reserve sorting, this paper does not implement reverse sorting constraint on all gallery images, while applies to several top gallery images in the forward sorting list. The top-ranked gallery images have a higher probability of containing true matches, so they are called content sets. In this paper, the reverse sorting constraint algorithm selects the content sets according to the dynamic method proposed by Jorge García et al. [2]. Figure 3 shows the relationship of the position of the gallery images in the forward sorting list of a probe image and the distance between the gallery image and the probe image: (1) at first ranks, the distance between the gallery image and the probe image increases abruptly, then flattens(first elbow); (2) from the first elbow, distances grow linearly till reaching high ranks, and at last, distances start increasing significantly again (second elbow). According to such trend, the gallery images can be divided into three classes: (1) the similar appearance class (C_{sa}), which corresponds to the gallery images whose positions are before the first elbow; (2) the difference appearance class (C_{da}), which corresponds to the gallery images whose positions are between the first elbow and the second elbow; (3) the opposite appearance class (C_{oa}), which corresponds to the gallery images whose positions are after the second elbow.

In order to find the positions of the three elbows and effectively divide the three types of the gallery set, this paper uses the k-means clustering algorithm to divide the original gallery set. The specific approach is as follows. First, defining the cluster mean value: $\mu_{sa} = d(p_i, g_1^i)$, $\mu_{da} = d(p_i, g_{(N/2)}^i)$ and $\mu_{oa} = d(p_i, g_N^i)$ where g_k^i denotes a gallery image ranked k-th in the forward sorting list of probe image p_i. Then the cluster mean is substituted into the following cost function:

$$\sum_{k \in \{sa,da,oa\}} \sum_{g_j^i \in C_k} ||d(p_i, g_j^i) - \mu_k||^2, j = 1, 2, ..., N \tag{1}$$

After continuously iterating and optimizing, the above cost function will converge. The similar appearance class will contain the first m gallery images of the forward sorting list, forming the content sets $B_i^{cn} = \{g_1^i, g_2^i, ..., g_m^i\}$. The reverse sorting constraint algorithm will be described in detail.

Given a probe image p_i and its forward sorting list $L(p_i, G)$. First of all, according to the method described above, we find the content set B_i^{cn} of the forward sorting list and then calculate the reverse sorting list $L(g_1^i, P)$ and $L(g_m^i, P)$ of gallery images g_1^i and g_m^i on the content set. Finally, we give certain "rewards" to the initial distance between p_i and g_1^i and the distance between p_i and g_m^i (decreasing the distance between them). The "reward" here is calculated by a reward function that gives more "rewards" to the distance between the top-ranked probe image and the gallery image of the reverse sorting list. According to the above properties, this paper proposes a reward function called Reward, as shown in the Eq. (2). It is worth noting that other reward functions that meet the above properties can also be used here.

$$Reward(p_i, g_j) = 2 - e^{0.01*rank(g_j, p_i)} \tag{2}$$

Algorithm 2: Reverse sorting constraint
Input: Initial distance matrix $dist(N, M)$
Output: Re-ranking distance matrix $outdist(N, M)$
for $i = 1, 2..., M$
    for $k = m, m - 1, ..., 2$
        $dist(p_i, g_k^i) -= Reward(p_i, g_k^i)$
        $dist(p_i, g_1^i) -= Reward(p_i, g_1^i)$
        Update sorting list.
    end
    $outdist = dist$
end

The $rank(g_j, p_i)$ indicates the position of the probe image p_i in the reverse sorting list of the gallery image g_j. Firstly, the "reward" is given to the distances between p_i and g_1^i, g_m^i, then we update the sorting list. Secondly, the "reward" operation is repeated on the distances between p_i and new g_1^i, g_{m-1}^i, then we update the sorting list again. Finally, we repeat the operation until "reward" the distances between p_i and g_1^i, g_2^i. The complete algorithm is summarized in Algorithm 2.

3 Experiments

3.1 Results of Different Datasets

We apply the proposed re-ranking algorithm on four different datasets of one baseline method to prove the universality of the method. The baseline method extracts hierarchical Gaussian descriptor [15] and uses cross-view quadratic discriminant analysis (XQDA) as distance metric. The four standard datasets include VIPeR [4], CUHK01 [8], PRID450S [18] and CUHK03 [9]. The test protocol of each dataset is the same as literature [15].

Table 1. Matching rates (%) of this algorithm on different datasets

Datasets	Method	r = 1	r = 5	r = 10
VIPeR	GOG	49.72	79.72	88.67
	GOG+OURS	**55.60**	**82.25**	**90.38**
	LOMO	40.00	68.13	80.51
	LOMO+OURS	**45.54**	**72.85**	**83.32**
CUHK01	GOG	57.91	79.14	86.27
	GOG+OURS	**63.68**	**83.60**	**89.29**
	LOMO	49.84	75.26	83.34
	LOMO+OURS	**55.16**	**78.16**	**85.66**
PRID450S	GOG	68.00	88.67	94.36
	GOG+OURS	**74.89**	**92.49**	**96.49**
	LOMO	57.42	81.07	88.31
	LOMO+OURS	**65.02**	**84.31**	**90.18**
CUHK03Labeled	GOG	68.47	90.69	95.84
	GOG+OURS	**77.08**	**94.59**	**97.20**
	LOMO	50.85	81.18	91.14
	LOMO+OURS	**56.90**	**84.83**	**93.69**
CUHK03 Detected	GOG	64.10	88.40	94.30
	GOG+OURS	**72.59**	**92.25**	**96.65**
	LOMO	44.10	78.70	87.70
	LOMO+OURS	**49.75**	**81.20**	**88.95**

The VIPeR dataset is a challenging dataset for person re-identification task, which contains 632 person image pairs, captured by different cameras in an outdoor environment. The CUHK01 dataset contains 971 persons and each person has two images in each camera. The PRID450S dataset is an extension of the PRID2011 dataset. It contains 450 pedestrian image pairs captured by two outdoor cameras. The background, lighting and viewpoints of two cameras are very different. The CUHK03 dataset contains 13164 images of 1360 identities captured by six surveillance cameras. Each pedestrian in the dataset has an average of 4.8 images per view. It provides bounding boxes manually labeled pedestrian and bounding boxes automatically detected by the pedestrian detector.

As shown in the Table 1, experimental results on above four datasets demonstrate that our approach achieves obvious improvements on the baseline method. Among them, the rank-1 matching rates increase by 5.32% at least and 8.85% at most. At the same time, there are also improvements on rank-5 and rank-10. It can be concluded that the proposed algorithm can be applied to different datasets and will not be affected by the acquisition environment and magnitude of datasets.

3.2   Results of Different Person Re-Identification Algorithms

In order to verify the scalability of the proposed algorithm, we apply the proposed re-ranking algorithm on four different person re-id methods of the VIPeR dataset.

Table 2. Matching rates (%) of different pedestrian re-identification algorithms

Method	VIPeR (%)			
	Rank-1	Rank-5	Rank-10	Rank-20
KCCA	37.25	71.39	84.56	92.81
KCCA+OURS	**43.39**	**75.06**	**85.76**	**93.04**
LOMO	40.00	68.13	80.51	91.03
LOMO+OURS	**45.54**	**72.85**	**83.32**	**92.94**
GOG	49.72	79.72	88.67	94.53
GOG+OURS	**55.60**	**82.25**	**90.38**	**95.51**
SCSP	53.54	82.59	91.49	96.65
SCSP+OURS	**57.75**	**83.01**	**91.23**	**98.92**

As shown in Table 2, our re-ranking algorithm can significantly improve performances for different person re-identification algorithms. And our method improves the rank-1 matching rates by 4% at least.

3.3   Comparison with Other Re-Ranking Methods

In this section, we compare our method with other re-ranking algorithms, Prototype-Specific Feature Importance (PSFI) [14], Individual-Specific Feature Importance (ISFI) [14], and Probe-Specific re-ranking (PSR) [19]. The results are copied from their papers and recorded in Table 3 for comparison.

It is shown in Table 3 that all the re-ranking methods can achieve higher matching accuracies compared with their baseline algorithms. Compared with our baseline method GOG, our result increases by 5.88% at Rank-1, which is the best improvement at rank1 among the compared methods. It is shown that our method outperforms these existing methods. Besides, our final results are better than those of the compared methods.

3.4   Effect of Major Components

From Sects. 2.2 and 2.3, it can be observed that the proposed re-ranking algorithm contains two parts: forward sorting constraint and reverse sorting constraint. The evaluation of each part of the algorithm is performed with the Cumulative Matching Characteristics (CMC) curve on the VIPeR dataset.

Table 3. Matching rates (%) of different re-ranking methods

Method	VIPeR (%)				
	Rank-1	Rank-5	Rank-10	Rank-20	Improvement of Rank-1
RDC	12.15	27.78	38.94	54.46	4.97
RDC+ISFI	**17.12**	**38.96**	**52.94**	**67.34**	
RSVM	12.93	31.46	43.91	59.64	2.83
RSVM+PSFI	**15.76**	**38.70**	**51.36**	**66.84**	
AML	43.04	72.28	83.96	93.54	2.15
AML+PSR	**45.19**	**73.58**	**85.35**	**93.99**	
GOG	49.72	79.72	88.67	94.53	5.88
GOG+OURS	**55.60**	**82.25**	**90.38**	**95.51**	

As shown in Fig. 4, both parts of the re-ranking algorithm (red and blue curves) have achieved good performances compared with the original results of baseline (black curve), and the combination of both parts will have better performance.

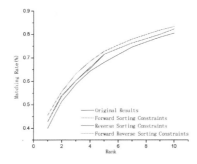

Fig. 4. Performance is obtained by separately considering the forward sorting constraint and reverse sorting constraint. Results have been computed considering the GOG baseline. (Color figure online)

3.5 Effect of Hyper-parameter on Algorithm Performance

Hyper-parameter λ is introduced to implement different levels of "punishment" for error matching in the algorithm 1. Figure 5 shows the influence of different λ values at rank-1 on three datasets. It is shown that the performances of person re-identification are best for VIPeR, CUHK01, and PRID450S, when λ values are 0.6, 0.5, and 0.5, respectively.

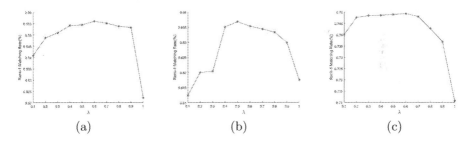

Fig. 5. The impact of the parameter λ on person re-identification performance on three datasets. (a) VIPeR dataset. (b) CUHK01 dataset. (c) PRID450S dataset.

4   Conclusions

In this paper, we use the implicit constraint information in the initial sorting list, which is formed by the existing person re-identification algorithm, and we conversely re-rank the initial sorting list to improve the performance of the original person re-identification algorithm. The experimental results show that the application of this algorithm on different datasets can significantly improve the performance of the original person re-identification algorithm, especially when the dataset is relatively large, the effect is even more gratifying. It is worth mentioning that the forward and reverse sorting constraints algorithm proposed in this paper is fully automatic and unsupervised, and it can be easily applied to existing person re-identification algorithms.

References

1. Chen, D., Yuan, Z., Chen, B., Zheng, N.: Similarity learning with spatial constraints for person re-identification. In: Computer Vision and Pattern Recognition, pp. 1268–1277 (2016)
2. Garcia, J., Martinel, N., Gardel, A., Bravo, I., Foresti, G.L., Micheloni, C.: Discriminant context information analysis for post-ranking person re-identification. IEEE Trans. Image Process. **26**(4), 1650–1665 (2017)
3. Garcia, J., Martinel, N., Micheloni, C., Gardel, A.: Person re-identification ranking optimisation by discriminant context information analysis. In: IEEE International Conference on Computer Vision, pp. 1305–1313 (2015)
4. Gray, D., Tao, H.: Viewpoint Invariant Pedestrian Recognition with an Ensemble of Localized Features. Springer, Heidelberg (2008)
5. Hirzer, M.: Large scale metric learning from equivalence constraints. In: IEEE Conference on Computer Vision and Pattern Recognition, pp. 2288–2295 (2012)
6. Leng, Q., Hu, R., Liang, C., Wang, Y., Chen, J.: Person re-identification with content and context re-ranking. Multimedia Tools Appl. **74**(17), 6989–7014 (2015)
7. Li, W., Wu, Y., Mukunoki, M., Minoh, M.: Common-near-neighbor analysis for person re-identification. In: IEEE International Conference on Image Processing, pp. 1621–1624 (2013)
8. Li, W., Zhao, R., Wang, X.: Human Reidentification with Transferred Metric Learning. Springer, Heidelberg (2013)
9. Li, W., Zhao, R., Xiao, T., Wang, X.: DeepReID: deep filter pairing neural network for person re-identification. In: Computer Vision and Pattern Recognition, pp. 152–159 (2014)
10. Liao, S., Hu, Y., Zhu, X., Li, S.Z.: Person re-identification by local maximal occurrence representation and metric learning. In: Computer Vision and Pattern Recognition, pp. 2197–2206 (2015)
11. Liao, S., Li, S.Z.: Efficient PSD constrained asymmetric metric learning for person re-identification. In: IEEE International Conference on Computer Vision, pp. 3685–3693 (2015)
12. Lisanti, G., Masi, I., Bimbo, A.D.: Matching people across camera views using kernel canonical correlation analysis, pp. 1–6 (2014)
13. Liu, C., Chen, C.L., Gong, S., Wang, G.: POP: person re-identification post-rank optimisation. In: IEEE International Conference on Computer Vision, pp. 441–448 (2014)
14. Liu, C., Gong, S., Chen, C.L.: On-the-fly feature importance mining for person re-identification. Patt. Recogn. **47**(4), 1602–1615 (2014)
15. Matsukawa, T., Okabe, T., Suzuki, E., Sato, Y.: Hierarchical Gaussian descriptor for person re-identification. In: Computer Vision and Pattern Recognition, pp. 1363–1372 (2016)
16. Nguyen, N.B., Nguyen, V.H., Ngo, T.D., Nguyen, K.M.T.T.: Person re-identification with mutual re-ranking. Vietnam J. Comput. Sci. **4**, 1–12 (2017)

17. Nguyen, V.H., Ngo, T.D., Nguyen, K.M.T.T., Duong, D.A., Nguyen, K., Le, D.D.: Re-ranking for person re-identification. In: Soft Computing and Pattern Recognition, pp. 304–308 (2015)
18. Roth, P.M., Hirzer, M., Köstinger, M., Beleznai, C., Bischof, H.: Mahalanobis distance learning for person re-identification, pp. 247–267 (2014)
19. Xie, Y., Yu, H., Gong, X., Levine, M.D.: Adaptive metric learning and probe-specific reranking for person reidentification. IEEE Sig. Process. Lett. **24**(6), 853–857 (2017)
20. Ye, M., et al.: Person reidentification via ranking aggregation of similarity pulling and dissimilarity pushing. IEEE Trans. Multimedia **18**(12), 2553–2566 (2016)
21. You, J., Wu, A., Li, X., Zheng, W.S.: Top-push video-based person re-identification, pp. 1345–1353 (2016)
22. Zheng, L., Wang, S., Tian, L., He, F., Liu, Z., Tian, Q.: Query-adaptive late fusion for image search and person re-identification. In: Computer Vision and Pattern Recognition, pp. 1741–1750 (2015)
23. Zhong, Z., Zheng, L., Cao, D., Li, S.: Re-ranking person re-identification with k-reciprocal encoding, pp. 3652–3661 (2017)

Content-Based Co-Factorization Machines: Modeling User Decisions in Event-Based Social Networks

Yilin Zhao[1], Yuan He[1], and Hong Li[2(✉)]

[1] HeFei University of Technology, Hefei 230009, Anhui, China
[2] HeFei University, Hefei 230601, Anhui, China
lihong@hfuu.edu.cn

Abstract. Event-based online social networks (EBSNs) have attracted millions of users to attend events and join event groups. However, the EBSNs are often overwhelmed with too many events and groups, making it is hard for users to attend events and join groups that interest them. Thus, it is natural to design recommender systems to recommend events and groups to users. One key challenge is that, though users have different kinds of behaviors (e.g., user-event behavior, user-word review behavior, and user-group behavior), these data are very sparse for prediction. To that end, in this paper, we propose a content-based co-factorization machines based method for the two recommendation tasks by co-relating users' different kinds of behaviors. Besides, to alleviate the data sparsity issue, we also model the content information in the co-factorization machines. Finally, experiments on three real-world datasets show the effectiveness of our proposed model on the two prediction tasks.

Keywords: Factorization machines · Event-based social networks
Recommender systems

1 Introduction

With the popularity of online social media, EBSNs have emerged in recent years. These EBSNs are online event platforms, in which users can create, distribute and share upcoming offline events. For instance, in the network platform Meetup (www.meetup.com), there are more than 16 million users with more than 300,000 events announced per month [17]. In order to help users quickly obtain his interests, a good social recommendation system is in demand according to his preference.

Unlike general online social media, an EBSNs platform is made up of a wide variety of behaviors: user-event behavior, user-word review behavior and user-group behavior and so on. The social events has its unique temporal and spatial characteristics, which make the recommendation is more challenging than

---

This work was supported in part by the National Natural Science Foundation of China (Grant No. 61632007), the Open Project Program of the National Laboratory of Pattern Recognition (NLPR Grant No. 201700017).

R. Hong et al. (Eds.): PCM 2018, LNCS 11165, pp. 780–791, 2018.
https://doi.org/10.1007/978-3-030-00767-6_72

traditional item recommendation, such as films recommendation, product recommendation. Recommendation system of EBSNs faces the following three challenges: (1) Sparsity: Few features can truly impact users with rich information in EBSNs, that the limited observed preference between users and events is not enough to reach satisfactory recommendation performance. (2) Cold-start: Events published in EBSNs are short-lived which means they are always organized in the future, having little or no trace of historical attendance, which raises the issue of new item cold-start problem. (3) Multi-tasks: Recommendation in EBSNs needs to solve multi-recommendation tasks for providing better services for users, for instance, recommending events, groups.

In the past decades, many EBSNs recommendation methods have been proposed, such as HeSig [10], MCLRE [7] and MF-EUN [5]. However, they just focus on event recommendation in EBSNs. In order to learn the group preferences of users, PTARMIGAN [17] and HeteRS [9] are proposed to learn the influence of weights for good understanding of user behaviors. In addition, there are few words used to describe events in EBSNs, and those short texts contain abundant useful knowledge. Short text is different from traditional documents, principally in its shortness. As a result, short text learning tasks are characterized by the high dimension of their feature space, in which most words have a low frequency, usually only appear once. Thus, the content cooperated event-group recommendation is still unsatisfactory.

To the end, we propose a *C*ontent-based *Co-F*actorization *M*achines(CC-FM) for event-group recommendation in EBSNs. In order to solve cold-start problem and provide an interpretable low-dimensional representation of content information, we combine content analysis based on probabilistic topic model [1] and factorization machine(FM) [11] which is a general approach that use factorized interaction to capture interaction between variables and is widely used in recommender systems. Our contributions can be summarized as follows:

- We propose a CC-FM model for the recommendation in EBSNs that combines the content information, user features, and event features to model user interests. The proposed model deals with the sparsity, cold-start and multi-tasks problem in the event-group recommendation by applying factorization machines and a probabilistic topic model to short text data as features for learning to rank.
- We conduct extensive experimental results on three real-world datasets with two recommendation tasks. The results clearly show that our method achieves better performance in event recommendation and group recommendation compared to the state-of-the-art baselines for modeling user decisions.

In the remainder of this paper is organized as follows. In Sect. 2, we review several relevant related works. Next, we describe proposed our recommendation models in detail in Sect. 3. In Sect. 4, we provide the learning problems of our model. The setup and the results of experiments are reported in Sect. 5, followed by the conclusion in Sect. 6.

2   Related Work

In this section, we briefly review the related work from two aspects: classic recommendation methods, event recommendation methods.

2.1   Classic Recommendation

Classic recommendation algorithms can be summarized into three categories: collaborative filtering, content-based recommendation, and hybrid recommender systems. In collaborative filtering, models based on latent factor perform well, such as matrix factorization [4], probabilistic matrix factorization [8] represent users and items in a shared latent low-dimensional space. Bayesian personalized ranking [12] learn from relative pairwise preferences by maximum posterior estimator derived from a Bayesian analysis. However, a simple latent factor model cannot easily capture the interaction of additional information in recommendation systems. Later, Steffen Rendle proposed factorization machines can naturally deal with features by parameterized the weight of a cross feature as the inner product of the embedding vectors [11]. In content-based recommendation, probabilistic topic models [1,2] can discover topics from a large collection of document to create event features, and realize recommendation. In hybrid recommender systems, NHPMF [15] use the tagging data to select neighbors of each user and each item, then add unique Gaussian distributions on each user's (item's) latent feature vector in the matrix factorization. [16] incorporate the underlying social theories to explain and model the evolution of users' behaviors. [13] leverage social influence to enhance temporal social recommendation performance remains pretty much open.

2.2   Event Recommendation

Recently event recommendation has garnered increased attention with the advent of EBSNs. As for the event recommendation in EBSNs, Xingjie Liu et al. defined the event social network which bridge online and offline social interactions, and conducted unique and highlighted characteristics of such networks for event participation prediction [6]. HeSig [10] is proposed for event recommendation by combining heterogeneous social and geographical information in a Bayesian matrix factorization. A hybrid model MCLRE [7] do recommendation by considering social-aware, content-aware and time-aware information in EBSNs. A collaborative filtering based algorithm MF-EUN [5] is proposed in 2016. It proposed an additional information based neighborhood discovery method to calculates the similarity between users based on three aspects, including user influences on topics, regions, and organizers. For the event-based group recommendation, Wei Zhang et al. [17] exploited matrix factorization based method PTARMIGAN for group recommendation in EBSNs by considering both explicit features to model interactions between users and groups. In 2015, a general graph-based model HeteRS [9] is proposed to solve three recommendation tasks in EBSNs in one framework, including recommending groups to users, recommending tags to groups, and recommending events to users.

3   Our Model

In this section, we first define the problem of recommending event and group. Then, we detail how factorization machines can be used for different types of responses. Lastly, we describe the content-based co-factorization machines.

3.1   Problem Definition

Given an event-based social network $N = (U, G, E)$, where U shows the set of users, G shows the set of groups, E shows the set of events. The words in event description documents are denoted as W. For user $u \in U$, the past response of user u to event $e \in E$ is denoted as y_{ue} and the past response of user u to group $g \in G$ is denoted as y_{ug}. While y_{wg} shows the count of word $w \in W$ in event description documents of group $g \in G$. Our goal is to recommend to user u a list of events, groups that will maximize his/her satisfaction and discover topics which truly influence users decisions based on his/her past preferences.

3.2   Modeling User Decisions with Content

In this paper, we leverage extended factorization machines to model EBSNs data from different aspects. We have three tasks to achieve for understanding and modeling user responses in EBSNs. First, we wish to uncover what kind of events that a user is truly interested and will attend. Second, we wish to find groups in which a user will be willing to join. In addition, we discover groups' topics in which users are interested and how these topics influence users' decisions. In other words, we wish to recommend events and groups as accurately as possible and discern topics from the huge amount of event description content at the same time.

Modeling User-Event Decisions. For the first task, we focus on a binary response y_{ue}: whether a target user u will attend the event e, and event e is organized by group g. For the convenience of discussion, we compose a simple feature vector consisting of one categorical feature to indicate the user, one categorical feature to indicate the event, and one categorical feature to indicate the group. Following the definition of factorization machines, we can use s_{ue} to represent an estimation of y_{ue}:

$$s_{ue} = \beta_0 + \beta_u + \beta_e + <\boldsymbol{\theta}_u, \boldsymbol{\theta}_e> + <\boldsymbol{\theta}_u, \boldsymbol{\theta}_g> + <\boldsymbol{\theta}_e, \boldsymbol{\theta}_g> \qquad (1)$$

where β_0 is the global bias, $\boldsymbol{\beta}$ denotes the weight of features of users, events, and groups, $\boldsymbol{\theta}$ are treated as latent factors for users, events and groups.

Modeling User-Group Decisions. The second task is similar to the first task, y_{ug} is a binary response: whether the target user u will join in the group g. We associate one categorical feature to indicate the user and another one to indicate the group. And s_{ug} is the estimation of y_{ug}:

$$s_{ug} = \alpha_0 + \boldsymbol{\alpha}_u + \boldsymbol{\alpha}_g + <\boldsymbol{\phi}_u, \boldsymbol{\phi}_g> \qquad (2)$$

where α_0 is the global bias, $\boldsymbol{\alpha}$ denotes the weight of features of users and groups, $\boldsymbol{\phi}$ are treated as latent factors for users and groups.

Auxiliary Content. For the third task, we will model words in events description documents of the group g. We denote s_{wg} to represent our estimation of the raw word count of word w in group g, y_{wg} which is the response in this task. We associate one categorical feature to indicate the word index and another one to indicate the group index. Following the definition of factorization machines, the estimation of y_{wg}:

$$s_{wg} = \gamma_0 + \boldsymbol{\gamma}_g + \boldsymbol{\gamma}_w + <\boldsymbol{\delta}_g, \boldsymbol{\delta}_w> \tag{3}$$

where γ_0 is the global bias, $\boldsymbol{\gamma}$ denotes the weight of features of groups and words, $\boldsymbol{\delta}$ are treated as latent factors for groups and words.

For $\boldsymbol{\delta}_g$, it is a K-dimensional vector and plays a similar role as $P(z \mid \theta)$ in traditional topic models like probabilistic latent semantic analysis (PLSA) [2] or latent Dirichlet allocation (LDA) [1]. For $\boldsymbol{\delta}_w$, it is a K-dimensional vector for word w and it can be treated as a $P(z \mid v)$. In order to recover a similar modeling power from topic model, follow [3], using construct matrix $\mathbf{M}$, $\mathbf{M} \in R^{K \times W}$ where $\mathbf{M}_{.w} = \boldsymbol{\delta}_w$:

$$\sum_\delta \delta_{kw} = 1 \text{ for all } k, \ \delta_{kw} \geq 0; \text{ and } \delta_{gk} \geq 0;$$

and restrict all γ to be non-negative, resulting in a non-negative decomposition of the term matrix.

3.3   Content-Based Co-Factorization Machines

In this subsection, we first review Co-Factorization Machines (Co-FM) to address the problem of modeling multiple aspects of data. Following the setting described in Eqs. 1, 2 and 3, we have three separate FM to model three aspects where the three aspects are not linked together. Notice that we have learned three different latent representations of the same group: $\boldsymbol{\theta}_g$, $\boldsymbol{\alpha}_g$ and $\boldsymbol{\delta}_g$. Linking three factorization machines might be possible if these three latent representations of the group can be coupled in certain ways.

We assume that the latent factor in different aspects is exactly the same. Therefore, some parts of the latent factors of the same group are shared across different aspects. We use $\mathbf{v}$ to indicate the shared latent factors, the three aspects of two tasks under this formalism are follows:

$$
\begin{aligned}
s_{ue} &= \beta_0 + \boldsymbol{\beta}_u + \boldsymbol{\beta}_e + \boldsymbol{\beta}_g + <\boldsymbol{\theta}_u, \boldsymbol{\theta}_e> + <\boldsymbol{\theta}_u, \mathbf{v}_g> + <\boldsymbol{\theta}_e, \mathbf{v}_g> \\
s_{ug} &= \alpha_0 + \boldsymbol{\alpha}_u + \boldsymbol{\alpha}_g + <\boldsymbol{\phi}_u, \mathbf{v}_g> \\
s_{wg} &= \gamma_0 + \boldsymbol{\gamma}_g + \boldsymbol{\gamma}_w + <\mathbf{v}_g, \boldsymbol{\delta}_w>
\end{aligned} \tag{4}
$$

This approach would be convenient when multiple factors will be shared. For instance, for each group, we can add one more categorical feature to indicate the historical events of the group and therefore obtain a latent representation of its events through content modeling:

$$s_{ue} = \beta_0 + \beta_u + \beta_e + \beta_g + <\boldsymbol{\theta}_u, \mathbf{v}_e> + <\boldsymbol{\theta}_u, \mathbf{v}_g> + <\mathbf{v}_e, \mathbf{v}_g>$$
$$s_{ug} = \alpha_0 + \alpha_u + \alpha_e + \alpha_g + <\boldsymbol{\phi}_u, \mathbf{v}_e> + <\boldsymbol{\phi}_u, \mathbf{v}_g> + <\mathbf{v}_e, \mathbf{v}_g>$$
$$s_{wg} = \gamma_0 + \gamma_e + \gamma_g + \gamma_w + <\mathbf{v}_g, \boldsymbol{\delta}_w> + <\boldsymbol{\delta}_w, \mathbf{v}_e> + <\mathbf{v}_e, \mathbf{v}_g> \qquad (5)$$

And we combine the topic model to discover group topic. The topic model [1] assume there are K topics, each represented by a word distribution. When a group creates an event, first a topic is chosen based on the group's topic distribution, then a bag of words is chosen one by one based the topic model. Then, we combine topic model and latent model:

$$\mathbf{v}_g = \mathbf{v}_g + \lambda_{topic}\boldsymbol{\theta}^g \qquad (6)$$

Factorization machines parameters use L2 regularization corresponds to Gaussian priors [11]. We can draw group latent factors:

$$\mathbf{v_g} \sim \mathcal{N}(\lambda_{topic}\boldsymbol{\theta}^g, \sigma)$$

where $\boldsymbol{\theta}^g$ is topic of group g, σ is a regularization parameter of latent factors. This will recover the formalism from [14].

4 Learning

4.1 Optimization

Optimization with Content. We view the words in all event description documents of a group as count data, and since words are sparse in documents where one word will most likely appear only once in all event description documents of a group, we can use Bernoulli distribution and thus $y_{wg} \sim Bernoulli\left(\delta\left(s_{wg}\right)\right)$. Therefore, we use logistic loss function to optimize the task of modeling content:

$$Opt\left(C\right) = \arg\min_{\gamma,\Delta} \sum_{w=1}^{W} \sum_{g=1}^{G} \left(l^C\left(y_{wg}, s_{wg}\right)\right) + \sum_{j=1}^{P_1} \sigma_{\delta,j} \|\delta_j\|_F^2$$
$$s.t.: \boldsymbol{\gamma} \geq 0, \delta_{kw} \geq 0, \forall k, w; \boldsymbol{\delta_k} \in P, \forall k \in K; \delta_{gk} \geq 0, \forall g, k \qquad (7)$$

where P_1 is the number of features used for each word w in group g, and P is a $(K-1)$-simplex.

Optimization with User-Event Response. For task of modeling user-event decision D_e, we use regression least-squares loss function.

$$Opt\left(D_e\right) = \arg\min_{\beta,\Theta} \sum_{e=1}^{E} \left(l^{LS}\left(y_{ue}, s_{ue}\right)\right) + \sum_{j=1}^{P_2} \sigma_{\theta,j} \|\theta_j\|_F^2 \qquad (8)$$

where P_2 is the number of features used for each event e.

Optimization with User-Group Response. For task of modeling user-group decision D_g, we use regression least-squares loss function.

$$Opt\left(D_g\right) = \underset{\alpha,\Phi}{\arg\min} \sum_{g=1}^{G} \left(l^{LS}\left(y_{wg}, s_{wg}\right)\right) + \sum_{j=1}^{P_3} \sigma_{\phi,j}\|\phi_j\|_F^2 \tag{9}$$

where P_3 is the number of features used for each group g.

4.2   Summary

In conclusion, a content-based factorization machines framework for learning a model for user-event decision, content understanding can be formalized as:

$$Opt\left(D_e\right) + \lambda_{ec}Opt\left(C\right) \tag{10}$$

For user-group decision and content understanding can be formalized as:

$$Opt\left(D_g\right) + \lambda_{gc}Opt\left(C\right) \tag{11}$$

where λ_{ec} is a parameter to balance optimization with user-event decisions and optimization with content, λ_{gc} is a parameter to balance optimization with user-group decisions and optimization with content. We adopt coordinate descent to optimize Eqs. 10 and 11.

5   Experiments

5.1   Experiments Setup

We conduct experiments on the public Meetup [7] dataset, which is crawled from Meetup.com. Based on different cities, the data set can be divided into Phoenix, Chicago, and San Jose. In the preprocessing phase of dataset, we only select the users who have attended at least 3 events, and the groups with more than 20 events for further process. After data preprocessing, the statistics of the data are given in Table 1. In order to simulate a realistic scenario in EBSNs, we select a timestamp, for user-event recommendation, data within six months before target timestamps are used for training, and we form candidate events which created before the target timestamps and take place after timestamps are used for testing to be recommended to the corresponding users. For user-group recommendation, we take 80% of the total user-group in each train-test partitions for training, the remaining part is used for testing.

   To assess the effectiveness of our model, we compare CC-FM with the following state-of-the-art baselines:

1. **MP:** It is a baseline which recommends the most popular events or groups.
2. **FM** [11]**:** It is a state-of-art framework for latent factor models with rich features.

Table 1. Statistics of the Meetup.com dataset

City	Groups	Users	Events
Chicago	2,528	12,234	10,253
Phoenix	1,667	8,645	19,458
San Jose	2,906	14,560	14,560

3. **Co-FM** [3]: It is an extending factorization machines that model user decisions and contents. We choose the best performance method, namely, Co-FM with shared latent space.

In our experiments, the evaluation metric we use Precision@K and MAP@K, which is widely used in evaluating recommendation results. It is defined as follows:

Precision@K computes the percentage of correctly recommended items in top K positions on a recommendation list.

$$Precision@K = \frac{top\ K\ recommendations \cap true\ items}{K} \tag{12}$$

MAP@K is mean average Precision@K.

$$MAP@K = \frac{\sum_{m=1}^{K} Precision@m \times l_m}{K} \tag{13}$$

where l_m is a binary indicator whether the position m has been response yes or not and K is the total number of positions evaluated.

5.2   Experimental Settings

Event Recommendation. The parameters for all models are tuned by grid search. We use coordinate descent to tune all models. The following model parameters were defined to San Jose, Chicago and Phoenix respectively: In event recommendation, for FM, we set its number of latent factors to 12, for Co-FM (shared latent space), we set its number of latent factors in 150, for CC-FM, we set its number of latent factors in 200, and $\lambda_{ec} = \{0.03, 0.04, 0.04\}$, $\lambda_{topic} = \{0.002, 0.002, 0.001\}$.

Figures 1 and 2 show the comparison of Precision@K and MAP@K of the CC-FM and other methods on event recommendation task. We can see $K = 5$ and $K = 10$ perform well, and $K = 15$ is general because of most user-event responses are lower than 15, the dataset is sparse.

Group Recommendation. In group recommendation, for FM, we set its number of latent factors to 12, for Co-FM (shared latent space), we set its number of latent factors in 200, and for CC-FM, we set its number of latent factors in 200, and $\lambda_{gc} = \{0.04, 0.03, 0.03\}$, $\lambda_{topic} = \{0.002, 0.002, 0.001\}$.

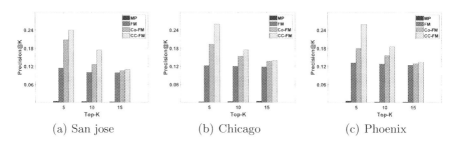

(a) San jose   (b) Chicago   (c) Phoenix

Fig. 1. Comparisons of event recommendation on Precision@K.

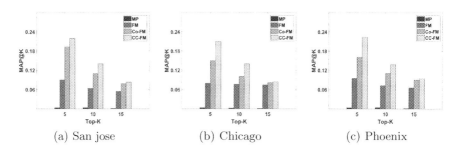

(a) San jose   (b) Chicago   (c) Phoenix

Fig. 2. Comparisons of event recommendation on MAP@K.

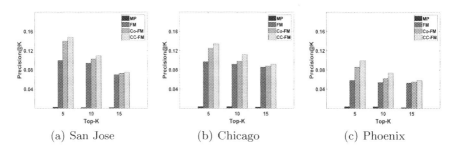

(a) San Jose   (b) Chicago   (c) Phoenix

Fig. 3. Comparisons of group recommendation on Precision@K.

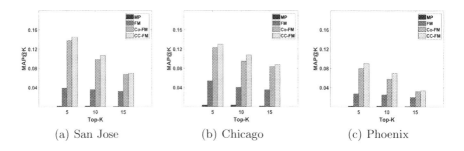

(a) San Jose   (b) Chicago   (c) Phoenix

Fig. 4. Comparisons of group recommendation on MAP@K.

From Figs. 3 and 4 show the comparison of Precision@K and MAP@K of the CC-FM and other methods on the group recommendation task. Note that since the Co-FM already achieves a good performer, the improvement of our algorithm is not large. But our method needs fewer iterations than other methods. In our experiment, it is usually 12 iterations, and Co-FM usually requires 23 iterations.

5.3   Parameter Sensitivity

We investigate the parameter sensitivity in this section. Figure 5 show topic parameter λ_{topic} and balance parameter λ affect the performance of CC-FM. When balance parameter is small in CC-FM, CC-FM behaves more like factorization machines where no content is considered. When balance parameter increases, CC-FM requires more influence for content. When topic parameter is small in CC-FM, CC-FM behaves more like Co-FM where no group topic is considered. When the topic parameter increases, CC-FM is penalized for v_g diverging from the topic proportions.

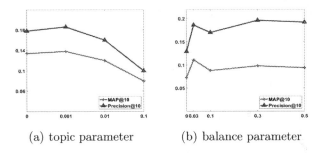

(a) topic parameter   (b) balance parameter

Fig. 5. How the parameter affects event recommendation performance of CC-FM in Phoenix.

6   Conclusion

In this paper, we propose the content-based co-factorization machines, which fully exploits the content influence for multi-recommendation tasks. In addition, we propose a new content-aware recommendation model using the latent factor model and topic model to simultaneously predict user decisions and modeling content in EBSNs by analyzing rich information. Extensive experiments on real-world datasets demonstrate that our proposed algorithm outperforms other recommendation algorithms and the experimental results show that the content factor plays important role in EBSNs.

References

1. Blei, D.M., Ng, A.Y., Jordan, M.I.: Latent dirichlet allocation. J. Mach. Learn. Res. **3**, 993–1022 (2003)
2. Hofmann, T.: Probabilistic latent semantic indexing. In: ACM SIGIR Forum, vol. 51, pp. 211–218. ACM (2017)
3. Hong, L., Doumith, A.S., Davison, B.D.: Co-factorization machines: modeling user interests and predicting individual decisions in Twitter. In: Proceedings of the Sixth ACM International Conference on Web Search and Data Mining, pp. 557–566. ACM (2013)
4. Koren, Y., Bell, R., Volinsky, C.: Matrix factorization techniques for recommender systems. Computer **42**(8), 30–37 (2009)
5. Li, X., Cheng, X., Su, S., Li, S., Yang, J.: A hybrid collaborative filtering model for social influence prediction in event-based social networks. Neurocomputing **230**, 197–209 (2017)
6. Liu, X., He, Q., Tian, Y., Lee, W.C., McPherson, J., Han, J.: Event-based social networks: linking the online and offline social worlds. In: Proceedings of the 18th ACM SIGKDD International Conference on Knowledge Discovery and Data Mining, pp. 1032–1040. ACM (2012)
7. Macedo, A.Q., Marinho, L.B., Santos, R.L.: Context-aware event recommendation in event-based social networks. In: Proceedings of the 9th ACM Conference on Recommender Systems, pp. 123–130. ACM (2015)
8. Mnih, A., Salakhutdinov, R.R.: Probabilistic matrix factorization. In: Advances in Neural Information Processing Systems, pp. 1257–1264 (2008)
9. Pham, T.A.N., Li, X., Cong, G., Zhang, Z.: A general graph-based model for recommendation in event-based social networks. In: 2015 IEEE 31st International Conference on Data Engineering (ICDE), pp. 567–578. IEEE (2015)
10. Qiao, Z., Zhang, P., Cao, Y., Zhou, C., Guo, L., Fang, B.: Combining heterogenous social and geographical information for event recommendation. In: AAAI, vol. 14, pp. 145–151 (2014)
11. Rendle, S.: Factorization machines with libFM. ACM Trans. Intell. Syst. Technol. (TIST) **3**(3), 57 (2012)
12. Rendle, S., Freudenthaler, C., Gantner, Z., Schmidt-Thieme, L.: BPR: Bayesian personalized ranking from implicit feedback. In: Proceedings of the Twenty-Fifth Conference on Uncertainty in Artificial Intelligence, pp. 452–461. AUAI Press (2009)
13. Sun, P., Wu, L., Wang, M.: Attentive recurrent social recommendation. In: The 41st International ACM SIGIR Conference on Research and Development in Information Retrieval, pp. 185–194. ACM (2018)
14. Wang, C., Blei, D.M.: Collaborative topic modeling for recommending scientific articles. In: Proceedings of the 17th ACM SIGKDD International Conference on Knowledge Discovery and Data Mining, pp. 448–456. ACM (2011)

15. Wu, L., Chen, E., Liu, Q., Xu, L., Bao, T., Zhang, L.: Leveraging tagging for neighborhood-aware probabilistic matrix factorization. In: Proceedings of the 21st ACM International Conference on Information and Knowledge Management, pp. 1854–1858. ACM (2012)

16. Wu, L., et al.: Modeling the evolution of users' preferences and social links in social networking services. IEEE Trans. Knowl. Data Eng. **29**(6), 1240–1253 (2017)

17. Zhang, W., Wang, J., Feng, W.: Combining latent factor model with location features for event-based group recommendation. In: Proceedings of the 19th ACM SIGKDD International Conference on Knowledge Discovery and Data Mining, pp. 910–918. ACM (2013)

Image-into-Image Steganography Using Deep Convolutional Network

Pin Wu[1] [iD], Yang Yang[1] [iD], and Xiaoqiang Li[1,2]([⊠]) [iD]

[1] School of Computer Engineering Science, Shanghai University, Shanghai, China
{adamcavendish,xqli}@shu.edu.cn
[2] Shanghai Institute for Advanced Communication and Data Science,
Shanghai University, Shanghai, China

Abstract. Raising payload capacity in image steganography without losing too much safety is a challenging task. This paper combines recent deep convolutional neural network methods with image-into-image steganography. We show that with the proposed method, the capacity can go up to 23.57 bpp (bits per pixel) by changing only 0.76% of the cover image. We applied several traditional steganography analysis algorithms and found out that the proposed method is quite robust.

The source code is available at: https://github.com/adamcavendish/Deep-Image-Steganography.

Keywords: Convolutional neural network · Image steganography
Steganography capacity

1 Introduction

Image steganography, aiming at delivering a modified cover image to secretly transfer hidden information inside with little awareness of the third-party supervision, is a classical computer vision and cryptography problem. Traditional image steganography algorithms go to their great length to hide information into the cover image while little consideration is tilted to payload capacity, also known as the ratio between hidden and total information transferred. The payload capacity is one significant factor to steganography methods because if more information is to be hidden in the cover, the visual appearance of the cover is altered further and thus the risk of detection is higher.

To maximize the payload capacity while still resistible to simple alterations, pixel level steganography is majorly used, in which LSB (least significant bits) method [11], BPCS (Bit Plane Complexity Segmentation) [9], and their extensions are in dominant. LSB-based methods can achieve a payload capacity of up to 50%, or otherwise, a vague outline of the hidden image would be exposed. However, most of these methods are vulnerable to statistical analysis, and therefore it can be easily detected.

Some traditional steganography methods with balanced attributes are hiding information in the JPEG DCT components. For instance, Almohammad's work

© Springer Nature Switzerland AG 2018
R. Hong et al. (Eds.): PCM 2018, LNCS 11165, pp. 792–802, 2018.
https://doi.org/10.1007/978-3-030-00767-6_73

(a)	(b)	(c)	(d)	(e)	(f)
Cover Image	StegNet Embedded	3-bit LSB Embedded	Hidden Image	StegNet Decoded	3-bit LSB Decoded

Fig. 1. StegNet and 3-bit LSB comparison (Embedded-Cover-Difference = 0.76%, Hidden-Decoded-Difference = 1.8%, Payload Capacity = 23.57 bpp)

[1] provides around 20% of payload capacity (based on the patterns) and still remains undetected through statistical analysis.

Most secure traditional image steganography methods recently have adopted several functions to evaluate the embedding localizations in the image, which enables content-adaptive steganography. HuGO [12], WOW and some other content-adaptive methods are quite robust against steganalysis, but they highly depend on the patterns of the cover image, and therefore the average payload capacity can be hard to calculate.

With our work, named after "StegNet", it is possible to raise payload capacity to an average of 98.2% or 23.57 bpp (bits per pixel), changing only around 0.76% of the cover image (See Fig. 1), and still robust to statistical analysis.

The payload capacity (23.57 bpp) is calculated from Eqs. 1–2, and 0.76% is calculated from Eq. 3.

$$\text{Decoded Rate} = 1 - \frac{\sum_{i=1}^{m} \sum_{j=1}^{n} |H_{i,j} - D_{i,j}|}{m \times n}, \tag{1}$$

$$\text{Payload Capacity} = \text{Decoded Rate} * 8 * 3 \tag{2}$$

$$\text{Cover Changing Rate} = \frac{\sum_{i=1}^{m} \sum_{j=1}^{n} |C_{i,j} - E_{i,j}|}{m \times n} \tag{3}$$

where C, H, E, D symbols stand for the cover image (C), the hidden image (H), the embedded image (E) and the decoded image (D) in correspondence, and "8, 3" stands for number of bits per channel and number of channels per pixel respectively.

This paper is organized as follows. Section 2 will describe some recent steganography related works based on deep learning. Section 3 will unveil the secret why neural network can achieve the amount of capacity encoding and decoding images. The architecture and experiments of our neural network is discussed in Sects. 4 and 5.

2   Related Work

Recently there're some works on applying neural networks for steganography, Volkhonskiy's [19] and Shi's [15] work focus on generating secure cover images for image steganography. Work [14] and work [13] applied deep neural networks for steganography analysis. Baluja [2] is working on the same field as StegNet, however, the hidden image is slightly visible on residual images of the generated embedded images, it requires more GPU memory than our work, and it takes longer time to embed.

3   Convolutional Neural Network for Image Steganography

3.1   High-Order Transformation

In image steganography, we argue that we should not only focus on where to hide information, which most traditional methods work on, but we should also focus on how to hide it.

Most traditional steganography methods usually directly embed hidden information into parts of pixels or transformed correspondings. The transformation regularly occurs in *where to hide*, either actively applied in the steganography method or passively applied because of file format. As a result, the payload capacity is highly related and restricted to the area of texture-rich part of the image detected by the *handcoded* patterns.

DCT-based steganography is one of the most famous transform domain steganograph. We can consider the DCT process in JPEG lossy compression process as a kind of one-level high-order transformation which works at a block level, converting each 8×8 or 16×16 block of pixel information into its corresponding frequency-domain representation. Even hiding in DCT transformed frequency-domain data, traditional works [1,4] embed hidden information in mid frequency band via LSB-alike methods, which eventually cannot be eluded.

While in contrast, deep convolution neural network makes multi-level high-order transformations possible for image steganography [21], where high-level features use less information to represent complex visual patterns. If the model stores only these high-level features, the steganography capacity can be considerably boosted.

3.2   Trading Accuracy for Capacity

Traditional image steganography algorithms mostly embed hidden information as it is or after applying lossless transformations. After decoding, the hidden information is extracted as it is or after the corresponding detransformations are applied. Therefore, empirically speaking, it is just as file compression methods, where lossless compression algorithms usually cannot outperform lossy compression algorithms in capacity.

We need to think in a "lossy" way in order to embed almost equal amount of information into the cover. The model needs to learn to compress the cover image and the hidden image into an embedding of high-level features and converts them into an image that appears as similar as the cover image, which comes to the vital idea of trading accuracy for capacity.

Trading accuracy for capacity means that we do not limit our model in reconstructing at a pixel-level accuracy of the hidden image, but aiming at "recreating" a new image with most of the features in it with a panoramic view, i.e. the spider in the picture, the pipes' position relatively correct, the outline of the mountain, etc.

In other words, the traditional approaches work in lossless ways, which after some preprocesses to the hidden image, the transformed data is crammed into the holes prepared in the cover image. However, StegNet approach decoded image has no pixelwise relationship with the hidden image at all, or strictly speaking, there is no reasonable transformation between each pair of corresponding pixels, but the decoded image as a whole can represent the original hidden image's meaning through neural network's reconstruction.

In the encoding process, the model needs to transform from low-level massive amount of pixelwise information into high-level limited sets of featurewise information with understanding of the image, and come up with a brand new image similar to the cover apparently but with hidden features embedded. In the decoding process, to the contrary, the model is shown only the embedded image, from which both cover and hidden high-level features are extracted, and the hidden image is rebuilt according to network's own comprehension.

As shown in Fig. 2 and StegNet part of Fig. 3, StegNet is not applying LSB-like or simple merging methods to embed the hidden information into the cover. The residual image is neither simulating random noise (LSB-based approach, see 3-bit LSB part of Fig. 3) nor combining recognizable hidden image inside. The embedded pattern is distributed across the whole image, and even magnified 5 to 10 times, the residual image is similar to the cover image visually which can help decreasing the abnormality exposed to human visual system and finally avoid to be detected.

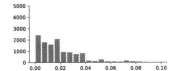 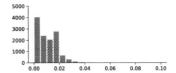

(a) Between Cover and Embedded   (b) Between Hidden and Decoded

Fig. 2. Residual image histograms

The residual image and the magnification operation is computed via

$$R(I_1, I_2) = \frac{|I_1 - I_2|}{\max |I_1 - I_2|}, \quad E(I, M) = \mathrm{clip}(I \cdot M, 0, 1). \tag{4}$$

I_1, I_2 are either cover image and embedded image, or hidden image and decoded image, which are all effectively normalized to $[0, 1]$. Operation $clip(*, 0, 1)$ is to limit the pixelwise enhanced image in range of $[0, 1]$ and M

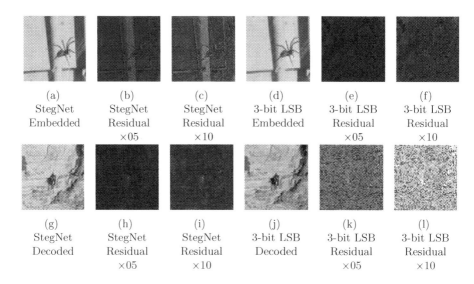

(a) StegNet Embedded	(b) StegNet Residual ×05	(c) StegNet Residual ×10	(d) 3-bit LSB Embedded	(e) 3-bit LSB Residual ×05	(f) 3-bit LSB Residual ×10
(g) StegNet Decoded	(h) StegNet Residual ×05	(i) StegNet Residual ×10	(j) 3-bit LSB Decoded	(k) 3-bit LSB Residual ×05	(l) 3-bit LSB Residual ×10

Fig. 3. StegNet and 3-bit LSB residual images ("×05" and "×10" are the pixelwise enhancement ratio)

is the magnification ratio, which 5 and 10 are chosen visualize the difference in this paper.

4   Architecture

4.1   Architecture Pipeline

The whole processing pipeline is shown in Fig. 4, which consists of two almost identical neural network structure responsible for encoding and decoding. The identical structures can help the neural network to model similar high-level features of images in their latent space. The details of embedding and decoding structure are described in Fig. 5. In the embedding procedure, the cover image and the hidden image are concatenated by channel while only the embedded image is shown to the network. Two parts of the network are both majorly made up of one lifting layer which lifts from image channels to a uniform of 32 channels, six 3×3 basic building blocks raising features into high-dimensional latent space and one reducing layer which transforms features back to image space.

The basic building block named "Separable Convolution with Residual Block" (abbreviated as "SCR" in following context) has the architecture as Fig. 5. Number of layers and channels are selected trading off between the use of GPU memory, the training time and the final effect. We adopt batch-normalization [8] and exponential linear unit (ELU) [6] for quicker convergence and better result.

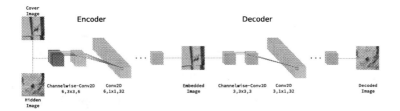

Fig. 4. StegNet processing pipeline

Cover Image	Hidden Image
Concatenation (by channel)	
SCR 6, 3x3, 32	
SCR 32, 3x3, 32	
SCR 32, 3x3, 64	
SCR 64, 3x3, 64	
SCR 64, 3x3, 128	
SCR 128, 3x3, 128	
Conv 128, 3x3, 32	
Batch Normalization	
Elu	
Conv 32, 3x3, 3	
Batch Normalization	
Elu	
Embedded Image	

Embedded Image
SCR 3, 3x3, 32
SCR 32, 3x3, 32
SCR 32, 3x3, 64
SCR 64, 3x3, 64
SCR 64, 3x3, 128
SCR 128, 3x3, 128
Conv 128, 3x3, 32
Batch Normalization
Elu
Conv 32, 3x3, 3
Batch Normalization
Elu
Decoded Image

(a) Embedding Structure   (b) Decoding Structure

x				
1, 3x3, 1	1, 3x3, 1	64 Paths ... Per Channel Convolution		1, 3x3, 1
Concatenation				
64, 1x1, 128				
Batch Normalization				
Elu				

(c) SCR Block

Fig. 5. StegNet network architecture

4.2   Separable Convolution with Residual Block

Our work adopt skip connections in Highway Network [16], ResNet [7] and ResNeXt [20], and separable convolution [5] together to form the basic building block "SCR".

The idea behind separable convolution [5] originated from Google's Inception models [17,18], and the hypothesis behind is that "cross-channel correlations and spatial correlations are decoupled". Further, in Xception architecture [5], it makes an even stronger hypothesis that "cross-channel correlations and spatial correlations can be mapped completely separately". Together with skip-connections [7] the gradients are preserved in backpropagation process via skip-connections to frontier layers and as a result, ease the problem of vanishing gradients.

4.3   Stride or Pooling Impact

StegNet does not use strides like ResNet [7] because strides or poolings have negative impact on embedded and decoded images. We notice that if we apply strides or pooling layers in the steganography model, the spatial information, or the appearance of the image, is gradually lost with more strided layers inserted. See Figs. 6 and 7.

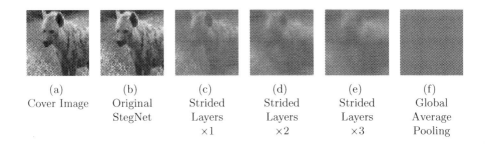

(a)	(b)	(c)	(d)	(e)	(f)
Cover Image	Original StegNet	Strided Layers ×1	Strided Layers ×2	Strided Layers ×3	Global Average Pooling

Fig. 6. Strided layers' impact on embedded images

The spatial information can be completely destroyed to the utmost with global average pooling applied.

We suppose the impact made by the strided layers is because the noise included by padding and normalization might weight more when the image is scaled down, so it would be a lot harder for the network to reconstruct the original signal through dirty signals.

4.4   Training

Learning the end-to-end mapping function from cover and hidden image to embedded image and embedded image to decoded image requires the estimation

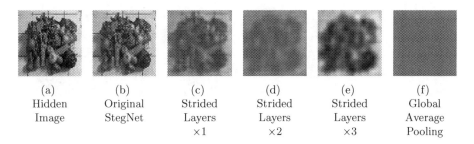

(a)	(b)	(c)	(d)	(e)	(f)
Hidden	Original	Strided	Strided	Strided	Global
Image	StegNet	Layers	Layers	Layers	Average
		×1	×2	×3	Pooling

Fig. 7. Strided layers' impact on decoded images

of millions of parameters in the neural network. It is achieved via minimizing the weighted loss of L_1-loss between the cover and the embedded image, L_1-loss between the hidden and the decoded image, and their corresponding variance losses (variance should be computed across images' height, width and channel). (See Eqs. 5, 6 and 7)

$$E_i = F_{CE}(C_i, H_i; \Theta_{CE}) \qquad\qquad D_i = F_{ED}(E_i; \Theta_{ED}) \qquad (5)$$

$$L_{CE} = \frac{1}{n} \sum_{i=1}^{n} |E_i - C_i| \qquad\qquad L_{HD} = \frac{1}{n} \sum_{i=1}^{n} |D_i - H_i| \qquad (6)$$

$$\text{Loss} = \frac{1}{4}(L_{CE} + L_{HD} + \text{Var}(L_{CE}) + \text{Var}(L_{HD})) \qquad (7)$$

Here F_{CE} and F_{ED} represents the embedder network and decoder network and their corresponding parameters are noted down as Θ_{CE} and Θ_{ED}.

L_{CE} is used to minimize the difference between the embedded image and the cover image, while L_{HD} is for the hidden image and the decoded image. Choosing only to decode the hidden image while not both the cover and the hidden images is under the consideration that the embedded image should be a concentration of high-level features apparently similar to the cover image whose dimension is half the shape of those two images, and some trivial information has been lost. It would have pushed the neural network to balance the capacity in embedding the cover and the hidden if both images are extracted at the decoding process.

Furthermore, adopting variance losses helps to give a hint to the neural network that the loss should be distributed throughout the image, but not putting at some specific position (See Fig. 8 for differences between. The embedded image without variance loss shows some obvious noise spikes (blue points) in the background and also some around the dog nose).

5   Experiments

5.1   Environment

Our work is trained on one NVidia GTX1080 GPU and we adopt a batch size of 64 using Adam optimizer [10] with learning rate at 10^{-5}. We use no image

| (a) Embedded Image with Variance Loss | (b) Embedded Image without Variance Loss | (c) Red Box Magnified (with Variance Loss) | (d) Red Box Magnified (without Variance Loss) |

Fig. 8. Variance loss effect on embedding results (Color figure online)

augmentation and restrict model's input image to 64 × 64 in height and width because of memory limit. We use 80% of the ImageNet dataset for training and the remaining for testing to verify the generalization ability of our model.

5.2   Statistical Analysis

The embedded images shown in Fig. 1 are visually quite similar, however LSB method embedded image is very fragile to histogram analysis. Figure 9 is a comparison of histogram analysis between LSB method and our work. It shows a direct view of robustness of StegNet against statistical analysis, which the StegNet embedded's histogram and the cover image's histogram are much more matching.

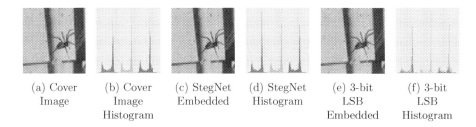

| (a) Cover Image | (b) Cover Image Histogram | (c) StegNet Embedded | (d) StegNet Histogram | (e) 3-bit LSB Embedded | (f) 3-bit LSB Histogram |

Fig. 9. Histogram comparison between StegNet and plain LSB

A more all-round test is conducted through StegExpose [3], which combines several decent algorithms to detect LSB based steganography. The detection threshold is its hyperparameter, which is used to balance true positive rate and false positive rate of the StegExpose's result. The test is performed with linear interpolation of detection threshold from 0.00 to 1.00 with 0.01 as step. Figure 10 is the ROC curve, where true positive stands for an embedded image correctly identified that there's hidden data inside while false positive means a clean figure falsely classified as an embedded image. The figure is plotted in red-dash-line-connected scatter data, showing that StegExpose can only work a little better than random guessing, the line in green. In other words, the proposed steganography method can better resist StegExpose attack.

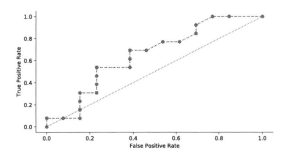

Fig. 10. ROC curves: detecting steganography via StegExpose

6   Conclusion and Future Work

We have presented a novel deep learning approach for image steganography, and the proposed method has achieved superior performance to traditional methods. We tested the proposed method with several steganography analysis methods and proves its robustness.

In spite of the good capacity StegNet provides, there's still some noise generated at non-texture-rich areas in the generated images, i.e., plain white or plain black parts. Therefore, the variance loss adopted by StegNet might not be the optimal solution to loss distribution.

References

1. Almohammad, A., Hierons, R.M., Ghinea, G.: High capacity steganographic method based upon JPEG. In: The Third International Conference on Availability, Reliability and Security (2008)
2. Baluja, S.: Hiding Images in Plain Sight: Deep Steganograph, pp. 2069–2079. Curran Associates, Inc. (2017). http://papers.nips.cc/paper/6802-hiding-images-in-plain-sight-deep-steganography.pdf
3. Boehm, B.: StegExpose - A Tool for Detecting LSB Steganography. arXiv e-prints (2014). arXiv:1410.6656
4. Chang, C.C., Chen, T.S., Chung, L.Z.: A steganographic method based upon JPEG and quantization table modification. Inf. Sci. **141**(1–2), 123–138 (2002)
5. Chollet, F.: Xception: Deep Learning with Depthwise Separable Convolutions. arXiv e-prints (2016). arXiv:1610.02357
6. Clevert, D.A., Unterthiner, T., Hochreiter, S.: Fast and Accurate Deep Network Learning by Exponential Linear Units (ELUs). arXiv e-prints (2015). arXiv:1511.07289

7. He, K., Zhang, X., Ren, S., Sun, J.: Deep Residual Learning for Image Recognition (2016)
8. Ioffe, S., Szegedy, C.: Batch Normalization: Accelerating Deep Network Training by Reducing Internal Covariate Shift. arXiv e-prints (2015). arXiv:1502.03167
9. Kawaguchi, E., Eason, R.: Principle and applications of BPCS-Steganography. In: Proceedings of SPIE, Multimedia Systems and Applications, vol. 3528, p. 464 (1998)
10. Kingma, D., Ba, J.: Adam: A Method for Stochastic Optimization. arXiv e-prints (2014). arXiv:1412.6980
11. Mielikainen, J.: LSB matching revisited. IEEE Signal Process. Lett. **13**(5), 285–287 (2006)
12. Pevný, T., Filler, T., Bas, P.: Using high-dimensional image models to perform highly undetectable steganography. In: Böhme, R., Fong, P.W.L., Safavi-Naini, R. (eds.) IH 2010. LNCS, vol. 6387, pp. 161–177. Springer, Heidelberg (2010). https://doi.org/10.1007/978-3-642-16435-4_13
13. Sedighi, V., Fridrich, J.: Histogram layer, moving convolutional neural networks towards feature-based steganalysis. Electron. Imaging **2017**(7), 50–55 (2017)
14. Sharifzadeh, M., Agarwal, C., Aloraini, M., Schonfeld, D.: Convolutional neural network steganalysis's application to steganography. arXiv:1711.02581 [cs], November 2017
15. Shi, H., Dong, J., Wang, W., Qian, Y., Zhang, X.: SSGAN: secure steganography based on generative adversarial networks. arXiv:1707.01613 [cs], July 2017
16. Srivastava, R.K., Greff, K., Schmidhuber, J.: Highway Networks. arXiv e-prints (2015). arXiv:1505.00387
17. Szegedy, C., Ioffe, S., Vanhoucke, V., Alemi, A.: Inception-v4, Inception-ResNet and the Impact of Residual Connections on Learning. arXiv e-prints (2016). arXiv:1602.07261
18. Szegedy, C., et al.: Going Deeper with Convolutions. arXiv e-prints (2014). arXiv:1409.4842
19. Volkhonskiy, D., Nazarov, I., Borisenko, B., Burnaev, E.: Steganographic generative adversarial networks. arXiv:1703.05502 [cs, stat], March 2017
20. Xie, S., Girshick, R., Dollár, P., Tu, Z., He, K.: Aggregated Residual Transformations for Deep Neural Networks (2017)
21. Zeiler, M.D., Fergus, R.: Visualizing and understanding convolutional networks **8689**, 818–833 (2013). https://doi.org/10.1007/978-3-319-10590-1_53

Deep Forest with Local Experts Based on ELM for Pedestrian Detection

Wenbo Zheng[1], Sisi Cao[1], Xin Jin[2], Shaocong Mo[3(✉)], Han Gao[4], Yili Qu[5(✉)],
Chengfeng Zheng[6], Sijie Long[6], Jia Shuai[6], Zefeng Xie[6], Wei Jiang[6],
Hang Du[6], and Yongsheng Zhu[6]

[1] School of Software Engineering, Xi'an Jiaotong University, Xi'an, China
[2] School of Management, Huazhong University of Science and Technology,
Wuhan, China
[3] College of Computer Science and Technology, Zhejiang University,
Hangzhou, China
yhmsc007@hotmail.com
[4] School of Cyber Security, University of Science and Technology of China,
Hefei, China
[5] School of Data and Computer Science, Sun Yat-sen University, Guangzhou, China
quwillpower@gmail.com
[6] School of Computer Science and Technology, Wuhan University of Technology,
Wuhan, China

Abstract. Despite recent significant advances, pedestrian detection continues to be an extremely challenging problem in real scenarios. Recently, some authors have shown the advantages of using combinations of part/patch-based detectors in order to cope with the large variability of poses and the existence of partial occlusions. In the beginning of 2017, deep forest is put forward to make up the blank of the decision tree in the field of deep learning. Deep forests have much less parameters than deep neural network and the advantages of higher classification accuracy. In this paper, we propose a novel pedestrian detection approach that combines the flexibility of a part-based model with the fast execution time of a deep forest classifier. In this proposed combination, the role of the part evaluations is taken over by local expert evaluations at the nodes of the decision tree. We first do feature select based on extreme learning machines to get feature sets. Afterwards we use the deep forest to classify the feature sets to get the score which is the results of the local experts. We tested the proposed method with well-known challenging datasets such as TUD and INRIA. The final experimental results on two challenging pedestrian datasets indicate that the proposed method achieves the state-of-the-art or competitive performance.

Keywords: Pedestrian detection · Extreme learning machines
Deep forest

© Springer Nature Switzerland AG 2018
R. Hong et al. (Eds.): PCM 2018, LNCS 11165, pp. 803–814, 2018.
https://doi.org/10.1007/978-3-030-00767-6_74

1   Introduction

Pedestrian detection is an important application domain in computer vision. It has been widely used in surveillance, robotic, intelligent vehicles, etc. However, owing to the large variations in pedestrians'pose and clothing, as well as the varying background and illumination. The seminal work of Dalal and Triggs [8] showed the importance of using rich block-based descriptors such as the Histograms of Oriented Gradients (HOG) representation, which provides both robustness and distinctiveness. Building upon this work, other authors [5] have proposed additional features that enrich the visual representation, including the use of texture through block-based Local Binary Patterns (LBP) [24], and the design of efficient gradient-based features via integral channels [10,13].

All of these approaches are holistic, in the sense that the whole pedestrian is described by a single feature vector and is classified at once. Some authors [7,15,25,33] have proposed successful methods which provides more flexibility in the spatial configuration of the different parts of the object, and these methods lead to higher adaptability to the different poses of the pedestrian. Felzenszwalb et al. [14] proposed local part-based approach which has shown state-of-the-art results in several challenging datasets, being consistently ranked among the top performers.

In recent years, deep learning has become the focus of research in artificial intelligence, computer vision and machine learning [18,19,30,31]. The beginning of 2017, Zhou [34] put forward a deep forest, make up the blank of the decision tree in the field of deep learning. Deep forests have much less parameters than deep neural network and the advantages of higher classification accuracy. In this paper, we propose a novel pedestrian detection approach that combines the flexibility of a part-based model with the fast execution time of a deep forest classifier. In this proposed combination, the role of the part evaluations is taken over by local expert evaluations at the nodes of the decision tree. As an image window proceeds down the tree, a variable configuration of local experts is evaluated on its content, depending on the outcome of previous evaluations. Thus, our proposed approach can flexibly adapt to different pedestrian viewpoints and body poses. In addition to this, the decision tree structure of the deep forest ensures that only a small number of local experts are evaluated on each detection window, resulting in fast execution. We tested the proposed method with well-known challenging datasets such as TUD and INRIA. The final experimental results on two challenging pedestrian datasets indicate that the proposed method achieves the state-of-the-art or competitive performance.

2   Related Work

2.1   HOG Feature and HOG-LBP Feaure

HOG is an excellent descriptor for capturing the edge or local shape information of objects. It has been applied successfully to detect the whole body of human by Dalal [8]. In the original approach [6,9,29], each $64 * 128$ sliding window is divided into cells of size $8 * 8$ pixels. The gradient magnitude of each pixel in the cell is voted into 9 bins according to the orientation of the pixel's gradient. For better performance, each $2 * 2$ cells is integrated into a block whose size is $16 * 16$ pixels. The histograms of the four cells in the block are concatenated into a 36-D feature vector that is normalized to an L2 unit length. Each $64 * 128$ detection window is represented by 105 overlapped blocks and thus concatenated into a 3780-D feature vector.

HOG is a good descriptor of representing the local edge shape information. Actually, there is also abundant texture information in human body, such as face, head-shoulder [32]. Therefore, the combination of the shape information and the texture information can describe the human body better and thus enhance the detection performance. Here, we use uniform LBP [24] as the texture descriptors.

$LBP^2{}_{8,1}$ uniform patter is used to calculate the histogram of each block. The process of dividing blocks is similar to what used in calculating the HOG features. However, in order to reduce the dimension of the final feature vector, all the blocks are not overlapped in calculating the uniform LBP feature vector. For each block, pixels in the block with different uniform patterns are voted into different bins and all of the non-uniform patterns are voted into one bin. Then a block is represented by a 59-D feature vector that is normalized to an L1-sqrt unit length. For a $64 * 128$ detection window, all the vectors of blocks in the window are concatenated to a uniform LBP feature vector.

In order to get a better detection performance, we extract the HOG feature and the uniform LBP feature and then combine both named the HOG-LBP feature.

2.2 Extreme Learning Machines

The Extreme Learning Machine (ELM) algorithm is proposed by Huang et al. [20,26], which is a new kind of single-hidden-layer structure of feed-forward neural networks, which has proved that ELM has the same global approximation ability as the neural network. Its network structure is divided into three layers: input layer, a single hidden layer and output layer, the network input weights and biases of the hidden nodes randomly generated, and once generated do not need to adjust, the output weight value calculated by the least squares estimation method directly. ELM not only has good generalization ability, but because do not need iteration to adjust the network weight value, avoid the local minima, the algorithm of gradient descent learning time is long, and the impact of vector, greatly improve the speed of training and testing, so the research and application of the ELM has received the widespread attention.

For N different training sample sets $(x_i, t_i), i = 1, 2, \cdots, N, x_i = [x_{i1}, x_{i2}, \cdots, x_{im}]^T \in R^m, t_i = a[t_{i1}, t_{i2}, \cdots, t_{in}]^T \in R^n$. Mathematically, the single hidden layer feed-forward neural network shown in Fig. 1 with L-hidden layer node and activation function $g(x)$ can be written as

$$\sum_{i=1}^{M} \beta_i g(w_i \cdot x_i + b_i) = o_j, j = 1, 2, \cdots, N \tag{1}$$

where $w_i = [w_{1i}, w_{2i}, \cdots, w_{mn}]$ is the input weight vector which means weight of the connect between the i-th hidden layer node and the input layer node, $\beta_i = [\beta_{i1}, \beta_{i2}, \cdots, \beta_{in}]^T$ is the input weight vector which means weight of the connect between the i-th hidden layer node and the output layer node and $o_j = [o_{j1}, o_{j2}, \cdots, o_{jn}]^T$ is the output weight vector which means weight of the output layer.

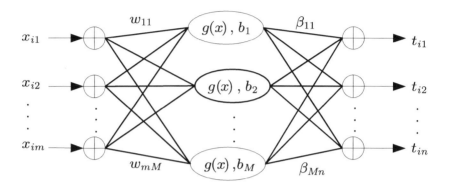

Fig. 1. The single hidden layer feed-forward neural network

Huang et al. [16] have proved that if the activation function $g(x)$ infinitely differentiable, then the single hidden layer of network input weights and bias values of hidden layer can be randomly generated, and do not need iterative adjustment once fixed, in this case, the matrix H is a constant matrix, and the learning process of the limit learning machine can be equivalent to the least squares solution of the least norm of the linear system $H\beta = T$, where $T = [t_1, t_2, \cdots, t_n]^T$ is the target matrix, and $H = \begin{bmatrix} g(w_1 \cdot x_1 + b_1) & \cdots & g(w_L \cdot x_1 + b_L) \\ \vdots & \ddots & \vdots \\ g(w_1 \cdot x_m + b_1) & \cdots & g(w_L \cdot x_m + b_L) \end{bmatrix}_{m \times L}$, where $b_i, i = 1, 2, 3, \cdots, M$ is the hidden layer node width value. Mathematically, the single hidden layer feed-forward neural network can be written as

$$\sum_{i=1}^{M} \beta_i g(w_i \cdot x_i + b_i) = t_j, j = 1, 2, \cdots, N \tag{2}$$

Usually the least squares method is used to determine the output weight of the linear system.

The learning process of ELM algorithm is mainly divided into three steps [3]:

(1) The input weight value and biases of the hidden nodes are randomly generated.
(2) Calculate the hidden layer output matrix H.
(3) Calculate the output weight β.

2.3 Deep Forest

Zhou and Feng [34] propose gcForest (multi-Grained Cascade Forest) shown in Fig. 2, a novel decision tree ensemble method. This method generates a deep forest(DF) ensemble, with a cascade structure which enables gcForest to do representation learning. Its representational learning ability can be further enhanced by multi-grained scanning when the inputs are with high dimensionality, potentially enabling gcForest to be contextual or structural aware. The number of cascade levels can be adaptively determined such that the model complexity can be automatically set, enabling gcForest to perform excellently even on small-scale data.

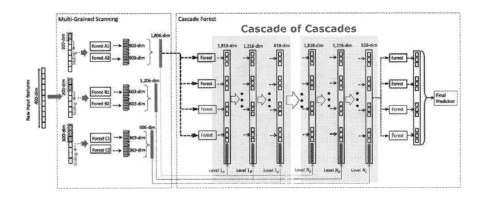

Fig. 2. Multi-Grained Cascade Forest [34]

3 Proposed Pedestrian Detection Algorithm

In this work we define a novel ensemble of local experts based on an averaged combination of deep forest, and the pipeline is shown in Fig. 3. We first do feature select based on extreme learning machines to get feature sets. Afterwards we use the deep forest to classify the feature sets to get the score which is the results of the local experts.

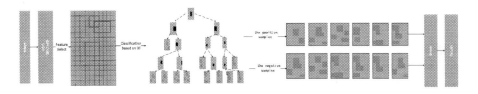

Fig. 3. The pipeline of proposed pedestrian detection algorithm

3.1   Feature Select Based on Extreme Learning Machines

We define our ensemble of local experts through the definition of an appropriate feature selector $\phi(\mathbf{v})$. Given an image window, a block based descriptor $\mathbf{v}$ such as HOG is extracted by partitioning the window into $N \times M$ blocks. Given this block based descriptor $\mathbf{v}$ each feature selector ϕ_k defines a rectangular region formed by contiguous blocks.

In particular, the k-th feature selector ϕ_k is generated by randomly selecting the coordinates (i, j) of the top-left block, and randomly generating the width W and height H of the rectangular area, where $1 \leqslant W \leqslant L$ and $1 \leqslant H \leqslant L$, with L the predefined maximum size. Given the previous definition of the feature selector ϕ_k, the k-th local expert is defined as $E_k(\mathbf{v}) = \psi^T_k \cdot \phi_k(\mathbf{v})$. The transformation ψ_k is learned by using ELM, using the transformed samples as training set $S^{\phi_k}{}_j$. The algorithm is mentioned in Algorithm 1.

Algorithm 1: Feature Select based on Extreme Learning Machines

---

Input: The local expert number k
Output: Feature select results h
1: Randomly generate a subset $\{\phi_1, \ldots, \phi_k\}$ of k feature selectors $\phi_k(\mathbf{v})$.
2: **for** $k = 1, 2, 3, \ldots, K$ **do**
    (1) Let $S^{\phi_k}{}_j$ be the transformed set of samples: $S^{\phi_k}{}_j = \{\phi_k(\mathbf{v}) : \mathbf{v} \in S_j\}$.
    (2) Obtain a discriminant linear transformation ψ_k by ELM classifier mentioned in 2.2 over the transformed samples $S^{\phi_k}{}_j$.
    (3) Find the threshold τ_k that maximizes the purity of the following partition:
$$S^L{}_j = \{\mathbf{v} \in S_j : \psi^T{}_k \cdot \phi_k(\mathbf{v}) \leqslant \tau_k\}$$
$$S^R{}_j = \{\mathbf{v} \in S_j : \psi^T{}_k \cdot \phi_k(\mathbf{v}) > \tau_k\}$$
    Note that the projected values $\psi^T{}_k \cdot \phi_k(\mathbf{v})$ are classification scores provided by the previously learned ELM classifier.
    (4) Let $P_k = I(\phi_k, \psi_k, \tau_k)$ be the maximum purity value obtained in the previous step.
3: Let $k^* = \arg\max_{k=1.2.3.\ldots,k} P_k$. Define the split function for node j as:
$$h(\mathbf{v}) = [\psi_{k^*} \cdot \phi_{k^*}(\mathbf{v}) \leqslant \tau_{k^*}]$$

4: **return** Feature select results h

This is equivalent to extracting a local block-based feature vector from the same rectangular area across the different image windows introduced into the node, and feeding them to a learner that obtains a model of this part of the window.

3.2   Pedestrian Feature Classification Based on Deep Forest

Regarding the output of the deep forest, let $p_t(c|\mathbf{v})$ be the probability that the window $\mathbf{v}$ belongs to class c, computed by the t-th tree of the forest. This probability

---

Algorithm 2: Pedestrian Feature Classification based on Deep Forest

---

Input: Training set $S = h$

Output: The final classification result score

1: Set the initial training set as $S = P \cup N$, where P is the set of cropped pedestrians, and N is an initial set of negative windows that are randomly sampled.

2: Set the initial forest as $F = \emptyset$

3: **for** $i = 1, 2, 3, ..., N$ **do**

> (1) Train M new trees using the training set S.
> Add the trees to the current forest F.
>
> (2) Use the current forest F for detecting false positives in the training images.
> Consider these false positives as negative samples and add them to the training set S.
>
> (3) Use the new training set S for updating the leaf probabilities $p_t(c|\mathbf{v})$ for all the trees in F.

4: $score = \frac{1}{M} \sum_{t=1}^{M} p_t(c|\mathbf{v})$

5: $t = M$

6: **while** $score > \eta$ and $t < T$ **do**

$$score = \frac{1}{t+1}(score * t + p_{t+1}(c = 1|\mathbf{v})$$

$$t = t + 1$$

7: **return** score

---

is obtained during the training stage. Every leaf stores the class distribution of the training samples that reach it, and then each leaf probability is set according to this distribution. Given this, we use the average as aggregation rule in order to compute the probability for the whole forest: $p_F(c|\mathbf{v}) = \frac{1}{T} \sum_{t=1}^{T} p_t(c|\mathbf{v})$, where T represents the number of trees in the DF F. The algorithm is mentioned in Algorithm 2.

4 Experimental Evaluation

This section will examine the classification and recognition performance of our algorithm by conducting experiments on TUD pedestrians [17] and INRIA pedestrians [22] datasets.

In order to investigate the effectiveness of the algorithm in this paper, in the experiment separately to realize the DCNN [23] algorithm, the MLC algorithm [4], KLCF algorithm [2], CAKSVM algorithm [1], MBN algorithm [21], SDH+MSRC algorithm [28] and SSM algorithm [27], TUD pedestrians and INRIA pedestrians datasets on the experiment and comparing with the result of the experiment.

All experiments were conducted using a 4-core PC with an Intel Core i7 6700HQ with the NVIDIA GTX 1080, 16 GB of RAM, and Ubuntu Linux in practice.

4.1   Experimental Results on TUD Database

The TUD pedestrian dataset is a widely used benchmark for human detection. This dataset includes 400 training images and 250 test images with 311 pedestrians. Because the background in this dataset is mainly street; moreover, the diversities of backgrounds are low; we suggest collecting negative samples from INAIA dataset. In addition, we randomly select some positive samples from INAIA dataset.

To evaluate the performance of different methods evaluated on TUD pedestrian test set, the Receiver Operating Characteristic (ROC) curves to describe the statistical comparison of different methods. But ROC is not more intuitive than the Equal Error Rate (EER) which is the point on the ROC curve that corresponds to have an equal probability of missclassifying a positive or negative sample. We compute the EER of different methods, as shown in Table 1.

Table 1. EER of different methods evaluated on TUD pedestrian dataset

Algorithm	Equal Error Rate (EER)
Ours	0.033
DCNN [23]	0.045
MLC [4]	0.077
KLCF [2]	0.089
CAKSVM [1]	0.111
MBN [21]	0.147
SDM+MSRC [28]	0.188
SSM [27]	0.206

From Table 1, we know our method is better than others. Regarding detection speed, we evaluate it on TUD pedestrian dataset. With cascade detection architecture which is the multi-grained cascade forest in our method, the mean detection time of one test image achieves 0.05 s which is faster than the DCNN's 0.047 s, the SSM's 0.68 s, the SDH+MSRC's 0.73 s, the CAKSVM's 0.98 s, the KLCF's 1.02 s, the MBN's 1.2 s, and the MLC's 1.3 s.

4.2   Experimental Results on INRIA Database

The INRIA pedestrian dataset is also a popular benchmark for pedestrian detection. This dataset is very challenging because of various intraclass variations and cluttered scenes. The training set includes 614 images with 1208 pedestrians and 1218 background images. In order to tolerate changes caused by poses, views, occlusions, and so forth, we flip the 1208 normalized pedestrian windows and get 2416 normalized positive samples. Negative training windows are sampled randomly from 1218 background images. The test set includes 288 images with 1126 pedestrians and 453 images without them.

During training the our method for pedestrian detection on these two datasets, all samples are normalized to 51×95 pixels.

The INRIA pedestrian test set contains pedestrians with large intraclass variability. The statistical comparison of different methods is defined by miss rate at 1 false positive per image (FPPI). We follow the definition in [22] that the miss rate is computed as

$$miss\ rate = \frac{FN}{nP}$$

where FN is the false negative during test and nP is the total number of positive samples in the test dataset.

For pedestrian detection, if the threshold of the classifier is low, the miss rate will decrease; at the same time, the number of false positive in each test image will increase. To make fair comparison, we should specify a statistical indicator and use another indicator to evaluate the performance of different methods. Generally, miss rate at $FPPI = 1$ is used, as shown in Table 2.

Table 2. Performances of different methods evaluated on INRIA pedestrian dataset at 1 FPPI

Algorithm	Miss rate at 1 FPPI
Ours	0.008
DCNN [23]	0.02
MLC [4]	0.07
CAKSVM [1]	0.08
KLCF [2]	0.12
SDH+MSRC [28]	0.14
MBN [21]	0.14
SSM [27]	0.22

From Table 2, we know our method is better than others. Regarding detection speed, the mean detection time of one test image in our method achieves 0.03 s which is faster than the DCNN's 0.047 s, the SSM's 0.70 s, the SDH+MSRC's 0.74 s, the CAKSVM's 1.01 s, the KLCF's 1.02 s, the MBN's 1.023 s, and the MLC's 1.3 s.

5   Conclusion

We propose a novel pedestrian detection approach that combines the flexibility of a part-based model with the fast execution time of a deep forest classifier. The key of our method is taken over by local expert evaluations at the nodes of the decision tree. As an image window proceeds down the tree, a variable configuration of local experts is evaluated on its content, depending on the outcome of previous evaluations, It is results that our proposed approach can flexibly adapt to different pedestrian viewpoints and body poses. The final experimental results on two challenging pedestrian datasets indicate that the proposed method achieves the state-of-the-art or competitive performance. In future, we will expand this method to the lager scale datasets like Caltech benchmark dataset [11,12], and we will expand this method to intelligent monitoring.

Acknowledgment. This paper was supported in part by Science & Technology Pillar Program of Hubei Province under Grant (#2014BAA146), Nature Science Foundation of Hubei Province under Grant (#2015CFA059), Science and Technology Open Cooperation Program of Henan Province under Grant (#152106000048) and Fundamental Research Funds for the Central Universities (#2018-JSJ-A1-01, #2018-JSJ-A1-02, #2018-JSJ-B1-12, #2018-JSJ-B1-05, #2018-JSJ-B1-06, #2018-JSJ-B1-07, #2018-JSJ-B1-08, WUT:2017II03XZ).

References

1. Baek, J., Kim, J., Kim, E.: Fast and efficient pedestrian detection via the cascade implementation of an additive kernel support vector machine. IEEE Trans. Intell. Transp. Syst. **PP**(99), 1–15 (2017)
2. Bilal, M.: Algorithmic optimization of histogram intersection kernel SVM-based pedestrian detection using low complexity features. IET Comput. Vis. **11**(5), 350–357 (2017)
3. Erik, C., et al.: Extreme learning machines [trends & controversies]. IEEE Intell. Syst. **28**(6), 30–59 (2013)
4. Cao, J., Pang, Y., Li, X.: Learning multilayer channel features for pedestrian detection. IEEE Trans. Image Process. **26**(7), 3210–3220 (2016)
5. Cao, S., et al.: Access control of cloud users credible behavior based on IPv6. In: 2018 IEEE 3rd International Conference on Big Data Analysis (ICBDA), pp. 303–308. IEEE (2018)
6. Creusen, I.M., Wijnhoven, R.G.J., Herbschleb, E., De With, P.H.N.: Color exploitation in hog-based traffic sign detection. In: IEEE International Conference on Image Processing, pp. 2669–2672 (2010)
7. Dai, S., Yang, M., Wu, Y., Katsaggelos, A.: Detector ensemble. In: IEEE Conference on Computer Vision and Pattern Recognition, pp. 1–8 (2007)
8. Dalal, N., Triggs, B.: Histograms of oriented gradients for human detection. In: IEEE Computer Society Conference on Computer Vision and Pattern Recognition, CVPR 2005, pp. 886–893 (2005)
9. Ding, S., Liu, Z., Li, C.: AdaBoost learning for fabric defect detection based on HOG and SVM. In: International Conference on Multimedia Technology, pp. 2903–2906 (2011)

10. Dollar, P., Tu, Z., Tao, H., Belongie, S.: Feature mining for image classification. In: IEEE Conference on Computer Vision and Pattern Recognition, CVPR 2007, pp. 1–8 (2007)
11. Dollár, P., Wojek, C., Schiele, B., Perona, P.: Pedestrian detection: a benchmark. In: CVPR, June 2009
12. Dollár, P., Wojek, C., Schiele, B., Perona, P.: Pedestrian detection: an evaluation of the state of the art. PAMI **34**(4), 743–761 (2012)
13. Dollár, P., Appel, R., Kienzle, W.: Crosstalk cascades for frame-rate pedestrian detection. In: Fitzgibbon, A., Lazebnik, S., Perona, P., Sato, Y., Schmid, C. (eds.) ECCV 2012. LNCS, pp. 645–659. Springer, Heidelberg (2012). https://doi.org/10.1007/978-3-642-33709-3_46
14. Felzenszwalb, P.F., Girshick, R.B., Mcallester, D., Ramanan, D.: Object detection with discriminatively trained part-based models. IEEE Trans. Pattern Anal. Mach. Intell. **47**(2), 6–7 (2014)
15. Gall, J., Lempitsky, V.: Class-specific hough forests for object detection. In: Criminisi, A., Shotton, J. (eds.) Decision Forests for Computer Vision and Medical Image Analysis. Advances in Computer Vision and Pattern Recognition, pp. 1022–1029. Springer, London (2013). https://doi.org/10.1007/978-1-4471-4929-3_11
16. Huang, Z., Yu, Y., Gu, J., Liu, H.: An efficient method for traffic sign recognition based on extreme learning machine. IEEE Trans. Cybern. **47**(4), 920–933 (2017)
17. Diplomarbeit Im Fach Informatik, Oliver Sander, Bioinformatik U Abteilung, and Angewandte Algorithmik. Max-planck-institut fr informatik (2013)
18. Jin, X., et al.: Pattern learning based parallel ant colony optimization. In: Ubiquitous Computing and Communications (ISPA/IUCC), 2017 IEEE International Symposium on Parallel and Distributed Processing with Applications and 2017 IEEE International Conference on, pp. 497–502. IEEE (2017)
19. Lecun, Y., Bengio, Y., Hinton, G.: Deep learning. Nature **521**(7553), 436–444 (2015)
20. Li, S., You, Z.H., Guo, H., Luo, X., Zhao, Z.Q.: Inverse-free extreme learning machine with optimal information updating. IEEE Trans. Cybern. **46**(5), 1229–1241 (2016)
21. Peng, P., Tian, Y., Wang, Y., Li, J., Huang, T.: Robust multiple cameras pedestrian detection with multi-view Bayesian network. Pattern Recognit. **48**(5), 1760–1772 (2015)
22. Taiana, M., Nascimento, J.C., Bernardino, A.: An improved labelling for the INRIA person data set for pedestrian detection. In: Sanches, J.M., Micó, L., Cardoso, J.S. (eds.) IbPRIA 2013. LNCS, vol. 7887, pp. 286–295. Springer, Heidelberg (2013). https://doi.org/10.1007/978-3-642-38628-2_34
23. Tom, D., Monti, F., Baroffio, L., Bondi, L., Tagliasacchi, M., Tubaro, S.: Deep convolutional neural networks for pedestrian detection. Signal Process. Image Commun. **47**, 482–489 (2016)
24. Wang, X., Han, T.X., Yan, S.: An HOG-LBP human detector with partial occlusion handling. In: IEEE International Conference on Computer Vision, pp. 32–39 (2010)
25. Bo, W., Nevatia, R.: Detection and segmentation of multiple, partially occluded objects by grouping, merging, assigning part detection responses. Int. J. Comput. Vis. **82**(2), 185–204 (2009)
26. Yang, Y., Wu, Q.M.J.: Extreme learning machine with subnetwork hidden nodes for regression and classification. IEEE Trans. Cybern. **46**(12), 2885–2898 (2016)
27. Zhang, S., Bauckhage, C., Cremers, A.B.: Efficient pedestrian detection via rectangular features based on a statistical shape model. IEEE Trans. Intell. Transp. Syst. **16**(2), 763–775 (2015)

28. Zhao, X., He, Z., Zhang, S., Liang, D.: Robust pedestrian detection in thermal infrared imagery using a shape distribution histogram feature and modified sparse representation classification. Pattern Recognit. **48**(6), 1947–1960 (2015)
29. Zheng, W., et al.: Image super-resolution reconstruction algorithm based on Bayesian theory. In: 2018 13th IEEE Conference on Industrial Electronics and Applications (ICIEA), pp. 1937–1941. IEEE (2018)
30. Zheng, W., et al.: Face recognition based on weighted multi-channel Gabor sparse representation and optimized extreme learning machines. In: 2017 3rd IEEE International Conference on Computer and Communications (ICCC), pp. 1616–1620. IEEE (2017)
31. Zheng, W., Mo, S., Duan, P., Jin, X.: An improved pagerank algorithm based on fuzzy c-means clustering and information entropy. In: 2017 3rd IEEE International Conference on Control Science and Systems Engineering (ICCSSE), pp. 615–618. IEEE (2017)
32. Zheng, W., et al.: Robust and high capacity watermarking for image based on DWT-SVD and CNN. In: 2018 13th IEEE Conference on Industrial Electronics and Applications (ICIEA), pp. 1237–1241. IEEE (2018)
33. Zheng, W., Wang, F., Wang, K.: An ACP-based approach to color image encryption using DNA sequence operation and hyper-chaotic system. In: 2017 IEEE International Conference on Systems, Man, and Cybernetics (SMC), pp. 461–466. IEEE (2017)
34. Zhou, Z., Feng, J.: Deep forest: towards an alternative to deep neural networks (2017)

AdvRefactor: A Resampling-Based Defense Against Adversarial Attacks

Jianguo Jiang[1], Boquan Li[1,2], Min Yu[1,2(✉)], Chao Liu[1], Jianguo Sun[3],
Weiqing Huang[1], and Zhiqiang Lv[1]

[1] Institute of Information Engineering, Chinese Academy of Sciences, Beijing, China
{jiangjianguo,liboquan,yumin,liuchao,huangweiqing,lvzhiqiang}@iie.ac.cn
[2] School of Cyber Security, University of Chinese Academy of Sciences,
Beijing, China
[3] School of Computer Science and Technology, Harbin Engineering University,
Harbin, China
sunjianguo@hrbeu.edu.cn

Abstract. Deep neural networks have achieved great success in many domains. However, they are vulnerable to *adversarial attacks*, which generate *adversarial examples* by adding tiny perturbations to legitimate images. Previous studies providing defense mostly focus on modifying DNN models to mitigate adversarial attacks. We propose a resampling-based defense, *AdvRefactor*, which aims at transforming the inputs of models, and thereby eliminates the adversarial perturbations. We explore two resampling algorithms, proximal interpolation and bilinear interpolation, which cost less, suit more models and are combinable with other defenses. Our evaluation results demonstrate that AdvRefactor can significantly mitigate the adversarial attacks.

Keywords: Deep neural networks · Image recognition
Adversarial examples · Defense against adversarial attacks
Resampling

1 Introduction

Deep neural networks (DNN) are known to achieve great performance in many domains of artificial intelligence, one representative example of which is image recognition [7]. Nevertheless, recent researches have discovered that neural networks are vulnerable to be cheated by *adversarial examples*, which are generated by *adversarial attacks*. The adversary crafts adversarial examples by adding tiny and purposeful perturbations to images, and thereby forces target models to produce unexpected outputs [20]. As an example shown in Fig. 1, the pixels required for perturbation can be so small to perceive, making it hard to distinguish adversarial examples from legitimate examples [19].

This work is supported by National Natural Science Foundation of China (No. 61601459).

R. Hong et al. (Eds.): PCM 2018, LNCS 11165, pp. 815–825, 2018.
https://doi.org/10.1007/978-3-030-00767-6_75

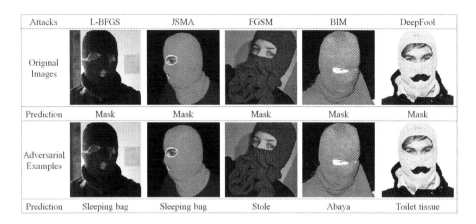

Fig. 1. Adversarial examples generated by five typical attacks we evaluated in Sect. 4. The form presents five original images which are all correctly predicted as masks. By adding imperceptible perturbations to each image, the model is fooled to recognize all images with incorrect predictions. If it happens in reality, the recognition system might fail to seek out malicious images due to such perturbation.

Adversarial attack is a serious threat particularly in security sensitive systems, including road sign recognition system of self-driving cars [4], face recognition system for access control [18], etc. Thus, it is necessary to apply efficient defenses to mitigating adversarial attacks. Defense of adversarial attacks is a emerging area researched in recent years. Most of the previous work, including gradient masking [6], defensive distillation [15] and adversarial training [5], aims at modifying or retraining the network to mitigate adversarial attacks, i.e. focusing on dealing with DNN model itself. Instead, our approach, which is named *AdvRefactor*, focuses on transforming the adversarial inputs into legitimate inputs. Compared with modifying the model, AdvRefactor requires less cost, suits more recognition model, and is easy to combine with other defenses.

In reality, the adversary merely perturbs a fraction of pixels in an image to generate an adversarial example. Hence the basic idea of AdvRefactor is to alter the perturbed pixels by refactoring the pixels of an image, and thereby eliminates the adversarial perturbations. The refactoring process is achieved via resampling. Concretely, we employ two basic algorithms of resampling: proximal interpolation and bilinear interpolation. In our evaluation, AdvRefactor significantly improves the accuracy on recognizing the perturbed images, preserves the accuracy on original images, and shows better performance than a benchmark strategy, which achieves the goal to mitigate the adversarial attacks.

We make the following contributions in this paper:

- We propose and evaluate a resampling-based defense by transforming the adversarial inputs rather than modifying models, which can significantly mitigate the adversarial attacks.

- We suggest a feasible direction towards refactoring the pixels to eliminate the adversarial perturbations of images.
- We perform our research in a specific scenario (as an example shown in Fig. 1), which attracts an attention of such security threat existing in recognizing malicious images.

The rest of the paper is structured as follows. We first give a brief introduction to existing adversarial attacks and defense strategies in Sect. 2. Next, we elaborate on our proposed approach in Sect. 3 and evaluate the effectiveness of our defense in Sect. 4. Finally, we summarize our work in Sect. 5.

2   Background

This Section provides a brief introduction to existing adversarial attacks and defense strategies.

2.1   Adversarial Attacks

Existing researches proposed two types of adversarial attacks [1]. In a targeted attack, the adversary's goal is to craft an adversarial example which causes the classifier to produce a particular incorrect output. In an untargeted attack, the adversary's goal is just to make an adversarial example be classified as any incorrect output. Szegedy et al. first observed that images with perturbations could fool deep learning models, and proposed a *box-constrained L-BFGS (Targeted)* attack to generate adversarial examples [20]. Further, they also demonstrated the robustness of DNN against adversarial examples could be improved by *adversarial training*. To provide training examples for adversarial training, Goodfellow et al. designed a *fast gradient sign method: FGSM (Untargeted)* to generate adversarial examples [5], as well as proposed the primary cause of neural networks' vulnerabilities was their linear nature. Based on FGSM, Kuarkin et al. designed a *basic iterative method: BIM (Untargeted)* to extend FGSM's one-step gradient ascent method [10], which took multiple small steps interactively when adjusting the direction after each step. Moreover, BIM explored the adversarial examples in the physical world as well. Moosavi-Dezfooli et al. proposed *DeepFool (Untargeted)* to compute a minimal-norm adversarial perturbation in an interactive pattern [11], which achieved similar fooling ratios whereas with smaller perturbations than FGSM. Papernot et al. created a *Jacobian-based saliency map attack: JSMA (Targeted)*, which achieved to perturb images with only a few pixels [13].

2.2   Defense Strategies

To mitigate the adversarial attacks, Gu et al. proposed to add a gradient penalty term in the training objective [6]. The approach was generally defined as *gradient masking*, which sought to reduce the sensitivity of DNN to small input changes.

However, gradient masking was queried to reduce the efficiency and accuracy of models on many tasks [14]. Papernot et al. proposed the notion of *defensive distillation* to enable DNN to be more robust against adversarial examples [15]. Different from making the classifier output class labels, defensive distillation made class confidences as outputs and leveraged these class confidences to train a identical-architecture model. The approach prevented the adversary to fool a new model by leveraging original attacks. It is worth noting that Carlini et al. have successfully attacked against the defensive distillation defense [2,3]. Some popular attacks synchronously utilized *adversarial training* as the first parclose against those attacks [5,11,20]. It focused on the training process of a model, which was trained with both adversarial and legitimate examples. Adversarial training increased the classification accuracy, enhanced the robustness of the model, and regularized the network to reduce over-fitting [5]. Nevertheless, expensive computation resources made adversarial training inefficient in practice.

3   Resampling-Based Defense

In this section, we present a resampling-based defense to mitigate adversarial attacks, named AdvRefactor. We elaborate on our approach, and make an illustration about why resampling can mitigate the adversarial attacks.

3.1   Mechanism of Defense

In our strategy, we preprocess the perturbed images before they are input into the recognition model. Namely, we attempt to transform the adversarial inputs into legitimate inputs. As illustrated in Fig. 2, AdvRefactor is akin to a filtration barrier between inputs and a recognition model. By employing AdvRefactor, we get an anticipative prediction of the presented image.

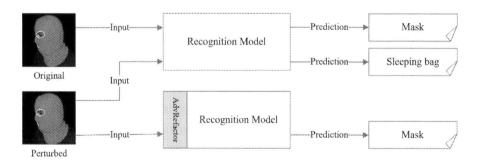

Fig. 2. Framework of AdvRefactor for mitigating adversarial attacks. AdvRefactor is akin to a filtration barrier between inputs and a recognition model. By employing the defense, the model could output a correct prediction though the input is perturbed.

In order to transform the perturbed inputs, we propose the method of resampling. In reality, the adversary merely perturbs a fraction of pixels in an image to generate an adversarial example (introduced in Sect. 2). Hence we donate our work on the perturbed pixels. We employ resampling for our defense, which attempts to alter the perturbed pixels in an adversarial example by refactoring the pixels of images, and thereby eliminating the adversarial perturbations. In addition, we must ensure such image processing will not degrade the accuracy on recognizing original images.

The idea of resampling is proved to be effective in Sect. 4. On account of the effectiveness of our work, our resampling-based defense suggests a feasible direction to the defense against adversarial attacks: towards refactoring the pixels to eliminate the adversarial perturbations of images.

3.2 Resampling Algorithms

Although the category of resampling algorithms is numerous, we employ two basic types: proximal interpolation and bilinear interpolation [9]. In this section, we denote $f(i, j)$ as the value of the pixel (i, j).

Proximal Interpolation. The proximal interpolation algorithm sets the value of the sampling pixel by replicating the pixel value located at the shortest distance, as shown in Eq. (1):

$$f(i + u, j + v) = f(i, j) \tag{1}$$

where $u, v \in (-0.5, 0.5)$, $f(i + u, j + v)$ represents the value of sampling pixel. Currently, although proximal interpolation requires little computation cost, it is not a widely-used resampling approach on account of the loss of image quality. Nevertheless, proximal interpolation algorithm performs well in mitigating the adversarial attacks, as demonstrated in Sect. 4.

Bilinear Interpolation. For a bilinear interpolation algorithm, the value of the sampling pixel is set to the weighted average of four neighboring pixels, whose weights are determined by the distances from the sampling pixel. The complete precess can be decomposed into two first-order linear interpolations processes in horizontal and vertical directions, respectively as shown in Eq. (2):

$$f(i + u, j + v) = (1 - u)(1 - v)f(i, j) + (1 - u)vf(i, j + 1) \\ + u(1 - v)f(i + 1, j) + uvf(i + 1, j + 1) \tag{2}$$

where $u, v \in [0, 1]$, and the value of sampling pixel $f(i + u, j + v)$ is determined by the values of four approximal pixels: $f(i, j)$, $f(i, j + 1)$, $f(i + 1, j)$, and $f(i + 1, j + 1)$. The bilinear interpolation algorithm is also proved to be effective in Sect. 4.

3.3   Feasibility of Resampling

It is a controversial topic about why the adversarial examples could fool the recognition model, thus we merely make an intuitive illustration about why resampling could eliminate the adversarial perturbations of images.

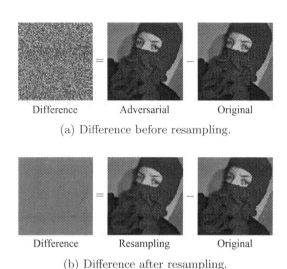

(a) Difference before resampling.

(b) Difference after resampling.

Fig. 3. Comparison of the difference between images. (a) shows the difference between an adversarial example (generated by FGSM attack) and an original image. (b) shows the weakened difference after resampling (with bilinear interpolation algorithm).

To intuitively illustrate the adversarial perturbations, we perform a difference operation on the image matrixs of an adversarial example and an original image, As shown in Eq. (3):

$$Matrix_{difference} = Matrix_{adversarial} - Matrix_{original} \tag{3}$$

here $Matrix_{difference}$ reflects the perturbations added to an original image. We output the difference matrix before and after resampling an adversarial example, as it is presented in Fig. 3.

Visually, the difference is weakened after resampling. Here we intuitively deem the change of the difference implies adversarial perturbations are indeed weakened. Further, it attests resampling takes effect on eliminating the adversarial perturbations of adversarial images.

4   Evaluation

In this section, we evaluate the effectiveness of AdvRefactor. Concretely, we execute five typical attacks on a pre-trained ResNet50 model [8] with excellent performance. Further, we evaluate the recognition accuracy on the perturbed images and the ability to preserve the accuracy on original images. For a fair comparison, we also employ a state-of-the-art strategy as a benchmark.

4.1 Experiment Settings

In the experiment, we leverage a set of specific-category images of people in masks from the popular dataset ImageNet [17], the total of which is 1105. It is noted that the 1105 images are from a complete synset without any manual intervention (synset refers to a set of synonymous images from a dataset). The dataset is not huge whereas enough for evaluating our work in the scenario of mask recognition.

We set up a pre-trained ResNet50 model as the target model, which achieves a top-5 accuracy of 94.75% on ImageNet, with the 50-layer residual nets [8]. Notably, we suppose that the adversary has full knowledge of the target model, but no ability to change it (i.e. the adversary generates adversarial examples with white-box attacks).

We evaluate AdvRefactor on five typical attacks introduced in Sect. 2.1, including two targeted attacks (L-BFGS and JSMA) and three untargeted attacks (FGSM, BIM and DeepFool). We implement these attacks based on a python toolbox: Foolbox [16], and conduct our experiments on a server with an i7-7820X 3.60 GHz CPU, a 32GiB system memory, and a GeForce GTX 1080Ti graphics card.

4.2 Evaluation of Attacks

We first evaluate the effect of five attacks, which is the first step to evaluate the performance of our defense. Conducting the attacks requires enormous computing resources and time, hence we merely leverage the first 300 original images in the dataset, which are enough for evaluation.

For each attack, we generate 300 adversarial examples and evaluate the top-1 accuracy on recognizing these images. In target attacks, we set *sleepingbag* as the target class. Noting that there are some errors when adding perturbations to several images (most of them occur on the DeepFool attack). Since this is out of the scope of this paper, we no more consider such case in our evaluation.

Table 1 reports the results on evaluating the five attacks. Before attacking, the ResNet50 model correctly recognizes 259 (86.33%) original images. By performing these attacks, all images are recognized incorrectly. It attests that all attacks are efficient in generating adversarial examples against the recognition model. For target attacks, we consider an attack success if and only if the image is predicted as *sleepingbag*. As the results, both of the two targeted attacks achieve a success rate of 100%.

4.3 Evaluation of Defense

For evaluating AdvRefactor, we merely leverage the 259 correctly predicted original images and corresponding adversarial examples obtained from previous experiment in this part, so as to ensure the recognition is caused by adversarial attacks rather than the low performance of the model itself.

We compare the top-1 accuracy on recognizing the adversarial examples with and without defense. We also verify whether the performance of the model is degraded by adding the defense, via investigating the accuracy on recognizing 259 original images by the adding-defense ResNet50 model.

Table 2 summarizes the evaluation results of AdvRefactor. Although the quantity of image is limited, such huge improvements can fully demonstrate the efficiency of our strategy. Observed from the results, with any of the two resampling methods, the accuracy on recognizing adversarial examples is significantly improved. Bilinear interpolation shows better performance, which increases the average accuracy from 0% to 77.22%, and nearly preserves the accuracy on original images (85.71%). The results prove that AdvRefactor achieves our goal of mitigating the adversarial attacks.

Table 1. Evaluation of attacks.

Attacks	Targeted		Untargeted			Original images
	L-BFGS	JSMA	FGSM	BIM	DeepFool	
Accuracy	0%	0%	0%	0%	0%	86.33%
SuccessRate	100%	100%	–	–	–	–

Table 2. Evaluation of defense with AdvRefactor.

Defense	Attacks					All attacks	Original images
	L-BFGS	JSMA	FGSM	BIM	DeepFool		
Without defense	0%	0%	0%	0%	0%	0%	100%
Proximal	74.90%	74.90%	67.57%	66.02%	70.66%	70.81%	80.69%
Bilinear	81.85%	81.85%	74.90%	71.81%	75.68%	**77.22%**	**85.71%**

4.4   Comparison with State-of-the-Art Work

We make a comparison with a state-of-the-art approach, DeepCAPTCHA, for which is representative and shows a similar technical route to AdvRefactor in changing the adversarial images. DeepCAPTCHA is proposed by Osadchy et al. [12], who utilizes a 5×5 median filter to preprocess inputs. Concretely, we reproduce DeepCAPTCHA with same parameters in their literature, and compare the effectiveness of median filter with the proposed two resampling methods.

To avoid the contingency of the comparison results, we leverage the first 1000 original images in the dataset. In addition, we merely generate adversarial examples by FGSM attack in this part, because it requires less computing resource and time than other attacks, which can be supported by our lab environment. Preparatively, we recognize these 1000 original images by the ResNet50 model, and obtain 832 correct-recognition original images for further evaluation.

We first evaluate whether the performance of the model is degraded after employing the three defenses. Here we recognize the 832 original images with three adding-defense ResNet50 models. The results is summarized in Fig. 4(a), it can be observed that model with bilinear interpolation preservers a best accuracy of 85.22%, and both of resampling methods perform better than the median filter method of DeepCAPTCHA (63.94%).

In Fig. 4(b), we compare the accuracy on recognizing 832 adversarial examples. As the results, the bilinear interpolation method performs best (76.92%), and both of our methods achieve better accuracies than DeepCAPTCHA (60.34%). Therefore, our defense of AdvRefactor achieve better performance in terms of the accuracy on both original images and adversarial examples.

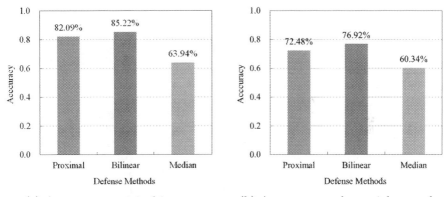

(a) Accuracy on original images.   (b) Accuracy on adversarial examples.

Fig. 4. Evaluation of three defense methods. The ability to preserve the accuracy on original images is summarized in (a), and (b) shows the accuracy on recognizing adversarial examples.

5 Conclusion

We proposed AdvRefactor to mitigate adversarial attacks on image recognition. AdvRefactor employed two resampling algorithms: proximal interpolation and bilinear interpolation, which refactored the pixels of perturbed images thereby eliminating the adversarial perturbations. From the results of evaluation, AdvRefactor improved the accuracy on recognizing adversarial examples from 0% to 77.22%, preserved the accuracy of 85.71% on original images, and showed better performance than a benchmark strategy. The results attested that AdvRefactor can significantly mitigate the adversarial attacks.

References

1. Akhtar, N., Mian, A.: Threat of adversarial attacks on deep learning in computer vision: a survey. arXiv preprint arXiv:1801.00553 (2018)
2. Carlini, N., Wagner, D.: Defensive distillation is not robust to adversarial examples. arXiv preprint arXiv:1607.04311 (2016)
3. Carlini, N., Wagner, D.: Towards evaluating the robustness of neural networks. In: 2017 IEEE Symposium on Security and Privacy (SP), pp. 39–57. IEEE (2017)
4. Evtimov, I., et al.: Robust physical-world attacks on deep learning models. arXiv preprint arXiv:1707.08945 (2017)
5. Goodfellow, I.J., Shlens, J., Szegedy, C.: Explaining and harnessing adversarial examples. In: International Conference on Learning Representations (ICLR) (2015)
6. Gu, S., Rigazio, L.: Towards deep neural network architectures robust to adversarial examples. arXiv preprint arXiv:1412.5068 (2014)
7. He, K., Zhang, X., Ren, S., Sun, J.: Delving deep into rectifiers: surpassing human-level performance on ImageNet classification. In: Proceedings of the IEEE International Conference on Computer Vision, pp. 1026–1034 (2015)
8. He, K., Zhang, X., Ren, S., Sun, J.: Deep residual learning for image recognition. In: Proceedings of the IEEE Conference on Computer Vision and Pattern Recognition, pp. 770–778 (2016)
9. Jain, A.K.: Fundamentals of Digital Image Processing. Prentice Hall, Englewood Cliffs (1989)
10. Kurakin, A., Goodfellow, I., Bengio, S.: Adversarial examples in the physical world. In: International Conference on Learning Representations (ICLR) Workshop (2015)
11. Moosavi Dezfooli, S.M., Fawzi, A., Frossard, P.: DeepFool: a simple and accurate method to fool deep neural networks. In: Proceedings of 2016 IEEE Conference on Computer Vision and Pattern Recognition (CVPR) (2016)
12. Osadchy, M., Hernandez-Castro, J., Gibson, S., Dunkelman, O., Pérez-Cabo, D.: No bot expects the deepcaptcha! IEEE Trans. Inf. Forensics Secur. $12(11)$, 2640–2653 (2017)
13. Papernot, N., McDaniel, P., Jha, S., Fredrikson, M., Celik, Z.B., Swami, A.: The limitations of deep learning in adversarial settings. In: 2016 IEEE European Symposium on Security and Privacy (EuroS&P), pp. 372–387. IEEE (2016)
14. Papernot, N., McDaniel, P., Sinha, A., Wellman, M.: Towards the science of security and privacy in machine learning. arXiv preprint arXiv:1611.03814 (2016)
15. Papernot, N., McDaniel, P., Wu, X., Jha, S., Swami, A.: Distillation as a defense to adversarial perturbations against deep neural networks. In: 2016 IEEE Symposium on Security and Privacy (SP), pp. 582–597. IEEE (2016)
16. Rauber, J., Brendel, W., Bethge, M.: Foolbox: a python toolbox to benchmark the robustness of machine learning models. arXiv preprint arXiv:1707.04131 (2017)
17. Russakovsky, O., et al.: ImageNet large scale visual recognition challenge. Int. J. Comput. Vis. (IJCV) $\mathbf{115}(3)$, 211–252 (2015). https://doi.org/10.1007/s11263-015-0816-y

18. Sharif, M., Bhagavatula, S., Bauer, L., Reiter, M.K.: Accessorize to a crime: real and stealthy attacks on state-of-the-art face recognition. In: Proceedings of the 2016 ACM SIGSAC Conference on Computer and Communications Security, pp. 1528–1540. ACM (2016)

19. Su, J., Vargas, D.V., Kouichi, S.: One pixel attack for fooling deep neural networks. arXiv preprint arXiv:1710.08864 (2017)

20. Szegedy, C., Zaremba, W., Sutskever, I., Bruna, J., Erhan, D., Goodfellow, I., Fergus, R.: Intriguing properties of neural networks. In: International Conference on Learning Representations (ICLR) (2014)

Author Index

Printed in the United States
By Bookmasters